Practical Distillation Control

Practical Distillation Control

William L. Luyben, Editor

VNR VAN NOSTRAND REINHOLD
New York

Library of Congress Catalog Card Number: 92-10642
ISBN: 0-442-00601-2

Manufactured in the United States of America.

Published by Van Nostrand Reinhold
115 Fifth Avenue
New York, New York 10003

Chapman and Hall
2-6 Boundary Row
London, SE 1 8HN, England

Thomas Nelson Australia
102 Dodds Street
South Melbourne 3205
Victoria, Australia

Nelson Canada
1120 Birchmount Road
Scarborough, Ontario M1K 5G4, Canada

16 15 14 13 12 11 10 9 8 7 6 5 4 3 2 1

Library of Congress Cataloging-in-Publication Data

Practical distillation control / edited by William L. Luyben
p. cm.
Includes bibliographical references and index.
ISBN 0-442-00601-2
1. Distillation apparatus. 2. Chemical process control.
I. Luyben, William L.
TP159.D5P73 1992
660'.28425—dc20 92-10642
CIP

Contributors

John E. Anderson
Hoechst Celanese
Corpus Christi, TX 78469
(Chapter 19)

Page S. Buckley
Consultant
Newark Delaware 19713
(Chapter 2)

James J. Downs
Advanced Controls Technology
Eastman Chemical Company
Kingsport, TN 37662
(Chapter 20)

James G. Gerstle
Amoco Corporation
Naperville, Illinois 60566
(Chapter 12)

Vincent G. Grassi II
Air Products and Chemicals, Inc.
Allentown, PA 18195-1501
(Chapters 3 and 18)

Kurt E. Haggblom
Process Control Laboratory
Åbo Akademi
20500 Åbo, Finland
(Chapter 10)

David A. Hokanson
Exxon Chemicals
H. R. Rotterdam No. 231768
The Netherlands
(Chapter 12)

Henk Leegwater
DSM
6190 AA Beek
The Netherlands
(Chapter 16)

William L. Luyben
Department of Chemical Engineering
Lehigh University
Bethlehem, PA 18015
(Chapters 1, 11, 22, 24, and 25)

Randy C. McFarlane
Amoco Corporation
Amoco Research Center
Naperville, IL 60566
(Chapter 7)

Charles Moore
Department of Chemical Engineering
University of Tennessee
Knoxville, TN 37996
(Chapter 8)

Cristian A. Muhrer
Air Products and Chemicals, Inc.
Allentown, PA 18195-1501
(Chapters 23 and 25)

Antonis Papadourakis
Rohm and Haas Co.
Bristol, PA 19007
(Chapter 4)

Ferdinand F. Rhiel
Corporate Division of Research & Development
Bayer AG
D-5090 Leverkusen, Germany
(Chapter 21)

John E. Rijnsdorp
University of Twente
7500AE Enschede
Netherlands
(Chapter 4)

Daniel E. Rivera
Department of Chemical Engineering
Arizona State University
Tempe, Arizona 85287
(Chapter 7)

F. Greg Shinskey
The Foxboro Co.
Foxboro, MA 02035
(Chapter 13)

Sigurd Skogestad
Chemical Engineering
University of Trondheim, NTH
N-7034 Trondheim, Norway
(Chapter 14)

Terry L. Tolliver
Monsanto Co.
St. Louis, MO 63167
(Chapter 17)

Bjorn D. Tyreus
Engineering Department
E. I. DuPont de Nemours & Co.
Wilmington, DE 19898
(Chapters 5 and 9)

Ernest F. Vogel
Advanced Control Technology
Tennessee Eastman Co.
Kingsport, TN 37662
(Chapter 6)

Kurt Waller
Process Control Laboratory
Åbo Akademi
20500 Åbo, Finland
(Chapters 10 and 15)

This book is dedicated to Bea, Gus, and Joanne Luyben and Bill Nichol, four of the most avid bridge players I have ever known. Run 'em out!

Preface

Distillation column control has been the subject of many, many papers over the last half century. Several books have been devoted to various aspects of the subject. The technology is quite extensive and diffuse. There are also many conflicting opinions about some of the important questions.

We hope that the collection under one cover of contributions from many of the leading authorities in the field of distillation control will help to consolidate, unify, and clarify some of this vast technology. The contributing authors of this book represent both industrial and academic perspectives, and their cumulative experience in the area of distillation control adds up to over 400 years! The collection of this wealth of experience under one cover must be unique in the field. We hope the readers find it effective and useful.

Most of the authors have participated at one time or another in the Distillation Control Short Course that has been given every two years at Lehigh since 1968. Much of the material in the book has been subjected to the "Lehigh inquisition" and survived! So it has been tested by the fire of both actual plant experience and review by a hard-nosed group of practically oriented skeptics.

In selecting the authors and the topics, the emphasis has been on keeping the material practical and useful, so some subjects that are currently of mathematical and theoretical interest, but have not been demonstrated to have practical importance, have not been included.

The book is divided about half and half between methodology and specific application examples. Chapters 3 through 14 discuss techniques and methods that have proven themselves to be useful tools in attacking distillation control problems. These methods include dynamic modelling, simulation, experimental identification, singular value decomposition, analysis of robustness, and the application of multivariable methods. Chapters 15 through 25 illustrate how these and how other methods can be applied to specific columns or important classes of columns.

Contents

1

Techniques and Methods

1

Introduction

William L. Luyben
Lehigh University

1-1 IMPORTANCE OF DISTILLATION IN INDUSTRY

Despite many predictions over the years to the contrary, distillation remains the most important separation method in the chemical and petroleum industries. Distillation columns constitute a significant fraction of the capital investment in chemical plants and refineries around the world, and the operating costs of distillation columns are often a major part of the total operating costs of many processes. Therefore, the availability of practical techniques for developing effective and reliable control systems for efficient and safe operation of distillation systems is very important.

These considerations are the basis for the development of this book. We hope that the generic techniques and the industrial case studies presented in these chapters will help working engineers in their important task of controlling distillation columns.

Distillation columns present challenging control problems. They are highly multivariable and usually quite nonlinear. They have many constraints and are subject to many disturbances. Therefore, their control is not a trivial task.

1-2 BASIC CONTROL

This section presents some of the fundamentals of distillation column control that are applicable to almost all distillation columns. Some of the ideas may seem obvious and trivial, but unless they are kept in mind, it is often easy to make some fundamental mistakes that will result in very poor control.

We will concentrate on the basic, "plain-vanilla" simple distillation column shown in Figure 1-1. The nomenclature is shown on the figure. There is a single feed, and two products are produced. Heat is added in the partial reboiler and removed in the total condenser. Reflux is added on the top tray. Trays are numbered from the bottom.

1-2-1 Degrees of Freedom

In the context of process control, the degrees of freedom of a process is the number of variables that can or must be controlled. It is always useful to be clear about what this number is for any process so that you do not attempt to over- or undercontrol any process.

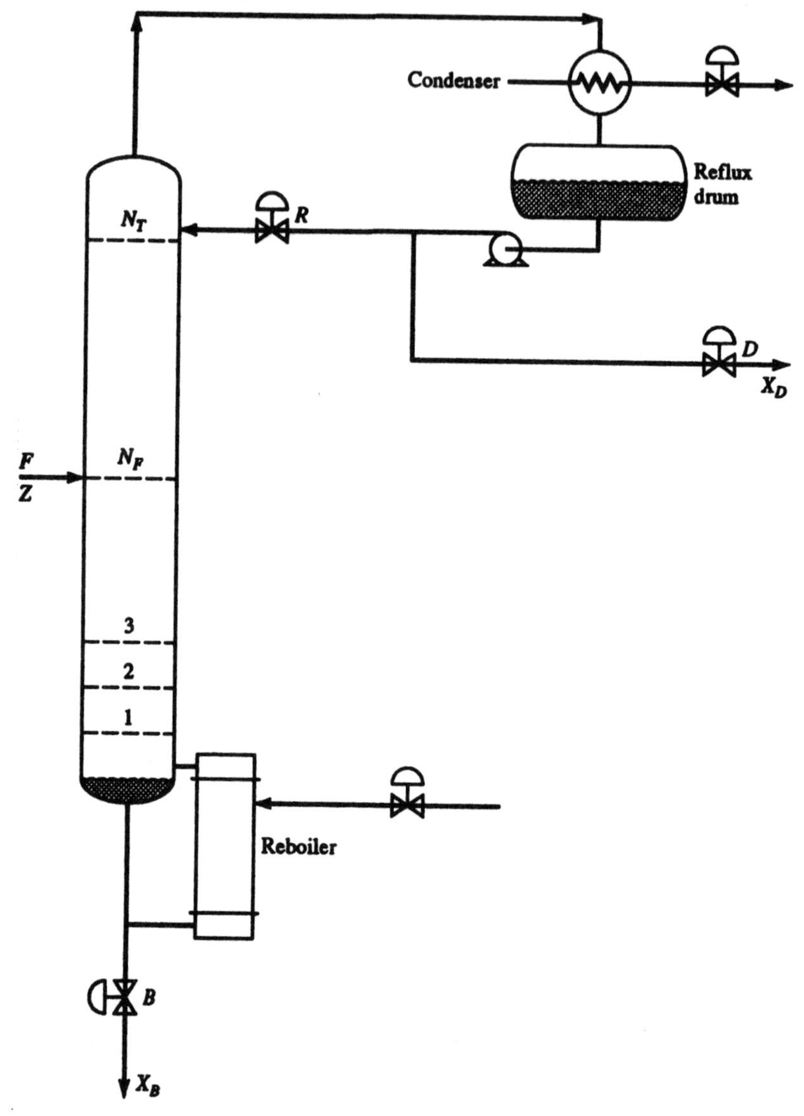

FIGURE 1-1. Basic column and nomenclature.

The mathematical approach to finding the degrees of freedom of any process is to total all the variables and subtract the number of independent equations. This is an interesting exercise, but there is a much easier approach. Simply add the total number of rationally placed control valves. The "rationally placed" qualification is to emphasize that we have avoided poorly conceived designs such as the placement of two control valves in series in a liquid filled system.

In Figure 1-1 we see that there are five control valves, one on each of the following streams: distillate, reflux, coolant, bottoms, and heating medium. We are assuming for the moment that the feed stream is set by the upstream unit. So this simple column has 5 degrees of freedom. But inventories in any process always *must* be controlled. Inventory loops involve liquid levels and pressures. In our simple distillation column example, this means that the liquid level in the reflux drum, the liquid level in the base of the column, and the column pressure must be controlled. When we say that the pressure is controlled, we do not necessarily mean that it is held constant. If we minimize it, we are "controlling" it.

If we subtract the three variables that must be controlled from 5, we end up with 2 degrees of freedom. Thus, there are two and only two additional variables that can (and must) be controlled in this distillation column. Notice that we have made no assumptions about the number or type of chemical components we are separating. So a simple, ideal, binary system has 2 degrees of freedom; a complex, multicomponent, nonideal distillation system also has 2 degrees of freedom.

The two variables that are chosen to be controlled depend on many factors. Some common situations are:

1. Control the composition of the light-key impurity in the bottoms and the composition of the heavy-key impurity in the distillate.
2. Control a temperature in the rectifying section of the column and a temperature in the stripping section of the column.
3. Control the flow rate of reflux and a temperature somewhere in the column.
4. Control the flow rate of steam to the reboiler and a temperature near the top of the column.
5. Control the reflux ratio (ratio of reflux flow to distillate flow) and a temperature in the column.

These examples illustrate that (a) only two things can be controlled and (b) normally at least one composition (or temperature) somewhere in the column must be controlled.

Once the five variables to be controlled have been specified (e.g., two temperatures, two levels, and pressure), we still have the problem of deciding what manipulated variable to use to control what controlled variable. This "pairing" problem is called determining the structure of the control system. It will be discussed in many chapters in this book.

1-2-2 Fundamental Variables for Composition Control

The compositions of the products from a distillation column are affected by two fundamental manipulated variables: feed split and fractionation. The feed split (or cutpoint) variable refers to the fraction of the feed that is taken overhead or out the bottom. The "fractionation" variable refers to the energy that is put into the column to accomplish the separation. Both of these fundamental variables affect both product compositions but in different ways and with different sensitivities.

Feed split: Taking more of the feed out of the top of the column as distillate tends to decrease the purity of the distillate and increase the purity of the bottoms. Taking more of the feed out of the bottom tends to increase distillate purity and decrease bottoms purity. On a McCabe–Thiele diagram, assuming a constant reflux ratio, we are shifting the operating lines to the left as we increase distillate flow and to the right as we decrease distillate flow.

Fractionation: Increasing the reflux ratio (or steam-to-feed ratio) produces more of a difference between the compositions of the products from the column. An increase in reflux ratio will reduce the impurities in both distillate and bottoms.

Feed split usually has a much stronger effect on product compositions than does fractionation. This is true for most distillation columns except those that have very low product purities (less than 90%).

One of the important consequences of the overwhelming effect of feed split is that it is usually impossible to control any composition (or temperature) in a column if the feed split is fixed, that is, if the distillate or bottoms flows are held constant. Any small changes in feed rate or feed composition will drastically affect the compositions of both products, and it will not be possible to change fractionation enough to counter this effect. A simple example illustrates the point. Suppose we are feeding 50 mol of component A and 50 mol of component B. Distillate is 49 mol of A and 1 mol of B; bottoms is 1 mol of A and 49 mol of B. Thus product purities are 98%. Now suppose the feed changes to 40 mol of A and 60 mol of B, but the distillate flow is fixed at a total of 50 mol. No matter how the reflux ratio is changed, the distillate will contain almost 40 mol of A and 10 mol of B, so its purity cannot be changed from 80%.

The *fundamental* manipulated variables (fractionation and feed split) can be changed in a variety of ways by adjustment of the control valves that set distillate, reflux, bottoms, steam, and cooling water flow rates. Fractionation can be set by adjusting reflux ratio, steam-to-feed ratio, reflux-to-feed ratio, and so forth. Feed split can be set *directly* by adjusting distillate or bottoms flows or *indirectly* by adjusting reflux or steam and letting level controllers change the product streams.

1-2-3 Pressure Control

Pressure in distillation columns is usually held fairly constant. In some columns where the difficulty of separation is reduced (relative volatility increased) by decreasing pressure, pressure is allowed to float so that it is as low as possible to minimize energy consumption. In any case it is important to prevent pressure from changing rapidly, either up or down. Sudden decreases in pressure can cause flashing of the liquid on the trays, and the excessive vapor rates can flood the column. Sudden increases in pressure can cause condensation of vapor, and the low vapor rates can cause weeping and dumping of trays.

There are a host of pressure control techniques. Figure 1-2 shows some common examples.

(a) *Manipulate coolant:* A control valve changes the flow rate of cooling water or refrigerant. If an air cooled condenser is used, fan speed or pitch is changed. Note that the liquid in the reflux drum is at its bubble point. Changes in coolant temperature are compensated for by the pressure controller.

(b) *Vent-bleed:* Inert gas is added or bled from the system using a dual split-ranged valve system so that under normal conditions both valves are closed. The reflux must be significantly subcooled in order to keep the concentration of product in the vent gas stream low. Changes in coolant temperature cause changes in reflux temperature.

(c) *Direct:* A control valve in the vapor line from the column controls column pressure. This system is only useful for fairly small columns.

(d) *Flooded:* Liquid is backed up into the condenser to vary the heat transfer area. Reflux is subcooled. If inerts are present, the condenser must be mounted horizontally to permit venting.

(e) *Hot-vapor bypass:* A blanket of hot vapor exists above a pool of cold liquid in the reflux drum. Heat transfer and condensation occur at this interface, and it is important to avoid disturbances that would change the interfacial area. The formation of ripples on the surface can result in the cold liquid "swallowing" the hot vapor and produce rapid drops in pressure. If the reflux drum is located below the condenser, a large vapor line connects the drum to the condenser inlet and a control valve in the liquid line below the condenser floods the condenser. If the reflux drum is located above the condenser, a control valve in the vapor line controls the pressure in the reflux drum (and indirectly the pressure in the column). Reflux is subcooled.

(f) *Floating pressure:* A valve-position controller is used to keep the cooling water

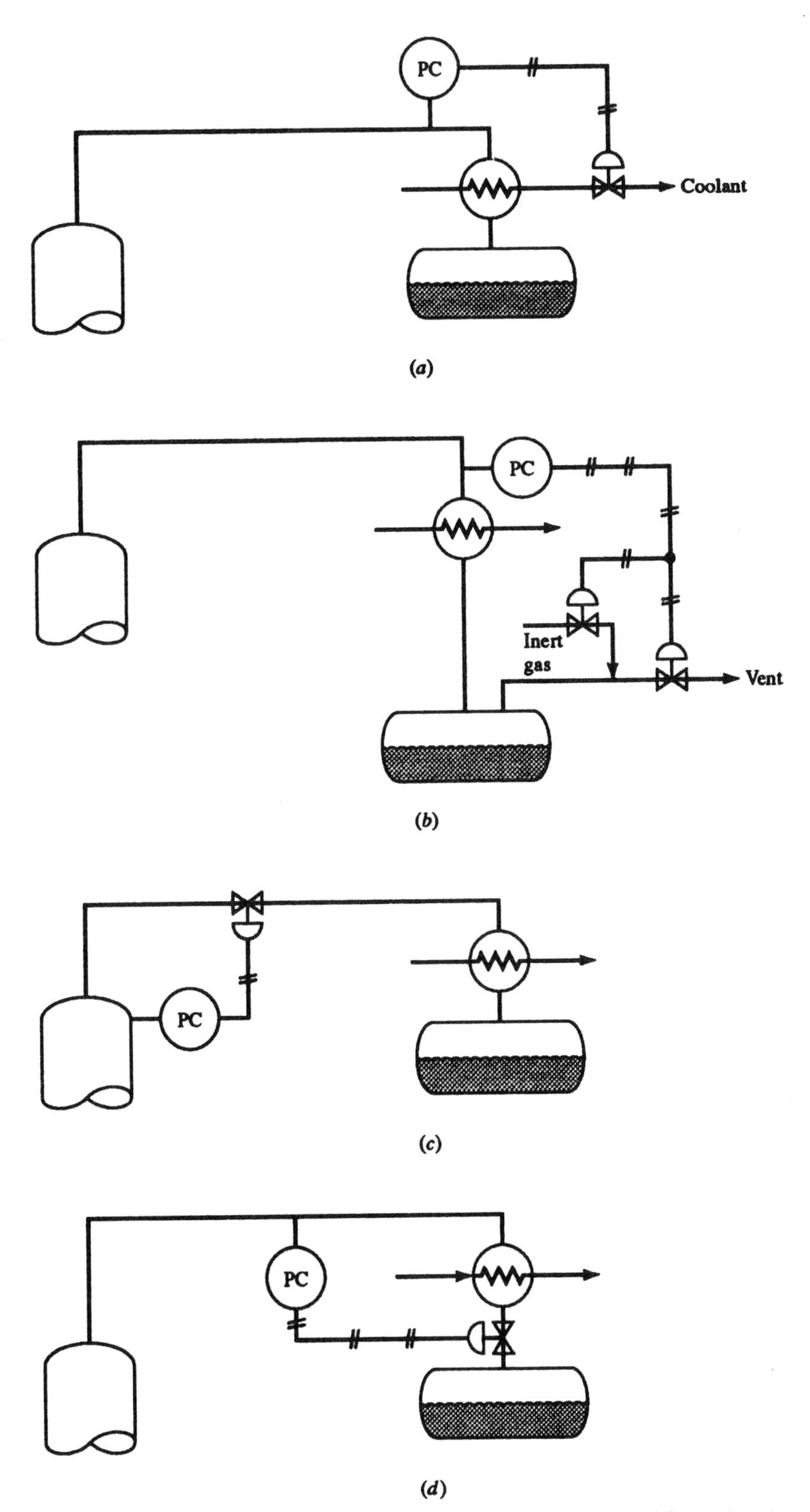

FIGURE 1-2. Pressure control schemes. (*a*) Coolant manipulation; (*b*) vent bleed; (*c*) direct; (*d*) flooded condenser; (*e*) hot vapor bypass; (*f*) floating pressure.

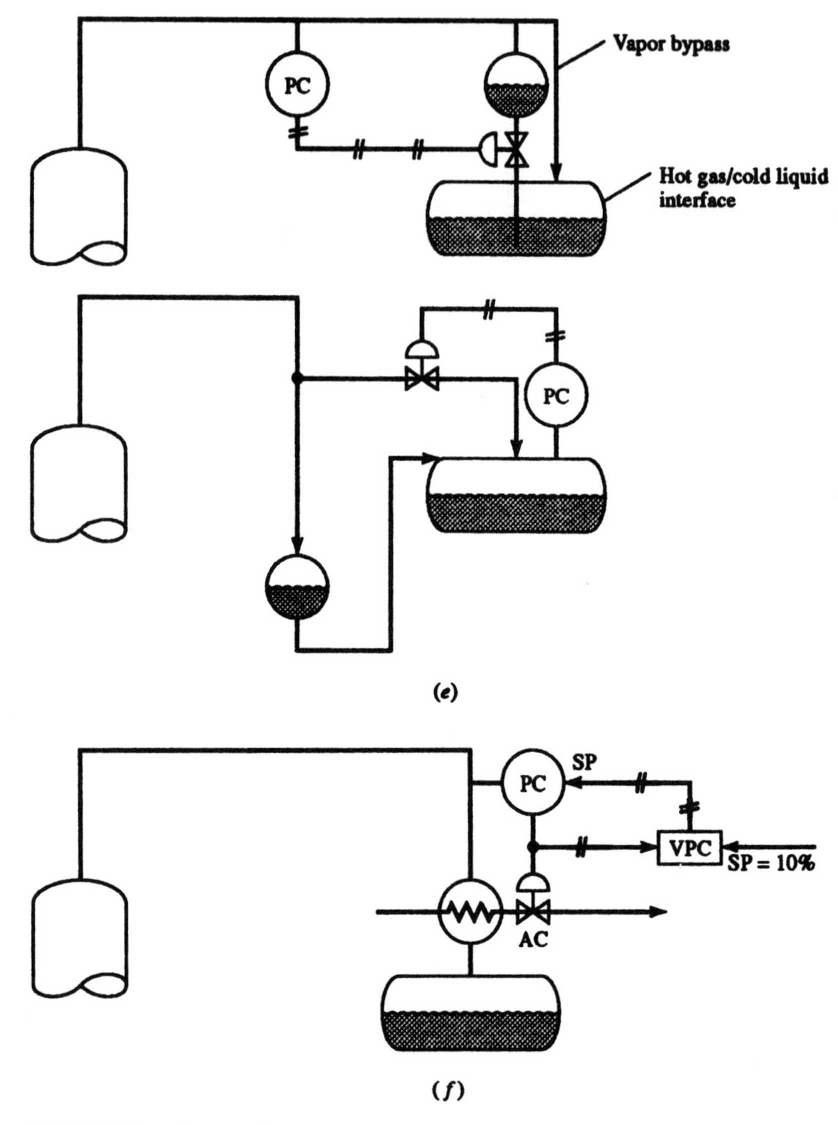

FIGURE 1-2. *Continued*

valve nearly wide open by slowly changing the setpoint of a fast pressure controller. Dual composition control (or some scheme to reduce reflux ratio as pressure is reduced) must be used to realize the energy savings.

1-2-4 Level Control

The two liquid levels that must be controlled are in the reflux drum and column base (or in the reboiler if a kettle reboiler is used). These levels are controlled in very different ways, depending on a number of factors.

If the column is part of a series of units in a plant, it is usually important from a plantwide control standpoint to use the liquid levels as surge capacities to dampen out disturbances. In such an environment, it is usually preferable to control base level with bottoms flow and reflux drum level with distillate flow, using a proportional-only controller. Proportional controllers minimize the flow disturbances that propagate to downstream units.

However, this is not always possible. One important example is a high reflux ratio column ($R/D > 5$). Using distillate to control level would require large changes in D for fairly small changes in R or V. Thus the disturbances would be amplified in the variations in the distillate flow rate. In columns with high reflux ratios, the reflux drum level should be controlled by reflux. This is an example of a very simple and useful rule that was proposed by Richardson (1990) of Union Carbide:

> Richardson's Rule: Always control level with the largest stream.

This heuristic works well in a remarkably large percentage of processes.

Application of the same logic would suggest that base level should be controlled by heat input in high reflux ratio columns. This is done in some columns, but you should be careful about potential problems with "inverse response" that can occur when this loop is closed. Increasing heat input will decrease base liquid level in the long term, but there may be a short transient period when the level momentarily *increases* instead of *decreases* with an increase in heat input. This inverse response can result from the "swell" effect in the reboiler and/or on the trays of the column itself.

In the reboiler, an increase in heat input can quickly increase the fraction of vapor. In a thermosiphon reboiler this can push liquid back into the base of the column, resulting in a momentary increase in the liquid level in the column base. In a kettle reboiler, the increase in vapor fraction causes the material in the reboiler to swell and more liquid flows over the outlet weir into the surge volume in the end of the reboiler. Therefore, the liquid level in this section momentarily increases.

On the trays in the column, an increase in vapor rate can cause the liquid on all the trays to swell. This will increase liquid rates flowing over the weir. Of course the increase in vapor rate will also increase the pressure drop through the trays, and this will require a higher height of liquid in the downcomer, which tends to decrease the liquid flow onto the tray. Thus, there are two competing effects: swell tends to increase internal liquid flows in the column and pressure drop tends to decrease them. If the former effect is larger (e.g., in valve tray columns where pressure drop does not change much with vapor flow), an increase in heat input can result in a momentary increase in liquid flow rate into the reboiler, which can make the base level increase for a short period of time.

This inverse response phenomena corresponds to a positive zero in the transfer function (i.e., a zero that is located in the right half of the s plane). Because the root-locus plot always goes to the zeros of the open-loop system transfer function, the presence of a positive zero pulls the root-locus plot toward the unstable region of the s plane. This clearly shows why inverse response results in poor control performance.

1-3 UNIQUENESS OF DISTILLATION COLUMNS

One of the often-quoted expression in the distillation area is: "There are no two distillation columns that are alike." This is true in the vast majority of cases. We will discuss briefly in succeeding text some of the reasons for this and its impact on both design and control.

It is true that there are many columns that fit into general classes:

Stabilizers: The distillate stream is only a small fraction of the feed and the relative volatility between the key components is large.

Superfractionators: The separation is difficult (relative volatilities < 1.2) and the columns have many trays and high reflux ratios (see Chapter 22).

High "K" columns: The separation is extremely easy, resulting in a very sharp temperature profile. This can lead to difficult control problems because of nonlinearity.

Sidestream columns: These complex configuration produce more than two products by removing sidestreams from the column. Sometimes sidestream strippers or sidestream rectifiers are also used.

The preceding partial list illustrates some of the generic columns that occur in distillation. However, despite the similarities, the individual columns within each of these classifications can be very different in both design and control for a number of reasons. Some conditions that vary from column to column are summarized:

Feed conditions: The number of components and the types of components in the feed can have a drastic effect on the type of column, condenser, and reboiler used and on the control system. The feed thermal condition can also strongly affect the column design and the control system. Subcooled liquid feed requires a different column and a different control structure that does superheated vapor feed.

Product specifications: High purity columns are more nonlinear and sensitive to disturbances. Larger feed tanks and temperature-composition cascade control systems may be required. If "on-aim" control is required (purity must be held within a narrow band), a blending system may be required.

Energy costs: Probably the major source of variability among distillation columns is the cost of energy in a particular plant environment. If excess low-pressure steam is available in the plant, the energy cost for any distillation column that can use it is very small.

For example, if you need a propylene–propane distillation column in a plant with excess low-pressure steam, you would probably build a conventional column with many trays (200) and high reflux ratio (14) and operate at 17 atm so that cooling water could be used in the condenser. The control system would maximize recovery of propylene by using as much heat input as possible (operating against a flooding or high-pressure constraint) and controlling only the distillate composition.

If, on the other hand, in your plant you have to produce incremental steam by burning more fuel, you would probably build a vapor recompression system with fewer trays (150), lower reflux ratio (11) and lower pressure (11 atm). The control system would try to keep the column at the optimum distillate and bottoms compositions (dual composition control) that represent the best tradeoff between compressor horsepower costs and propylene recovery.

1-4 INTERACTION BETWEEN DESIGN AND CONTROL

As with any process, in distillation systems there are interesting and challenging interactions between the basic design of the process, which historically has been based on only steady-state economics, and its controllability. Discussion of some examples follows.

1-4-1 Increasing Column Size (Trays and Diameter)

Traditional designs yield columns that operate at reflux ratios of 1.1 to 1.2 times the minimum reflux ratio. Usually this corresponds to a column with about twice the minimum number of trays (which can be calculated from the Fenske equation for constant relative volatility systems). Column diameters and heat exchangers are designed to handle flow rates that are 10 to 20% above design.

These designs leave little excess capacity that can be used to handle the inevitable disturbances that occur. It is usually good engineering practice to increase the number of trays by 10% and increase the capacity of the column, reboiler, and condenser by 20% to make it easier to control the column.

1-4-2 Holdups in Column Base and Reflux Drum

From a steady-state economic standpoint, these holdups should be as small as possible so as to minimize capital investment. Increasing base holdup means that the column shell must be longer and requires that the column be positioned higher above grade. Both of these effects increase capital costs. Bigger reflux drums mean high costs for both the drum and its supporting superstructure. Large holdups of liquids may also be undesirable because of safety considerations (if the material is toxic, explosive, thermally sensitive, etc.).

However, from the standpoint of dynamics and control, the ability of the column to ride through disturbances is usually improved by having more liquid holdup available. Thus there is a conflict between steady-state economics and controllability, and an engineering trade-off must be made.

Liquid holdup times of about 5 min are fairly typical. This holdup time is based on the total material entering and leaving the column base or reflux drum, not just the stream that is being used to control the level. If the column is very large, subject to many large disturbances, or forms part of a series of operation units, somewhat larger holdup times may be appropriate. If small liquid holdups are required because of safety or thermal degradation considerations, somewhat smaller holdup time may be appropriate.

But remember, one of the most common reasons why columns give control difficulties is insufficient liquid holdups, so make sure the process designers do not squeeze too hard for the sake of a few dollars. Safety and thermal degradation problems can often be handled by using external tankage for surge capacity so that liquid levels will not be lost or flow rate to downstream units will not vary excessively during disturbances.

1-4-3 Effects of Contacting Devices

The type of contact devices used in the column can significantly affect the dynamics of the column. Most columns with trays have more liquid holdup and therefore slower dynamic responses than columns with packing. The structured packing that has become more widely used in recent years in processes where pressure drop is important has faster dynamics than tray columns. These columns respond more quickly to changes in manipulated variables, but they also respond more quickly to disturbances. These considerations should be kept in mind when designing control systems.

1-4-4 Sensors

It is important for the operation of the column to provide adequate sensors: enough thermocouples located so that the temperature profile can be determined, differential pressure measurements to sense flooding, flow measurements of all streams, and so forth. It is important to include these measurements in the original equipment design when it is quite inexpensive and easy. Having to make field modification to an operating column can be very expensive and may take a long time to accomplish.

1-5 SPECIAL PROBLEMS

Throughout the chapters in this book there will be much discussion and many examples of various types of columns and systems. Several commonly encountered situations that present special control problems are discussed briefly in the following text.

1-5-1 High-Purity Products

As noted earlier, distillation columns that produce high-purity products are very nonlinear and sensitive to disturbances. Very small changes in the feed split can produce drastic changes in product concentrations at steady state. The theoretical linear time constants of these high-purity column are very large.

These columns can be effectively controlled if disturbances can be detected quickly and dynamic corrective action be taken in time to keep the column near the desired operating point. The system is somewhat analogous to the "inverted pendulum" process that is studied by mechanical engineers (i.e., balancing a stick on the palm of your hand): If the position of the stick can be detected quickly, a vertical position can be maintained.

Thus it is vital that measurement and sampling times be minimized. It is also important to slow down the effects of disturbances to these columns. Large upstream feed tanks may be required to filter disturbances.

1-5-2 Small Temperature Differences

Difficult separations occur when boiling point differences between key components are small. This results in a temperature profile in the column that is quite flat and can make the use of temperature to infer composition quite ineffective. Sometimes more sensitivity can be obtained by using differential or double-differential temperatures, but nonmonotonic behavior can sometimes occur, which can crash the system. If the separation is quite difficult (relative volatilities less than 1.2), temperature becomes essentially useless and some type of composition analyzer is required to control composition.

1-5-3 Large Temperature Differences

Extremely easy separation (e.g., peanut butter and hydrogen) yields temperature profiles that are very sharp and lead to control problems because of high process gains and nonlinearity (saturation of the measurement signal). This problem can be handled effectively by using a "profile position" control system: several (four to five) temperatures are measured at tray locations in the column below and above the tray where the temperature break occurs under steady-state design conditions. These temperatures are averaged, and the average temperature is controlled, typically by manipulation of heat input. This technique reduces the process gain and avoids the measurement saturation problem that would be experienced if only a single tray temperature were measured.

1-5-4 Gravity-Flow Reflux

Instead of pumping reflux back to the column from a reflux drum located somewhat above grade (the most common configuration), it is sometimes desirable to locate the condenser and reflux drum above the column in the superstructure. Then gravity can be used to overcome the pressure differential between the reflux drum and the top of the column. This avoids a pump, which can be a real advantage in highly corrosive, toxic, or dangerous chemical systems.

However, the design of the gravity-flow reflux system requires some explicit consideration of dynamics. The condenser must be located high enough above the column so that there is enough head to get liquid back into the column in the worst-case situation. The head must overcome pressure drop through the vapor line, the condenser, the liquid line, the reflux flow measurement device, and the reflux control valve. Because these pressure drops vary as the square of the flow rates, a 30% increase in flow increases the pressure drops through the fixed resistances by a factor of $(1.3)^2 = 1.69$. This 69% increase in pressure drop can only come from some combination of an increase in liquid height and a decrease in control valve pressure drop.

Gravity-flow reflux systems are often not designed with control considerations in mind, and insufficient height is provided between the condenser and the top of the column. This can lead to very poor control.

1-5-5 Dephlegmators

An extension of the gravity-flow-reflux scheme is to eliminate not only the pump but also the external condenser. The condenser is mounted directly on the top of the column. These systems are particularly popular in low-temperature systems where heat losses must be minimized.

The usual installation has the vapor flowing upward through the vertical tubes in the condenser. The condensed liquid flows countercurrent down the tubes and refluxes the column. The uncondensed vapor leaving the top of the condenser is the vapor distillate product.

From a control engineer's point of view, these dephlegmator systems can be nightmares. The lack of any liquid holdup makes the system very sensitive to disturbances. For example, a slug of noncondensibles can drop the rate of condensation drastically and the reflux flow will drop immediately. Fractionation on the trays in the column is immediately adversely affected. Columns do not recover quickly from changes in liquid flow rates because they propagate down the column fairly slowly (3 to 6 s per tray).

I recommend that a total trap-out tray be installed under the condenser to provide reflux surge capacity. This type of tray has a chimney to let the vapor pass through to the condenser, and should be designed with sufficient depth so that about 5 min of liquid holdup is provided. Reflux from this reflux-surge tray to the top of the column should be controlled by an external control valve and the flow rate measured. Clearly the reflux-surge tray has to be located high enough to provide the required pressure drops.

1-6 CONCLUSION

This chapter has attempted to present some of the basic principles of distillation column control. Some of the problems and constraints have been briefly discussed. Many of these topics will be treated in more detail in later chapters.

Reference

Richardson, R. (1990). 1990 Lehigh Distillation Control Short Course. Lehigh, PA: Lehigh University.

2

Historical Perspective

Page S. Buckley

Process Control Consultant

2-1 INTRODUCTION

History is normally approached in a chronological fashion, starting at an earlier date and proceeding stepwise and logically to a later date. To discuss distillation control in this manner is difficult. Most important developments have been evolutionary, occurring gradually over a period of time. Consequently, we will be discussing this subject primarily in terms of eras, rather than specific dates. Furthermore, the time period of most interest for distillation control history is very short, about 40 years. As every historian knows, the closer one comes to the present, the more difficult it is to achieve a valid perspective. Bias worsens when the historian has been a participant in the events under discussion.

2-2 WHAT IS CONTROL?

Before talking about distillation control, let us define the meaning of *control*, at least as it is used in this chapter. A chemical plant or refinery must produce a product or products that meet certain quality specifications. Preferably, there should be enough inventory to ship upon receipt of orders, although this is not always feasible, and some orders call for shipping a certain amount continuously or intermittently over a period of time. Plant operation must meet production requirements while observing certain constraints of safety, environmental protection, and limits of efficient operation. We do not sell flow rates (usually), liquid levels, temperatures, pressures, and so forth. Contrary to popular belief, these variables need not be held constant; primarily they should be manipulated to achieve the following operational objectives:

Material balance control: An overall plant material balance must be maintained. Production rate must, on the average, equal rate of sales. Flow rate changes should be gradual to avoid upsetting process equipment.

Product quality control: Final product or products must meet sales specifications. Product quality needs to be constant at only one point in a process—final inventory.

Constraints: If there is a serious malfunction of equipment, interlocks may shut the plant down. Otherwise, override controls may nudge a process away from excessive pressure or temperature, excessively high or low liquid levels, excessive pollutants in waste streams, and so on.

An important implication of the preceding is that for optimum operation, one must let all variables vary somewhat except final product quality, and, sometimes, even that may vary within prescribed limits.

Classical control theory emphasizes (a) rapid response to setpoint changes and (b) rapid return to setpoint in the face of disturbances. These performance requirements really apply only to product quality controls and to some constraint controls. Some of the successes of computer control have resulted from restoring control flexibility that was originally lost by excessive reliance on fixed setpoint control of individual variables.

2-3 COLUMN DESIGN METHODS

The reader is presumably familiar with steady-state column design, so little more will be said about it.

In the 1930s and 1940s, what chemical engineers call "unit operations theory" made tremendous strides. The objectives were twofold:

1. To ensure that design of piping, heat exchangers, distillation columns, and so forth, would have at least flow sheet capacity.
2. To ensure that this equipment would be as little overdesigned as possible to minimize capital investment.

For distillation columns, the McCabe–Thiele procedure for estimating the required number of theoretical trays was published in 1925. It did not, however (at least in my experience), achieve wide usage until the early 1940s. During World War II a group of us trying to increase production rates in a heavy chemicals plant found that existing columns designed before 1930 occasionally had tremendous overdesign factors: they were 10 to 15 times larger than necessary.

By 1948, we were using a safety factor of 2 for designing distillation columns and heat exchangers, and 25 to 50% for piping.

Today, overdesign safety factors are very small. Columns are designed to run much closer to flooding and thermosyphon reboilers often operate close to choked-flow instability. Older columns with more safety factor in design, and particularly with bubble-cap trays, were easier to control.

2-4 COLUMN TRAY AND AUXILIARY DESIGN

2-4-1 Tray Design

Circa 1950 we began to design columns with sieve trays instead of bubble-cap trays. For a given load, a sieve tray column is smaller and cheaper. However, sieve tray columns have more limited turndown: about 2 : 1 instead of about 8 : 1. In addition they are subject to weeping and dumping. These characteristics, together with tighter design practices, increase control problems.

More recently, valve trays have become popular. They have more turndown capabilities than sieve trays, say 4 : 1, but have the same potential for weeping and dumping. In addition, they commonly have another problem: inverse response.

As noted by Rijnsdorp (1961), with an increase in boilup some columns show a momentary increase in internal reflux, followed eventually by a decrease in internal reflux. This is termed *inverse response* because it causes a temporary increase in low boiler composition at the bottom of the column, followed by an eventual decrease. The mechanism was elucidated by Buckley, Cox, and Rollins (1975), who found that most sieve tray columns demonstrated this effect at low boilup rates, whereas most valve tray columns demonstrated it over the entire turndown. Thistlethwaite (1980) also studied inverse response and worked out more details. Composition control via boilup at the base of a column is vastly more difficult when a column is afflicted with inverse response. Momentarily, the controller seems to be hooked up backwards. For the control engineer this can be even worse than deadtime.

Since the mid-1980s packed columns, particularly those with structured packing, have been used more extensively. They have about the same turndown as valve trays, say 4:1, but do not suffer from inverse response. On the other hand, changes in reflux flow can change column pressure drop significantly.

2-4-2 Reboiler Design

Thermosyphon reboilers, particularly of the vertical type, are popular because of low cost. Several kinds of dynamic problems, however, can result from their use (Buckley, 1974). One is choked flow instability (Shellene et al., 1967), which occurs when the reboiler is sized too snugly with excessive heat flux.

Another problem is "swell." At low heat loads there is little vapor volume in the tubes; at high heat loads there is much more vapor in the tubes. Hence, an increase in steam flow will cause liquid to be displaced into the column base, causing a temporary increase in base level. If base level is controlled by steam flow, the controller momentarily acts like it is hooked up backwards. When "swell" is accompanied by inverse response of reflux, controlling the base level via heating medium may be impossible.

Recent trends in thermosyphon design seem to have had at least one beneficial effect: critical ΔT is rarely a problem. Prior to 1950, it could be observed fairly often. This improvement apparently is due to much higher circulating rates.

2-4-3 Flooded Reboilers

Particularly with larger columns, it is common to use steam condensate pots with level controllers instead of traps. In recent years some reboilers have been designed to run partially flooded on the steam side.

The large steam control valve may be replaced by a much smaller valve for steam condensate. Also, it is usually easier to measure condensate flow rather than steam flow. Flooded reboilers are, however, more sluggish than the nonflooded ones, typically by a factor of 10 or more. Flooded reboilers sometimes permit use of low pressure, waste steam. There may be little pressure drop available for a steam supply valve.

2-4-4 Column Base Designs

The size of the column base, the design of the internals (if any), and the topological relationship to the reboiler are all important factors in column control. Column base holdups may range from a few seconds to many hours (in terms of bottom product flow rate). If there are no intermediate holdups between columns or process steps, the column base must serve as surge capacity for the next step.

Originally, column bases were very simple in design, but in recent years elaborate internal baffle schemes have often been used. The objective is to achieve one more stage of separation. Many of these schemes make column operation more difficult and limit the use of the holdup volume as surge capacity. Overall, their economics is questionable; increasing the number of trays in a column by one is relatively cheap.

Elimination of intermediate tanks can save a lot of money, both in fixed investment (tanks) and working capital (inventory).

2-4-5 Overhead Design

In the chemical industry, particularly for smaller columns, it is common to use gravity flow reflux and vertical, coolant-in-shell condensers. Instrumentation and controls can be much simpler than with pumped-back reflux. Regardless, however, of condenser design, gravity flow reflux systems have a dynamics problem if distillate is the controlled flow and reflux is the difference flow. This is the well known problem of "reflux cycle." Its mechanism and several corrective measures are discussed in a paper by Buckley (1966).

When horizontal, coolant-in-tube condensers are used, control is sometimes simplified by running the condensers partially flooded. This technique is especially useful when the top product is a vapor. Flooded condensers seem to be becoming more popular, but like flooded reboilers, are more sluggish. For both, however, we have worked out mathematical models that enable us to calculate the necessary derivative compensation (Buckley, Luyben, and Shunta, 1985).

2-5 INSTRUMENTATION

Before automatic controls and automatic valves became prevalent, control was achieved by having operators observe local gages and adjust hand valves to control process variables to desired values. According to stories told to me by old timers, some applications required such careful attention and frequent adjustment that an operator would be kept at a single gage and hand valve. To help keep the operator alert, a one-legged stool was provided.

Gradually, this practice diminished as manufacturers developed automatic controls in which measurement, controller, and valve were combined in one package. Sometimes the valve itself, usually powered by air and a spring and a diaphragm actuator, was a separate entity. Slowly, the practice of separating measurement, controller, and valve evolved.

The development of measurement devices that could transmit pneumatic or electric signals to a remote location was well underway by 1940. This permitted the use of central control rooms and, together with more automatic controls, a reduction in the required number of operators.

Prior to 1940 most controllers were either proportional-only or had manual rather than automatic reset. A few vendors, however, had begun to use primitive versions of automatic reset.

By 1945, the chief instruments available for distillation control were:

- *Flow:* Orifice with mercury manometer plus pneumatic transmission, or rotameters.
- *Temperature:* Thermocouple or mercury-in-bulb thermometers.
- *Pressure:* Bourdon tube plus pneumatic transmission.
- *Control valve:* Air-operated with spring-and-diaphragm or electric motor operator. Most valves, except for small ones, were of the double-seated variety; small ones were single-seated. Valve positioners occasionally were used.

By 1950 the picture had begun to change dramatically. The so-called force balance principle began to be used by instrument designers for pneumatic process variable transmitters and controllers. This feature, together with higher capacity pneumatic pilots, provides more sensitivity, freedom from drift, and speed of response. Mercury-type flow transmitters gave way to "dry" transmitters and mercury-in-bulb temperature transmitters were replaced by gas-filled bulb transmitters.

Piston-operated valves with integral valve positioners appeared on the market. The practicality of larger sized single-seated valves permitted less expensive valves for a given flow requirement and also provided increased sensitivity and speed of response.

With improved actuators and positioners, reliability improved, and many users abandoned control valve bypasses except for a few cases. This saves money (investment) and in many cases reduces maintenance.

In 1974, the Instrument Society of America (ISA) published new standards for calculating flow through control valves. This was one of the most valuable developments of the past 25 years. It permitted much more accurate valve sizing, even for flashing, cavitation, or incipient cavitation. In a parallel development, valve vendors began to make available more accurate plots or tables relating C_v, the valve flow coefficient, to valve stem position. Together with nonlinear functions available in new microprocessor controls, these permitted the design of valve flow compensators (Buckley, 1982), which in turn, permits much more constant stability over a wide range of flows.

New pneumatic controllers with external reset feedback appeared, which permitted the use of anti-reset windup schemes. Circa 1952, one vendor marketted an extensive line of pneumatic computing relays, which permitted addition, subtraction, and multiplication by a constant, high- and low-signal selection, and so-called impulse feedforward. Decent pneumatic devices for multiplying or dividing two signals did not appear until the mid-1960s. Taken together, these various devices permitted on-line calculation of heat and material balances. Calculation of internal reflux from external reflux and two temperatures also became practical, and simple pressure-compensated temperature schemes became feasible.

The improved ΔP transmitters, originally developed for orifice flow measurements, permitted sensitive specific gravity and density measurements. They could also be used to measure column ΔP, increasingly important for columns designed to run closer to flooding.

Parallel to developments in conventional instruments, there was a much slower development in on-line analyzers. In the late 1940s, Monsanto developed an automatic refractometer, which was used in a styrene monomer distillation train. In the early 1950s, several prototypes of on-line infrared and optical analyzers were available. Problems with reliability and with sample system design plus high cost severely limited their applications. pH measurements, on the other hand, expanded more rapidly, as the result of improved cell design, better cable insulation, and high input impedance electronic circuitry.

The first analog electronic controls appeared in the mid-1950s. They had good sensitivity and speed of response, but due to the use of vacuum tubes, had somewhat high maintenance. In addition, they were far less flexible than pneumatics. Subsequent switching to solid state circuitry improved reliability, but flexibility still suffered. It was not until the early 1970s that one manufacturer produced an electronic line with all of the computation and logic functions available in existing pneumatics. In the meantime, the cost competitiveness of pneumatics was improved by a switch from metallic transmission tubing to black polyethylene tubing (outdoors) and soft vinyl tubing (in the control room).

In the late 1970s a potentially momentous change got underway—the introduction of so-called distributed digital control systems. By using a number of independent microprocessors, manufacturers were able to "distribute" the control room hardware and achieve better reliability than would be obtained with a single, large time-shared computer (avoid "all eggs in one basket"). Early versions were quite inflexible compared with the best pneumatic and electronic hardware, but this situation has now changed. Some of the newest systems have more flexibility than any analog system and lend themselves well to the design of advanced control systems. Transmitters and control valves in the field usually may be either pneumatic or electronic.

No discussion of instrumentation is complete without some mention of computers. The first process control computer was put on the market in the early 1950s. A well-publicized application was a Texaco refinery, where some optimization was performed. Another optimization application of that era was a Monsanto ammonia plant at Luling, LA. Other machines came on the market in the next few years, but all of the early machines suffered from lack of reliability, slow operation, and insufficient memory. In the 1960s and 1970s applications were chiefly as data loggers and supervisory controls. Production personnel found that CRT displays, computer memory, and computer printouts were a big improvement over conventional gages and recorders. As the memory and speed of computers have increased and as the expertise of users has improved, we have seen a resurgence of interest in optimization.

The latest development in instrumentation began in mid-1983 when one vendor put a line of "smart" transmitters on the market. These featured built-in microprocessors to permit remote automatic calibration, zeroing, and interrogation. These features, together with reduced drift and sensitivity to ambient conditions, and with more accuracy and linearity, improve plant operation. They also facilitate and reduce maintenance. However, test data I have seen on such instruments indicate that they are slower (poorer frequency response) than their pneumatic counterparts. Eventual extension to instruments with multiple functions such as temperature and pressure is foreseen. Present indications are that within the next several years a number of manufacturers will put "smart" instruments on the market.

2-6 CONTROL SYSTEM DESIGN METHODS

Prior to 1948, distillation control systems as well as other process control systems were designed by what might be called the instrumentation method. This is a qualitative approach based principally on past practice and intuition. Many "how-to" papers have appeared in the literature. A good summary of the state-of-the-art in 1948 is given by Boyd (1948).

In the period 1948 to 1952, there appeared a large number of books based on control technology which had been developed for military purposes during World War II. Commonly the title mentioned "servomechanisms theory." Two of the best known books are by Brown and Campbell (1948) and Chestnut and Mayer (1951). A wartime text by Oldenbourg and Sartorius (1948) was translated from German into English and became well known. Although the theory required significant modifications to adapt it to process control, for the first time it permitted us to analyze many process control situations quantitatively.

The technique involves converting process equations, including differential equations, into linear perturbation equations. By perturbations we mean small changes around an average value. Linear differential equations with constant coefficients are easily solved by use of the Laplace transformation. It is usually easy to combine a number of system equations into one. This equation, which relates output changes to input changes, is called the *transfer function*. Time domain solutions are readily calculated, and if the uncontrolled system is stable, the transfer function may be expressed in terms of frequency response.

The transfer function approach permits convenient study of system stability, permits calculation of controller gain and reset time, and permits calculation of system speed of response to various forcing functions. In the precomputer era, these calculations were made with a slide rule and several graphical aids. Today, the transfer function techniques are still useful but the calculations usually may be made more rapidly with a programmable calculator or a computer. Early applications were in flow, pressure, and liquid level systems. Although frequency response methods of calculation were most commonly used, some users prefer root locus methods.

Throughout the 1950s and 1960s the transfer function approach was used extensively for piping, mixing vessels, heat exchangers, extruders, weigh belts, and, occasionally, distillation columns. Two books that appeared in the 1950s illustrated the application of transfer function techniques to process control (Campbell, 1958; Young, 1957).

The relatively simple modelling techniques developed earlier were of limited use for systems with significant nonlinearities or those describable by either partial differential equations or by a large number of ordinary differential equations. Gould (1969) explored some more advanced modelling techniques for such systems.

The limitations of the simple transfer function approach for distillation columns led some workers to try another approach—computer simulation. The work of Williams (1971) and Williams, Harnett, and Rose (1956) in applying this approach is well known. Originally, simulations were mostly run on analog computers. Then, when improved digital computers and FORTRAN programming became available, there was a shift to digital techniques. Basically, simulation consists of writing all of the differential equations for the uncontrolled system and putting them on the computer. Various control loops can then be added, and after empirically tuning the controllers, the entire system response can be obtained. In some cases real controllers, not simulated ones, have been connected to the computer.

In 1960, Rippin and Lamb (1960) presented a combined transfer function–simulation approach to distillation control. The computer simulation was used to develop open loop transfer functions. Then standard frequency response methods were used to calculate controller gain and reset as well as desired feedforward compensation. In our own work, we have found this combination approach to be quite fruitful.

Since about 1980, there has been enormous progress in small or personal computers (PCs) accompanied by a more gradual evolution in software. The machines have more memory and speed, and both machines and software are more user friendly. Simple distillation columns are readily modeled and programmed; see, for example, the program illustrated in Luyben (1990). Modelling, including newer approaches, is discussed by other authors in this book.

As the speed and memory of technical computers increased, distillation models have become more detailed and elaborate. Tolliver and Waggoner (1980) have presented an excellent discussion of this subject.

Many newer control techniques have been devised. Among these are "internal model control" and "dynamic matrix control." These and other techniques are summarized and reviewed by Luyben (1990). There has been some application to distillation control, as discussed in Chapter 12.

Although most process measurements are continuous, some are not, as, for example, once per day sampling and laboratory analysis of column product streams. This led to an interest in the techniques of sampled-data control. This permits the quantitative design of discontinuous control systems. There have been few applications to distillation control in spite of the fact that many composition measurements are discontinuous. One application has been the use of Shannon's data reconstruction theorem to establish the relationship between a specified sampling frequency and required process holdup, or between a specified holdup and required sampling frequency.

These newer techniques require a lot of training beyond that required for transfer function techniques, and industrial applications have so far been limited. There is no question, however, that there exist ample incentives and opportunities for this newer control technology.

An interesting trend in recent years has been toward the development of standardized programs for control system design. This is commonly called computer aided design (CAD). Although originally aimed at larger machines, many useful programs exist for small computers and even programmable calculators. The portability and permanent memory features of many of the latter render them very useful in a plant or maintenance shop.

Since about 1970, there has been increasing interest in newer approaches to control.

The incentives have been twofold:

1. Because flow sheet data are often incomplete or not exact, it is desirable to measure process characteristics of the real plant after startup. Furthermore, process dynamics change as operating conditions change. The increasing availability of on-line computers permits the use of "identification" techniques. Information thus obtained may be used to retune controllers either manually or automatically. If a column is not equipped with analyzers, identification techniques may be used to deduce compositions from a combination of temperature, pressure, and flow measurements. A particular approach that was developed by Brosilow and Tong (1978) is termed *inferential control*. For binary or almost binary distillation, pressure-compensated temperature usually suffices.
2. Many processes are plagued by interactions. There are a variety of techniques for dealing with this that may be lumped under the heading "modern control theory" (Ray, 1981). That aspect relating to time-optimal control will likely be helpful in future design of batch distillation columns. Implementation of this technology usually requires an on-line computer.

Some other methods of dealing with interactions include relative gain array (RGA), inverse Nyquist array (INA), and decouplers. See Luyben (1990) for a good short summary.

2-7 PROCESS CONTROL TECHNIQUES

In 1945 there existed no overall process control strategy. As one process design manager put it, "instruments are sprinkled on the flow sheet like ornaments on a Christmas tree." Not until 1964 was there published an overall strategy for laying out operational controls from one end of a process to the other (Buckley, 1964). This strategy, as mentioned earlier, has three main facets: material balance control, product quality control, and constraints. What kind of controls do we use to implement them?

Material balance controls are traditionally averaging level and averaging pressure controls. Although the basic concept dates back to 1937, averaging controls are still widely misunderstood. The original theory was expanded in 1964 (Buckley, 1964) and again quite recently (Buckley, 1983). To minimize required inventory and achieve maximum flow smoothing we use quasilinear or nonlinear proportional integral (PI) controllers cascaded to flow controllers. If we can use level control of a column base to manipulate bottom product flow and level control of the overhead receiver to adjust top product flow, then we can usually avoid the use of surge tanks between columns or other process steps. This saves both fixed investment and working capital.

Product quality controllers are usually also of the PI type. Overhead composition is usually controlled by manipulating reflux and base composition by manipulating heat flow to the reboiler. It is also increasingly common to provide feedforward compensation for feed rate changes. This minimizes swings in boilup and reflux, and in top and bottom compositions.

One of the most important developments in control techniques is that of variable configuration and variable structure controls. Although it is much more common to use fixed configuration and fixed structures, in reality either or both should change as process conditions change. For example, the steam valve for a distillation column reboiler may, depending on circumstances, respond to controllers for:

Steam flow rate
Column Δp
Column pressure
Base temperature
Column feed rate
Column base level
Column bottom product rate

The seven variables listed may also exert control on five or six other valves. We call this *variable configuration*.

For level control, the quasilinear structure referred to earlier consists normally of PI control, but proportional-only control is used when the level is too high or too low. We call this *variable structure*. To accomplish this automatically, it was proposed in 1965 (Buckley, 1968) to use a type of control called *overrides*. In ensuing years a number of other publications, (Buckley and Cox, 1971; Buckley, 1969b; Cox, 1973) described the applications to distillation. The math and theory are presented in a two-part article (Buckley, 1971).

Fairly recently, there has been a growing interest in "self-tuning regulators." One of the most prolific contributors has been Åström (1979) in Sweden. So far, applications to industrial distillation columns are not known, but they have been tried on pilot columns.

Although little has been published, there has been a certain amount of work with nonlinear controllers. Unlike linear theory, which has a cohesive, well-developed methodology, nonlinear theory is essentially a bag of tricks. The best approach I have found is to run a large number of simulations with different algorithms and for various types and sizes of forcing functions.

For averaging level control we found (Buckley, 1986) that with linear reset and gain proportional to the error squared, we got flow smoothing far superior to that achievable with any linear controller, whether P or PI, and experiences where P-only controllers were replaced with this type of controller confirmed simulation results.

For composition control, we tried nonlinear gain to filter out noise. Sometimes it worked well, sometimes not.

Our experience with trial-and-error tuning of nonlinear controllers in a plant has been negative. These applications should be simulated. For those interested in nonlinear control, we recommend a book by Oldenburger (1966).

2-8 INFLUENCE ON DISTILLATION CONTROL

Some years ago I was present when one of our most experienced distillation experts was asked "What is the best way to control a column?" His answer: "Gently." When approaching column control one should keep in mind that conditions in a column cannot be changed rapidly. Rapid changes in boilup or rate of vapor removal, for example, may cause momentary flooding or dumping. The developments since 1945 have therefore not been aimed at rapid control but rather at smoother operation with better separation at lower cost.

If we now look back at 1945 and move forward we can see how the various developments previously discussed have interacted to influence distillation control.

1. *Switch to sieve and valve trays and tighter column design.* These have encouraged the use of minimum and maximum boilup overrides. A high Δp override on steam is now universally employed by some engineering organizations. We also frequently provide maximum and minimum overrides for reflux.
2. *Increased use of packed columns.* As mentioned earlier, packed columns, particularly those with structured packing, have become more popular.
3. *Tighter design of heat exchangers, especially reboilers.* To prevent choked flow instability we sometimes provide maximum steam flow limiters.
4. *Increased use of side-draw columns.* One side-draw column is cheaper than two conventional columns with two takeoffs. Turndown is less, however, and the column is harder to control (Buckley, 1969a; Doukas and Luyben, 1978; Luyben, 1966). Such columns usually require more over-

rides, particularly to guarantee adequate reflux below the side draw or draws.

5. *Energy crunch of 1973.* This has had two major effects:
   a. Increased energy conservation of conventional columns. After improved insulation, one of the big items has become double-ended composition control. A column runs more efficiently and has more capacity if composition is controlled at both ends. Another consequence has been more interest in composition measurements instead of just temperature. Temperature is a nonspecific composition measurement except for a binary system at constant pressure. This, together with the common practice of having temperature control of only one end of a column, usually leads production personnel to run a column with excess boilup and reflux to ensure meeting or exceeding product purity specifications. This wastes steam and limits column capacity. Experience indicates that 10 to 30% savings are often achievable. To get an idea of the numbers involved, consider a column that uses 1000 pph of steam. If we can save 100 pph, if average operating time is 8000 h/yr, and if steam costs $5 per 1000 lb, then annual savings are $4000. This means we can invest $8000 to $12,000 in improved controls to achieve the desired steam saving.
   b. Heat recovery schemes. Much larger energy savings are often possible with energy recovery schemes, usually involving the condensation of vapor from one column in the reboiler of another column (Buckley, 1981). Our experience has been that designing controls for energy recovery schemes is much more difficult and time-consuming than for conventional columns. This extra cost, however, is usually a small fraction of savings. An interesting feature of some heat recovery schemes is that the pressure in the column supplying heat is allowed to float. Without going into details I will simply say that we have found that this simplifies process design (usually no auxiliary condensers are needed) and simplifies instrumentation and control (no pressure controller is needed). Because maximum operating pressure occurs when boilup is a maximum, there is no problem with flooding. There are other problems, however. For example, the feed valve to a column with floating pressure has a much larger ΔP variation with feed rate than is commonly encountered. Additionally, the column bottom product valve has a low ΔP at low rates and a high ΔP at high rates. To obtain reasonably constant stability, one may need to provide "flow characteristic compensation" (Buckley, 1982). This need is increased by a trend toward lower ΔP for valves in pumped systems. There are some differences in control requirements between the chemical and petroleum industries. Tolliver and Waggoner (1980) have presented an excellent discussion of this subject.

Other approaches to heat recovery are discussed in Chapter 24.

2-9 CONCLUSION

From the preceding discussions we can see that since 1945 we have moved from generously designed columns with bubble-cap trays and simple controls to tightly designed columns with sieve or valve trays or packing and with more sophisticated controls. For new design projects with difficult applications it may be appropriate in some cases to consider going back to bubble-cap trays and larger design safety factors to avoid the necessity of very complex and perhaps touchy controls.

The literature on distillation control, particularly that from academic contributors, emphasizes product composition control. Our experience has been that many operating difficulties, in addition to those due to tight design, are due to improperly designed column bases and auxiliaries, including piping. In my own experience, the single factor that contributes most to low cost operation and good composition control has been properly designed averaging liquid level controls.

As of mid-1990, lack of adequate product quality measurements is probably the Achilles' heel of distillation control. With adequate quality measurements and composition control of each product stream, the next problem area will be that of interactions.

Although a great deal of progress has been made in quantitative design of control systems, I would say that today we are about where distillation design was in 1948.

2-10 COMMENTS ON REFERENCE TEXTS

There are now several texts on distillation control. One of them, which has the most complete technical treatment, is that by Rademaker, Rijnsdorp, and Maarleveld (1975). Another, by Shinskey (1977) is notable for its treatment of energy conservation. A third, by Nisenfeld and Seemann (1981), has a well-organized treatment of distillation fundamentals and types of columns. Neither of these last two books deals with control in a technical sense: there is no stability theory, Laplace transforms, or frequency response. But each contains many gems of practical details about column control. There are two more recent books. One, by Buckley, Luyben and Shunta (1985), has an initial section devoted to practical details and a second half that discusses distillation dynamics and control with modest theory. A second book, by Deshpande (1985), has a more advanced treatment.

References

Åström, K. J. (1979). Self-tuning regulators. Report LUTFD2/(TRFT-7177)/1-068, Lund Institute of Technology, Sweden.

Boyd, D. M. (1948). Fractionation instrumentation and control. *Petr. Ref.*, October and November.

Brosilow, C. B. and Tong, M. (1978). *Inferential control of processes: Part II. AIChE J.* 24, 485.

Brown, G. S. and Campbell, D. P. (1948). *Principles of Servomechanisms*. New York: John Wiley.

Buckley, P. S. (1964). *Techniques of Process Control*. New York: John Wiley.

Buckley, P. S. (1966). Reflux cycle in distillation columns. Presented at IFAC Conference, London, 1966.

Buckley, P. S. (1968). Override controls for distillation columns. *InTech*. 15, August, 51–58.

Buckley, P. S. (1969a). Control for sidestream drawoff columns. *Chem. Eng. Prog.* 65, 45–51.

Buckley, P. S. (1969b). Protective controls for sidestream drawoff columns. Presented at AIChE Meeting, New Orleans, March 1969.

Buckley, P. S. (1971). Designing override and feedforward controls. Part 1. *Control Eng.* August, 48–51; Part 2, October, 82–85.

Buckley, P. S. (1974). Material balance control in distillation columns. Presented at AIChE Workshop, Tampa, FL, November 1974.

Buckley, P. S. (1981). Control of heat—integrated distillation columns. Presented at Engineering Foundation Conference, Sea Island, GA, January 1981.

Buckley, P. S. (1982). Optimum control valves for pumped systems. Presented at Texas A&M Symposium, January 1982.

Buckley, P. S. (1983). Recent advances in averaging level control. Presented at Instrument Society of America meeting, Houston, April 1983.

Buckley, P. S. (1986). Nonlinear averaging level control with digital controllers. Presented at Texas A&M Instrumentation Symposium, January 1986.

Buckley, P. S. and Cox, R. K. (1971). New developments in overrides for distillation columns. *ISA Trans.* 10, 386–394.

Buckley, P. S., Cox, R. K., and Rollins, D. L. (1975). Inverse response in a distillation column. *Chem. Eng. Prog.* 71, 83–84.

Buckley, P. S., Luyben, W. L., and Shunta, J. P. (1985). *Design of Distillation Column Control*

Systems. Research Triangle Park, NC.: Instrument Society of America.

Campbell, D. P. (1958). *Process Dynamics*. New York: John Wiley.

Chestnut, H. and Mayer, R. W. (1951). *Servomechanisms and Regulating System Design*, Vols. I and II. New York: John Wiley.

Cox, R. K. (1973). Some practical considerations in the application of overrides. Presented at ISA Symposium, St. Louis, MO, April 1973.

Deshpande, P. B. (1985). *Distillation Dynamics and Control*. Research Triangle Park, NC: Instrument Society of America, 1985.

Doukas, N. and Luyben, W. L. (1978). Control of sidestream columns separating ternary mixtures. *InTech*. 25, 43–48.

Gould, L. A. (1969). *Chemical Process Control*. Reading, MA: Addison-Wesley.

Luyben, W. L. (1966). 10 schemes to control distillation columns with sidestream drawoff. *ISA J*. July, 37–42.

Luyben, W. L. (1990). *Process Modeling, Simulation and Control for Chemical Engineers*, 2nd ed. New York: McGraw-Hill, 1990.

Nisenfeld, A. E. and Seemann, R. C. (1981). *Distillation Columns*. Research Triangle Park, NC: Instrument Society of America.

Oldenburger, R. (1966). *Optimal Control*. New York: Holt, Reinhart, and Winston.

Oldenbourg, R. C. and Sartorius, H. (1948). *The Dynamics of Automatic Controls*. New York: ASME.

Rademaker, O., Rijnsdorp, J. E., and Maarleveld, A. (1975). *Dynamics and Control of Continuous Distillation Columns*. New York: Elsevier.

Ray, W. H. (1981). *Advanced Process Control*. New York: McGraw-Hill.

Rijnsdorp, J. E. (1961). *Birmingham University Chemical Engineer* 12, 5–14, 1961.

Rippin, D. W. T. and Lamb, D. E. (1960). *A Theoretical Study of the Dynamics and Control of Binary Distillation*. Newark, DE: University of Delaware.

Shellene, K. R., Stempling, C. V., Snyder, N. H., and Church, D. M. (1967). Experimental study of a vertical thermosyphon reboiler. Presented at Ninth National Heat Transfer Conference AIChE–ASME, Seattle, WA, August 1967.

Shinskey, F. G. (1977). *Distillation Control*. New York, McGraw-Hill.

Thistlethwaite, E. A. (1980). Analysis of inverse response behavior in distillation columns. M.S. Thesis, Department of Chemical Engineering, Louisiana State University.

Tolliver, T. L. and Waggoner, R. C. (1980). Distillation column control: A review and perspective from the CPI. *ISA Adv. Instrum*. 35, 83–106.

Williams, T. J. (1971). Distillation column control systems. Proceedings of the 12th Annual chemical and Petroleum Industries Symposium, Houston, April 1971.

Williams, T. J., Harnett, B. T., and Rose, A. (1956). *Ind. Eng. Chem*. 48, 1008–1019.

Young, A. J., Ed. (1957). *Plant and Process Dynamic Characteristics*. New York: Academic.

2

Methods

3

Rigorous Modelling and Conventional Simulation

Vincent G. Grassi II

Air Products and Chemicals, Inc.

3-1 OVERVIEW

The aim of this chapter is to show the reader how to develop a dynamic simulation for a distillation tower and its control from first principles. The different classes of dynamic distillation models and various approaches to solving these models will be presented. The author hopes to dispel the myth that modelling and simulation of distillation dynamics must be difficult and complex.

Dynamic modelling and simulation has proven to be an insightful and productive process engineering tool. It can be used to design a distillation process that will produce quality products in the most economic fashion possible, even under undesirable process disturbances. Working dynamic models provide a process engineering tool that has a long and useful life.

Dynamic simulation can be used early in a project to aid in the process and control system design. It ensures that the process is operable and can meet product specifications when the process varies from steady-state design. Later in the project the simulation can be used to complete the detailed control system design and solve plantwide operability problems. After the project, the same simulation is useful for training. Years later, as product and economic conditions change, the simulation can be used for plant improvement programs.

The authors of this book offer different approaches to distillation modelling. Chapter 4 discusses reduced and simplified models. Chapter 5 presents a novel concept in object orientated simulation. Chapter 6 presents concepts necessary to develop a plantwide simulator. These chapters on modelling and simulation will provide readers with a solid framework so the methods and ideas presented in the remaining chapters of this book can be implemented.

This chapter will be devoted to developing process models that realistically predict plant dynamics and their formulation in algorithms suitable for digital computer codes. Computer source codes are readily available from many sources (Franks, 1972; Luyben, 1990), so they will not be repeated here. I intend to share my experience in properly selecting models that accurately predict the dynamics of real distillation columns one is likely to find in the plant.

A dynamic model is needed to study and design composition controls. To do this we will develop a sufficiently rigorous tray-by-tray model with nonideal vapor–liquid and stage equilibrium. Proportional-integral feedback controllers will control product compositions or tray temperatures. Vapor flow and pressure dynamics often can be assumed negligible. I will discuss how systems with vapor hydraulic pressure dynamics can be modelled and simulated. My approach is based on fundamental process engineering principles. Only the relationships that are necessary to solve the problem should be modelled. Most importantly, these are models that any chemical engineer can easily simulate. They are suitable for a small personal computer, in whatever programming language you prefer.

3-1-1 Conventional Simulation

Let me begin by defining what I mean by *conventional* simulation. I view conventional simulation as a technique used by many practicing engineers over the past 30 years based on a *physical approach*. I consider nonconventional techniques to be simulation methods based on *mathematical approaches*. The conventional approach usually follows the process flows, solving each physical phenomenon in an organized sequence of steps. The nonconventional approach writes the governing equations for each unit operation and solves the entire system as a set of simultaneous differential and algebraic equations. Each method has advantages and disadvantages. The conventional method usually results in a model that is numerically easier to solve, but specific to process. Mathematically, a conventional simulation is easier to initialize and more robust when dealing with discontinuities. The nonconventional method is numerically more difficult to solve, but is flexible and less specific. Flexibility is important when developing a general purpose dynamic simulator.

Chemical engineering models can be classified by type. Models can be classified in four categories: algebraic, ordinary differential, differential algebraic, and partial differential. Table 3-1 contains a general classification for chemical engineering models.

Our models for distillation towers use the equilibrium stage approach. Differential-algebraic equations (DAE) govern this tower model. Differential-algebraic models are characterized by their *index* (Brenan, Campbell, and Petzold, 1989). The index is equal to the number of times we must differentiate the set of algebraic equations to obtain continuous differential equations for all unknown variables. An index of zero is a set of differential equations without any algebraic equations. If the algebraic equations are solved separately from the differential equations, the system is of index 0. Higher levels of index result when the algebraic and differential equations are solved simultaneously. Figure 3-1*a* contains three examples of various index problems. The unknown variables are y_1, y_2, and z in these examples.

Index 0 problems are much easier to solve numerically than those with index 1. Indices greater than 1 are very difficult to solve, if solvable at all. There are commercial integration packages available for solving index 1 problems reliably, such as DASSL (Brenan, Campbell, and Petzold, 1989). Currently reliable algorithms to solve problems with index 2 or greater do not exist. If you try to simulate a differential-

TABLE 3-1 Classification of Chemical Engineering Models

Model Type	Characteristics
Algebraic	Linear
	Nonlinear
Ordinary differential	Linear
	Nonlinear
	Nonstiff
	Stiff
Differential algebraic	Index = 0
	Index = 1
	Index > 1
Partial differential	Parabolic
	Hyperbolic
	Elliptical

INDEX = 1

$$\dot{y}_1 = a_{11}y_1 + a_{12}y_2 + z$$
$$\dot{y}_2 = a_{21}y_1 + a_{22}y_2 + z$$
$$y_1 + y_2 + z = 0$$

Differentiate once to get $\dot{y}_1$, $\dot{y}_2$, and $\dot{z}$

INDEX = 2

$$\dot{y}_1 = a_{11}y_1 + a_{12}y_2 + z$$
$$\dot{y}_2 = a_{21}y_1 + a_{22}y_2 + z$$
$$y_1 + y_2 = 0$$

Differentiate twice to get $\dot{y}_1$, $\dot{y}_2$, and $\dot{z}$

INDEX = 3

$$\ddot{y}_1 = y_3 y_1$$
$$\ddot{y}_2 = y_3 y_2 - g$$
$$y_1^2 + y_2^2 - L^2 = 0$$

Differentiate three times to get $\dot{y}_1$, $\dot{y}_2$, and $\dot{y}_3$

(a)

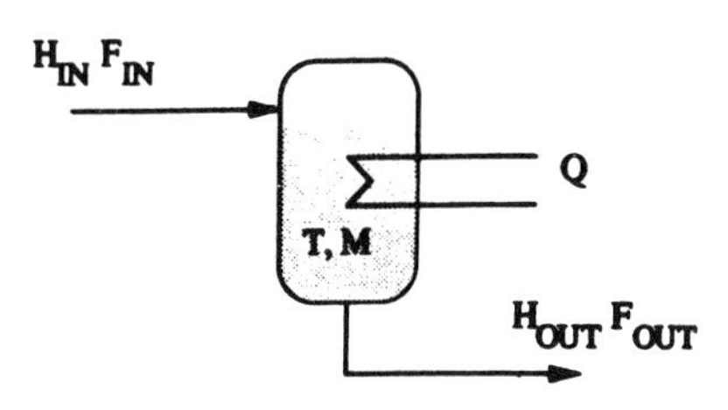

$$\frac{dM}{dt} = F_{\text{in}} - F_{\text{out}}$$
$$\frac{dE}{dt} = H_{\text{in}}F_{\text{in}} - H_{\text{out}}F_{\text{out}} + Q$$
$$E = H_{\text{out}}M$$
$$H_{\text{out}} = AT + BT^2$$

Specify Q, calculate $T \Rightarrow$ Index = 1
Specify T, calculate $Q \Rightarrow$ Index = 2

where F = molar flow rate
H = specific molar enthalpy
E = total energy
Q = heat addition rate
A, B = enthalpy correlation constants
t = time
T = temperature
in = inlet stream
out = outlet stream

(b)

FIGURE 3-1. (a) Index. The *index* is the number of differentiations required to obtain all derivatives explicitly. (b) Index example.

algebraic distillation model, be very careful to pose the model so that the index does not exceed 1.

Figure 3-1b contains an example of the index problem similar to the type we find in distillation dynamics. The example is a simple tank. A known pure component feed enters a tank. Heat is added through a heating coil. The outlet flow is set downstream from the tank.

We write the total differential material and heat balances. We can calculate the outlet enthalpy by a correlation of temperature. Our model has two differential and two algebraic equations that can be solved using the DAE approach.

The index of this problem can change depending on what variables we set and what variables we calculate. For example, we could set the heat addition rate and solve the problem to calculate the temperature. We would need to differentiate the algebraic equation with respect to time once to get the temperature differential. All the unknown variables are then given in differential form, resulting in a problem of index 1.

Alternatively, we could set the temperature and solve the problem to calculate the heat addition rate. This problem is index greater than 1. We would have to differentiate the energy balance once to get the heat addition differential. This differentiation creates a second derivative of the total energy, resulting in an index 2 problem. We can reduce the problem to index 1 by adding another algebraic equation for the heat. This equation could be a heat transfer equation for the heating coil.

Another numerical difficulty that appears in the solution of differential equations is *stiffness*. When we model the dynamics of a chemical process, we are seeking the time response of the process to changes in its inputs. The process is modelled by many differential equations. Each differential equation takes a characteristic amount of time to reach steady state. Some equations can take hours whereas others might take seconds. Stiffness can be measured as the ratio of the longest to the shortest time for these differential equations to approach steady state. We solve the system of differential equations by stepping forward in time, but we cannot take steps larger than those necessary to solve the shortest differential equation. As stiffness increases we must take more steps to solve the equations, and the simulation runs slower.

The conventional simulation methods that we will develop in this chapter result in index 0 problems. We avoid the index problem by solving the algebraic equations as separate modules before integrating the differential equations. In this way we do not have to solve the differential-algebraic system simultaneously. These models can be stiff or not, depending on what problem we need to solve.

3-2 DISTILLATION PROCESS FUNDAMENTALS

3-2-1 Continuity Equations

The development of a dynamic distillation model can be simple if you wisely use your process knowledge and engineering judgment. It can be difficult if you try to model details that are not needed to solve the problem at hand. You might find it interesting that it can be easier to simulate the dynamics of a distillation column than its steady state. A steady-state model must solve all the algebraic equations for mass and energy simulation. In a dynamic simulation we compute the change in mass and energy and step forward in time. A dynamic simulation is more intuitive with the way in which we think about distillation operation than a steady-state simulation.

To develop our dynamic model we begin by writing dynamic continuity equations of mass and energy for each unit operation where mass or energy can accumulate. For a distillation tower we develop the continuity equations for the stages, condenser, reflux accumulator, bottom sump, reboiler, thermowells, integrating controllers, and so forth. Let us start by discussing the dynamic material and energy balances for a general system.

The dynamic continuity equations state that the rate of accumulation of material (mass or energy) in a system is equal to the amount of material entering and generated, less the amount leaving and consumed within the system:

$$\begin{bmatrix}\text{rate of accumulation}\\ \text{of mass (energy)}\end{bmatrix} = \begin{bmatrix}\text{mass (energy) flow}\\ \text{into the system}\end{bmatrix} - \begin{bmatrix}\text{mass (energy) flow}\\ \text{out of the system}\end{bmatrix} + \begin{bmatrix}\text{mass (energy) generated}\\ \text{within the system}\end{bmatrix} - \begin{bmatrix}\text{mass (energy) consumed}\\ \text{within the system}\end{bmatrix}$$

The accumulation term is a first order time derivative of the total mass or energy. The flow terms are algebraic. This results in a first order ordinary differential equation that is usually nonlinear. We write material balances for each component (or the total flow combined with all but one component flow) and one energy balance. The component balances are modelled as

$$\frac{dx_i M}{dt} = z_i^{\text{in}} F^{\text{in}} - x_i F_{\text{out}} + \dot{M}_i^{\text{gen}} - \dot{M}_i^{\text{con}} \tag{3-1}$$

where z_i = feed composition (mole fraction)
x_i = composition of material in system (mole fraction)
M = material in system (lb-mol)
F = flow rate (lb-mol/h)
$\dot{M}_i^{\text{gen}}$ = rate of generation of component i in system (lb-mol/h)
$\dot{M}_i^{\text{con}}$ = rate of consumption of component i in system (lb-mol/h)

We will not consider reactive distillation systems in our model, so there are no chemical components generated or consumed in any part of the distillation tower. The material generation and consumption terms are zero.

A general energy balance is represented as

$$\frac{d(E^{\text{out}}M)}{dt} = E^{\text{in}}F^{\text{in}} - E^{\text{out}}F^{\text{out}} + Q - W \quad \text{(3-2a)}$$

$$E = U + \text{PE} + \text{KE} \quad \text{(3-2b)}$$

$$U = H - \text{PV} \quad \text{(3-2c)}$$

where M = material in system (lb-mol)
F = material flow in/out of the system (lb-mol/h)
E = specific total energy (Btu/lb-mol)
Q = heat added to system (Btu/h)
W = work produced by the system (Btu/h)
U = specific internal energy (Btu/lb-mol)
PE = specific potential energy (Btu/ lb-mol)
KE = specific kinetic energy (Btu/lb-mol)
H = specific enthalpy (Btu/ lb-mol)
PV = pressure–volume work (Btu/lb-mol)

The PE, KE, and PV terms of total energy are negligible in distillation towers. We will not consider vapor recompression systems in this chapter, so there is no work produced by the system. Vapor recompression will be considered in Chapter 23.

Heat can be added or removed from many unit operations in distillation, such as reboilers, condensers, or interstage heat exchangers. This means that the energy terms considered in our distillation model are equal to enthalpy. Our overall energy balance can be written as

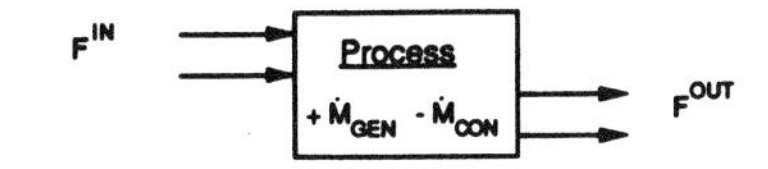

Rate of Accumulation = In - Out + Generation - Consumption

$$\frac{dM_I}{dt} = \sum_{i=1}^{m} F_i^{IN} - \sum_{i=1}^{n} F_i^{OUT} + \sum_{i=1}^{o} \dot{M}_{GEN,I} - \sum_{i=1}^{p} \dot{M}_{CON,I}$$

Initial Condition: $M_I(t=0) = M_I^0$

$I = \{1, 2, 3 \ldots n_{EQ}\}$

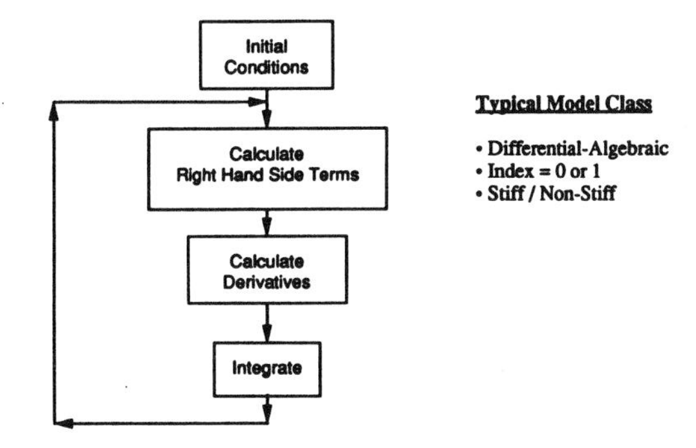

FIGURE 3-2. Integration block diagram.

$$\frac{d(H^{\text{out}}M)}{dt} = H^{\text{in}}F^{\text{in}} - H^{\text{out}}F^{\text{out}} + Q \quad \text{(3-2d)}$$

We solve the system of differential equations, for all units in the process, by integrating forward in time. First we specify a set of initial conditions for the state variables at the starting time. Next, all differentials are calculated from the material and energy flows at the current time. Each equation is integrated to the next time step. The current time is increased by the time step, and the integration process continues. Figure 3-2 contains a representation of steps in this process.

The integration procedure must be started with a set of initial conditions for each state variable. Initial conditions may or may not correspond to a steady-state solution. If a steady-state solution is not available for the first run, a set of consistent values may be chosen. The solution will usually converge, dynamically, to a bona fide steady state as the integration proceeds. This method of initialization is useful to get a dynamic simulation running quickly. In fact, often it is numerically more stable to converge to the steady state dynamically than to solve for the steady state algebraically.

The following simple first order Euler integrator will be used to integrate the model:

$$\chi_{t+\Delta t} = \chi_t + \frac{d\chi_t}{dt}\Delta t \qquad (3\text{-}3a)$$

$$\chi(t = 0) = \chi_0 \qquad (3\text{-}3b)$$

where χ is a function of time (t). The Euler integrator is very easy to implement and is very effective at solving a system of nonstiff differential equations. The Euler integrator works well for solving the distillation model that we develop in this chapter. Models that include vapor hydraulics and pressure dynamics are stiff. It is usually more efficient to use an implicit integrator such as LSODE (Hindmarsh, 1980) for such stiff systems. It is important to remember the difficulty with stiffness is that it slows down the simulation. Because the computational speed of computers has dramatically increased over a short period, systems that were considered a few years ago to be too stiff to solve explicitly are now quickly solved by present day computers. Explicit methods are generally easier to implement than implicit integrators and the solution behavior of an explicit integrator is easier to follow than an implicit integrator. This is a big benefit when you are debugging your simulation.

We have discussed how to develop and simulate dynamic models from continuity balances. In the next few sections, we will develop the principles necessary to compute the terms of the differential equations specifically for distillation towers. We will look at the algorithms that we need to solve the algebraic models and we will develop the complete differential models of the various subsections within the distillation tower. Finally, we will discuss the performance of these models on various computer platforms and find how well the models match real plant towers.

3-2-2 Vapor-Liquid Equilibrium

It is essential to have an accurate model or correlation for the vapor–liquid equilibrium (VLE) and physical properties of the components in the distillation tower. There are many useful sources for obtaining literature data and correlating equations to represent your system (Dauber and Danner, 1988; Gmehling and Onken, 1977; Reid, Prausnitz, and Sherwood, 1977; Walas, 1985; Yaws, 1977). The phase equilibrium of some systems is essentially ideal. Raoult's law,

$$y_i P = x_i P_i^s \qquad (3\text{-}4a)$$

represents ideal systems. Others may have constant relative volatility:

$$\alpha_{ij} = \frac{K_i}{K_j} = \frac{y_i/x_i}{y_j/x_j} \qquad (3\text{-}4b)$$

Some can be correlated simply as a polynomial in temperature:

$$K_i = \frac{y_i}{x_i} = A_i + B_i T + C_i T^2 + \cdots \qquad (3\text{-}4c)$$

where y_i, x_i = vapor, liquid composition (mole fraction)
P = total pressure (psia)
P_i^s = vapor pressure (psia)
α_{ij} = relative volatility
K_i = vapor–liquid composition ratio
A, B, C = correlation constants

Many chemical systems have significant nonidealities in the liquid phase, and sometimes also the vapor phase. A more rigorous VLE model is required. The *fugacities* of the vapor and liquid phases are equal at equilibrium:

$$f_i^V = f_i^L \qquad (3\text{-}5a)$$

Expressions can be written for the fugacity of each phase:

$$\phi_i^V y_i P = \phi_i^L x_i \qquad (3\text{-}5b)$$

$$\phi_i^V y_i P = \gamma_i x_i P_i^s \exp\left(\frac{\nu_i^c (P - P_i^s)}{RT}\right) \qquad (3\text{-}5c)$$

where f_i^V, f_i^L = vapor, liquid fugacity
ϕ_i^V, ϕ_i^L = vapor, liquid fugacity coefficient
P_i^s = vapor pressure (psia)
ν_i^c = specific critical volume (ft^3/ lb-mol)
γ_i = liquid activity coefficient
R = gas constant (ft^3 psia/lb-mol °R)
T = temperature (°F)

If an equation of state is available that accurately represents both the vapor and liquid phases, use Equation 3-5b. If the liquid phase cannot be modelled by an equation of state, which is usually the case, use Equation 3-5c, which models the nonidealities of the liquid phase by a correlation of the liquid phase activity coefficient γ. The exponential term corrects the liquid fugacity for high pressure (greater than 150 psia). Most of the nonideal effects for low pressure (less than 150 psia) chemical systems can be represented by a simplified form of Equation 3-5c, given as

$$y_i P = \gamma_i x_i P_i^s \qquad (3\text{-}5d)$$

The vapor phase is very close to ideal and the exponential correction is negligible. It should be noted that Equation 3-5d reduces to Raoult's law if the liquid phase is ideal, that is, the liquid phase activity coefficient is unity.

The liquid activity coefficient can be modelled for a component system by a correlating equation (Gmehling, 1977; Walas, 1985) such as the Van Laar, Wilson, NRTL, or UNIQUAC equations. Pure component vapor pressure is modelled by the Antoine equation,

$$\ln(P_i^s) = C_1 + \frac{C_2}{C_3 + T} \qquad (3\text{-}5e)$$

where C_1, C_2, C_3 are Antoine coefficients.

3-2-3 Murphree Vapor Phase Stage Efficiency

Tray vapor composition can be calculated from the relationships developed in the previous section. Mass transfer limitations prevent the vapor leaving a tray from being in precise equilibrium with the liquid on the tray. This limitation can be modelled as a deviation from equilibrium. Three types of stage efficiencies are commonly used:

1. Overall efficiency.
2. Murphree efficiency.
3. Local efficiency.

Overall efficiency pertains to the entire column, relating the total number of actual to ideal stages. Murphree efficiency pertains to the efficiency at a specific stage whereas local efficiency pertains to a specific location on a single stage. We will use Murphree stage efficiency in our model. We base the Murphree stage efficiency on the vapor composition's approach to phase equilibrium. Figure 3-3 contains a geometric interpretation of the Murphree vapor phase stage efficiency for a binary system. The Murphree stage efficiency is the ratio between the actual change in vapor composition between two stages and the change that would occur if the vapor was in equilibrium with the liquid leaving the stage.

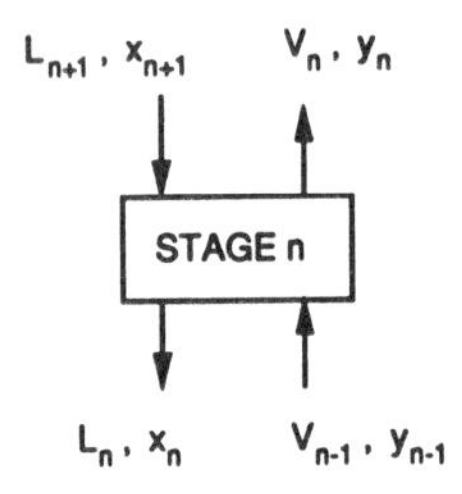

Assume the liquid and vapor are in thermal equilibrium, but not in phase equilibrium

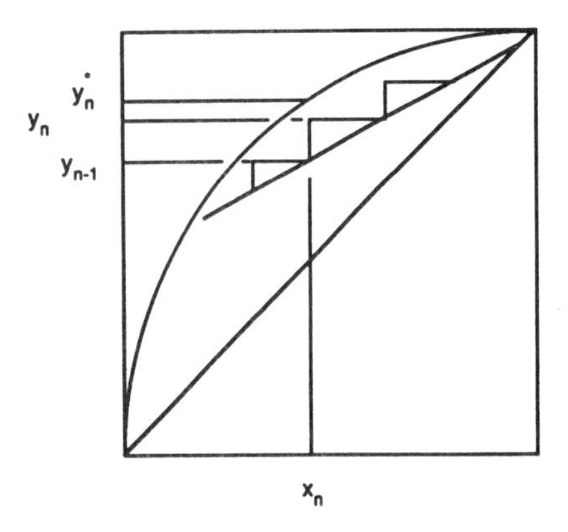

$$E_n = \frac{y_n - y_{n-1}}{y_n^* - y_{n-1}} \Rightarrow y_{i,n} = (y_{i,n}^* - y_{i,n-1})E_n + y_{i,n-1}$$

FIGURE 3-3. Murphree vapor phase stage efficiency.

3-2-4 Enthalpy

Total energy reduces to enthalpy in our distillation model. We must therefore calculate the enthalpies of the vapor and liquid streams as functions of temperature, pressure, and composition. Liquid enthalpy is not a function of pressure because liquids are incompressible. Pressure effects on vapor enthalpy are negligible for low to moderate pressure systems. Enthalpy can be easily correlated with composition and temperature, based on a linear fit of heat capacity with temperature. Selecting 0°F as the reference temperature results in the correlations

$$h_i^L = A_i^L T + B_i^L T^2 \tag{3-6a}$$

$$h_i^V = A_i^V T + B_i^V T^2 + \Delta H_i^V\,(0°\text{F}) \tag{3-6b}$$

where h_i^V, h_i^L = vapor, liquid pure component specific enthalpy
A, B = correlation constants
ΔH_i^V (0°F) = pure component heat of vaporization at 0°F

We can obtain the correlation constants for Equations 3-6a and 3-6b from heat capacity data. Because we used 0°F for our enthalpy reference temperature, the heat of vaporization in Equation 3-6b must be that at 0°F. Mixing rules are used to find multicomponent properties. The vapor phase enthalpy is mixed ideally as the molar average of the pure component enthalpies. The liquid phase can contain significant nonidealities (heat of mixing) unless the liquid activity coefficient is unity. Heat of mixing can be modelled with

$$h_{\text{mix}}^L = -RT^2 \sum_i x_i \frac{\partial \ln \gamma_i}{\partial T} \tag{3-6c}$$

The liquid activity coefficient in Equation 3-6c is modelled with the same correlation we used for vapor–liquid equilibrium in Section 3-1-2. Good results can be obtained often, even with significant liquid nonidealities, neglecting the heat of mixing or substituting a constant value.

Vapor and liquid enthalpy for mixtures can be represented as

$$h^V = \sum_i y_i h_i^V \tag{3-7a}$$

$$h^L = \sum_i x_i h_i^L + h_{\text{mix}}^L \tag{3-7b}$$

3-2-5 Liquid and Froth Density

Liquid density can be calculated from a correlation or an equation of state. A useful correlation for the pure component liquid density (Yaws, 1977) is

$$\rho_i = AB^{-(1-T_r)^{2/7}} \tag{3-8a}$$

Yaws gives constants for many common components, or you can regress data to obtain the parameters. To compute the liquid density of a mixture use

$$\rho_L = M_L \bigg/ \left(\sum_i \frac{x_i M_i}{\rho_i} \right) \tag{3-8b}$$

where ρ_i = pure component liquid density (lb/ft^3)
ρ_L = mixture liquid density (lb/ft^3)
M_L = liquid mixture molecular weight (lb/lb-mol)
M_i = pure component molecular weight (lb/lb-mol)
T_r = reduced temperature
A, B = pure component density correlation constants

Vapor passing through a distillation tower tray aerates the liquid, creating a froth. The clear liquid density, Equation 3-8b, is corrected for frothing by correlations, usually obtained from the tray vendor design guides or generalized correlations (Van Winkel, 1967). Computing the froth density as a function of vapor rate is necessary if you want to study flow hydraulics. Changes in froth density with vapor rate can lead to inverse response behavior in bottom sump composition and level dynamics. There is no need to compute the froth density continuously if you are not interested in froth dynamics. Omitting this calculation at each time step will improve the speed of the simulation.

3-3 COMPUTER SIMULATION

3-3-1 Algebraic Convergence Methods

Our conventional simulation of a distillation tower requires solving algebraic equations before we integrate the system of ordinary differential equations. We will use two simple and intuitive methods to solve implicit algebraic relations. The first of these is the Wegstein, or secant, convergence method. Figure 3-4 presents this method graphically. The Wegstein method requires the objective function to be written in the form $x = f(x)$. The method requires two initial estimates of the solution (x_1 and x_2), and the function is evaluated at these two points (f_1 and f_2). A secant is then drawn between the two values. A third point (f_3) is calculated where the secant intersects the line where $f(x) = x$, and the function is evaluated at this third point (f_3). If f_3 is sufficiently close to x_3, the solution has been found; if not, the last two points are used as the next set of initial guesses and the procedure is repeated. The Wegstein method is very easy to implement in a computer program.

The second convergence method that is very useful for distillation problem is the Newton–Raphson method. The Newton–Raphson method requires the objective function to be written in the form $f(x) = 0$. This method requires the first derivative of the function to be evaluated, preferably analytically. This requirement adds some complication that the Wegstein method does not have, but improves the rate of convergence. Newton–Raphson is a very powerful numerical method for thermodynamic calculations, such as the bubble point. The analytical derivative is readily available for

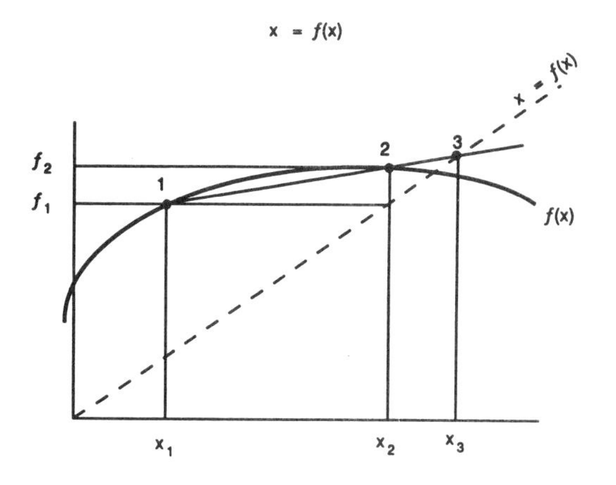

$$\frac{f_2 - f_1}{x_2 - x_1} = \frac{x_3 - f_2}{x_3 - x_2}$$

$$\boxed{x_3 = \frac{f_2 x_1 - f_1 x_2}{f_2 - f_1 - x_2 + x_1}}$$

FIGURE 3-4. Wegstein convergence.

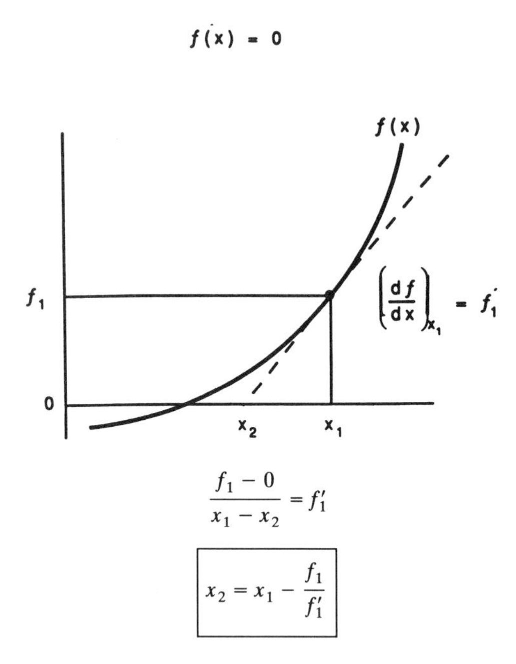

FIGURE 3-5. Newton–Raphson convergence.

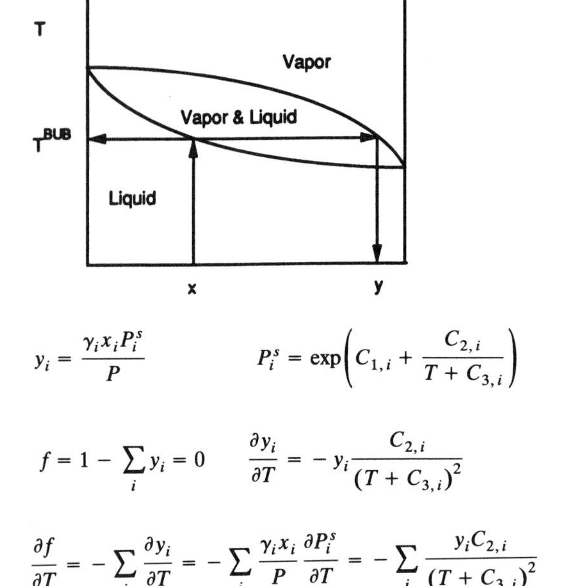

Converge T: $T_{k+1} = T_k + f \Big/ \left(\sum_i \frac{\partial y_i}{\partial T}\right)$

FIGURE 3-6. Equilibrium bubble point calculation (binary system at constant pressure).

these calculations. Figure 3-5 presents the Newton–Raphson method graphically.

The Newton–Raphson method requires an initial estimate of the solution (x_1). An analytical expression gives the derivative of the objective function at the estimate. The derivative is the tangent to the curve at this point. The next estimate (x_2) is computed where the tangent intersects the x axis at $f = 0$. This solution procedure is very similar to the Wegstein method but a tangent is used to compute the next estimate instead of a secant.

These convergence methods require initial estimates for the solution. During the dynamic simulation, the solution at the previous time step is used for these estimates. This is very close to the solution at the current time step. The algebraic equations are solved very quickly, in only a few iterations.

3-3-2 Equilibrium Bubble Point Calculation

The development of an equilibrium stage model requires the calculation of the vapor composition and temperature in equilibrium with a known liquid composition at a known pressure. This is a *bubble point* calculation. Equation 3-5d will be used for our thermodynamic model. A starting temperature is guessed. The Antoine equation, Equation 3-5e, computes the vapor pressures. A single dimensional Newton–Raphson method is used to converge temperature and compute the equilibrium vapor composition. The relations and a graphical representation of the solution are presented in Figure 3-6.

The function that we choose to solve is unity minus the sum of the calculated vapor compositions. The derivative of this function with respect to temperature is given in Figure 3-6. The liquid activity coefficient is a weak function in temperature and can be excluded from the analytic derivative. This method to compute the tray bubble point temperature and equilibrium vapor composition is very fast and robust.

3-3-3 Equilibrium Dew Point Calculation

The reverse of the bubble point is the dew point. Here we would like to calculate the temperature and liquid composition in equilibrium with a vapor. This calculation is necessary to model a partial condenser. The entire model for a partial condenser is not developed in this chapter, but the dew point calculation is presented in Figure 3-7.

3-3-4 Distillation Stage Dynamic Model

We will now develop a general model for a single distillation stage. We assume single-flow pass tray hydraulics. Liquid enters the tray through the downcomer of the tray above. Vapor enters the tray from the tray below. The vapor and liquid completely mix on the tray. Vapor leaves the tray, in equilibrium with the tray liquid composition, and passes to the tray above. Liquid flows over the outlet weir into the downcomer and the tray below. The tray may contain a feed or a sidestream. A dynamic model for the tray will contain N_c differential material balances, where N_c is the number of components in the system, and one overall energy balance. I like to model the material balances as $N_c - 1$ component balances and one overall material balance. You can model the system as N_c component balances if you prefer; both are equivalent. Do not forget to include the liquid in the downcomer as material in the stage.

If the tower pressure is high (greater than 150 psi), the vapor can represent a significant fraction of the total stage mass. The liquid mass should be increased to account for the vapor if it is up to 30% of the total material. If the vapor mass is more than 30 or 40% of the total stage material and variable, the vapor phase must be modelled independently.

The change in specific liquid enthalpy is usually very small compared to the total tray enthalpy. Therefore, the energy balance may often be reduced to an algebraic relation that can be used to calculate the vapor rate leaving the stage. A schematic of this model is given in Figure 3-8.

Given this structure, the following procedure can be used to solve the stage model. The procedure is started from the bottom and proceeds up through the column.

1. Calculate the equilibrium vapor composition and temperature from pressure and liquid composition by a bubble point calculation.
2. Calculate the actual vapor composition from the Murphree vapor phase stage efficiency.
3. Calculate the vapor and liquid enthalpy from their composition and stage temperature.
4. Calculate the clear liquid density from composition and temperature.
5. Calculate the liquid froth density at the stage vapor rate.
6. Calculate the liquid rate leaving the stage from the Francis weir formula.
7. Calculate the vapor flow leaving the stage from the energy balance.
8. Calculate the component and total mass derivatives.

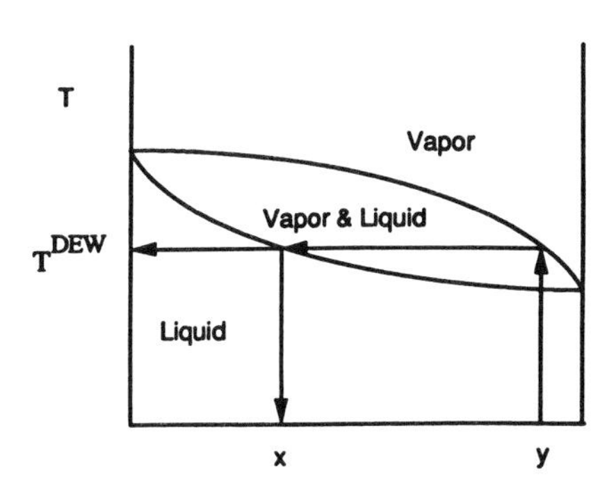

$$x_i = \frac{y_i P}{\gamma_i P_i^s} \qquad P_i^s = \exp\left(C_{1,i} + \frac{C_{2,i}}{T + C_{3,i}}\right)$$

$$f = 1 - \Sigma x_i = 0 \qquad \frac{\partial x_i}{\partial T} = x_i \frac{C_{2,i}}{(T + C_{3,i})^2}$$

$$\text{Converge } T\text{:} \quad T_{k+1} = T_k + f \Big/ \left(\Sigma \frac{\partial x_i}{\partial T}\right)$$

FIGURE 3-7. Equilibrium dew point calculation (binary system at constant pressure).

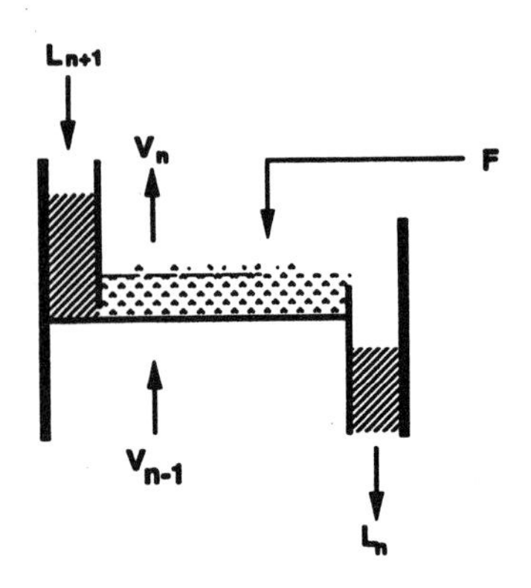

Material Balance $$\frac{dM_n}{dt} = F + L_{n+1} + V_{n-1} - L_n - V_n$$

Component Balance $$\frac{d(x_i M_n)}{dt} = z_{i,f}F + x_{i,n+1}L_{n+1} + y_{i,n-1}V_{n-1} - x_{i,n}L_n - y_{i,n}V_n$$

Energy Balance $$\frac{d(h_n^L M_n)}{dt} = h_F F + h_{n+1}^L L_{n+1} + h_{n-1}^V V_{n-1} - h_n^L L_n - h_n^V V_n$$

$$\frac{dh_n^L}{dt} \approx 0$$

Liquid Hydraulics (Francis weir) $$L_n = C\rho_n^L w_{\text{len}} H_{\text{ow}}^{1.5} \quad (C \text{ is a constant for units conversion})$$

Vapor–Liquid Equilibrium $$y_{i,n} = K_{i,n}x_{i,n} \quad \Leftrightarrow \quad K_{i,n} = \frac{\gamma_{i,n}P_{i,n}^s}{P_n}$$

FIGURE 3-8. Distillation tray model (2a).

The stage model is the keystone of the dynamic distillation tower model. For each stage in the column, we repeat it in a loop. The stage model presented here can be modified easily to meet your specific modelling needs. For example, if precise froth density is not important for your needs, the correlation can be replaced easily with a constant frothing factor. Sidestream products are added easily by an additional term for the sidestream flow leaving the stage.

We have treated stage pressure as a constant in this model. Variable pressure drop requires the addition of a pressure drop relation expressed in terms of vapor flow. This requires the energy balance to be solved as a differential equation. The change in specific enthalpy is not neglected.

1. Calculate the specific enthalpy and liquid composition by integration.
2. Calculate the tray temperature from enthalpy and tray composition.
3. Calculate the tray pressure from temperature and liquid composition.
4. Calculate the vapor rate leaving the tray from the tray pressure and tray vapor flow hydraulics.

Vapor hydraulics are orders of magnitude faster than liquid composition dynamics. This results in a very stiff set of differential equations. It will require a small time step, a fraction of a second, to integrate the equations. It probably will be more efficient to use a stiff integrator, such as LSODE, to solve this model. Fortunately, vapor and pressure dynamics are not usually important for product composition control problems and we do not have to resort to this model too often.

We can use the stage model we have developed here to help classify different types of distillation stage models. Five models are given in Figure 3-9. The simplest model (1) is appropriate when the molar latent heats of the system components are very close, producing equimolar overflow. The vapor rate through each section of the column is constant, eliminating the energy balance. We integrate the differential mass continuity equations, and the Francis weir formula gives the liquid rate.

Models 2a, 2b, and 3 are the most widely applicable dynamic distillation tower models used by the chemical industry. We have developed model 2a here. I have found this model sufficient to solve 95% of the industrial distillation problems that I have worked on. The remaining 5% were solved using models 2b or 3. Use model 2b for high pressure (greater than 150 psia) systems where the vapor holdup is constant. Use model 3 if there is significant variation in tray pressure drop, or if pressure dynamics are important. Distillation columns operating under vacuum might require model 3. The differential equations become stiff when solving model 3, so model 3 is best solved using an implicit integrator, such as LSODE.

Model 4 is rarely applicable to industrial columns. You need to use this model when there is significant vapor holdup, that is, pressure greater than 150 psia, and variable tray pressure drop is important. Variable tray pressure drop is usually only important in vacuum columns. Yet vacuum columns have insignificant vapor holdup, so model 3 is usually used. Applications for model 4 are very rare, but should the need for it arise, you will be ready. This model is probably solved best using a nonconventional simulation, that is, solving the system of differential and algebraic equations simultaneously. The differential-algebraic integration program DASSL can be used to solve the model.

Model 1

- Equimolal Overflow
- Negligible Vapor Holdup
- Constant Pressure
- Single Liquid Phase

Model 2a

- Negligible Vapor Holdup
- Negligible Specific Enthalpy Change
- Constant Pressure or Tray Pressure Drop

Model 2b

- Significant, but Constant, Vapor Holdup
- Negligible Specific Enthalpy Change
- Constant Pressure or Tray Pressure Drop

Model 3

- Negligible Vapor Holdup
- Significant Specific Enthalpy Change
- Variable Tray Pressure Drop

Model 4

- Significant Vapor Holdup
- Variable Tray Pressure Drop

FIGURE 3-9. Distillation model classifications.

3-3-5 Bottom Sump

The model of the distillation tower bottom sump is similar to that of the tower stage. The bottom sump is modelled as an equilibrium stage. There is heat input from the reboiler. Material holdup is large and variable. Changes in specific enthalpy will not be neglected. The dynamic model is presented in Figure 3-10.

The time derivative of the specific enthalpy can be approximated numerically by backward difference. The specific enthalpy is saved at each time step. The current value is subtracted from the previous value and the result is divided by the time step. It is necessary to compute the specific enthalpy derivative so that the vapor rate leaving the sump can be calculated from the energy balance.

3-3-6 Condenser

The energy dynamics of the condenser are small relative to column composition dy-

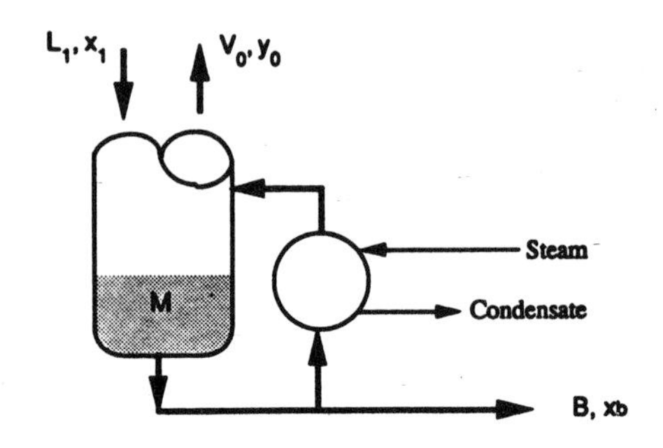

Material Balance $$\frac{dM}{dt} = L_1 - V_0 - B$$

Component Balance $$\frac{d(x_{b,i}M)}{dt} = x_{1,i}L_1 - y_{0,i}V_0 - x_{b,i}B$$

Energy Balance $$\frac{d(h_b^L M)}{dt} = h_1^L L_1 - h_0^V V_0 - h_b^L B + Q_R$$

$$V_0 = \left(h_b^L B + h_b^L \frac{dM}{dt} + M\frac{dh_b^L}{dt} - h_1^L L_1 + Q_R\right)\Big/ h_0^V \quad \text{where } \frac{dh_b^L}{dt} \approx \frac{\Delta h_b^L}{\Delta t}$$

Vapor–Liquid Equilibrium $$y_{0,i} = \frac{\gamma_i P_i^s}{P_0} x_{b,i}$$

FIGURE 3-10. Tower bottom sump model.

namics. We will specify the condensate temperature of the liquid leaving the condenser to allow for subcooling. The condenser duty is equal to the latent heat required to condense the overhead vapor to its bubble point plus the sensible heat for any subcooling. Our simulation reduces the enthalpy of the overhead stream by the condenser duty before the stream is passed to the reflux accumulator:

$$Q_c = H_{\text{in}}F_{\text{in}} - H_{\text{out}}F_{\text{out}} \tag{3-9}$$

where H = stream specific molar enthalpy
F = stream molar flow rate
in = vapor inlet stream into the condenser
out = liquid outlet stream from the condenser

3-3-7 Reflux Accumulator

The reflux accumulator contains a significant and variable quantity of material. There is no heat into or leaving the accumulator and there is no vapor leaving the accumulator. The total enthalpy is computed by integrating the energy balance directly. Dividing the total enthalpy by the total mass gives us the specific enthalpy of the liquid contents. Integrating the material and energy balances produces the liquid composition and specific enthalpy. We must compute the temperature of the accumulator contents from composition and enthalpy. Equation 3-10 can be solved, using Wegstein convergence, to compute the temperature of the contents given specific enthalpy and composition:

$$T = \frac{h^L}{\Sigma_i x_i A_i^L + T\Sigma_i x_i B_i^L} \tag{3-10}$$

where A, B are correlation constants.

3-3-8 Feedback Controllers

Column inventories and product compositions are controlled with proportional (P) and proportional-integral (PI) controllers. I do not like to use proportional-integral-derivative (PID) controllers in dynamic distillation simulations. Derivative action performs better in noise-free simulations than in the real plant. If you use PI controllers in the simulation, your expectations for the plant will be conservative. Derivative action can always be added in the field to improve the response further.

Proportional controllers are used for simple level loops that do not require tight setpoints. Proportional control will allow the steady-state level to move as disturbances pass through the process and smooth the exit flow. Downstream disturbances are minimized. This is very important if this stream is fed to another unit operation. The control law for a proportional controller is

$$C_0 = K_c(P_v - S_p) + B_0 \qquad \text{(3-11a)}$$

where C_0 = controller output (%)
K_c = controller gain (%/%)
P_v = process variable (%)
S_p = setpoint (%)
B_0 = valve bias (%)

All the controller variables are scaled to instrument ranges. The process variable and setpoint are divided by the transmitter span. The controller output is a percentage of the total range of the control valve. The actual flow through the control valve is a function of the controller output and valve flow characteristics. The bias is equal to the desired controller output when the difference between the process and setpoint is zero. Because the bias term is constant, there will be a steady-state offset when the controller moves the output to a different steady state.

A PI controller eliminates steady-state offset. This is desirable for product composition control. The control law for a PI controller is

$$C_0 = K_c\left[(P_v - S_p) + \frac{1}{\tau_i}\int_0^t (P_v - S_p)\,dt\right] \qquad \text{(3-11b)}$$

where C_0 = controller output (%)
K_c = controller gain (%/%)
τ_i = controller reset time (h)
P_v = process variable (%)
S_p = setpoint (%)
t = time (h)

Again, the controller variables are appropriately scaled to instrument ranges. The integral term is evaluated by integrating the operand with the differential continuity equations. The bias term is the error time integral. This term drives the error to zero at all steady states because it varies.

3-4 WRITING A DYNAMIC DISTILLATION SIMULATOR

With the ideas and references I have presented in this chapter you now have enough information to write a dynamic distillation simulator. You can write your simulator using the *conventional* or *nonconventional* approach; it is entirely up to you. Both methods work. Each has benefits over the other depending on your background and the nature of your model. I have used the following procedure to write a *conventional* simulator using model 2a.

1. Create a data file with the tower mechanical, instrument, and physical property data.
2. Specify the initial conditions of the state variables. If you have made previous runs, a snapshot of the process variables is saved in a data file that can be used as the initial conditions. If this is the first run, specify a set of consistent initial conditions.
3. Calculate the derivatives of each stage starting with the bottoms sump and working up the column.

4. Calculate the derivatives for the condenser and reflux accumulator.
5. Calculate the derivatives for the controllers containing integral action.
6. Integrate all the derivatives.
7. Increment time by the integration time step.
8. If necessary, print the display variables and save a snapshot of the current process variables.
9. Go back to step 3 and continue.

In step 2 I suggest that if this is a new simulation consistent initial conditions be chosen. To do this I ask the user to input the feed stream, reflux, heat input, and product compositions. Tray liquid compositions are then computed as a linear profile from top to bottom. A constant vapor rate is used based on the heat input. The liquid rate is set using equimolar overflow and assuming the feed is a liquid. Bubble point, enthalpy, liquid density, and froth density are calculated for each tray. The dynamic simulation is then run from these conditions until it converges to a bona fide steady state.

A method that can be used to simplify programming of the simulator is to integrate the differential equations at each stage instead of integrating the entire tower. The way I have presented these models and simulation approach is well suited to this technique. This avoids creating arrays for all the differential equations. If you use this technique, be very careful to make sure that your variables do not get out of time step. Remember, once you integrate a differential equation, the state variable is at the next time step.

3-5 PLANT-MODEL VERIFICATION

Other chapters in this book discuss methodologies on how to design, control, and operate distillation towers. The dynamic models developed in this chapter may be used as a tool to apply or test these methods. Before you begin to apply your simulation to a real

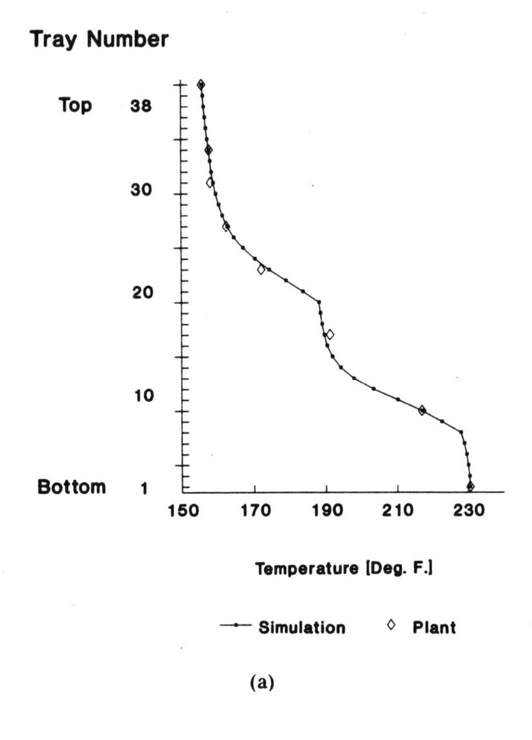

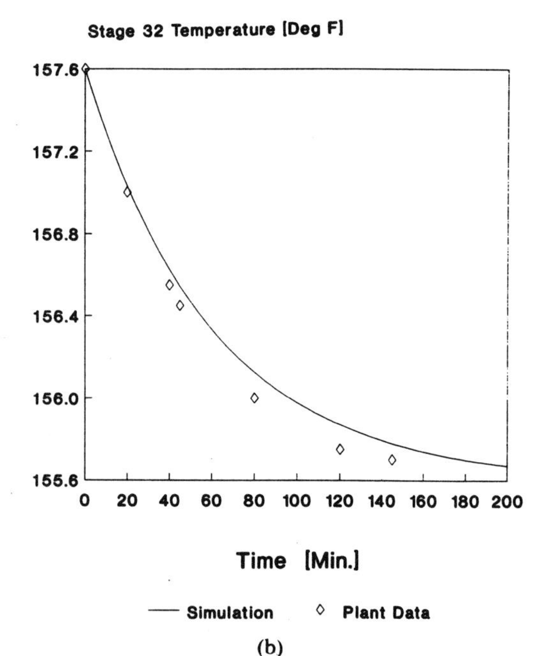

FIGURE 3-11. (*a*) Alcohol tower steady-state temperature profile verification and (*b*) dynamic response verification.

tower, you need to verify that the model does mimic plant operation.

A good model will match the plant at steady state and accurately track the process during dynamic upsets. A verification of the model includes a check of the overall material and energy balances, steady-state temperature profile, and open loop dynamic disturbance tests. A final test of closed loop performance will ensure that your model is accurate. Once you have verified many different columns, you will have confidence in modelling a new design that cannot be verified until startup.

Figures 3-11*a* and 3-11*b* are two verification plots completed for a real 38-tray alcohol–water column. Figure 3-11*a* is the steady-state temperature profile computed using the simulation with plant data overlaid. Figure 3-11*b* is a dynamic open loop test of a tray temperature for a step change in reflux flow. These curves are illustrative of my experience in successfully verifying dynamic distillation models with plant data. The simulation can now be used to develop new and improved process control strategies.

3-6 COMPUTATIONAL PERFORMANCE

Computational performance and productivity have progressed well beyond our expectations over the last 30 years. It is likely to exceed our current expectations in the next few decades.

A decade ago, complex engineering problems were only discussed qualitatively. Today, these problems are quickly quantified and solutions put into production.

We now have the power of supercomputing at an affordable price, either on the desk or tied to it through a network of computers. System software and services have reached the point that engineers without extensive training in computer technology can program steady-state and dynamic models. The modelling activity itself gives the engineer improved process insight and solves very significant problems.

TABLE 3-2 Computer System Execution Speed

Computer	MFLOPS[a]	Real Time Factor[b]
Cray Y-MP (1 processor)	161	860
HP 9000/720	10	580
IBM RS/6000-530	14	390
DECstation 5000/200	4	250
VAX 8700	1	90
Compaq 386/25 MHz	0.2	27

[a] Dongara (1991).
[b] Simulation based on a 7.5 ft, 38 tray alcohol–water column with a setpoint change:

$$\text{real time factor (RTF)} = \frac{\text{process real time}}{\text{elapsed simulation time}}$$

Table 3-2 contains the run time performance, from various computing platforms, of a dynamic distillation simulation. These data correspond to the 38-tray binary alcohol–water distillation tower that we verified in the previous section.

We measure execution performance in terms of a *real time factor*. This is the number of times faster than real time the simulation runs. Computing performance is growing geometrically each year. The desktop, laptop, or fingertop computer of tomorrow will be fast enough that entire plants of five or more towers will be added to the top of Table 3-2.

3-7 CONCLUSIONS

I have presented an approach to modelling distillation dynamics that builds on engineering experience and judgment. The approach is based on modelling the physical elements as modules and connecting the components together in the body of a simulation. The modules chosen are those that are important to solve our engineering objectives for the problem at hand.

We developed a rigorous model for a general distillation tower. The model is based on first principles. We implemented the model in a computer simulation using an explicit integrator. The simulation accurately predicted operating data from a real plant. The simulation runs very fast on a personal computer. Written in FORTRAN 77, it is portable to many other computer platforms. We will use this simulator later in Chapter 18. We will use model 2a to understand the process dynamics and design a control system of an extractive distillation system.

3-8 NOMENCLATURE

B	Controller bias	%
C	Correlation coefficient	
E	Murphree stage efficiency	
E	Specific total energy	Btu/lb-mol
F	Feed flow rate	lb-mol/h
f	Fugacity	psia
H	Specific enthalpy	Btu/lb-mol
h	Specific enthalpy	Btu/lb-mol
K_c	Controller gain	%/%
K	Vapor–liquid composition ratio	
KE	Specific kinetic energy	Btu/lb-mol
L	Liquid flow rate	lb-mol/h
M	Total material	lb-mol
N_c	Number of components	
P_v	Controller process variable	%
P	Pressure	psia
PE	Specific potential energy	Btu/lb-mol
PV	Pressure–volume work	Btu/lb-mol
Q	Heat input	Btu/h
R	Gas constant	
S_P	Controller setpoint	%
T	Temperature	°F
T_r	Reduced temperature (actual/critical)	
t	Time	h
U	Specific internal energy	Btu/lb-mol
V	Vapor flow rate	lb-mol/h
W	Work produced	Btu/h
w	Weir length	ft
x	Liquid composition	mole fraction
y	Vapor composition	mole fraction
z	Feed composition	mole fraction

Greek

ε	Error	
ϕ	Vapor fugacity coefficient	
γ	Liquid activity coefficient	
ν	Liquid specific volume	ft^3/lb-mol
ρ	Liquid density	lb/lb-mol
τ_i	Controller reset time	h

Subscripts

con	Material consumed within system
gen	Material generated within system
in	Flow in system
L	Liquid
len	Length
mix	Heat of mixing
out	Flow from system

Superscripts

*	Equilibrium
L	Liquid
s	Saturated (vapor pressure)
V	Vapor
·	Time rate of change

References

Brenan, K. E., Campbell, S. L., and Petzold, L. R. (1989). *Numerical Solution of Initial-Value Problems in Differential-Algebraic Equations*. New York: North-Holland.

Daubert, T. E. and Danner, R. P. (1988). Data compilation, tables of properties of pure compounds. DIPPR, AIChE, New York.

Dongarra, J. T. (1991). Performance of various computers using standard linear equations software. University of Tennessee, Knoxville, TN.

Franks, R. G. E. (1972). *Modeling and Simulation in Chemical Engineering*. New York: Wiley-Interscience.

Gmehling, J. and Onken, U. (1977). *Vapor–Liquid Equilibrium Data Collection*. Frankfurt, Germany: DECHEMA.

Hindmarsh, A. C. (1980). LSODE: Livermore solver for ordinary differential equations. Lawrence Livermore National Laboratory.

Luyben, W. L. (1990). *Process Modeling Simulation and Control for Chemical Engineers*, 2nd ed. New York: McGraw-Hill.

Reid, R. C., Prausnitz, J. M., and Sherwood, T. K. (1977). *The Properties of Gases and Liquids*. New York: McGraw-Hill.

Van Winkel, M. (1967). *Distillation*. New York: McGraw-Hill.

Walas, S. M. (1985). *Phase Equilibrium in Chemical Engineering*. Boston: Butterworth.

Yaws, C. L. (1977). *Physical Properties*. New York: McGraw-Hill.

4

Approximate and Simplified Models

Antonis Papadourakis
Rohm and Haas Company
John E. Rijnsdorp
University of Twente

4-1 INTRODUCTION

Although the basic principle of distillation is simple, modelling columns with many trays leads to large models with complex overall behavior. In the past, this has encouraged the development of many shortcut models for process design purposes. However, with the present availability of cheap and powerful computer hardware and software it has become possible to utilize rigorous static and dynamic models (see preceding chapter) for off-line use.

Why do we then still need approximate and simplified models? First, we can better understand the behavior of a distillation process. Without understanding it is almost impossible to design good control structures or to improve process controllability. Second, simple models are needed in process computer systems for optimizing and advanced regulatory control, where strict real-time and reliability conditions prohibit complex ones. Third, advanced plant design brings about strongly integrated process flow schemes, which cannot be segregated into individual process units. Then the overall complexity of integrated rigorous dynamic models is even beyond the possibilities of present-day work stations.

In what follows, an attempt is made to look at simple static and dynamic models in a systematic way, classify them according to their major characteristics, and demonstrate the application of the best among them.

4-2 CLASSIFICATION OF SIMPLE MODELS

In general, the model structure has to be in line with the purpose for which it will be used. Table 4-1 shows some common uses of models.

For continuous distillation processes, highly simplified static models can be used to track relatively small changes in the optimum operation point, provided the model is validated by means of a rigorous model or by means of test run data (Hawkins, Tolfo, and Chauvin, 1987). Combining optimizing control with model adaptation requires nonlinear ARMA-type models (Bamberger and Isermann, 1978, have used Hammerstein models for this purpose). In the case of batch distillation the major dynamics are in the kettle, but the rectifier dynamics may not be completely ignored (Betlem, 1991). Here dynamic optimization techniques are only feasible in practice if the overall model is of low order, say of order 10 or less.

TABLE 4-1 Purposes for Simple Distillation Models

Understanding process behavior
Optimization of process operation
Startup, switchover, or shutdown
Assessing process controllability
Designing regulatory control structures
Composition estimation
Model-based control
Simulation, particularly of integrated plants (for operability and controllability analysis and for operator training)

Startup, shutdown, and major switchovers can only be analyzed by nonlinear models. For complete plants, consisting of many process units, order reduction is mandatory, but not very easy. For an approximate analysis of minor switchovers, linear models are sufficient and high order does not impose any great difficulties.

There is a need to assess process controllability in early design phases, when the flowsheet has not yet been frozen. Here singular value decomposition (see Skogestad and Morari, 1987a, for a brief description and a worthwhile extension) has proven to be a simple and valid method (Perkins, 1990). This method requires a (linear) frequency response matrix between process inputs and the variables to be controlled.

The same model can be used for designing the basic control structures, which includes a separation between single-input–single-output (SISO) and multiple-input–multiple output (MIMO) subsystems. For the latter, the block relative gain array method (Arkun, 1987) has proven to be quite useful.

Real-time composition estimation and model-based control are only practical when the model order is low. In addition, in case of high-purity separations, nonlinearities play an important role, which brings us again to the problem of order reduction in nonlinear models (see Section 4-6 for more detail).

Finally, in later process design phases, simulation of complete plants is desirable for a more thorough analysis of plant operability, controllability, and switchability, and for operator training. In the latter case, speed of computation is more critical than in the former case, which warrants model simplification.

4-3 SIMPLE STEADY-STATE MODELS

As has already been indicated in the preceding section, simple models can be used for tracing optimum operation points, provided they are validated for the actual operating region. A very simple model has been proposed by Shinskey (1967):

$$V/F = \beta_c \ln S_c \tag{4-1}$$

where V = vapor flow rate (mol/s)
F = feed flow rate (mol/s)
β_c = column characterization factor
S_c = separation factor, equal to

$$S_c = (LK_D/LK_B)(HK_B/HK_D) \tag{4-2}$$

where LK = mole fraction of the light key component
HK = mole fraction of the heavy key component
D = top product
B = bottom product

The column characterization factor is a function of the number of trays, the tray efficiency, and the relative volatility of the two key components. It is the parameter to be fitted to the operating region.

Equation 4-1 does not contain the influences of feed enthalpy and feed tray location. The first type influence can be accounted for by defining separation factors for each column section (Rijnsdorp, 1991):

$$L_R/D = \beta_R \ln S_R \tag{4-3}$$

$$V_S/B = \beta_S \ln S_S \tag{4-4}$$

where L_R = liquid flow in the rectifying section (mol/s)
V_S = vapor flow in the stripping section (mol/s)
S_R, S_S = characterization factors for the rectifying and the stripping section, respectively, defined by

$$S_R = (LK_D/LK_F)(HK_F/HK_D) \quad (4\text{-}5)$$

$$S_S = (LK_F/LK_B)(HK_B/HK_F) \quad (4\text{-}6)$$

where LK_F = mole fraction of the light key component at the feed tray
HK_F = mole fraction of the heavy key component at the feed tray

Equations 4-3 to 4-6 can be combined, with substitution of 4-2, into

$$\ln S_c = (L_R/D)/\beta_R + (V_S/B)/\beta_S \quad (4\text{-}7)$$

Most column design methods implicitly assume that the feed enters at the optimum location. However, in actual operation the feed tray location is rarely adapted to the prevailing conditions (although this could be quite worthwhile for large deviations from the design conditions). The effect on the column performance can be accounted for by comparing the ratio of the key component mole fractions at the feed tray according to Equations 4-3 and 4-4 to the optimal mole fractions, which are given by the intersection of the operating lines for the two column sections:

$$y_F = (L_R/V_R)x_F + (D/V_R)x_D \quad (4\text{-}8)$$

$$y_F = (L_S/V_S)x_F - (B/V_S)x_B \quad (4\text{-}9)$$

where x_F = component mole fraction in the liquid on the feed tray
y_F = component mole fraction in the vapor above the feed tray

Solution yields

$$x_F = \frac{(D/V_R)x_D + (B/V_S)x_B}{L_S/V_S - L_R/V_R} \quad (4\text{-}10)$$

The difference between the actual (according to Equations 4-3 and 4-4) and the optimal (according to Equation 4-10) ratio of the two key component fractions is now used to increase the values of the characterization factors in a suitable way, for instance,

$$\beta_R = \beta_{R,\text{opt}}\left[1 + c_R\left(\text{pos}\{(LK_F/HK_F)_{\text{opt}} - (LK_F/HK_F)_{\text{act}}\}\right)^2\right] \quad (4\text{-}11)$$

$$\beta_S = \beta_{S,\text{opt}}\left[1 + c_S\left(\text{pos}\{(LK_F/HK_F)_{\text{act}} - (LK_F/HK_F)_{\text{opt}}\}\right)^2\right] \quad (4\text{-}12)$$

where c_R, c_S = parameters to be determined by comparison with tray-to-tray calculations
pos{ } = its argument if the latter is positive, and equals zero if the latter is nonpositive

4-4 PARTITIONING OF THE OVERALL DYNAMIC MODEL

4-4-1 Introduction

At first sight, a complete dynamic model of distillation column behavior offers a bewildering interaction between many variables. For every tray there are pressures, vapor and liquid flows, and vapor and liquid concentrations. Similar variables also appear for the reboiler, the condenser, and the top and bottom accumulators. It would be very worthwhile to partition these variables into a number of clusters, so that the overall dynamics can be represented by less complicated submodels, each associated with a particular control issue.

As will be shown in the next subsection, this can be realized under relatively mild assumptions, with submodels for the following clusters of process variables (see Figure 4-1):

1. Pressure and vapor flow rates, associated with pressure and pressure drop control.
2. Liquid holdups and flow rates, associated with liquid level control in the top and bottom accumulators.
3. Vapor and liquid concentrations, associated with temperature and composition control.
4. Condenser temperatures and heat flow rates.
5. Reboiler temperatures and heat flow rates.

The first three submodels have no interactions, but there is an important action from the first to the second and third, and

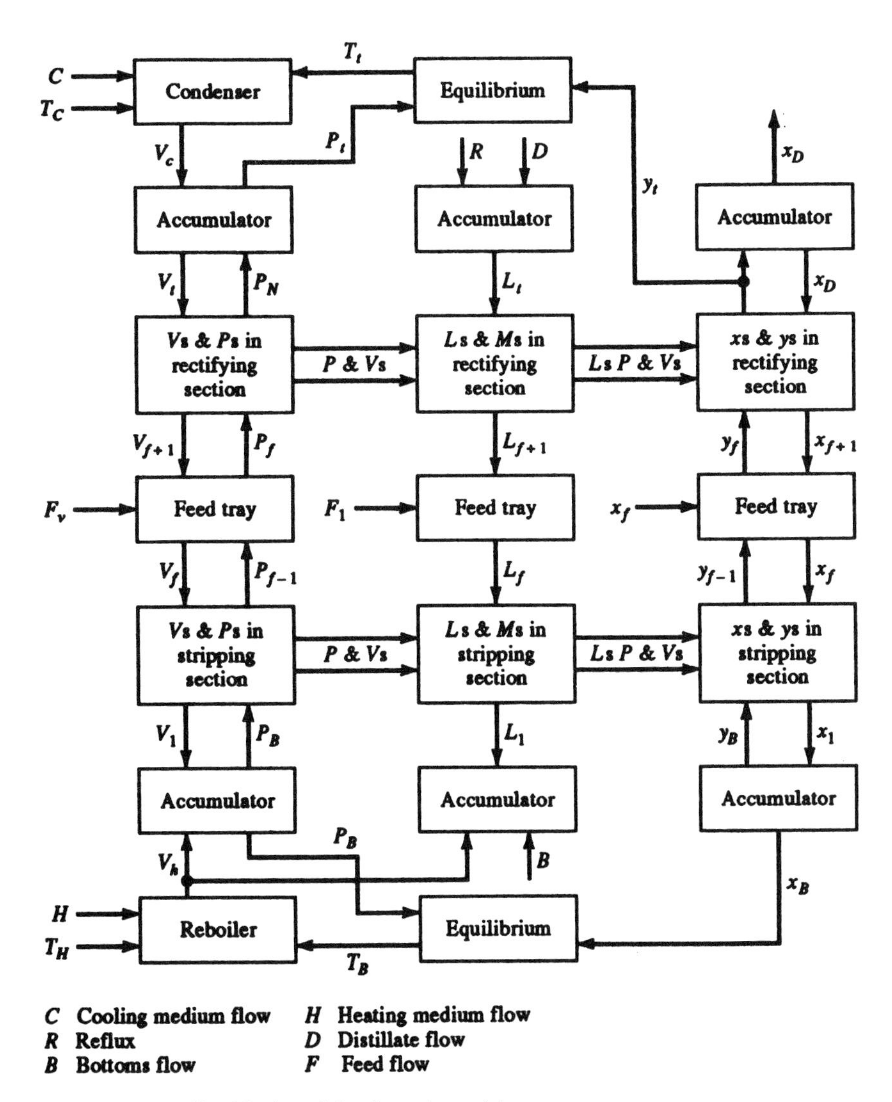

FIGURE 4-1. Partitioning of the dynamic model.

from the second to the third. In fact, the influences from the latter on the former mainly go through the reboiler and the condenser submodels, which cause inherent regulation and deregulation, respectively.

4-4-2 Assumptions

The assumptions are:

- Equal (pseudo) molal overflow.
- Small influence from liquid tray holdup on tray pressure drop.
- Tray pressure drop small compared to the absolute value of the pressure.
- Negligible heat loss.

In distillation column design, a popular assumption is that of equal molal overflow. In terms of the relations between specific enthalpies and compositions this assumption is equivalent to (where the summation is over all except the least volatile component)

$$H_L(s,t) = H_0(s,t) - \Sigma a_j x_j(x,t) \quad (4\text{-}13)$$

and

$$H_v(s,t) = H_0(s,t) + H_c(s,t) - \Sigma a_j y_j(s,t) \quad (4\text{-}14)$$

where $x_j(s,t)$ = liquid mole fraction for component j on tray s
$y_j(s,t)$ = vapor mole fraction for component j on tray s
a_j = sensitivity coefficient for component j, assumed to be equal for vapor and liquid
$H_0(s,t)$ = specific liquid enthalpy of the least volatile component (J/mol)
$H_c(s,t)$ = heat of vaporization of the least volatile component (J/mol)

Evidently, in the enthalpy–composition hyperspace this corresponds to parallel hyperplanes for liquid and vapor enthalpy. Robinson and Gilliland (1950) have extended the scope by introducing pseudomolecular weights. In this way, unequal sensitivity coefficients for vapor and liquid can be equalized. The accuracy of the method has been demonstrated for the design of a column separating the nonideal mixture of ammonia and water. Chen (1969) has shown that this method is also consistent for the dynamic case.

In actual trays, changes in liquid holdup will influence the pressure drop across the tray. For trays with downcomers within the normal operating range (where most of the liquid holdup variations appear in the downcomer), the liquid height on the tray proper does not vary much, so the influence of the vapor flow is most important. Of course this no longer holds when the trays are overloaded, that is, when the downcomers start to overflow.

The third assumption even holds for vacuum operation, possibly with exception for trays near the top where the absolute pressure is very low. Finally, the fourth assumption is warranted for well insulated big columns and for small columns provided with a compensatory heating system.

On the basis of the assumptions, the energy balance of the tray can be reduced to the following, very simple balance for the vapor flows (Rademaker, Rijnsdorp, and Maarleveld, 1975):

$$B(s,t)\frac{dP(s,t)}{dt} = V(s-1,t) - V(s,t) \quad (4\text{-}15)$$

where $B(s,t)$ = effective vapor capacity of the tray
$P(s,t)$ = pressure on the tray

The physical interpretation is that an increasing pressure requires compression of the vapor holdup and condensation of vapor for heating the tray to the new equilibrium temperature, both resulting in a difference between the vapor flow entering the tray and that leaving the tray.

The effect of resistance to mass transfer between vapor and liquid can be accounted for by a more refined version of Equation 4-15:

$$B^*(s,t)\frac{dP^*(s,t)}{dt} + B_V(s,t)\frac{dP(s,t)}{dt} = V(s-1,t) - V(s,t) \quad (4\text{-}16)$$

where $B_V(s,t)$ = vapor compression term of $B(s,t)$

$B^*(s,t)$ = remainder of $B(s,t)$

$P^*(s,t)$ = pressure in equilibrium with the liquid temperature, determined by

$$B^*(s,t)\frac{dP^*(s,t)}{dt} = \frac{[P(s,t) - P^*(s,t)]}{R_p(s,t)} \quad (4\text{-}17)$$

where $R_p(s,t)$ is the resistance to mass transfer between the phases on tray s (Pa s/ mol).

4-4-3 Propagation of Vapor Flow and Pressure Responses

The equations for the vapor flow balances across the trays (see Equation 4-15, or Equation 4-16 together with Equation 4-17) form the basis of the submodel for the pressures and the vapor flows. We just have to extend it with an equation for the tray pressure drop. According to the second assumption, the latter only depends on the vapor flow rate. Now Equation 4-16 can be approximated by

$$B^*(s,t)\frac{dP^*(s,t)}{dt} + B_V(s,t)\frac{dP(s,t)}{dt} = f_V\{P(s-1,t) - P(s,t)\} - f_V\{P(s,t) - P(s+1,t)\} \quad (4\text{-}18)$$

where $f_V\{\ \}$ represents the relationship between vapor flow rate and pressure drop.

For n trays, Equations 4-17 and 4-18 lead to a set of state equations with the $2n$ tray pressures [the $P(s,t)$s and the $P^*(s,t)$s] as state variables. In analogy to electrical networks it can also be interpreted as a resistance–capacitance ladder (see Figure 4-2) with nonlinear resistances. Note that this network does not tell the complete story. In fact, the condenser reacts to the change in top vapor flow and so provides a feedback to the submodel. A similar influence occurs at the bottom end via the reboiler.

4-4-4 Propagation of Liquid Flow and Liquid Holdup Variations

The responses of liquid holdups and flow rates are determined by the total mass bal-

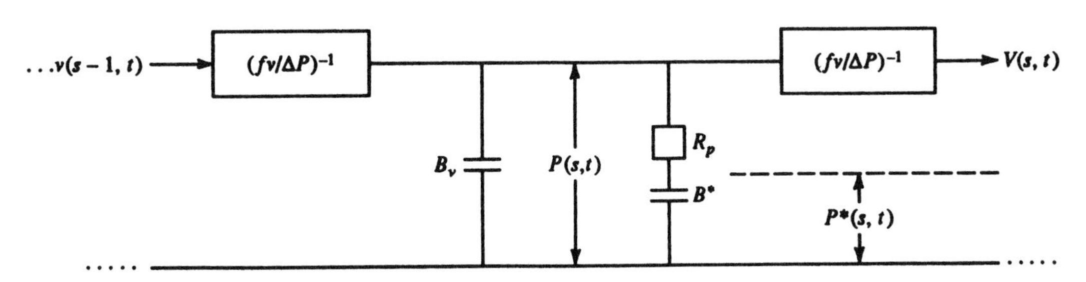

FIGURE 4-2. Electrical network interpretation of propagation of vapor flow and pressure responses.

ance of the tray:

$$\frac{dM_L(s,t)}{dt} = L(s+1,t) - L(s,t) + V(s-1,t) - V(s,t) \quad (4\text{-}19)$$

With Equation 4-15 and with the hydraulic equation for the tray,

$$M_L(s,t) = f_M[L(s,t), V(s-1,t)] \quad (4\text{-}20)$$

we find

$$\tau_L(s,t)\frac{dL^*(s,t)}{dt} = L^*(s+1,t) - L^*(s,t) + [1 - K_{VL}(s,t)]B(s,t)\frac{dP(s,t)}{dt} \quad (4\text{-}21)$$

where, by definition,

$$L^*(s,t) = L(s,t) - K_{VL}(s,t)V(s-1,t) \quad (4\text{-}22)$$

$$\tau_L(s,t) = \left(\frac{\partial M_L(s,t)}{\partial L(s,t)}\right)_{V(s-1,t)} \quad (4\text{-}23)$$

$$K_{VL}(s,t) = \left(\frac{\partial L}{\partial V}\right)_{M_L} \quad (4\text{-}24)$$

As K_{VL} is poorly known, we have taken its average value, which also helps to keep the model simple. When the total pressure drop across the trays is small compared to the average (absolute) pressure, the propagation of pressure variations is fast compared to the propagation of liquid flow variations (Rademaker, Rijnsdorp, and Maarleveld, 1975). Then $P(s,t)$ can be replaced by its average $P(t)$.

4-4-5 Propagation of Vapor and Liquid Concentrations

Finally, we come to the concentration responses. These are determined by the partial mass balances of the trays:

$$\frac{d[M_L(s,t)x_j(s,t)]}{dt} + \frac{d[M_V(s,t)y_j(s,t)]}{dt} = L(s+1,t)x_j(s+1,t) - L(s,t)x_j(s,t) + V(s-1,t)y_j(s-1,t) - V(s,t)y_j(s,t) \quad (4\text{-}25)$$

where $x_j(s,t)$ = mole fraction of component j in the liquid on tray s

$y_j(s,t)$ = mole fraction of component j in the vapor on tray s

Subtracting the total mass balance, multiplied by $x_j(s,t)$, yields

$$M_L(s,t)\frac{dx_j(s,t)}{dt} + M_V(s,t)\frac{dy_j(s,t)}{dt} = L(s+1,t)[x_j(s+1,t) - x_j(s,t)] + V(s-1,t)[y_j(s-1,t) - y_j(s,t)] + [y_j(s,t) - x_j(s,t)] \times B^*(s,t)\frac{dP(s,t)}{dt} \quad (4\text{-}26)$$

where use has been made of Equation 4-15.

Note that the liquid flow from above, the vapor flow from below, and the rate of change of the tray pressure constitute the inputs to this tray model for the concentrations. In their turn the top concentration influences the top pressure by way of the condenser, and the bottom concentration influences the bottom pressure by way of the reboiler.

For the case of theoretical trays, the vapor concentrations are related to the liquid composition, so Equation 4-26 constitutes a set of $(n_j - 1)$ (the number of components minus 1) times n (the number of trays) simultaneous differential equations. In the binary case, there is just one differential equation per tray.

4-5 LINEAR MODELS

The linear models for distillation columns are essentially linearized versions of the original nonlinear models that are developed based on mass and energy balances around each equilibrium stage. In the vast majority of published work, only composition dynamics are considered, and in very few cases the flow dynamics have been superimposed to obtain a more accurate description. Usually, the nonlinear model is linearized around some steady state (i.e., the operating state) of the column, and the study of the resulting linear model reflects the dynamics of the process around this steady state. For a more comprehensive examination of the existing linear models of distillation columns, we distinguish them according to the domain they are presented in:

1. Linear models in the time domain
2. Linear models in the Laplace domain
3. Linear models in the frequency domain

Because a model presented in one domain can be transformed to a model of another domain, there is an ambiguity in such a classification. However, as we will see in the following paragraphs, there are certain domain-dependent properties of these models.

4-5-1 Linear Models in the Time Domain

Modal reduction techniques can be used for the approximation of high-order linear systems in the time domain (Bonvin and Mellichamp, 1982a). Several techniques are available in the literature, and a critical review of six of them is given by Bonvin and Mellichamp (1982b). These techniques can be applied to chemical engineering processes in general, and to distillation in particular.

A modal reduction technique especially suited for staged processes was developed by Georgakis and Stoever (Georgakis and Stoever, 1982; Stoever and Georgakis, 1982). Their method of nonuniform lumping (and uniform lumping as a special case) can be used for the order reduction of tridiagonal dynamic models of staged processes. The advantages of their method are good retention of the dominant time constants, small deviation of the eigenrows, physical significance of the states of the reduced model, and flexibility in the choice of lumping. The disadvantages of the method include the inaccurate simulation of the deadtimes, as well as that the initial response of the reduced model can be fundamentally different from that of the original model. Also, the requirement for a tridiagonal form of the Jacobian matrix of the system limits the applicability of their method.

4-5-2 Linear Models in the Laplace Domain

It has been observed by many distillation control practitioners that the nonlinear dynamic composition response of distillation columns often resembles a linear first-order response. It is also known from theoretical studies (Levy, Foss, and Grens, 1969) that the dominant time constant of a distillation column is well separated from the rest and is nearly the same regardless of where a disturbance is introduced or where composition is measured (Skogestad and Morari, 1987b). In addition, if constant molar flows are assumed, then the flow dynamics are decoupled from the composition dynamics, and the dominant part of the dynamics can be captured by modelling only the composition dynamics.

The preceding realizations have had a twofold effect. First, they have led many investigators to attempt representation of the composition dynamics of distillation columns using models of the simple forms

$$\frac{Ke^{-ds}}{(\tau_1 s + 1)} \quad \text{or} \quad \frac{K(\tau_0 s + 1)e^{-ds}}{(\tau_1 s + 1)(\tau_2 s + 1)}$$

where τ_1 is the dominant time constant of the column. Many investigators focused their attention on developing such transfer functions for distillation columns, assuming that a linearized model for the column is available. Kim and Friedly (1974) used the fact that distillation columns exhibit a dominant eigenvalue well separated from the rest (i.e., $\lambda_1 \gg \lambda_2 > \cdots > \lambda_n$) to approximate a normalized transfer function of the form

$$\frac{1}{\prod_{j=1}^{n}(1 - s/\lambda_j)} \tag{4-27}$$

by a second-order-plus-deadtime model, where the two dominant eigenvalues are used as the time constants and the rest of the eigenvalues are lumped as the deadtime:

$$\frac{e^{-ds}}{(\tau_1 s + 1)(\tau_2 s + 1)} \tag{4-28}$$

with

$$\tau_1 = -\frac{1}{\lambda_1}, \qquad \tau_2 = -\frac{1}{\lambda_2},$$
$$d = \sum_{j=3}^{n}\left(-\frac{1}{\lambda_j}\right) \tag{4-29}$$

The transfer function given by Equation 4-27 has no numerator dynamics. Such transfer functions arise in staged systems when considering the response of a variable at one end of a cascade to a disturbance introduced at the opposite end. In cases where the transfer function of the column contains numerator dynamics (i.e., zeros), the numerator and denominator dynamics of the column can be approximated separately (Kim and Friedly, 1974; Celebi and Chimowitz, 1985; Papadourakis, Doherty, and Douglas, 1989). The approximation of the denominator dynamics provides the dominant time constants (poles) and a contribution to the deadtime, whereas the approximation of the numerator dynamics provides the zero(s) and the remainder of the deadtime. In other words, a normalized transfer function of the form

$$\frac{\prod_{i=1}^{m}(1 - s/\mu_i)}{\prod_{j=1}^{n}(1 - s/\lambda_j)} \tag{4-30}$$

can be approximated by

$$\frac{(\tau_0 s + 1)e^{-ds}}{(\tau_1 s + 1)(\tau_2 s + 1)} \tag{4-31}$$

where

$$\tau_0 = -\frac{1}{\mu_1}, \qquad \tau_1 = -\frac{1}{\lambda_1}, \qquad \tau_2 = -\frac{1}{\lambda_2},$$
$$d = \sum_{j=3}^{n}\left(-\frac{1}{\lambda_j}\right) - \sum_{i=2}^{m}\left(-\frac{1}{\mu_i}\right) \tag{4-32}$$

with $\lambda_1 \gg \lambda_2 > \cdots > \lambda_n$ and $\mu_1 > \mu_2 > \cdots > \mu_m$.

Celebi and Chimowitz (1985) provide a method for obtaining the rigorous transfer function (Equation 4-30) for a distillation column, but their method is useful only when considering the response of the column to changes in feed composition.

In general, if the full linearized model of a distillation column is available in analytical form in the Laplace domain, many order-reduction methods can be used to produce an approximate model similar to the one given by Equation 4-31. Surveys of these methods (i.e., moment matching methods, cumulant matching methods, Routh approximations, impulse energy approximations, least squares approximations, etc.) are given by Genesio and Milanese (1976) and by Papadourakis, Doherty, and Douglas (1989). However, the applicability of the preceding methods is limited because the full transfer function model of a distillation column is seldom available in analytical form.

The second effect of the realization that the nonlinear distillation dynamics are dominated by a time constant well separated from the rest is a large number of attempts to relate the dominant time constant to steady-state parameters. Many investigators (Moczek, Otto, and Williams, 1965; Bhat and Williams, 1969; Wahl and Harriot, 1970; Toijala (Waller) and his co-workers, 1968–1978; Weigand, Jhawar, and Williams, 1972; Skogestad and Morari, 1987b, 1988) have tried to estimate this time constant, which appears to be equal to "the change in column holdup of one component over the imbalance in supply of this component," that is,

$$\tau_c = \frac{\text{change in total column holdup of one component (mol)}}{\text{imbalance in supply of this component (mol/min)}}$$

Skogestad and Morari (1987b) developed an analytical expression for the calculation of the dominant time constant based on steady-state parameters only. Such a formulation also had been suggested by Rijnsdorp (see discussion in Moczek, Otto, and Williams, 1963). According to Skogestad and Morari's work, the dominant time constant of the composition dynamics of a two-product distillation column is given by

$$\tau_c = \frac{\Sigma_{i=1}^{N_T} M_{L_{if}} \Delta x_i + M_{L_{B_f}} \Delta x_B + M_{L_{D_f}} \Delta x_D}{D_f \, \Delta y_D + B_f \, \Delta x_B} \tag{4-33}$$

where $\Delta x = x_{\text{initial}} - x_{\text{final}}$
N_T = total number of trays (excluding the condenser and the reboiler)
M_{L_i} = liquid holdup of tray i
M_{L_B} = liquid holdup of the reboiler
M_{L_D} = liquid holdup of the condenser

the subscript f denotes the final steady state reached by the column after a step change has been introduced and subscript initial refers to the initial steady state of the column.

The derivation of τ_c is based on two assumptions:

Assumption I
The flow dynamics are immediate (i.e., the holdups and the flows change instantaneously).

Assumption II
All the trays have the same dynamic response, which corresponds to treating the column as a large mixing tank.

The following comments apply on the derivation of τ_c:

1. The column model, on which the derivation of τ_c is based, is not linearized.
2. τ_c depends on the magnitude and direction of change.
3. Equation 4-33 applies to any component in a multicomponent mixture.
4. Equation 4-33 applies to any change that changes the external material balance of the column:

$$D_f \, \Delta y_D + B_f \, \Delta x_B \neq 0$$

5. To compute τ_c, a steady-state model of the column is needed in order to calculate the compositions on all stages for both the initial and the final steady states.
6. For small perturbations to columns with both products of high purity, very large time constants are found.
7. There are three contributions to τ_c (caused by the holdup inside the column, in the condenser, and in the reboiler).

As follows from the first comment, the time constant given by Equation 4-33 is the nonlinear time constant of the composition dynamics of the column. The main advantages of this method of evaluating the time constant is that it depends on steady-state data only and that it can be applied to multicomponent separations. The disadvantage of the method is that it requires the availability of a nonlinear steady-state model (not a great requirement in today's world) and the calculation of compositions on all trays.

A simple formula for a linearized time constant limited to binary separations was also developed by Skogestad and Morari (1987b). The derivation of this formula was based on three more assumptions, in addition to the two assumptions made for the derivation of τ_c:

Assumption III
All the trays (except the condenser and the reboiler) have equal and constant holdup M_L.

Assumption IV
The average composition inside the column is given by

$$\overline{x_I} = \left(1 + \frac{\ln((1 - x_B)/x_B)}{\ln(y_D/(1 - y_D))}\right)^{-1} \tag{4-34}$$

Assumption V
The steady-state gains may be estimated by assuming that the separation factor $S = y_D(1 - x_B)/(1 - y_D)x_B$ is constant for any given change.

Based on these assumptions, the linearized value of τ_c is given by

$$\tau_{sc} = \frac{M_I}{I_s \ln S} + \frac{M_{L_D}(1 - y_D)y_D}{I_s} + \frac{M_{L_B}(1 - x_B)x_B}{I_s} \tag{4-35}$$

where M_I is the total holdup of liquid inside the column,

$$M_I = N_T * M_L \tag{4-36}$$

I_s is the "impurity sum,"

$$I_s = Dy_D(1 - y_D) + Bx_B(1 - x_B) \tag{4-37}$$

and S is the separation factor,

$$S = \frac{y_D(1 - x_B)}{(1 - y_D)x_B} \tag{4-38}$$

τ_{sc} is expected to be in very good agreement with τ_c for cases where Assumptions IV and V hold (i.e., for high-purity columns with high reflux). The advantages of the analytical formula for τ_{sc} is that it depends only on readily available steady-state data of the initial operating state of the column, it does not require the compositions on all stages, and it does not require the availability of a nonlinear steady-state model.

The main value of τ_{sc} is that it can provide a good understanding of the nonlinear dynamics of the column and their dependence on steady-state parameters. For example, and taking into account the fact that the contribution to τ_c from the holdup inside the column ($M_I/I_s \ln S$) is dominant, it is easy to see that the time constant is determined mainly by I_s (because ln S does not change significantly with the operating conditions).

The usefulness and simplicity of the method is illustrated by the following example.

Example 4-1

Consider the binary column shown in Figure 4-3 with steady-state data given in Table 4-2. A disturbance was introduced by reducing the distillate flow rate by 0.01%. The nonlinear column response (obtained by applying material balances on each equilibrium stage) was fitted by eye to a second-order linear response $1/(1 + \tau_1 s)(1 + \tau_2 s)$. Both responses are shown in Figure 4-4. The dominant time constant τ_1 of the fitted response is 4.3 h, whereas the time constant

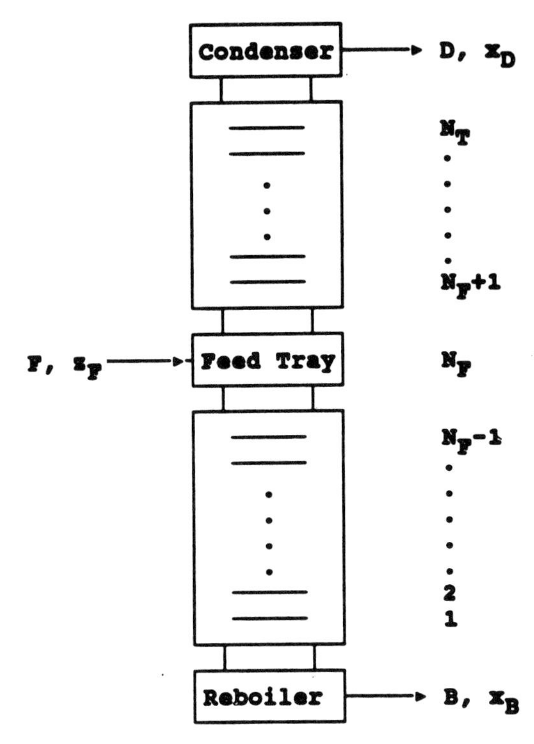

FIGURE 4-3. Binary distillation column used in Example 4-1.

TABLE 4-2 Steady-State Data for Column of Example 4-1

Parameter	Symbol	Value	Units
Feed composition	z_F	0.4	
Distillate composition	x_D	0.99285	
Bottoms composition	x_B	0.00048	
Feed quality	q	1.0	
Relative volatility	α	2.0	
Number of trays	N_T	23	
Feed tray location	N_F	12	
Distillate/feed flow	D/F	0.4	
Reflux/feed flow	L/F	1.6	
Tray holdup/feed flow	M_T/F	0.01	hours
Cond. holdup/feed flow	M_D/F	0.1	hours
Reb. holdup/feed flow	M_B/F	0.1	hours

τ_c calculated by Equation 4-33 was found to be 3.94 h (i.e., an error of only 8%). The value of τ_{sc} calculated by Equation 4-35 for the same column was found to be 4.15 h (i.e., an error of only 3.5%).

4-5-3 Linear Models in the Frequency Domain

For small variations about an equilibrium point, Equation 4-18 can be linearized. Us-

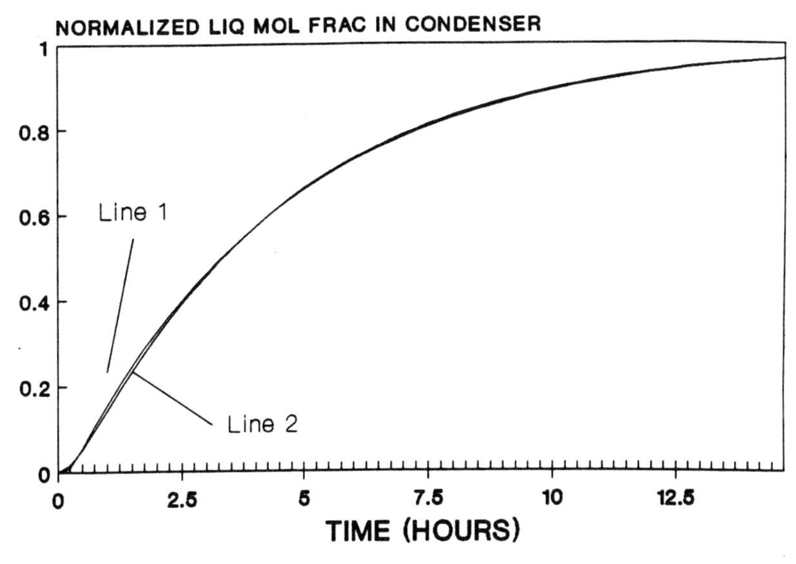

FIGURE 4-4. Transient response of the condenser of the distillation column of Example 4-1 to a step decrease of the distillate flow rate by 0.01%. Line 1: full model; line 2: fitted model $1/(1 + \tau_1 s)(1 + \tau_2 s)$.

ing average values for $B^*(s,t)$, $B_V(s,t)$, $R_c(s,t)$, and the resistance value $R_{V'}(s,t)$, and transforming to the frequency domain gives

$$j\omega\tau_{P,V} + \frac{j\omega\tau_{P,L}}{1 + j\omega\tau_c}\,\delta P(s, j\omega) = \delta P(s-1, j\omega) - 2\,\delta P(s, j\omega) + \delta P(s+1, j\omega) \tag{4-39}$$

where

$$\tau_{P,V} = B_V R_V \tag{4-40}$$

$$\tau_{P,L} = B^* R_V \tag{4-41}$$

$$\tau_c = R_c B^* \tag{4-42}$$

the average time constant for the condensation–vaporization phenomenon. This formula represents a set of linear difference equations, which can be solved analytically. The result is

$$\delta P(s, j\omega) = \frac{(J_1^s - J_2^s)\,\delta P(m, j\omega)}{J_1^m - J_2^m} + \frac{(J_1^{m-s} - J_2^{m-s})\,\delta P(0, j\omega)}{J_1^m - J_2^m} \tag{4-43}$$

where $\delta P(m, j\omega)$ = pressure variation above the top tray (Pa)

δP_0 = pressure variation below the bottom tray (Pa)

J_1, J_2 = roots of the characteristic equation

$$J^2 - \left(2 + j\omega\left[\tau_{P,V} + \frac{\tau_{P,L}}{1 + j\omega\tau_c}\right]\right)J + 1 = 0 \tag{4-44}$$

The transfer functions for the vapor flow responses can be found simply from the linearized equation for the pressure drop across the tray:

$$\delta V(s-1, j\omega) = \frac{\delta P(s-1, j\omega) - \delta P(s, j\omega)}{R_V} \tag{4-45}$$

For small deviations from equilibrium and equal parameter values for the various trays, the equation for the liquid flow variations, Equation 4-21, is linearized to

$$j\omega\tau_L\,\delta L^*(s, j\omega) = \delta L^*(s+1, j\omega) - \delta L^*(s, j\omega) + (1 - K_{VL})\,j\omega B^*\,\delta P(j\omega) \tag{4-46}$$

Repeated application for a series of trays yields

$$\delta L^*(1, j\omega) = \frac{\delta L^*(M+1, j\omega)}{(1 + j\omega\tau_L)^m} + \delta P(j\omega)(1 - K_{VL}) \times \left(\frac{B^*}{\tau_L}\right)\left[1 - \frac{1}{(1 + j\omega\tau_L)^m}\right] \tag{4-47}$$

with Equation 4-41;

$$\delta L(1, j\omega) = \frac{\delta L^*(m+1, j\omega)}{(1 + j\omega\tau_L)^m} + K_{VL}\left[\delta V(0, j\omega) - \frac{\delta V(m, j\omega)}{(1 + j\omega\tau_L)^m}\right] + \delta P(j\omega)(1 - K_{VL})\left(\frac{B}{\tau_L}\right) \times \left[1 - \frac{1}{(1 + j\omega\tau_L)^n}\right] \tag{4-48}$$

The tray parameter K_{VL} has a strong impact on the response of the bottom accumulator level to the vapor flow. If it equals 1, then the term $K_{VL}\,\delta V(0, j\omega)$ just cancels the direct effect of the vapor flow variation, so the bottom level has to wait for the term with $\delta V(m, j\omega)$, which is delayed by n times the tray time constant τ_L. This makes bottom level control by vapor flow as slow as by reflux flow. In practice this can force column designers to increase the size of the bottom holdup, and so make the column more expensive.

Similar to the preceding submodels, one can develop a linearized version of the concentration submodel with average values for tray holdups, capacities, and pressures. Approximating the equilibrium relationship between vapor and liquid by a straight line,

$$y = e_0 + e_1 x \tag{4-49}$$

Substitution into Equation 4-26 yields

$$\begin{aligned}&[M_L(s,t) + e_1 M_V(s,t)]\frac{dx(s,t)}{dt}\\ &= L(s+1,t)[x(s+1,t) - x(s,t)]\\ &\quad + e_1 V(s-1,t)[x(s-1,t) - x(s,t)]\\ &\quad + [(e_1 - 1)x(s,t) + e_0]B(s,t)\frac{dP(s,t)}{dt}\end{aligned} \tag{4-50}$$

By linearization and transforming Equation 4-26 to the frequency domain,

$$\begin{aligned}&j\omega\tau_x\,\delta x(s,j\omega)\\ &= [\delta x(s+1,j\omega) - \delta x(s,j\omega)]\\ &\quad + E[\delta x(s-1,j\omega) - \delta x(s,j\omega)]\\ &\quad + j\omega[(e_1 - 1)\,\delta x(s) + e_0]B^*\,\delta P(s,j\omega)\\ &\quad + [x(s+1) - x(s)]\\ &\quad \times\left[\frac{\delta L(s+1,j\omega)}{L} - \frac{\delta V(s-1,j\omega)}{V}\right]\end{aligned} \tag{4-51}$$

where

$$E = e_1\frac{V}{L} \tag{4-52}$$

$$\tau_x = \frac{M_L + e_1 M_V}{L} \tag{4-53}$$

and use has been made of the static partial mass balance of the tray,

$$\begin{aligned}&L[x(s+1) - x(s)]\\ &\quad + V[y(s-1) - y(s)] = 0\end{aligned} \tag{4-54}$$

The pressure-dependent term in Equation 4-53 is rather awkward to handle analytically (Rademaker, Rijnsdorp, and Maarleveld, 1975, state the formidable analytical expression for its effect on the concentration), but solving for the vapor and the liquid flow is straightforward.

The analytical expressions for the responses of tray pressures, vapor flows, liquid flow, liquid holdups, and liquid and vapor concentrations are very convenient for the design of regulatory control structures.

4-6 NONLINEAR MODELS

The number of equations constituting a nonlinear model for a distillation column depends primarily on three factors:

1. Simplifying assumptions
2. Number of components
3. Number of stages

In what follows, we will try to examine the effect of these three factors on the complexity and size of the models.

4-6-1 Simplifying Assumptions

The most rigorous model for describing the transient behavior of stagewise distillation processes should include a large number of variables as state variables (i.e., tray liquid compositions, liquid enthalpies, liquid holdups, vapor enthalpies, vapor holdups, etc.). By employing certain assumptions, which we will call *simplifying assumptions*, some of the state variables can be eliminated, and the order of the model can be significantly reduced. Such simplifying assumptions are:

a. Negligible vapor holdup
b. Fast vapor enthalpy dynamics
c. Fast flow dynamics
d. Fast liquid enthalpy dynamics

Assumptions a and b are more or less standard in the literature of distillation dynamics because interest is focused on composition dynamics and control. Assumption c has been employed by many investigators and is only reasonable if the measurement lags in quality analysis are dominant (that is, the flow dynamics are fast compared with the composition dynamics). In case the flow dynamics are significant, they can be treated separately from the composition dynamics, and then they can be superimposed on them (see Section 4-4). Assumption d implies fast enthalpy changes, turning the energy balances on each tray into a system of algebraic equations that must be satisfied at all times (see Section 4-4).

4-6-2 Number of Components

After some simplifying assumptions have been made, the resulting model can still be of high order, due to the possible presence of a very large number of components. Two approaches have been suggested in the literature to overcome this problem. The first consists of lumping some components together (pseudocomponents) and modelling the column dynamics by applying energy and material balances on these pseudocomponents only. This approach has been criticized as crude and arbitrary (Kehlen and Ratzsch, 1987), mainly because there are no good rules to guide the component-lumping decisions.

The second approach is more recent and is related to the development and application of continuous thermodynamics (Cotterman, Bender, and Prausnitz, 1985; Cotterman and Prausnitz, 1985; Cotterman, Chou, and Prausnitz, 1986) to chemical process modelling. Application of these methods to steady-state multicomponent distillation calculations was performed by Kehlen and Ratzsch (1987); it is expected that this kind of work will extend the application of the method to the modelling of distillation dynamics. Recently, Shibata, Sandler, and Behrens (1987) combined the method of pseudocomponents with continuous thermodynamics for phase equilibrium calculations of "semicontinuous mixtures." Their approach could be used for distillation modelling applications.

In general, the methods of continuous thermodynamics are suited for mixtures containing very many components (i.e., petroleum fractions, shale oils, etc.), but not for mixtures with a moderate number of components (i.e., less than 10). For the latter cases, the lumping of some components to pseudocomponents can be used to reduce the number of equations, but a lot remains to be done (i.e., development of lumping rules) before such an approach can be successful.

4-6-3 Number of Stages — Orthogonal Collocation

The third factor that affects the number of equations that describe the dynamic model of a distillation column is the number of stages. There are several methods for reducing the order of the system in this context, mainly by attempting to approximate the dynamics of a number of stages by the dynamics of a fewer number of "pseudostages." There are two approaches for such an approximation. The first is the method of compartmental models (Benalou, Seborg, and Mellichamp, 1982). These investigators considered a distillation column as a compartmental system, in which a number of stages are lumped to form an equivalent stage. The resulting model is low order, nonlinear, and preserves both material balances and steady states for arbitrary changes in the input variables. The advantages of this model are that it retains the nonlinear form and the gain of the original model, the model parameters are related to the process parameters, and changing the order of the reduced model does not require any additional calculations. The disadvantages are that there are no rules on how to divide a given column into compartments and that the simplified model can exhibit dynamic behavior fundamentally different from that of the original model (i.e., the compartmental model can exhibit inverse response whereas the full model does not).

The second approach involves the reduction of the models using orthogonal collocation methods. There are several papers on this topic (Wong and Luus, 1980; Cho and Joseph, 1983a, 1983b, 1984; Stewart, Levien, and Morari, 1985; Srivastava and Joseph, 1987a, 1987b). Their differences are mainly in the selection of the polynomials, the zeros of which are used as the grid points, and in the selection of the polynomials used as the weighting functions. One advantage that the model of Stewart, Levien, and Morari seems to have over the rest is the fact that it uses the zeros of the Hahn polynomials, which are more suited for discrete (staged) systems, as the location of the collocation points. Most of the other models use the zeros of the Jacobi polynomials, which are more suited for continuous descriptions. Stewart, Levien, and Morari's model converges to the full stagewise solution as the number of collocation points approaches the number of stages.

The collocation models are, in a sense, compartmental models, and they are good for simulation purposes because they reduce the order of the system significantly. The advantages of these models (especially of the one proposed by Stewart, Levien, and Morari) is that they retain the nonlinear form of the original model, they can be used for multicomponent distillation, they allow free choice of thermodynamic subroutines, they can be implemented without a full order solution, and they can reduce the computational time significantly. Their main disadvantage is the nonretention of the original model's gain in an exact manner. However, the gain predicted by the collocation models is usually in very good agreement with that of the full model. The basic method for using orthogonal collocation to obtain a reduced-order model for a distillation column is illustrated in succeeding text and follows along the lines of the work of Stewart, Levien, and Morari (1985).

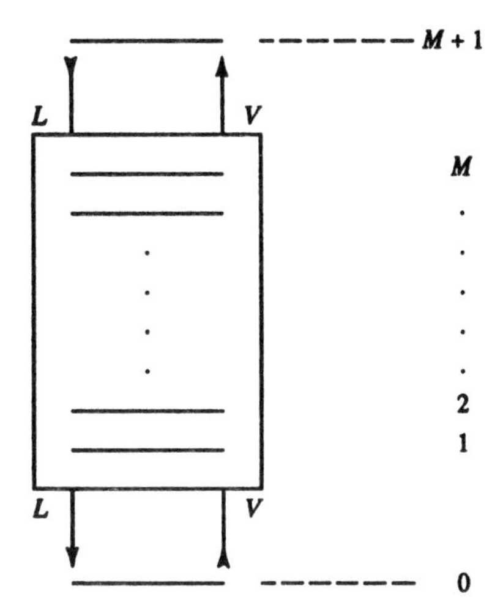

FIGURE 4-5. A sequence of M process stages used for the separation of a multicomponent mixture.

Consider a sequence of M process stages used for a multicomponent separation as depicted in Figure 4-5. Assuming constant molar overflow, the dynamic component material balances on each tray s are given by

$$M_L \frac{dx(s,t)}{dt} = Lx(s+1,t) - Lx(s,t) + Vy(s-1,t) - Vy(s,t) \tag{4-55}$$

where $s = 1, \ldots, M$. The compositions in the module can be approximated by polynomials, using $n \leq M$ interior grid points $s_1, s_2, \ldots, s_n$ plus two entry points $s_{n+1} = M + 1$ for the liquid and $s_0 = 0$ for the vapor compositions. (The reader should be aware of the fact that in the original paper of Stewart, Levien, and Morari the stages are numbered from top to bottom, whereas in this book the stages are numbered from the bottom to the top. The equations presented here comply fully with the notation and numbering used in this book and not in the paper.)

TABLE 4-3 Collocation Points for a Module Consisting of M Stages

Number of Collocation Points n	Points s_j
1	$\dfrac{M+1}{2}$
2	$\dfrac{M+1}{2} \pm \sqrt{\dfrac{M^2-1}{12}}$
3	$\dfrac{M+1}{2}, \dfrac{M+1}{2} \pm \sqrt{\dfrac{3M^2-7}{20}}$
4	$\dfrac{M+1}{2} \pm \sqrt{\dfrac{3M^2-13}{28} \pm \sqrt{\dfrac{6M^4-45M^2+164}{980}}}$

The corresponding equations of the collocation model for the module of Figure 4-5 are

$$M_L \frac{d}{dt}\tilde{x}(s_j, t) = L\tilde{x}(s_j + 1, t) - L\tilde{x}(s_j, t) + V\tilde{y}(s_j - 1, t) - V\tilde{y}(s_j, t) \tag{4-56}$$

for $j = 1, \ldots, n$, where the tilde denotes an approximate value. Therefore, the number of equations describing the component material balances for the module is reduced from M (number of stages in the module) to n (number of collocation points in the module). The locations of the collocation points $s_1, \ldots, s_n$ are the zeros of the Hahn polynomials (Hahn, 1949) with $(a, b) = (0, 0)$. The zeros of the Hahn polynomials for $n \leq 4$ are given in Table 4-3.

Srivastava and Joseph (1985) have shown that using Hahn polynomials with $a = 0$ and $b > 0$ leads to better approximations where the error is more evenly distributed throughout the column.

The compositions in each stage of the module are given by

$$\begin{aligned} \tilde{x}(s, t) &= \sum_{j=1}^{n+1} w_{j_L}(s)\tilde{x}(s_j, t) \\ \tilde{y}(s, t) &= \sum_{j=0}^{n} w_{j_V}(s)\tilde{y}(s_j, t) \end{aligned} \tag{4-57}$$

where the w functions are the Lagrange polynomials

$$\begin{aligned} w_{j_L}(s) &= \prod_{\substack{k=1 \\ k \neq j}}^{n+1} \frac{s - s_k}{s_j - s_k}, \qquad j = 1, \ldots, n+1 \\ w_{j_V}(s) &= \prod_{\substack{k=0 \\ k \neq j}}^{n} \frac{s - s_k}{s_j - s_k}, \qquad j = 0, \ldots, n \end{aligned} \tag{4-58}$$

Note again that $s_0 = 0$ and $s_{n+1} = M + 1$.

The method is applicable for the solution of both steady-state and dynamic models of multicomponent distillation columns with

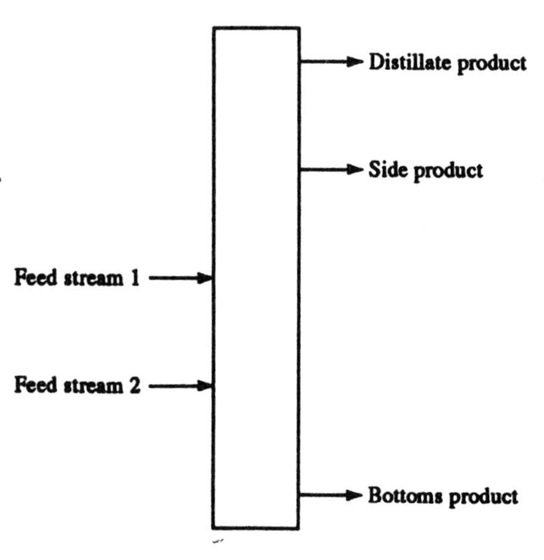

FIGURE 4-6. A complex distillation column with two feed streams and three product streams.

multiple feeds and/or product streams, and it can accommodate columns with tray-dependent temperatures, liquid and vapor flows, mixtures with nonideal thermodynamics, trays with efficiencies less than 1, and so on.

One issue related to the implementation of the method is associated with the way in which a given column is divided into modules. Consider the column shown in Figure 4-6; it consists of two feed streams and three product streams. A modular decomposition of this column is shown in Figure 4-7*a* (decomposition scheme A). An alternative decomposition (scheme B) is shown in Figure 4-7*b*.

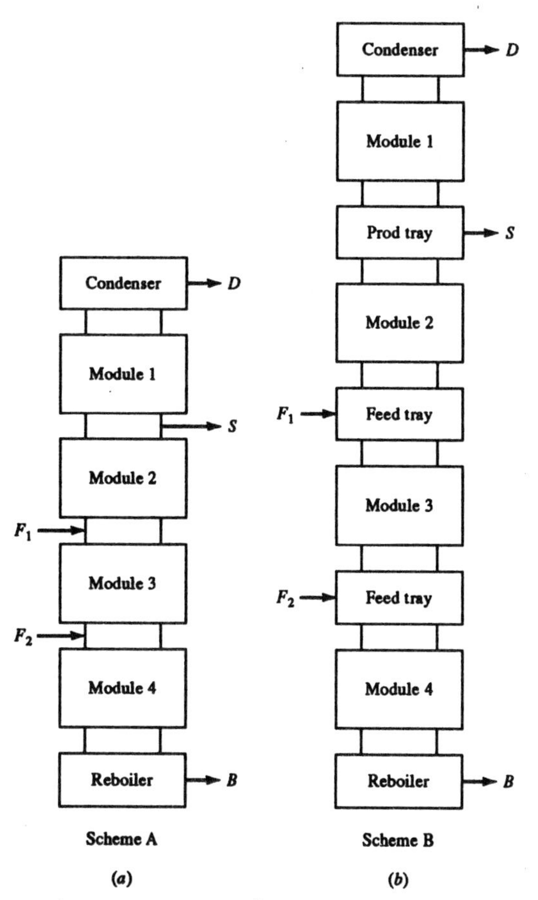

FIGURE 4-7. Modular decomposition schemes for the column shown in Figure 4-6. (a) Decomposition scheme A in which the feed and intermediate product trays are lumped inside a module. (b) Decomposition scheme B in which all feed and product trays are used as collocation points.

There is a trade-off between the two decomposition schemes. Scheme A leads to a smaller number of equations (i.e., equations for modules plus equations for condenser and reboiler), but the Jacobian of the resulting model is a full matrix (i.e., very few nonzero elements). In contrast, scheme B requires more equations (i.e., equations for modules plus equations for condenser and reboiler plus equations for feed–product trays), but the corresponding Jacobian has a block diagonal form that facilitates the solution of the system (Papadourakis, 1985). The superiority of scheme B, which corresponds to treating all the feed–product trays as collocation points, was also recognized (for different reasons) by Pinto and Biscaia (1988). In addition, the modular decomposition of a column according to scheme B is easier to implement. The application of the preceding method is illustrated in the following example.

Example 4-2

Consider again the binary column described in Example 4-1. Applying the decomposition scheme B, the stripping section trays (excluding the feed tray) constitute one module, the rectifying section trays constitute a second module, and the feed tray, the condenser, and the reboiler are used as collocation points by themselves. Applying this decomposition scheme and the collocation method of Stewart, Levien, and Morari (1985), a reduced-order model of the column was developed.

The first equation is the mass balance around the reboiler:

$$M_{L_B}\frac{d}{dt}\tilde{x}(s_1,t) = (L+F)\tilde{x}(s_1+1,t) - B\tilde{x}(s_1,t) - V\tilde{y}(s_1,t) \tag{4-59}$$

where s_1 is the collocation point corresponding to the reboiler and from Table 4-2 for $n = 1$ and $M = 1$, we get $s_1 = 1$.

Assuming that the dynamics of the stripping section can be described by using only two collocation points, the corresponding mass balance equations become

$$M_L \frac{d}{dt}\tilde{x}(s_2, t)$$
$$= (L+F)\tilde{x}(s_2+1, t) - (L+F)\tilde{x}(s_2, t)$$
$$+V\tilde{y}(s_2-1, t) - V\tilde{y}(s_2, t) \quad (4\text{-}60)$$

$$M_L \frac{d}{dt}\tilde{x}(s_3, t)$$
$$= (L+F)\tilde{x}(s_3+1, t) - (L+F)\tilde{x}(s_3, t)$$
$$+V\tilde{y}(s_3-1, t) - V\tilde{y}(s_3, t) \quad (4\text{-}61)$$

The locations of the collocation points s_2 and s_3 are found from Table 4-2 for $n = 2$ (number of collocation points in the section) and for $M = 11$ (number of trays in the section). It turns out that $s_2 = 2.84$ and $s_3 = 9.16$. However, because there is an additional stage (the reboiler) below the stripping section, then $s_2 = 3.84$ and $s_3 = 10.16$.

The mass balance for the feed tray is given by

$$M_L \frac{d}{dt}\tilde{x}(s_4, t)$$
$$= L\tilde{x}(s_4+1, t) - (L+F)\tilde{x}(s_4, t) + Fx_F$$
$$+V\tilde{y}(s_4-1, t) - V\tilde{y}(s_4, t) \quad (4\text{-}62)$$

where s_4 is the collocation point for the feed tray. Again, from Table 4-2 we get (for $n = 1$ and $M = 1$) that $s_4 = 1$. However, because there are 13 stages below the feed tray (reboiler + 12 trays of the stripping section), we set $s_4 = 14$.

Assuming that the dynamics of the rectifying section can also be described by using only two collocation points, the corresponding mass balance equations become

$$M_L \frac{d}{dt}\tilde{x}(s_5, t)$$
$$= L\tilde{x}(s_5+1, t) - L\tilde{x}(s_5, t)$$
$$+ V\tilde{y}(s_5-1, t) - V\tilde{y}(s_5, t) \quad (4\text{-}63)$$

$$M_L \frac{d}{dt}\tilde{x}(s_6, t)$$
$$= L\tilde{x}(s_6+1, t) - L\tilde{x}(s_6, t)$$
$$+V\tilde{y}(s_6-1, t) - V\tilde{y}(s_6, t) \quad (4\text{-}64)$$

The location of the collocation points s_5 and s_6 is found from Table 4-2 for $n = 2$ (number of collocation points in the section) and for $M = 11$ (number of trays in the section). It turns out that $s_5 = 2.84$ and $s_6 = 9.16$. Again, because there are 14 trays below the rectifying section, we set $s_5 = 16.84$ and $s_6 = 23.16$.

Finally, the mass balance for the total condenser takes the form

$$M_{L_D} \frac{d}{dt}\tilde{x}(s_7, t) = -L\tilde{x}(s_7, t) - D\tilde{x}(s_7, t)$$
$$+ V\tilde{y}(s_7 - 1, t) \quad (4\text{-}65)$$

where $s_7 = 25$.

Equations 4-59 to 4-65 constitute a system where the unknowns are $\tilde{x}(s_i, t)$, $i = 1, 7$. Note that a full scale model for the column (using the same assumptions) would consist of 25 equations (i.e., number of equations = number of stages).

In Equations 4-59 to 4-65 the corresponding $\tilde{y}(s_i, t)$ can be found from the equilibrium relationship

$$\tilde{y}(s_i, t) = \frac{\alpha\tilde{x}(s_i, t)}{1 + (\alpha - 1)\tilde{x}(s_i, t)} \quad (4\text{-}66)$$

The remaining variables in the preceding equations are functions of the liquid and vapor mole fractions of the collocation points as described by Equations 4-57 and 4-58. For example, the variable $\tilde{x}(s_3 - 1, t) = \tilde{x}(9.16, t)$ is given by

$$\tilde{x}(9.16, t) = \sum_{j=2}^{4} w_{j_L}(9.16)\tilde{x}(s_j, t) \quad (4\text{-}67)$$

where the w functions are the Lagrange polynomials

$$w_{j_L}(9.16) = \prod_{\substack{k=2 \\ k \neq j}}^{4} \frac{9.16 - s_k}{s_j - s_k}, \qquad j = 2, \ldots, 4 \tag{4-68}$$

Note that $\tilde{x}(9.16, t)$ is the liquid mole fraction of a fictitious tray that is inside the stripping section (module). Therefore, and according to Equations 4-57 and 4-58, it will depend on the liquid mole fractions of the collocation points of this module only, plus on the liquid mole fraction of the stage immediately above the module. Using the decomposition scheme shown on Figure 4-7b, a tray immediately above or below a module is a feed or product stage and, therefore, it constitutes the location of a lone collocation point.

Note also that the vapor mole fraction of a fictitious (or real) tray that belongs to a certain module depends only on the vapor mole fractions of collocation points of the module and on the vapor mole fraction of the stage (collocation point) immediately below the module.

By setting the derivatives of the collocation dynamic model equal to zero, an approximate steady-state model for the column was obtained. The steady-state composition profile of the column was calculated by first solving the full model, and also by solving the reduced-order model. The results are shown in Figure 4-8. In the same figure, the results are shown from approximate models developed by using 3 and 4 collocation points in each module (resulting in 9 and 11 collocation points for the entire column). As can be seen in the figure, the agreement between the full model and the reduced-order model is excellent when 11 collocation points are used in the column, very good when 9 points are used, and poor when only 7 points are used for the entire column. It should be noted here that the solution that exhibits very good agreement with the full solution (i.e., when 9 collocation points were used for the entire column) results from the solution of 9 equations, whereas the full model required the solution of 25 equations.

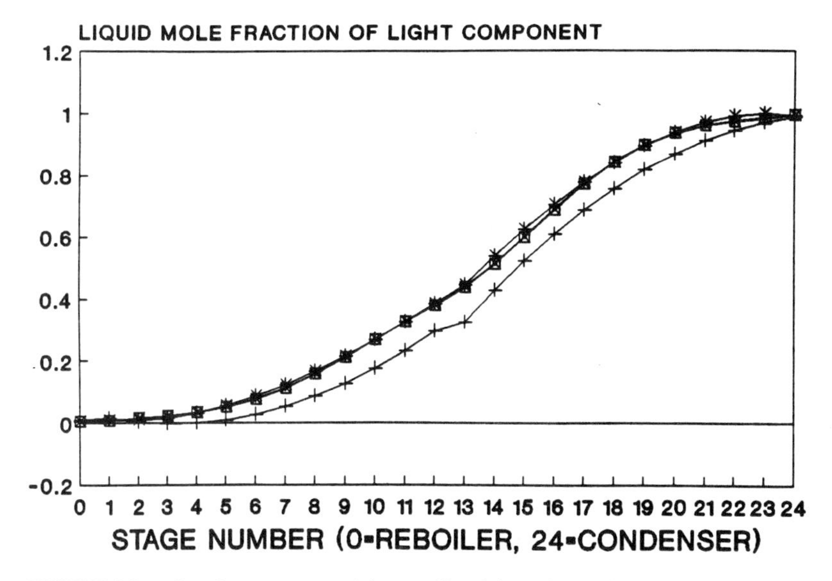

FIGURE 4-8. Steady-state composition profile of the column of Example 4-2. ■ full model; + reduced model with 7 collocation points; * reduced model with 9 collocation points; □ reduced model with 11 collocation points.

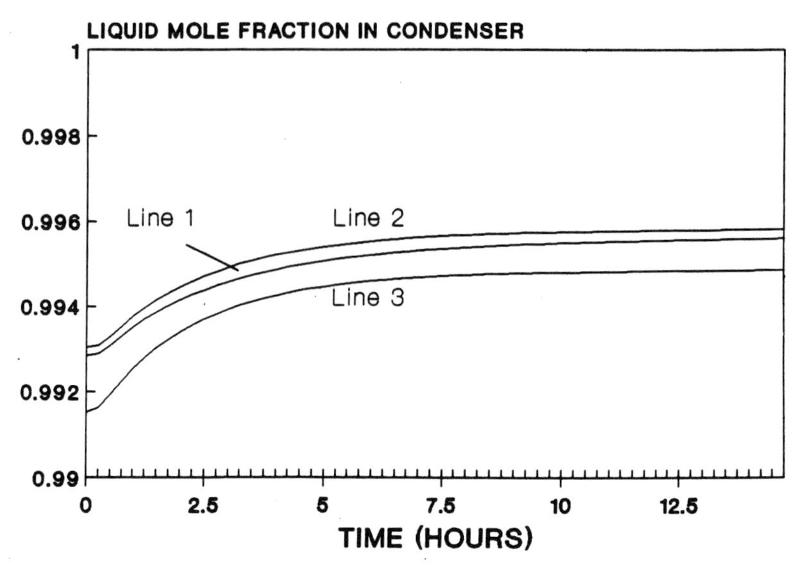

FIGURE 4-9. Transient response of the condenser of the distillation column of Example 4-2 to a step decrease of the distillate flow rate by 1%. Line 1: full model; line 2: reduced model with 9 collocation points; line 3: reduced model with 11 collocation points.

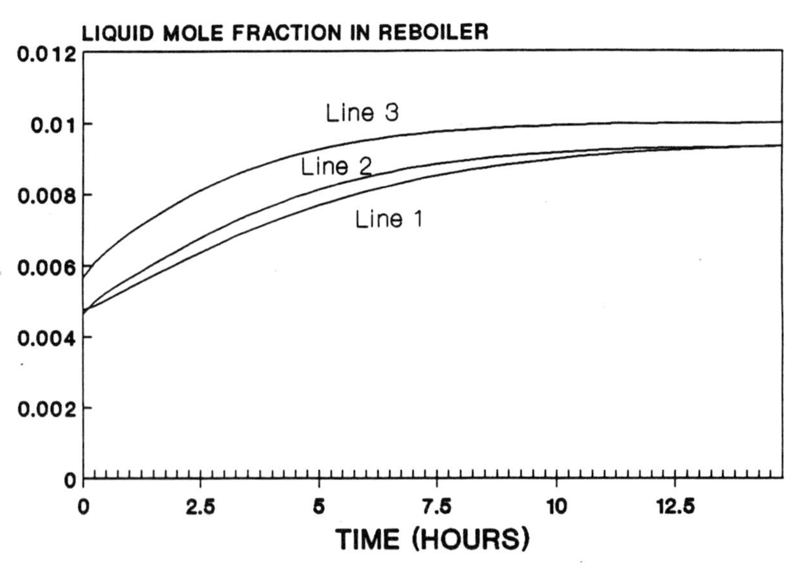

FIGURE 4-10. Transient response of the reboiler of the distillation column of Example 4-2 to a step decrease of the distillate flow rate by 1%. Line 1: full model; line 2: reduced model with 9 collocation points; line 3: reduced model with 11 collocation points.

The method was also applied to obtain a reduced-order dynamic model for the column. The predicted column response for a 1% reduction in the distillate flow rate is shown in Figures 4-9 and 4-10. Again, the agreement between the reduced-order models and the full model ranges from excellent (model with 11 collocation points in the column) to very good (model with 9 collocation points in the column).

The degree of order reduction achievable for a given column (using orthogonal collocation) was dealt with in a paper by Srivastava and Joseph (1985). They developed an index called *order-reduction parameter* that can be used to predict the extent of order reduction achievable and to provide guidelines for choosing an appropriately sized reduced-order model.

In general, reduced-order models obtained from the application of the previously mentioned orthogonal collocation methods can be used for both steady-state and dynamic simulation purposes with excellent results. Use of these approximate models can lead to reductions in computational times proportional to $(N_c/N)^2$ where N_c is the total number of collocation points in the column and N is the total number of equilibrium stages in the column (Pinto and Biscaia, 1988).

References

Arkun, Y. (1987). Dynamic block relative gain array and its connection with the performance and stability of decentralized control structures. *Int. J. Control* 46, 1187–1193.

Bamberger, W. and Isermann, R. (1978). Adaptive steady-state optimization of slow dynamic processes. *Automatica* 14, 223–230.

Benalou, A., Seborg, D. E., and Mellichamp, D. A. (1982). Low order, physically lumped dynamic models for distillation column control. Presented at the AIChE Annual Meeting, November 1982, Los Angeles, CA.

Betlem, D. H. L. (1991). Personal communication.

Bhat, P. V. and Williams, T. J. (1969). Approximate dynamic models of distillation operations and their relation to overall column control methods. *Int. Chem. Eng. Symposium Series* 32, 22–32.

Bonvin, D. and Mellichamp, D. A. (1982a). A generalized structural dominance method for the analysis of large-scale systems. *Int. J. Control* 35, 807–827.

Bonvin, D. and Mellichamp, D. A. (1982b). A unified derivation and critical review of modal approaches to model reduction. *Int. J. Control* 35, 829–848.

Celebi, C. and Chimowitz, E. H. (1985). Analytic reduced order dynamic models for large equilibrium staged cascades. *AIChE J.* 31, 2039–2051.

Chen, C. H. (1969). M.Sc. thesis, Polytechnic Institute of New York.

Cho, Y. S. and Joseph, B. (1983a). Reduced-order steady-state and dynamic models for separation processes Part I. Development of the model reduction procedure. *AIChE J.* 29, 261–269.

Cho, Y. S. and Joseph, B. (1983b). Reduced-order steady-state and dynamic models for separation processes Part II. Application to nonlinear multicomponent systems. *AIChE J.* 29, 270–276.

Cho, Y. S. and Joseph, B. (1984). Reduced-order models for separation columns—III. Application to columns with multiple feeds and sidestreams. *Comput. Chem. Eng.* 8, 81–90.

Cotterman, R. L. and Prausnitz, J. M. (1985). Flash calculations for continuous or semicontinuous mixtures using an equation of state. *Ind. Eng. Chem. Proc. Des. Dev.* 24, 434–443.

Cotterman, R. L., Bender, R., and Prausnitz, J. M. (1985). Phase equilibria for mixtures containing very many components. Development and application of continuous thermodynamics for chemical process design. *Ind. Eng. Chem. Proc. Des. Dev.* 24, 194–203.

Cotterman, R. L., Chou, G. F., and Prausnitz, J. M. (1986). Comments on "Flash calculations for continuous or semicontinuous mixtures using an equation of state." *Ind. Eng. Chem. Proc. Des. Dev.* 25, 840–841.

Genesio, R. M. and Milanese, M. (1976). A note on the derivation and use of reduced-order models. *IEEE Trans. Automat. Control* AC-21, 118–122.

Georgakis, C. and Stoever, M. A. (1982). Time domain order reduction of tridiagonal dynamics of staged processes—I. Uniform lumping. *Chem. Eng. Sci.* 37, 687–697.

Hahn, W. (1949). Uber Orthogonalpolynome die q-Differenzengleichungen genugen. *Math. Nachrichten* 2, 4–34.

Hawkins, D. E., Tolfo, F., and Chauvin, L. (1987). Group methods for advanced column control. In *Handbook of Advanced Process Control Systems and Instrumentation*, L. Kane, ed. Houston: Gulf Publishing Co.

Kehlen, H. and Ratzsch, M. T. (1987). Complex multicomponent distillation calculations by continuous thermodynamics. *Chem. Eng. Sci.* 42, 221–232.

Kim, C. and Friedly, C. F. (1974). Dynamic modeling of large staged systems. *Ind. Eng. Chem. Proc. Des. Dev.* 13, 177–181.

Levy, R. E., Foss, A. S., and Grens, E. A., II (1969). Response modes of a binary distillation column. *Ind. Eng. Chem. Fundam.* 8, 765–776.

Moczek, J. S., Otto, R. E., and Williams, T. J. (1963). Proceedings of the Second IFAC Congress, Basle. New York: Pergamon.

Moczek, J. S., Otto, R. E., and Williams, T. J. (1965). Approximation models for the dynamic response of large distillation columns. *Chem. Eng. Prog. Symp. Ser.* 61, 136–146.

Papadourakis, A. (1985). *Stability and Dynamic Performance of Plants with Recycle*. Ph.D. Dissertation, University of Massachusetts, Amherst, MA.

Papadourakis, A., Doherty, M. F., and Douglas, J. M. (1989). Approximate dynamic models for chemical process systems. *Ind. Eng. Chem. Res.* 28, 546–552.

Perkins, J. D. (1990). Interactions between process design and process control. In *Dynamics and Control of Chemical Reactors, Distillation Columns and Batch Processes. IFAC Symposium Series 1990*, No. 7, J. E. Rijnsdorp et al., eds. Oxford: Pergamon, 195–203.

Pinto, J. D. and Biscaia, E. C., Jr. (1988). Order reduction strategies for models of staged separation systems. *Comput. Chem. Eng.* 12, 812–831.

Rademaker, O., Rijnsdorp, J. E., and Maarleveld, A. (1975). *Dynamics and Control of Continuous Distillation Units*. Amsterdam: Elsevier Science Publishers.

Rijnsdorp, J. E. (1991). *Integrated Process Control and Automation*. Amsterdam: Elsevier Science Publishers.

Robinson, C. S. and Gilliland, E. R. (1950). *Elements of Fractional Distillation*. New York: McGraw-Hill.

Shibata, S. K., Sandler, S. I., and Behrens, R. A. (1987). Phase equilibrium calculations for continuous and semicontinuous mixtures. *Chem. Eng. Sci.* 42, 1977–1988.

Shinskey, F. G. (1967). *Process Control Systems*. New York: McGraw-Hill.

Skogestad, S. and Morari, M. (1987a). Effect of disturbance directions on closed loop performance. *Ind. Eng. Chem. Res.* 26, 2029–2035.

Skogestad, S. and Morari, M. (1987b). The dominant time constant for distillation columns. *Comput. Chem. Eng.* 11, 607–617.

Skogestad, S. and Morari, M. (1988). Understanding the dynamic behavior of distillation columns. *Ind. Eng. Chem. Res.* 27, 1848–1862.

Srivastava, R. K. and Joseph, B. (1985). Reduced-order models for separation columns—V. Selection of collocation points. *Comput. Chem. Eng.* 9, 601–613.

Srivastava, W. E. and Joseph, B. (1987a). Reduced-order models for staged separation columns—IV. Treatment of columns with multiple feeds and sidestreams via spline fitting. *Comput. Chem. Eng.* 11, 159–164.

Srivastava, W. E. and Joseph, B. (1987b). Reduced-order models for staged separation columns—VI. Columns with steep and flat composition profiles. *Comput. Chem. Eng.* 11, 165–176.

Stewart, W. E., Levien, K. L., and Morari, M. (1985). Simulation of fractionation by orthogonal collocation. *Chem. Eng. Sci.* 40, 409–421.

Stoever, M. A. and Georgakis, C. (1982). Time domain order reduction of tridiagonal dynamics of staged processes—II. Nonuniform lumping. *Chem. Eng. Sci.* 37, 699–705.

Toijala, K. V. (1968). An approximate model for the dynamic behavior of distillation columns. Part I. General theory. *Acta Acad. Aboensis (Math. Phys.)* 28(8), 1–18.

Toijala, K. V. (1969). An approximate model for the dynamic behavior of distillation columns. Part II. Composition responses to changes in feed composition for binary mixtures. *Acta Acad. Aboensis (Math. Phys.)* 29(10), 1–34.

Toijala, K. V. (1971a). An approximate model for the dynamic behavior of distillation columns. Part IV. Comparison of theory with experimental results for changes in feed composition. *Acta Acad. Aboensis (Math. Phys.)* 31(3), 1–25.

Toijala, K. V. (1971b). An approximate model for the dynamic behavior of distillation columns. Part V. Comparison of theory with experimental results for changes in reflux flow rate. *Acta Acad. Aboensis* (*Math. Phys.*) 31(5), 1–19.

Toijala, K. V. (1978). Simple models for distillation dynamics. Paper presented at the 86th National AIChE Meeting, April 1–5, 1978, Houston, Texas.

Toijala, K. V. and Fagervik, K. (1972). An approximate model for the dynamic behavior of distillation columns. Part VII. Comparison of model with experimental results for controlled columns by the use of digital simulation. *Acta Acad. Aboensis* (*Math. Phys.*) 32(1), 1–17.

Toijala, K. V. and Gustafsson, S. (1971). Process dynamics of multi-component distillation. *Kemian Teollisuus* 28, 113–120.

Toijala, K. V. and Gustafsson, S. (1972a). Process dynamics of multi-component distillation. Comparison of theoretical model with published experimental results. *Kemian Teollisuus* 29, 95–103.

Toijala, K. V. and Gustafsson, S. (1972b). On the general characteristics of component distillation dynamics. *Kemian Teollisuus* 29, 173–184.

Toijala, K. V. and Jonasson, D. (1970). An approximate model for the dynamic behavior of distillation columns. Part III. Composition response to changes in reflux flow rate, vapour flow rate and feed enthalpy for binary mixtures. *Acta Acad. Aboensis* (*Math. Phys.*) 30(15), 1–21.

Toijala, K. V. and Jonasson, D. (1971). An approximate model for the dynamic behavior of distillation columns. Part VI. Composition response for changes in feed flow rate for binary mixtures. *Acta Acad. Aboensis* (*Math. Phys.*) 31(11), 1–27.

Wahl, E. F. and Harriot, P. (1970). Understanding and prediction of the dynamic behavior of distillation columns. *Ind. Eng. Chem. Proc. Des. Dev.* 9,396.

Weigand, W. A., Jhawar, A. K., and Williams, T. J. (1972). Calculation method for the response time to step inputs for approximate dynamic models of distillation columns. *AIChE J.* 18, 1243–1252.

Wong, K. T. and Luus, R. (1980). Model reduction of high-order multistage systems by the method of orthogonal collocation. *Can. J. Chem. Eng.* 58, 382.

5

Object-Oriented Simulation

B. D. Tyreus

E. I. Du Pont de Nemours & Co.

5-1 INTRODUCTION

Steady-state process simulation is used routinely in the evaluation, selection, and design of new processes (Biegler, 1989). In the past, all the pertinent design factors could be addressed with steady-state programs. Today, however, there are new factors to be considered, and some of these have more to do with the operation and control of the plant than with steady-state aspects. For example, product quality demands, increasing safety concerns, stringent environmental requirements, process flexibility, minimization of capital expenditure through process simplification and integration, and elimination of overdesign are all important factors that tend to influence process operations. However, the operational aspect of process design is difficult or impossible to glean from steady-state simulations, and we must, therefore, consider dynamic simulations for answers. How then are dynamic simulations used in process design? Traditionally, they are used in control system evaluation and tuning but also to find the appropriate controller structure for a particular process. To date, such simulations have mostly involved one or two unit operations with little emphasis on plantwide control.

The operational factors cited in the foregoing text suggest a broader use for dynamic simulations than for just control studies of unit operations. For example, with current integrated plant designs, the greatest benefit of a dynamic analysis might be in solving the plantwide control problem. This idea is further reinforced in Chapter 6. In that context, the management of inventories, recycle paths, product transitions, startup, and shutdown can be studied as they impact safety, environment, and product quality. Furthermore, when a plantwide dynamic simulator has been developed for process design, it can later be used for operator training and process improvement studies.

Are we ready then to meet an increased demand of dynamic simulations? Unfortunately not, due to the prevailing paradigm used when considering such simulations. Under this paradigm, dynamic simulations are viewed as control system evaluation tools, and because process control is a specialized field mastered by control engineers, we let the specialists create and use the simulations according to their needs. Sometimes this means that the model is a set of transfer functions used in a linear analysis; sometimes a more detailed nonlinear dynamic model is written from first principles and implemented in FORTRAN or BASIC (see Chapter 3 for example). The models are almost always tailored to the specific problem at hand; seldom extensible to other problems and almost never usable by others.

The tools we know today for creating dynamic simulations are consistent with the prevailing paradigm. Whether they are integrators like LEANS, libraries of FORTRAN subroutines like DYFLO (Franks, 1972), or equation-based numerical solvers like SPEEDUP (Perkins, 1986), these tools are aimed at the specialist who has considerable experience in using the tool, who knows how to model various processes in terms of their fundamental equations, and who is willing to spend considerable time in entering code and data into input files, which are compiled, edited, and debugged before they yield results in terms of time plots of prespecified variables over fixed time periods.

I feel that in order to meet the broader need for dynamic simulations, a paradigm shift is required. Under a different paradigm, dynamic simulations would go beyond control system studies, and all process engineers should be able to create and use such simulations. I also feel that today's tools and techniques for generating dynamic simulations will be inadequate under a different paradigm. Instead, I suggest that new software approaches, such as object-oriented programming (OOP), will replace current tools in order to support a new paradigm (Stephanopoulos, Henning, and Leone, 1990). I am also aware of a couple of simulation software companies that have reached the same conclusion.

The purpose of this chapter is thus to familiarize you with an emerging software technology for dynamic simulations. I will describe the concepts of objects and messages and contrast them to their more familiar counterparts: data structures and subroutines. The important features of inheritance and polymorphism are also covered. All this is put in the context of creating dynamic simulation models for distillation systems.

This chapter is, however, not detailed enough to show you how to write your own object-oriented dynamic simulator, nor does it give enough insight into how I have written mine. Instead I merely hope to remove some of the mystique around object-oriented programming and to spark enough interest that you might consider experimenting with an object-oriented compiler next time you write a simulation model. In case you do not like writing your own simulations from scratch but rely on vendor tools, this chapter will hopefully interest you a little more in emerging simulation packages built on object-oriented principles.

5-2 IDEAL DYNAMIC SIMULATOR

Before delving into the principles of OOP, we will take a closer look at a vision of a suitable tool for creating dynamic simulations.

In order to shift to a different paradigm, one has to ask the question: "What is it that we cannot do now that if it could be done would have a significant impact on our business?" When I asked this question about dynamic models and the simulation tools we use, I found the following to be true:

- Most process engineers cannot create or use a dynamic simulation because the tools and techniques are designed for specialists.
- Dynamic simulators are not user-friendly, preventing the casual user from considering them.
- Dynamic simulations are expensive to develop because it is generally difficult to reuse previously written code in new applications, thereby forcing the developer to rewrite substantial sections of the new model.
- Dynamic simulations are not readily accessible because the development envi-

ronments and final models require relatively large, expensive computers.

- Many simulators are not configurable and require extensive programming and almost always compilation before the model can be run.
- Dynamic simulations are often slow.
- The user cannot interact with the simulation during a run.
- Even for the sophisticated user, most dynamic simulation packages are limited because they are not portable, extensible, or maintainable.

If instead we had simulation tools that were configurable, user-friendly, flexible, portable, readily accessible, fast, and inexpensive, we could dramatically change the way we consider dynamics in process design and control. Such an ideal dynamic simulator would allow process engineers and control engineers to routinely develop rigorous, nonlinear, dynamic simulations for most processes. It would allow us to study the impact of process design and controller structure on the operation of not just different unit operations, but entire processing units and even plants (see also Chapter 6). The ideal dynamic simulator would allow process engineers to explore and understand existing processes so the processes could be simplified and improved. It would allow us to quickly validate our control systems and to train operators with realistic, validated models. In short, the ideal dynamic simulator would change the way we do business; it would help create a paradigm shift.

Once the vision of an ideal dynamic simulator has been created, we need to know how such a tool could be developed. Object-oriented programming provides an excellent vehicle to take us to our goal.

5-3 OBJECT-ORIENTED PROGRAMMING

To help explain the concepts of object-oriented programming, I will contrast this new software technique with traditional programming approaches supported by procedural languages, such as BASIC, FORTRAN, Pascal, and C. In procedural programming, the focal point is data. Data are transformed to results by letting procedures or subroutines operate on them. For example, in a simulation package, such as DYFLO (Franks, 1972), there are data structures defined that specify how component compositions, temperature, and pressure are stored, how parameters enter the simulation, and how the results of chemical reactions are made available when they are needed. DYFLO then consists of a library of subroutines that either take the data structures as parameters or use the data directly from COMMON areas to produce results. Similarly, when writing a Pascal program, the first thing that has to be done is to define the data types and variables that will be used by the procedures and functions that get invoked in a predefined, procedural fashion.

If in procedural programming the key words are data and subroutines, then in object-oriented programming the key words are *objects* and *messages* (Pascoe, 1986). Although there are some similarities between data and objects on the one hand and subroutines and messages on the other, there are several significant differences. The most significant difference is between data and objects. Data structures are passive storage places for information that are transformed by subroutine calls in the program. Objects, on the other hand, are active entities that exhibit behavior when they receive a message. Objects are data structures and subroutines packaged into one unit (*encapsulations*) where the data are globally known to all the subroutines that belong to that object. A message is then the mechanism to trigger the appropriate subroutine within the object. Because objects send messages to each other and because the particular object to receive a particular message can change during execution of the program, there is no counterpart to the sequence of subroutine calls found in procedural programs. Here lies part of the difficulty in transitioning from the procedural paradigm to an object-oriented approach, but it also contains the real power of object-oriented programming.

5-3-1 Classes and Objects

Sometimes terminology can be an obstacle when describing a new concept. Object-oriented programming has its share of special nomenclature such as classes, encapsulations, data abstraction, objects, instantiations, inheritance, polymorphism, and dynamic binding. If these words were mere synonyms for well known concepts, I could simply provide a glossary and be done. Unfortunately, this is not the case. Each of these words stands for a concept that is unique to object-oriented programming, and for that reason, it is probably a good idea that they do not sound like things we are familiar with. I will describe each concept as I introduce it and give some examples, but for now, we will focus on classes, instances, and objects. A *class* is the template for an object. A class defines what data the object contains, what behavior the resulting objects can exhibit (what subroutines are contained in the object), and what messages the object responds to. The declaration and definition of classes look and feel like programming in just about any language; there are data declarations and subroutine definitions. Given the declaration of a class (the template), one or more objects can be instantiated (cloned) from that class anywhere and anytime in the program or during program execution. This does not have a simple analog in conventional programming. The working program then revolves around active objects that send and receive messages that cause desirable actions and produce results.

5-3-2 Modelling a Column Tray

In order to illustrate some of the important concepts in object-oriented programming, we will model the dynamics of the fluid on a tray in a distillation column. We will compare an object-oriented solution to the more familiar procedural approach. Figure 5-1 shows a section of a distillation column with a single tray. The liquid hydraulics and multicomponent composition dynamics can be described by the following equations (Luyben, 1990; see also Chapter 3 for more details on modelling a distillation column):

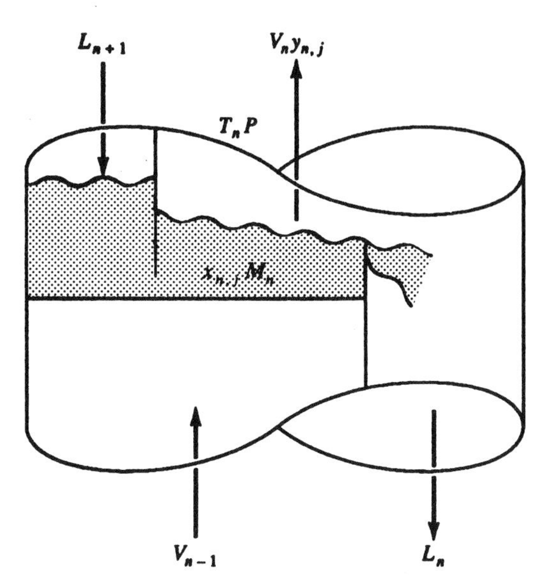

FIGURE 5-1. *n*th tray of multicomponent column.

$$\frac{dM_n}{dt} = L_{n+1} - L_n \tag{5-1}$$

$$\frac{d(M_n x_{n,j})}{dt} = L_{n+1}x_{n+1,j} - L_n x_{n,j} + V_{n-1}y_{n-1,j} - V_n y_{n,j} \tag{5-2}$$

$$M_n = f(L_n) \tag{5-3}$$

$$y_{n,j} = \frac{\alpha_j x_{n,j}}{\sum_{k=1}^{NC} \alpha_k x_{n,k}} \tag{5-4}$$

Procedural Approach

We will first study a procedural approach to implementing the tray model. The equations are conveniently coded just as they appear in a procedural language like FORTRAN or Pascal. In writing the code, we would, however, be responsible for defining data areas or storage locations for the *x*s,

```
C****************** COLUMN MODULE************************************
                    SUBROUTINE COLUMN(NT,NF,DT)
                    COMMON NC
                    COMMON DM,DXM,XM,X,Y,T,P
                    DIMENSION DM(50),DXM(50,5),XM(50,5),X(50,5),Y(50,5),
                    T(50),XX(5),YY(5)
                    DO 15 N=1,NT
                    DO 5 J=1,NC
    5               XX(J)=X(N,J)
                    CALL BUBPT(T(N),XX,YY,P)
                    DO 10 J=1,NC
   10               Y(N,J)=YY(J)
   15               CONTINUE
C........CODE TO CALCULATE VAPOR AND LIQUID TRAFFIC HERE
         :
         :
                    DO 20 N=2,NF-1
   20               DM(N)=L(N+1)+V(N-1)-L(N)-V(N)
C........DM FOR FEED TRAY AND REST OF COLUMN HERE
         :
         :
                    DO 100 J=1,NC
         :
                    DO 50 N=2,NF-1
   50               DXM(N,J)=X(N+1,J)*L(N+1)+Y(N-1,J)*V(N-1)-X(N,J)*L(N)
                    -Y(N,J)*V(N)
C........DXM FOR FEED TRAY AND REST OF COLUMN HERE
         :
         :
  100               CONTINUE
C........INTEGRATE A LA EULER
                    DO 200 N=1,NT
  200               M(N)=M(N)+DM(N)*DT
                    DO 400 J=1,NC
                    DO 300 1,NT
                    XM(N,J)=XM(N,J)+DXM(N,J)*DT
                    X(N,J)=XM(N,J)/M(N,J)
                    IF (X(N,J) .GT. 1.0)X(N,J)=1.0
                    IF (X(N,J) .LT. 0.0)X(N,J)=0.0
  300               CONTINUE
  400               CONTINUE
                    RETURN
                    END
C************************* VLE MODULE *************************
                    SUBROUTINE BUBPT(T,X,Y,P)
                    COMMON NC
                    COMMON ALPHA
                    DIMENSION ALPHA(5),X(5),Y(5)
                    SUM=0.0
                    DO 1 K=1,NC
    1               SUM=SUM+ALPHA(K)*X(K)
                    DO 2 J=1,NC
                    Y(J)=ALPHA(J)*X(J)/SUM
                    RETURN
                    END
```

FIGURE 5-2. Procedural implementation of distillation column dynamics.

*y*s, alphas, *L*s, and *M*s. Typically, these variables would be stored in single or multidimensional arrays.

As convenient as it might be to directly code the equations into a program, we would introduce a serious limitation in our code by mixing the tray equations with the details of the equilibrium calculations. If we later want to change the way we calculate equilibrium, we would have to go right to the heart of the program and make changes. Such editing is always dangerous because there is a risk of breaking or invalidating a perfectly fine program. We could, of course, make a copy of the program and keep the original as *version1* and edit the copy and call it *version2*. Such proliferation of versions has its own problems because it quickly leads to an unmanageable set of programs that not even the owner can keep track of. A better approach is to separate the programs into different modules by use of subroutines. Equations 5-1 through 5-3, which describe the composition dynamics on the tray, are not going to change much and should be kept in their own module. The equilibrium calculation (Equation 5-4), which is likely to change more, should be in a different place. In order to communicate between the two modules, we use a subroutine call such as: CALL BUBPT (T,X,Y,P) in place of Equation 5-4. Strictly speaking, we only need X and Y as arguments when we use relative volatilities, but it might be a good idea to provide temperature (T) and pressure (P) in case we later plan to perform bubble point calculations based on vapor pressures.

The procedural implementation of the tray model now seems to be in good shape. We have one module that describes the tray dynamics and another module that describes the equilibrium (see Figure 5-2). We have eliminated the need to make changes in the tray section code and we have removed the multiple version syndrome. Or have we? There are still some loose ends. In return for removing Equation 5-4 from the tray section code, we had to permanently introduce the call to BUBPT (T,X,Y,P). We made the call general enough to cover many cases, but how do we keep track of the cases? If we decide to always use relative volatilities (Equation 5-4) in the body of BUBPT, then that means that all simulations using the tray section module will use relative volatilities for equilibrium. That is not what we had in mind. We could of course implement a different equilibrium method in BUBPT, but even then, all simulations using the tray module would be forced to use the same equilibrium calculation. For dynamic simulations, this is most undesirable because different columns have different modelling requirements. For example, columns separating components with low relative volatilities usually require many trays (80 to 120). These systems are almost always ideal, meaning that relative volatilities adequately describe their equilibrium. Using Equation 5-4 in dynamic simulations of such columns is very desirable because the calculation of equilibrium is explicit and fast. However, not all systems are ideal and they should therefore not be modelled by relative volatilities. Wide boiling fluid systems are best described by vapor pressures, possibly in combination with activity coefficients. The bubble point calculation for liquids modelled by vapor pressures and liquid activity coefficients is extremely time consuming and can easily make dynamic simulations of columns with many trays impractical. Fortunately, the separation of wide boiling systems requires only a few trays, making these columns as feasible to simulate as the low relative volatility columns provided we use the appropriate bubble point calculation in each case. However, we just finished saying that all our simulations will use the same version of BUBPT, which implies that some of our applications using the tray module will be incorrect or hopelessly slow depending on which method we use in BUBPT. To resolve this dilemma, we could of course create a couple of different tray modules that make calls to different BUBPT routines (BUBPT1, BUBPT2, etc.), but now we are back where we started in version management and proliferation of similar looking code.

Problems like these, which only magnify as we go beyond a single tray and into more complex models, have prevented development of the ideal dynamic simulator. It is virtually impossible to come to grips with these problems when using a procedural approach. On the other hand, object-oriented programming elegantly removes these obstacles. To see how this is possible I need to describe how a tray is modelled in an object-oriented style.

Object-Oriented Approach

In the object-oriented approach we start by creating two different classes. One class would model the tray itself and its behavior, whereas another class would model the properties and behavior of the fluid system on and around the tray. For example the fluid class might look like this:

```
class Fluid
{
      :
      :
         bubblePoint();
         getX(j);
         getY(j);
         moles(j);
         getNumberOfComponents();
         :
protected:
         :
         //data area
};
```

where the messages *bubblePoint()*, *getX(j)*, *getY(j)*, *moles(j)*, and *getNumberOfComponents()* would invoke specific behavior from all objects instantiated from class **Fluid**. The nomenclature I am using to describe the classes is similar to that of the object-oriented programming language C++. The data area where compositions, component quantities, temperature, pressure, and physical property parameters are stored is protected from outside use or inspection; only the message routines have complete access to these data even though the data are not formally passed as parameters to the routines. It is as if the data were present in invisible COMMON blocks in each of the message routines.

Next we specify a class for the tray itself. It might look like this:

```
class Tray
{     :
         update(dt);
         Fluid* liquidIn;
         Fluid* liquidOut;
         Fluid* vaporIn;
         Fluid* vaporOut;
protected:
         :
         Fluid* holdup;
         //more data
};
```

The **Tray** class would typically have several messages that tray objects react to, but only one of these messages, *update*, is shown here. In the protected data section, tray is referring to a fluid object called *holdup*. In the C++ nomenclature **Fluid*** designates the variable type and *holdup* the variable name. Our object-oriented model closely mimics the real world; a tray is a physical entity that holds a fluid phase. A tray also knows about its environment such as the liquid coming in from the tray above and the vapor leaving the tray. These are all fluid objects that the tray object can refer to.

Now let us see what the *update(dt)* message might cause a tray object to do.

```
update(dt)
{
holdup → bubblePoint();
    :
    :
n = holdup → getNumberOfComponents();
for all n do:                                        (5-5)
holdup → moles(j) = holdup → moles(j) +
    (liquidIn → moles(j) + vaporIn → moles(j) −
    holdup → getX(j)*L −
    holdup → getY(j)*V)*dt;
    :
}
```

The message *update(dt)* takes one external parameter *dt*. The tray object receiving the *update* message responds by first sending the message *bubblePoint()* to its *holdup* object [holdup → bubblePoint() means "send the message bubblePoint to the object identified as holdup"]. The *bubblePoint* message requests *holdup* to calculate its bubble point. Then *update* calculates the liquid (L) and vapor (V) flow rates leaving the tray (code not shown here). Next, *update* asks *holdup* how many components it has and in return informs *holdup* how each of these components will change as a result of integrating the component balance. The component balance is derived from Equation 5-2 integrated with a simple Euler integration step *dt*.

Comparison of Approaches

Equation 5-5 implements the object-oriented form for tray component dynamics, whereas Figure 5-2 shows the procedural version. There are several important differences between these implementations. In Figure 5-2, we are explicitly declaring and referring to the mole fractions X and Y that are stored as part of the tray module. In Equation 5-5, the composition data are not stored as part of the tray object but instead as part of each fluid object (e.g., *holdup*). However, even in the fluid object, the data are not explicit because we are not directly referring to the mole fractions in holdup (e.g., holdup.x[j]). Instead, we are referring to the mole fractions in an abstract way by sending a message to *holdup* asking it to get the appropriate mole fraction for us [e.g., holdup → getX(j)]. This concept is called *data abstraction* and is another key idea in object-oriented programming. Data abstraction allows us to change the data structures in classes without affecting any code that uses objects instantiated from those classes. For example, initially we might have chosen to store the mole fractions X and Y as arrays in the fluid class. Later we can change our mind and decide not to store the mole fractions explicitly but to store the number of moles of each component and calculate X and Y as we need them. With data abstraction such a change does not impact users of the **Fluid** class (e.g., tray objects).

5-3-3 Inheritance and Polymorphism

We are now ready to tackle the problem of different equilibrium subroutines for different column applications. First, assume that fluid objects calculate their bubble points from relative volatilities (Equation 5-4). Furthermore, we have built a general purpose distillation column from the tray class. Now, we consider an application where vapor pressures and activity coefficients should be used to correctly calculate the bubble point. This is where we got stuck before with procedural programming. Object-oriented programming allows us to proceed. First we use the object-oriented concept of *inheritance* to derive a new fluid class from **Fluid**. Call the new fluid class **VaporPressureFluid**:

```
class VaporPressureFluid : public Fluid
{
    bubblePoint();
};
```

Creating this new class was simple enough: All we had to do was to give the class a distinct name and mention from what other class it should inherit properties and behavior. A **VaporPressureFluid** object will thus have all the properties and behavior of a **Fluid** object. They will only differ in one aspect; they will calculate their bubble points using different methods. A **Fluid** object still calculates bubble points from relative volatilities, whereas the **VaporPressureFluid** will use vapor pressures and activity coefficients to complete its bubble point calculation. The interesting part is that in both cases, the message is the same: *bubblePoint()*. We can thus keep our tray class and its *update* message intact because *update* was sending that very same message *bubblePoint()* to its holdup. We simply have to provide tray objects with **Fluid** objects when we want to continue calculating bubble points with relative volatilities or provide the tray objects with **VaporPressureFluid** objects if we want the bubble point to be calculated using vapor pressures. This statement might seem puzzling at first because the **Tray** class is declared with references to **Fluid** and not to **VaporPressureFluid**. How then can we arbitrarily substitute a **Fluid** object with a **VaporPressureFluid** object? This is where the most powerful concept of object-oriented programming enters, namely, *polymorphism*.

Polymorphism means many shapes and refers to the principle that any object instantiated from a derived class (e.g., **VaporPressureFluid**) can replace an object from a more basic class (e.g., **Fluid**) without telling the user class (**Tray**) about the substitution. **Tray** objects send the messages as before assuming that they are referring to **Fluid** objects when in reality all those objects receiving the messages are **VaporPressureFluid** objects and as such respond in their own way to the same messages. Not only can we perform this powerful substitution of base classes with derived classes when we link our modules together, but we can also make the substitution while the program is

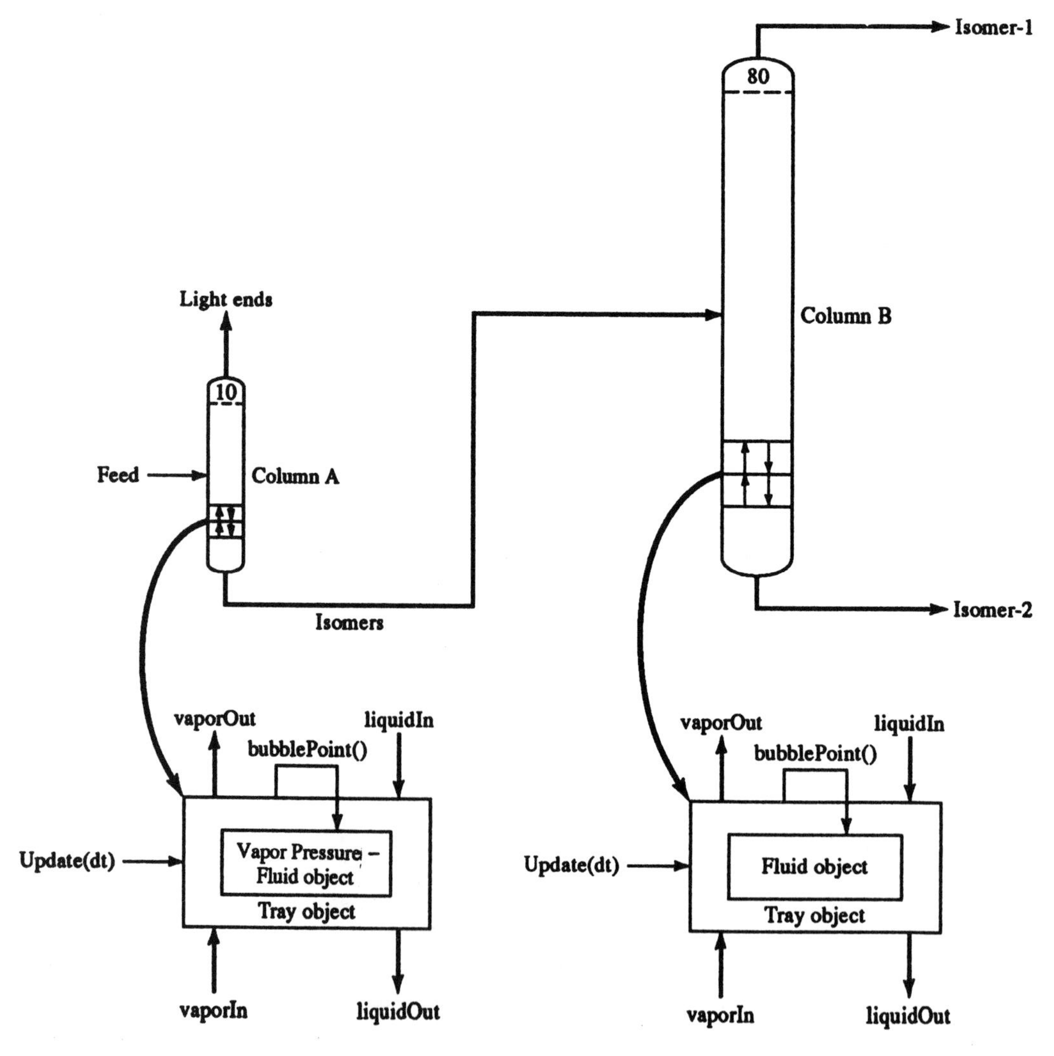

FIGURE 5-3. Graphical representation of polymorphism.

executing. The run time assignment of objects from derived classes is called *dynamic binding* and is probably the most intriguing aspect of object-oriented programming (Cornish, 1987).

Figure 5-3 summarizes the concepts of inheritance, polymorphism, and dynamic binding as applied to the fluid objects around the trays in a distillation column. Two different columns are part of a simulation separating several components. Both columns are instantiated from a generic **Column** class built from the **Tray** class by connecting **Tray** objects together. When a **Column** object is instantiated from a **Column** class we specify how many stages it will have. This can be done by the user during execution of the program and does not require separate compilation. Column A is removing light ends from a mixture of mostly isomers and requires only 10 stages. Column A should be simulated with a rigorous bubble point calculation involving vapor pressures and activity coefficients. Column B on the other hand separates the isomers and needs 80 stages to do the job. For column B, relative volatilities are adequate to describe the vapor–liquid equilibrium. The two columns can be simulated together using the same **Column** class with different fluid objects on the trays. The different fluid objects can be specified by the user, at run time, when the user requests two different instances of the same **Column** class.

5-4 DISTILLATION COLUMN SIMULATION WITH OBJECT-ORIENTED PROGRAMMING

I have demonstrated how inheritance, polymorphism, and dynamic binding help build a generic distillation column tray section that can be used with many different fluid systems and physical property calculation methods. I will now describe how these same concepts and the notions of encapsulation and data abstraction help us build complete simulation systems that are very flexible. Before doing this, it is necessary to review the principles of model structuring (Astom and Kreutzer, 1986; Mattsson, 1988).

5-4-1 Structured Models

The first principle of model structuring is that of *hierarchical submodel decomposition*, which suggests that a complex system should be built from a set of small, self-contained units. The small units should in turn be built from yet smaller units until there is a smallest logical unit that corresponds to some physical entity in the real world. In the other direction, a complex unit can be part of an even more complex system. A distillation column is an excellent example of the potential use of hierarchical submodels because a complete column with trays and auxiliaries can be considered a complex system. According to the hierarchical submodel principle, the column should be built from small pieces, in this case a tray section module, a reboiler module, and a condenser module. Furthermore, the tray section module should be a collection of individual tray modules, which in turn contain different fluid phase modules. In a different hierarchy, a column can be part of a larger, more complex system, say a refining train or a complete plant.

The second modelling principle we need to consider is that of *parametrization* (Mattsson, 1988; Nilsson, 1989). In design of the modules for a hierarchical subsystem, it is important that these modules be flexible and modifiable. If we can modify a module's behavior based on parameters we give it, we are using parametrization. A simple example is the number of trays in a distillation column tray section. A generic tray section of say 20 trays might be useful for many simulations but not nearly as useful as a generic tray section with a user-specified number of trays. The substitution of different fluid objects in the tray class can be viewed as another example of parametrization (Nilsson, 1989). Here the tray is the flexible module using the fluid system as a parameter to modify its behavior.

The concept of *terminals* is the third modelling principle we should consider (Mattsson, 1989). A terminal on a software module is analogous to the antenna terminal on a regular television set or the serial port on a personal computer. These terminals provide agreed upon formats for communicating information to or from the unit. The serial port on a personal computer might be the best example of a terminal. From the computer's point of view, we need not know what device is at the other end of the cable attached to the serial port. We only need to know how to read from and write to the serial port itself. Examples of terminals on a distillation column module would be the feed nozzle, distillate nozzle, and bottoms nozzle. If these terminals have the characteristics of a fluid, we can model and simulate a column without knowing what is attached to the other ends of the pipes connected to those nozzles.

The last modelling concept we will consider is that of *connections* (Mattsson, 1989). Connections are modules that connect units together via their terminals. For a distillation column, the connections correspond to the pipes that connect various submodules together. For example, the reflux pipe on a real column would be represented by a connection module in a model of the column.

5-4-2 Structured Models and Object-Oriented Programming

Hierarchical submodules, parametrization, terminals, and connections are concepts that have been used successfully in many process simulation systems. They do not imply nor are they synonymous with object-oriented programming. For example, both DYFLO and SPEEDUP have subunits that can be connected together with terminals and streams. Parametrization is also used in these programs to modify structure and behavior. However, neither DYFLO nor SPEEDUP uses an object-oriented paradigm. So why do I advocate object-oriented programming as the preferred way of implementing dynamic simulation systems? For the simple reason that object-oriented programming makes it extremely easy to implement and fully exploit model structuring principles.

For example, there is a very natural mapping between hierarchical submodels and classes. I have already indicated how the **Tray** class was used to build a tray section. Similarly, **Condenser** and **Reboiler** classes can be used with the tray section to make a complete column. The object-oriented principles of encapsulation and data hiding encourage making these units self-contained and autonomous. Their only communication with the outside world is through terminals and connections that can be made very versatile through data abstraction and polymorphism.

In this fashion I have used object-oriented programming to build an extremely flexible dynamic simulator for distillation columns. One generic class represents the tray section, one class represents a total condenser, and one class represents a thermosiphon reboiler. These three classes are then connected to form a generic **Column** class. From the **Condenser** class, several other types of condensers, such as partial condensers, condensers handling inerts, and spray condensers are derived. Because all the derived condensers can take the place of a total condenser by way of polymorphism, different types of overhead systems can be realized with the generic column without ever having to modify or add to the code describing the column. This way I did not have to concern myself up front about all the possible configurations for a column because if I want to modify any part of the column, I simply derive a new class or classes from the basic ones, slip them into the generic column class, and the new behavior appears almost as magic. There is never a risk of modifying and possibly breaking existing proven code or subunits. Because I use the same principle to model the fluid streams in the column, I can use the generic column to process literally every possible fluid system with any physical property methods I care to consider. This even allows me to simulate reactive distillation systems with one and the same generic column model. After all, chemical reactions on the trays of a column are not a function of what column we use but merely depend on the properties and the state of the fluid we process in the column.

Flexible dynamic simulators do not come without a price. The major disadvantage in working with object-based systems is the lack of standard design principles for creating classes. This is bothersome because the whole concept of object-oriented programming critically hinges on the existence of a few powerful base classes such as the **Fluid** class and the **Tray** class described earlier. Without these well thought out base classes, object-oriented programming can turn into a jungle of many small, specialized classes that are hard to use and keep track of. It takes considerable time and effort, including many design iterations, to come up with a good set of base classes. However, once this front end loading effort is completed, the downstream possibilities are virtually endless.

5-5 EXPERIENCE IN USING OBJECT-ORIENTED SIMULATION FOR DISTILLATION

The vision of the ideal dynamic simulator was created independently of any perceived relation to object-oriented programming. So when I originally started to explore object-oriented programming for dynamic simulations, I did it for very specific and less visionary reasons. The first reason was to provide a better framework for reusing the code in the many custom simulations I had developed over time. Often new simulation models were sufficiently different from previous ones such that the old code could not be used without substantial error prone surgery. This forced me to start from scratch many times. On the other hand, as I developed a new simulation, I always felt that there were more than 50% common code and concepts from simulation to simulation. By using model structuring principles and object-oriented programming, I have made it possible to reuse substantial amounts of proven code. For example, when it used to take me a couple of months to write, debug, and run a dynamic model of a distillation column, I can today put together and evaluate the same simulation in a few days. Furthermore, I hand over an interactive model to the plant personnel so they can make their own assessment of what needs to be done in the process. The second reason I started to look at object-oriented programming was the desire to have an interactive simulator that engineers and operators could use to learn the principles of process dynamics and process control. The object-oriented features I have described make it relatively easy to write programs that simulate different unit operations or even a whole process. After all, the concept of object-oriented programming grew out of the language SIMULA, which had been specifically developed in the 1960s to deal with large scale dynamic simulations (Magnusson,1990).

In relation to languages, it might be of interest to mention which ones I have tried and what I believe works. There are many different high-level languages that can be classified as object-oriented. Smalltalk is considered the purest object-oriented language. In Smalltalk everything is an object and all computing is done by message passing. Because Smalltalk is an interactive language, there is no distinction between the development environment and the run time simulation. New classes can therefore be defined at run time. This makes Smalltalk ideal for model building because hierarchical models can be built and tested without having to explicitly compile and link the program. Although run time interaction is the strength of Smalltalk, it is also the major weakness when it comes to simulation. I have found Smalltalk to be unacceptably slow for distillation simulation purposes.

Lisp is another language often mentioned in an object-oriented context. Although Lisp itself is not object-oriented, it is easily extended to be so. Flavors is one of the most used extensions to Common Lisp whereas CLOS has been accepted by ANSI as the standardized extension. Lisp and its object-oriented extensions offer faster alternatives to Smalltalk when it comes to simulation because they support numeric calculations without message passing. Lisp is, however, difficult to learn and master and is therefore not something I recommend for practical simulation applications.

C++ is an extension to C that supports object-oriented programming. It is also intended to be a better C. C++ is a compiled language and is therefore very fast during execution, which is a most desirable feature for large distillation simulations. Although C++ requires more work than Smalltalk to provide interactive model building it is clearly superior when it comes to program size and execution speed. This has enabled me to develop large scale simulations in C++ that run many times faster than real time even on a personal computer. I feel that C++ is currently the most suitable language for object-oriented simulation of chemical systems.

Run time speed is extremely important for dynamic simulations. That is primarily why C++ is preferable to Smalltalk and Lisp. However, because C++ is comparable to FORTRAN and Pascal in run time speed, you might ask why I advocate C++ over these older and more common languages? The answer is that run time speed is only one issue in dynamic simulation and the total time from problem formulation to solution is what really counts. Here is where object-oriented techniques have significant advantages in that they encourage generic module building. Once the generic modules are available, complex dynamic models can be constructed in extremely short time.

5-6 CONCLUSION

By using object-oriented programming I have already realized the two major goals of code reuse and a simple configurable training simulator as described. However, with recent advances in computer hardware and compilers I could use object-oriented programming to move closer to the ideal simulator. Needless to say this task is difficult to achieve because such a simulator must not only have a large dose of chemical process simulation capability, but must also have a flexible and user-friendly interface. Without doubt, I see object-oriented programming as the right way to get there.

Commercial products will soon be available to meet the need for the ideal simulator. Some of these products will be based on object-oriented principles. One could perhaps argue that it is irrelevant what language was used in creating a simulator that is run time configurable; the user and the computer do not care if the source code was FORTRAN, Pascal, or C++. However, there is always a need to make new simulation units or to modify existing ones. The user should be able to do so with inheritance and dynamic binding features that encourage building libraries of small, efficient, and well-tested modules that can be shared across an organization. This extremely powerful aspect of a good simulator has, to date, been fully explored only in a few commercial products (Bozenhart and Etra, 1989).

References

Åstrom, K. J. and Kreutzer, W. (1986). System representations. *IEEE Third Symposium on Computer-Aided Control System Design, Arlington, Virginia*. New York: IEEE.

Biegler, L. T. (1989). Chemical process simulation. *Chem. Eng. Prog.* 85, 50–61.

Bozenhart, H. and Etra, S. (1989). Intelligent simulation (Mercury ISIM). Product publication, Artificial Intelligence Technologies, Inc., Hawthorne, NY.

Cornish, M. (1987). What would you do with object-oriented programming if you had it? *FYAI-TI Technical Tips* 1(10), 3–12.

Franks, R. G. E. (1972). *Modeling and Simulation in Chemical Engineering*. New York: Wiley Interscience.

Luyben, W. L. (1990). *Process Modeling, Simulation and Control for Chemical Engineers*, 2nd ed. New York: McGraw-Hill.

Magnusson, B. (1990). European perspective. *J. Object-Oriented Program*. 2, 52–55.

Mattsson, S. E. (1988). On model structuring concepts. 4th IFAC Symposium on Computer-Aided Design in Control Systems, P. R. China.

Mattsson, S. E. (1989). Modeling of interactions between submodels. 1989 European Simulation Multiconference, ESM'89, Rome, Italy.

Nilsson, B. (1989). Structured modeling of chemical processes with control systems. AIChE fall meeting, San Francisco, Nov. 1989.

Pascoe, G. A. (1986). Elements of object-oriented programming. *BYTE* Aug., 139–144.

Perkins, J. D. (1986). Survey of existing systems for dynamic simulation of industrial process. *Modeling, Identification and Control* 7, 2.

Stephanopoulos, G., Henning, G., and Leone, H. (1990). MODEL.LA., a modeling language for process engineering. *Comput. Chem. Eng.* 14(8), 813–869.

6

Plantwide Process Control Simulation

Ernest F. Vogel
Tennessee Eastman Company

6-1 INTRODUCTION

Process simulation is a valuable tool for the design of process control strategies, particularly for distillation columns. Typically, control strategies use conventional single-input–single-output (SISO) PID controllers. Therefore, designing a control strategy normally includes determining which process variables should be measured and controlled and which control valve should be paired with each controlled variable. Control strategy design also includes identifying locations in the process where conventional single-input–single-output PID controllers may not work well.

Distillation columns are relatively complex unit operations and it is often difficult for an engineer to develop a satisfactory distillation control strategy without the assistance of a distillation column simulator. In many cases, distillation columns operate somewhat independently of other unit operations and the columns can be studied one at a time. For example, for a process where all unit operations are in series with no recycle (Figure 6-1), a control strategy can be developed for each column one at a time. Additionally, tanks on column feed or product streams isolate columns, allowing column control strategies to be developed independently (Figure 6-2). However, for processes with recycles (Figure 6-3) or processes with coupled distillation columns (Figure 6-4), a distillation column control strategy often must be developed in consideration of associated unit operations. Chapter 20 provides additional examples.

Studying the control and operation of distillation columns and associated equipment requires a plantwide process control simulator. A plantwide process control simulator can simulate multiple unit operations of various types (a flowsheet simulator) and can perform both steady-state and dynamic simulation.

This chapter describes the use of a plantwide process control simulator for control strategy design and for other problems as well. The chapter also discusses model accuracy, simulation environments, and necessary features of a plantwide process control simulator.

6-2 APPLICATIONS OF A PLANTWIDE PROCESS CONTROL SIMULATOR

In addition to the development of process control strategies, there are a number of other related problems where a plantwide process control simulator can be applied.

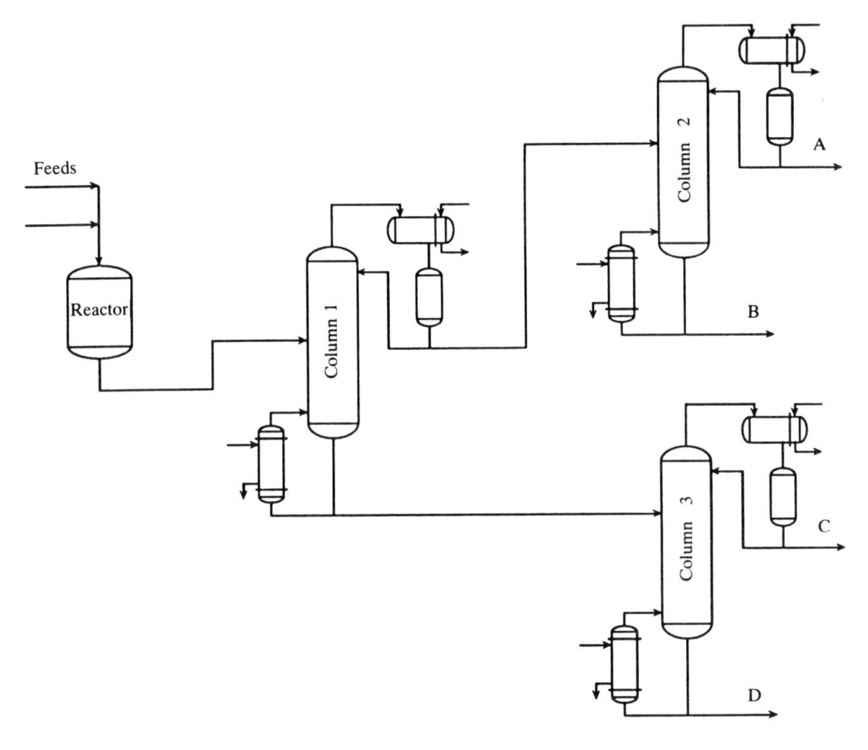

FIGURE 6-1. Process with units in series.

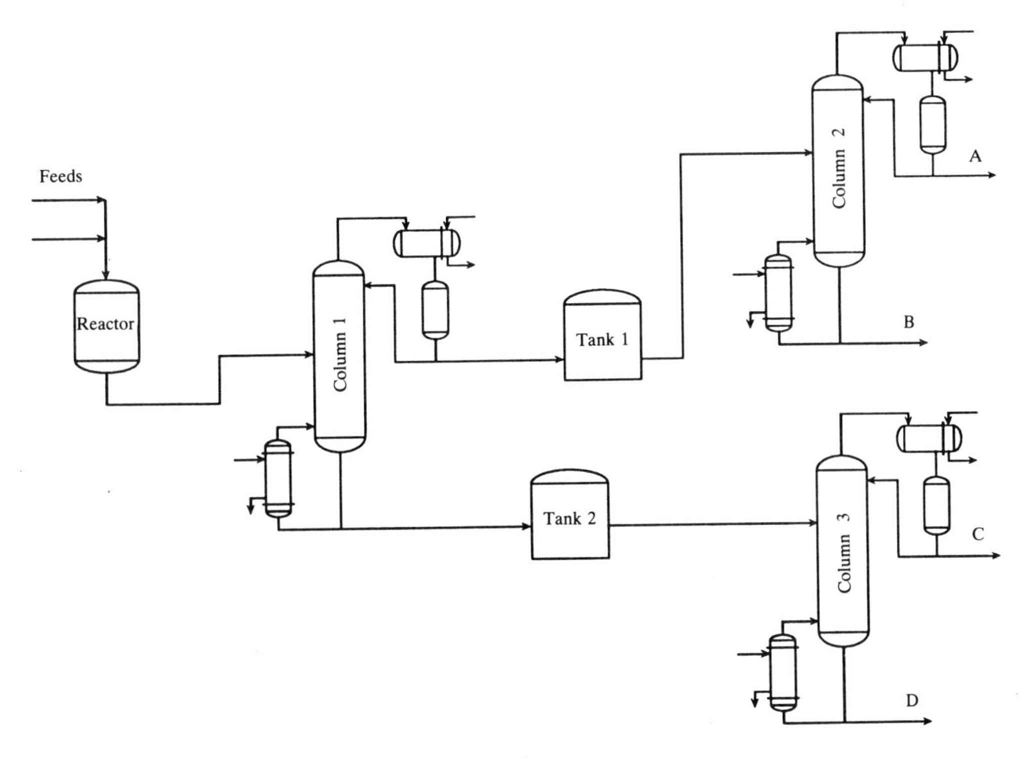

FIGURE 6-2. Process with intermediate storage tanks.

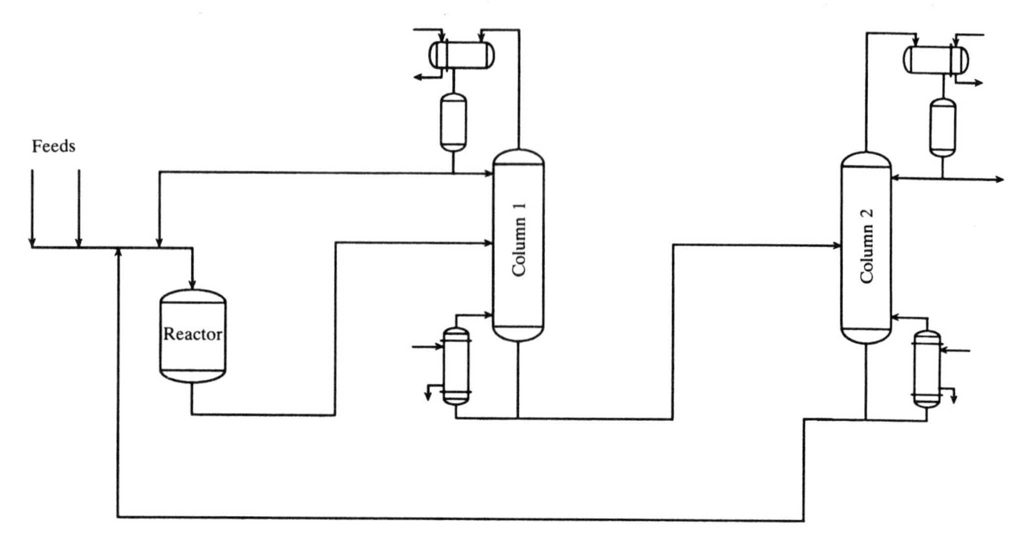

FIGURE 6-3. Process with recycles.

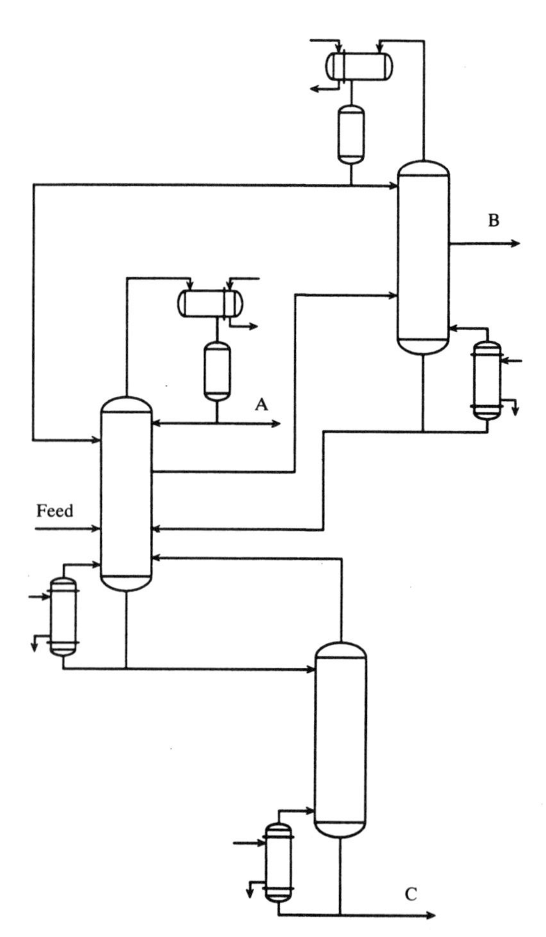

FIGURE 6-4. Coupled distillation columns.

The applications fall into three major categories: process control, process design, and process safety. The following sections discuss the application of a plantwide control simulator in all three of these areas.

6-2-1 Process Control

Control Strategy Design
Control strategy design is the procedure of determining which process measurements need to be controlled and which valves should be used to control them. This procedure assumes that conventional single-input–single-output PID controllers will work satisfactorily, which is normally the case. This procedure also identifies cases where the conventional control algorithm is not sufficient. Steady-state simulation with sensitivity analysis and interaction analysis is valuable for identifying good potential control strategies. Dynamic simulation provides a demonstration of how the candidate control strategies will actually perform for expected disturbances and setpoint changes.

Justification of Control Strategy Modifications
Changing the control strategy in an existing plant typically requires convincing plant engineers to agree to the change. Changing the strategy also requires justifying any ex-

pense of implementation, process downtime or hardware and software expenses. Dynamic simulations showing how both the old and the new strategies perform for typical process disturbances often can accomplish both of these tasks.

On-line Analyzer Justification
On-line analyzers, which provide process composition measurements either continuously or frequently (every 10 min or less), can provide significant improvements in process control. However, many on-line analyzers are more expensive and require more maintenance compared to more common measurements such as pressure and temperature. Dynamic simulation can demonstrate how much an on-line analyzer can improve control through use in a feedforward or feedback control scheme. Dynamic simulation can also be used to determine the frequency at which new composition measurements must be available.

Advanced Control Algorithm Evaluation
Advanced control algorithms, such as predictive control, require significantly more effort to implement and maintain than conventional PID control. Implementation of advanced control algorithms on realistic process models provides an evaluation of their performance compared to conventional control and justification of their implementation.

Batch Process Control and Operation
Dynamic simulation provides a method to evaluate strategies and procedures for controlling and operating batch processes. Examples of batch control and operation issues include tracking temperature trajectories and determining cycle times and operating sequences.

6-2-2 Process Design

Surge Capacity Sizing
Intermediate holdup tanks provide attenuation of flow rate or composition disturbances before they reach downstream equipment. Such surge tanks are essential for interfacing batch and continuous processes. Simulation provides a way to determine how large surge tanks need to be and where they should be located to sufficiently attenuate harmful disturbances.

Flowsheet Design
Some process flowsheet designs that produce the same product are inherently easier to control and operate than others. The dynamic performance and operability of alternate flowsheet designs can be compared using a plantwide dynamic simulation.

Heat Integration Systems
Although heat integration can result in significant energy savings, it can also cause process disturbances to propagate to parts of a process that would not be affected without heat integration. Plantwide process control simulation with and without heat integration provides an evaluation of the consequences of additional disturbance propagation.

6-2-3 Process Safety

Emergency Shutdown Evaluation
Most processes have emergency shutdown systems to prevent the occurrence of a catastrophe if the process moves to a dangerous operating condition. Unfortunately, the performance of these systems can only be truly tested if an event occurs. However, dynamic simulation provides a way to test and compare emergency shutdown strategies and to quantify how well they would work.

Accident Evaluation
When a process accident occurs, there are often many proposed explanations and scenarios for what happened. Usually, it is not clear what actually caused the problem. With dynamic simulation, the proposed explanations can be tested to evaluate their feasibility. The dynamic simulations often rule out several of the possibilities and help point to the real cause.

Operator Training
Good operator training is important with processes that are relatively complex or potentially hazardous. In these cases, operator training simulators are very helpful. With an operator training simulator, the operator's interface to the "process" is the same control system operating console as used for the real plant, but the process measurements are provided by a real-time, dynamic process simulator and the operator's commands go to the simulator rather than the actual process. A plantwide process control simulator facilitates the building of such operator training simulators.

Pre-Start-up Knowledge
Running simulations of new processes provides a feel for how they will operate, how quickly measurements respond, and how sensitive they are to the manipulated variables. This insight is helpful during actual process start-ups.

6-2-4 Example

Figure 6-5 shows an example process used here to illustrate the role of a plantwide process control simulator in the process and control strategy design. This process has two feeds, components A and B. A and B react together to form the desired product R. B reacts with itself to form the undesired by-product S. The reactions are exothermic. The process consists of the reactor and the two distilltion columns to separate the reactant and products. The reactor includes internal cooling coils to remove the heat of reaction. The first distillation column separates the reactants (A and B) from the products (R and S). The reactants are recycled back to the reactor. The second distillation column separates the products (R and S). In the design of the process, the control, and the safety system for this example, there are several issues where simulation with a plantwide process control simulator is helpful.

Process Control
What are possible control strategies? How well will each perform for expected disturbances? Where should the production rate be set: the feeds to the reactor, the feed to the first column, another valve in the process?

What variables need to be controlled? Should reactor temperature be controlled? Which temperature or temperatures should be controlled in the distillation columns?

Is temperature control on the columns sufficient, or are composition measurements

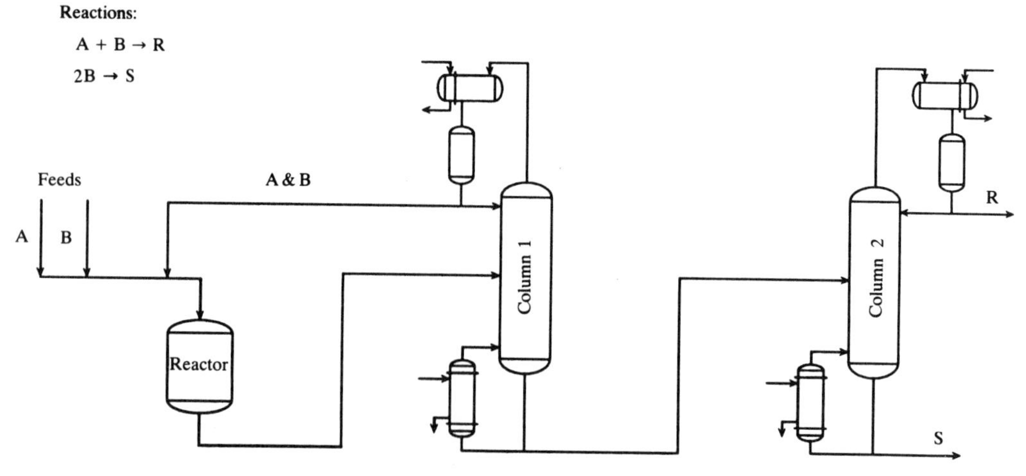

FIGURE 6-5. Example process for plant-wide simulation.

needed? How much might they help? How frequently are composition measurements needed?

Are SISO PID controllers satisfactory? Are there any severe interactions? Are they bad enough to justify a more advanced algorithm? How much will an advanced algorithm help?

Which disturbances have the most effect? What happens if feed A or feed B is in excess?

Process Design

Is a surge tank needed between the columns? The simulation can justify it if needed or prove that one is not necessary if that is the case.

Is the reflux drum on the first column big enough so that disturbances in the column are sufficiently attenuated and do not upset the reactor through the recycle?

Process Safety

Is the exothermic reaction potentially dangerous? If so, will stopping the feeds and maximizing the cooling water flow stop a temperature excursion? Could a change in the recycle flow or composition initiate a temperature excursion?

6-3 BENEFIT FROM PLANTWIDE PROCESS CONTROL SIMULATION

The primary result or benefit from process simulation is improved process insight and understanding. The thinking process the user goes through in developing and running a simulation is what leads to the improved process insight and understanding. Better process understanding results in the formulation of better control strategy designs, as well as better flowsheet designs and better safety systems. A process simulator does not explicitly provide such solutions, but as a thinking tool it assists in the development of solutions. A simulation reveals which process parameters are important and which ones are not. Sometimes the simulated dynamic behavior is easily explained, but counterintuitive. The mind does not easily predict dynamic responses by superimposing all of the simultaneous dynamic effects in a chemical process, particularly if the process contains one or more recycles.

The explicit results from process control simulation tend to be more qualitative than quantitative. Application of process control simulation often results in a decision as illustrated in the example above, for example, use control strategy B rather than strategy A, buy a continuous analyzer for use in a feedback or feedforward strategy, a surge tank is necessary, and so forth. To contrast, process design simulations often provide absolute numbers, for example, reboiler heat duty, distillation column diameter, and so forth.

6-4 DEFINING THE SCOPE OF A PLANTWIDE PROCESS SIMULATION

The first step in performing a plantwide process control simulation is to define the scope of the simulation. The scope includes the size and the accuracy of the simulation. The size of a simulation refers to how much of the process is simulated. The accuracy of a simulation refers to how well the steady-state and dynamic responses of the simulator match that of the actual process. To minimize time and effort in building and running a simulation, the engineer determines the size and accuracy actually needed in order to avoid simulating more of the process than necessary or simulating the process more accurately than necessary. Both the size and accuracy needed for a simulation depend on the problem being solved and normally must be determined on a case-by-case basis.

In defining the size of the simulation, for example, which unit operations to simulate, the engineer looks for logical breaks in the process. Large interprocess inventories make good boundaries for simulations because most disturbances are significantly attenuated at those points. Also, portions of a process that are not connected by recycles can be segregated.

The accuracy of a simulation is not easily measured. In some cases, the plant does not yet exist. For existing processes, perturbing the plant to observe dynamic responses is usually not desirable. Generally, to obtain more accuracy, the simulation must use more rigorous or complex models of unit operations, valves, and so on.

In control strategy design, dynamic simulation is typically used for making relative performance comparisons between alternate control strategies. When making relative comparisons, a high degree of accuracy is not as important as when determining absolute values. If two control schemes are significantly different, simulations will demonstrate that fact, even if the model response does not exactly match the process response. If the differences between two schemes are small, other criteria should be used to choose between them, for example, ease of operator understanding and ease of implementation and support.

As an accuracy guideline for control strategy design, match the simulation's steady-state base case to the plant operation or to the steady-state process design conditions to within 5%. These conditions include flows, temperatures, pressures, compositions, levels (inventories), and vessel volumes. Note that with good estimates of vessel volumes, and flow rates, residence times (time constants) will be similar to those of the process. Thus, validation of dynamic responses is generally not necessary for control strategy design. If the process conditions are matched as previously described, the process dynamics are usually sufficiently accurate for control strategy comparisons.

Note, however, that for other applications model accuracy is much more important. For instance, when evaluating an emergency shutdown system on an exothermic reactor and determining how high the temperature will go during an upset and how quickly it will rise, accuracy of the model is very important.

6-5 BUILDING A PLANTWIDE PROCESS SIMULATION

There are several simulation environments an engineer can use to build a dynamic flowsheet simulation. Three possibilities are described in the following subsections.

6-5-1 Programming Environment

In this case, the engineer writes source code for the differential and algebraic equations that describe the process to be simulated. The programmer must provide numerical methods and interface routines as well. Simulations built in this environment tend to be customized for simulating one particular process and are not easily modified for other problems. Writing and debugging simulations is likely to be time intensive. This approach offers the most flexibility, but requires a high level of expertise in both modelling and programming.

6-5-2 Equation Solving Environment

The equation solving environment is a higher level than the programming environment. Similar to the programming environment, the engineer must provide the differential and algebraic equations that describe the process to be simulated. However, the numerical method and interface routines are provided. This approach is similar to the programming approach in that simulations tend to be customized and writing and debugging the simulation can be time consuming. Although this environment may not be quite as flexible as the programming environment, the total time required to build the simulation and the programming expertise required will likely be less.

6-5-3 Steady-State – Dynamic Flowsheet Simulation Environment

A steady-state–dynamic flowsheet simulator offers a significantly higher level simulation environment compared to the previous two options. A steady-state–dynamic flowsheet simulator has preprogrammed unit operation models and control algorithms and the engineer builds the simulation by interactively linking the necessary pieces together. The simulator also provides all the needed interfaces and numerical methods. Thus, this approach requires a lower level of expertise in the modelling and programming areas. Also, because models are preprogrammed, significantly less time is necessary for building and debugging simulations. However, this approach offers less flexibility for unusual cases because the engineer is limited to the provided library of process and controller models. To address this potential problem, the simulator may offer a way to build custom models, but again the use of custom models is limited by the engineer's time and modelling expertise.

6-5-4 Practical Considerations

The preceding three options offer a trade-off between flexibility and ease of use. The first two options offer more flexibility for creating custom models that may more closely represent the actual process. Realistically, however, these two choices are of little value to most control strategy design engineers. The time required to build custom simulations starting from the point of writing differential equations is often prohibitive except in critical cases. The control strategy designer is likely to be in an environment where the schedules and workload make quick turnaround of projects essential. Design schedules allow little time for process simulation. Process control trouble shooting type problems often need answers as soon as possible, typically within two to three days. Further, the control strategy design engineer may not have the necessary modelling or programming expertise. Thus, the interactive flowsheet simulator option appears to be the best choice for most control strategy designers.

Given the control strategy designer's limitations of time and expertise, it is important that the simulator be interactive. An interactive simulator includes a menu driven or graphical interface for building and changing the simulation. Such an interface allows the user to quickly and easily examine or change any parameter. Interactive means that the user can monitor the progress of a simulation and change parameters during the simulation or that the user can interrupt a simulation, change a parameter or configuration, and resume the simulation from where it was interrupted. Interactive also means that building or changing a simulation never requires recompiling or relinking the simulator. Interactive does not mean: editing an input file with a screen editor, searching for and changing a parameter, exiting the editor, and then issuing a command that recompiles and relinks the simulator and then starts the dynamic simulation. The effort and time such an environment requires discourages use.

Eliminating the need for programming greatly speeds the building of a simulation. In order for a user to avoid programming, the simulator must have the needed capabilities readily available to solve typical problems: physical properties for common chemicals, unit operations, and control algorithms should all be available from data bases or libraries.

Although not all processes can be simulated using a unit operation library, most control strategy design problems can be satisfactorily simulated with a library of traditional chemical engineering unit operations. For instance, most liquid phase reactors can be modeled as simple continuous stirred tank reactors (CSTRs). This type of assumption is possible because a high level of accuracy is generally not needed for control strategy design. As discussed previously, less accuracy will suffice because control strategy design simulations involve relative performance comparisons. However, some processes contain very unique unit operations and no matter how complete the unit operation library in a flowsheet simulator, there will be problems that require custom modelling.

6-6 FEATURES OF A PLANTWIDE PROCESS SIMULATOR FOR CONTROL STRATEGY DESIGN

There are several tools in a plantwide process control simulator that set it apart from other more general flowsheet simulators. A key feature is the capability to perform both steady-state and dynamic simulation. Steady-state simulation facilitates the development and initial screening of candidate control strategies by providing a quick way to calculate several steady-state performance measures.

A common first step in control strategy design is the calculation of steady-state gains for selected controlled (c) and manipulated (m) variables. Steady-state gain may be defined as $\partial c/\partial m$. To calculate the steady-state gains, each manipulated variable is perturbed one at a time from an initial base case condition. After each manipulated variable is perturbed, the simulation is reconverged to steady state and the changes in the controlled variables are calculated. The derivative is then approximated as $\Delta c/\Delta m$.

From steady-state gains, the simulator can then calculate several interaction measures as described in Chapter 8: relative gain array (Bristol, 1966; McAvoy, 1983), singular value decomposition (Forsythe, Malcolm, and Moler, 1977), Niederlinski's stability index (Niederlinski, 1971), and block relative gain (Manousiouthakis, Savage, and Arkun, 1986).

Also from steady-state simulation, sensitivity plots as illustrated by Luyben (1975) can be generated. These sensitivity plots have information similar to the steady-states gains, but the plots also reveal the nonlinearities of the process. Figure 6-6 shows an example of a sensitivity plot for the first column of the example process (Figure 6-5). This plot shows how the bottoms composition varies for three disturbances while a proposed control strategy holds the controlled variables constant. The curves are generated by varying each disturbance variable one at a time and reconverging the simulation to steady state with each variation. In this case, the study shows that the

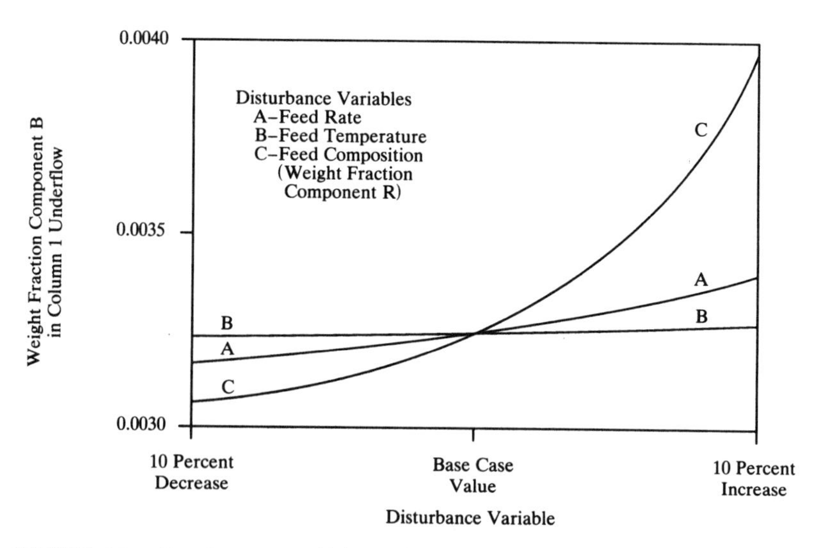

FIGURE 6-6. Steady-state sensitivity analysis.

composition is relatively insensitive to the feed temperature whereas it is more sensitive to the feed composition, particularly for increasing amounts of component R in the feed. If this amount of variation in the B composition is acceptable, then the proposed control strategy may merit further evaluation with dynamic simulation. However, if this amount of composition variation exceeds acceptable limits, there is no point in further considering this control strategy.

After developing a set of candidate control strategies using steady-state analysis, dynamic simulation provides a demonstration of how the control strategies will actually perform. Often, two strategies that appear to have comparable steady-state performance greatly differ dynamically. Further, the strategy that appears to have the best steady-state performance may not perform well at all dynamically.

Integrating the steady-state and dynamic simulation capabilities and the control system analysis tools into one plantwide simulation package greatly facilitates control strategy design. Testing control strategies using representative, nonlinear process models provides good indications of their actual performance.

References

Bristol, E. H. (1966). On a new measure of interaction for multivariable process control. *IEEE Trans. Automatic Control* AC-11, 133–134.

Forsythe, G. E., Malcolm, M. A., and Moler, C. B. (1977). *Computer Methods for Mathematical Computations*. Englewood Cliffs, NJ: Prentice-Hall.

Luyben, W. L. (1975). Steady-state energy conservation aspects of distillation column control system design. *Ind. Eng. Chem. Fundam.* 14, 321–325.

McAvoy, T. J. (1983). *Interaction Analysis, Principles and Applications*. Research Triangle Park, NC: Instrument Society of America.

Manousiouthakis, V. Savage, R. and Arkun, Y. (1986). Synthesis of decentralized process control structures using the concept of block relative gain. *AIChE J.* 32, 991–1003.

Niederlinski, A. (1971). A heuristic approach to the design of linear multivariable control system. *Automatica* 7, 691.

7

Identification of Distillation Systems

R. C. McFarlane
Amoco Corporation

D. E. Rivera[1]
Arizona State University

7-1 INTRODUCTION

In this chapter we address the problem of identification of distillation systems for the purpose of obtaining models for process control. Much of the control literature has focused on controller synthesis procedures that are derived under the assumption that suitable models are available. Significantly less attention has been paid to the specific problem of defining the requirements of models for process control purposes and how to make the best choices of design variables in identification to obtain them. Among the objectives of this chapter is to survey the available literature in this area and present some ideas and procedures that, when incorporated into the well established methodology for system identification, make the identification more relevant to the needs of process control.

The identification of distillation systems has many features that are common to most chemical engineering systems. Although distillation processes are inherently nonlinear, if operated over a sufficiently small region, control systems based on linear input–output models often perform satisfactorily on them. The literature on linear system identification is a mature one, and it would be redundant to provide detailed descriptions here of methods that are well defined in the literature. Rather, we present an overview of the general methodology of linear system identification, and provide details of specific methods that are "control-relevant" and useful for identifying distillation systems. For distillation systems that are highly nonlinear (for example, high purity columns), a number of different identification approaches are possible. We do not review nonlinear system identification because the control structures that have proven most useful for distillation, even in cases where nonlinearity is pronounced, are based on linear input–output representations. Therefore, we suggest adaptations of the basic linear identification approach (e.g., use of linearizing transformations) to obtain the required linear models.

The class of multivariable predictive controllers based on linear step response or truncated impulse response models has proven useful in a wide range of industrial distillation systems (see Chapter 12). Direct identification using these models poses

[1]Also affiliated with the Control Systems Engineering Laboratory, Computer-Integrated Manufacturing Systems Research Center, Arizona State University.

problems because they are not parsimonious (i.e., compact) system representations. Overparametrization and lack of independence among the model parameters results in a poorly conditioned estimation problem and estimates with high variance result when conventional least-squares estimators are used. To compensate for this problem, practitioners are forced to perform long dynamic process tests to generate large quantities of data for the estimation. In some plants, performing very long dynamic tests may be acceptable. Nevertheless, an overall goal of system identification is to obtain the required models with the minimum of disruption to normal process operation. Therefore, we present two alternative approaches for obtaining step or impulse models, namely, estimation of parsimonious low-order transfer function models (that are converted after estimation to the required step or impulse form) and the use of biased least-squares estimators (for example, ridge regression and partial least squares). The use of biased estimators is one way to provide more efficient estimation if impulse or step response models are to be identified directly from process data.

We assume in this chapter that model estimation is performed off-line using batch data collected from a dynamic process test conducted over a fixed period of time for the purpose of designing a fixed-structure controller. We do not consider on-line recursive estimation because the resulting adaptive controllers, particularly for multivariable constrained systems, have not proven reliable enough to be considered as practical tools for distillation control at this time.

Finally, we would like to emphasize that the daunting programming task previously associated with getting started in system identification has been eliminated with the availability of a number of commercial software packages for system identification. These packages provide a wide range of data analysis and identification tools, implemented with efficient numerical procedures, allowing a user to concentrate on making the design decisions and judgments involved in system identification.

The chapter is organized as follows. Section 7-1-1 describes the general class of transfer function models on which the identification methods covered in this chapter are based, whereas Section 7-1-2 presents the overall iterative methodology of system identification. More detailed descriptions of each step in the process are provided in subsequent sections: Section 7-2 examines the design of perturbation signals to obtain information for single-input–single-output (SISO) and multi-input–multi-output (MIMO) systems; Section 7-3 covers topics associated with identifying model structure and estimation of model parameters, which includes a discussion of bias–variance trade-offs in system identification (Section 7-3-1), nonparametric and parametric identification (Sections 7-3-2 and 7-3-3), control-relevant identification (Section 7-3-4), identification in the closed-loop (Section 7-3-5), and the treatment of nonlinearity in distillation systems (Section 7-3-6). Model validation is discussed in Section 7-4, some practical aspects of conducting dynamic tests on a process are discussed in Section 7-5, and an example is provided in Section 7-6.

7-1-1 Discrete Transfer Function Models for Distillation Systems

We assume that the system to be identified is linear and time-invariant and can be represented by a transfer function model of the form

$$y(t) = G_0(q)u(t) + \nu(t) \qquad (7\text{-}1)$$

where

subscript 0 = the true (unknown) system

$y(t)$ = a system output (e.g., a temperature or composition measurement from an on-line analyzer, or *transformations* of these measurements) measured at equally spaced intervals of time (the sampling interval T_s)

$u(t)$ = a manipulated system input (or a measured disturbance variable)

$G_0(q)$ = the transfer function representing the deterministic process dynamics

The disturbance $\nu(t)$ represents the effect of all unmeasured disturbances acting on the system, including measurement noise, and is modelled as

$$\nu(t) = H_0(q)e(t) \tag{7-2}$$

where $e(t)$ = a sequence of random shocks with a specified probability density function (PDF)

$H_0(q)$ = the transfer function of the disturbance process

Equation (7-2) is used to model serially correlated stochastic disturbances (as described in Box and Jenkins, 1976), as well as *piecewise deterministic* disturbances, as discussed by Åström and Wittenmark (1984) and MacGregor, Harris, and Wright (1984). For piecewise deterministic disturbances, $e(t)$ represents a signal that is zero except at isolated points. A more in-depth discussion of disturbances that typically affect distillation systems is made in Section 7-3-3-1.

The object of identification is to find a parametrized representation of the true system in the form

$$y(t) = G(q)u(t) + H(q)e(t) \tag{7-3}$$

Determining specific structural forms for $G(q)$ and $H(q)$ is part of the identification problem, as well as designing perturbations on $u(t)$ to be applied to the process to obtain data with sufficient information to perform the identification. In the case of distillation, the identification problem includes determining whether transformations on certain measured outputs are necessary to allow use of the linear model Equation 7-3. As pointed out in Chapter 12, even with transformations, highly nonlinear columns cannot always be adequately represented by linear models of this form.

Multivariate systems are represented using extensions of Equation 7-3. For example, for a system with k outputs and m inputs, a convenient form for the MIMO transfer function model is

$$y_i(t) = G_{i1}(q)u_1(t) + G_{i2}(q)u_2(t) + \cdots + G_{im}(q)u_m(t) + H_i(q)e_i(t), \quad i = 1, 2, \ldots, k \tag{7-4}$$

Note that this model assumes that disturbances act independently on each output. This assumption is usually made to simplify the estimation problem. More general multivariable forms are available in the literature (e.g., see Gevers and Wertz, 1987; Ljung, 1987, Appendix 4.A).

7-1-2 Iterative Methodology of System Identification

System identification has traditionally been considered as an iterative process, consisting of four main steps:

Step 1. Experimental planning. An engineer must decide on the type of input [e.g. pulse, step, relay, or pseudo-random binary sequence (PRBS)], design parameters associated with each signal type, the test duration, whether the plant will be kept in open or closed loop during testing, and, in the case of multivariable systems, how the test will be conducted (i.e., perturbations applied to inputs simultaneously or one at a time).

Step 2. Selection of a model structure. Within the class of linear transfer function models described by Equation (7-3), a specific structure must be selected [e.g., AutoRegressive with eXternal input (ARX), Box–Jenkins (BJ), output error (OE), or finite impulse response (FIR)] as well as the orders of the polynomials and time delays.

Step 3. Parameter estimation. The parameter estimation step involves the selection of a number of design variables: the numerical procedure to obtain estimates for the model parameters, the form of the objective function (e.g., sum of squared prediction errors), the number of steps ahead in the prediction-error equation, and whether to use a prefilter, and if so, its form. A complicating factor in this step is the potential for numerical problems in the parameter estimation algorithm (e.g., ill-conditioning, multiple local minima).

Step 4. Model validation. Having obtained a model, its adequacy must be assessed. Among the issues to consider are:

What information about unmodelled dynamics is resident in the residuals (the time series resulting from the difference between the predicted model output and measured output) and how this information assists in the iterative identification procedure?

Does the step response of the estimated model $G(q)$ agree with physical intuition?

Although these issues are important, the ultimate test of model adequacy is whether it meets the requirements of the intended application (which in this case is process control).

In practice, it is desirable to iterate on Steps 2 to 4 only, because repetition of dynamic tests is unduly disruptive to process operation. Because the ability to perform Steps 2 to 4 efficiently depends to a great extent on the quality of information available in the dynamic test data, the design of a dynamic test to obtain the necessary information in one attempt (and often with only approximate a priori information about the system) is perhaps the most challenging aspect of process identification. In the following sections we discuss each step in more detail. Much of the discussion is general to the problem of how to perform identification when the end-use of a model is for process control. Topics specific to distillation are discussed where they arise.

Again, we emphasize that commercial software packages for system identification are available today that greatly facilitate Steps 2 to 4.

7-2 PERTURBATION SIGNAL DESIGN

7-2-1 Discussion

A primary goal of system identification is to be able to discriminate among competing models within a specified model set. This discrimination (and the related operation of parameter estimation) will be efficient only when performed using data obtained from a process test in which certain identifiability conditions have been satisfied. These conditions are usually expressed as the requirement that a perturbation signal be persistently exciting (e.g., see Ljung, 1987, Chapter 14). In practice, this requires a perturbation signal that is sufficiently rich in frequency content to excite a system up to some specific frequency, above which the dynamics are considered unimportant. For example, for control purposes, we are often not interested in modelling the fastest modes of a system (e.g., the dynamics of valve motion). A number of questions arise:

1. What is the frequency range important for control?
2. How can one design a perturbation signal that excites a system over this frequency range without having complete a priori knowledge of the system frequency response (and controller tuning)?

The first question, as it relates to the requirements of system identification, is discussed in Section 7-3-4. The second question is addressed in this section. As described in Section 7-1-2, the design of a perturbation signal is part of the iterative process of identification. In theory, the design of a control system should also be incorporated in this iterative procedure because controller design and tuning affect the modelling requirements for process control and therefore influence signal design, that is, the frequency range of interest. However, in practice, including controller design as an iteration is not necessary; sufficient approximate system information is usually available to allow informative dynamic tests to be designed.

The perturbation signal most often used to satisfy the preceding identifiability conditions (as well as closed-loop identifiability conditions; see Section 7-3-5) is the pseudo-random binary sequence (PRBS) signal. The design of PRBS signals for perturbation of single-input–single-output (SISO) and multi-input–multi-output (MIMO) systems for control-relevant identification is discussed in the following sections.

Discrete sampling of a continuous system leads to inevitable information loss, and therefore selection of sampling interval is an important topic in identification for discrete-time control models. Sampling too slowly can distort the spectrum of a sampled signal in the frequency range of interest as a result of aliasing (or foldover), leading to faulty identification. The selection of sampling interval and design of anti-aliasing filters to avoid this problem is discussed in most texts dealing with system identification (e.g., Ljung, 1987, Section 14.5).

7-2-2 Pseudo-random Binary Sequence Signals

7-2-2-1 PRBS Signals for SISO Systems

As described by Davies (1970), a PRBS signal can be generated using shift registers, and such signals can be characterized by two parameters: n, the number of shift registers used to generate the sequence, and T_{cl}, the clock period (i.e., the minimum time between changes in level of the signal, as an integer multiple of the sampling period T_s). The sequence repeats itself after $T = NT_{cl}$ units of time, where $N = 2^n - 1$. For T_{cl} = one sampling period, the power spectrum of a PRBS signal is approximately equal to 1 over the entire frequency spectrum. For values of $T_{cl} > 1$, the spectrum drops off at higher frequency and has a specific bandwidth useful for system excitation. The power spectrum of a PRBS signal generated using $n = 7$ and $T_{cl} = 3$ (with $T_s = 1$ s) is shown in Figure 7-1 (plot generated numerically using the *spa* function in MATLAB, 1989).

An analytical expression for the power spectrum of a PRBS signal generated using shift registers is given by (Davies, 1970)

$$S(\omega) = \frac{a^2(N+1)T_{cl}}{N}\left[\frac{\sin(\omega T_{cl}/2)}{\omega T_{cl}/2}\right]^2 \tag{7-5}$$

where a = signal amplitude. At low frequencies the power spectrum has the approximate value $a^2(N+1)T_{cl}/N$. At $\omega = 2.8/T_{cl}$, the spectrum is reduced by half, defining the upper limit of its useful range. Due to the periodicity of the autocovariance function of a PRBS signal, its power spectrum is discrete (Eykhoff, 1974) with points in the spectrum separated by $2\pi/T$ and the first point at $2\pi/T$. Therefore, the frequency range of a PRBS signal useful for system excitation is

$$\frac{2\pi}{T} \le \omega \le \frac{2.8}{T_{cl}} \text{ rad/s} \tag{7-6}$$

Selection of n and T_{cl} allows a user to tailor this frequency range for excitation of a particular system. If a priori estimates of the system dominant time constant τ_{dom} and deadtime τ_d are available, then the following can be used as a guideline for designing a PRBS signal:

$$\frac{1}{\beta(\tau_{dom} + \tau_d/2)} \le \omega \le \frac{1}{(\tau_{dom} + \tau_d/2)/\alpha} \tag{7-7}$$

where α is specified to ensure that suffi-

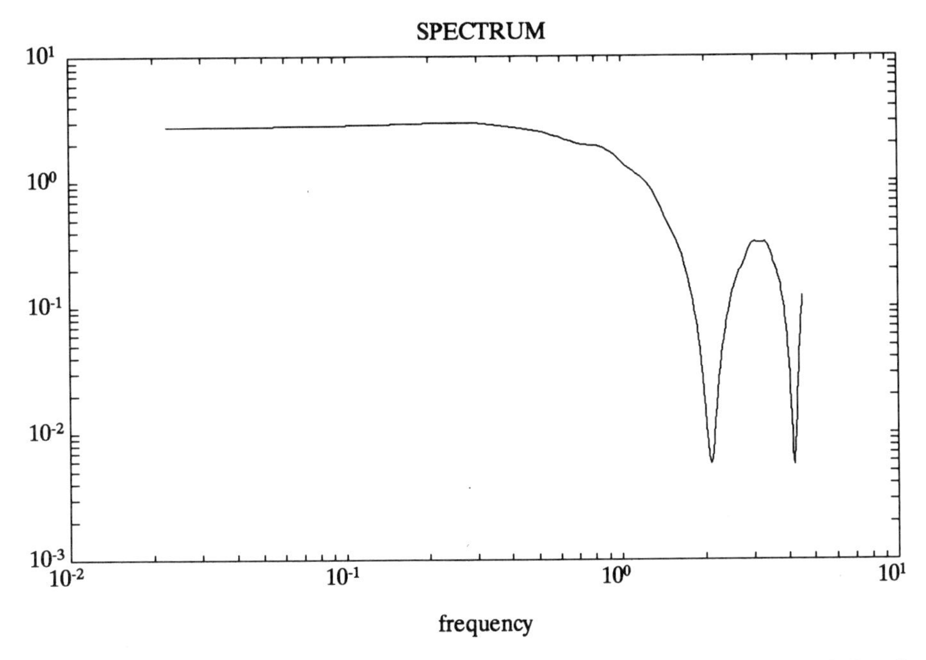

FIGURE 7-1. Power spectrum of a PRBS signal generated using $n = 7$ and $T_{\mathrm{cl}} = 3$ ($T_s = 1$ s).

ciently high frequency content is available in the perturbation signal, commensurate with how much faster the closed-loop response is expected to be relative to the open-loop response:

$$\tau_{\mathrm{cl}} = \frac{\tau_{\mathrm{ol}}}{\alpha}$$

For example, with $\alpha = 2$, the designer expects the closed-loop time constant to be one-half that of the open-loop response (i.e., twice as fast). As discussed in more detail in Section 7-3, the accuracy of the frequency response of a fitted model, relative to that of the real process, is greatest in those frequency ranges where the perturbation signal has most energy. A more aggressively tuned model-based controller (larger α) will require an identified model that is accurate of higher frequencies. Therefore, the perturbation signal should be designed (e.g. using Equation 7-8) to ensure sufficient energy at higher frequencies.

β is specified to tailor how much low-frequency information will be present in the perturbation signal (i.e., how low in frequency does one wish to model the plant). Choosing higher values of β results in lower-frequency information: $\beta = 3$ will provide information down to a frequency roughly corresponding to the 95% settling time of process, $\beta = 4$ for the 98% settling time, and $\beta = 5$ for the 99% settling time.

Comparing Equations (7-6) and (7-7) gives

$$T_{\mathrm{cl}} = \frac{2.8(\tau_{\mathrm{dom}} + \tau_d/2)}{\alpha} \tag{7-8}$$

and

$$N = \frac{2\pi\alpha\beta}{2.8} \tag{7-9}$$

Consider the design of a PRBS signal for a first-order plus dead-time process where $\tau = \tau_{\mathrm{dom}} = 100$ s and $\tau_d = 20$ s. Choosing $\alpha = 1$ (i.e., only the open-loop dynamics are of interest and the controller is not required to be tightly tuned) gives $T_{\mathrm{cl}} = 308$ s. Thus, the *minimum* time between changes in level of the PRBS signal is roughly 3 time constants, that is, very nearly what can be considered conventional step tests where the process is allowed to reach steady state after each step. In this particular example, because the control system is expected to be tuned so that the closed-loop bandwidth is not significantly wider than the open-loop one (a situation that arises frequently in industrial control problems), the perturbation signal need only emphasize low-frequency excitation. How few step changes (i.e., the duration of the test) will suffice depends on the presence of noise. In the absence of noise or corrupting disturbances, a single step (followed by a step in the opposite direction to ensure that the system returns to its original level) in this example would provide the required information. Higher-frequency information will be necessary only when the system to be identified is known to have faster modes (i.e., lower dominant time constant) or the control system is expected to be tuned aggressively so that frequency response in the higher ranges is important.

7-2-2-2 PRBS Signals for MIMO Systems

When a dynamic test is performed by perturbing only one input at a time, the estimation problem is greatly simplified because SISO models can be estimated one at a time (assuming that a noise model is not estimated). This approach has advantages and disadvantages from an operational point of view. By varying only one input, the behavior of process outputs is more predictable, reducing the probability of entering an undesirable operating region and/or violating constraints. The probability of operator intervention becomes more likely with a test specifically designed to perturb all inputs simultaneously (for example, with independent PRBS signals applied to each input), because it is difficult to predict when certain combinations of input levels might drive a process into an undesirable operating region. This is a particular problem for multivariable systems that exhibit strong gain directionality, that is, the gain of a particular output to simultaneous changes in input levels varies widely with different combinations of input moves.

As described by Eykhoff (1974) and Briggs and Godfrey (1966), PRBS signals applied to multiple inputs simultaneously must satisfy certain conditions to allow identification of multivariable systems. Most important, for a system with p inputs, the cross-correlation functions $\phi_{u_i u_j}(\tau)$ for each pair of inputs $u_i u_j$, $i, j = 1, 2, \ldots, p$, must be negligible for τ less than the maximum system settling time (this can be taken as the largest 99% settling time among the input–output pairs, $T_{99\%,\max}$). PRBS signals generated independently for each input would satisfy this requirement. However, a more practical approach is to generate multiple PRBS signals as delayed versions of a single PRBS. To satisfy the requirement on the cross-correlation functions, the minimum allowed delay between any two applied sequences is given by D:

$$D = \frac{NT_{\text{cl}}}{p} > T_{99\%,\max} \tag{7-10}$$

N can be increased as necessary to satisfy the inequality.

Another approach to conduct a multivariable test is to conduct one long test in which PRBS signals are applied consecutively to the inputs, one at a time. During a period in which a PRBS signal is applied to one particular input, the other inputs are held constant if possible, but allowed to vary if necessary (e.g., due to automatic or operator initiated control action). Data for the entire test are collected as a single batch, from which multiple-input–single-output (MISO) models of the form described by Equation 7-4 are identified for each process output variable (that are then converted to SISO models if required for the control system design). Thus, the effect of one or more inputs wandering during a period in which another input is perturbed is accounted for by the estimation procedure.

Design of perturbation signals for multi-input–multi-output systems that are ill-conditioned (e.g., high purity distillation systems with composition to be controlled at both ends of the tower) is a more complex undertaking. These processes exhibit strong gain directionality and the stability of any control system (regardless of its sophistication) is very sensitive to identification errors. In these situations, input perturbations should emphasize directions in which output gain is least, that is, independent PRBS signals will not suffice. For more information on the design of input signals for ill-conditioned multivariable systems, the reader is referred to Koung and MacGregor (1991) and the references therein. Reducing the effect of nonlinearity in identification of linear input–output models for high purity distillation systems is discussed in Section 7-3-6.

7-3 MODEL STRUCTURE SELECTION AND PARAMETER ESTIMATION

7-3-1 Bias – Variance Trade-Offs in System Identification

One of the most fundamental concepts in system identification is the trade-off between bias and variance. Generally speaking, errors in system identification are the result of bias and variance:

$$\text{error} = \text{bias} + \text{variance}$$

Bias is often considered in the context of *systematic* error in parameter estimates or model predictions. However, as discussed in more detail later, examining bias in the frequency domain (i.e., the error between the frequency response of the real system and that of an estimated model) provides significant insight when identifying models for control. Bias effects are present in identification even if the data are free of noise or the data length is arbitrarily long. Among the factors contributing to bias are:

Choice of model structure. An incorrect choice of model structure (or a low-order structure that is an approximation of the real system) for $G(q)$ will lead to bias. Frequency-domain analysis also shows that the disturbance model $H(q)$ affects bias in the estimated $G(q)$. Different forms of $H(q)$ (i.e., autoregressive, moving average, and the orders of the polynomials) affect the frequency domain bias of $G(q)$ in different ways.

System excitation. The spectrum of the input signal can significantly affect bias in the estimated model. Frequency-domain bias is reduced in those frequency ranges where spectral power is strong.

Mode of operation. Whether the estimation is carried out on data generated from open-loop or closed-loop identification is important. The presence of a control system can either amplify or attenuate the frequency content of the input signal in certain frequency ranges, thus affecting frequency-domain bias (that may be undesirable) in those ranges.

These factors are explored in more detail in subsequent sections.

Variance in the parameter estimates and predicted outputs, on the other hand, is affected by such phenomena as noise in the data, the number of model parameters, and the duration of the experiment. Variance effects exist in identification even when unbiased estimation is possible.

The phenomenon of bias–variance trade-off occurs regardless of the type of parameter estimator. Analysis in the frequency domain is an effective means for understanding bias–variance effects. This does not mean that the identification has to be carried out in the frequency domain, but rather that frequency-domain arguments provide very useful insight into the identification problem when the end use of the model is for process control. The information for analysis is resident in the power spectrum of the prediction errors, as described in succeeding text.

Consider the true system represented by Equations 7-1 and 7-2 that is to be identified using the model in Equation (7-3). Prefiltering of the input and output data,

$$y_F(t) = L(q)y(t) \quad (7\text{-}11)$$

$$u_F(t) = L(q)u(t) \quad (7\text{-}12)$$

where $L(q)$ is the specified filter, may be performed to remove nonstationarities and to influence the goodness-of-fit (in the frequency domain), as will be described. In this case the power spectrum $\Phi_{e_F}(\omega)$ of the prefiltered prediction error $[e_F(t) = L(q)e_P(t)]$, where the prediction error $e_P(t)$ is given by

$$e_P(t) = y(t) - \hat{y}(t) \quad (7\text{-}13)$$

and $\hat{y}$ is the one-step-ahead output prediction

$$\hat{y}(t) = H^{-1}(q)G(q)u(t) + (1 - H^{-1}(q))y(t) \quad (7\text{-}14)$$

is given by Ljung (1987):

$$\Phi_{e_F}(\omega) = \left(\left|G_0(e^{j\omega}) - G(e^{j\omega})\right|^2\Phi_u(\omega) + \Phi_v(\omega)\right)\frac{\left|L(e^{j\omega})\right|^2}{\left|H(e^{j\omega})\right|^2} \quad (7\text{-}15)$$

$\Phi_u(\omega)$ and $\Phi_v(\omega)$ are the power spectral densities of the input signal u and the disturbance v, respectively, computed from the expression

$$\Phi_x(\omega) = \sum_{k=-\infty}^{\infty} c_x(k)e^{-j\omega Tk} \quad (7\text{-}16)$$

$c_x(k)$ is the covariance of a time series x at lag k, evaluated using

$$c_x(k) = \frac{1}{N}\sum_{i=1}^{N-k} x(i)x(i+k), \qquad k = 0, N-1 \quad (7\text{-}17)$$

When dealing with a deterministic signal, the power spectrum is equivalent to the squared magnitude

$$\Phi_x(\omega) = \left|x(j\omega)\right|^2 \quad (7\text{-}18)$$

It can be shown (Ljung, 1987) that the parameter estimation criterion based on minimizing the sum of prediction errors (see Equation 7-35) can be equivalently expressed in the frequency domain using Parseval's theorem, that is,

$$\lim_{N\to\infty}\sum_{i=1}^{N} e_F^2(i) = \frac{1}{2\pi}\int_{-\pi}^{\pi}\Phi_{e_F}(\omega)\,d\omega \quad (7\text{-}19)$$

From Equations 7-15 and 7-19 we can discern four factors that significantly affect frequency-domain bias in the estimated $G(q)$ in the parameter estimation problem:

The input signal power spectral density $\Phi_u(\omega)$. Because the input spectrum directly weights the term $|G_0(e^{j\omega}) - G(e^{j\omega})|^2$ (i.e., the frequency-domain bias), error between the true system and the estimated model is reduced at those frequencies at which the input power spectrum is highest. Therefore, it is imperative to define the frequency range of importance to the control problem and ensure that the input signal is rich in frequency content in this range.

Prefilter $L(q)$. The prefiltering also acts as a frequency dependent weight, providing another way to selectively emphasize the frequency range of importance for control.

The structure of $G(q)$ and $H(q)$. The choice of model structure plays a significant role in frequency-domain bias in the parameter estimation problem. Note that the noise model $H(q)$ weights the objective function in a manner similar to the prefilter, and therefore has a direct effect on frequency-domain bias in the estimated $G(q)$.

The disturbance power spectral density $\Phi_v(\omega)$. The parameter estimation problem, when examined in the frequency domain, is a multiobjective minimization whenever a disturbance model $H(q)$ is present [because both $G(q)$ and $H(q)$ contain parameters to be estimated]. In this case there is trade-off between reducing bias between $G_0(q)$ and $G(q)$, and fitting the spectrum of the disturbance $\nu(t)$. This problem becomes particularly pronounced when autoregressive terms that are common to both $G(q)$ and $H(q)$ are included in the parameter estimation problem.

Variance issues can be examined in the frequency domain using an expression of the form (Ljung, 1987)

$$\text{variance} \sim \left(\frac{n}{N} \frac{\Phi_v(\omega)}{\Phi_u(\omega)} \right) \tag{7-20}$$

where n = the number of parameters in the model
N = the length of the data set

Equation 7-20 clearly shows that reducing the number of model parameters, increasing the length of the data set, and increasing the magnitude of the input signal all contribute to variance reduction in system identification.

7-3-2 Nonparametric Methods

We present nonparametric techniques because they can serve as useful precursors to parametric estimation, providing approximate information about the plant dynamics that can be useful later in the identification process.

7-3-2-1 Correlation Analysis

Correlation analysis is often used to obtain an initial guess of the process impulse response. The kth impulse parameter v_k is given by (Box and Jenkins, 1976)

$$v_k = \frac{c_{uy}(k)}{\sigma_u^2} \tag{7-21}$$

where

$$c_{uy}(k) = \frac{1}{N} \sum_{i=1}^{N-k} u'(i) y'(i+k), \qquad k = 0, N-1$$

$$\sigma_u^2 = \frac{1}{N} \sum_{i=1}^{N} u'(i)^2$$

N is the number of observations, whereas the prime denotes that detrending and prewhitening operations were performed on the time series; under these conditions, $c_{uy}(k)$, the estimated cross-covariance function, can be used to obtain an estimate of the kth impulse response of a system. Other uses of correlation analysis include discerning the deadtime of a system, the 95% settling time, and also, most importantly, determining if an input signal u has a significant effect in describing the dynamics of the process. This latter use involves evaluating the estimated impulse response coefficients against the hypothesis that the data reflect only white noise. The standard deviation of v_k is given by

$$\sigma_v(k) = \frac{\sigma_y}{\sigma_u} \frac{1}{\sqrt{N-k}} \tag{7-22}$$

where σ_u and σ_y are the standard deviations of the input and output sequences, respectively (Box and Jenkins, 1976). Two standard deviations constitute 95% confidence bounds for the estimated impulse parameters.

7-3-2-2 Spectral Analysis

The frequency response of a system is the Fourier transform of its impulse response, and provides information about the dominant time constant and the order of the transfer function to be estimated. The estimated frequency response, $\hat{G}(j\omega)$ can be obtained directly from plant data using spectral analysis (Jenkins and Watts, 1969) with the equation

$$\hat{G}(j\omega) = \frac{\hat{\Phi}_{uy}(\omega)}{\hat{\Phi}_{u}(\omega)} \tag{7-23}$$

where $\hat{\Phi}_{uy}$ and $\hat{\Phi}_{u}$ denote the "smoothed" cross- and autospectra between u and y, described by the formulas

$$\hat{\Phi}_{uy}(\omega) = \sum_{k=-M}^{M} c_{uy}(k)w(k)e^{-j\omega k} \tag{7-24}$$

$$\hat{\Phi}_{u}(\omega) = \sum_{k=-M}^{M} c_{u}(k)w(k)e^{-j\omega k} \tag{7-25}$$

$w(k)$ is the lag "window" that accomplishes smoothing. A common window choice is the Tukey (Hamming) lag window

$$w(k) = \begin{cases} \frac{1}{2}(1+\cos(\pi k/M)), & |u| \le M \\ 0, & |u| > M \end{cases} \tag{7-26}$$

where M is the truncation parameter for the window.

Spectral analysis can also be applied to the residual time series to obtain norm-bound uncertainty descriptions that are useful for robust control. A more detailed analysis of this is explained in Section 7-4.

7-3-3 Parametric Models

7-3-3-1 Prediction-Error Model (PEM) Structures

For prediction-error methods, the general model structure Equation 7-3 is represented using polynomials in the backward shift operator q^{-1} [defined as $q^{-k}x(t) = x(t-k)$]:

$$A(q)y(t) = \frac{B(q)}{F(q)}u(t-n_k) + \frac{C(q)}{D(q)}e(t) \tag{7-27}$$

where A, B, C, D, and F are polynomials in q^{-1}:

$$A(q) = 1 + a_1 q^{-1} + \cdots + a_{n_a} q^{-n_a}$$
$$B(q) = b_1 + b_2 q^{-1} + \cdots + b_{n_b} q^{-n_b+1}$$
$$C(q) = 1 + c_1 q^{-1} + \cdots + c_{n_c} q^{-n_c}$$
$$D(q) = 1 + d_1 q^{-1} + \cdots + d_{n_d} q^{-n_d}$$
$$F(q) = 1 + f_1 q^{-1} + \cdots + f_{n_f} q^{-n_f}$$

The A polynomial is an autoregressive (AR) term; B corresponds to the external (X) input u, whereas C is a moving average (MA) term. D is an autoregressive term applied exclusively to the disturbance model. n_k is the system dead-time. Comparing Equations 7-27 and 7-3 gives the transfer functions

$$G(q) = \frac{B(q)}{A(q)F(q)}q^{-n_k} \tag{7-28}$$

$$H(q) = \frac{C(q)}{A(q)D(q)} \tag{7-29}$$

Note that the A polynomial contains poles that are common to both the disturbance and process transfer functions.

Equation 7-27 is written for multi-input systems by adding terms on the right hand side

$$\frac{B_j(q)}{F_j(q)}u_j(t-n_{kj}).$$

A number of factors must be considered in the decision to identify a disturbance model [i.e., the polynomials $C(q)$ and $D(q)$, and possibly $A(q)$] as part of the overall system model to be used for control of a distillation system. For processes that are affected by unmeasured disturbances that can be modelled adequately by Equation (7-29), and where the disturbance structure is fixed (i.e., time-invariant), then inclusion of the disturbance model into a fixed-structure controller design can be expected to deliver superior performance over a controller lacking a disturbance model. As mentioned previously, Equation 7-29 is capable of modelling stochastic autoregressive integrated moving average (ARIMA) disturbances [as described by Box and Jenkins (1976)] or randomly occurring deterministic disturbances. However, when performing off-line batch identification and off-line controller design, it is assumed that the disturbance process does not change with time (i.e., its structure is fixed). This assumption is very often not valid in typical processing systems. Disturbances affecting a distillation column, as well as most other process units, are typically random in *structure*. Process units within a plant are highly interconnected and integrated through process and utility streams. A particular process unit may be disturbed at any instant from an upset or change in operating condition from any number of other process units that interact with it. These upsets arise from a multitude of different sources; for example, equipment failure, scheduled or unscheduled maintenance operations, abrupt changes in ambient conditions due to passage of fronts, changes in raw material type or composition, operator control moves motivated by any number of causes (e.g., shift changes, optimization moves, response to disturbances), and so on. In addition, disturbances from different sources propagate through a unit in different ways, again suggesting that each will have a different structure if modelled with a linear disturbance model in the form of Equation 7-29. Control structures that employ an identified model for $G(q)$ only, and ignore disturbance dynamics, have been found in practice to deliver acceptable control performance.

Again, we do not consider model-based adaptive schemes (particularly for multivariable constrained systems) to be viable at this time. Thus the identification problem as discussed in this chapter is to obtain the best possible estimate for $G(q)$ from the given batch of data. Estimators that are capable of giving unbiased estimates for the parameters in $G(q)$, even when structural inadequacies are present in $H(q)$, including the case where no disturbance model is present [i.e. $H(q) = 1$], are particularly useful in this situation. The prediction error estimates implemented by Ljung (1988) fall into this category.

Another consideration is frequency domain bias in the estimated $G(q)$ that is introduced by including a disturbance model in the estimation problem. From Equations 7-15 and 7-19 it is clear that $H(q)$ acts as a frequency domain weight on the estimation of $G(q)$. Including the term $D(q)$ [and $A(q)$] emphasizes high frequencies in the estimation of $G(q)$; the model will more accurately predict fast dynamics and hinder the low-frequency and steady-state predictive ability of the model. Including $C(q)$ has the opposite effect; high-frequency fit is deemphasized. These effects are a current area of research in the literature.

Various substructures arising from Equation 7-27 and the resulting parameter estimation problems are now described.

ARMAX ARMAX identification uses the structure

$$A(q)y(t) = B(q)u(t - n_k) + C(q)e(t) \tag{7-30}$$

This structure has been popular in the control literature, particularly for designing adaptive controllers.

ARX The ARX structure is

$$A(q)y(t) = B(q)u(t - n_k) + e(t) \tag{7-31}$$

Using this structure, the problem of determining the right order for A and B is often circumvented by overparametrization, that is, the orders of the A and B polynomials ($n = n_a = n_b$) are selected to be high. The theoretical justification for this choice of variables is that for an infinite number of observations, u white noise, and e a zero-mean stationary sequence, the estimate of the first n impulse response coefficients of

$G(q)$ will be unbiased. A high-order ARX model is therefore capable of approximating any linear system arbitrarily well.

Although high-order ARX modelling possesses a number of attractive theoretical properties, the conditions under which they apply are seldom seen in practice. Furthermore the presence of the autoregressive term $A(q)$ results in an emphasis on the high-frequency fit, which is not necessarily good for control system design. The bias problem with autoregressive terms becomes worse if low-order ARX models are used.

FIR The FIR (finite impulse response) model structure

$$y(t) = B(q)u(t - n_k) + e(t) \quad (7\text{-}32)$$

is convenient because some predictive controllers are based on this model structure [and it is easily converted to step-response form for others that require it, e.g., dynamic matrix control (DMC)]. Although the appropriate order of the finite impulse response depends on the selected sampling time and the settling time of the process, the result is usually high (30 to 100). When u is persistently exciting, e is stationary, and u and e are uncorrelated, the estimated impulse response coefficients will be unbiased. Again, as with ARX, these conditions are seldom fully observed in practice. As mentioned in the introduction, direct estimation of the impulse response model is inefficient, resulting in parameter estimates with high variance. Direct identification of the FIR model requires special test and estimation procedures, as described in subsequent text.

Box–Jenkins Box and Jenkins (1976) proposed a model of the form

$$y(t) = \frac{B(q)}{F(q)}u(t - n_k) + \frac{C(q)}{D(q)}e(t) \quad (7\text{-}33)$$

One advantage of this structure over the ARMAX structure in Equation 7-30 is that it separately parametrizes the input and disturbances, avoiding transfer functions that have common poles. The estimation problem is nonlinear, but reliable and fast estimation methods are available. Nonstationary disturbances are handled by forcing $D(q)$ to have one or more roots on the unit circle [i.e., write as $(1 - q^{-1})^d D(q)$, where d is typically equal to 1].

Output Error The output error model is a simplified form of the Box–Jenkins model structure

$$y(t) = \frac{B(q)}{F(q)}u(t - n_k) + e(t) \quad (7\text{-}34)$$

The output error structure retains the advantage of separate parametrizations for the input and disturbance, and does not require a choice of the structure for the disturbance model.

7-3-3-2 Prediction-Error Parameter Estimation

The objective minimized in prediction-error methods is the sum of the squared prediction errors

$$V = \frac{1}{N}\sum_{t=1}^{N}\frac{1}{2}(y(t) - \hat{y}(t))^2 = \frac{1}{N}\sum_{t=1}^{N}\frac{1}{2}e_P^2(t) \quad (7\text{-}35)$$

In the case of FIR and ARX models, the predictor structure is

$$\hat{y}(t) = B(q)u(t) + (1 - A(q))y(t) \quad (7\text{-}36)$$

which can be written in regression form as

$$\hat{y}(t) = \varphi^T(t)\theta \quad (7\text{-}37)$$

φ is the regression vector

$$\varphi = [-y(t-1) \quad \cdots \quad -y(t-n_a) \quad u(t-1) \quad \cdots \quad u(t-n_b)]$$

and θ is the vector of parameters to be estimated

$$\theta = \begin{bmatrix} a_1 & \cdots & a_{n_a} & b_1 & \cdots & b_{n_b} \end{bmatrix}$$

Because the prediction equation is linear in the parameters, estimates are obtained in the usual manner for linear least-squares problems:

$$\hat{\theta} = \left[\frac{1}{N} \sum_{t=1}^{N} \varphi(t)\varphi^T(t) \right]^{-1} \frac{1}{N} \sum_{t=1}^{N} \varphi(t) y(t) \tag{7-38}$$

For the ARMAX structure, the predictor structure becomes

$$\hat{y}(t) = B(q)u(t) + (1 - A(q))y(t) + (C(q) - 1)(y(t) - \hat{y}(t)) \tag{7-39}$$

In this case the regression vector is

$$\varphi = \begin{bmatrix} -y(t-1) & \cdots & -y(t-n_a) \\ u(t-1) & \cdots & u(t-n_b) \\ e_P(t-1) & \cdots & e_P(t-n_c) \end{bmatrix}$$

and the parameter vector is

$$\theta = \begin{bmatrix} a_1 & \cdots & a_{n_a} \\ b_1 & \cdots & b_{n_b} & c_1 & \cdots & c_{n_c} \end{bmatrix}$$

For ARMAX estimation, the regression problem is no longer linear with respect to the parameters and a simple least-squares solution is no longer applicable. Again, nonlinear least-squares algorithms are readily available for solution of this estimation problem.

Output error and Box–Jenkins estimation are also nonlinear least-squares problems. A summary of explicit search techniques (e.g., Newton–Raphson, Levenberg–Marquardt, Gauss–Newton) is found in Ljung (1987). Although these techniques have favorable theoretical properties (i.e., separate parametrization of input vs. disturbance effects), numerically they represent more challenging problems than their linear counterparts. The possibility of lack of convergence or multiple local minima must be considered when using these methods.

Estimation algorithms for all of the preceding model structures, including the case where there are multiple inputs, are available in commercial identification packages.

7-3-4 Identification for Control System Design

From the previous discussion it is clear that a myriad of design variables exist in system identification. Consequently, it is important to determine how these design variables can be manipulated for the purpose of identification for control design. In this section, our objective is to present some recent techniques that have been described in the literature that consider this problem.

7-3-4-1 Use of Auxiliary Information in Identification

The trade-off between performance and robustness of a control system in the presence of model uncertainty can be analyzed in the frequency domain using the Nyquist stability criterion (Morari and Zafirou, 1989). For robust stability, each member of the family of Nyquist curves resulting from a characterization of the model uncertainty (see Chapter 2 of Morari and Zafiriou for a more detailed discussion) must satisfy the Nyquist criterion. For a given level of model uncertainty, a controller must be detuned so that the Nyquist curve corresponding to the worse case model does not encircle the point $(-1, 0)$. Thus higher model uncertainty leads to poorer performance. Conversely, less uncertainty in the fitted model, particularly around the frequency where the Nyquist curve most closely approaches the point $(-1, 0)$, allows tighter controller tuning and thus a greater closed-loop bandwidth and better control performance. Therefore, reducing the bias (and uncertainty) of the fitted model around this frequency is an important objective when the model is to be used for control system design.

From previous discussions, one way to improve the accuracy of the fitted process model $G(q)$ in a particular frequency range is to ensure that the perturbation signal is rich in these frequencies. Designing a perturbation signal to emphasize system excitation around the frequency where the Nyquist curve most closely approaches the point $(-1, 0)$ poses a problem because a priori knowledge of the closed-loop frequency response for a given controller tuning would be required. Although an iterative design procedure could be adopted, an alternative is the use of a simple relay test (Åström and Hägglund, 1984) that determines the point of intersection of the Nyquist curve with the negative real axis of the open-loop system $G_0(q)$ (i.e., the critical point, or the ultimate gain and frequency). In an industrial setting, a typically tuned PID controller (or other controller design) provides only a small amount of phase lead, and the frequency at which the Nyquist curve most closely approaches the point $(-1, 0)$ (as well as the crossover frequency of the compensated system, of equal importance) is not often not far from the region of the critical point of the open-loop process. In these situations, knowledge of the open-loop critical point is of practical value in system identification for control purposes.

Åström and Hägglund used the relay test to provide information for PID controller tuning. The use of such auxiliary information (i.e., knowledge of one or more points on the system Nyquist curve, obtained from experiment or from other process knowledge) to improve the quality of models identified for control purposes has been described by Wei, Eskinat, and Luyben (1991) and Eskinat, Johnson, and Luyben (1991). The basic idea is as follows. Data are collected from the process to be identified using a conventional perturbation procedure (e.g., step or PRBS tests). A model of specified structure is fit to these data, subject to the constraint that the Nyquist curve of the fitted model passes exactly through the known point(s) on the Nyquist diagram, thus reducing the bias in the fitted model at this frequency to zero. As shown in the preceding references, the constraint can be expressed as a set of linear equations, resulting in a quadratic optimization problem when the objective function for fitting the model is quadratic (as is the case with prediction-error methods). For certain model structures, analytical solutions to the quadratic optimization are available; otherwise a numerical procedure must be used.

To perform a relay test to obtain auxiliary information for fitting a discrete time model, a relay is connected in a sampled feedback loop, as shown in Figure 7-2. The sampling frequency for the relay test must be the same as the sampling period used to collect the data from the perturbation test. If the process has a phase lag of at least $-\pi$ radians (which will always be the case in practice because all real processes exhibit some deadtime), under relay feedback it will enter a sustained oscillation with period P_u (the critical period), giving the ultimate frequency $\omega_u = 2\pi/P_u$. The ultimate gain is given approximately by $K_u = 4h/(a\pi)$, where h is the height of relay and a is the amplitude of the principal harmonic of the output. The relay test is easier and safer to apply than the procedure suggested by Ziegler and Nichols because it does not require adjustment of controller gain in order to generate sustained oscillation. It only requires approximate knowledge of the pro-

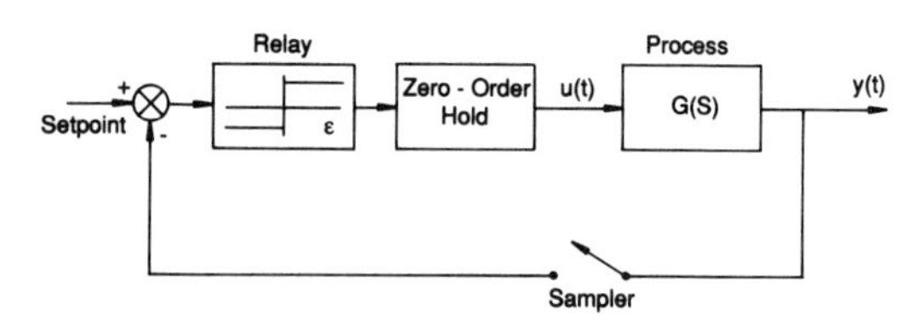

FIGURE 7-2. Configuration for a relay test to estimate crossover frequency for use as auxiliary information in identifying a discrete-time model.

cess gain (ideally at the critical frequency, but because this is not likely available, steady-state gain can be used) in order to specify an appropriate relay amplitude.

Because the relay is a nonlinear device, when applied to a linear process the measured critical point is an approximation to the real one. However, as shown by Wei, Eskinat, and Luyben (1991), the information is sufficiently accurate to be of value in system identification for control purposes.

The presence of noise may require the relay test to be implemented with a dead band within which the response variable can move without causing a jump in level of the input. In a very noisy or highly disturbed process, it may be difficult to obtain useful results from a relay test.

7-3-4-2 Control-Relevant Prefiltering

The previous section described a procedure for reducing the impact of modelling bias on control design by emphasizing the fit near one frequency, namely, the crossover frequency. Prefiltering, on the other hand, allows the engineer to emphasize the goodness of fit over a *range* of frequencies, and thus is a useful technique for identifying models that are not only to be used for feedback control but for other control structures, such as combined feedback–feedforward control, decentralized control, and cascade control (Rivera, 1991a).

As described in Section 7-3-1, the prefilter $L(q)$ acts as a frequency-dependent weight in the parameter estimation problem, and as such can be used to emphasize those frequency regions that are most important for control system design. Assuming that a sufficiently informative dynamic test has been performed, a prefilter can be used to arbitrarily influence the goodness-of-fit in the frequency domain without having to modify the model structure or repeat the dynamic test. For example, prefiltering can be used to obtain a good steady-state fit without requiring a step test input (provided that sufficient low-frequency information is available from the test data).

The literature remains vague about prefilter design; questions regarding the systematic choice of structure (low pass, high pass, bandpass) and the choice of filter parameters remain unanswered issues. However, recent work in this area has begun to provide some answers. In Rivera et al., 1990 and Rivera, Pollard, and Garcia, 1990, the derivation of a control-relevant prefilter for SISO feedback control is described that explicitly incorporates the model structure, the desired closed-loop response, and the setpoint–disturbance characteristics of the control problem into the statement of the prefilter:

$$L(q) = H(q)G^{-1}(q)(1 - \tilde{\eta}(q)) \times \tilde{\eta}(q)(r(q) - d(q)) \quad (7\text{-}40)$$

$\tilde{\eta} = GK(1 + GK)^{-1}$ is the complementary sensitivity function that describes the desired closed-loop response. Two algorithms for implementation of the prefilter, one iterative and the other "single-pass", are described by Rivera et al. (1990) and Rivera, Pollard, and Garcia (1990). The single-pass implementation is described here.

The statement of the single-pass prefilter is obtained from Equation 7-40 following some assumptions and simplifications. This prefilter assumes that the essential dynamics of the system can be reasonably approximated by a first-order with deadtime model

$$G(q) \approx \frac{q^{-n_k+1}}{(q - \alpha)} \quad (7\text{-}41)$$

where $\alpha = e^{-T_s/\tau_{\text{dom}}}$ and τ_{dom} is a user-provided estimate of the dominant time constant of the system. An estimate of the steady-state gain is not necessary because the gain simply appears as a constant in Equation 7-40.

Regarding the structure for $\tilde{\eta}$, one must recognize that it is usually dictated by the control system design procedure. In quadratic dynamic matrix control (QDMC), for example, the effect of move suppression results in $\tilde{\eta}(q)$ that is an order greater than the plant (Balhoff and Lau, 1985). In internal model control (IMC) (Morari and Zafiriou, 1989), $\tilde{\eta}$ is specified directly via the factorization of nonminimum phase elements and the choice of a filter. In accordance with this philosophy for the model in Equation 7-41 we can use the structure

$$\tilde{\eta}(q) = q^{-n_k} f(q) \tag{7-42}$$

$f(q)$ is a low-pass filter that either corresponds to the filter used in the IMC design procedure or approximates the effect of move suppression in QDMC. For model structures such as output error (OE) and finite impulse response (FIR), $H(q) = 1$, which leads to the definition of the prefilter as

$$L(q) = (q - \alpha)(1 - q^{-n_k} f(q)) \times q^{-1} f(q)(r(q) - d(q)) \tag{7-43}$$

For autoregressive with external input (ARX) models, we can approximate $\tilde{p}_e$ with the same dominant time constant guess made for $\tilde{p}$:

$$H(q) \approx \frac{q}{(q - \alpha)} \tag{7-44}$$

which leads to

$$L(q) = (1 - q^{-n_k} f(q)) f(q) \times (r(q) - d(q)) \tag{7-45}$$

Specific forms for these prefilters are shown via an example. Consider the first-order plant model according to

$$G(q) = \frac{k}{q - \alpha} \tag{7-46}$$

which will be controlled using QDMC. The resulting structure for $\tilde{\eta}$ is second order, so we define $f(q)$ as second order:

$$f(q) = \frac{(1 - \delta)^2 q^2}{(q - \delta)^2} \tag{7-47}$$

where $\delta = e^{-1.555 T_s / \tau_{cl}}$
τ_{cl} = the desired closed-loop time constant, specified by the engineer

Assuming step setpoint and disturbance changes for the control system, the resulting prefilter for FIR and OE estimation is fourth order,

$$L(q) = \frac{(1 - \delta)^2 q^2 (q - \alpha)(q - \delta^2)}{(q - \delta)^4} \tag{7-48}$$

and (for the ARX structure)

$$L(q) = \frac{(1 - \delta)^2 q^3 (q - \delta^2)}{(q - \delta)^4} \tag{7-49}$$

Note how Equations 7-48 and 7-49 meet the requirements for a practical prefilter. Having defined a model structure and the nature of the controller design problem, the choice of the prefilter is reduced to simply providing estimates for the closed-loop speed-of-response (τ_{cl}) and the open-loop time constant (τ_{dom}). This information can be readily obtained in most situations faced by process control engineers.

Restrictions regarding the benefits of prefiltering can be gathered through an understanding of bias–variance trade-offs, as noted in Rivera (1990a). In output error estimation, control-relevant prefiltering was shown to substantially improve the estimation for low-order models. Prefiltering in the ARX estimation problem, however, worsens the quality of the plant estimate if the ratio $\Phi_v(\omega)/\Phi_u(\omega)$ is high over the bandwidth of the prefilter because this causes the estimator to favor the fit of $A(q)$ to noise. For the case of FIR, a significant reduction in the low-frequency uncertainty is obtained using prefiltering; variance effects, however, become more pronounced, and the resulting step responses may look increasingly more jagged. One way to overcome this variance problem is to estimate the FIR parameters using a biased least-squares estimator, as discussed in Section 7-3-4-3.

Work is currently being carried out (Rivera, 1991b) that unifies the control-relevant prefiltering approach with the crossover frequency approach described in Section 7-3-4-1. A fundamental analysis of modelling requirements for control shows that there are four factors that determine the adequacy of a model for feedback control purposes:

1. The choice of plant and noise model structure.
2. The desired closed-loop speed-of-response.
3. The *shape* of the closed-loop response (i.e., underdamped or overdamped).
4. The nature of the setpoints and disturbances that will be encountered by the closed-loop system.

Under certain circumstances (the presence of an integrator in the model and control requirements demanding very fast, underdamped responses, for example), the power spectrum of the prefilter shows a very strong amplification of the crossover frequency, in which case both approaches yield equivalent results. However, when considering the case of a first-order system with substantial deadtime (ratio of time delay to time constant equal to 1), the requirement of an overdamped closed-loop response with a correspondingly smooth manipulated variable response results in modelling requirements that emphasize the intermediate and final segments of the plant's step response *without* demanding a close fit of the plant's ultimate frequency. This occurs because for the specified control requirements, the frequency responses of the plant and the loop transfer function $G_0(s)K(s)$ differ substantially at the crossover point; the assumption that the controller does not introduce substantial phase shift (a key assumption for the method described in Section 7-3-4-1) is violated in this case.

7-3-4-3 Biased Least-Squares Estimators

Direct estimation of a truncated impulse response model results in an estimation in which the bias–variance trade-off is shifted toward reduction in bias at the expense of variance, due to the large number of parameters (and their lack of independence). Therefore, an estimated impulse model might give an excellent fit to the data that were used in the estimation, but its use as a prediction tool on a different set of data may be extremely limited because of the resulting high variance of the parameter estimates (and thus predictions). There are a number of ways to reduce the variance of estimates to improve the predictive ability of models:

1. Estimate a parsimonious transfer function model (low number of parameters) and convert the nominal estimated model to impulse or step form (with no loss of information).
2. Perform a longer dynamic test to increase the size of the data set, (which, as mentioned previously, may be quite acceptable for some processes).
3. Use a biased estimator that shifts bias–variance trade-off toward allowing some bias in order to reduce variance.

Biased estimators, as the name implies, produce estimates that are biased and not least squares, but are more desirable from a prediction point of view because the reduction in prediction mean square error (MSE) they achieve more than compensates for the minimal bias that is introduced. We briefly describe a few approaches that have been used in the literature for biased estimation of truncated impulse response models and we provide references for further information.

Ridge Regression Consider the finite impulse response (FIR) model Equation 7-32. The prediction equation for this model structure is given by Equation 7-36 [with $A(q) = 1$], and the least-squares parameter estimates are given by Equation 7-38, which can be written in matrix notation as

$$\hat{\theta} = [X^T X]^{-1} X^T y \tag{7-50}$$

With a typical order of 30 to 100 for $B(q)$, the parameters will not be independent, resulting in poor conditioning of $X^T X$ and an inversion that is sensitive to small variations in the data. In the simplest form of ridge regression, Equation 7-50 is replaced by

$$\hat{\theta} = [X^T X + cI]^{-1} X^T y \tag{7-51}$$

where I = an identity matrix of appropriate dimension and
c = a specified positive number

For more information, the reader is referred to Draper and Smith (1966) and Hoerl and Kennard (1970).

Partial Least-Squares Estimation Another biased estimator that has been used to estimate FIR models (Ricker, 1988) is partial least squares (PLS), a general method for solving poorly conditioned least-squares problems. A brief explanation of the method follows. For more information, the reader is referred to Ricker (1988), Lorber, Wangen, and Kowalski (1987), Hoskuldsson (1988), Geladi (1988), and a tutorial paper by Geladi and Kowalski (1986).

A data matrix X (dim $m \times n$), can always be decomposed into two matrices, $\tilde{T}$ $(m \times r)$ (the "scores") and $\tilde{P}$ $(n \times r)$ (the "loadings"):

$$X = \tilde{T}\tilde{P}^T = t_1 p_1^T + t_2 p_2^T + \cdots + t_r p_r^T \tag{7-52}$$

where $r = \text{rank}(X)$
t_i, p_i = are the ith columns of $\tilde{T}$ and $\tilde{P}$, respectively

Because of the correlation that exists between the parameters in an FIR model, not all of the t_i and p_i are required to explain the total variation that exists in the data matrix X. For the same reason, not all the t_i and p_i are required to develop a satisfactory predictive relationship between X and the response vector y. Therefore, in PLS, X is written as

$$X = t_1 p_1^T + t_2 p_2^T + \cdots + t_q p_q^T + E$$
$$= TP^T + E$$

where q is referred to as the number of latent variables, and is usually very much less than r. E is a residual matrix that contains the random errors in X as well as the discarded components of X that have no significant predictable effect on y. In PLS, the regression is performed between y and the t_i in Equation 7-53 and is an integral part of the numerical procedure that performs the decomposition (in the general PLS problem where there are multiple response variables that may be correlated, the numerical procedure includes decomposition of the response matrix Y; see for example, Kresta, MacGregor, and Marlin, 1991).

PLS requires the user to select the number of latent variables used in the decomposition. The object in PLS is to find the minimum number of latent variables that results in a satisfactory predictor of the response variable. This involves evaluating the predictive ability of fitted models with increasing number of latent variables, usually using cross-validation (the model is fit on one segment of data, and evaluated for predictive ability on another segment). The number of latent variables is increased until no further significant decrease in the sum of squares of residuals between the predictions from the fitted model and the validation data set is observed. The number of latent variables will be far less than the number of FIR parameters in a typical FIR model. With fewer parameters to estimate, a more acceptable balance between bias and variance in the estimated model is obtained. As pointed out by Rivera et al., 1990 and

Rivera, Pollard, and Garcia (1990), prefiltering prior to PLS estimation gives the engineer a further handle on the bias–variance trade-off in a frequency-dependent sense.

7-3-5 Identifiability Conditions for Closed-Loop Systems

When data are collected with a process in closed-loop operation, certain identifiability conditions must be satisfied to ensure that proper identification can be performed. Historical plant data collected under closed-loop conditions do not generally satisfy these conditions and cannot be used for dynamic model identification. The pitfalls associated with ignoring these requirements have been well documented in the literature (e.g., Box and MacGregor, 1974). Two issues are involved:

Identifiability. Identifiability refers to whether a suitable model describing the plant dynamics can be extracted from the available closed-loop data. When specific identifiability conditions are satisfied (for example, see Ljung, 1987, Chapter 8, specifically Theorem 8.3) prediction-error estimators can be applied to closed-loop data as though they were obtained under open-loop conditions, and no other special consideration is required (except that the model to be identified be sufficiently parametrized that it is capable of representing the true system dynamics).

Closed-loop frequency-domain bias. Even if identifiable, the quality of the estimated model obtained from closed-loop data is an issue (discussed in Section 7-3-5-2).

7-3-5-1 Identifiability

Söderström, Ljung, and Gustavsson (1976) and Gustavsson, Ljung, and Söderström (1977) provide detailed analyses of the closed-loop identifiability problem. Depending on the conditions imposed during the collection of closed-loop data, two levels of identifiability can be achieved:

System identifiability (SI). Given that certain strict conditions are met with regard to the model structure and the order of the compensator, model parameter estimates that correctly represent the plant can potentially be obtained from the data.

Strong system identifiability (SSI). This means that no special restrictions need be imposed on the model structure or the order of the compensator in order to obtain reasonable estimates. Frequency-domain bias remains as an issue, however.

It is clearly desirable to obtain experimental conditions that are SSI because regression techniques can be applied in a direct fashion with no other special considerations. Conditions that lead to SSI are

- Closed-loop data generated by shifting between different feedback controllers.
- Closed-loop data generated by adding an external signal to the feedback loop that is persistently exciting (typically a PRBS signal). Possible injection points for this signal include the controlled variable setpoint (r) and the manipulated variable (u).

From a process operations viewpoint, applying an external signal to the feedback loop is preferable to switching between regulators.

One important result of the work of Söderström, Ljung, and Gustavsson and Gustavsson, Ljung, and Söderström is that data are collected from a closed-loop system responding strictly to disturbances is SI but *not* SSI. Only if certain restrictive conditions are met with regard to the plant and the compensator will useful models be obtained from data. These conditions are unattainable from a practical standpoint, however, because they depend upon a priori knowledge of the true plant structure; hence it is not possible to determine a posteriori if the system was really identifiable. One is better off carrying out closed-loop identification under circumstances that generate SSI, which implies that a carefully designed external signal should be used.

It should be noted that nonparametric identification techniques (such as correlation analysis and spectral analysis) cannot be applied directly to closed-loop data from normal plant operation. This has been previously pointed out by Box and MacGregor (1974), among others. The standard application of these techniques results in model estimates that are the inverse of the controller.

7-3-5-2 Closed-Loop Frequency Domain Bias
An expression similar to Equation 7-15 can be derived for closed-loop systems to analyze frequency-domain bias when a regulator is present. Such an analysis is more complicated than its open-loop counterpart. We will thus focus on the effect of the feedback controller C on the power spectrum of the input signal, in order to provide some insight into the problem of frequency-domain bias generated from closed-loop dynamic tests. If an external signal u_d is added onto the manipulated variable u, the power spectrum of the input is given by

$$\Phi_u(\omega) = \left|\varepsilon(e^{j\omega T})\right|^2 \Phi_{u_d}(\omega) \tag{7-53}$$

where $\varepsilon = (1 + G_0 K)^{-1}$ is the sensitivity function of the closed-loop system that acts as a weight to the parameter estimation problem in a manner similar to prefiltering. The effect of the controller is to attenuate low-frequency information, as shown in Example 7-1. For controllers with integral action, the sensitivity function is 0 at $\omega = 0$, which means that the controller attenuates the low-frequency portion of the external signal. Significant detuning of K, the feedback controller, may be required in order to obtain an appropriate steady-state fit.

It is possible to avoid detuning by introducing the signal at an alternate point in the loop, the controlled variable setpoint. The expression for the input power spectrum in this case is

$$\Phi_u(\omega) = \left|K(1 + G_0 K)^{-1}\right|^2 \Phi_{u_d}(\omega)$$
$$= |G_0^{-1}\eta|^2 \Phi_{u_d}(\omega) \tag{7-54}$$

where $\eta = G_0 K(1 + G_0 K)^{-1}$ is the complementary sensitivity function. Note that the effect of an external setpoint change on the estimation of $G(q)$ is weighted by $|G_0^{-1}\eta|^2$, as opposed to $|\varepsilon|^2$, which does not attenuate the low frequencies when the controller is tightly tuned. Furthermore, a tightly tuned controller may also lead to amplification of the higher frequencies, as Example 7-1 shows.

Example 7-1

Consider the first-order plant given by:

$$G_0(q) = \frac{k}{q - \alpha} = \frac{0.096}{q - 0.904} \tag{7-55}$$

This plant corresponds to a first-order transfer function with an open-loop time constant of 10 min. Assume that the desired closed-loop response is also a first-order transfer function, represented by the expression

$$\eta(q) = 1 - \varepsilon(q) = \frac{(1 - \delta)}{q - \delta} \tag{7-56}$$

where

$$\delta = e^{-T_s/\tau_{\mathrm{cl}}}$$

T_s is the sampling time (and control interval) and τ_{cl} is the closed-loop time constant.

A PI controller tuned using the rules described by Prett and Garcia (1988) would result in such a closed-loop system.

In this example we set $T_s = 1.0$ min and examine the amplitude ratios of $G_0^{-1}\eta$ and ε for the three cases:

$\tau_{cl} = 1$ min (closed-loop speed faster than open-loop).
$\tau_{cl} = 10$ min (closed-loop speed equivalent to open-loop).
$\tau_{cl} = 100$ min (closed-loop speed slower than open-loop).

Figures 7-3 and 7-4 compare the normalized amplitude ratios of $G_0^{-1}\eta$ and ε. ε resembles a high-pass filter with bandwidth (the frequency at which the amplitude ratio first reaches $1/\sqrt{2}$) defined by $1/\tau_{cl}$. $G_0^{-1}\eta$ is a low-pass filter for slow closed-loop speed-of-response; however, its ability to attenuate the high-frequency range decreases with increasing τ_{cl}, and when $\tau_{cl} < \tau_{ol}$, the high frequencies are amplified. In this example, $G_0^{-1}\eta$ is an allpass when $\tau_{cl} = \tau_{ol}$, which represents an ideal situation. We thus recommend that when injecting an external signal at the controlled variable setpoint, the controller should be tuned such that the closed-loop speed of response equals the open-loop speed of response.

7-3-6 Treatment of Nonlinearity

Distillation columns are inherently nonlinear. Severe nonlinearity is most often seen in high purity columns, but a number of other factors can contribute as well, for example, nonideal vapor–liquid equilibrium. Although columns operated above approximately 98% purity often exhibit severe nonlinearity, behavior varies from column to column and it is difficult to define a specific purity above which severe nonlin-

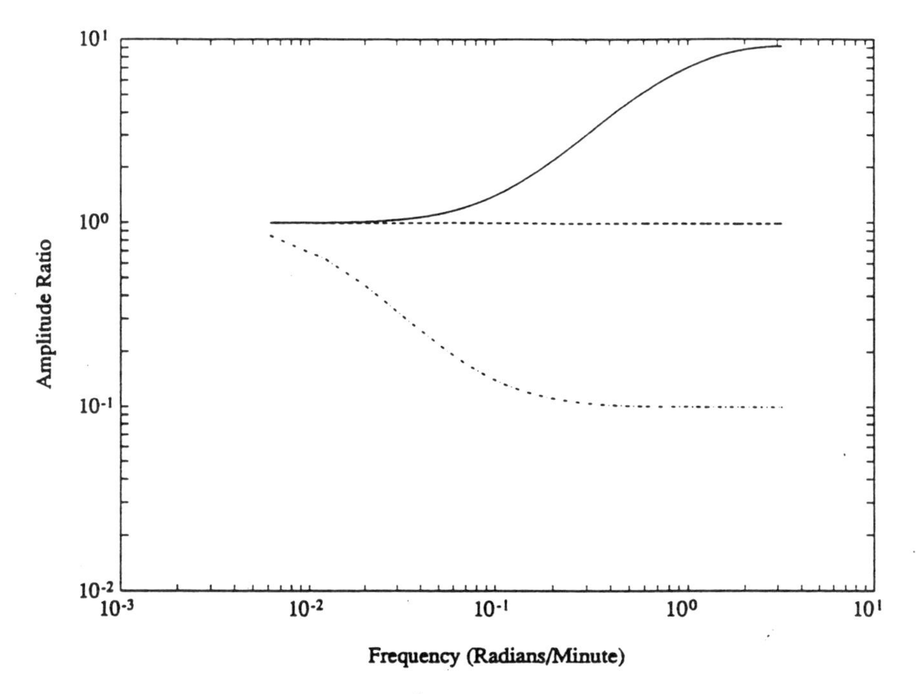

FIGURE 7-3. Amplitude ratio for $G_0^{-1}\eta$. ——: $\tau_{cl} = 1$ min; - - -: $\tau_{cl} = 10$ min; —·—: $\tau_{cl} = 100$ min.

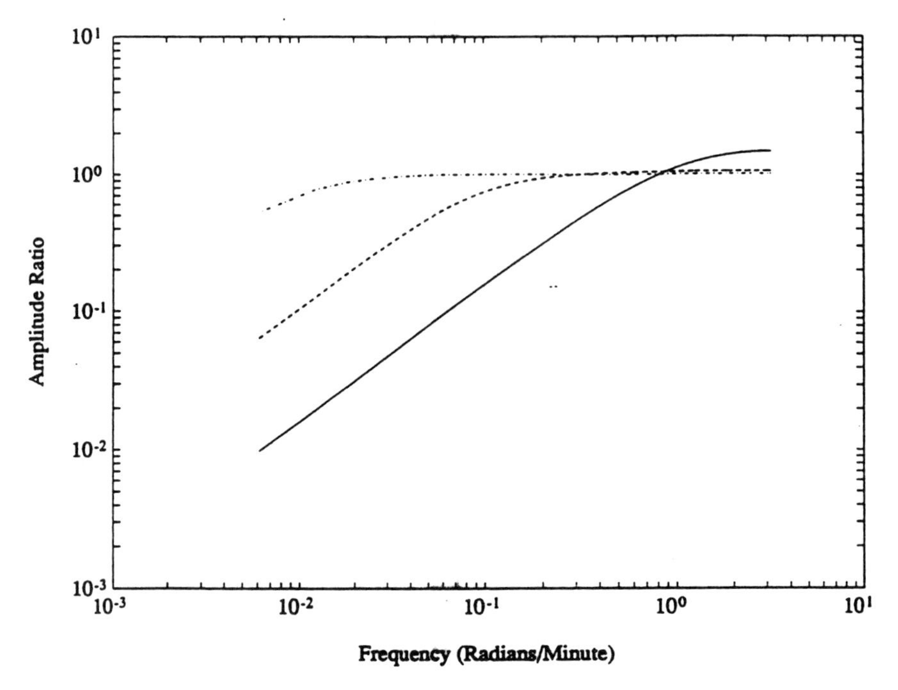

FIGURE 7-4. Amplitude ratio for ε. ——: τ_{cl} = 1 min; - - -: τ_{cl} = 10 min; —·—: τ_{cl} = 100 min.

earity can be expected. As with other nonlinear systems, if operated over a sufficiently small region, linear models often provide an adequate system description. Although large setpoint changes are uncommon in distillation systems, disturbances often drive response variables far enough from steady-state conditions that the performance of linear model-based control systems can deteriorate significantly. This section presents a number of tools for adapting the basic linear identification approach when significant nonlinearity is present in distillation systems.

The most widely applied tool for identifying nonlinear distillation systems is the use of linearizing transformations on composition measurements (Alsop and Edgar, 1987; Koung and Harris, 1987). The Koung and Harris transformation is based on an analysis of the fundamental performance equations described by Eduljee (1975), which are particularly accurate for columns operated near minimum reflux. The transformed composition measurements are given by

$$\tilde{X}_D = \ln \frac{1 - x_D}{1 - x_D^{\text{set}}} \tag{7-57}$$

$$\tilde{X}_B = \ln \frac{x_B}{x_B^{\text{set}}} \tag{7-58}$$

where x = composition measurement
set = setpoint

Georgiou, Georgakis, and Luyben (1988) applied dynamic matrix control (DMC) based on transformed composition measurements to three simulated distillation systems: a moderate purity methanol–water system (99% purity), the same system at higher purity (99.9%), and a very high purity system (10 ppm impurity). As purity increased, significant performance improvements were obtained by using nonlinear DMC (i.e., variables transformed) over linear DMC and a conventional LV control strategy using multiple PI loops. However, with the very high purity column, the performance of the conventional PI-based control system was superior to that of the nonlinear DMC, suggesting that the simple Koung and Harris transformation was inadequate at this purity level. The more complex Alsop and Edgar transformations were not investigated in the Georgiou, Georgakis, and Luyben study.

A number of different identification approaches were used in the Georgiou, Georgakis, and Luyben study. Linear transfer function models were readily obtained for the moderate purity column using conventional pulse testing techniques (Luyben, 1990). However, at higher purities, nonlinearities make it difficult to obtain a suitable linear transfer function model. Estimated transfer function parameters can vary widely, depending on the size and direction of the perturbations to the inputs. Luyben (1987) proposed an identification procedure to overcome this difficulty in high purity distillation systems. In the proposed procedure, steady-state gains are obtained from detailed and highly accurate steady-state simulation (rating) programs. Relay tests (as described in Section 7-3-4-2) are then applied to the actual system to obtain the critical gains and frequencies for the diagonal elements of the plant transfer function matrix. Individual transfer functions are obtained by finding the best transfer function fit to the zero frequency and ultimate frequency data. The utility of this identification procedure for very high purity columns when combined with linearizing transformations was demonstrated by Georgiou, Georgakis, and Luyben (1988).

As discussed in Chapter 12, it is difficult to predict in advance whether a linearizing transformation will be helpful for a given column. A useful procedure for testing the value of a linearizing transformation for a particular column is to put competing models on line and compare the variance and autocorrelation structure of their prediction errors over an extended period.

Further information on the use of linearizing transformations can be found in Shinskey (1988).

7-4 MODEL VALIDATION

7-4-1 Classical Techniques

Having performed parameter estimation, the remaining problem in identification is to assess the validity of the model. As noted previously, the issue of importance here is whether the model meets the intended purpose, which is control system design. We briefly describe some of the commonly used validation techniques.

Simulation Simulation of the measured (y) versus predicted ($\hat{y}$) output is the most common validation tool in identification. Adequately determining how "close" the predicted output is to the actual output, however, is a more challenging task than it appears.

Impulse or Step Responses Examining the model's response to a standard input such as a step or a ramp is a commonly used technique that allows an engineer to compare the results of identification versus personal process understanding. A model with wrong gain sign or one with dynamic behavior that contradicts the engineer's own intuitive understanding of the process will result in repeating some step(s) of the identification procedure.

Cross-Validation Generally speaking, if a model is validated using the same data that were used for parameter estimation, the fit can be expected to improve with increasing number of parameters in the model. To judge the quality of a fitted model in this circumstance requires a measure that accounts for both the decrease in the fit's loss function and the potential loss in predictive power of a model with increasing number of parameters. One commonly used measure is the Akaike information theoretic criterion (AIC) (see Ljung, 1987, Section 16.4). An alternate approach is to perform cross-validation, in which a portion of the experimental data set is retained for validation purposes and not used in parameter estimation. Simulating the model output over the cross-validation data set allows an engineer to assess the true predictive ability of the model.

Residual Analysis Residual analysis consists of testing the residual time series against the hypotheses that the noise series is Gaussian and uncorrelated with the input series. If a nonunity disturbance model is chosen, that is, $H(q) \neq 1$, then residual analysis is performed on the calculated prediction errors. Residual analysis is discussed in detail by Box and Jenkins (1976). The measures that are used in residual analysis are the calculated autocorrelation

$$\rho_{e_P}(k) = \frac{c_{e_P}(k)}{\sigma_{e_P}^2} \tag{7-59}$$

and for the cross-correlation between residual and the plant input

$$\rho_{ue_P}(k) = \frac{c_{ue_P}(k)}{\sqrt{\sigma_u^2 \sigma_{e_P}^2}} \tag{7-60}$$

If the auto- or cross-correlation functions exhibit a recognizable pattern (i.e., autocorrelation at certain lags fall outside their estimated 95% confidence intervals), model inadequacy is suggested. If examination of the cross-correlation function(s) suggests no inadequacy of the process transfer function $G(q)$ (i.e., no cross-correlations outside their estimated 95% confidence limits), then the inadequacy is probably in the disturbance model. An overestimated deadtime (n_k too large) shows up as significant cross-correlation between residuals and the input as lags corresponding the missing b_k terms in the polynomial $B(q)$.

7-4-2 Control-Relevant Techniques

A fundamental problem with the model validation techniques described in the previous section is that they are open-loop measures for the goodness-of-fit. An increased understanding of modelling for control design has led to the conclusion that a model that may appear to be a good fit in the open-loop can result in very bad closed-loop performance and vice versa (see Åström and Wittenmark, 1989, Chapter 2). Clearly, there is a need for *control-relevant* validation measures that determine the adequacy of the model for control purposes. To this end, current research draws from robust control theory and the structured singular value paradigm to develop these validation measures (see Rivera, Webb, and Morari, 1987 and Smith and Doyle, 1989). The key challenge here, however, is obtaining appropriate estimates of the plant uncertainty from the residual time series.

The most conventional means for obtaining model uncertainty is to use $100(1-\alpha)\%$ confidence intervals generated from the solution of the linear least-squares problem Equation 7-38. These confidence intervals can be computed from the identification data set and an arbitrary plant input $\hat{u}$ (such as a step or an impulse) as follows:

$$t\left(N-n, 1-\frac{\alpha}{2}\right) \times \sigma_{e_P} \sqrt{\hat{u}^T \left[\frac{1}{N} \sum_{t=1}^{N} \varphi(t) \varphi^T(t)\right]^{-1} \hat{u}} \tag{7-61}$$

$t(N-n, 1-\alpha/2)$ is the t statistic obtained from Student's distribution. These confidence intervals must be viewed with caution because they imply that only Gaussian noise affects the process. Neither the effects of undermodelling nor the problems associated with autocorrelated residuals are considered.

An alternative is to compute frequency-domain uncertainty descriptions from the residual time series, as shown by Kosut (1987). Kosut uses spectral analysis to obtain an estimate of unmodelled dynamics. This result is applied by Rivera et al. (1990) and Rivera, Pollard, and Garcia (1990) to generate uncertainty bounds that capture both bias and variance effects in the identification. A brief summary follows.

Consider plant residuals represented by the symbol υ, which reflect both unmodelled dynamics and the effect of noise $\nu(t)$:

$$\begin{aligned} \upsilon(t) &= (G_0(q) - G(q))u(t) + \nu(t) \\ &= \Delta(q)u(t) + \nu(t) \end{aligned} \tag{7-62}$$

Estimating uncertainty, then, requires obtaining an estimate of the unmodelled dynamics $\Delta(q)$ and realizing that the quality of this estimate is affected by the presence of noise and length of the data set. The additive uncertainty bound l_a is obtained from

$$l_a(\omega) = |\hat{\Delta}(e^{j\omega})| + |\Delta(e^{j\omega}) - \hat{\Delta}(e^{j\omega})| \tag{7-63}$$

where the first term is a *bias* term whereas the second is a *variance* term. To obtain the bias term, one uses spectral analysis to obtain $\hat{\Delta}$:

$$\hat{\Delta}(j\omega) = \frac{\hat{\Phi}_{\upsilon u}(j\omega)}{\hat{\Phi}_u(\omega)} \tag{7-64}$$

where $\hat{\Phi}_{\upsilon u}$ and $\hat{\Phi}_u$ denote the "smoothed" estimates of the cross- and autospectra between υ and u. A result presented by Jenkins and Watts (1969) is then used to obtain the variance term

$$|\Delta - \hat{\Delta}|^2 \leq \frac{2}{\varpi - 2} \frac{\hat{\Phi}_\nu(\omega)}{\hat{\Phi}_u(\omega)} f_{2,\varpi-2} \tag{7-65}$$

where $\hat{\Phi}_\nu$ = an estimate of the disturbance power spectral density

$$\hat{\Phi}_\nu(\omega) = \hat{\Phi}_\upsilon(\omega) - |\hat{\Delta}|^2 \hat{\Phi}_u(\omega)$$

$\hat{\Phi}_\upsilon(\omega)$ = computed power spectrum for υ

ϖ = degrees of freedom of the spectral estimator (which is proportional to the length of the data set)

$f_{2,\varpi-2}$ = the two-way Fisher statistic for a user-specified confidence level.

These results are used by Rivera et al. and Rivera, Pollard, and Garcia (1990) to analyze the effects of different design variables on diverse prediction-error model structures.

The usefulness of uncertainty modelling from residuals for robust control purposes must be clarified. To begin with, both the confidence interval and frequency domain approaches assume that the true plant is linear. Hence their usefulness when applied to nonlinear plants must be qualified, as is done by Webb, Budman, and Morari (1989). Because of operating restrictions in identification testing, however, many data sets gathered from process plants fall under this assumption. Thus the appropriate question to ask in evaluating these approaches is, even if the estimated uncertainty does not cover all true plant uncertainty, is this information still useful for controller design? The answer here is yes. Performing control over varying operating regions is a task that requires a careful combination of robust control design, process monitoring tools, and adaptation mechanisms. The usefulness of uncertainty modelling from residuals must be viewed within this context.

7-5 PRACTICAL CONSIDERATIONS

In the previous sections we discussed selection of design parameters for experimental design, model structure selection, and parameter estimation that are pertinent when the end use of the identified models is for process control. Ideally, one would consider the model requirements and trade-offs presented by different choices of these design parameters and then proceed with the dynamic process test and identification based on the particular design parameters selected. In practice, however, constraints (process as well as management) sometimes make it impossible to conduct the desired process test. Prejudice for selection of a certain model structure also affects the process test, as described in the following text.

The presence of disturbances and measurement noise directly affects the variance of estimated parameters. In general, the size of perturbations applied to the inputs should be as large as permitted (to maximize the signal-to-noise ratio) in order to reduce the required duration of the experiment. In situations where measurement noise is low and a process operates relatively smoothly between disturbances that occur only infrequently, a shorter duration experiment can be designed. In these situations, a small number of step changes (perhaps combined with a relay test) will provide sufficient information to obtain an acceptable control model. However, this is only true if (low-order) transfer functions are to be identified. If an a priori decision is made to statistically estimate truncated impulse (or step response) models, then this dictates that a long duration experiment be performed, to improve the variance of the resulting estimates. As mentioned previously, a biased least-squares estimator might also be performed with this model structure to obtain a more favorable bias–variance trade-off.

Other practical considerations may also influence the decision to identify truncated impulse models directly over transfer function forms: dimensionality (number of inputs and outputs) and the ability to conduct tests on a multivariable process one input at a time. If the process is sufficiently well behaved to allow one input to be perturbed while holding the others constant, then this greatly reduces the complexity of the estimation problem, because SISO models can be estimated one at a time (assuming the multivariable model Equation 7-4 is deemed adequate and no disturbance model is to be identified). Determining the structure, deadtime, and polynomial orders for a transfer function model Equation 7-27 can be performed readily. The resulting transfer function forms can be converted numerically to truncated impulse or step response form without any loss of information.

Regardless of whether inputs are perturbed one at a time or simultaneously, it may become necessary to allow an operator to make moves in other inputs in order to keep a process within an acceptable operating region. In either case the structural and parameter estimation problem becomes more complex if a multi-input model rather than a single-input model must be identified, particularly if the number of inputs becomes large. The trade-offs that occur between duration of experiment, signal-to-noise ratio (input amplitudes), model structure (impulse vs. low-order transfer function) must now also include the experience level of the person performing the identification. As dimensionality increases, the problem of satisfactorily determining deadtimes, structures, and polynomial orders for transfer function models becomes more complex (particularly if the system exhibits dynamic behavior that is more complex than simple first or second order with deadtime, as very typically happens). In these situations the probability of obtaining an incorrect structure becomes higher. Depending on the training and experience of the user, direct identification of impulse or step models (that require only one parameter to be specified—the process settling time) may be the best alternative. However, it must be recognized that as dimensionality increases, experiments of increasing duration are required. At some point, even the most accommodating operations superintendent may not permit such a long test, and a more efficient approach will be required.

The following are some suggestions for increasing the probability of performing a successful dynamic test. The first suggestion is to perform a "pre-test" on a system some time before the actual dynamic test. The pre-test allows one to identify potential problems that may hamper (or postpone) the actual test. Typical things to look for are equipment problems (valves, transmitters, analyzers, etc.), inadequate tuning of regulatory controllers, equipment in "nonnormal" operating mode or range (e.g., bypass valves), and functioning of the data-logging program(s), to name a few. The pre-test also allows one to explain to the operations staff what the dynamic test needs to accomplish (and to agree on how it should be performed to ensure that quality and operating constraints are not violated). In addition to information obtained from process instrument diagrams, the pre-test can help to identify points that should be logged during the test. In addition to the points for which models will be required, any point that may potentially be used to verify normal operation (or conversely to later explain an unusual operating occurrence) should be logged. Points on the unit under test as well as points on upstream (or interconnected) units that have a direct effect on the test unit should be logged. Redundant or useless data can always be discarded, but missing data can, in the worst scenario, invalidate a test. Data from the pre-test can also provide information on the system dynamics, which allows one to design the perturbation signals for the subsequent dynamic test.

During the actual test it is important to have an operations engineer monitor the system to record any events (e.g., opening of a bypass valve, upset or shutdown of an upstream unit, flushing of a heat exchanger, tuning of a controller, putting a controller in manual, unscheduled maintenance operation, etc.) that might invalidate a segment of test data. The data logger should record points with a time stamp so that segments of data corresponding to periods of upsets or unscheduled events can be identified later.

7-6 EXAMPLE

As an example, an identification is performed on a moderate purity refinery depropanizer. Feed to the column is the distillate flow from a debutanizer, which is manipulated by the debutanizer pressure controller. Flow rate of this stream is measured, but its composition is not. Analyzers on the overhead and bottom streams of the depropanizer provide composition measurements every 3 min. A multivariable control design resulted in three inputs: feed flow rate (a feedforward variable), reflux, and reboiler steam flow rates (manipulated variables) and two controlled outputs: overhead and bottoms compositions. Discrete impulse models were required for the control package, one for each input–output pair.

All data analysis, model structure determination, parameter estimation, and model validation was performed using the System Identification Toolbox (Ljung, 1988) in MATLAB (1989).

The data set for the identification consisted of 3 min samples taken over approximately 60 h. Discrete impulse response models relating reflux flow, reboiler steam flow, and feed flow to overhead and bottoms composition were obtained as follows. The entire data set was used initially to obtain least-squares estimates of truncated impulse models relating each input–output pair, which were converted to step response models for examination. From plots of the step response models, approximate deadtimes and model orders were determined. A smaller portion of the data (the portion considered to be most reliable) was then used to estimate low-order transfer function models [of MISO output error (OE) structure], using trial and error search of values of deadtime and model polynomial orders around those determined from the estimated impulse models. The best model was selected as the one that minimized the Akaike information criterion (AIC) and that satisfied model adequacy criteria based on cross-correlation analysis (between the residuals of the fitted OE model and the corresponding inputs), cross-validation using a segment of the original data set that was held back from the estimation, and examining the poles and zeros of the fitted model (if the 95% confidence region of a pole of the transfer function for a particular input–output pair overlaps with a zero of the same transfer function, it is overparametrized). Disturbance models were not identified because the control package used assumed disturbance models (randomly occurring deterministic steps), and it was not considered necessary to modify it for this particular application. Nevertheless, simultaneous identification of process and disturbance transfer function models can be readily accomplished using the available fitting routines in the toolbox.

All data were normalized by subtracting off sample means (see function *detrend* in the toolbox). Variance normalization (to unit standard deviation) can also be performed (for an example, see Section 5.4 of the MATLAB manual).

Only the identification of MISO models for bottoms composition is shown here. Similar procedures were followed to obtain overhead composition models.

Normalized test data are shown in Figure 7-5 (*a*: bottoms composition (% C3); *b*: reflux flow rate; *c*: reboiler steam flow rate; *d*: feed flow rate).

Early in the test it was realized that the column was overrefluxed (and reboiled). Over roughly one shift (sample number 125 to sample number 275—approximately 7.5 h), both reflux and reboiler steam flow rates were gradually reduced to bring the column back to a more normal operating region. These data were retained for preliminary estimation of the impulse models (to obtain rough estimates of deadtimes and model orders) and for cross-validation (with a certain level of skepticism due to the abnormal operating condition).

Due to exceptional operating problems and disturbances in the debutanizer, feed flow to the depropanizer experienced greater fluctuation than normal during the dynamic test period. This was not entirely undesirable because it excited the column for identification of models relating feed flow rate and the response variables. It sometimes can be difficult to arrange intentional perturbation of feedforward variables (because this usually means upsetting an upstream unit). Because of the wide swings in feed flow rate during the test, operators did occasionally have to manipulate reflux and reboiler steam rates to maintain operation within quality constraints as much as possible. As described earlier in the chapter, when data for identification are obtained when feedback is present (inadvertently, as may have occurred in this case, or due to a controller that is on-line), it is recommended that external perturbations be applied somewhere in the feedback loop. During the depropanizer test, random perturbations were applied to reflux and reboil steam flow rates to as great a degree as possible, considering the unusual operating conditions. However, these conditions did preclude applying preprogrammed PRBS signals to the inputs. As well, it was impossible to perform a dynamic test in which only one input at a time was perturbed (while holding the others constant), which would have simplified the model estimation stage of the identification.

Several additional operating problems arose during the dynamic test that were not apparent in the pre-test, which had been conducted several weeks earlier. As shown in Figure 7-5*b*, a problem in the reflux control loop caused high-frequency oscillations in the reflux flow. Tuning of the PID flow controller did not resolve the problem. Troubleshooting for mechanical problems in the loop hardware also did not immediately solve the problem. Because the oscillations were high frequency, and would sim-

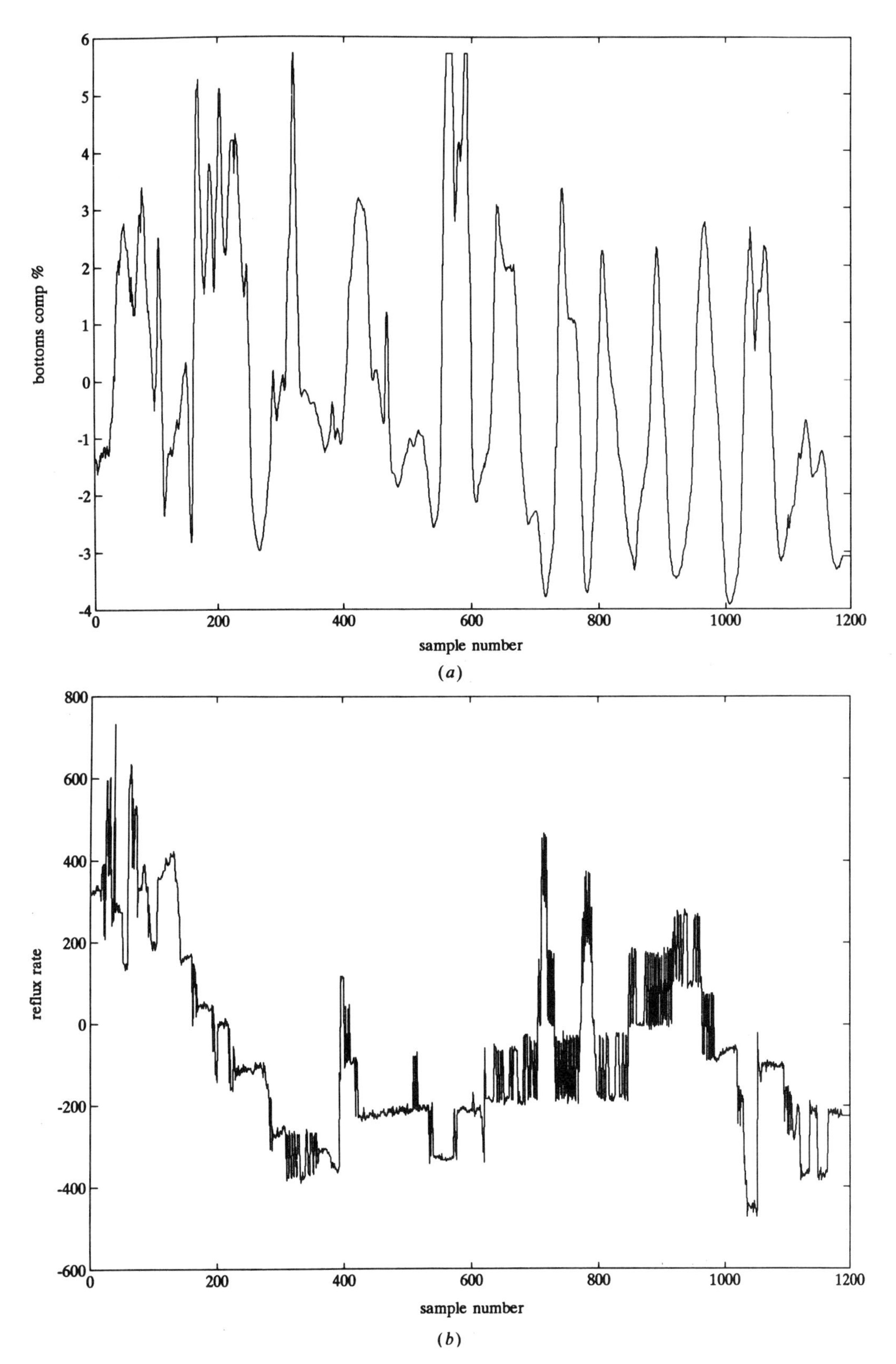

FIGURE 7-5. Depropanizer test data. (*a*) bottoms composition; (*b*) reflux flow rate; (*c*) reboiler steam flow rate; (*d*) feed flow rate.

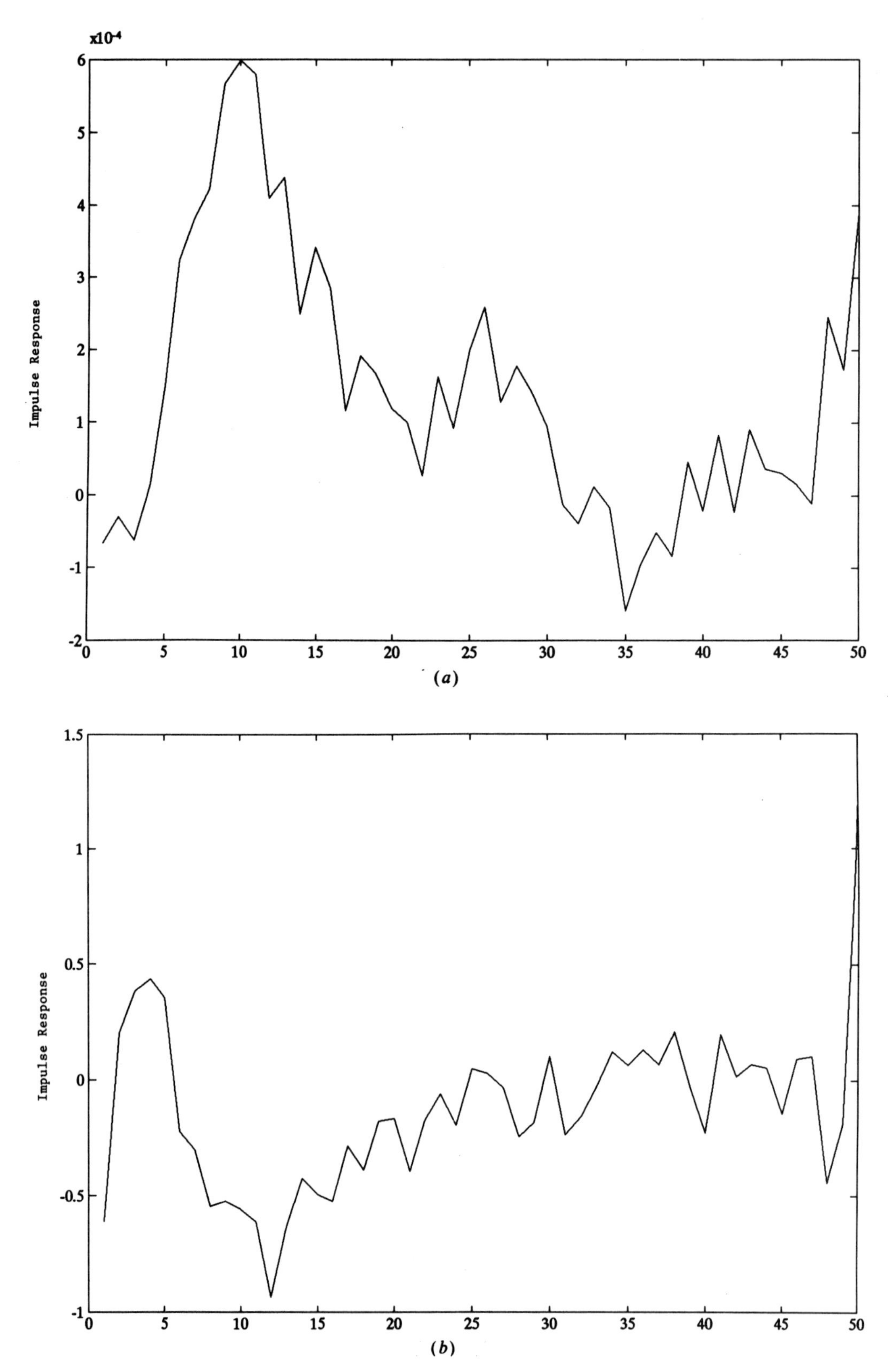

FIGURE 7-7. Estimated impulse models. (*a*) bottoms composition/reflux; (*b*) bottoms composition/reboiler; (*c*) bottoms composition/feed.

ply be filtered by the process, and it was still possible to make setpoint changes to the reflux flow controller, it was decided to proceed with the test while troubleshooting continued. A mechanical problem was finally resolved near the end of the test (around sample number 975).

Power spectra were generated for all inputs using the *spa* function in the toolbox to determine the useful frequency range of the input perturbations. The spectrum for reflux flow rate is shown in Figure 7-6. Examination of the frequency functions (i.e. Bode plots) obtained from spectral analysis (generated by the *spa* function) and the Bode plots of the final estimated models indicated that the input perturbations did not contain sufficiently high frequency information. The resulting control models would be considered adequate only if relatively sluggish controller tuning proved to be adequate. Power spectra for the other inputs were similar.

Low-frequency drift in the data should also be removed by prefiltering (for example, using the filter function *idfilt* in the toolbox as a high-pass filter). Because any controller with integral action will be able to regulate effectively against very slow disturbances, models identified for use in process control need not be accurate at low frequencies (i.e., near steady state). The need to obtain a high quality estimate of steady-state gain for control models is overemphasized in the identification folklore. Differencing data will also remove nonstationarity (low-frequency drift), but will result in fitted models in which bias at higher frequencies is reduced (at the expense of accuracy at lower frequencies that arc of interest for process control) because taking the derivative of a signal distorts its spectrum (i.e., amplifies power) at higher frequencies.

For systems that do not exhibit strong gain directionality (as is the case with this moderate purity tower), input perturbations should be independent (noncorrelated). Cross-correlation functions for the input sequences were generated and plotted (not shown) using the routine *xcov* in the Signal Processing Toolbox in MATLAB. No significant cross-correlation between the inputs was found.

Preliminary least-squares estimates of the impulse responses for each input–output pair were obtained by fitting a three-input–one-output ARX model [Equation

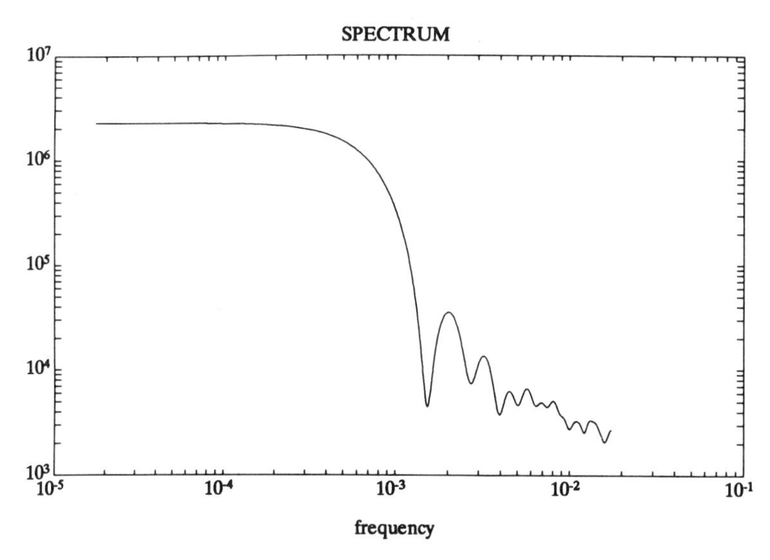

FIGURE 7-6. Power spectrum of reflux flow rate.

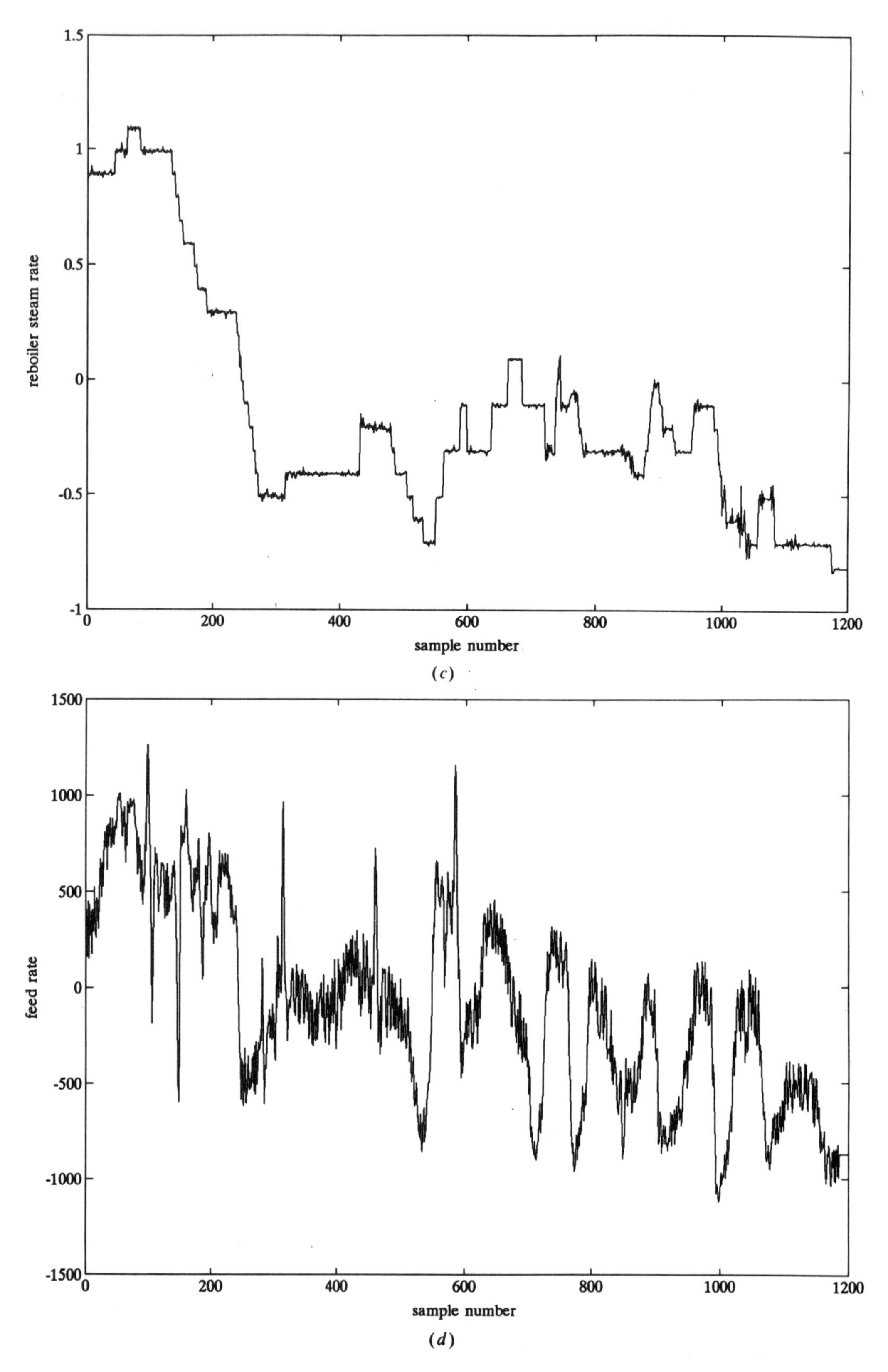

FIGURE 7-5. *Continued.*

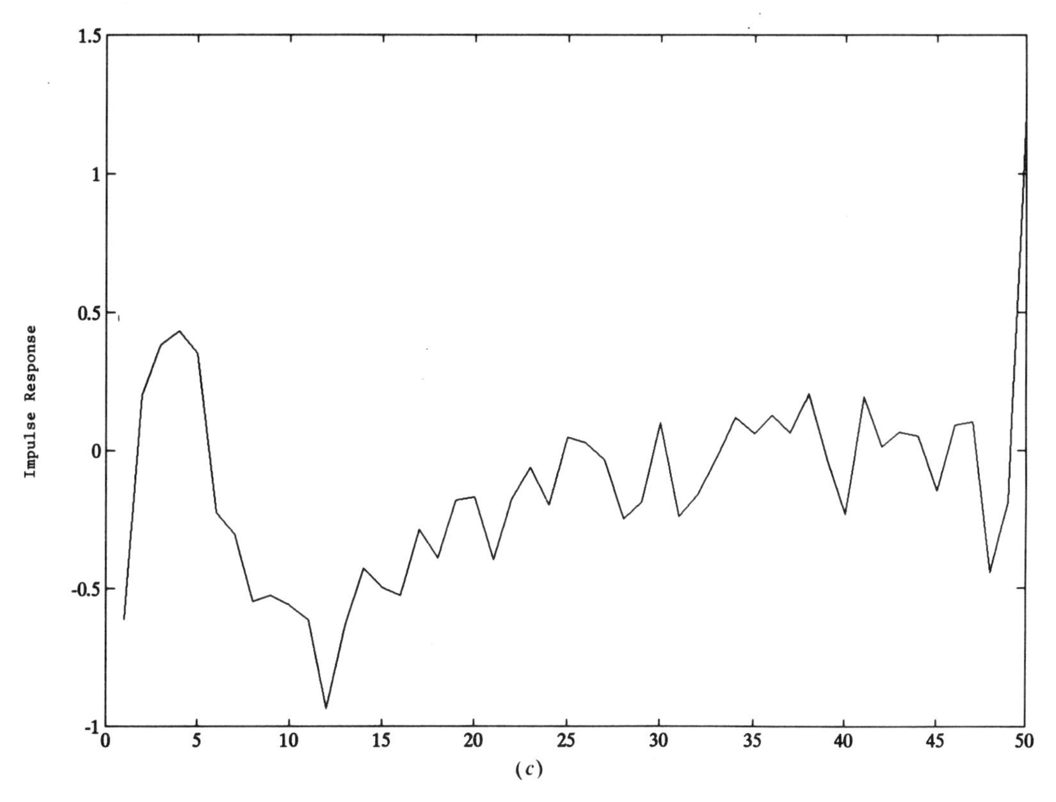

FIGURE 7-7. *Continued.*

7-31, with $n_a = 0$, i.e., $A(q) = 1$ and $n_{bi} = 50$, the order of $B_i(q)$ for input i, $i = 1, 2, 3$] using the toolbox routine *arx*. Alternately the toolbox routine *cra*, which obtains impulse models by a cross-correlation technique, could have been used. The estimated impulse models are shown in Figure 7-7. The toolbox routine *idsim* was used to obtain the corresponding step response models (by simulating the response of the estimated three-input–one-output impulse model to a unit step change applied to each input, one at a time).

The unit step response models are shown in Figure 7-8. From these plots it was determined that the responses of bottoms composition to reflux, reboiler, and feed flow rate each has a deadtime of approximately five sampling periods (15 min). Each of the responses appears to be low order [the order of $F(q)$ for a transfer function model will be low]. The response of bottoms composition to reboiler steam flow rate rate exhibits inverse behavior, suggesting that the order of $B_i(q)$, $i = 2$ (reboil), for a transfer function model should be at least 2.

Low-order transfer function models were identified using sample numbers 630 to 1193 (approximately 28 h). Because no disturbance models were to be estimated, an output error (OE) structure was selected for identification. An ARX structure could also have been used (Equation 7-31), but as shown by Equation 7-29 this structure implicitly assumes a disturbance model of the form $H(q) = 1/A(q)$. This is undesirable mainly because $H(q) = 1/A(q)$ results in a fit in the frequency domain that emphasizes reducing bias at high frequencies, as shown by Equations 7-15 and 7-19 [with $H(q) = 1/A(q)$, the quantity $1/|H(e^{j\omega T})|^2$ becomes larger at higher frequencies, thereby result-

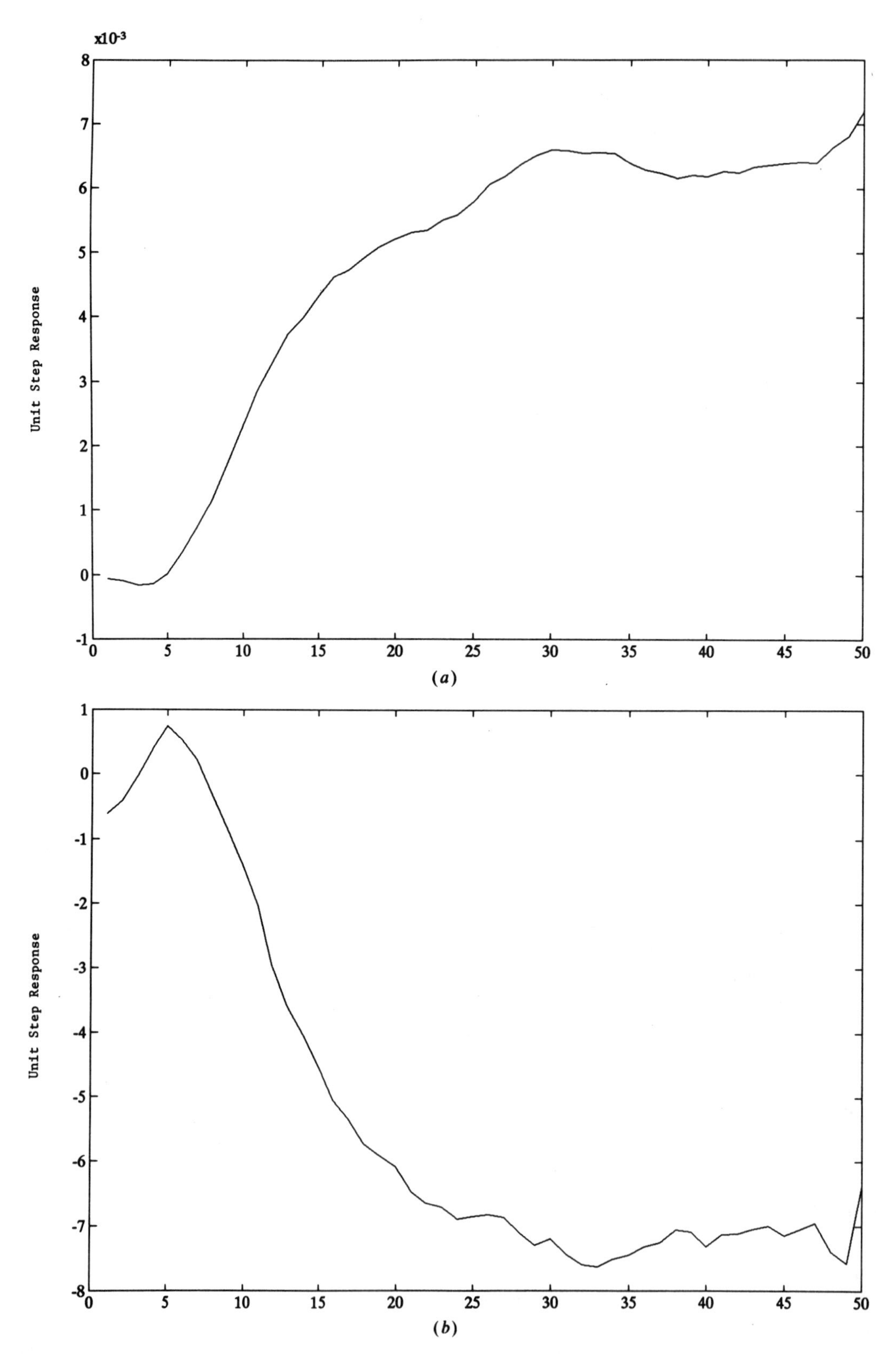

FIGURE 7-8. Unit step responses obtained from estimated impulse models. (*a*) bottoms composition/reflux; (*b*) bottoms composition/reboiler; (*c*) bottoms composition/feed.

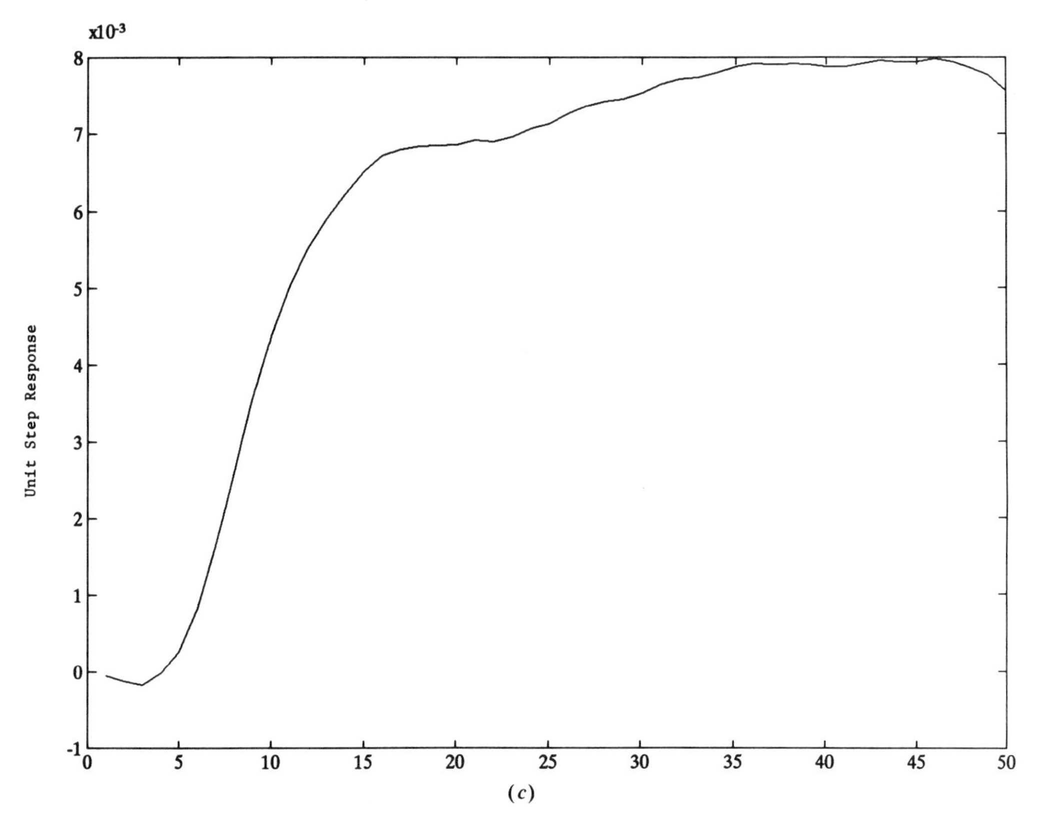

FIGURE 7-8. *Continued.*

ing in greater weight being placed on reducing bias in the fitted model at higher frequencies].

The toolbox routine *oe* for estimating output error models allows only one input and so the general routine for prediction-error models, *pem*, was used instead. With three inputs and the polynomial orders for $A(q)$, $C(q)$, and $D(q)$ set to zero, Equation 7-27 becomes

$$y(t) = \frac{B_1(q)}{F_1(q)}u_1(t - n_{k1}) + \frac{B_2(q)}{F_2(q)}u_2(t - n_{k2}) + \frac{B_3(q)}{F_3(q)}u_3(t - n_{k3}) + e(t) \quad (7\text{-}66)$$

where y = bottoms composition
u_1 = reflux flow rate
u_2 = reboiler steam flow rate
u_3 = feed flow rate

The following estimated OE model minimized the AIC (and satisfied all other validation criteria, as shown below). Parameter estimates (denoted by carets) are given with their 95% confidence intervals in parentheses.

Bottoms composition/reflux flow rate:

$n_{k1} = 5$

$n_{b1} = 1 \quad \hat{b}_{11} = 0.57 \times 10^{-3} \, (\pm 0.11 \times 10^{-3})$

$n_{f1} = 1 \quad \hat{f}_{11} = -0.92 \, (\pm 0.014)$

Bottoms composition/reboil steam flow rate:

$n_{k2} = 4$

$n_{b2} = 2 \quad \hat{b}_{21} = 1.03\ (\pm 0.64)$

$\hat{b}_{22} = -1.34\ (\pm 0.74)$

$n_{f2} = 2 \quad \hat{f}_{21} = -1.59\ (\pm 0.15)$

$\hat{f}_{22} = 0.634\ (\pm 0.14)$

Bottoms composition/feed flow rate:

$n_{k3} = 5$

$n_{b3} = 1 \quad \hat{b}_{31} = 0.98 \times 10^{-3}\ (\pm 0.07 \times 10^{-3})$

$n_{f3} = 1 \quad \hat{f}_{31} = -0.86\ (\pm 0.015)$

None of the confidence intervals encompasses zero, indicating that all the estimates are statistically significant.

The poles and zeroes of the transfer function for each input–output pair were calculated using the toolbox routine *th2zp* and plotted (along with their 95% confidence regions, which are not shown) using the toolbox routine *zpplot*. Possible redundancy in poles and zeros was only a concern for the transfer function relating bottoms composition and reboil steam rate because this input–output pair had two poles. The pole–zero plot obtained from *zpplot* for this input–output pair showed no overlapping of the 95% confidence regions for the poles and the zero, and therefore overparametrization was not suspected.

A plot of the residuals for the fitted three-input–one output OE model is shown in Figure 7-9. The autocorrelation function for these residuals (Figure 7-10*a*) shows significant values (i.e., outside the 95% confidence limits given by the dotted lines), which is expected because a disturbance model was not estimated. At this point a disturbance model could be identified by adding the polynomials $C(q)$ and $D(q)$ (Equation 7-27) in the toolbox routine *pem* and deter-

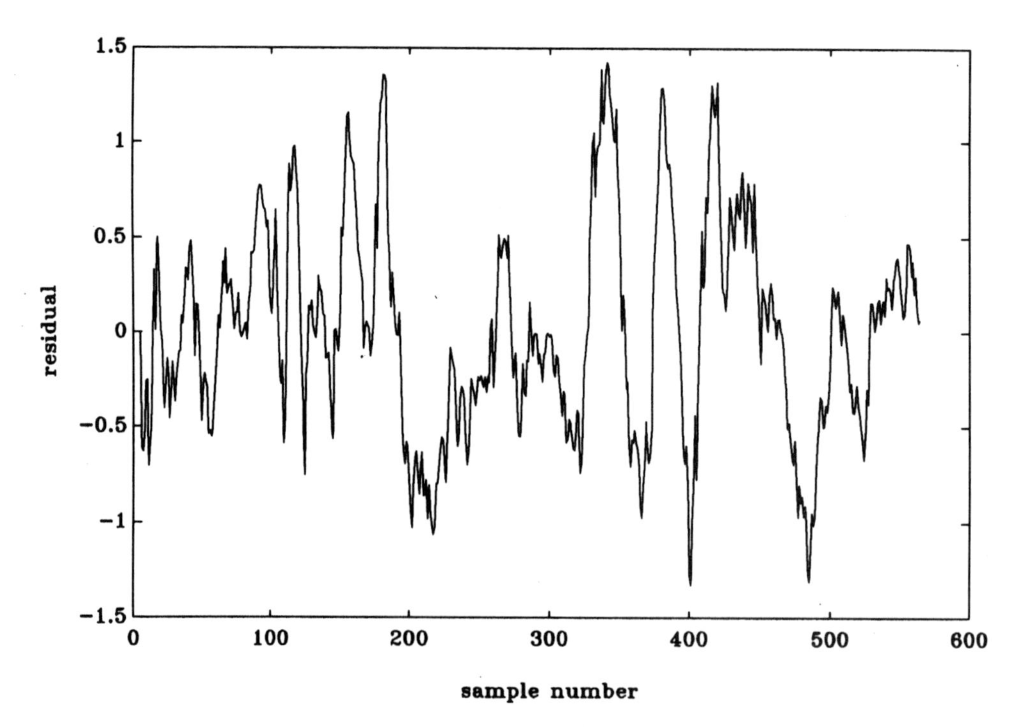

FIGURE 7-9. Residual plot for estimated OE models.

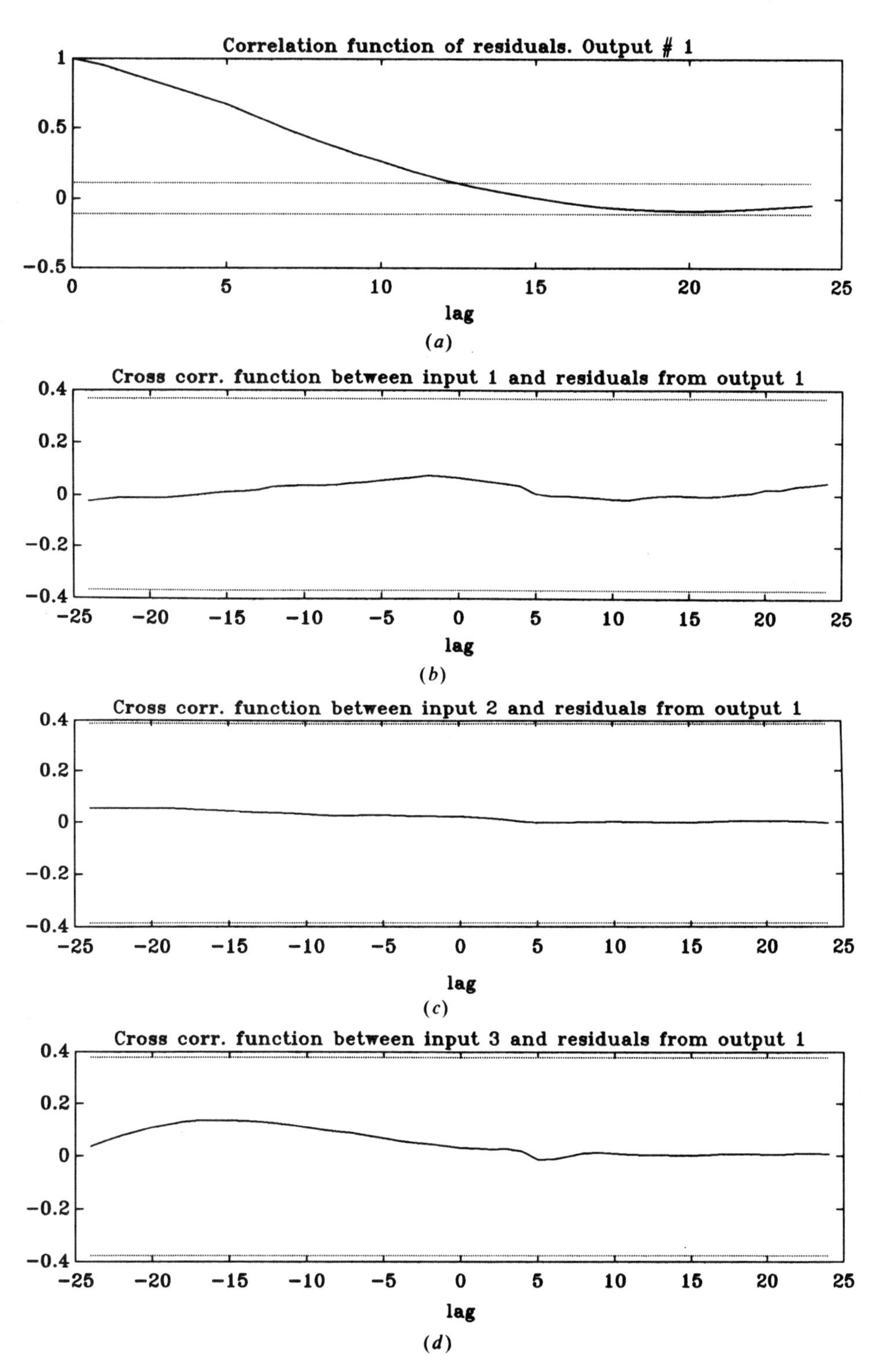

FIGURE 7-10. Auto- and cross-correlation functions for estimated OE models. (*a*) Residual autocorrelation function; cross-correlation functions between residuals and inputs; (*b*) input 1 (reflux); (*c*) input 2 (reboiler); (*d*) input 3 (feed).

mining an adequate structure as described by Box and Jenkins (1976).

The cross-correlation functions between the residuals of the fitted three-input–one-output model and each of the inputs are shown in Figure 7-10*b–d*. None of the cross-correlation functions shows significant values at positive lags, indicating that no structural deficiencies are present in the model. Also, there is no significant cross-correlation at negative lags, indicating that no feedback existed between the output (bottoms composition) and the inputs during the test period.

Residual and correlation functions were generated and plotted using the toolbox routine *resid*.

Figure 7-11 is a plot of the simulated model output for bottoms composition (generated with the toolbox routine *idsim*) and the measured bottoms composition. The simulated output obtained from *idsim* is the response of the model to changes in the inputs (reflux, reboil steam, and feed flow rates) only. Because no disturbance model was identified, unmodelled disturbances affecting the actual measured bottoms composition will show up in this plot as offset between the simulated model output and the measurements. Therefore, model adequacy must be judged with this plot by examining overall dynamic behavior of the simulated output against the actual measurements (i.e., is the model capturing essential dynamic response despite the presence of offset, due to unmodelled disturbances, between the simulated and measured output?). This way of plotting model response is better than calculating the model as a one-step-ahead predictor (i.e., to predict the output at time t, the model has all measured input and output information up to time $t - 1$) because the model is required to predict only one step ahead, mak-

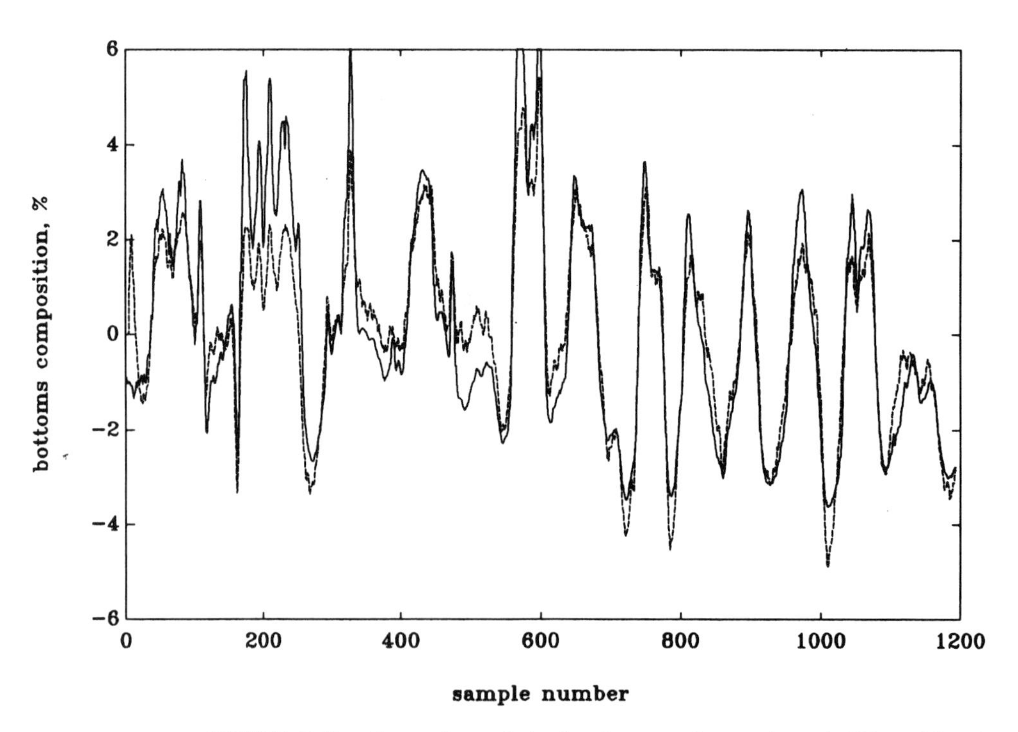

FIGURE 7-11. Comparison of simulated output from estimated OE model and measured data for bottoms composition. ——: measured bottoms composition; – – –: simulated model output.

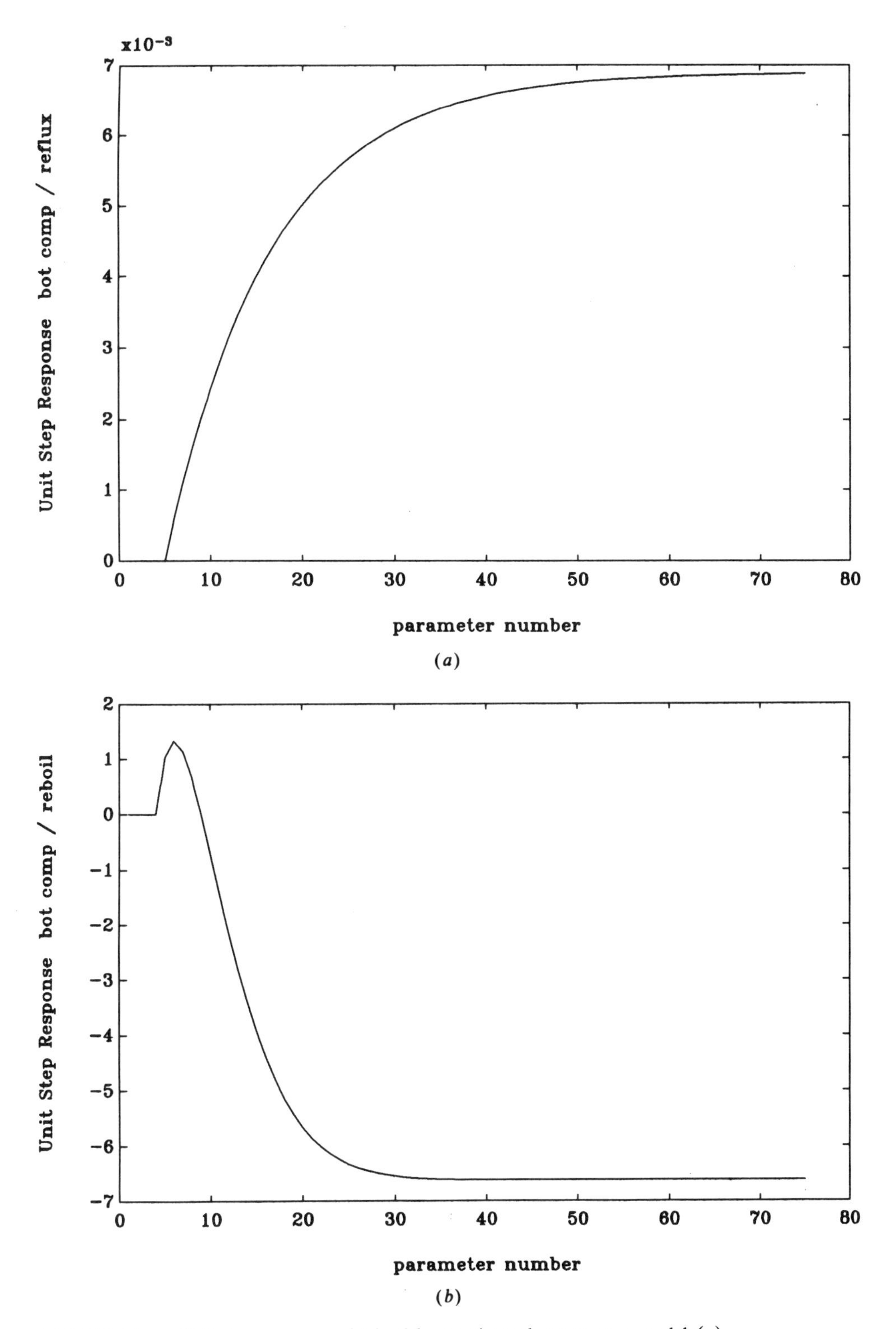

FIGURE 7-12. Unit step responses obtained from estimated output error model. (*a*) bottoms composition/reflux; (*b*) bottoms composition/reboiler; (*c*) bottoms composition/feed.

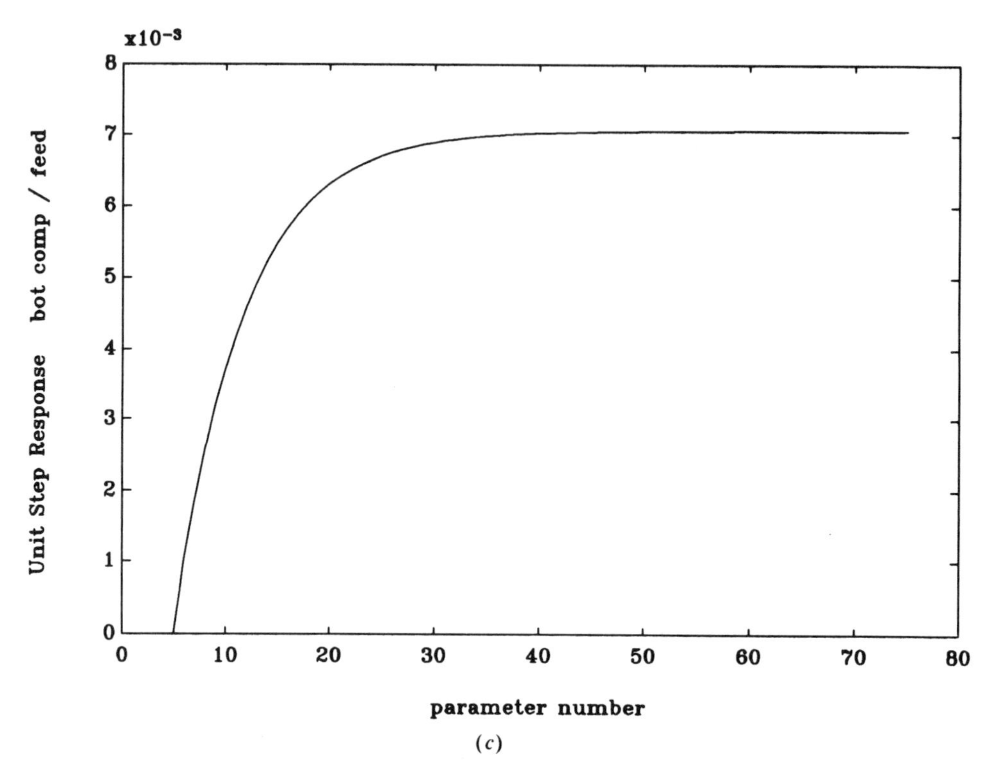

(c)

FIGURE 7-12. *Continued.*

ing it more difficult to detect model inadequacy.

Recall that the model was estimated using sample numbers 630 to 1193. For cross-validation, we are interested in the predictive ability of the model on a segment of data that was not involved in the estimation (in this case, sample numbers 1 to 629, Figure 7-11). The model appears to predict dynamic response very well for these data, despite the abnormal operating condition, over a range of frequencies (fast dynamic changes are predicted as well as slower changes), even during the initial period when reflux and reboiler steam rates were stepped down to a more normal operating region. As mentioned previously, due to the lack of higher frequency excitation in the input perturbations, these simulation results do not provide useful information on the predictive ability of the model at higher frequency.

Discrete impulse and step response models for each input–output pair were obtained from the estimated three-input–one-output OE model using the toolbox routine *idsim*. For brevity, only the unit step response models are shown here, in Figures 7-12 (*a*: bottoms composition/reflux; *b*: bottoms composition/reboiler; *c*: bottoms composition/feed). These model responses agree with our physical intuition of the process (settling times, sign of the steady-state gains, and dynamics, particularly the inverse response present in the transfer function relating bottoms composition to reboiler steam flow rate, which was attributed to the dynamic response of liquid level at the bottom of the tower).

7-7 NOMENCLATURE

$\|\cdot\|_2$ two-norm objective function, $\|e\|_2 = (\Sigma_{k=0}^{\infty} e_k^2)^{1/2}$

α Speed-of-response increase, PRBS design; $e^{-T_s/\tau_{\text{dom}}}$, used in prefilter design;

Symbol	Description
	Used to determine confidence intervals
β	PRBS design parameter for desired low-frequency range
$\Delta(q)$	Uncertainty from undermodelling
$\hat{\Delta}(q)$	Estimated uncertainty from undermodelling
ε	Sensitivity operator $(1 + G_0(q)K(q))^{-1}$
$\tilde{\varepsilon}$	Sensitivity operator based on estimated model $(1 + G(q)K(q))^{-1}$
η	Complementary sensitivity operator $G_0(q)K(q)(1 + G_0)(q)K(q))^{-1}$
$\tilde{\eta}$	Complementary sensitivity operator based on the estimated model $G(q)K(q)(1 + G(q)K(q))^{-1}$
θ	Vector of model parameters
$\hat{\theta}$	Vector of estimated model parameters
$\nu(t)$	Unmeasured disturbance
ρ_x	Autocorrelation function for a signal x
ρ_{xy}	Cross-correlation function between signals x and y
σ_x	Standard deviation of a signal x
τ_d	Deadtime
τ_{cl}	Closed-loop speed-of-response
τ_{ol}	Open-loop speed-of-response
τ_{dom}	System dominant time constant
Φ_x	Power spectrum of a signal x
$\hat{\Phi}_x$	Smoothed power spectrum of a signal x
φ	Regression vector
ω	Frequency (radians/time)
ϖ	Degrees of freedom, spectral estimator
a	Signal magnitude, PRBS signal
$A(q)$	Autoregressive polynomial
$B(q), F(q)$	Polynomials describing the model structure for inputs
$C(q), D(q)$	Polynomials describing the disturbance model
c	Design parameter in ridge regression
c_x	Covariance function, signal x
c_{xy}	Cross-covariance function between x and y
D	Delay in multi-input PRBS sequences
E	Residual matrix in partial least-squares estimation
$e_F(t)$	$L(q)e(t)$, prefiltered prediction error
$e_P(t)$	Prediction error
$f_{2,\varpi-2}$	Fisher statistic
$f(q)$	Filter function for IMC/QDMC analysis
G	Identified process model
G_0	True process input–output transfer function
H	Identified disturbance transfer function model
H_0	True disturbance transfer function
$K(q)$	Feedback controller
$L(q)$	Prefilter
l_a	Additive uncertainty norm bound
n	Number of shift registers, PRBS sequence; Number of model parameters
n_k	Model deadtime
N	PRBS signal length; Number of observations in an identification
p	Number of inputs in a multi-input PRBS signal
$\tilde{P}$	Matrix of loadings in partial least-squares estimation
p_i	ith column of $\tilde{P}$
q^{-1}	Backshift operator $x(t-k) = q^{-k}x(t)$
r	Reference setpoint
$\mathrm{Re}(\cdot)$	Real part of a complex number
$S(\omega)$	Power spectrum of PRBS signal
t	time

T	Period of repetition of PRBS signal
t_i	ith column of $\tilde{T}$
T_{cl}	Clock period of PRBS signal
T_s	Sampling period
$T_{99\%,\max}$	Largest 99% settling time among input–output pairs of a multivariable system
$\tilde{T}$	Matrix of scores in partial least-squares estimation
v_k	Estimated impulse response coefficients
$w(k)$	Lag window for spectral analysis
x_B	Composition measurement on bottoms stream
x_D	Composition measurement on distillate stream
$\tilde{X}_B$	Transformed composition measurement on bottoms stream
$\tilde{X}_D$	Transformed composition measurement on distillate stream
y	Output of true system
$\hat{y}$	One-step-ahead prediction
y_F	Prefiltered output time series

References

Alsop, A. and Edgar, T. F. (1987). Nonlinear control of a high purity distillation column by the use of partially linearized control variables. AIChE Meeting, Paper 10b.

Åström, K. J. and Hägglund, T. (1984). Automatic tuning of simple regulators with specification on phase amplitude margins. *Automatica* 20, 645–651.

Åström, K. J. and Wittenmark, B. (1984). *Computer-Controlled Systems: Theory and Design.* Englewood Cliffs, NJ: Prentice-Hall.

Åström, K. J. and Wittenmark, B. (1989). *Adaptive Control*. Reading, MA: Addison-Wesley.

Balhoff, R. and Lau, H. K. (1985). A transfer function form of dynamic matrix control and its relationship with some classical controllers. American Control Conference, Boston.

Box, G. E. P. and Jenkins, G. (1976). *Time Series Analysis*. San Francisco: Holden-Day.

Box, G. E. P. and MacGregor, J. F. (1974). The analysis of closed-loop dynamic stochastic systems. *Technometrics* 16, 391.

Briggs, P. A. N. and Godfrey, K. R. (1966). Pseudorandom signals for the dynamic analysis of multivariable systems. *Proc. IEE* 113, 1259–1267.

Davies, W. D. T. (1970). *System Identification for Self-Adaptive Control.* London: Wiley Interscience.

Draper, N. R. and Smith, H. (1966). *Applied Regression Analysis*. New York: Wiley.

Eduljee, H. E. (1975). Equations replace Gilliland plot. *Hydroc. Process.* 54, 120.

Eskinat, E., Johnson, S. H., and Luyben, W. L. (1991). The use of auxiliary information in system identification. Submitted to Ind. Eng. Chem. Research.

Eykhoff, P. (1974). *System Identification: Parameter and State Estimation*. New York: Wiley.

Geladi, P. (1988). Notes on the history and nature of partial least-squares modeling. *J. Chemometrics* 2, 231–246.

Geladi, P. and Kowalski, B. R. (1986). Partial least-squares regression: A tutorial. *Anal. Chim. Acta* 185, 19–32.

Georgiou, A., Georgakis, C., and Luyben, W. L. (1988). Nonlinear dynamic matrix control for high-purity distillation columns. *AIChE J.* 34, 1287–1298.

Gevers, M. and Wertz, V. (1987). Techniques for the selection of identifiable parametrizations for multivariable linear systems. *Control and Dynamic Systems* 26, 35–86.

Gustavsson, I., Ljung, L., and Söderström, T. (1977). Identification of processes in closed-loop: Identifiability and accuracy aspects. *Automatica* 13, 59.

Hoerl, A. E. and Kennard, R. W. (1970). Ridge regression: Biased estimators for nonorthogonal problems. *Technometrics* 12, 55–67.

Hoskuldsson, A. (1988). PLS regression methods. *J. Chemometrics* 2, 211–228.

Jenkins, G. and Watts, D. (1969). *Spectral Analysis and its Applications*. San Francisco: Holden-Day.

Kosut, R. L. (1987). Adaptive uncertainty modeling: On-line robust control design. 1987 American Control Conference, Minneapolis.

Koung, C. W. and Harris, T. J. (1987). Analysis and control of high-purity distillation columns using nonlinearity transformed composition measurements. Canadian Engng. Centennial Conf., Montreal.

Koung, C. W. and MacGregor, J. F. (1991). Design of experiments for robust identification of MIMO systems. 1991 AIChE Annual Meeting, Los Angeles, CA.

Kresta, J. V., MacGregor, J. F., and Marlin, T. E. (1991). Multivariate statistical monitoring of process operating performance. *Can. J. Chem. Eng.* 69, 35–47.

Lorber, A., Wangen, L. E., and Kowalski, B. (1987). A theoretical foundation for PLS algorithm. *J. Chemometrics* 1, 19.

Ljung, L. (1987). *System Identification: Theory for the User*. Englewood Cliffs, NJ: Prentice-Hall.

Ljung, L. (1988). *System Identification Toolbox, MATLAB*. South Natick, MA: The MathWorks, Inc.

Luyben, W. L. (1990). *Process Modeling, Simulation and Control for Chemical Engineers*. 2nd Ed. New York: McGraw-Hill.

Luyben, W. L. (1987). Derivation of transfer functions of highly nonlinear distillation columns. *Ind. Eng. Chem. Res.* 26, 2490–2495.

MacGregor, J. F., Harris, T. J., and Wright, J. D. (1984). Duality between the control of processes subject to randomly occurring deterministic disturbances and ARIMA disturbances. *Technometrics* 26, 389.

MATLAB (1989). South Natick, MA: The MathWorks, Inc.

Morari, M. and Zafiriou, E. (1989). *Robust Process Control*. Englewood Cliffs, NJ: Prentice-Hall.

Prett, D. M. and Garcia, C. E. (1988). *Fundamental Process Control*. Boston: Butterworths.

Ricker, N. L. (1988). The use of biased least-squares estimators for parameters in discrete-time pulse response models. *Ind. Eng. Chem. Res.* 27, 343.

Rivera, D. E. (1991a). Control-relevant parameter estimation: A systematic procedure for prefilter design. 1991 American Control Conference, Boston.

Rivera, D. E. (1991b). A re-examination of modeling requirements for process control. Unpublished.

Rivera, D. E., Pollard, J. F., and Garcia, C. E. (1990). Control-relevant parameter estimation via prediction-error methods: implications for digital PID and QDMC control. Paper 4a, AIChE Annual Meeting, Chicago.

Rivera, D. E., Pollard, J. F., Sterman, L. E., and Garcia, C. E. (1990). An industrial perspective on control-relevant identification. 1990 American Control Conference, San Diego.

Rivera, D. E., Webb, C., and Morari, M. (1987). A control-relevant identification methodology. Paper 82b, AIChE Annual Meeting, New York.

Shinskey, F. G. (1988). *Process Control Systems*, 3rd ed. New York: McGraw-Hill.

Smith, R. S. and Doyle, J. C. (1989). Model validation: A connection between robust control and identification. American Control Conference, Pittsburgh.

Söderström, T., Ljung, L., and Gustavsson, I. (1976). Identifiability conditions for linear multivariable systems operating under feedback. *IEEE Trans. Autom. Control* AC-21, 837.

Webb, C., Budman, H., and Morari, M. (1989). Identifying frequency domain uncertainty bounds for robust controller design: Theory with application to a fixed-bed reactor. American Control Conference, Pittsburgh.

Wei, L., Eskinat, E., and Luyben, W. L. (1991). An improved autotune identification method. Ind. Eng. Chem. Res. 30, 1530–1541.

8

Selection of Controlled and Manipulated Variables

Charles F. Moore
University of Tennessee

8-1 INTRODUCTION

Distillation control philosophy and practice vary greatly. Some control systems are designed using simple PID controllers whereas others are designed around sophisticated model based control algorithms. Some strategies are implemented in state-of-the-art distributed control systems (DCS), whereas others use a collection of more conventional electronic and pneumatic control devices. Distillation control strategies vary in cost and complexity, yet the performance of all strategies, large and small, are at the mercy of the sensor and valve systems to which they are connected. Poor valve choices and/or confused sensor information will render ineffective even the best of control systems.

The purpose of this chapter is to present key sensor and valve issues to be considered by the control engineer in the design and/or analysis of a column control strategy. Methodologies and procedures will be presented that provide systematic ways to address these important but seldom discussed issues.

8-2 SENSOR AND VALVE ISSUES

There are many valve and sensor issues to be considered in the design of a control strategy (see Figure 8-1). Some issues are general and some are specific. All are important to the overall performance of a column control strategy.

Distillation control systems characteristically include two levels of focus: A first-level focus, concerned with stabilizing the basic operation of the column, and a second-level focus, addressing the separation taking place in the column. The sensor and valve issues that should be considered by the control engineer are different at each level.

8-2-1 Inventory Control Concerns

The first level is the more mundane aspect of any distillation control system. It typically includes flow controls on the material and energy streams, inventory controls on the reboiler and accumulator, and pressure controls at the accumulator and/or base of the column. The sensor issues at this level are minimal. Measuring flows, levels, and pressures are straightforward and have a local focus. Valves are a much larger concern. Specifically, the control engineer must decide which valves will be used to control the column inventories. These decisions are critical to the overall performance of the column control system in two ways: They define what valves are available to adjust (*a*) Summary of Column Control Sensor Issues

Sensor Location: Distillation columns are distributed processes. Temperature, pressure, and composition vary throughout the column. Choosing the proper location of control sensors is critical to the performance of the column control system. The locations that best facilitate the control system are highly specific to the column and to the separation it is designed to maintain.

Analyzer Type: Information that reflects the state of the separation is necessary for all column control strategies. Compositions are sometimes measured directly using process analyzers but they are most frequently measured indirectly using temperature. The effectiveness of temperature-based sensors in controlling composition varies greatly from column to column. It is important for the control system designer to realize that controlling temperature does not always solve the underlying composition control problem.

Analyzer Focus: The degrees of freedom of most multicomponent distillation columns suggest that it is not possible to control the composition of all the components. The choice of analyzer locations and which component composition to determine at each is important to the success of the column control strategy.

Sensor Sensitivity: The zero and span of the sensor are important to the performance of the control systems. Sensors with too broad a span lack the resolution and sensitivity to meet the control objectives. Sensors with too narrow a span tend to operate off scale and function in the control system more like two position relays.

Sensor Consistency: One of the main purposes of a column control system is to stabilize the operation of the column. If the control sensor is not stable in its reading such local variation can be propagated by the control system to other parts of the column operation. This problem is particularly severe in complex multivariable control strategies in which errors in a single sensor measure can affect the movement of a number of manipulated variables.

Sensor Reliability: Sensors that fail can cause very severe problems in a control system. It is important for the control system designer to know the failure probability of key sensors. It may be critical to design a control structure that degrades gracefully. It may also be important to design fault detection and sensor validation schemes that alert the control system to the occurrence of such failures.

Non-Column Sensors: It should be realized that some of the most useful information in controlling the column may come from sensors not located in the column. In designing a column control system it is important to take a more complete plantwide perspective. The problems with the operation and control of the column may best be corrected by better control elsewhere in the plant. Care should also be given that the local column control strategy does not inadvertently cause serious problems to other unit operations downstream.

(*b*) Summary of Column Control Valve Issues

Range: The valve range is the maximum flow attainable with the valve fully opened. It represents a limitation on how large an adjustment a control system can make. Such physical limitations need to be considered in both the column design and in the design of its control systems.

Resolution: The valve resolution represents a physical limitation on how small an adjustment a control system can make. Typically, valve resolutions are 2% or greater. If more subtle adjustments are necessary the control system will struggle to find a steady state.

Linearity: Typically, installed valves are not linear with respect to the relationship between flow and signal from controller. Highly nonlinear valves can cause problems in control loops. The effect of the nonlinearity is especially difficult in control loops that have slow dynamics.

Size: The size of a control valve is set by its CV. Typically, control valves are sized too large for the flow service. In such cases the problems of linearity and valve resolution are greatly exaggerated.

Dynamics: The dynamics of the distillation column is an important issue. The nature and speed of response is different for each valve and is a function of the role of the valve in the column control strategy.

Role: The local role of each control valve is a major issue. Is it to be set by the inventory control system? Is it to be set by the separation control system? Is it to be set manually by the control room operator?

Plantwide Implications: A distillation column is typically only one of many unit operations that make up a processing plant. Manipulation of column control valves affects more than the local operation of the column. The plantwide implications of changes in each control valve need to be considered thoroughly. Such implications should strongly influence the design of the local control system.

FIGURE 8-1. Summary of sensor and valve issues.

the separation in the column and they help establish how the column interacts with other unit operations on a plantwide basis.

8-2-2 Separation Control Concerns

The second level of the column control addresses controlling the separation taking place in the column. At this level the main valve issue is deciding the role of the remaining valves (i.e., valves not used in the first level strategy). The control engineer must decide which of the valves available for separation control are to be manipulated by the control system and which are to be set manually by the control room operator.

Perhaps the most confusing aspect of the second-level control concerns the sensors used to monitor the state of the separation taking place in the column. There are three fundamental sensor questions to be addressed early in the design of the separation control strategy:

What kind of composition sensor?
How many measurements?
Where in the column should they be located?

These sensor questions are highly specific to the column and to the separation taking place in the column. They must be answered before the control strategy can be properly designed.

8-2-3 Loop Sensitivity Issues

Loop sensitivity issues are important considerations in selecting valves and locating sensors. Although such issues are fundamental to both levels of column control, they are frequently overlooked (or ignored) in the design of control strategies. Failing to properly understand the importance of sensitivity issues can lead to serious operational and control problems.

Loop sensitivity is a measure of how the control sensor responds at steady state to changes in the manipulated variable. Mathematically, the loop sensitivity can be defined as the partial derivative of the sensor signal with respect to changes in the signal to the valve:

$$\text{loop sensitivity} = \frac{\partial S_i}{\partial M_j}$$

where S_i = signal from control sensor i, expressed as a percentage of the maximum signal
M_j = signal to control valve j, expressed as a percentage of the maximum signal

Both valve and sensor have a range beyond which they cannot operate. The units of the loop sensitivity should be chosen to clearly reflect these physical limitations. Such limitations are included if the loop sensitivity is expressed in percentage units indicated in the preceding text. The numerator is the change in the sensor reading expressed in percent of sensor span. The denominator is the change in manipulated variable expressed in percent of the range of the signal to the valve.

Measured in such percentage units, the ideal loop sensitivity is 1% per percent. In such cases, if the control system is well balanced a 1% change in the signal to the control valve will result in a 1% change in the sensor. If the sensitivity is either significantly larger or smaller than 1, control problems are likely to develop. These problems can seriously diminish the effectiveness of a feedback control loop and should be understood and considered in the selection and location of all column control sensors.

High Sensitivity

High loop sensitivity can be a problem in distillation control. Control systems highly sensitive to changes in the manipulated variables tend to overcompensate. The control system is inclined to "hunt" (without much success) for a valve position to satisfy the setpoint. In such cases, steady-state operation is seldom achieved.

There are two aspects of the high sensitivity problem: one concerns the sensor and the other concerns the actuator. The sensor problem occurs when the control sensor has a narrow span compared to the range over which the process variable can be manipulated. In such cases, even small changes in the manipulated variable will cause the sensor to saturate. During disturbances, most of the time the sensor will present to the control system either the maximum or the minimum value of its output. In effect, it will function in the control loop much like a two position relay.

The problems of high sensitivity also can be associated with the manipulated variable. Valves and other actuators all have a minimum resolution with respect to positioning. These limitations restrict the fine adjustments often necessary for high gain processes to reach a steady operation. If the fine adjustment necessary for steady state is less than the resolution of the valve, sustained oscillation will likely occur. Consider, for example, a steam valve with a resolution of $\pm 1.0\%$. If a valve position of 53.45% is necessary to meet the target temperature, then the valve will, at best, settle to a limit cycle that hunts over a range from about 55 to 53%. If the process gain is 10, the hunting of the valve will cause a limit cycle in the control temperature of 20%.

There are many considerations that should be balanced in trying to improve a poor loop sensitivity. The problem can be addressed to some extent by resizing the sensors and/or valves. For example, a common problem in distillation control is oversized control valves. The actuation characteristics of a typical oversized valve are shown in Figure 8-2. Note that 90% of the maximum flow is achieved when the signal to the valve is only 10% of its maximum. The result is high sensitivity through most of the flow range of the valve. Also, the effective valve resolution is greatly reduced. If the flow for a linear valve can be positioned $\pm 1.0\%$ of the maximum, the oversized valve can be positioned to only $\pm 10.0\%$ of the same maximum. The solution to the problem of an oversized valve can be easily found by changing the valve trim to a more appropriate size.

The problem of high sensitivity may well be the result of the separation. High purity separations typically have temperature profiles that are flat at both ends but break sharply in the middle of the column. The sharp break indicates where the primary separation of the components takes place; the flat portions represent where the products are being purified. The sensitivity is very low except on the few trays where the separation is taking place. In such cases the problem can be reduced by using multiple sensors. Temperature sensors, well placed throughout the sensitive section of the column, can be averaged to provide a net feedback signal that has lower gain and better linearity.

Low Sensitivity

Low loop sensitivity is a more common problem in distillation control. In such cases, even large changes in the signal to the control valve will have little effect on the signal from the control sensor. For example, a loop sensitivity of 0.05% per percent indicates a change in the valve position of 10% will result in only a 0.5% change in the sensor. The most obvious problem caused by such low loop sensitivity is leverage. The control system is severely limited with respect to the magnitude of errors that can be corrected by the valve. The magnitude of change necessary to correct for an error may well exceed the capability of the control valve. When valves saturate (fully open or fully closed), the control objective cannot be met until the error returns to a more manageable magnitude. For processes with low sensitivities, the range of manageable errors may be very small.

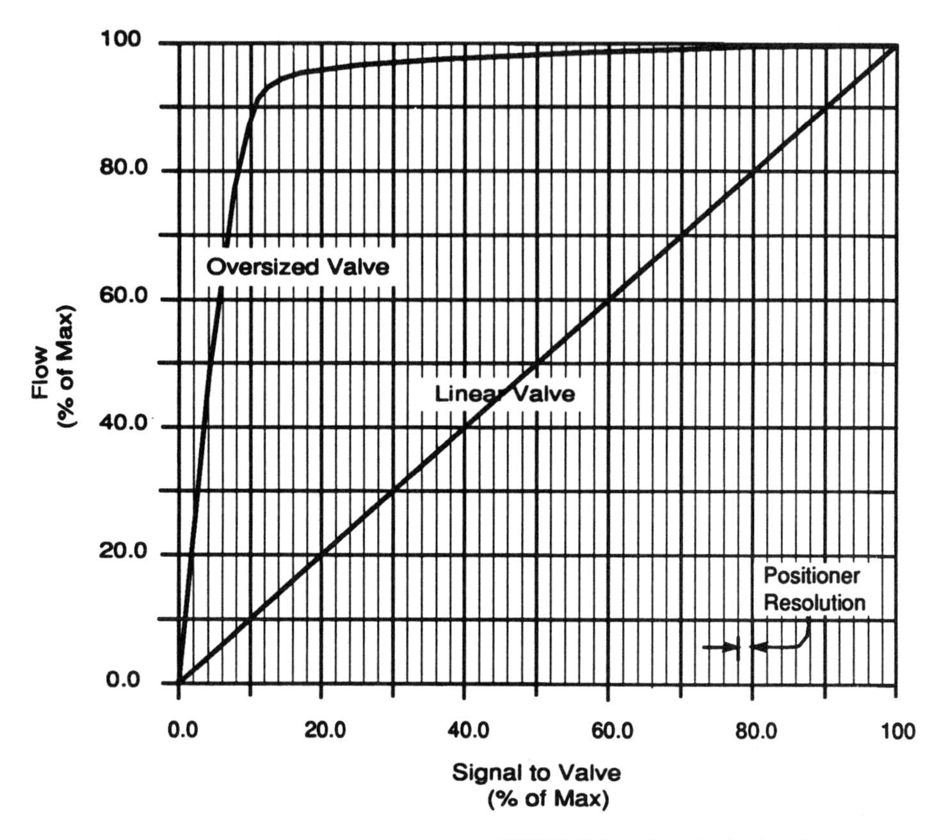

FIGURE 8-2. Oversized valve characteristics.

The second problem caused by low process sensitivity concerns the effect sensor noise has on loop performance. In order to compensate for the low process sensitivity, the feedback controller gains are typically set high. Such high controller gains will cause problems if the control system is connected to a sensor containing noise in its signal. (The only sensors the author can think of that do not contain noise are the ones used in typical academic simulations.) A high gain controller amplifies and reflects sensor noise to its manipulated variables. Consider, for example, a process sensitivity of 0.01. For such a process, a reasonable value of controller gain might be 100.0 (this value provides a closed loop gain of 1.0). In such cases, sensor noise as small as $\pm 0.5\%$ will cause the valve to swing wildly between its saturation limits. From an operations perspective, such valve action is universally unacceptable.

It should be noted that the problem caused by noise for processes with low sensitivity is even more pronounced in control systems using multivariable control algorithms. In such systems, the noise is amplified and reflected to more than one manipulated variable. Problems that were only of local concern in a conventional control system can be inadvertently propagated throughout the entire system by the multivariable algorithm.

The problem caused by the noise can be reduced by either lowering the controller gain or adding filters. Lowering the controller gain reduces the sensitivity of the control loop to noise; but, it also reduces the sensitivity to those errors the control system was intended to correct. Adding a filter between the sensor and the controller can also make the control system less sensitive to noise. Filters can be tuned for the noise; nevertheless they also tend to reduce the sensitivity of the control system at all frequencies.

The biggest leverage in terms of improving low sensitivities is through sensor selection. The type of sensors used and the location of those sensors are critical to the sensitivity problem. The practice of selecting location and type using rule-of-thumb and historical precedent often leads to problems of low sensitivity. The next sections discuss analytical techniques that can be used to effectively make these critical sensor selections.

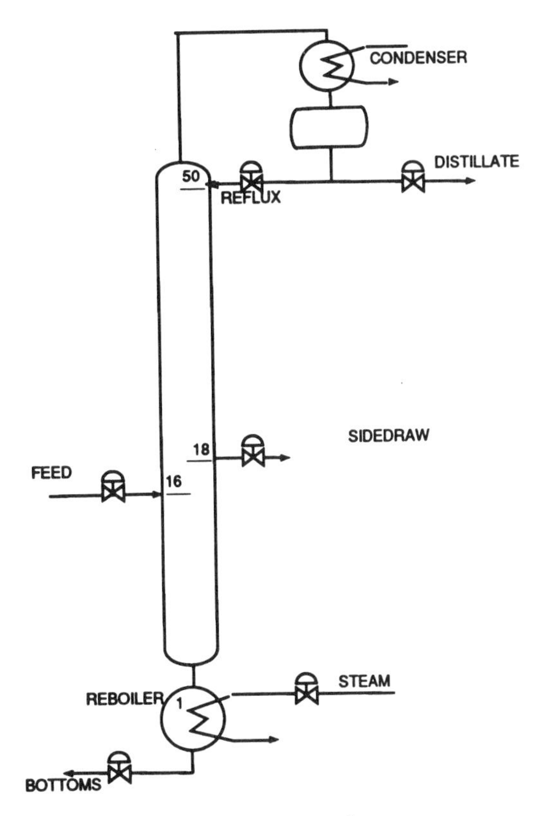

FIGURE 8-3. Ethanol—water column.

8-3 LOCATION OF TEMPERATURE SENSORS

One of the most important issues in column control is to establish an appropriate location for the sensors used to control the column separation. The question of sensor location cannot be completely separated from the question of what kind of sensor. For the sake of clarity, let us first consider only temperature sensors. Temperature is a fast, inexpensive sensor. It provides, with varying degrees of accuracy, an indirect measurement of composition. The procedures developed in this section for determining the appropriate locations for temperature sensors will be extended in later sections for more direct measurements of composition.

The distributed nature of distillation provides many possible choices for locating temperature sensors. Consider the ethanol–water column shown in Figure 8-3. In this example, there are 50 stages in the column. Each provides a possible location for a unique measurement to be made. The objective of the control system designer is to make sensor choices that result in a well behaved column. The fundamental idea is to control one or more well selected temperatures and to reduce the effect the disturbances have on the purity of the various product streams.

8-3-1 Determining Temperature Sensitivities

The first step in establishing the location of the temperature sensor(s) is to determine the sensitivity of the column with respect to temperature. This should be done for each tray in the column and with respect to each variable potentially manipulated by the control system. (It should be noted that the first-level inventory control scheme must be established before the potential manipulated variables for the separation control system can be identified.) The sensitivity information can be conveniently summarized in matrix form as

$$K_{n \times m} = \begin{bmatrix} \left.\dfrac{\partial T_1}{\partial M_1}\right| & \cdots & \left.\dfrac{\partial T_1}{\partial M_j}\right| & \cdots & \left.\dfrac{\partial T_1}{\partial M_m}\right| \\ \vdots & & \vdots & & \vdots \\ \left.\dfrac{\partial T_i}{\partial M_1}\right| & \cdots & \left.\dfrac{\partial T_i}{\partial M_j}\right| & \cdots & \left.\dfrac{\partial T_i}{\partial M_m}\right| \\ \vdots & & \vdots & & \vdots \\ \left.\dfrac{\partial T_n}{\partial M_1}\right| & \cdots & \left.\dfrac{\partial T_n}{\partial M_j}\right| & \cdots & \left.\dfrac{\partial T_n}{\partial M_m}\right| \end{bmatrix}$$

where T_i = temperature on stage i
M_j = manipulated variable j
$K_{n \times m}$ = steady-state temperature gain matrix for an n stage column with m manipulated variables

Such sensitivity information can be determined from perturbation studies using any distillation simulation based on detailed stage-to-stage calculations. It is important, however, to use extended precision and only very small perturbations to insure the resulting changes in the temperature profile are not too large to approximate the local derivatives.

The numerical values of the full 50 × 2 gain matrix for the ethanol–water column are listed in Figure 8-4. These numbers can be used directly in establishing a sensitive sensor location. The graphical presentation in Figure 8-5 provides a clearer picture of the procedure. In Figure 8-5 the elements of each vector of the gain matrix are plotted. The top line indicates the sensitivity of each stage to changes in the distillate rate whereas the bottom line indicates the corresponding sensitivities for changes in steam flow. The analysis clearly points to the trays around tray 19 as the strongest place to locate the control sensors.

$K_{\text{temperature}}$ =

$\left.\dfrac{\delta T_i}{\delta D}\right\|_Q$	$\left.\dfrac{\delta T_i}{\delta Q}\right\|_D$	Tray Number
0.0014726	−0.000128	50 Top
0.0018209	−0.000159	49
0.0022324	−0.000194	48
0.0027183	−0.000233	47
0.0032927	−0.000277	46
0.0039730	−0.000327	45
0.0047806	−0.000383	44
0.0057430	−0.000446	43
0.0068946	−0.000517	42
0.0082802	−0.000597	41
0.0099576	−0.000687	40
0.0120030	−0.000790	39
0.0145190	−0.000907	38
0.0176430	−0.001041	37
0.0215650	−0.001196	36
0.0265520	−0.001376	35
0.0329830	−0.001587	34
0.0414110	−0.001838	33
0.0526460	−0.002139	32
0.0679110	−0.002506	31
0.0890670	−0.002960	30
0.1190000	−0.003535	29
0.1622500	−0.004275	28
0.2260400	−0.005251	27
0.3220100	−0.006569	26
0.4691400	−0.008394	25
0.6988000	−0.010988	24
1.0635000	−0.014779	23
1.6535000	−0.020511	22
2.6323000	−0.029640	21
4.3028000	−0.045441	20
6.6139000	−0.068335	19
6.0330000	−0.053858	18
2.4169000	0.000620	17
0.3651600	0.023556	16
0.3494200	0.038175	15
0.3160100	0.046997	14
0.2708000	0.049842	13
0.2210700	0.047769	12
0.1731700	0.042482	11
0.1311600	0.035678	10
0.0967280	0.028655	9
0.0698550	0.022208	8
0.0496110	0.016712	7
0.0347400	0.012260	6
0.0240050	0.008786	5
0.0163460	0.006158	4
0.0109200	0.004219	3
0.0070880	0.002821	2
0.0028392	0.001246	1 Bottom

FIGURE 8-4. Temperature gain matrix for the ethanol–water column.

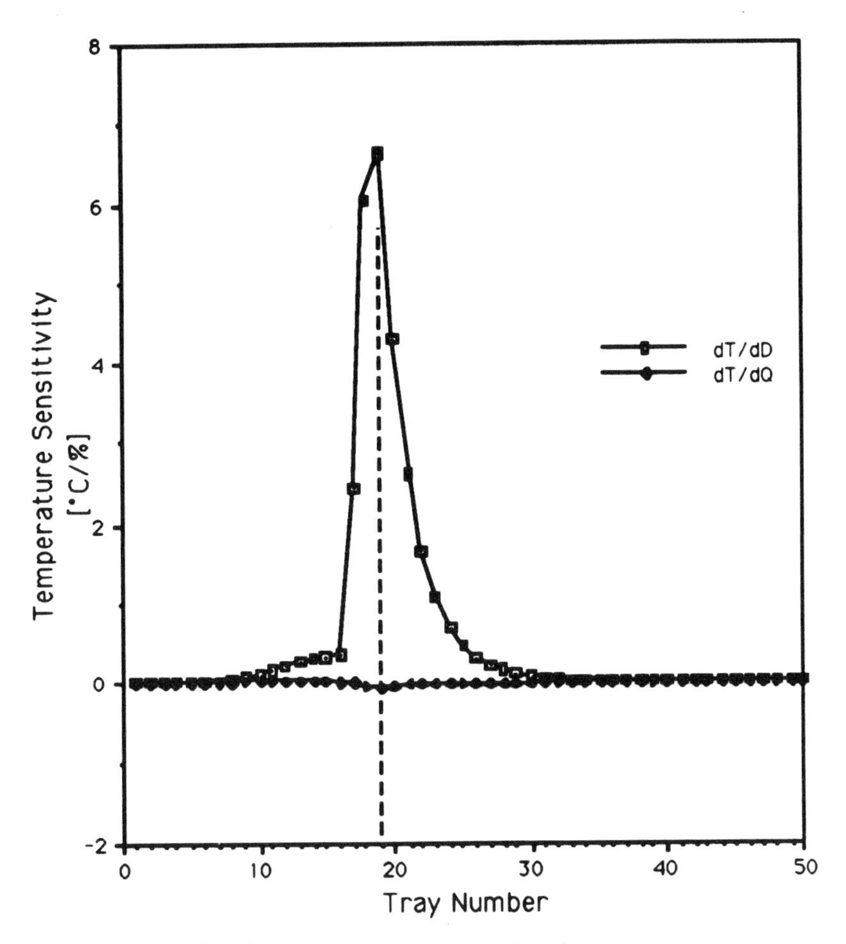

FIGURE 8-5. Ethanol–water temperature gain matrix plot.

8-3-2 Selecting a Temperature Sensor for Single-Ended Control

Before the sensor question can be completely answered, the dimension of the control system must be addressed. Will the control system manipulate one or both of the manipulated variables? Will the control system attempt to control one or more temperatures? On a practical level, there are many difficulties with control systems attempting more than one control objective. The implications of a more ambitious control strategy on sensor location will be discussed in Section 8-3-3. For now, let us assume only one temperature sensor and only one manipulated variable will be used by the control system.

For a single temperature control system manipulating only one variable, the gain matrix provides insight into two important control questions: sensor location and selection of an appropriate manipulated variable. In this example, tray 19 is clearly the most sensitive sensor location. Note also the distillate flow is by far the strongest manipulated variable. A change in distillate flow affects the column about 10 times more than a similar change in steam flow. (Recall the unit of change for both are expressed in terms of percentage of the maximum flow.) There may be other considerations in selecting a valve–sensor pairing (i.e., plantwide concerns, nature and source of the disturbance, complexity of the control system, process dynamics, etc.). On the basis of this steady-state analysis, a strategy that uses distillate flow to regulate the temperature on tray 19 would be expected to yield the strongest loop.

This example considers only the control configuration where the distillate and steam flows are available as manipulated variables for the separation control strategy. In general, other control configurations should be studied as well. Common control configurations and a rationale for evaluating various alternatives are discussed in Section 8-3.

In practice, temperature sensors are often located at or near the top or the bottom of a column. The rationale for such practice is to place the temperature control sensor close to the product. Figure 8-4 demonstrates the problem using such practice. In this example, as in most columns, the temperature sensitivities at the column extremes are very weak. The performance of any control system (simple or complex) attempting control based on such insensitive temperature information will be poor at best.

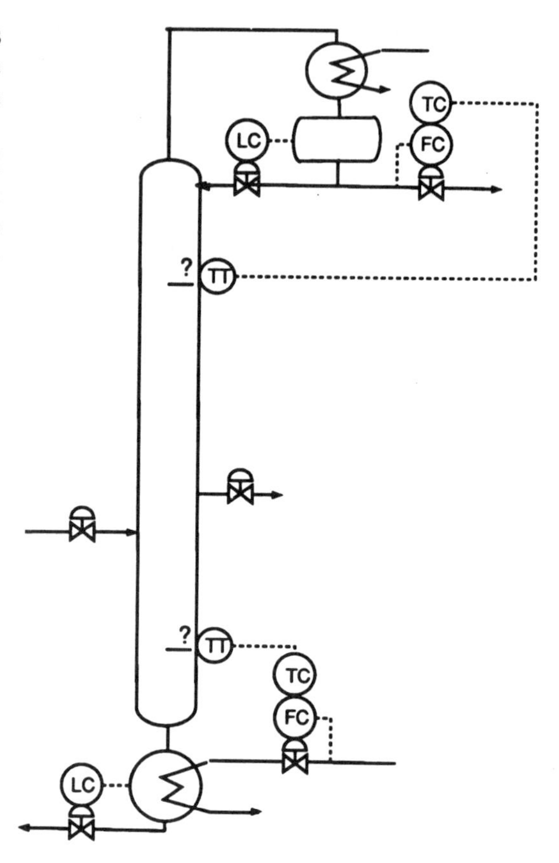

FIGURE 8-6. A dual-ended control for the ethanol–water column.

8-3-3 Selecting Temperature Location for Dual-Ended Control

Distillation columns have at least two product streams: a distillate product and bottom product. Some columns may also have one or more side-draw streams (i.e., product or purge streams). (In the ethanol–water example the purpose of the side draw is to purge the buildup of fusel oil intermediates.) For the sake of clarity, the following discussion considers only conventional two product columns. The ideas and analysis procedures presented can be extended to more complex separations.

Designing a control system to regulate both ends of a column is appealing. Potentially, such schemes provide tighter control over both product qualities as well as reduce significantly the amount of energy used in the column. The problem is to design a scheme that actually works. Selecting the proper sensor locations is a major part of the design often overlooked.

The fundamental sensor problem affecting performance in single-ended control is loop sensitivity. For dual-ended strategies, an additional problem concerns the interaction between the control sensors. Both poor loop sensitivity and sensor interaction can limit severely the performance of the best column control strategies. Both problems need to be considered in the selection of sensor locations for use in a dual-ended strategy.

Consider selecting temperature sensors for the dual-ended control system shown in Figure 8-6. The sensitivity plots (shown in Figure 8-5) indicate that trays 18 and 19 are the two most sensitive locations. However, these locations would not be good choices. Temperatures on adjacent (or nearly adjacent) trays will be highly correlated and therefore inadequate choices of control sensors. With respect to the problem of interaction, a better sensor location is near the column extremes. For example, locating sensors on tray 1 and tray 50 would yield much less interaction. The problem with this choice is sensitivity. As can be seen in Figure 8-5, the sensitivities of all trays near the top and the bottom of the column are extremely low.

Selecting sensor locations for a multivariable control strategy requires balancing the problem of sensor interaction with the problem of loop sensitivity. The gain matrix can provide insight into both problems; however, further analysis is required. Singular-value decomposition (SVD) provides a basis for further analysis. The SVD analysis and an approach to select multiple sensor locations are described in the following sections.

SVD Analysis of the Gain Matrix

Singular-value decomposition (SVD) is a numerical algorithm useful in analyzing the multivariable aspect of the gain matrix. It decomposes a matrix into three unique component matrices as

$$K = U \Sigma V^T$$

where K = an $n \times m$ matrix

U = an $n \times n$ orthonormal matrix, the columns of which are called the left singular vectors

V = an $m \times m$ orthonormal matrix, the columns of which are called the right singular vectors

Σ = an $n \times m$ diagonal matrix of scalars called the singular values and organized in descending order such that $\sigma_1 \geq \sigma_2 \geq \cdots \geq \sigma_m \geq 0$

SVD and a closely related statistical algorithm called principal component analysis are discussed in many text books (Forsythe and Moler, 1967; Golub and Reinsch, 1970; Lawson and Hanson, 1978), and software is readily available through a number of standard numerical packages (for example, MATLAB, 1985).

Singular-value decomposition was developed to minimize computational errors involving large matrix operations. When applied to a steady-state gain matrix the component matrices provide insight into the multivariable control problem. To understand the use of SVD in addressing the sensor placement problem, it is helpful to consider the physical interpretation of these component matrices presented in Figure 8-7.

Sensor Placement Based on U-Vector Plots

The U and K matrices are both measures of sensor sensitivity. The geometric properties of the U matrix provides subtle insights. They are not obvious in the gain matrix alone. The U vectors present tray sensitivities in a orthogonal coordinate system. These are useful in considerations of both sensor interaction and sensor sensitivity. The analysis and interpretation of the U vector is as follows:

U_1 indicates the most sensitive combination of tray temperatures in the column. The principal component of the U_1 vector is the most sensitive location for a temperature sensor in the column.

U_2 indicates the next most sensitive combination of tray temperatures, orthogonal to the tray temperatures represented in U_1. The principal component of the U_2 vector represents location sensitivity but exhibits the least possible interactions with the primary sensor.

It should be noted that although the U vectors are noninteracting (by definition), the locations indicated by the principal components of each of the U vectors may exhibit considerable interaction. Unfortunately, it is seldom possible to select two

$$K = U\Sigma V^T$$

$K = \delta T_i/\delta M_1 : \delta T_i/\delta M_2 : \cdots : \delta T_i/\delta M_m$

The gain matrix is composed of a set of physically scaled vectors that describe the steady-state temperature sensitivity on each of the n trays to changes in each of the m manipulated variables.

$U = U_1 : U_2 : \cdots : U_n$

The left singular vectors are the most important component matrix. It provides a sensor coordinate system for viewing the sensitivity of the column to changes in the manipulated variables. This coordinate system is such that U_1 indicates the vector direction that is most responsive to changes in the manipulated variables. U_2 indicates the next most responsive vector direction. The third vector is the next most responsive, and so on. The properties of U provide that each vector direction is orthogonal to the other vector directions.

$V = V_1 : V_2 : \cdots : V_m$

The right singular vectors are not directly used in the analysis of the sensor placement problem; however, its interpretation is helpful in completing the physical picture provided by SVD. The right singular vector provides a manipulated variable coordinate system for viewing the sensitivity of the column. This coordinate system is such that V_1 indicates the combination of control actions that most affect the column, V_2 indicates the next most effective combination of control actions, etc. The properties of V are such that the vector directions are orthogonal.

$\Sigma = \text{diag}(\sigma_1, \sigma_2, \ldots, \sigma_m)$

The scalar singular value provides an indication of the multivariable gains in a decoupled process. This information is useful in evaluating the difficulty of a proposed dual-ended control system.

FIGURE 8-7.

sensors totally free of interaction. This procedure does identify two sensitive locations with little interaction relative to other choices.

Consider the results of the singular-value decomposition of the ethanol–water gain matrix as shown in Figure 8-8. The U matrix can be considered directly, but a graphic analysis presents a clearer picture of the sensor selection problem. In Figure 8-9 the elements of the U vectors are plotted versus tray locations. The maximum absolute values of the U vectors indicate the locations of the principal components. These are generally good locations to consider in placing

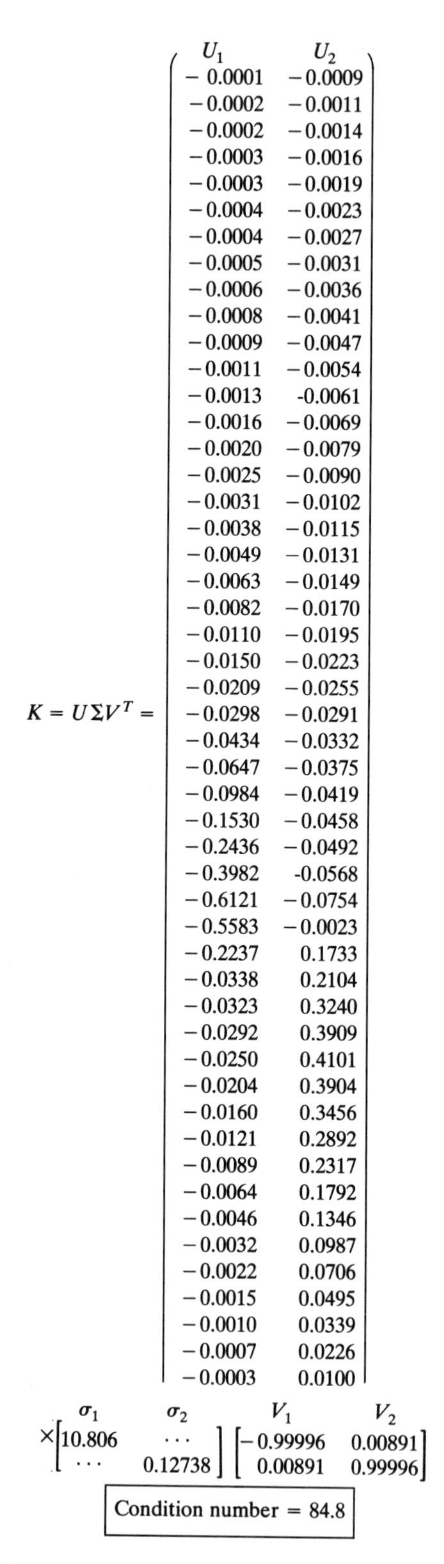

$K = U\Sigma V^T =$

U_1	U_2
−0.0001	−0.0009
−0.0002	−0.0011
−0.0002	−0.0014
−0.0003	−0.0016
−0.0003	−0.0019
−0.0004	−0.0023
−0.0004	−0.0027
−0.0005	−0.0031
−0.0006	−0.0036
−0.0008	−0.0041
−0.0009	−0.0047
−0.0011	−0.0054
−0.0013	-0.0061
−0.0016	−0.0069
−0.0020	−0.0079
−0.0025	−0.0090
−0.0031	−0.0102
−0.0038	−0.0115
−0.0049	−0.0131
−0.0063	−0.0149
−0.0082	−0.0170
−0.0110	−0.0195
−0.0150	−0.0223
−0.0209	−0.0255
−0.0298	−0.0291
−0.0434	−0.0332
−0.0647	−0.0375
−0.0984	−0.0419
−0.1530	−0.0458
−0.2436	−0.0492
−0.3982	-0.0568
−0.6121	−0.0754
−0.5583	−0.0023
−0.2237	0.1733
−0.0338	0.2104
−0.0323	0.3240
−0.0292	0.3909
−0.0250	0.4101
−0.0204	0.3904
−0.0160	0.3456
−0.0121	0.2892
−0.0089	0.2317
−0.0064	0.1792
−0.0046	0.1346
−0.0032	0.0987
−0.0022	0.0706
−0.0015	0.0495
−0.0010	0.0339
−0.0007	0.0226
−0.0003	0.0100

$$\times \begin{bmatrix} 10.806 & \cdots \\ \cdots & 0.12738 \end{bmatrix} \begin{bmatrix} -0.99996 & 0.00891 \\ 0.00891 & 0.99996 \end{bmatrix}$$

(columns labelled σ_1, σ_2 and V_1, V_2)

Condition number = 84.8

FIGURE 8-8. SVD analysis of the ethanol–water temperature gain matrix.

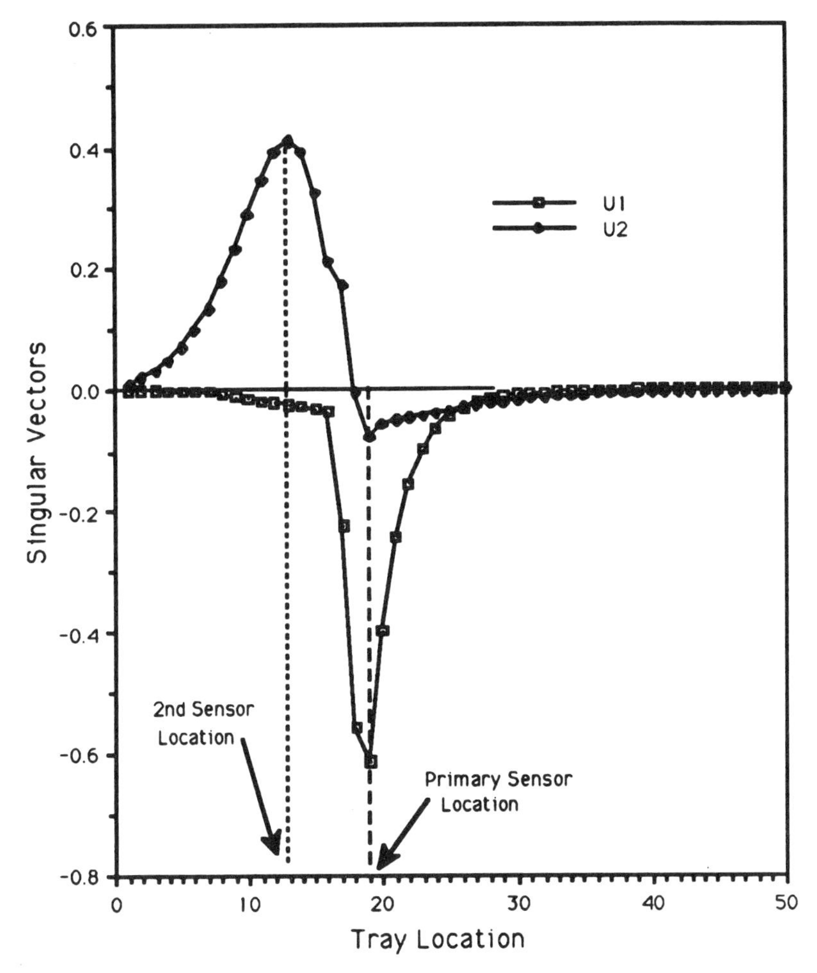

FIGURE 8-9. *U* vector plots for the ethanol–water column.

the control sensors. Likewise, in this example trays 19 and 13 can be chosen for placing temperature sensors.

A modified version of the principal component analysis is shown in Figure 8-10. It consists of only one function defined by the difference between the absolute values of the elements of the U vectors:

$$Z_i = |U_{1i}| - |U_{2i}|$$

The maximum of this function suggests a tray location for the primary sensor and the minimum of this function suggests a tray location for the secondary sensor. This method offers a slight advantage over the principal component method because it considers the individual sensor interactions in more detail.

Note that in this example the results of the two methods are similar except for the location of the primary sensor. In the modified analysis the primary sensor was selected to be at tray 18 to reduce the interaction that exists between the two vector components at tray 19.

Degrees of Freedom and the Condition Number

A fundamental question to be considered in the design of any dual-ended control system (and the placement of its sensors) is the question of degrees of freedom of the column. Does the column have 2 degrees of freedom or only 1? Unfortunately, when applied to the control problem, the answer is seldom clear. Nearly all columns have 2 degrees of freedom, but they may be too narrow for a control system to successfully regulate. Singular-value decomposition provides insight into the pragmatic side of the degrees-of-freedom question through the condition number.

The condition number (CN), in general, is defined as the ratio of the largest singular value to the smallest singular value. For the dual-ended control problem, it is defined as the first singular value divided by the second singular value:

$$\mathrm{CN} = \frac{\sigma_{\text{largest}}}{\sigma_{\text{smallest}}} = \frac{\sigma_1}{\sigma_2}$$

It provides a numerical indication of the balance of sensitivities in the multivariable system. Low condition numbers indicate the multivariable gains are well balanced and should result in a system with sufficient degrees of freedom to meet the dual-ended control objectives. On the other hand, high condition numbers evidence an imbalance in the multivariable gains. Imbalance signifies the column to be much more sensitive in one vector direction than it is in the other. Consequently it may have difficulty meeting the dual-ended control objectives.

At the extreme values of the condition number the results are unambiguous: A condition number of 1 (very unlikely) clearly indicates two full degrees of freedom, whereas a condition number of infinity indicates only 1 degree of freedom. What about the condition numbers between 1 and infinity? Consider, for example, the ethanol–water column. The SVD analysis yielded a condition number of 84.8. This indicates that the column is 84.8 times more respon-

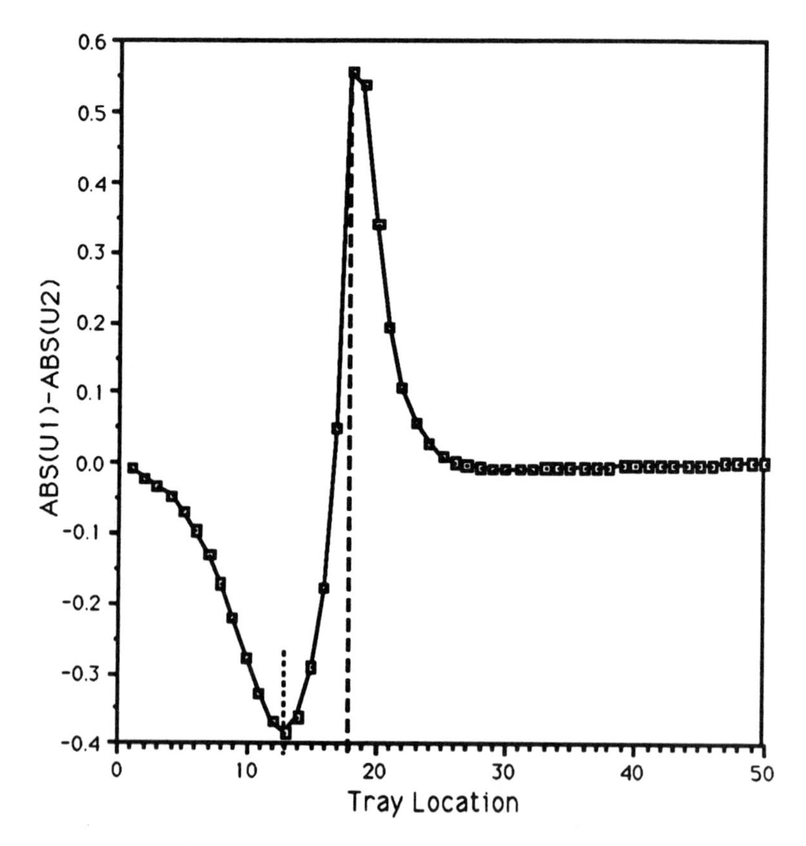

FIGURE 8-10. Modified principal component analysis.

sive in the $U1$ sensor direction than in the $U2$ sensor direction. (Recall, $U1$ and $U2$ are orthogonal vectors indicating the most sensitive sensor directions in the column.) The degree of freedom for the ethanol–water column is 2. It should be qualified as a relatively narrow 2 degrees of freedom.

The practical control problems resulting from narrow degrees of freedom (high condition numbers) are directly related to the problems of high and low sensitivities discussed earlier. Such sensitivity problems are even more pronounced in a multiloop control system. A dual-ended control system will have to overcome these problems in an environment where the control loops interact. Problems in one loop can easily propagate through the entire system.

Overall Versus Control Specific Condition Numbers With respect to the question of multiple sensor placement there are two levels of SVD analysis: an overall analysis and a control specific analysis. The analysis described in the preceding text is applied to the full gain matrix. The primary purpose of this overall analysis is to locate the specific trays (or stages) to locate control sensors. The condition number in this overall analysis is helpful to gage the magnitude of the multivariable control problem; however, it does not completely represent the process a dual-ended control system would actually experience. The next level of the SVD analysis focuses only on the 2×2 matrix specific to the two control temperatures. Considering only trays 18 and 13 in the ethanol–water column, the overall gain matrix is reduced to

$$K_{2\times 2} = \begin{vmatrix} 6.614 & -0.0683 \\ 0.271 & 0.0498 \end{vmatrix}$$

The singular values and condition number for this control specific matrix are presented in Figure 8-11 along with the corresponding values for the overall analysis. Note that for the control specific analysis the values of the singular values are lower and the condition number is higher. Both

	Overall (50 × 2)	Control Specific (2 × 2)
σ_1	10.806	6.61
σ_2	0.1274	0.0526
CN	84.8	125.89

FIGURE 8-11. Overall versus control specific condition numbers for the ethanol–water column.

suggest the dual temperature control problem to be somewhat more difficult than indicated in the initial overall analysis.

Critical Condition Numbers What is the critical value of the control specific condition number? This is a reasonable question but it does not have a specific numerical answer. Large condition numbers indicate only the hidden presence of unbalanced multivariable gains. The practical effect of unbalanced gains is to amplify and to exaggerate the effect of local control problems in a multivariable control system. Control problems resulting from sensor and valves (i.e., resolution, constraints, noise, etc.) as well as problems with process dynamics and multivariable interaction become increasingly pervasive in multivariable control systems for processes with large condition numbers.

Successful dual-ended control of columns with control specific condition numbers as high as 100 have been observed but this has been on columns well behaved with respect to other problems. If the local problems are extreme, the dual-ended control of processes, even with small condition numbers, may not be possible.

Global Methods of Dual Sensor Placement

The difference between the overall and the control specific analysis can also be considered in the placement of sensors. The method of sensor placement outlined previously is based on the SVD analysis of the overall gain matrix. The principal components of the U vectors are used to indicate tray locations sensitive in each of the multivariable dimensions of the column. The results from these analyses lead to a single control specific subsystem to be studied further. The global approaches that will be presented systematically consider all possible control specific subsystems.

There are two problems with a global approach to sensor placement. The first concerns the large number of control specific subsystems. (For the ethanol–water column there are 1200 unique combinations of dual sensor locations.) This can be a large but manageable computational problem. The second problem is the selection of a suitable basis for comparing the subsystems. Three different formulations of the global optimization problem are described in the following text along with a rationale for using SVD to measure the strengths and weaknesses of the various control specific subsystems.

Minimum of the Condition Numbers Consider first the control specific condition number as a basis of comparing sensor pair subsystems. Using the subsystem condition numbers, the sensor placement problem can be formulated as

$$\frac{\min}{i,j} \mathrm{CN}_{i,j} \qquad \left[\text{or } \frac{\max}{i,j} \frac{1}{\mathrm{CN}_{i,j}}\right]$$

where $\mathrm{CN}_{i,j}$ = control specific condition number

i, j = tray location of the control sensors

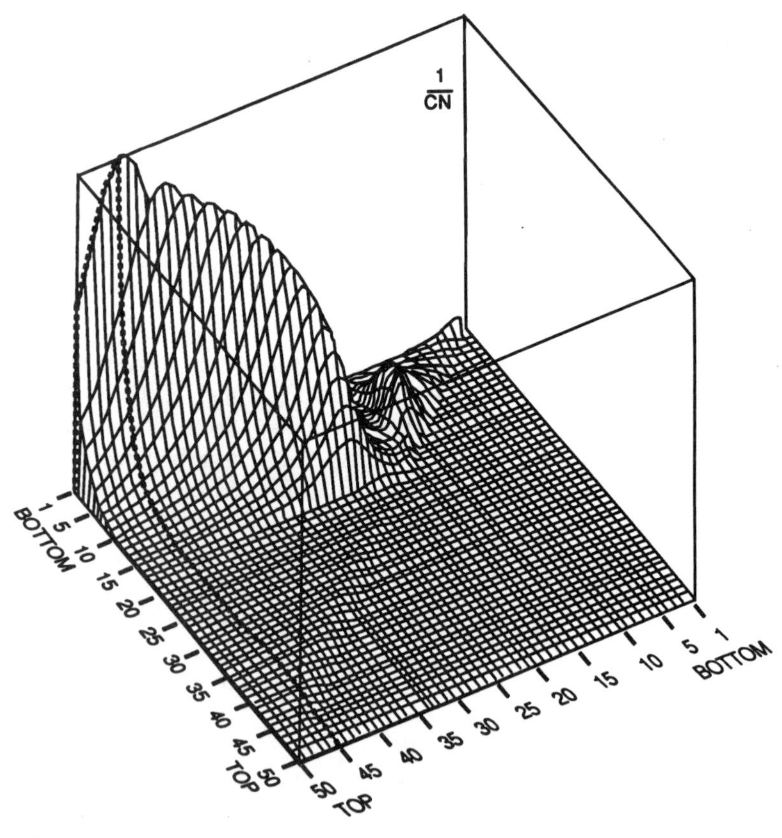

FIGURE 8-12. Control specific condition number versus sensor locations for the ethanol–water column.

For the sake of scale and clarity the surface is plotted as the reciprocal of the condition number. In the reciprocal condition number plots, the global optimum appears as a maximum. The surface defined by the control specific condition numbers for the ethanol–water column is shown in Figure 8-12. Note, that the "best" sensor locations suggested by this optimization strategy are curiously at tray 1 and tray 44. The problem with these two locations is loop sensitivity. The individual sensitivities shown in Figure 8-4 indicate a 10% change in either the distillate or the steam flow affects the two temperatures by less than 0.03° C. This is far too weak a response for either a dual- or single-ended control system to be effective.

In general, the problem with basing sensor locations solely on control specific condition numbers is the neglected issue of sensor sensitivity. Balanced gains are important to the dual-ended control problem, but even a perfectly balanced system that has low sensitivities cannot be controlled.

2. *Maximum of the Smallest Singular Values* Another perspective of the global problem to consider is the multivariable sensitivity of each of the control specific subsystems. The singular values of the subsystem are a measure of the multivariable sensitivity. In terms of placing sensors for maximum sensitivity the focus typically should be on the weakness of the subsystem. The weakness of the system is reflected in the smaller of the singular values.

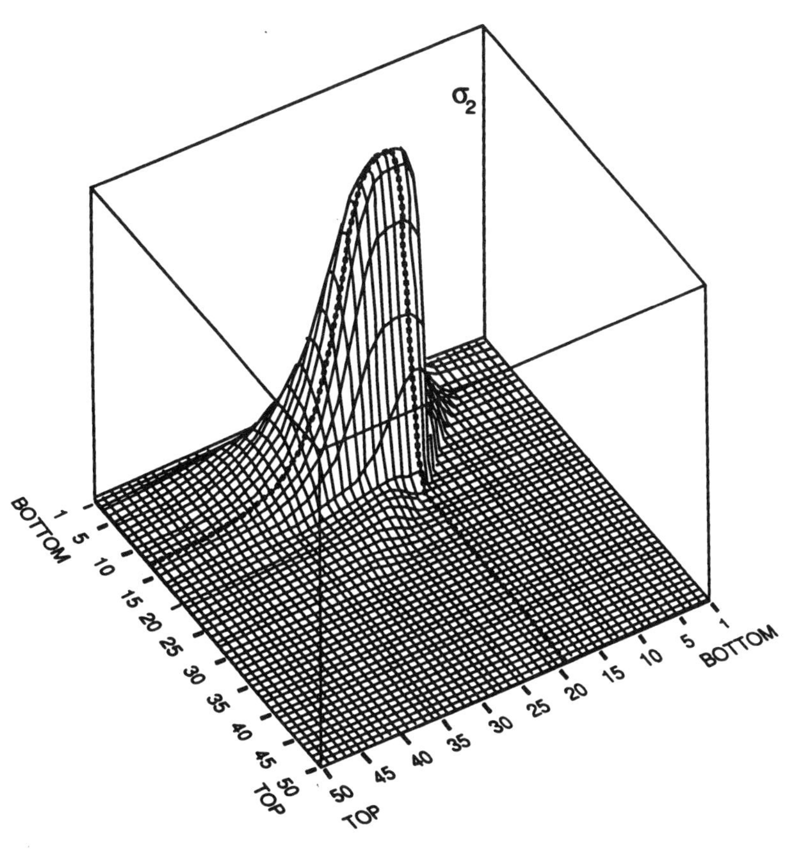

FIGURE 8-13. Minimum singular value versus sensor location for the ethanol–water column.

(For the dual-ended control problem, σ_2 is the smallest singular value.) Based on this rationale the global optimization problem can be stated as

$$\max_{i,j} \sigma_{2i,j}$$

where $\sigma_{2i,j}$ = smallest control specific singular value

i, j = location of the two control sensors

Figure 8-13 illustrates the surface of this global optimization for the ethanol–water column. The optimum occurs when the temperature sensors are located at trays 19 and 13. It represents the sensor pair that yields the strongest multivariable sensitivity and is very similar to the results found in the principal component analysis of the overall gain matrix.

3. Maximum "Intersivity Index" A global optimization based on minimum condition number considers balance but does not consider sensitivity. Conversely, the global optimization based on the maximum of the smallest singular values considers sensitivity but does not consider balance. Let us explore an optimization considering both aspects of the sensor placement problem.

$$\max_{i,j} I_{i,j}$$

where $I_{i,j} = \sigma_{2i,j}/\mathrm{CN}_{i,j} = (\sigma_{2i,j})^2/\sigma_{1i,j}$ = intersivity index

$\sigma_{2i,j}$ = smallest control specific condition singular value

$\sigma_{1i,j}$ = largest control specific singular value

i, j = location of the two control sensors

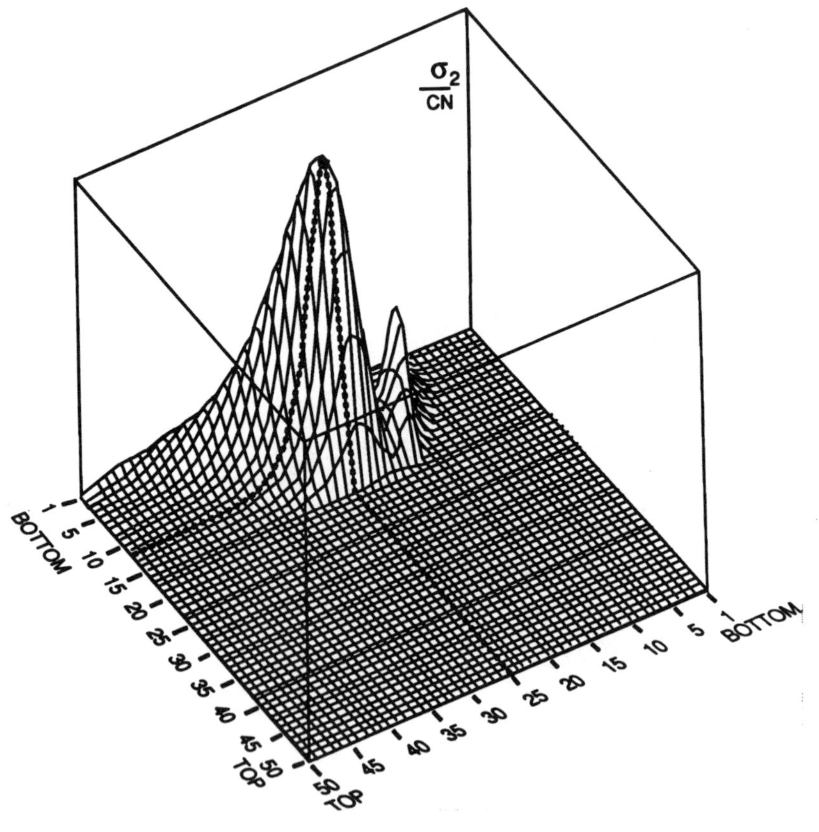

FIGURE 8-14. Intersivity versus sensor location for the ethanol–water column.

The surface defined by the intersivity index is the product of the two surfaces discussed earlier. The result for the ethanol–water column is shown in Figure 8-14. The maximum intersivity occurs when the sensor pairs are located at trays 25 and 12. This choice results in a control specific subsystem with both good sensitivity and balance.

Method 1: Principal Component Analysis. Requires only a single SVD analysis of the full column gain matrix. Results in sensors that represent a good balance between interaction and sensitivity. Can be applied to a control system of any reasonable dimension.

Method 2: Modified Principal Component Analysis. Requires only a single SVD analysis of the full column gain matrix. Weighs more heavily the interaction between sensors than does Method 1. Can only be applied to the analysis of a 2×2 multivariable control problem.

Method 3: Global Minimum Condition Number. Requires an SVD analysis of all possible sensor combinations. Considers only the problem of sensor interaction. Because it does not consider the serious problem of sensor sensitivity it can yield very misleading results.

Method 4: Global Maximum of the Minimum Singular Value. Requires an SVD analysis of all possible sensor combinations. Directly considers sensor sensitivity but forces the sensor interaction also to be considered by looking at the multivariable weakness. Yields good results that are similar to those obtained by Methods 1 and 2. Can be applied to any dimension control problem; however, for systems greater than 2×2 the number of sensor combinations that should be considered becomes very large.

Method 5: Global Maximum of the "Intersivity." Intersivity is a function defined to explicitly include both sensor sensitivity and sensor interaction. Requires an SVD analysis of all possible sensor combinations. Weighs more heavily the interaction between sensors than do Methods 1, 2, and 4. As a result the secondary sensor is the same but the primary sensor is moved to a weaker location for the sake of interaction. Can with some difficulty be applied to control problems with dimensions greater than 2×2.

FIGURE 8-15. Summary and comparison of methods for locating sensors for dual-ended control.

The results are similar to those found using the principal component analysis of the overall gain matrix with one important difference. The position of the weak sensor is essentially the same; yet the position of the strong sensor is adjusted (moved up the column several trays) for the sake of a multivariable balance.

4. Choosing a Method for Locating Sensors for Dual-Ended Control Five methods have been described for selecting sensor locations in dual-ended temperature control. A brief summary of the characteristic strengths and weakness of each method is presented in Figure 8-15. Generally, Methods 1 and 2 are the most commonly used because they involve a single SVD analysis. They also yield similar results to the more extensive procedures of Methods 4 and 5.

8-4 SELECTING SENSOR TYPE: TEMPERATURE VERSUS COMPOSITION

Historically, temperature has been the sensor of choice in monitoring and controlling the separation. Temperature measurement is simple, fast, inexpensive, reliable, and does not need the complex sampling systems that many analyzer-based schemes require. In many columns, however, controlling temperature does not meet the separation control objectives. In such cases more direct measurements of composition should be considered.

8-4-1 Limitation of Temperature Sensors

Before considering the use of process analyzers, perhaps it is appropriate to consider the limitations and failings of conventional temperature-based control. Temperature is useful in controlling column separation only to the extent that it is a reliable indicator of composition. The use of temperature sensors to indirectly measure composition is loosely based on the physics of the liquid–vapor phase transition of mixtures. In general, the composition, temperature, and pressure of a boiling mixture are thermodynamically related as indicated by

$$T = f(x_1, x_2, \ldots, x_i, \ldots, x_n, P)$$

where T = temperature of the boiling mixture
x_i = composition of component i in a n component mixture
P = vapor pressure of the boiling mixture

The effectiveness of a temperature-based composition control scheme is largely related to the complexity of the relationships defined by the thermodynamics of the mixture. Such schemes are the most effective for binary mixtures if the column pressure is constant. In such cases composition is only a function of temperature. Therefore, if the control system is able to hold temperature constant, then the composition will also be held constant. For more complex (and more realistic) systems, holding temperature constant does not imply that composition will also be constant.

Pressure Variations

One problem of temperature-based column control is pressure variations. It is important to realize all separation control systems manipulate (either directly or indirectly) the liquid and vapor rates inside of the column. As these internal flows change, so will the pressure distribution in the column. Even in columns where pressure is formally controlled, variations are likely to occur at the tray locations where temperature measurements are made.

The potential problems caused by such pressure variation are demonstrated by the following relationship for a simple binary mixture:

$$dT = \left.\frac{\delta T}{\delta x}\right| dx + \left.\frac{\delta T}{\delta P}\right| dP$$

When a temperature control has been implemented to insure temperature remains constant, the following problem develops:

$$dT = 0$$

$$dx = \frac{(\delta T/\delta P)}{(\delta T/\delta x)} dP$$

In such cases, variations in pressure are forced by the temperature control system to propagate to the composition. The magnitude of the problem is defined by the magnitude of pressure variation and by the physical properties of the mixture.

It is relatively easy to estimate the pressure problem by considering the thermodynamics of the mixture at the control tray. If the ratio of the two partial derivatives is even modestly large, temperature control may actually cause more problems than it solves. It should be noted that it is relatively easy to develop a strategy that compensates for pressure variations. Such strategies do need, however, a pressure measurement for each of the temperature control points.

Nonideal and Multicomponent Mixtures

The nonideal and multicomponent nature of mixtures can cause problems in temperature-based control systems more difficult to solve than pressure variations. In thermodynamically complex systems, temperature can be a very poor indicator of composition. Controlling temperature in such systems may be counterproductive to the fundamental compensation control objectives of most systems. Consider, for example, the azeotropic column shown in Figure 8-16. An extensive simulation-based study was made for the Department of Energy (Roberts et al., 1991) in which three different sensor-based control systems were evaluated with respect to the effect various levels of variation in feed composition have on product purity. The study results (summarized in Figure 8-17) indicate a control system using a well placed internal composition measurement would be expected to yield results

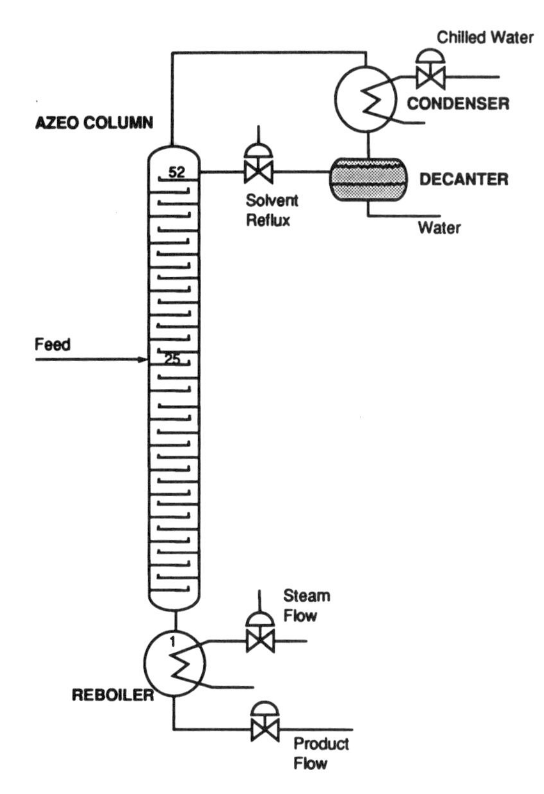

FIGURE 8-16. Azeotropic distillation column.

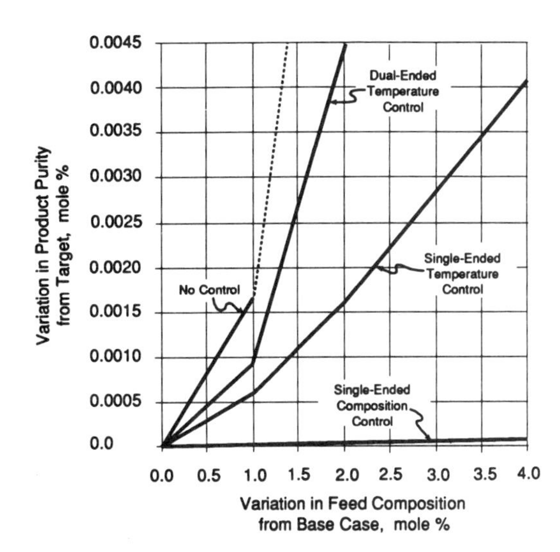

FIGURE 8-17. Disturbance rejection of various sensor-based control systems for an acetic acid recovery column.

orders of magnitude better than the conventional temperature-based schemes.

The acetic acid study also showed an interesting result with respect to single-ended versus dual-ended temperature control. The conventional wisdom is dual-ended control, if possible, should give better performance than single-ended control. The rationale is one of degrees of freedom. If a column exhibits two controllable degrees of freedom, tying down those degrees of freedom with a dual-ended control system should reduce variation throughout the column. The study indicates, however, fixing the degrees of freedom of the temperature profile actually results in higher variation in composition.

Temperature-based control systems can yield good (perhaps even exciting) results in terms of temperature performance. Nevertheless, the underlying control objective is composition and not temperature. The performance of all column control systems (be they single-ended, dual-ended, or even more complex) should always be evaluated with respect to product compositions. Unfortunately, this is seldom applied in practice.

8-4-2 Operational Concerns with Using Process Analyzer

Generally, a column separation control based on direct composition measurements will always perform better than one based on temperature. The problem is obtaining the composition measurements. Such measurements typically require the service of an on-line analyzer. Such devices can provide more direct indications of composition than do temperature measurements; yet they have a number of practical problems that tend to limit their application (see Figure 8-18).

Process analyzers are expensive and can be the source of many operational headaches; nevertheless, the benefits often outweigh the disadvantages. When industry becomes more aware of these benefits and new analyzer technologies are developed,

Costs: On-line process analyzers are typically expensive both in terms of the cost of the device and in terms of the cost of maintenance.

Auxiliary Equipment: On-line analyzers typically require considerable auxiliary equipment such as samplers, calibration standards, environmental enclosures, etc., which must also be designed, purchased, and maintained.

Temperamental Nature: Process analyzers and the necessary auxiliary equipment tend to demand a lot of attention to insure reliable results.

Speed of Response: Most analyzers are relatively slow, particularly if they are multiplexed to a number of sample points.

Application Specific: Unlike temperature sensors in which only the span needs to be specified, on-line analyzers have to be selected and tailored in considerable detail for each application.

FIGURE 8-18. Headaches and operational concerns in using on-line process analyzers for control.

the use of process analyzers will undoubtedly increase.

8-4-3 Schemes for Using Analyzers in Distillation Control

Analyzers can have a broader role in distillation control than can temperature. They can be applied to feed and product streams as well as to stages inside the column. (Recall, the temperature sensor can be used to infer composition only when a liquid and vapor are in thermal equilibrium. This condition seldom exists in product and feed streams.)

In discussing the role of analyzers for distillation control it is important to distinguish between off-line and on-line analyzers. Off-line analyzers are typically laboratory-based equipment and procedures. On-line analyzers are directly connected to the automatic control system. The role of these on-line analyzer will be discussed first.

Figure 8-19 illustrates three basic analyzer control schemes using reflux as a manipulated variable. The first scheme (Figure 8-19*a*) uses the analyzer to measure the product composition and a composition controller to directly adjust reflux flow. The second scheme (Figure 8-19*b*) also uses product composition, but the role of the composition controller is to make feedback adjustments to a temperature controller. In the third scheme (Figure 8-19*c*) the analyzer monitors an internal tray instead of the product. The composition controller directly manipulates the reflux flow based on the internal composition in much the same way as the temperature schemes discussed in earlier sections.

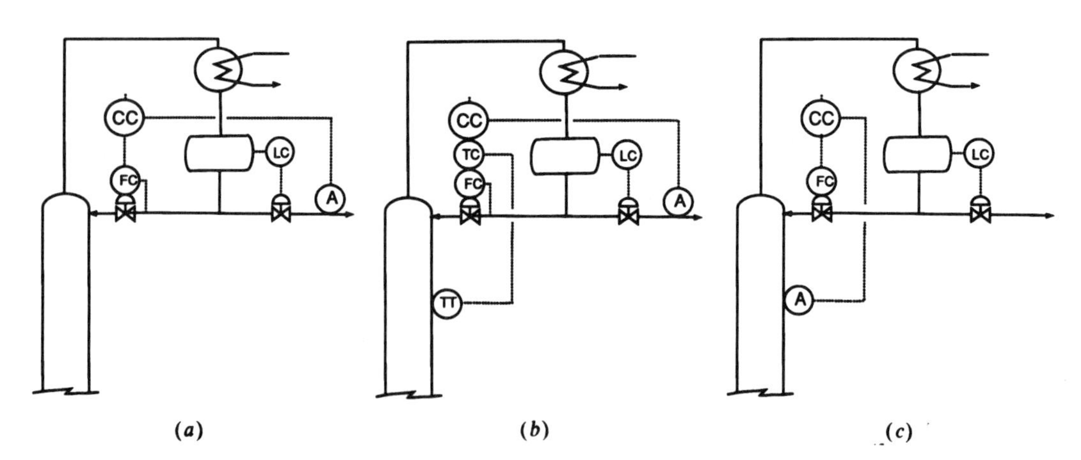

FIGURE 8-19. Analyzer control schemes. (*a*) Scheme 1: product composition control. (*b*) Scheme 2: temperature composition cascade control. (*c*) Scheme 3: internal composition control.

The first two schemes historically represent the most common use of analyzers in the control of column separation. The third scheme, however, can offer advantage if loop sensitivity is a problem. The rationale for placing analyzers directly in the product stream is based on a desire to directly control the product purity. This rationale may not always result in the best control system. The sensitivity issues discussed for temperature-based control also can be a problem in control systems using analyzers. The least sensitive location to measure composition is typically at the product streams. Inside the column is where most of the action is taking place and it should be considered in selecting a location for the control analyzer.

8-4-4 Analyzer Resolution Requirements versus Location

Locating separation control analyzers inside the column as indicated in Figure 8-19*c* can

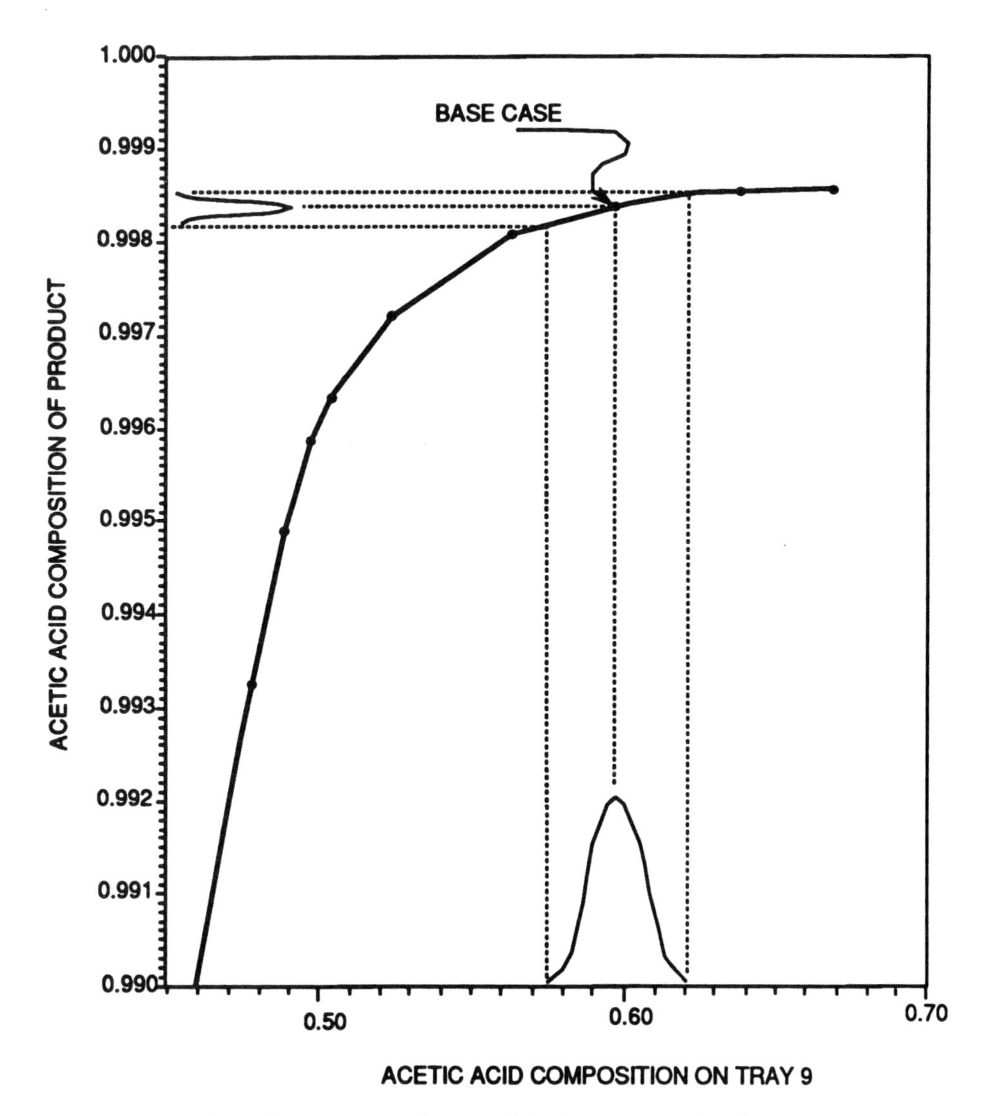

FIGURE 8-20. Control tray versus product sensitivity for an azeotropic column.

provide advantages other than increasing loop sensitivity. Selecting a location inside the column can relax the extreme accuracy and resolution requirements placed on the product analyzer. Consider the control sensitivity studies shown in Figure 8-20 (Roberts et al., 1991). It demonstrates the relationship between the variation in the acetic acid composition at tray 9 and the corresponding variations in product purity for the azeocolumn shown in Figure 8-16. A control loop capable of controlling the composition of acetic acid on tray 9 within a 6σ standard deviation of $\pm 2.0\%$ will result in control of the product composition of $\pm 0.015\%$. This indicates a leverage greater than 2 orders of magnitude in terms of analyzer resolution and accuracy. An analyzer located at the product would need to measure acetic acid compositions to less than 1 part per 10,000, whereas an analyzer located at tray 9 would need only 1 part per 100. In general, reduced analyzer resolution requirements means that faster and less expensive analyzers can be used.

8-4-5 Determining Composition Sensitivities

The sensitivities for composition can be determined in much the same way as described for the temperature sensitivities. A separate gain matrix is required for each composition investigated:

$$K_k = \begin{bmatrix} \left.\dfrac{\partial X_{k,1}}{\partial M_1}\right| & \cdots & \left.\dfrac{\partial X_{k,1}}{\partial M_j}\right| & \cdots & \left.\dfrac{\partial X_{k,1}}{\partial M_m}\right| \\ \vdots & & \vdots & & \vdots \\ \left.\dfrac{\partial X_{k,i}}{\partial M_1}\right| & \cdots & \left.\dfrac{\partial X_{k,i}}{\partial M_j}\right| & \cdots & \left.\dfrac{\partial X_{k,i}}{\partial M_m}\right| \\ \vdots & & \vdots & & \vdots \\ \left.\dfrac{\partial X_{k,n}}{\partial M_1}\right| & \cdots & \left.\dfrac{\partial X_{k,n}}{\partial M_j}\right| & \cdots & \left.\dfrac{\partial X_{k,n}}{\partial M_m}\right| \end{bmatrix}$$

where $X_{k,i}$ = composition of component k on tray i
M_j = manipulated variable j
K_k = steady-state composition gain matrix for an n stage column with m manipulated variables

Again, such sensitivity information can be determined using any distillation simulation based on detailed stage-to-stage calculations. As with the temperature gain matrix, the proper scaling of the composition gain matrices is important to the sensor selection and location analysis. It is particularly important that the units of the manipulated variables be expressed as a percentage of their maximum values. Also, it is important to recall that all gain matrices (composition as well as temperature) are a function of the first-level inventory control scheme. A separate study must be conducted for each first-level strategy considered in the control system design.

8-4-6 Selecting an Analyzer Location for Single-Ended Control

The composition gain matrix can be used to select analyzer locations. Plotting the gain vectors versus tray location can be used to help establish a tray location as well as a component focus. For example, consider again the acetic acid example illustrated in Figure 8-16. The tray sensitivity plots for the column are shown in Figure 8-21. Note, the most sensitive location to measure either acetic acid or solvent is tray 7. The most sensitive location to measure water is tray 10.

The gain analysis also provides an indication of the relative effect of each of the manipulated variables considered in the first-level control configuration. (The first-level control configuration indicates which variables are available for manipulation by the separation control scheme. For details

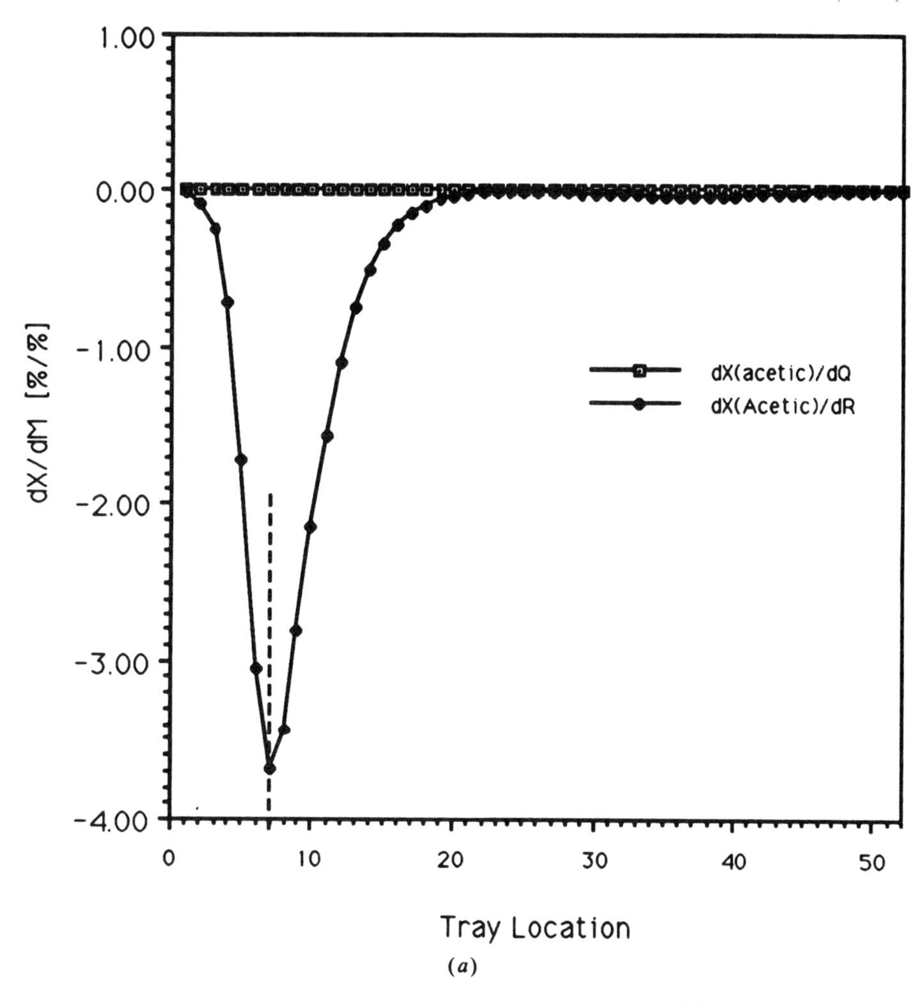

(*a*)

FIGURE 8-21. Sensitivities for an azeotropic column. (*a*) Acetic acid; (*b*) water; (*c*) solvent.

on selecting a first-level configuration see Section 8-6.)

Azeotropic columns are different from conventional columns in that a decanter is used to separate and return the solvent to the column as reflux. Except for losses at the decanter, the solvent is trapped in a recycle loop between the column and the decanter. For such systems the role of the reflux valve is very important. It can be used to control either the inventory of solvent in the decanter or the inventory of the solvent in the column. Conventional schemes typically use the reflux valve to control the level in the decanter. Note, however, that such schemes force the inevitable variation in solvent inventory to occur inside the column. The control configuration considered in this example (LQ) uses reflux to control the inventory in the column. The inventory in the decanter is allowed to float uncontrolled.

Although the solvent losses at the decanter are not great, they are continuous and the solvent inventory in the system will gradually decrease. Eventually, an operator will need to add fresh solvent to make up for the loss of inventory in the system. It is

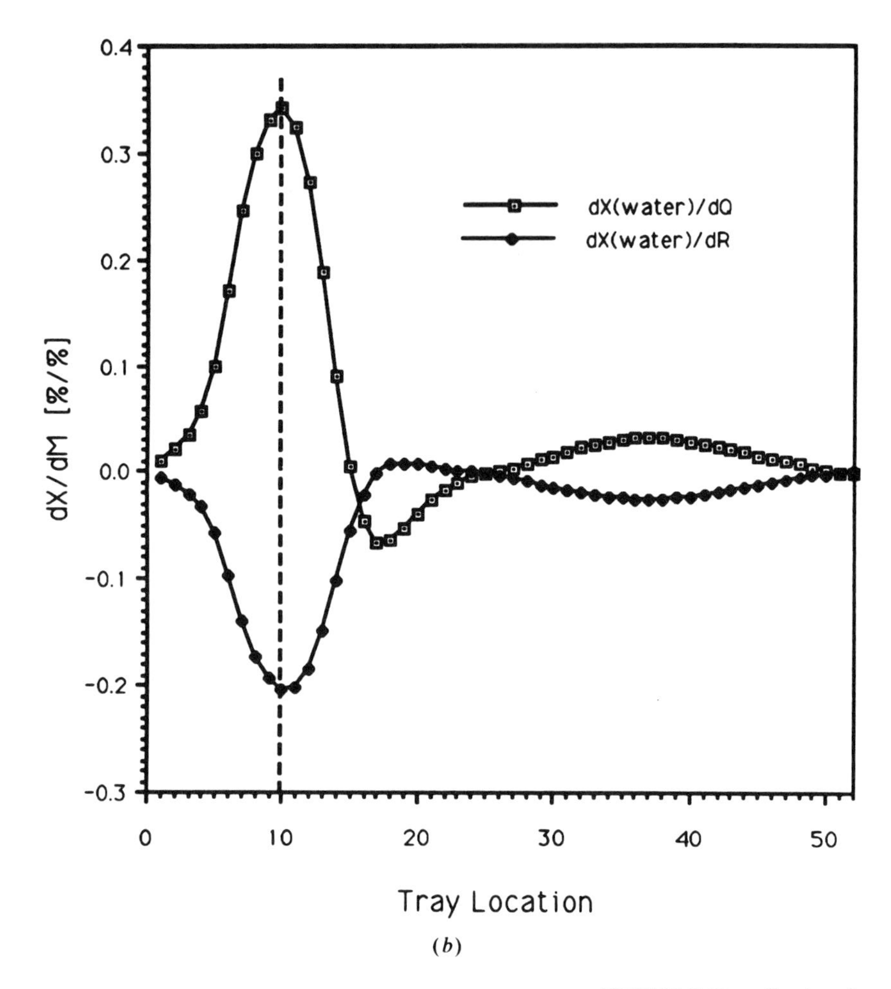

(*b*)

FIGURE 8-21. *Continued.*

better if such variations (both the gradual loss and the typically sudden correction) are forced to occur at the decanter instead of inside the column. If the decanter is large enough to maintain the short term variation in inventory, a control room operator can easily maintain the overall solvent inventory without disturbing the operation of the column. Control adjustment can be made on simple high and low alarms on the decanter level.

In the azeotropic example only the *LQ* control configuration is considered. For this configuration the acetic acid composition is responsive to the reflux but shows little sensitivity to changes in the steam. Water responds to both the reflux and the steam but the response is relatively weak. Of the three components the solvent composition is the most responsive to changes in both the steam and the reflux.

This example clearly shows the problem that often exists with analyzers located in the product streams. In this case, composition at the top and the bottom of the column essentially are not responsive to changes in either of the two manipulated variables. Control loops based on either the bottom acetic acid product or on the solvent water overhead would be very weak. Consequently, performance of such loops would be poor, at best.

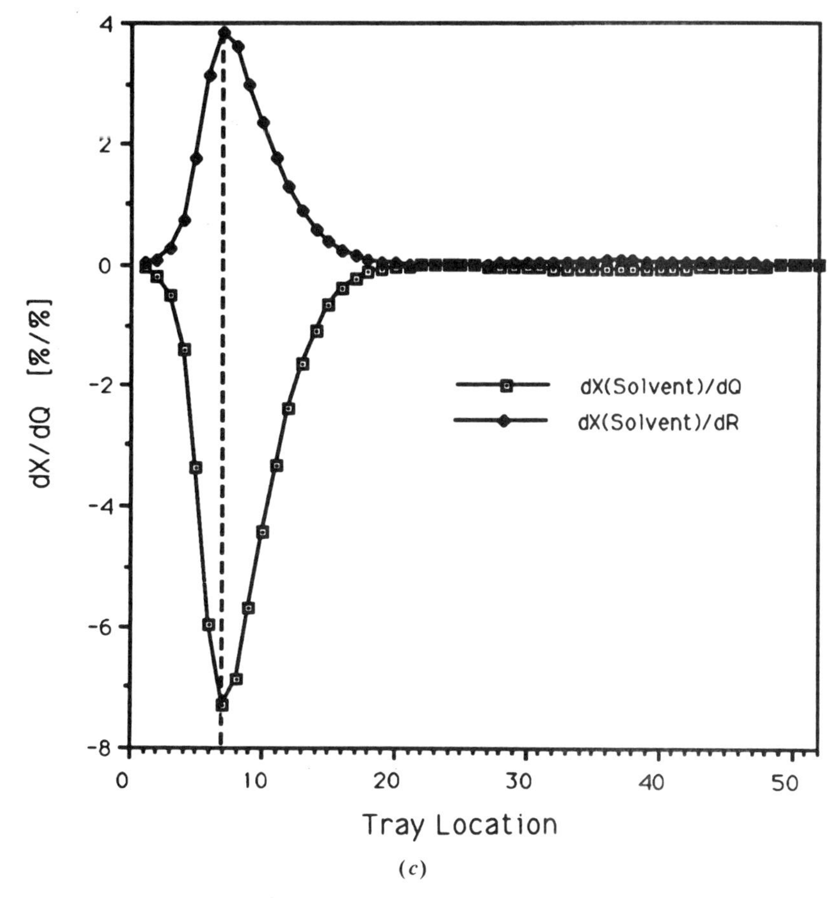

(*c*)

FIGURE 8-21. *Continued.*

Poor analyzer sensitivity in the product streams is not always the case. Consider the analysis of the ethanol–water column. The composition gain matrix and corresponding sensitivity plots are shown in Figures 8-22 and 8-23. Note that ethanol is the more sensitive component and its peak sensitivity is different for each manipulated variable. The gain analysis indicates the strongest composition control loop would be one controlling the ethanol composition on tray 18 by manipulating the distillate rate. If steam is to be used also as a manipulated variable, the composition of ethanol on tray 1 (the reboiler) would result in the strongest secondary loop.

Note that for this example it would not be advisable to consider a single-ended separation control strategy that manipulates only steam. Disturbances in either the feed rate or the feed composition cannot be compensated for by manipulating the steam rate in any *DQ* control configuration. All strategies that require the distillate flow to be set manually are very sensitive to material balance disturbances in the feed.

8-4-7 Selecting Analyzer Locations and Focus for Dual-Ended Control

The use of analyzers can present intriguing possibilities for multivariable column control systems. Analyzers better represent product purity than do temperature sensors. They also can provide a much better conditioned multivariable problem for the control

$K_{ethanol}$: $\left.\frac{\delta X_i}{\delta D}\right\|_Q$	$K_{ethanol}$: $\left.\frac{\delta X_i}{\delta Q}\right\|_D$	K_{water}: $\left.\frac{\delta X_i}{\delta D}\right\|_Q$	K_{water}: $\left.\frac{\delta X_i}{\delta Q}\right\|_D$
−0.222	0.019	1.361	−0.119
−0.249	0.022	1.500	−0.131
−0.278	0.024	1.647	−0.143
−0.310	0.027	1.801	−0.155
−0.344	0.029	1.964	−0.166
−0.382	0.032	2.138	−0.176
−0.425	0.034	2.323	−0.186
−0.471	0.037	2.521	−0.196
−0.524	0.039	2.735	−0.205
−0.583	0.042	2.967	−0.214
−0.651	0.045	3.220	−0.223
−0.728	0.048	3.496	−0.231
−0.817	0.051	3.801	−0.239
−0.922	0.055	4.138	−0.247
−1.044	0.059	4.514	−0.254
−1.190	0.063	4.935	−0.261
−1.366	0.067	5.411	−0.267
−1.581	0.072	5.950	−0.274
−1.847	0.078	6.566	−0.280
−2.182	0.085	7.274	−0.286
−2.612	0.093	8.091	−0.292
−3.175	0.103	9.039	−0.298
−3.928	0.115	10.142	−0.304
−4.962	0.131	11.427	−0.311
−6.423	0.152	12.921	−0.318
−8.555	0.180	14.641	−0.326
−11.785	0.222	16.586	−0.335
−16.901	0.286	18.694	−0.343
−25.442	0.389	20.755	−0.350
−40.548	0.568	22.156	−0.348
−67.802	0.887	21.168	−0.317
−105.170	1.283	14.578	−0.213
−110.770	1.073	5.543	−0.075
−63.662	−0.048	1.367	−0.009
−10.864	−0.718	0.177	0.009
−12.297	−1.373	0.158	0.015
−13.611	−2.071	0.135	0.018
−14.752	−2.787	0.110	0.018
−15.696	−3.499	0.087	0.017
−16.446	−4.190	0.066	0.014
−17.020	−4.852	0.049	0.012
−17.449	−5.481	0.036	0.009
−17.763	−6.077	0.026	0.007
−17.989	−6.642	0.018	0.005
−18.149	−7.177	0.013	0.003
−18.263	−7.685	0.009	0.002
−18.342	−8.168	0.006	0.002
−18.398	−8.627	0.005	0.001
−18.437	−9.067	0.003	0.001
−18.465	−9.604	0.001	0.000

FIGURE 8-22. Composition gain matrices for the ethanol–water column.

system to solve. In order for these potential advantages to be realized, it is important to determine the proper location and focus for the analyzer system.

Properly locating analyzers is somewhat more complicated than locating temperature sensors. There are two design questions that need to be addressed concerning composition-based control: Where should the analyzers be located and what component composition should be presented to the control system? The composition gain matrices and their singular value decompositions can provide insight into answering these critical questions.

Identifying Locations for Similar Composition Measurements

Consider an analysis of the ethanol–water column. Figure 8-24 shows the U-vector plots that result for the SVD analysis of the two composition gain matrices shown in Figure 8-22. Each plot indicates two analyzer locations that might yield a promising balance between sensitivity and interaction. The results suggest two possible measurement schemes: one that measures the ethanol composition on trays 1 and 18 and another that measures the water composition on trays 21 and 37.

A control specific SVD analysis of each scheme provides further insight and a basis for comparison. This will be discussed after the rationale for identifying measurement systems for dual-ended composition has been more fully developed.

Identifying Locations for Different Composition Measurements

Identifying locations using the U vectors assumes the same component composition determination at each measurement point. Measuring different components at each location should also be considered. The composition sensitivity plots illustrated in Figure 8-23 provide guidance in this direction.

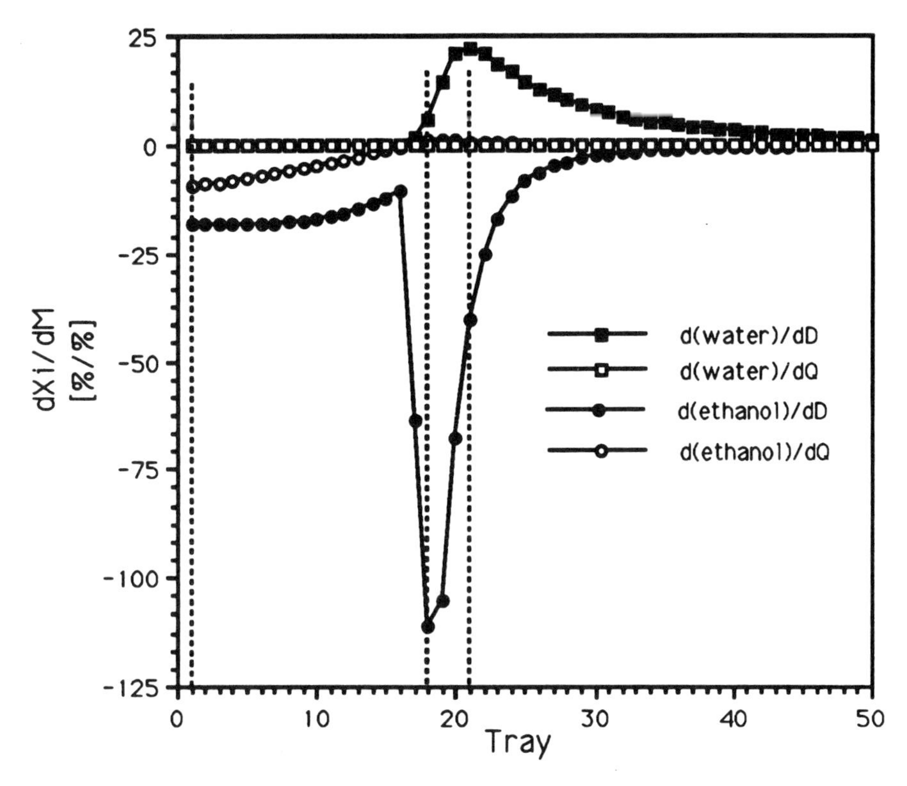

FIGURE 8-23. Composition sensitivity plots for the ethanol–water column.

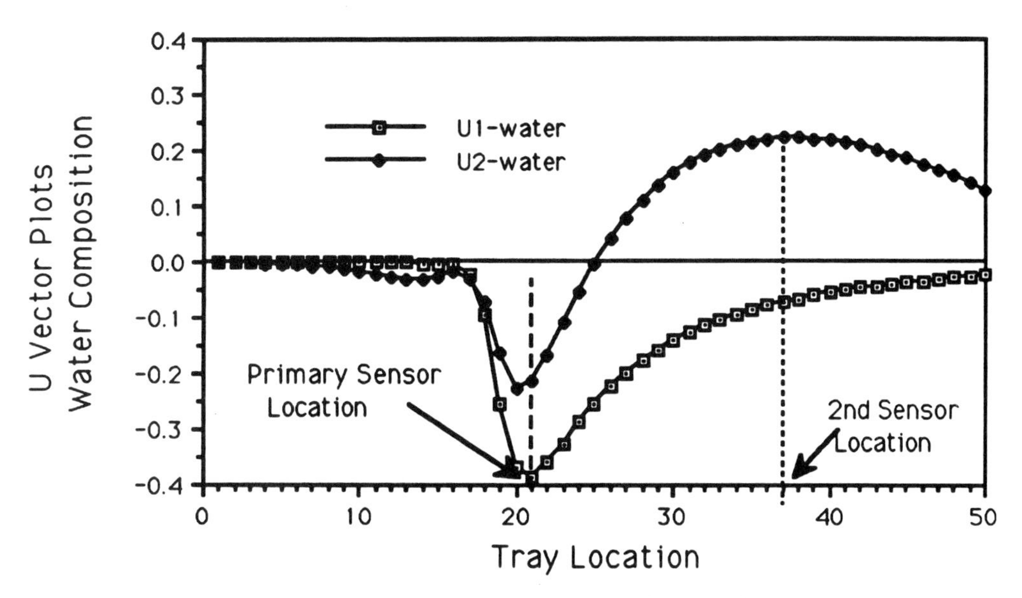

FIGURE 8-24. *U* vector plots for the water composition of the ethanol–water column.

The most sensitive location for each component is one obvious combination to consider. In this example, the maximum ethanol sensitivity is at tray 18 and the maximum water sensitivity occurs at tray 21. The close proximity of these locations, however, hints of interaction problems. Figure 8-25 suggests the next most sensitive location to measure ethanol is at the bottom. Locating an ethanol sensor at the bottom and the water sensor at tray 21 is another measurement scheme that should be considered.

Comparing the Different Measurement Schemes

The rationale just presented identified four possible measurement schemes: one based on ethanol measurements, one based on water measurements, and two schemes based on both ethanol and water measurements. To properly evaluate these schemes the multivariable problem each presents to a control scheme needs to be considered. Figure 8-26 reports the singular values and condition numbers resulting from the control specific SVD analysis of each proposed measurement system. (Recall, the control specific SVD analysis is of the 2×2 gain matrix defined by the location and type of sensors selected.)

In this example two composition-based schemes look acceptable. Scheme 2 (ethanol at trays 1 and 18) has a condition number of 11. Scheme 5 (ethanol at tray 1 and water at tray 21) with a condition number of 4 is even better. The singular values of both schemes also indicate strong sensitivity.

It is important to note that composition-based measurement schemes 2 and 5 are far better from a control point of view than the temperature-based measurement scheme. The temperature-based system has a condition number of 126. Such large condition numbers indicate a very narrow degree of freedom. In this case, a dual-ended control system would face a problem 126 times more

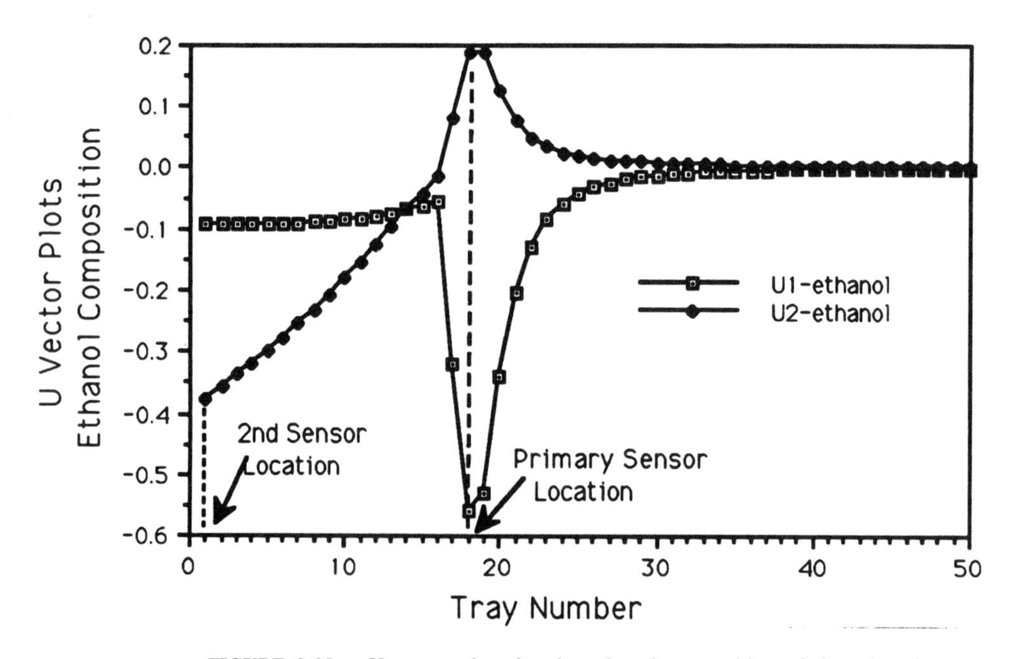

FIGURE 8-25. *U* vector plots for the ethanol composition of the ethanol–water column.

SENSOR SCHEME	SENSOR SELECTION & LOCATION	SINGULAR VALUES		CONDITION NUMBER (Control Specific)
		σ_1	σ_2	
Scheme 1	Temp-19 Temp-13	6.61	0.0526	125.87
Scheme 2	Ethanol-18 Ethanol-1	106.8	9.68	11.03
Scheme 3	Water-21 Water-37	22.54	0.1789	126.01
Scheme 4	Ethanol-18 Water-21	112.9	0.130	864.1
Scheme 5	Ethanol-1 Water-21	294.7	7.436	3.963

FIGURE 8-26. Comparison of various sensor schemes for the ethanol–water column.

sensitive in one multivariable direction than in the other. Another problem with the temperature-based control can be seen in the small magnitude of the second singular value. A loop sensitivity of 0.05 can cause difficulties even in a single loop control system. The problems caused by low loop sensitivity are even greater in a multiloop control system.

8-5 OTHER ROLES FOR COLUMN ANALYZERS

The preceding discussion concerns locating process analyzers to measure and control the column separation. It should be noted that analyzers can (and should) play a much broader role than providing sensor information for automatic feedback control systems. Analyzers also can provide sensor information important to coordinate the column operation with more plantwide concerns. In the following text we will list several other analyzer applications that should be considered in the overall design of the column control systems.

8-5-1 Feedforward Control

Column operations are typically very sensitive to changes in the material balance. Variation in the feed will propagate to the product purities if proper adjustments are not made. The purpose of a column control system is to make such adjustments. If the control system is based entirely on feedback information, such adjustments are late, at best. Some feedforward action with respect to feed disturbances can enhance the performance of any column control system.

Material balance disturbances associated with the feed stream can be of two types: changes in feed flow rate and changes in feed composition. Of the two, feed flow rate changes are the most commonly considered in a feedforward design. It is, however, the feed composition changes that are the most insidious with respect to column operations (i.e., require the most subtle control adjustments).

Subtle feedforward adjustments are quite possible in today's computer-based control systems. It is important to remember, however, they cannot be made without an analyzer in the feed stream. Locating an analyzer in the feed of the column should always be considered if the plantwide environment results in significant composition variations in the column.

8-5-2 Recycle Inventory Control

Distillation columns are frequently part of plantwide recycle systems. Their purpose is often to separate solvents and intermediates from salable products so they can be recycled for further processing. From a plantwide control perspective it is very important to monitor and control the component inventories in these recycle streams (see Chapter 20). Such sensor information is necessary for the control of these plantwide inventories.

An analyzer placed in the column feed can often serve a dual role: It can be used by the control room operator in the management of the plantwide inventories; it also can provide information necessary for the feedforward control discussed previously.

8-5-3 Measuring and Documenting Variation

Another important role of the analyzer is to measure and document variation. Part of this role is related to quality assurance. Today, customer contracts typically require statistical control charts be maintained on the product quality as well as other key process variables. The purpose of such charts is to document the state of control (in the statistical sense) of the process as well as the product.

The role of the analyzer in measuring and documenting variation should be much broader than providing information for quality assurance control charts. In a more general sense, such information can be used to improve process operations. Control systems that respond to disturbances are important; however, it is also important to understand where composition disturbances originate and how they propagate through a complex process system. Such understanding can often lead to rather simple changes in the process. Those changes will reduce or eliminate the need for complex column control systems.

In the role of monitoring and documenting variation, it is not essential to have an on-line measurement. In such applications versatility and accuracy are far more important than speed. These studies are typically conducted off-line by collecting and submitting samples to an analytical laboratory.

8-6 SELECTING MANIPULATED VARIABLES

The foregoing sections primarily concern selecting and locating sensors to be used by the column control systems. An equally important concern is the selection of the variables to be manipulated by those control systems. The selection of the controlled and manipulated variables fundamentally defines the structure of the control system. The selection of an appropriate control structure is critical to the success of the control system. It establishes the difficulty of the local column control problems and defines how the operation of the column interacts with the more global plantwide concerns.

The first step in selecting manipulated variables is to consider which valves will be used for inventory control and which valves will be used for controlling the separation. Designating the role of each valve and establishing the manipulated variables for the separation control system defines the first-level structure of the column control system.

Consider the 73 tray acid recovery column shown in Figure 8-27. In this case two valves must be used to control the inventories in the accumulator and reboiler, and the remaining two are available for the control of the column separation. A list of choices commonly considered are given in Figure 8-28. Note that the inventory control system manipulates the valves directly. The separation control system also includes

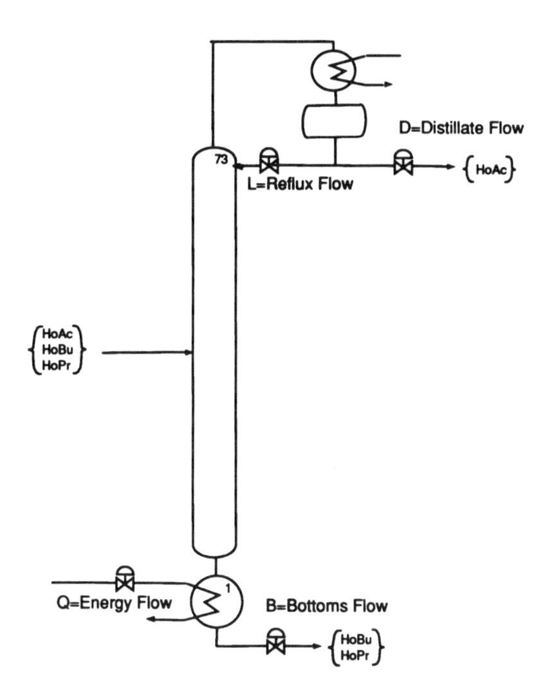

FIGURE 8-27. Acid splitter column.

schemes that indirectly manipulate the valves. In this example, two such indirect schemes are considered: reflux ratio (L/D) and boilup ratio (Q/B). In such cases a ratio controller provides the signal to the control valve. The setpoint of the ratio controller serves as the manipulated variable for the separation control strategy.

Selecting a control structure is a complex problem with many facets. It requires looking at the column control problem from several perspectives:

A local perspective considering the steady-state characteristics of the column.
A local perspective considering the dynamic characteristics of the column.
A global perspective considering the interaction of the column with other unit operations in the plant.

Presented in the following text are several analysis procedures that are helpful in sorting through the various considerations necessary to select a control structure appropriate for the column and the plantwide function it serves.

	CONTROL STRUCTURE	Role of Valve				Manipulated Variables	
		D	L	Q	B	Inventory Control	Separation Control
1	D-Q	SC	IC	SC	IC	L, B	D, Q
2	L-Q	IC	SC	SC	IC	D, B	L, Q
3	L-B	IC	SC	IC	SC	D, Q	L, B
4	L/D-Q	IC	SC	SC	IC	D, B	L/D, Q
5	L/D-B	IC	SC	IC	SC	D, Q	L/D, B
6	D-Q/B	SC	IC	SC	IC	L, B	D, Q/B
7	L-Q/B	IC	SC	SC	IC	D, B	L, Q/B
8	L/D-Q/B	IC	SC	SC	IC	D, B	L/D, Q/B

D= Distillate Flow Rate
L= Reflux Flow Rate
B = Bottoms Flow Rate
Q = Steam Flow Rate

SC= Separation Control
IC= Inventory Control

FIGURE 8-28. Typical column control structures.

8-6-1 Steady-State Considerations

The aggressive nature of the separation control strategy must be considered in the selection of the manipulated variables. Will the control system adjust both ends of the column or only one? If dual-ended control is to be considered, decisions have to be made about sensor–valve pairing. If only single-ended control is considered, decisions have to be made about which valve (or local control system) the control system will manipulate and which valves (or local control systems) will be supervised by a control room operator.

Note that the various control structures assume that all valves are manipulated either by a local flow controller or by a local flow ratio controller. The separation control systems, both manual and automatic, manipulate the local flow or flow ratio setpoints instead of the actual signals to the valves.

SVD Screening

The SVD analysis discussed in the sensor selections can be applied also to the problem of selecting manipulated variables. Each control structure considered in the design can be characterized by a unique gain matrix. The gain matrix describes the steady-state sensitivity of the column to the implied separation control strategy. The SVD analysis of each gain matrix calculates condition numbers and singular values for each control structure. These numbers can provide insight into the difficulty of the separation control problem and can be used to screen the control structures under consideration.

The condition number provides a measure of the degrees of freedom of the separation control problem. High condition numbers are an indication that dual-ended control may not be a practical consideration. Singular values also can be used to identify another practical aspect of the dual-ended control. Low singular values indicate low sensitivity. To compensate for the low sensitivity the controller gain must be set high. High controller gains typically result in saturated valves. Also, high controller gains can make the entire multivariable system sensitive to even low levels of sensor noise.

High singular values are less common; however, they suggest another practical problem that needs to be considered in the question of dual- versus single-ended control. High singular values indicate multivariable directions of high sensitivity. Such high sensitivity may place demands on the control system that cannot be met within the resolution of the valve. In a dual-ended control the hunting typical of such loops can affect the entire multivariable system.

Consider the SVD analysis of each temperature control structure for the acid recovery column. Figure 8-29 indicates the results of both overall SVD and control specific analyses. The overall analysis is of the full 73 × 2 temperature gain matrix for each control structure. (The gain matrices were determined for each control structure using a steady-state simulation of the column operating at its base case.) The control specific analysis for each structure is of the 2 × 2 gain matrix defined by the location of the two control sensors. The location of the control sensors was determined using the principal component method outlined in the sensor selection section.

Control Structure	Condition Number: Overall (73 × 2)	Condition Number: Control Specific[a]	Singular Values (CS)[a]: σ_1	Singular Values (CS)[a]: σ_2	λ_{max} RGA[a]
1 DQ	46.9	53.8	7.63	0.142	0.595
2 LQ	249.0	287.0	18.37	0.064	19.36
3 LB	46.9	53.9	3.87	0.072	0.607
4 L/DQ	123.0	142.0	8.95	0.063	9.486
5 L/DB	48.7	56.3	3.95	0.072	0.621
6 DQ/B	42.1	48.0	7.33	0.153	0.569
7 LQ/B	82.7	95.0	6.18	0.065	6.79
8 L/DQ/B	62.9	72.4	4.62	0.064	5.12

[a]The control specific analysis is based on temperature sensors on trays 16 and 44. These locations were chosen using the principal component analysis of the full U vectors for each control structure outlined in Section 8-3-3. The result of the study indicated identical sensor locations for all eight first-level control structures.

FIGURE 8-29. Condition numbers and singular values for various control structures in an acid recovery column.

Typically, the choice of sensor locations is not a strong function of control structure. For this example, sensor location analysis for all eight control structures indicated tray locations 16 and 44 to be the best choice.

In this example, the control structures using either distillate and steam or reflux and bottom as valves in the separation control strategy are better conditioned than the schemes using reflux and steam. The two schemes with condition numbers greater than 100 (LQ and L/DQ) can be discounted as candidates for dual-ended control.

Considering the singular values can further narrow the list of structures to be considered for dual-ended control. In this example, six of the eight structures have singular values less than 0.10. (For most systems, singular values less than 0.1% per percent indicate that the multivariable system will be prone to saturate the control valves and will be sensitive to noise.) The two structures with the largest second singular value are DQ and DQ/B. (It is interesting to note that these are the only structures that use the steam and distillate valves in the separation control strategy.)

Sensor–Valve Pairing

The SVD analysis also can be used to help in selecting the manipulated variables for the separation control strategy. The columns of the U matrix are orthogonal vectors indicating the sensor directions in which the column is most likely to respond to changes in the manipulated variables. The V matrix are orthogonal vectors indicating the combinations of manipulated variable that have the greatest effect on the sensors. A simple principal component analysis of the U and V matrices can be used to identify sensor–valve pairings. The principal component of the first column vector of U is paired with the principal component of the first column vector of V. Pairing the principal components of the second column vectors of each matrix defines the next pairing. (The pairing rationale is based on providing the least interaction between conventional SISO control loops.)

Consider the result of the SVD analysis of the acid separation column. Presented in Figure 8-30 are the results for the basic DQ and LQ control structures. For the DQ structure the principal component of the first U vector indicates the temperature on tray 16 is the most responsive. The principal component of the first row in the V^T matrix indicates that D is the strongest manipulated variable. The pairing suggested by this analysis is D with $T(16)$ and Q with $T(44)$. A similar analysis of the LQ structure indicates Q with $T(16)$ is the strongest pairing and L with $T(44)$ is the next strongest pairing. In a dual-ended control both pairings would be used. In a less ambitious single-ended control only the first pairing would be used.

It should be emphasized that using SVD to determine loop pairing considers only the open loop sensitivities. Other pairing analyses should be considered also. The SVD analysis cannot predict closed loop stability

SVD Analysis

	DQ-Structure			LQ-Structure		
K		D	Q		L	Q
	T(44)	1.2957	-0.0856	T(44)	-1.408	2.6446
	T(16)	7.5174	0.33854	T(16)	-8.174	16.179
U	T(44)	-0.1691	-0.9856	T(44)	0.1631	-0.9866
	T(16)	-0.9856	0.1691	T(16)	0.9866	0.1631
V^T		D	Q		L	Q
		-0.9991	-0.0418		-0.4515	0.8923
		-0.0418	0.9991		0.8923	0.4515
Σ		7.6349			18.373	
			0.1418			0.0640
CN		53.8			287.	

RGA Analysis

		D	Q		L	Q
Λ	T(44)	0.4051	0.5949	T(44)	19.36	-18.36
	T(16)	0.5949	0.4051	T(16)	-18.36	19.36

FIGURE 8-30. SVD and RGA analyses for the DQ and LQ control structures for the acid recovery column.

problems that other indexes such as relative gain array (RGA) and Niederlinski can provide. Note that the RGA for these two structures is presented in Figure 8-30. In this case the RGA and SVD suggest the same pairing. Both also suggest that the LQ structure would be a poor basis for dual-ended control.

8-6-2 Dynamic Considerations

The steady-state characteristics of each control structure are an important consideration in selecting a control structure. Good steady-state sensitivities are necessary; however, they are not a sufficient condition for good control system performance. The dynamics of each control structure also should be considered. Dynamic problems can play havoc even on systems that have good steady-state sensitivities.

The dynamics of distillation columns are complex. The relationship between valves and sensors characteristically involves a mixture of effects. These effects differ in direction and speed of response. The net result can seldom be described adequately by simple first order lag plus deadtime models. The characteristics of these relation-

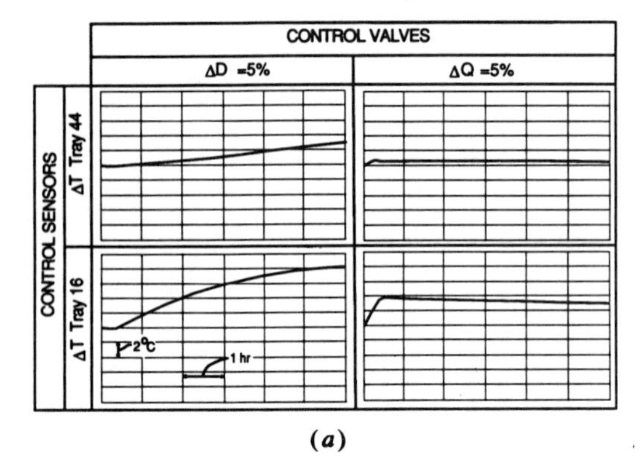

(*a*)

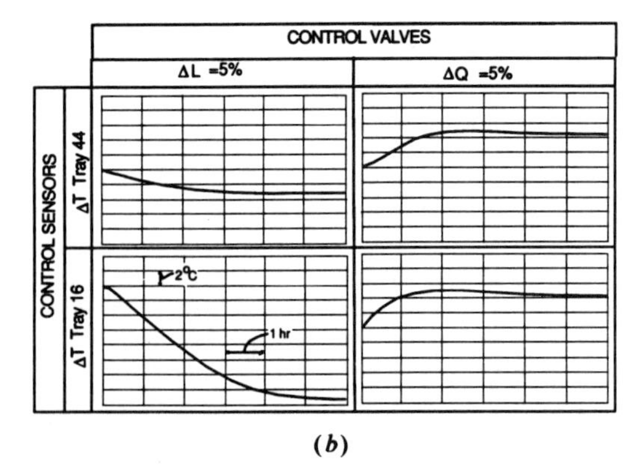

(*b*)

FIGURE 8-31. Dynamic response in the acid recovery column. (*a*) *DQ* control structure; (*b*) *LQ* control structure.

ships vary between control structures and between the various valves and sensors within each structure.

The dynamics of the acid separation example are illustrated in Figure 8-31. Presented are the open loop dynamics of the *DQ* and *LQ* schemes. Note that in this case the speed of response of both sensors is about the same for a specific manipulated variable. The speed of the response varies, however, between the various manipulated variables and between the two control structures. The distillate and reflux are the manipulated variables with the slowest speed. (Distillate flow has a time constant of about 3.5 h relative to $T(16)$ and appears to be integrating with respect to $T(44)$. Reflux flow has a time constant of about 2.5 h with respect to both locations.) Control systems that manipulate either of these variables will be facing very long sensor response times. Steam, on the other hand, is relatively fast. Its speed is a function of control structure. The response of both sensors to changes in Q is about three times faster in the *DQ* structure ($\tau \sim 20$ min) than in the *LQ* structure ($\tau \sim 60$ min).

This example illustrates some of the subtle but important complexities of distillation dynamics. Note that manipulating Q results in open loop overshoot. The initial effect of changing the steam is caused by the change in vapor rate, which occurs relatively fast. A secondary effect results from changes in reflux and distillate at the top of the column. These secondary effects are slower and in the opposite direction. Such effects are not extremely serious; however, they can cause trouble in tuning the feedback control.

Dynamics consisting of multiple cause and effects can manifest more serious dynamic problems. For example, when the minor effect is in the opposite direction of the major effect and is considerably faster, an inverse response occurs. Such characteristics are common in distillation columns. (Note that in the acid recovery example, manipulating D causes a slight inverse response in both temperatures). When they are severe, such characteristics can present very difficult problems for any feedback control system.

8-6-3 Plantwide Considerations

The results of the preceding steady-state and dynamics analyses are not by themselves sufficient to make wise decisions about the control structure. The plant environment in which the column is to operate also must be considered. Plantwide as well as local questions need to be asked during the selection of the control structure and the pairing of sensors and valves within that structure: What are the disturbances that are affecting the column? What are their magnitudes, frequencies, and durations? How will the different disturbances propagate through the column and its control systems to other unit operations? What are the overall plantwide objectives and how do decisions about the column control systems affect those objectives?

Disturbance Analysis

It is important to consider the nature of the disturbances in the selection of a control structure. What are the anticipated column disturbances and how will the more promising control systems react to each? Such information is necessary in order to fully interpret the results of the steady-state and dynamic analyses.

Again, consider the acid recovery column. The steady-state analysis indicates, from the point of view of condition number, that DQ should be a good control structure to consider. The dynamic analysis further indicates that although the dynamics for changes in Q are good, the dynamics for changes in D are extremely slow. One logical conclusion might be to implement a single-ended control system within the DQ structure that does not use D as a manipulated variable. Considering only the results of the steady-state and dynamics analyses, the most logical single-ended control structure is one that controls the temperature on tray 16 using the steam flow. Although this rationale is clever, it would obviously lead to a problem if feed disturbances occur. Feed flow or composition changes require adjustments in the distillate flow that are simply not provided for in this design.

A disturbance rejection study of the acid recovery column illustrates the problem more clearly. The study considered three common disturbances: overhead pressure, feed flow rate, and feed composition. Each disturbance was varied from its base operating condition. The results are presented as steady-state variations in purity of the acetic acid product stream as seen in Figure 8-32.

	Control Structure	D-Q	D-Q	L-Q	L/D-Q
	Valve-Sensor Pairing	D - T44	Q - T16	Q-T16	Q-T16
Disturbance Study	$\Delta P = \pm 15\%$	0.00517	0.00171	0.00118	0.0007
	$\Delta F_F = \pm 15\%$	0.00808	0.00363*	0.00517	0.00422
	$\Delta X_F = \pm 10\%$	0.00431	0.25636	0.02092	0.00813
		ΔX_d Change in Overhead Purity			

* Includes a feedforward controller that adjust the distillate flow for changes in feed flow

FIGURE 8-32. Change in overhead purity for various acid recovery column disturbances.

The schemes selected for this study were based on the results of the steady-state and dynamic analyses. Because of the high condition numbers, only single-ended schemes were considered. Because of the dynamic advantage offered in manipulating steam, only DQ, LQ, and RQ control structures were considered ($R = L/D$). For all three structures a sensor–valve pairing of T(16) with Q was considered. For the DQ scheme a sensor–valve pairing of T(44) with D was considered also. The temperature control loop in all four schemes also included pressure compensation.

Of the four schemes considered, the two using the DQ structure showed the poorest performance. The DQ strategy using D as a manipulated variable is sensitive to all three disturbances. The DQ strategy using Q is extremely sensitive to the feed composition. (This strategy would also be very sensitive to feed flow disturbances if the feedforward controller were not present to adjust the distillate flow for changes in the feed flow rate.) The LQ and RQ strategies demonstrate about the same disturbance rejection capabilities for feed flow disturbances. However, RQ shows a marked advantage with respect to changes in feed composition and changes in column pressure.

Propagation of Variation to the Plant:

The preceding discussion considers how disturbances affect product purity. It is important to be aware of other issues concerning how the column and its control system react to these disturbances. These other issues concern the relationship between the column and other unit operations in the plant. A column is typically connected to other unit operations through its feed and product streams. In highly heat-integrated plants it may also be coupled through heat exchangers and economizers. All these connections suggest that variation in the column will propagate to other parts of the plant.

There are several related questions to be considered during the selection of a control structure: What plantwide disturbances are caused by the actions of the column control system. What units are disturbed and how does that affect the plantwide objectives? A larger perspective in some cases may suggest using a different column control strategy. One that propagates its variation to more benign locations in the plant should be considered.

Component Inventories in Recycle Systems

Another question concerns the problem of plantwide inventories. Specifically, this concerns component inventories in recycle systems. A distillation column designed to consider only local performance may not behave well in such systems. Closely controlling the purity of the product streams may inadvertently serve as a dam to one or more of the components. If the component flow into the system is greater than the rate the column control system allows it to leave, accumulation occurs. The component buildup will continue until the column control system cannot handle it anymore or until an operator intervenes. In such cases the best solution is a less ambitious column control strategy. Some situations may not suggest using any separation control strategy. It is important to remember that the local column control system should conform to the plantwide concerns and goals.

A more extensive detailed discussion of plantwide concerns that should be considered in the design of the local column control system is included in Chapter 20.

Plant and Column Design

The easiest column to control is one that has no disturbances. The calmest column, in the plantwide sense, is one that has no disturbances. During the control system design (and after) the control engineer should become aware of where and how disturbances originate and propagate through the plant to the column. Understanding the disturbance paths may suggest simple process changes to eliminate or reduce the magnitude of such disturbances. For example, well placed inventory tanks can help reduce propagation of many disturbances from unit to unit. Such inventories can often be justified on the basis of improved quality and/or productivity.

The control engineer also should consider the control effectiveness of the local column design. For example, internal condensers are often used in column design. Such designs cost less to fabricate; however, they limit the control possibilities. An internal condenser-based design allows only control structures that vary the distillate to be implemented. If the column analysis indicates that a reflux-based control structure is appropriate, the internal condenser may not be cost effective after all.

It is obvious that design changes necessary to improve the control system are best identified before the process design phase is completed. Once fixed in concrete and steel, even simple modifications may be difficult unless the control engineer builds a very convincing case. Nevertheless, if such changes are to occur, the control system designer must also be involved in the column design as well as the overall plant design.

8-7 SUMMARY AND CONCLUSIONS

This chapter attempts to discuss a number of technical as well as philosophical issues related to the preliminary design of a column control system. These issues relate primarily to the selection of control structures, placement of sensors, and pairing those sensors with available manipulated variables within that structure. Proper choices at this level typically have a far greater impact on the overall performance of the column control systems than does the choice of control algorithms.

The more philosophical issues concern the role and responsibility of the control strategy specialists. As distillation control strategies become more complex there is a natural tendency for the control engineer to become more specialized. There is also a tendency for the role of the control engineer to be limited to solving control problems. A larger role should include working with other specialists to minimize the difficulty of the control problem that must be solved. This is particularly true with respect to process design. Developing a clearer understanding of the fundamental cause of column disturbances can lead to process improvements that reduce the need for complex control systems.

References

Cantor, D. L. (1987). A detailed steady-state control analysis of an ethanol–water distillation column. M.S. thesis, The University of Tennessee.

Deshpande, P. B., ed. (1988). Multivariable control methods. Research Triangle Park, NC: Instrument Society of America.

Downs, J. J. (1982). The control of azeotropic distillation columns. Ph.D. thesis, The University of Tennessee.

Forsythe, G. E. and Moler, C. B. (1967). *Computer Solution of Linear Algebra*. Englewood Cliffs, NJ: Prentice-Hall.

Golub, G. H. and Reinsh, C. (1970) Singular value decomposition and least squares solutions. *Numer. Math.* 14, 403–420.

Roberts, J. J., Muly, E. C., Garrison, A. A., and Moore, C. F. (1991). On-line chemical composition analyzer development, phase II, Final Report DOE DE FC07-88ID 12691, Department of Energy.

Lawson, C. L. and Hanson, R. J. (1978). *Solving Least Squares Problems*. Englewood Cliffs, NJ: Prentice-Hall.

MATLAB User's Guide (1985). The Mathworks, Inc., South Natick, MA.

Moore, C. F. (1985). A reliable distillation control analysis procedure for use during initial column design. In 1985 Annual Meeting of the AIChE.

Moore, C. F. (1991). A new role for engineering process control focused on improving quality. In *Competing Globally Through Customer Value—The Management of Strategic Suprasystems*, M. J. Stahl and G. M. Bounds, eds. New York: Quorum Books.

Moore, C. F. (1987). Determining analyzer location and type for distillation column control. In 14th Annual Meeting of the Federation of Analytical Chemistry and Spectroscopy Societies.

Moore, C. F., Hackney, J. E., and Canter, D. L. (1986). Selecting sensor location and type for multivariable processes. In *The Shell Process Control Workshop*. Boston: Butterworths.

Roat, S. D., Downs, J. J., Vogel, E. F., and Doss, J. E. (1986). The integration of rigorous dynamic modeling and control system synthesis for distillation columns: An industrial approach. In *Chemical Process Control—CPCIII*, 99–138. Elsevier, Amsterdam, 1986.

9

Selection of Controller Structure

B. D. Tyreus

E. I. Du Pont de Nemours & Co.

9-1 INTRODUCTION

In working with control system design for a multivariable process the terms control strategy, control structure, and controller structure are sometimes used interchangeably. They then mean the selection and pairing of controlled and manipulative variables to form a complete, functional control system. However, the three terms can also be taken to have their own individual meaning. In this case control strategy describes how the main loops in a process are to be set up in order to meet the overall objectives. For example, if the objective of a column is to produce a bottoms product stream, the control strategy to insure the purity of the product might be to use reboiler heat to keep the impurities out. Control structure, on the other hand, is the selection of controlled and manipulated variables from a set of many choices. For example, the choice of reflux flow and reboiler heat (R-V) to control both distillate and bottoms compositions in a column gives one possible control structure. Alternatively using distillate flow and reboiler heat (D-V) gives another control structure. Controller structure finally means the specific pairing of the controlled and manipulative variables by way of feedback controllers. For example, a controller structure that specifies reflux flow to control distillate composition and reboiler heat to control bottoms composition is possible in an R-V control structure.

The title of this chapter is selection of controller structure. However, I am not going to stick to the fairly narrow interpretation given above. Instead I will describe a methodology for designing a multivariable control system that includes elements of control strategy considerations, control structure selection, and variable pairing.

This methodology is largely empirical based on general principles for distillation control such as those outlined in Chapter 1. However, the methods presented here fit well into more formal approaches for selecting and tuning control structures as described in Chapters 8, 10, and 11.

The methodology for control system design I will present assumes that the process is fixed and process changes are not permitted in order to come up with the final control scheme. Although this assumption could severely limit the possibilities for effective control (Downs and Doss, 1991) it is realistic in practice because control engineers are often asked to design the control system for existing process configurations. It could either be for an existing plant or for a new one well into the design phase. The process equipment including buffer tanks and control valves are then already designated and the task is to select the appropriate variables to be controlled and design controllers that will tie the controlled variables to the control valves in such a way that the resulting controller structure meets the desired objectives.

I will also assume that the final controller structure will be built up around conventional control elements such as PID controllers, ratio, feedforward, and override control blocks as found in all commercial distributed control systems. I will not address the design of multivariable controllers such as DMC, which is treated in Chapter 12.

9-1-1 Control Design Principles

The process control field is characterized by algorithms and mathematical procedures for analysis, design, and control; but if you are looking for another algorithm for selection of controller structures you will not find it here. Not that I am against algorithms per se, I just happen to believe that they compete poorly against a combination of other activities in process control design, namely, definition of the control system objectives, process understanding and rigorous dynamic simulation. Others have expressed a similar view (Downs and Doss, 1991). Because these three activities are central to the design principles expressed in this chapter, I will say a few more words about each of them.

Defining and understanding the control system objectives should be a collaborative effort between process engineers, control engineers, and plant personnel. Left to any one of these contributors alone, the objectives can be severely biased. For example, a control engineer might be tempted to make the control system too complex in order for it to do more than is justified based on existing disturbances and possible yield and energy savings, whereas a process engineer might underestimate what process control can achieve and thus make the objectives less demanding. It is crucial to define what the control system should do as well as to understand what disturbances it has to deal with.

Process understanding is another key activity for successful control system design. Nobody seems to disagree with this notion, but in practice more time is usually spent on designing and implementing algorithms and complex controllers than on analyzing process data and figuring out how the process really works. Modelling and simulation are key parts in the process understanding step.

Rigorous dynamic simulation is the third important activity in control system design. A flexible dynamic simulation tool allows rapid evaluation of different control structures and their response to various disturbances. The attributes of a dynamic simulator ideal for control system work are outlined in Chapter 5.

The methodology I advocate for controller structure selection is the following:

- Define the objectives of the control system and the nature of the disturbances.
- Understand the principles of the process in terms of its dynamic behavior.
- Propose a control structure consistent with the objectives and the process characteristics.
- Assign controllers and evaluate through simulation the proposed control structure with the anticipated disturbances.

Before we embark on the details of this methodology it will be useful to first examine the nature and roles of the manipulative variables that will be used by the controllers in the proposed control structures.

9-2 MANIPULATIVE VARIABLES

When a process engineer works with a flowsheet or a detailed steady-state simulation of a distillation column a certain number of variables have to be specified or "controlled" in order to get a solution. For example, for a two-product column two specifications have to be given (e.g., product

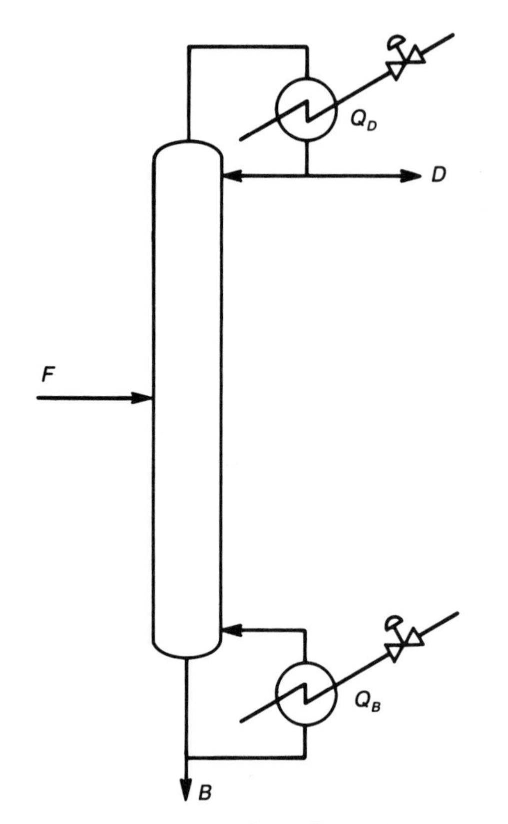

FIGURE 9-1. A two-product column with 2 steady-state degrees of freedom.

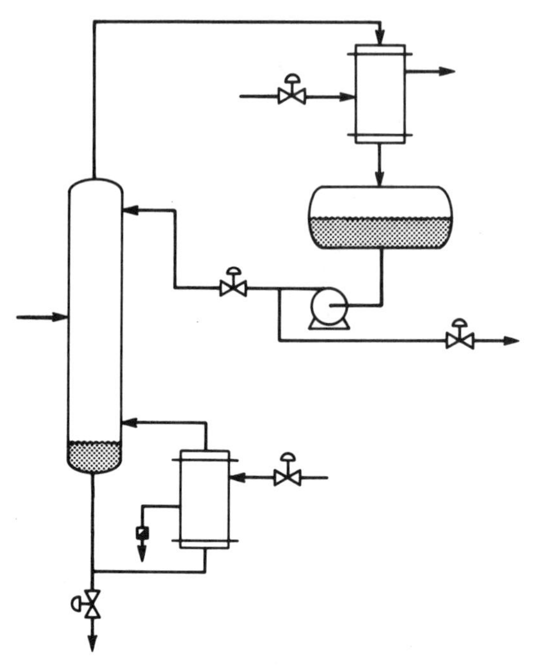

FIGURE 9-2. A two-product column with five manipulative variables for control.

composition and a component recovery). The program will then manipulate two variables, such as reboiler and condenser duties, in order to satisfy the specifications and close the steady-state material and energy balances (see Figure 9-1). We say that a two-product column has 2 steady-state degrees of freedom requiring two steady-state manipulative variables for control.

When the same column is examined by a control engineer the number of manipulative variables as counted by the valves has increased from two to five (see Figure 9-2). The three new manipulative variables are needed to control integrating process variables (e.g., levels and column pressure) that are not fixed by steady-state material and energy balances alone. The three additional manipulative variables that are needed to control the real process can be classified as "dynamic" manipulative variables providing 3 "dynamic" degrees of freedom.

When we consider the five manipulative variables in Figure 9-2 we see that the two variables of Figure 9-1 are a subset of the five. However, there is nothing magic about the heat duties that make them exclusive as steady-state manipulators and prevent them from being used for inventory control. As a matter of fact, in many control schemes the condenser duty is used for pressure control rather than for composition control. For the same reason we can assume that any three of the five valves are suitable for level and pressure control as well.

Although I have talked about manipulative variables as control valves there are many more choices for manipulative variables than just the individual valves. For instance, many columns have reflux ratio as a manipulative variable for either inventory control or composition control. If we allow ratios and linear combinations of variables, the choice of a manipulator for a given loop broadens considerably for a simple two-product column. However, the steady-state and dynamic degrees of freedom remain unchanged as 2 and 3, respectively.

9-2-1 Manipulative Variables and Degrees of Freedom

The relationship between control valves in the process and degrees of freedom for control is important for the proposed design methodology, and I will therefore illustrate the key ideas with a few more examples.

Consider first the extractive distillation system in Figure 9-3. It has 10 control valves as we might expect from combining two columns. However, the process has 5 steady-state degrees of freedom and not 4 as we initially would assume (four compositions and the recycle rate of extractive agent). Figure 9-4 shows one possible combination of values or manipulators that could be used to reach a solution on a steady-state simulator. Because the total degrees of freedom for actual plant operation equals the number of control valves in the process, we are only left with five dynamic degrees of freedom. This means that only five levels and pressures can be controlled with the control valves provided. A close look at Figure 9-3, however, shows that there are six integrating variables that we might attempt to regulate; two column pressures and four levels. How do we resolve this dilemma? The answer is that we should not attempt to control all levels and pressures at the same time with the control valves provided. Qualitatively it is not hard to see that the level in the base of the second column primarily depends on the total amount of extractive agent in the system and not on the bottoms flow rate or boilup for that column. An attempt to control the level in the base of the second column with say bottoms flow, while all other levels are also controlled, would lead to an unstable system.

For another example of the relationship between degrees of freedom and control, consider Figure 9-5, which shows an azeotropic column with an entrainer. The entrainer separates from the light compo-

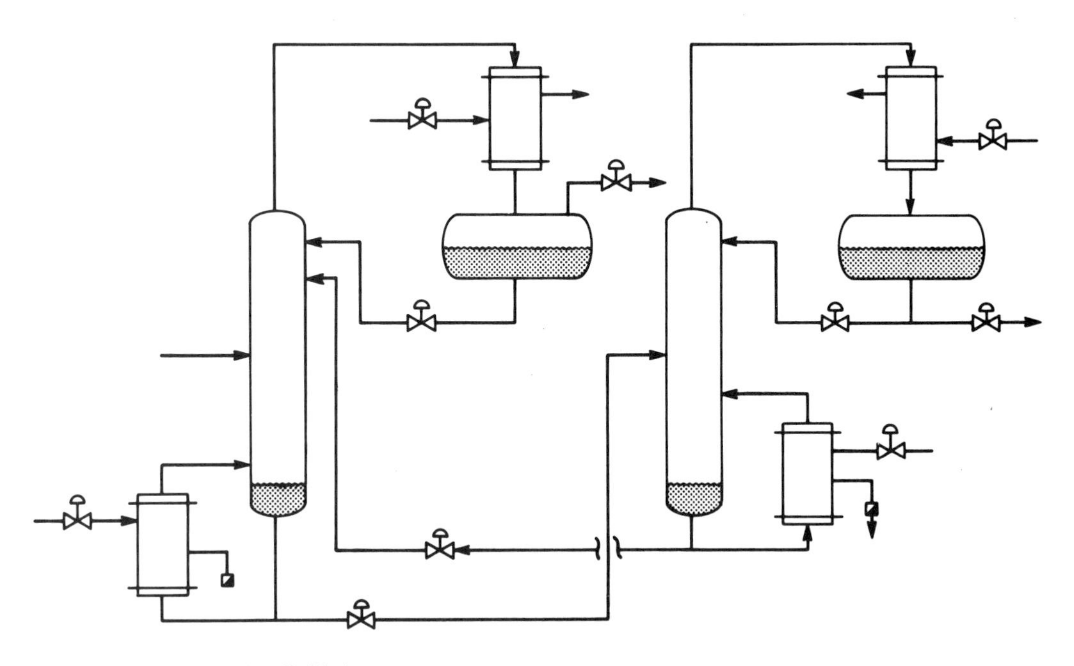

FIGURE 9-3. Extractive distillation system.

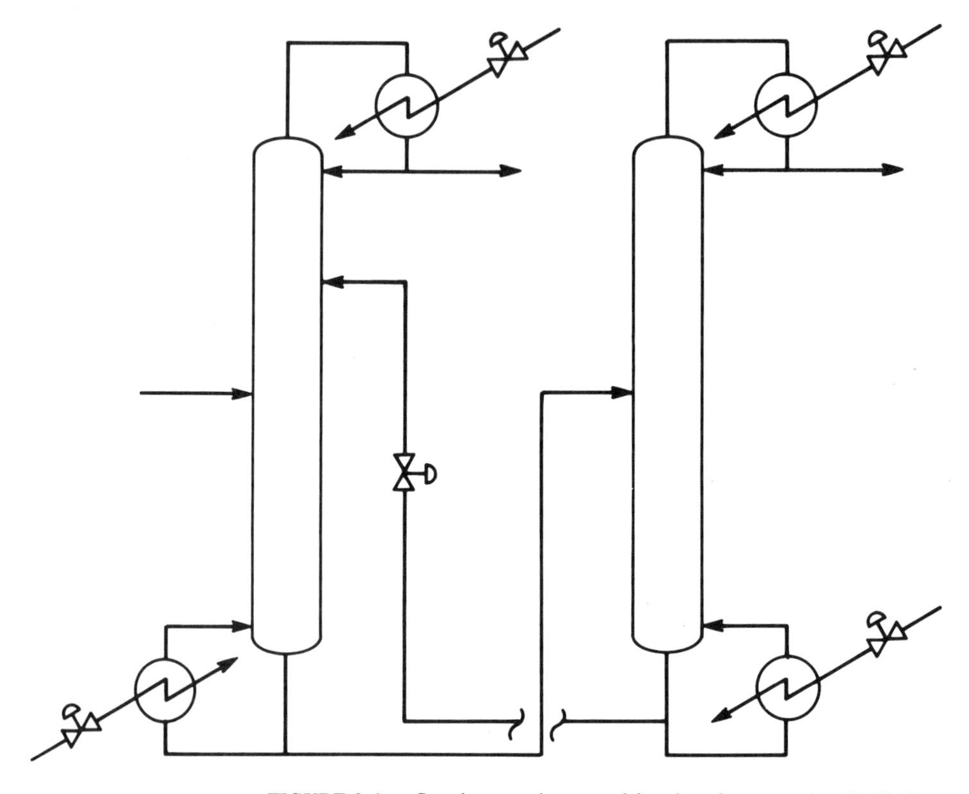

FIGURE 9-4. Steady-state degrees of freedom for extractive distillation.

nent in the overhead decanter and is then typically refluxed back into the column. To simulate such a column at steady state we need three manipulators. Two of these could be the condenser and reboiler duties, but the third has to be the entrainer reflux rate because this would be the only way of introducing that component to the balance equations. Of the six control valves in Figure 9-5 we would thus need three for steady-state control, leaving only 3 dynamic degrees of freedom. However, we have four integrating variables we might want to control; column pressure, the base level, the decanter interface level, and the decanter total level. We have thus lost 1 dynamic degree of freedom. In practice it means that the level of entrainer in the decanter should not be controlled by the reflux of entrainer. Instead the entrainer reflux should be flow controlled or set by some other variable, and the amount of entrainer in the reflux drum should be allowed to float. This way any loss of entrainer will show up in the reflux drum instead of in the column where it affects the separation. In some installations I have seen this loss of a dynamic degree of freedom was not recognized and the entrainer was level controlled with reflux, which caused the amount of entrainer to vary indeterminately in the column. This resulted in column instabilities and confusion for the operators. If there is a need to make up entrainer on a continuous basis, the entrainer level in the drum can be controlled with the entrainer makeup valve.

To summarize this section, we can say that the total degrees of freedom for actual plant operation equals the number of valves available for control in that section of the plant. To find out how many integrating variables (pressures and levels) we can control with the available valves, we first have to subtract the degrees of freedom required

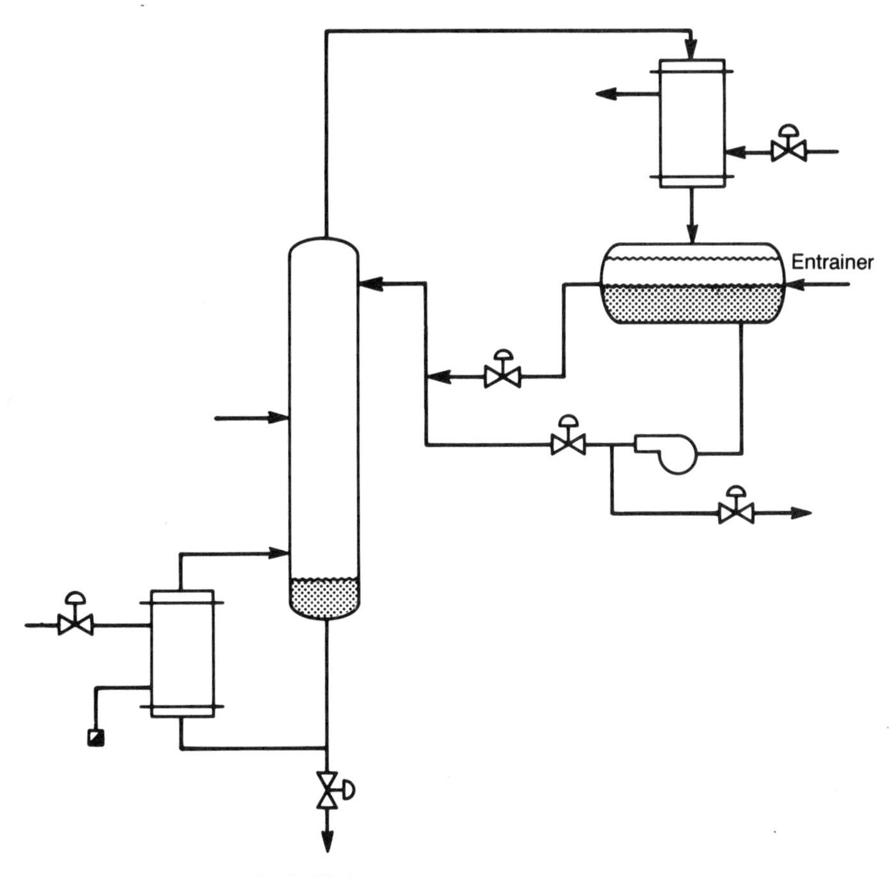

FIGURE 9-5. Azeotropic distillation.

for steady-state control from the total degrees of freedom. Recycle streams will reduce the dynamic degrees of freedom if the recycle contains a component that neither enters nor leaves the process. If it is important to control the inventory of such "trapped" components, it is necessary to provide extra control valves to account for the loss in degrees of freedom.

9-3 A METHODOLOGY FOR SELECTION OF CONTROLLER STRUCTURE

The economic performance of a distillation system is linked to its steady-state degrees of freedom. In other words the economic benefits of a column control scheme depend on how well it controls compositions, recovery, or yield and not on how well it holds integrating variables such as levels and pressure. The integrating variables must obviously be controlled too, but their control performances do not directly translate into profits. However, as all practitioners know, inventory controls can be the most troublesome of all loops and can preoccupy the operators to the point where the economically important composition and recovery loops are almost neglected. This problem was recognized early by industrial practitioners (Buckley, 1964) and has been resolved by designing the level and pressure controls before dealing with the compositions controls. I too have found this to be a very workable approach. However, I have also found that the choice of manipulators for inventory control significantly impacts the control performance of the composition loops (Tyreus, 1987). Others have made the same observation (Waller et al., 1988a).

I am now ready to summarize a methodology for selection of controller structure:

1. Count the control valves in the process to determine the overall degrees of freedom for control.
2. Determine from a steady-state analysis the steady-state degrees of freedom.
3. Subtract the steady-state degrees of freedom from the overall degrees of freedom to determine how many inventory loops can be closed with available control valves.
4. Design pressure and level controls and test for disturbance rejection.
5. Design composition controls based on product stream requirements.
6. Design optimizing controls with remaining manipulative variables.

9-3-1 Level and Pressure Controls

The assignment of manipulative variables for level and pressure control is, in principle, quite simple: choose the stream with the most direct impact. For example, in a column with a reflux ratio of 100, we have 101 units of vapor entering and 100 units leaving the reflux drum for every unit of distillate leaving. We should therefore use reflux flow or vapor boilup to control the drum level. If we were to violate this assignment principle and select distillate flow for level control, it only takes a change of slightly more than 1% in either vapor boilup or reflux flow to overwhelm the drum level controller and saturate the distillate valve.

Finding manipulative variables for pressure control obeys the same general principle as for level control. Column pressure is generated by boilup and is relieved by condensation and venting. To find an effective variable for pressure control we need to determine what affects pressure the most. For example, on a column with a total condenser, either the reboiler heat or the condenser cooling will be good candidates for pressure control. On a column with a partial condenser we need to determine whether removing the vapor stream affects pressure more than condensing the reflux. Sometimes the dominating effect is not immediately obvious. If the vent stream is small we might assume that condenser cooling should be manipulated for pressure control. However, if the vent stream contains noncondensables, these will blanket the condenser and affect condensation significantly. In this situation the vent flow, although small, is the best choice for pressure control.

There is only one caution in applying the principle of using the stream with the most impact for inventory control and it has to do with the loss of dynamic degrees of freedom discussed in Section 9-2-1. If the largest stream is a recycle stream, we need to examine if it contains components that either should stay completely within the process (e.g., an extractive agent) or enter and leave at a relatively small rate. I have already discussed the case of a component that remains in the process and the process loses a dynamic degree of freedom. The other case where a component enters and leaves at a small rate compared to what is recycled in the process presents a similar problem. If we use the larger recycle stream for inventory control without satisfying the overall material balance for the component, we also lose a dynamic degree of freedom. I will illustrate this point further in the examples.

9-3-2 Composition Controls

The composition control loops on a column are the most important steady-state controls. The purpose of composition control is to satisfy the constraints defined by product quality specifications. The constraints must be satisfied at all times, particularly in the face of disturbances.

The objective of a composition control loop is usually clear: Hold the controlled composition as close as possible to the imposed constraint without violating the constraint. This objective translates to on-aim, minimum variance control.

To achieve good composition control we need to examine two things: process dynamics and disturbance characteristics. Process dynamics includes the measurement, the process itself, and the control valve. Tight process control is possible if the equivalent deadtime in the loop is small compared to the shortest time constant of a disturbance with significant amplitude. To ensure a small overall deadtime in the loop, we always seek a rapid measurement along with a manipulative variable that gives an immediate and appreciable response. In distillation such a measurement often translates to a tray temperature, and a good manipulative variable is vapor flow, which usually has a significant gain on tray temperatures and compositions and also travels quickly through the column.

In situations where the apparent deadtime in the composition loop cannot be kept small compared to significant disturbances we need to focus our attention on these disturbances. Sometimes the important disturbances can be anticipated or measured, in which case direct feedforward control is appropriate. In other situations we might be able to affect the way disturbances affect our composition variable by rearranging other control loops, particularly the pressure and inventory loops (Shinskey, 1985; Tyreus, 1987; Waller et al., 1988a). Although several researchers have proposed algorithms for determining the disturbance sensitivity of different control structures (Waller et al., 1988b; Skogestad, Jacobsen, and Lundstrom, 1989), I have found that direct dynamic simulation of the strategies resulting from assignment of manipulative variables for pressure and level control gives the best insight to the viability of a proposed composition control scheme.

9-3-3 Optimizing Controls

After the pressure, level, and composition loops have been assigned, there typically remain a few manipulative variables. These variables can be used for process optimization. Because process optimization should be done on a plantwide scale (Fisher, Doherty, and Douglas, 1988), it is difficult to provide guidelines for how the manipulative variables should be used on a local, unit operation basis. In many cases the best strategy might be to operate the manipulative variables at their maximum values (e.g., full cooling load on a condenser) or to set them to a near optimum value such as constant reflux flow (Tyreus, 1987).

9-4 EXAMPLES

I will illustrate the methodology for selection of controller structure with two examples. Because most discussions of distillation control in the literature use conventional, two-product columns I will expand the scope and use multiproduct, sidestream columns in my examples.

9-4-1 A Column with a Stripping Section Sidestream

Figure 9-6 shows the flow sheet configuration of a column with a stripping section vapor sidestream. The multicomponent feed comes from an upstream unit in the process and the sidestream goes downstream for further processing. The distillate is the product stream. It has a certain purity specification in terms of sidestream material. The feed contains small quantities of heavy boilers that are removed with a purge from the reboiler.

The control scheme objective of this column is to operate close to the quality constraint of the distillate product while preventing high boilers from entering the sidestream. The major disturbances will be feed rate (due to overall changes in production rates) combined with feed composition increases of sidestream material.

At this stage, we consider all prior insights to the process and its dynamic behavior. Particularly relevant here is the control of base level. We know that manipulation of

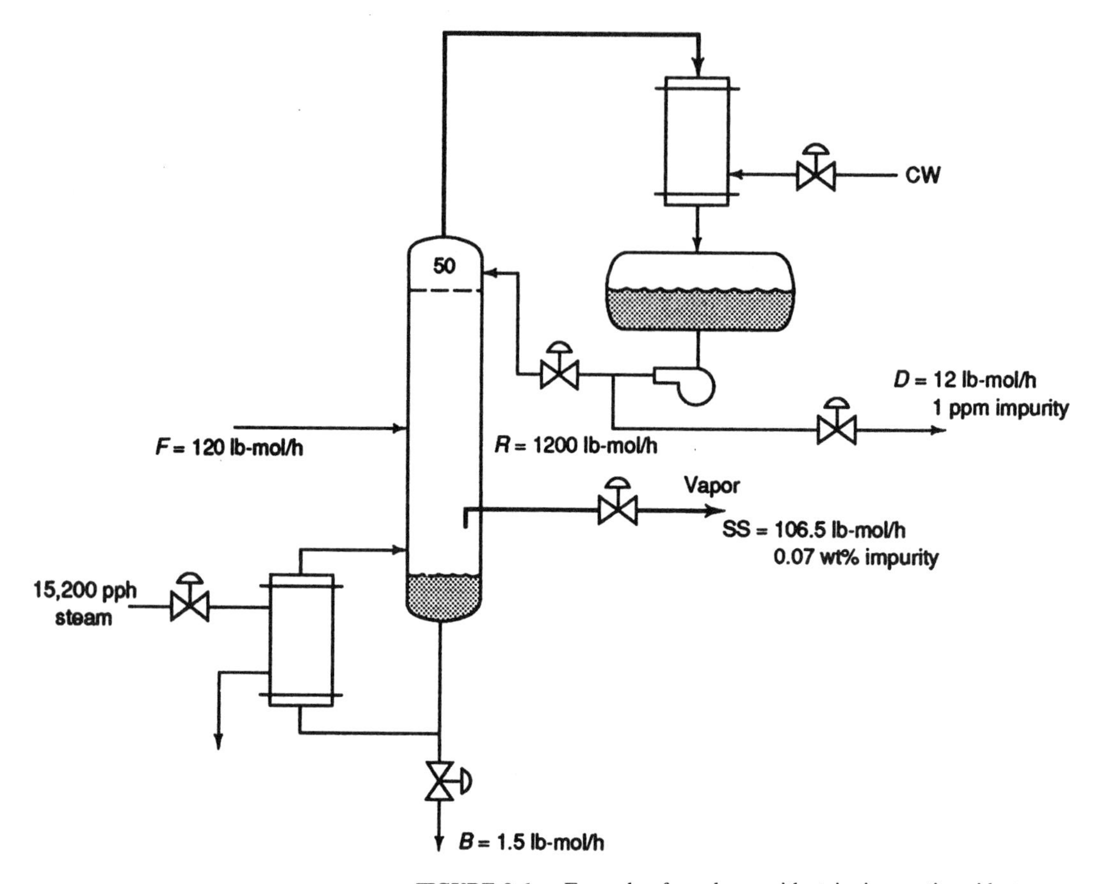

FIGURE 9-6. Example of a column with stripping section sidestream.

the vapor sidestream has a very indirect effect on the base level. For example, if the level increases and we attempt to reverse this trend by removing more sidestream vapor, our actions will not be felt by the level until the reduction in vapor flow up the column has resulted in a decrease in reflux flow and the liquid has worked itself down the column. Furthermore, this scheme assumes that reflux flow controls drum level or that there is a tray temperature controller manipulating reflux. Because the column has many trays, it will take appreciable time for reflux changes to be felt in the base. For this same reason reflux flow is also a poor manipulator for base level control.

We next assign manipulative variables for pressure and level controls. We have a total of six control valves giving 6 degrees of freedom. Three of these are needed to control the process at steady-state, leaving three manipulative variables for pressure and level controls. The column has a total condenser free from noncondensables. We therefore select condenser cooling as the best choice for pressure control. Control of the reflux drum level is also straightforward. Because the reflux ratio is 100:1, reflux flow is the only reasonable manipulator for drum level. However, we need to worry about the potential loss of a dynamic degree of freedom unless we ensure that the material balance for the distillate product is satisfied. One simple way of forcing distillate component to leave the column when it accumulates is to ratio the distillate flow to the reflux flow. The effective manipulator for reflux drum

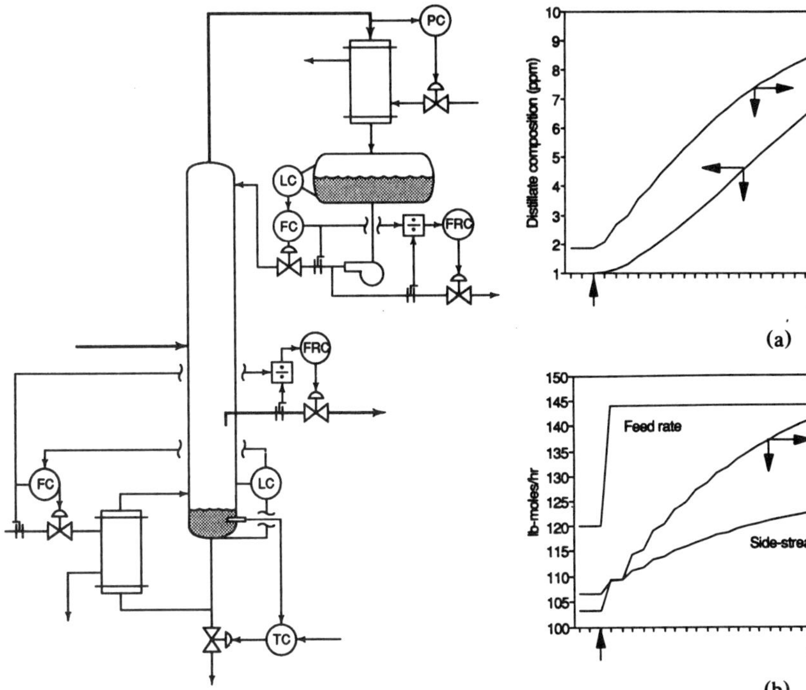

FIGURE 9-7. Inventory controls for the example column with stripping section sidestream.

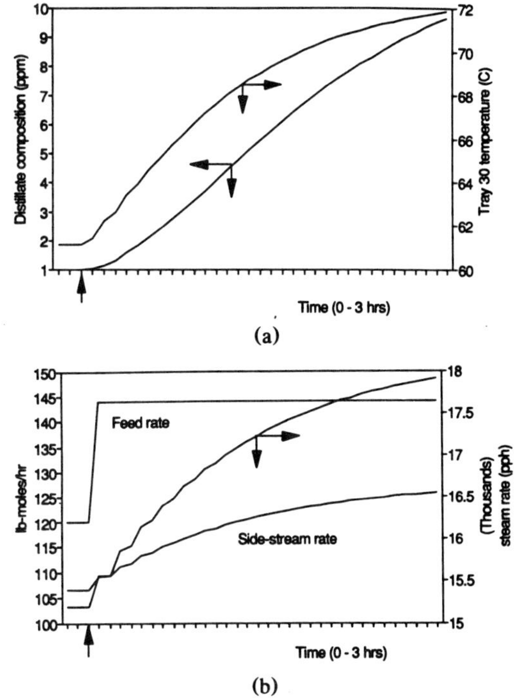

FIGURE 9-8. Inventory controller response to a 20% feed rate increase.

level is now reflux and distillate instead of just reflux flow.

Base level control is a little trickier. From process insights we have already eliminated a couple of potential manipulators. The only reasonable remaining variable is steam flow to the reboiler. However, vapor boilup recycles material back into the column, which could cause problems with loss of degrees of freedom unless we satisfy the material balances for both the sidestream material and the high boilers. We can prevent sidestream material from accumulating by ratioing the sidestream flow to the reboiler steam. To prevent high boilers from building up in the reboiler we could monitor the reboiler temperature and purge the column based on this temperature.

Figure 9-7 shows the proposed inventory control scheme for the column. With only the inventory loops closed, Figure 9-8 shows the response in distillate composition after a step change in feed rate to the column.

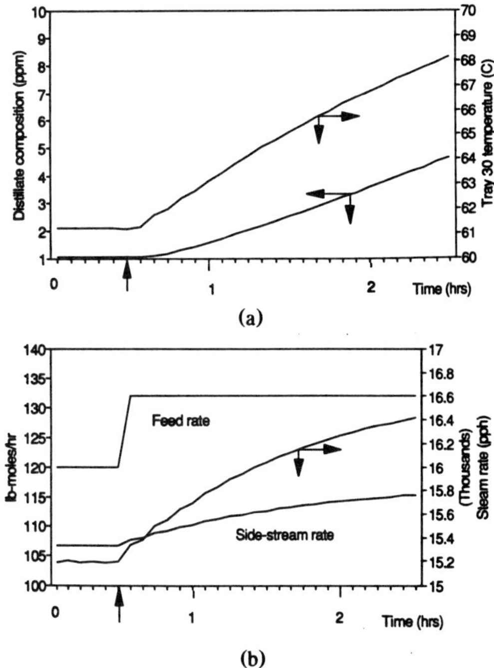

FIGURE 9-9. Inventory controller response to a 10% increase in feed rate and a 1% increase in feed composition.

Figure 9-9 shows the response to a disturbance where both the feed rate and the feed composition of sidestream material increase simultaneously. In both cases the basic inventory controls do an excellent job in rejecting the disturbances, making the job for the composition controllers simple.

There are three manipulators left for steady-state control: the overhead reflux ratio, the side-stream to steam ratio and the reboiler temperature controller setpoint. Manipulation of the sidestream to reboiler ratio has a direct impact on the vapor traffic in the column and ought to be a good candidate for tray temperature control in the rectifying section.

The remaining steady-state manipulative variables can be used for optimization. Setting the reflux ratio will determine the recovery of product from the sidestream whereas setting of the reboiler temperature controller determines the loss of sidestream material in the purge. There are several different ways of setting these remaining manipulators. For example, a third temperature controller could be used for multiple temperature control. Such multivariable composition control schemes are appropriate for energy minimization. However, if energy is less important than yield, one or more of the manipulators could be adjusted to maximize energy flow, which thereby maximizes recovery. In some situations it is obvious what constitutes an optimum strategy at the unit operation level, but in most cases the optimum policy for the remaining degrees of freedom must be determined from a plantwide optimization.

9-4-2 A Column with a Rectifying Section Sidestream

Figure 9-10 shows the flowsheet configuration for the second example column. The multicomponent feed comes from an upstream unit and the bottoms flow goes to further processing. In this case the liquid sidestream is the final product that has a purity specification in terms of bottoms material. The column has a gravity reflux, vent condenser where a small overhead purge stream of noncondensables is removed.

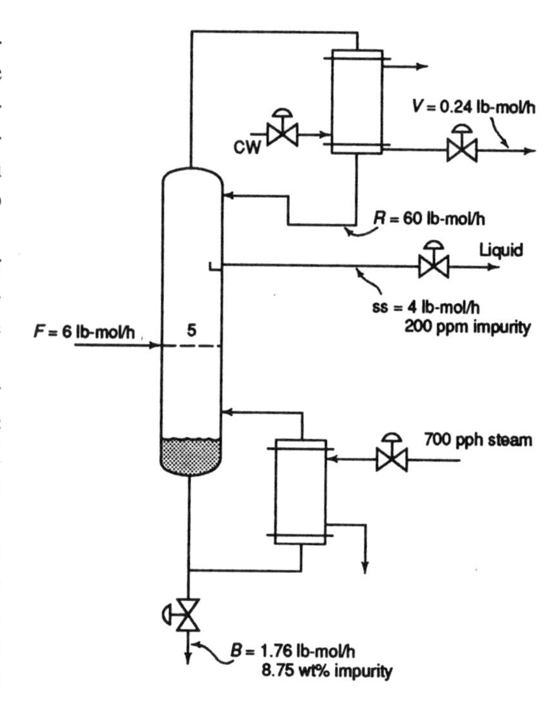

FIGURE 9-10. Example column with rectifying section sidestream.

The objective of the control system is to maintain the purity of the sidestream product despite changes in feed flow rate and feed composition of bottoms material.

This column has only five control valves and requires 3 degrees of freedom for steady-state control. However, all condensed liquid is returned with gravity to the column so there is no separate liquid inventory control needed in the condenser. Thus only two inventory loops have to be controlled: the column pressure and the base level. Given a vent flow of noncondensables we should use this stream for pressure control. Noncondensables in a column tend to accumulate in the condenser and significantly reduce the dew point of incoming vapors. The low dew point reduces heat transfer because of a small temperature driving force. Because the vent stream is rich in noncondensables, vent flow rate is effective in removing the noncondensables and thereby quickly increasing heat transfer whenever needed.

Control of base level is pretty much restricted to the use of reboiler steam due to the large vapor boilup to tails flow ratio. If we used sidestream flow for base level control we would introduce deadtime in the level loop from reflux moving down the column. If we used tails flow we would have a rangeability problem with the tails flow control valve. To ensure that bottoms flow material is provided a way out from the column we could ratio tails flow to steam. Alternatively, we could manipulate bottoms flow directly by way of a tray temperature controller.

At this point it might seem that we are done with the inventory controls and ready to tackle the composition controllers. However, we need to take a closer look at how sidestream material leaves the system. The partial condenser separates the noncondensables from sidestream material and recycles essentially all sidestream component back into the column. The base level controller will also send most of this component back into the column. To prevent sidestream component material from getting trapped in the column and cause a loss of a dynamic degree of freedom we need to provide a route for it to escape. This can be accomplished by ratioing the sidestream to the partial condenser reflux.

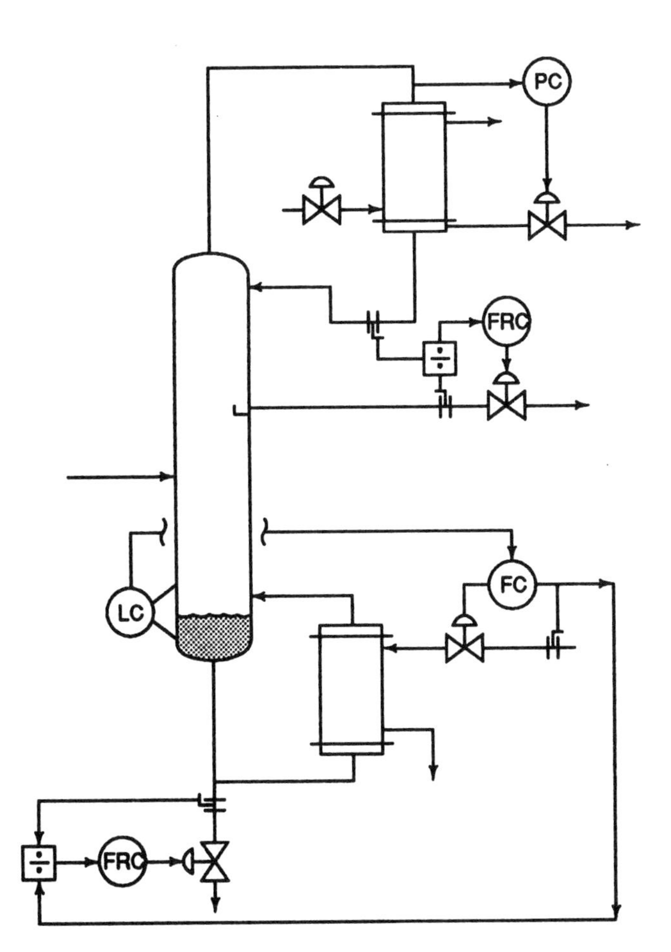

FIGURE 9-11. Inventory controls for the example column with rectifying section sidestream.

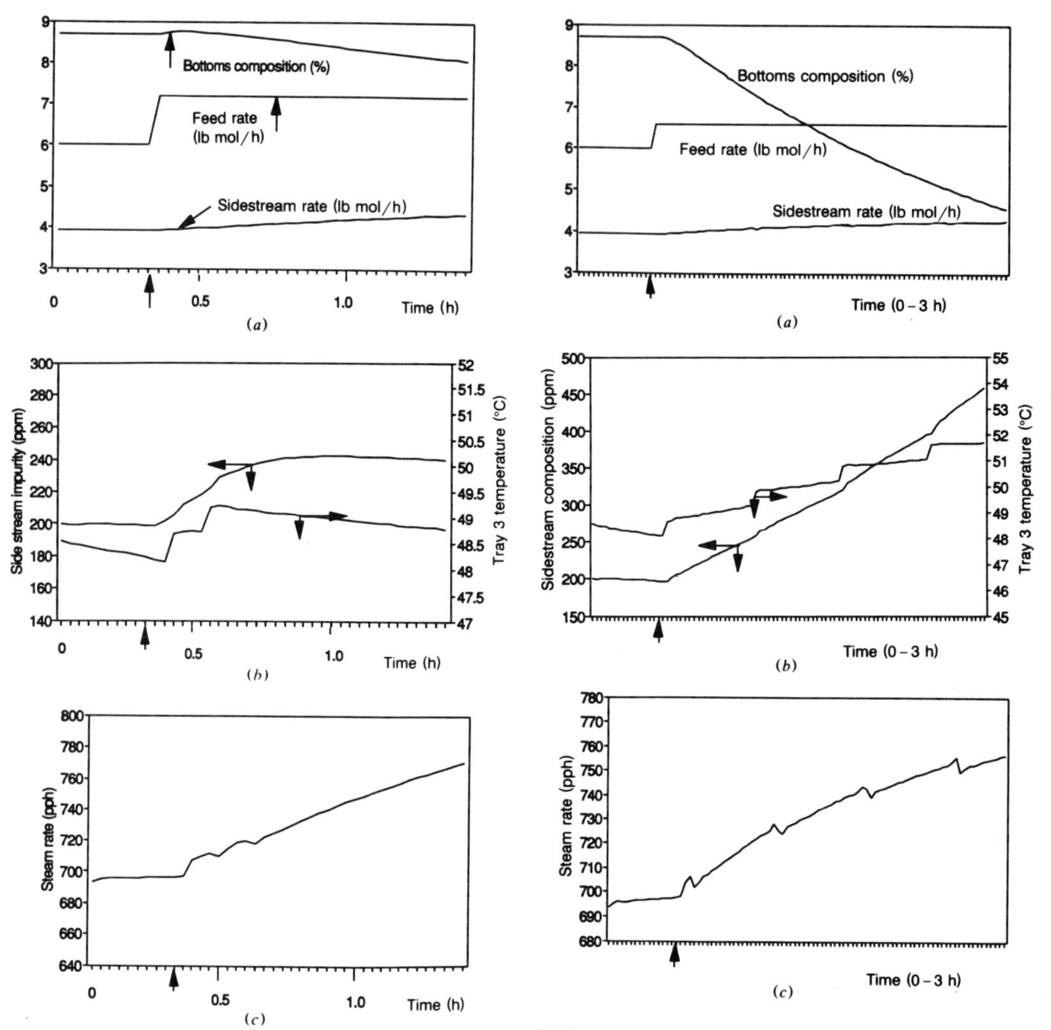

FIGURE 9-12. Inventory controller response to a 20% feed rate increase.

FIGURE 9-13. Inventory controller response (*a*) to a 10% increase in feed rate and a 7% feed composition increase of high boiler; (*b*) to a simultaneous feed rate and feed composition increase; (*c*) to feed rate and feed composition increase.

Figure 9-11 shows the proposed inventory control scheme for this column. Figure 9-12 shows how the inventory controllers react to a feed rate disturbance whereas Figure 9-13 shows the response to a combined feed rate and feed composition disturbance. In both cases the basic inventory controls handle the disturbances well, making composition control fairly simple.

The remaining steady-state manipulative variables are tails flow to steam ratio, sidestream to reflux ratio, and condenser cooling. For composition control I would choose the tails flow to steam ratio because this affects the vapor rate in the column. The condenser cooling and the sidestream to reflux ratio are then left for optimizing controls. Again there are several options for setting these remaining manipulative variables. We could use dual or triple tray temperature controls to minimize energy consumption or we could maximize utilities for best recovery. As before the optimum policy for these variables should come from a plantwide optimization but for this column chances are that maximum cooling in the condenser could be justified to minimize vent losses of product material. The sidestream to reflux ratio will then determine the recovery of product from the bottoms stream.

9-5 CONCLUSION

An empirical method for selection of distillation column controller structures has been presented. We first determine how many degrees of freedom exist for closing inventory loops on the process. These inventory loops are then assigned by using manipulative variables with the most impact on the controlled variables. Ratio controls are provided at this step to prevent a near loss of dynamic degree of freedom from internal recycles. The resulting inventory control scheme with appropriate ratio controls is then tested through dynamic simulation to verify its disturbance rejection capability. The ratio controllers usually provide excellent disturbance rejection toward feed rate changes, eliminating the need for separate feedforward controllers.

Composition controllers are assigned next. Here the objective is to find fast loops that measure composition or a surrogate for composition. This usually amounts to measuring a tray temperature and controlling it with a manipulator that directly or indirectly affects the vapor rate in the column. After the composition loops are assigned, any remaining manipulative variables can be used for steady-state optimization. This could amount to something as simple as holding a reflux rate constant or maximizing condenser cooling, but it could also be the result of running a plantwide optimizer.

References

Buckley, P. S. (1964). Overall process control. *Techniques of Process Control*. New York: Wiley, Chapter 13.

Downs, J. J. and Doss, J. E. (1991). Present status and future needs—A view from North American industry. Paper presented at the Fourth International Conference on Chemical Process Control, CPC IV, South Padre Island, 1991.

Fisher, W. R., Doherty, M. F., and Douglas, J. M. (1988). The interface between design and control. 3. Selecting a set of controlled variables. *Ind. Eng. Chem. Res.* 27, 611.

Shinskey, F. G. (1985). Disturbance-rejection capabilities of distillation control systems. Proceedings of the American Control Conference, Boston, MA, June 19–21.

Skogestad, S., Jacobsen, E. W., and Lundstrom, P. (1990). Selecting the best distillation control structure. *AIChE J.* 36, 753–764.

Tyreus, B. D. (1987). Optimization and multivariable control of distillation columns. *Proceedings of the ISA-87, Anaheim, California.*

Waller, K. V., Finnerman, D. H., Sandelin, P. M., Haggblom, K. E., and Gustafsson, S. E. (1988a). An experimental comparison of four control structures for two-point control of distillation. *Ind. Eng. Chem. Res.* 27, 624.

Waller, K. V., Haggblom, K. E., Sandelin, P. M., and Finnerman, D. H. (1988b). Disturbance sensitivity of distillation control structures. *AIChE J.* 34, 853–858.

10

Control Structures, Consistency, and Transformations

Kurt E. Häggblom and Kurt V. Waller

Åbo Akademi

10-1 INTRODUCTION

The operation of a multivariable process like a distillation column has to satisfy several control objectives. Typical objectives are to ensure the stability of the process, to produce specified products, and to optimize the operation economically. As the various objectives may be of quite different importance and normally require control actions at different time rates, it is usually desirable to decompose the full system into a number of subsystems according to the objectives.

For a distillation column it is natural to decompose the control problem into at least two subsystems: inventory control and product control. Inventory control is necessary in order to make continuous operation possible, but apart from that, the inventory control is of minor importance compared to product control. In practice, the subsystems are usually further decomposed into single-input–single-output (SISO) systems. The main reason for this is that a multiloop SISO system is easier to design and operate than a true multi-input–multi-output (MIMO) system.

The decomposition can be done in many different ways. For a two-product distillation column, three common control structures are the *energy balance structure* (L, V) and the *material balance structures* (D, V) and (L, B). Most industrial distillation columns on two-point control are probably operated by one of these control structures. In general, however, control structures with better properties in terms of control loop interaction, disturbance sensitivity and input–output linearity can be obtained by using compound input and/or output variables that are functions of the basic process variables.

The use of various flow ratios as manipulated variables has been recommended for a long time as a means of reducing interaction and disturbance sensitivity (for an early summary, see Rademaker, Rijnsdorp, and Maarleveld 1975; see also Ryskamp, 1980; Takamatsu, Hashimoto, and Hashimoto, 1982; Shinskey, 1984; Kridiotis and Georgakis, 1986; Häggblom and Waller, 1989; Skogestad, Lundström, and Jacobsen, 1990). Also control of sums and differences of compositions or temperatures instead of the compositions and temperatures themselves has been suggested as a way of reducing interaction (for a review, see Waller and Finnerman, 1987). The use of logarithmic concentrations as controlled variables has been proposed as a simple way of reducing the nonlinearity of the plant (Ryskamp, 1982; Skogestad and Morari, 1988).

In principle, the basic variables can be transformed in an unlimited number of ways resulting in an unlimited number of possible control structures with different sets of input and output variables. Even if the above-mentioned suggestions only are considered, the number of different control structures for a distillation column is huge. Because the properties of the control structures and distillation columns vary considerably, the selection of an appropriate control structure is undoubtedly one of the most important tasks—perhaps the most important one—in the design of distillation control systems.

A prerequisite for a rational selection of control structure is that one can assess the feasibility of different structures. In practice, this assessment requires that all control structures under consideration be described by simple analytical models. It is possible to determine such models from simulations or experiments with the control structures in question, but obviously this is not an attractive approach. Furthermore, it is not a trivial task even to determine an adequate linear steady-state model from simulated data for a nonlinear process (Häggblom, 1988b). Therefore, it would be desirable to have methods by which models for different control structures could be derived analytically from a single process model. Such an approach would be time saving and it would also make the whole procedure of control structure selection suitable for computer implementation.

Actually, methods for modelling control structures have been proposed and used in the literature. McAvoy (1983) and Shinskey (1984), for example, use approximate analytical models in the derivation of linear models for different distillation control structures. Other attempts at deriving control structure models have been made, for example, by Tsogas and McAvoy (1981), Kridiotis and Georgakis (1986), Skogestad and Morari (1987a), Takamatsu, Hashimoto, and Hashimoto (1987), and Koung and MacGregor (1991). All these methods are based on the assumption of "constant flows" in the column sections. As this assumption requires that the heat of vaporization of the mixture distilled is independent of its composition, it may be a very inaccurate assumption for real columns. Another drawback with the methods based on this assumption is that the boilup rate usually is used as a manipulated variable. In reality, however, the boilup rate is always a dependent variable that cannot be manipulated directly.

Häggblom (1986) and Häggblom and Waller (1986, 1988a) have developed rigorous analytical methods, which do not require any limiting assumptions, for the modelling of control structures based on steady-state models. Also Skogestad and Morari (1987b) have presented similar methods. By these methods, the gains of any control structure that can be described by a steady-state model can be determined from the gains of any other adequately modelled control structure. It has also been shown that the process gains of a control structure have to satisfy certain consistency relations, which follow from the external material balances for the column. These relations can, for example, be used to show that the assumption of "constant flows" in the column sections generally is incorrect and even unnecessary.

Recently the methods for modelling of control structures have been extended to dynamic models (Yang et al., 1990; Häggblom and Waller, 1991). The general dynamic case is much more complicated than the steady-state case. There are two situations, however, when dynamic models can be treated essentially as steady-state models (Häggblom and Waller, 1988a, 1991). One situation is when the reflux drum and reboiler holdups can be considered perfectly controlled. The other situation is when the same inventory control system is used in different control structures. This is possible, for example, if the reflux drum holdup is controlled by $L + D$ and the reboiler holdup by $V' + B$, where V' is an estimate of the boilup rate. Such a "generalized" inventory control system would in many cases be superior to the inventory control systems more commonly used.

For pedagogical reasons we will mainly deal with steady-state models in this chapter. As previously implied, this is not necessarily a severe limitation. Furthermore, most methods for rating control structures require only linear steady-state models. The most popular of these methods is the traditional relative gain analysis (Bristol, 1966; McAvoy, 1983; Shinskey, 1984). It has also been shown that many important closed-loop properties can be derived from the steady-state behavior of a process (Grosdidier, Morari, and Holt 1985; Koppel, 1985). Such properties are closed-loop stability, sensor and actuator failure tolerance, feasibility of decentralized control, and robustness with respect to modelling errors.

In addition to a number of numerical illustrations of the basic modelling principles, we will also present some more advanced applications. One application is the identification of control structure models from numerical (simulated or experimental) data. The parameters of the models should be determined subject to known consistency and other relations; otherwise the resulting models will almost surely violate these relations. This could, for example, mean that they violate the external material balances (i.e., the law of mass conservation).

Another application is relative gain analysis. We will show that the relative gains of different control structures are related to each other through exact analytical relationships. These relationships can be used, for example, to calculate relative gains for control structures for which no model has been determined.

Two other applications are concerned with the synthesis of control structures with specified properties. In one of the applications, decoupled control structures are synthesized by means of transformations of output variables. For the material balance structures (D, V) and (L, B) a very interesting result is obtained: It is possible to get perfect one-way steady-state decoupling even without a process model; it is sufficient to know some (nominal) steady-state data. In the other application, transformations of input variables are used to synthesize such a control structure that, in steady state, the primary control loops are completely decoupled and the primary outputs are insensitive to certain disturbances even in open-loop operation.

10-2 SOME BASIC PROPERTIES OF DISTILLATION CONTROL STRUCTURES

In this section we will discuss some fundamental properties of distillation control structures, using three commonly used control structures as examples. Somewhat surprisingly, these properties are largely unrecognized even among workers in the field. We will also briefly discuss some applications where these properties are important.

The basic notation and the form of the control structure models used in this chapter are also introduced.

10-2-1 Energy Balance Structure (L, V)

The *energy balance structure* can be considered to be the standard control structure for dual composition control of distillation. In this control structure, the reflux flow rate L and the boilup manipulator V (which can be any variable that directly affects the boilup rate, e.g., the steam flow to the reboiler) are used to control the outputs associated with the product specifications (the "primary" outputs, which usually are concentrations or temperatures in the column). The control structure is often also called the (L, V) structure after the primary manipulators. The reflux drum and reboiler holdups (the "secondary" outputs) are usually controlled by the distillate flow rate D and the bottoms flow rate B, but there are also other possibilities, as will be discussed. Figure 15-3 in Chapter 15 shows an implementation of this control scheme with D and B as inventory control manipulators. (To simplify the treatment, the pressure control loop is neglected.)

A steady-state model (i.e., the steady-state part of a dynamic model) for the (L, V) structure can be written

$$\begin{bmatrix} \Delta y \\ \Delta x \end{bmatrix} = \begin{bmatrix} K_{yL}^{LV} & K_{yV}^{LV} \\ K_{xL}^{LV} & K_{xV}^{LV} \end{bmatrix} \begin{bmatrix} \Delta L \\ \Delta V \end{bmatrix} + \begin{bmatrix} K_{yF}^{LV} & K_{yz}^{LV} \\ K_{xF}^{LV} & K_{xz}^{LV} \end{bmatrix} \begin{bmatrix} \Delta F \\ \Delta z \end{bmatrix} \tag{10-1a}$$

$$\begin{bmatrix} \Delta D \\ \Delta B \end{bmatrix} = \begin{bmatrix} K_{DL}^{LV} & K_{DV}^{LV} \\ K_{BL}^{LV} & K_{BV}^{LV} \end{bmatrix} \begin{bmatrix} \Delta L \\ \Delta V \end{bmatrix} + \begin{bmatrix} K_{DF}^{LV} & K_{Dz}^{LV} \\ K_{BF}^{LV} & K_{Bz}^{LV} \end{bmatrix} \begin{bmatrix} \Delta F \\ \Delta z \end{bmatrix} \tag{10-1b}$$

where y and x are the primary outputs. Two disturbance variables, the feed flow rate F and the feed composition z, are also included. Δ denotes a deviation from a nominal steady-state value (e.g., $\Delta y = y - \bar{y}$, where $\bar{y}$ is the nominal steady-state value for y). The superscript LV is used to distinguish the gain between two variables (denoted by subscripts on K) in the (L, V) structure from the gain between the same two variables in another control structure (see Equations 10-2 and 10-3).

This model of the (L, V) structure may not look familiar. When transfer function models are used in the literature, usually only the first part, Equation 10-1a, is considered or even recognized. However, when the inventory control loops are closed, the inventory control manipulators become dependent variables, and Equation 10-1b shows how they depend on L, V, F, and z. The numerical values of the flow gains in this equation are far from irrelevant. They mainly depend on the properties of the column itself, not on the inventory control system. Therefore, both the primary and secondary gains in any control structure where D or B, or some function of them, is used as a primary manipulator will depend on these flow gains. This means that the flow gains are needed if a model for such a control structure is to be derived from the model for the (L, V) structure. See Sections 10-4 and 10-5.

The flow gains are also important in certain "generalized" inventory control schemes (Häggblom and Waller, 1991). For example, it would be optimal from a steady-state point of view to control the reflux drum holdup by $\Delta D - K_{DL}^{LV} \Delta L - K_{DV}^{LV} \Delta V$ if F and z are not measured. If ν_D is the controller output, D would then be determined from $\Delta D = \Delta \nu_D + K_{DL}^{LV} \Delta L + K_{DV}^{LV} \Delta V$. This means that changes in L and V are fed forward to D before they affect the holdup in the reflux drum. A suboptimal approach would be to use feedforward only from L, that is, to calculate D from $\Delta D = \Delta \nu_D + K_{DL}^{LV} \Delta L$. Note that D is still a dependent variable in these generalized inventory control schemes. This means that Equations 10-1 may be used to describe the (L, V) structure irrespectively of how the inventory control system is implemented.

Most works on control structure modelling reported to date are based on the assumption of "constant flows" in the column sections. Basically, this assumption requires that the heat of vaporization (expressed in units consistent with the flow rates) of the mixture distilled is independent of its composition and that the reflux stream is saturated liquid. According to this assumption, the flow gains in Equation 10-1b would have the following values: $K_{DL}^{LV} = -K_{BL}^{LV} = -1$, $K_{DV}^{LV} = -K_{BV}^{LV} = 1$ (if V denotes boilup), $K_{DF}^{LV} = 1 - K_{BF}^{LV} = q_F$ (where q_F expresses the degree of vaporization of the feed such that $q_F = 0$ for saturated liquid and $q_F = 1$ for saturated vapor), and $K_{Dz}^{LV} = K_{Bz}^{LV} = 0$. These flow gains mean that the relationship $\Delta D = -\Delta L + \Delta V + q_F \Delta F$, or $D = -L + V + q_F F$ (+ a constant) would hold.

In Section 10-3 we will show that the flow gains of the (L, V) structure are related to the composition gains in Equation 10-1a through so-called consistency relations that can be rigorously derived from the external material balances for the column. This means that the preceding assumptions can

TABLE 10-1 Distillate Flow Gains for Some Distillation Columns

Separation	Column	K_{DL}^{LV}	K_{DV}^{LV}	K_{DF}^{LV}	K_{Dz}^{LV}
Methanol–water	Pilot-plant[a]	−0.25	0.49	0.37	—
Ethanol–water	Pilot-plant[b]	−0.61	1.35	0.06	0.94
Methanol–ethanol	Simulation[c]	−0.70	0.020	0.13	0.017
Benzene–toluene (col. A)	Simulation[d]	−0.81	0.81	—	—
n-, sec-, tert-butanol	Simulation[e]	−0.78	0.82	0.038	−1.20

[a]Wood and Berry, 1973; Häggblom and Waller, 1988b (mass units, V is steam).
[b]Waller et al., 1988a; Häggblom, 1989 (mass units, V is steam).
[c]Häggblom, 1988b; Häggblom and Waller, 1989 (molar units, V is heat).
[d]McAvoy and Weischedel, 1981; Häggblom and Waller, 1988a (molar units, V is boilup).
[e]Yang et al., 1990 (molar units, V is boilup, z is n-butanol)

be checked if the composition gains and some steady-state data are known. Table 10-1 shows calculated flow gains for some distillation columns taken from the literature. As can be seen, the flow gain K_{DL}^{LV}, for example, is often in the range -0.6 to -0.8, but a value as low as -0.25 has also been found. These values are even more remarkable considering that the reflux in experimental columns with a total condenser (as the column for which $K_{DL}^{LV} = -0.25$) usually is subcooled, in which case one would expect $K_{DL}^{LV} < -1$ according to the assumption of constant flows. This means that the "standard" assumption of constant flows and the corresponding values of the flow gains given previously generally are incorrect. Moreover, the assumption is unnecessary when the flow gains can be determined from the composition gains.

Because of the practical importance of flow gains like K_{DL}^{LV}, an explanation as to why K_{DL}^{LV} often is significantly different from -1 is appropriate. In practice, all mixtures are nonideal and the heat of vaporization, for example, is dependent on the composition of the mixture. This is especially true if the composition is not expressed in molar units. When the reflux flow rate L is increased, the concentration of light component in the column will also increase (if other inputs are constant). This usually means that the heat of vaporization of the mixture in the column will decrease, which has the effect that the amount of vaporized liquid in all parts of the column will increase, even if the heat input to the column (or the vapor boilup in the bottom of the column) remains unchanged. As a result of this, the overhead vapor flow rate will increase, which in turn means that the distillate flow rate D does not decrease as much as L is increased. Hence, $K_{DL}^{LV} > -1$, and in practice, $-1 < K_{DL}^{LV} < 0$. A more rigorous treatment of the flow dynamics is given in Häggblom (1991).

10-2-2 Material Balance Structures (*D*, *V*) and (*L*, *B*)

Two other frequently used control structures are the *material balance structures* (D, V) and (L, B). In the (D, V) structure, D and V are used as primary manipulators whereas L and B usually are used as inventory control manipulators. (Figure 15-4 shows an implementation of this control scheme.) The (D, V) structure may thus be described by the steady-state model

$$\begin{bmatrix} \Delta y \\ \Delta x \end{bmatrix} = \begin{bmatrix} K_{yD}^{DV} & K_{yV}^{DV} \\ K_{xD}^{DV} & K_{xV}^{DV} \end{bmatrix} \begin{bmatrix} \Delta D \\ \Delta V \end{bmatrix} + \begin{bmatrix} K_{yF}^{DV} & K_{yz}^{DV} \\ K_{xF}^{DV} & K_{xz}^{DV} \end{bmatrix} \begin{bmatrix} \Delta F \\ \Delta z \end{bmatrix} \tag{10-2a}$$

$$\begin{bmatrix} \Delta L \\ \Delta B \end{bmatrix} = \begin{bmatrix} K_{LD}^{DV} & K_{LV}^{DV} \\ K_{BD}^{DV} & K_{BV}^{DV} \end{bmatrix} \begin{bmatrix} \Delta D \\ \Delta V \end{bmatrix} + \begin{bmatrix} K_{LF}^{DV} & K_{Lz}^{DV} \\ K_{BF}^{DV} & K_{Bz}^{DV} \end{bmatrix} \begin{bmatrix} \Delta F \\ \Delta z \end{bmatrix} \tag{10-2b}$$

In the (L, B) structure, the opposite decomposition into subsystems is made, that is, the primary outputs are controlled by L and B, and the column inventory is usually controlled by D and V. The steady-state model for the (L, B) structure is thus

$$\begin{bmatrix} \Delta y \\ \Delta x \end{bmatrix} = \begin{bmatrix} K_{yL}^{LB} & K_{yB}^{LB} \\ K_{xL}^{LB} & K_{xB}^{LB} \end{bmatrix} \begin{bmatrix} \Delta L \\ \Delta B \end{bmatrix} + \begin{bmatrix} K_{yF}^{LB} & K_{yz}^{LB} \\ K_{xF}^{LB} & K_{xz}^{LB} \end{bmatrix} \begin{bmatrix} \Delta F \\ \Delta z \end{bmatrix} \qquad \text{(10-3a)}$$

$$\begin{bmatrix} \Delta D \\ \Delta V \end{bmatrix} = \begin{bmatrix} K_{DL}^{LB} & K_{DB}^{LB} \\ K_{VL}^{LB} & K_{VB}^{LB} \end{bmatrix} \begin{bmatrix} \Delta L \\ \Delta B \end{bmatrix} + \begin{bmatrix} K_{DF}^{LB} & K_{Dz}^{LB} \\ K_{VF}^{LB} & K_{Vz}^{LB} \end{bmatrix} \begin{bmatrix} \Delta F \\ \Delta z \end{bmatrix} \qquad \text{(10-3b)}$$

As for the (L, V) structure, Equations 10-2b and 10-3b are valid irrespectively of how the inventory control system actually is implemented (provided that the implementation is compatible with the choice of primary manipulators in the control structure).

These control structures may seem very similar to the (L, V) structure. The only difference between the (L, V) and the (D, V) structure is that the roles of L and D are switched. However, this has a profound effect on the properties of the control structures; the properties are generally very different (Waller et al., 1988a, 1988b). This also applies to the disturbance gains, although they are gains between the same input and output variables. This has to do with the fact that the assumption of constant flows does not generally hold. All these differences are disclosed by the analytical relations existing between the gains in the different control structures (see Section 10-4).

A generalized inventory control scheme similar to the ones discussed for the (L, V) structure has been used with the (D, V) structure for a long time in industrial operation. Van Kampen (1965) (see Shinskey, 1984, pages 128–129) used an inventory control scheme for the reflux drum where a change in D was fed forward to a corresponding change in L. Van Kampen found it beneficial to increase the factor k by which variations in the distillate was amplified to above unity. The reflux flow rate was thus determined from $\Delta L = \Delta \nu_D - k\,\Delta D$, where ν_D is the output from the inventory controller. From Equation 10-2b it follows that k should ideally be chosen as $k = -K_{LD}^{DV}$. Because $K_{LD}^{DV} = 1/K_{DL}^{LV}$ (see Section 10-4), van Kampen's suggestion that $k > 1$ is in agreement with $-1 < K_{DL}^{LV} < 0$, which is the normal situation in "real" columns (see Table 10-1 and Section 10-3).

An interesting property of the two material balance structures is that the primary gains in each control structure are not independent of each other if the primary outputs are product compositions. For example, for the (D, V) structure it can be shown (see Section 10-3) that $\bar{D}K_{yV}^{DV} + \bar{B}K_{xV}^{DV} = 0$. Similar relations exist between other primary gains both in the (D, V) and the (L, B) structure. This has important practical consequences when models for these control structures are determined through identification from numerical (experimental or simulated) data: the gains should be determined in such a way that these consistency relations are satisfied; otherwise the external material balances (i.e., the law of mass conservation) will be violated.

For the material balance structures it is not possible to determine all flow gains from consistency relations. If the model is determined from numerical data, this means that some of the flow gains have to be explicitly determined from such data. It is then possible to design, for example, a generalized inventory control scheme. It is also possible to derive models for other control structures, for example the (L, V) structure. Without the flow gains, this is not possible. If a complete steady-state model, including all flow gains, is known for a control structure, it is possible to derive the model for any other control structure (see Section 10-5).

10-3 CONSISTENCY RELATIONS

We will show that the gains of a control structure have to satisfy certain consistency relations, which follow from the external material balances for the column. In principle, consistency relations can also be derived from an external energy balance, but their usefulness is quite limited, because the energy balance contains variables that usually are not (or cannot be) measured (e.g., energy losses) or that are not important enough to be explicitly included in a model (e.g., stream enthalpies).

In this section we will only consider the (L, V), (D, V), and (L, B) structures. We will handle the general case in Section 10-5.

10-3-1 (L, V) Structure

Consider a simple distillation column with one feed stream (F) and two product streams (D and B). Recall the steady-state model for the (L, V) structure:

$$\begin{bmatrix} \Delta y \\ \Delta x \end{bmatrix} = \begin{bmatrix} K_{yL}^{LV} & K_{yV}^{LV} \\ K_{xL}^{LV} & K_{xV}^{LV} \end{bmatrix} \begin{bmatrix} \Delta L \\ \Delta V \end{bmatrix} + \begin{bmatrix} K_{yF}^{LV} & K_{yz}^{LV} \\ K_{xF}^{LV} & K_{xz}^{LV} \end{bmatrix} \begin{bmatrix} \Delta F \\ \Delta z \end{bmatrix} \quad (10\text{-}1a)$$

$$\begin{bmatrix} \Delta D \\ \Delta B \end{bmatrix} = \begin{bmatrix} K_{DL}^{LV} & K_{DV}^{LV} \\ K_{BL}^{LV} & K_{BV}^{LV} \end{bmatrix} \begin{bmatrix} \Delta L \\ \Delta V \end{bmatrix} + \begin{bmatrix} K_{DF}^{LV} & K_{Dz}^{LV} \\ K_{BF}^{LV} & K_{Bz}^{LV} \end{bmatrix} \begin{bmatrix} \Delta F \\ \Delta z \end{bmatrix} \quad (10\text{-}1b)$$

where y, x, and z now specifically denote the (mole or weight) fraction of the same component in the distillate, bottoms, and feed stream, respectively. This model has to satisfy the external material balances for the column, that is,

$$D + B = F \quad (10\text{-}4a)$$

$$Dy + Bx = Fz \quad (10\text{-}4b)$$

Linearization of the material balances at the nominal steady state $(\bar{D}, \bar{B}, \bar{F}, \bar{y}, \bar{x}, \bar{z})$ gives

$$\Delta D + \Delta B = \Delta F \quad (10\text{-}5a)$$

$$\bar{D}\,\Delta y + \bar{y}\,\Delta D + \bar{B}\,\Delta x + \bar{x}\,\Delta B = \bar{F}\,\Delta z + \bar{z}\,\Delta F \quad (10\text{-}5b)$$

The input variables in Equations 10-1 are independent of each other. Mathematically this means that Equations 10-1 and 10-5 have to hold regardless of the values of the input variables. Suppose, for example, that $\Delta L \neq 0$ and that all other inputs $(\Delta V, \Delta F, \Delta z)$ are 0. Elimination of ΔD and ΔB from Equation 10-5a by Equation 10-1b then gives Equation 10-6a, whereas elimination of ΔD, ΔB, Δy, and Δx from Equation 10-5b by Equations 10-1a and 10-1b gives Equation 10-7a. In a similar way, the consistency relations associated with the other inputs can be derived, giving

$$K_{DL}^{LV} + K_{BL}^{LV} = 0 \quad (10\text{-}6a)$$

$$K_{DV}^{LV} + K_{BV}^{LV} = 0 \quad (10\text{-}6b)$$

$$K_{DF}^{LV} + K_{BF}^{LV} = 1 \quad (10\text{-}6c)$$

$$K_{Dz}^{LV} + K_{Bz}^{LV} = 0 \quad (10\text{-}6d)$$

and

$$\bar{D}K_{yL}^{LV} + \bar{y}K_{DL}^{LV} + \bar{B}K_{xL}^{LV} + \bar{x}K_{BL}^{LV} = 0 \quad (10\text{-}7a)$$

$$\bar{D}K_{yV}^{LV} + \bar{y}K_{DV}^{LV} + \bar{B}K_{xV}^{LV} + \bar{x}K_{BV}^{LV} = 0 \quad (10\text{-}7b)$$

$$\bar{D}K_{yF}^{LV} + \bar{y}K_{DF}^{LV} + \bar{B}K_{xF}^{LV} + \bar{x}K_{BF}^{LV} = \bar{z} \quad (10\text{-}7c)$$

$$\bar{D}K_{yz}^{LV} + \bar{y}K_{Dz}^{LV} + \bar{B}K_{xz}^{LV} + \bar{x}K_{Bz}^{LV} = \bar{F} \quad (10\text{-}7d)$$

Equation 10-6a, for example, simply states that a change in L affects D and B so that the sum of the resulting changes in D and B is zero at steady state, which is completely in accordance with Equation 10-5a ($\Delta F = 0$ because L does not affect F). Similarly, the change in L will also, due to the component material balance in Equation 10-5b, affect y, x, D, and B in such a way that Equation 10-7a holds.

If the primary outputs are product compositions expressed as fractions of the same component, the gains in Equations 10-1 have to satisfy Equations 10-6 and 10-7. If the primary outputs are any other kinds of variables (e.g., temperatures), or if they are fractions of different components, only Equations 10-6 apply.

Equations 10-7 can be rearranged by means of Equations 10-6 to give

$$K_{DL}^{LV} = -K_{BL}^{LV} = -\frac{\bar{D}K_{yL}^{LV} + \bar{B}K_{xL}^{LV}}{\bar{y} - \bar{x}} \tag{10-8a}$$

$$K_{DV}^{LV} = -K_{BV}^{LV} = -\frac{\bar{D}K_{yV}^{LV} + \bar{B}K_{xV}^{LV}}{\bar{y} - \bar{x}} \tag{10-8b}$$

$$K_{DF}^{LV} = 1 - K_{BF}^{LV} = \frac{\bar{z} - \bar{x} - \bar{D}K_{yF}^{LV} - \bar{B}K_{xF}^{LV}}{\bar{y} - \bar{x}} \tag{10-8c}$$

$$K_{Dz}^{LV} = -K_{Bz}^{LV} = \frac{\bar{F} - \bar{D}K_{yz}^{LV} - \bar{B}K_{xz}^{LV}}{\bar{y} - \bar{x}} \tag{10-8d}$$

These equations show that if nominal steady-state data and the gains in Equation 10-1a are known, the gains in Equation 10-1b can be calculated when the primary outputs are product compositions expressed as fractions of the same component. Unless the primary gains and the nominal steady-state data are related in a very specific way, the popular assumption of constant flows is obviously incorrect and, in fact, inconsistent with the external material balances. [Recall that this assumption gives the flow gains $K_{DL}^{LV} = -K_{BL}^{LV} = -1$, $K_{DV}^{LV} = -K_{BV}^{LV} = 1$ (if V denotes boilup), $K_{DF}^{LV} = 1 - K_{BF}^{LV} = q_F$, and $K_{Dz}^{LV} = K_{Bz}^{LV} = 0$.] Moreover, the assumption is unnecessary if the primary outputs are fractions of the same component. If the primary outputs do not fulfill the requirements (e.g., because they are temperatures), Equations 10-8 do not apply and the assumption of constant flows cannot be checked, but this does not make the assumption correct, of course.

Table 10-1 shows values of the distillate flow gains for five columns studied in the literature. The listed references are the original source of the model and the work where the gains were determined. The unit for the disturbance in the feed composition is weight or mole percent of the light component, except for the ternary butanol column, where it is mole percent of the least volatile component. It should also be noted that the boilup manipulator varies in the five cases. For the methanol–water and the ethanol–water column it is the flow rate of steam to the reboiler, for the methanol–ethanol column it is the flow rate of heat from the reboiler to the column, and for the benzene–toluene and the ternary butanol column it is the boilup rate. As can be seen, the flow gains of "real" columns can drastically differ from the flow gains following from the assumption of constant flows.

Note that although the flow gains in Equation 10-1b often can be calculated from consistency relations (Equations 10-6 and 10-7), it is advisable when a model is determined from numerical data first to estimate all gains, including the flow gains, then to reconcile them subject to the consistency relations. This will reduce the uncertainty (variance) of the parameter estimates (Häggblom, 1988b).

10-3-2 (D, V) Structure

Let us next consider the (D, V) structure, which is described by the steady-state model

$$\begin{bmatrix} \Delta y \\ \Delta x \end{bmatrix} = \begin{bmatrix} K_{yD}^{DV} & K_{yV}^{DV} \\ K_{xD}^{DV} & K_{xV}^{DV} \end{bmatrix} \begin{bmatrix} \Delta D \\ \Delta V \end{bmatrix} + \begin{bmatrix} K_{yF}^{DV} & K_{yz}^{DV} \\ K_{xF}^{DV} & K_{xz}^{DV} \end{bmatrix} \begin{bmatrix} \Delta F \\ \Delta z \end{bmatrix} \tag{10-2a}$$

$$\begin{bmatrix} \Delta L \\ \Delta B \end{bmatrix} = \begin{bmatrix} K_{LD}^{DV} & K_{LV}^{DV} \\ K_{BD}^{DV} & K_{BV}^{DV} \end{bmatrix} \begin{bmatrix} \Delta D \\ \Delta V \end{bmatrix} + \begin{bmatrix} K_{LF}^{DV} & K_{Lz}^{DV} \\ K_{BF}^{DV} & K_{Bz}^{DV} \end{bmatrix} \begin{bmatrix} \Delta F \\ \Delta z \end{bmatrix} \tag{10-2b}$$

Consistency relations for this control structure may be derived in a similar way as those for the (L, V) structure, except that now the independent variables are D, V, F, and z. When independent changes in these variables are considered, one at a time, Equation 10-2b and the total material balance, Equation 10-5a, give the expressions

$$K_{BD}^{DV} = -1, \qquad K_{BV}^{DV} = 0,$$
$$K_{BF}^{DV} = 1, \qquad K_{Bz}^{DV} = 0 \qquad (10\text{-}9)$$

When these expressions are taken into account, Equations 10-2 and the component material balance, Equation 10-5b, give the consistency relations

$$\bar{D}K_{yD}^{DV} + \bar{B}K_{xD}^{DV} = \bar{x} - \bar{y} \qquad (10\text{-}10a)$$

$$\bar{D}K_{yV}^{DV} + \bar{B}K_{xV}^{DV} = 0 \qquad (10\text{-}10b)$$

$$\bar{D}K_{yF}^{DV} + \bar{B}K_{xF}^{DV} = \bar{z} - \bar{x} \qquad (10\text{-}10c)$$

$$\bar{D}K_{yz}^{DV} + \bar{B}K_{xz}^{DV} = \bar{F} \qquad (10\text{-}10d)$$

Equations 10-9 hold regardless of the type of primary outputs, whereas Equations 10-10 require the primary outputs to be product compositions.

As is evident from Equations 10-10, the primary gains in the (D, V) structure, that is, the gains in Equation 10-2a, are not independent of each other when the primary outputs are product compositions. If a model for the (D, V) structure is determined from numerical data, the primary gains should thus be determined subject to Equations 10-10. It is not necessary to estimate the secondary gains associated with the inventory control manipulator B because they are exactly given by Equations 10-9. If they are estimated, however, they give an indication of the reliability of the flow rate measurements.

The consistency relations do not give any information about the secondary gains associated with the inventory control manipulator L. Therefore, these gains always have to be included in a complete model for the (D,V) structure. However, in practice this never seems to be done, although the gains are needed, for example, if a model for the (L, V) structure is to be derived from a model for the (D, V) structure. Normal practice is to apply the generally incorrect assumption of constant flows in the column sections, in which case the gains would have the values $K_{LD}^{DV} = -1$, $K_{LV}^{DV} = 1$ (if V denotes boilup), $K_{LF}^{DV} = q_F$, and $K_{Lz}^{DV} = 0$. The fact that the assumption of constant flows cannot be checked from the consistency relations for the (D, V) structure is not a justification for using it.

10-3-3 (L, B) Structure

Consistency relations for the (L, B) structure can be derived in a similar way as those for the (D, V) structure. In the (L, B) structure, described by the model

$$\begin{bmatrix} \Delta y \\ \Delta x \end{bmatrix} = \begin{bmatrix} K_{yL}^{LB} & K_{yB}^{LB} \\ K_{xL}^{LB} & K_{xB}^{LB} \end{bmatrix} \begin{bmatrix} \Delta L \\ \Delta B \end{bmatrix} + \begin{bmatrix} K_{yF}^{LB} & K_{yz}^{LB} \\ K_{xF}^{LB} & K_{xz}^{LB} \end{bmatrix} \begin{bmatrix} \Delta F \\ \Delta z \end{bmatrix} \qquad (10\text{-}3a)$$

$$\begin{bmatrix} \Delta D \\ \Delta V \end{bmatrix} = \begin{bmatrix} K_{DL}^{LB} & K_{DB}^{LB} \\ K_{VL}^{LB} & K_{VB}^{LB} \end{bmatrix} \begin{bmatrix} \Delta L \\ \Delta B \end{bmatrix} + \begin{bmatrix} K_{DF}^{LB} & K_{Dz}^{LB} \\ K_{VF}^{LB} & K_{Vz}^{LB} \end{bmatrix} \begin{bmatrix} \Delta F \\ \Delta z \end{bmatrix} \qquad (10\text{-}3b)$$

L, B, F, and z are independent variables.

When independent changes in these variables are considered, Equations 10-3b and 10-5a give the consistency relations

$$K_{DL}^{LB} = 0, \qquad K_{DB}^{LB} = -1,$$
$$K_{DF}^{LB} = 1, \qquad K_{Dz}^{LB} = 0 \qquad (10\text{-}11)$$

Similarly, Equations 10-3 and 10-5b together with Equation 10-11 give the consistency relations

$$\bar{D}K_{yL}^{LB} + \bar{B}K_{xL}^{LB} = 0 \qquad (10\text{-}12a)$$

$$\bar{D}K_{yB}^{LB} + \bar{B}K_{xB}^{LB} = \bar{y} - \bar{x} \qquad (10\text{-}12b)$$

$$\bar{D}K_{yF}^{LB} + \bar{B}K_{xF}^{LB} = \bar{z} - \bar{y} \qquad (10\text{-}12c)$$

$$\bar{D}K_{yz}^{LB} + \bar{B}K_{xz}^{LB} = \bar{F} \qquad (10\text{-}12d)$$

Equations 10-11 hold regardless of the type of primary outputs, whereas Equations 10-12 require the primary outputs to be product compositions.

Similar conclusions can be drawn for the (L, B) structure as for the (D, V) structure. For example, when process gains are numerically determined, they should be determined subject to the preceding consistency relations. Furthermore, the flow gains K_{VL}^{LB}, K_{VB}^{LB}, K_{VF}^{LB}, and K_{Vz}^{LB} must always be given in a complete model; these gains cannot be calculated from consistency relations.

10-4 TRANSFORMATIONS BETWEEN CONTROL STRUCTURES

A multivariable process is generally controllable with several different control structures. The control variables in these control structures may be linear or nonlinear functions of the true physical variables that affect the process. Usually it is also possible to control a distillation column by completely different sets of manipulators in different control structures. A further possibility is to control linear or nonlinear transformations of the basic output variables.

In this section we will treat only transformations of input variables. An application of output variable transformations is given in Section 10-10. Furthermore, we will only consider transformations between the (L, V), (D, V), and (L, B) structures in this section. We will deal with the general case in Section 10-5.

10-4-1 Transformation from (L, V) to (D, V)

Recall the steady-state model for the (L, V) structure,

$$\begin{bmatrix} \Delta y \\ \Delta x \end{bmatrix} = \begin{bmatrix} K_{yL}^{LV} & K_{yV}^{LV} \\ K_{xL}^{LV} & K_{xV}^{LV} \end{bmatrix} \begin{bmatrix} \Delta L \\ \Delta V \end{bmatrix} + \begin{bmatrix} K_{yF}^{LV} & K_{yz}^{LV} \\ K_{xF}^{LV} & K_{xz}^{LV} \end{bmatrix} \begin{bmatrix} \Delta F \\ \Delta z \end{bmatrix} \qquad (10\text{-}1a)$$

$$\begin{bmatrix} \Delta D \\ \Delta B \end{bmatrix} = \begin{bmatrix} K_{DL}^{LV} & K_{DV}^{LV} \\ K_{BL}^{LV} & K_{BV}^{LV} \end{bmatrix} \begin{bmatrix} \Delta L \\ \Delta V \end{bmatrix} + \begin{bmatrix} K_{DF}^{LV} & K_{Dz}^{LV} \\ K_{BF}^{LV} & K_{Bz}^{LV} \end{bmatrix} \begin{bmatrix} \Delta F \\ \Delta z \end{bmatrix} \qquad (10\text{-}1b)$$

Equation 10-1b shows how all the variables that can be used to affect the process are related to each other when the column inventory is controlled. For continuous distillation, these steady-state relations have to hold regardless of the control structure. Therefore, they can be rearranged into any form suitable to represent a given control structure.

Consider now the (D, V) structure, where the holdup in the reflux drum instead of by D is controlled by L, which thus becomes a dependent variable. Explicitly solved for this variable, the upper part of Equation 10-1b gives

$$\Delta L = \frac{1}{K_{DL}^{LV}} \Delta D - \frac{K_{DV}^{LV}}{K_{DL}^{LV}} \Delta V - \frac{K_{DF}^{LV}}{K_{DL}^{LV}} \Delta F - \frac{K_{Dz}^{LV}}{K_{DL}^{LV}} \Delta z \quad (10\text{-}13)$$

Elimination of ΔL from Equations 10-1 by means of Equation 10-13 now gives the model for the (D, V) structure:

$$\begin{bmatrix} \Delta y \\ \Delta x \end{bmatrix} = \begin{bmatrix} K_{yD}^{DV} & K_{yV}^{DV} \\ K_{xD}^{DV} & K_{xV}^{DV} \end{bmatrix} \begin{bmatrix} \Delta D \\ \Delta V \end{bmatrix} + \begin{bmatrix} K_{yF}^{DV} & K_{yz}^{DV} \\ K_{xF}^{DV} & K_{xz}^{DV} \end{bmatrix} \begin{bmatrix} \Delta F \\ \Delta z \end{bmatrix} \quad (10\text{-}2a)$$

$$\begin{bmatrix} \Delta L \\ \Delta B \end{bmatrix} = \begin{bmatrix} K_{LD}^{DV} & K_{LV}^{DV} \\ K_{BD}^{DV} & K_{BV}^{DV} \end{bmatrix} \begin{bmatrix} \Delta D \\ \Delta V \end{bmatrix} + \begin{bmatrix} K_{LF}^{DV} & K_{Lz}^{DV} \\ K_{BF}^{DV} & K_{Bz}^{DV} \end{bmatrix} \begin{bmatrix} \Delta F \\ \Delta z \end{bmatrix} \quad (10\text{-}2b)$$

where

$$\begin{bmatrix} K_{yD}^{DV} & K_{yV}^{DV} \\ K_{xD}^{DV} & K_{xV}^{DV} \end{bmatrix} = \begin{bmatrix} \dfrac{K_{yL}^{LV}}{K_{DL}^{LV}} & K_{yV}^{LV} - K_{yL}^{LV} \dfrac{K_{DV}^{LV}}{K_{DL}^{LV}} \\ \dfrac{K_{xL}^{LV}}{K_{DL}^{LV}} & K_{xV}^{LV} - K_{xL}^{LV} \dfrac{K_{DV}^{LV}}{K_{DL}^{LV}} \end{bmatrix} \quad (10\text{-}14a)$$

$$\begin{bmatrix} K_{yF}^{DV} & K_{yz}^{DV} \\ K_{xF}^{DV} & K_{xz}^{DV} \end{bmatrix} = \begin{bmatrix} K_{yF}^{LV} - K_{yL}^{LV} \dfrac{K_{DF}^{LV}}{K_{DL}^{LV}} & K_{yz}^{LV} - K_{yL}^{LV} \dfrac{K_{Dz}^{LV}}{K_{DL}^{LV}} \\ K_{xF}^{LV} - K_{xL}^{LV} \dfrac{K_{DF}^{LV}}{K_{DL}^{LV}} & K_{xz}^{LV} - K_{xL}^{LV} \dfrac{K_{Dz}^{LV}}{K_{DL}^{LV}} \end{bmatrix} \quad (10\text{-}14b)$$

$$\begin{bmatrix} K_{LD}^{DV} & K_{LV}^{DV} \\ K_{BD}^{DV} & K_{BV}^{DV} \end{bmatrix} = \begin{bmatrix} \dfrac{1}{K_{DL}^{LV}} & -\dfrac{K_{DV}^{LV}}{K_{DL}^{LV}} \\ \dfrac{K_{BL}^{LV}}{K_{DL}^{LV}} & K_{BV}^{LV} - K_{BL}^{LV} \dfrac{K_{DV}^{LV}}{K_{DL}^{LV}} \end{bmatrix} \quad (10\text{-}14c)$$

$$\begin{bmatrix} K_{LF}^{DV} & K_{Lz}^{DV} \\ K_{BF}^{DV} & K_{Bz}^{DV} \end{bmatrix} = \begin{bmatrix} -\dfrac{K_{DF}^{LV}}{K_{DL}^{LV}} & -\dfrac{K_{Dz}^{LV}}{K_{DL}^{LV}} \\ K_{BF}^{LV} - K_{BL}^{LV} \dfrac{K_{DF}^{LV}}{K_{DL}^{LV}} & K_{Bz}^{LV} - K_{BL}^{LV} \dfrac{K_{Dz}^{LV}}{K_{DL}^{LV}} \end{bmatrix} \quad (10\text{-}14d)$$

When the consistency relations in Equation 10-6 between the flow gains in the (L, V) structure are taken into account, the gains associated with the output B in Equations 10-14c and 10-14d become

$$K_{BD}^{DV} = -1, \qquad K_{BV}^{DV} = 0,$$
$$K_{BF}^{DV} = 1, \qquad K_{Bz}^{DV} = 0 \quad (10\text{-}9)$$

In Section 10-3-2 we derived the same expressions without using the consistency relations for the (L, V) structure.

The preceding relationships between the gains in the (D, V) structure and the (L, V) structure are valid in general, regardless of the type of primary output variables. If the primary outputs are product compositions, the secondary gains in Equations 10-14 can be eliminated by means of Equations 10-8. Naturally, the resulting expressions for the gains in the (D, V) structure satisfy the consistency relations given in Equations 10-10.

Equations 10-14 illustrate the previously stated importance of flow gains like K_{DL}^{LV} in the modelling of distillation control structures. This generally has not been recognized: Published models are usually given as Equation 10-1a or 10-2a with no flow gains included. If needed, the flow gains are usually chosen according to the assumption of constant flows. As was shown in Section 10-3, this assumption is generally incorrect.

Equation 10-14b clearly shows that disturbances generally affect the outputs differently in two structures, that is, generally the disturbance sensitivity of different structures is different. This has also largely been overlooked in the distillation control literature. One reason for this is the simplifying assumptions commonly used in simulation studies (e.g., the assumption of constant flows), which, for example, result in the flow gains $K_{DF}^{LV} = 0$ and $K_{Dz}^{LV} = 0$. According to Equation 10-14b, the disturbance gains of the (L, V) structure and the (D, V) structure would then be equal. However, from practice it is known that generally this is not true (Waller et al., 1988a, 1988b).

Equations 10-14 are convenient for calculation of the gains in the (D, V) structure if a model for the (L, V) structure is known. If the model for the (D, V) structure is known, the same equations can be used to calculate the gains in the (L, V) structure.

10-4-2 Transformation from (L, V) to (L, B)

A model for the (L, B) structure can be derived from a model for the (L, V) structure in the same way as the model for the (D, V) structure. In the (L, B) structure, the reboiler holdup is controlled by V instead of by B as done in the (L, V) structure. For this situation, the lower part of Equation 10-1b gives

$$\Delta V = \frac{1}{K_{BV}^{LV}} \Delta B - \frac{K_{BL}^{LV}}{K_{BV}^{LV}} \Delta L - \frac{K_{BF}^{LV}}{K_{BV}^{LV}} \Delta F - \frac{K_{Bz}^{LV}}{K_{BL}^{LV}} \Delta z \quad (10\text{-}15)$$

Elimination of ΔV from Equations 10-1 by means of Equation 10-15 gives the model for the (L, B) structure:

$$\begin{bmatrix} \Delta y \\ \Delta x \end{bmatrix} = \begin{bmatrix} K_{yL}^{LB} & K_{yB}^{LB} \\ K_{xL}^{LB} & K_{xB}^{LB} \end{bmatrix} \begin{bmatrix} \Delta L \\ \Delta B \end{bmatrix} + \begin{bmatrix} K_{yF}^{LB} & K_{yz}^{LB} \\ K_{xF}^{LB} & K_{xz}^{LB} \end{bmatrix} \begin{bmatrix} \Delta F \\ \Delta z \end{bmatrix} \quad (10\text{-}3a)$$

$$\begin{bmatrix} \Delta D \\ \Delta V \end{bmatrix} = \begin{bmatrix} K_{DL}^{LB} & K_{DB}^{LB} \\ K_{VL}^{LB} & K_{VB}^{LB} \end{bmatrix} \begin{bmatrix} \Delta L \\ \Delta B \end{bmatrix} + \begin{bmatrix} K_{DF}^{LB} & K_{Dz}^{LB} \\ K_{VF}^{LB} & K_{Vz}^{LB} \end{bmatrix} \begin{bmatrix} \Delta F \\ \Delta z \end{bmatrix} \quad (10\text{-}3b)$$

where

$$\begin{bmatrix} K_{yL}^{LB} & K_{yB}^{LB} \\ K_{xL}^{LB} & K_{xB}^{LB} \end{bmatrix} = \begin{bmatrix} K_{yL}^{LV} - K_{yV}^{LV}\dfrac{K_{BL}^{LV}}{K_{BV}^{LV}} & \dfrac{K_{yV}^{LV}}{K_{BV}^{LV}} \\ K_{xL}^{LV} - K_{xV}^{LV}\dfrac{K_{BL}^{LV}}{K_{BV}^{LV}} & \dfrac{K_{xV}^{LV}}{K_{BV}^{LV}} \end{bmatrix} \quad (10\text{-}16a)$$

$$\begin{bmatrix} K_{yF}^{LB} & K_{yz}^{LB} \\ K_{xF}^{LB} & K_{xz}^{LB} \end{bmatrix} = \begin{bmatrix} K_{yF}^{LV} - K_{yV}^{LV}\dfrac{K_{BF}^{LV}}{K_{BV}^{LV}} & K_{yz}^{LV} - K_{yV}^{LV}\dfrac{K_{Bz}^{LV}}{K_{BV}^{LV}} \\ K_{xF}^{LV} - K_{xV}^{LV}\dfrac{K_{BF}^{LV}}{K_{BV}^{LV}} & K_{xz}^{LV} - K_{xV}^{LV}\dfrac{K_{Bz}^{LV}}{K_{BV}^{LV}} \end{bmatrix} \quad (10\text{-}16b)$$

$$\begin{bmatrix} K_{DL}^{LB} & K_{DB}^{LB} \\ K_{VL}^{LB} & K_{VB}^{LB} \end{bmatrix} = \begin{bmatrix} K_{DL}^{LV} - K_{DV}^{LV}\dfrac{K_{BL}^{LV}}{K_{BV}^{LV}} & \dfrac{K_{DV}^{LV}}{K_{BV}^{LV}} \\ -\dfrac{K_{BL}^{LV}}{K_{BV}^{LV}} & \dfrac{1}{K_{BV}^{LV}} \end{bmatrix} \tag{10-16c}$$

$$\begin{bmatrix} K_{DF}^{LB} & K_{Dz}^{LB} \\ K_{VF}^{LB} & K_{Vz}^{LB} \end{bmatrix} = \begin{bmatrix} K_{DF}^{LV} - K_{DV}^{LV}\dfrac{K_{BF}^{LV}}{K_{BV}^{LV}} & K_{Dz}^{LV} - K_{DV}^{LV}\dfrac{K_{Bz}^{LV}}{K_{BV}^{LV}} \\ -\dfrac{K_{BF}^{LV}}{K_{BV}^{LV}} & -\dfrac{K_{Bz}^{LV}}{K_{BV}^{LV}} \end{bmatrix} \tag{10-16d}$$

When the consistency relations in Equation 10-6 between the flow gains in the (L, V) structure are taken into account, the gains associated with the output D in Equations 10-16c and 10-16d become

$$K_{DL}^{LB} = 0, \qquad K_{DB}^{LB} = -1,$$
$$K_{DF}^{LB} = 1, \qquad K_{Dz}^{LB} = 0 \tag{10-11}$$

In Section 10-3-2 we derived the same expressions without using the consistency relations for the (L, V) structure.

Conclusions similar to those drawn for the (D, V) structure from Equations 10-14 can be drawn for the (L, B) structure from Equations 10-16. Equations 10-16 can be used, for example, to calculate the gains for the (L, B) from the gains for the (L, V) structure or vice versa.

10-5 CONTROL STRUCTURE MODELLING — THE GENERAL CASE

In the last sections we have been dealing with the most basic distillation control structures, the (L, V), (D, V), and (L, B) structures. There is also a fourth "simple" control structure, the (D, B) structure. This control structure has generally been considered infeasible, because it is so from a steady-state point of view (Häggblom and Waller, 1988a). However, Finco, Luyben, and Polleck (1989) have demonstrated experimentally that the (D, B) structure can be used for certain columns. The most important reason why the (D, B) structure can work seems to be the flow dynamics in the column (Häggblom, 1991; Häggblom and Waller, 1991). Because this control structure cannot be modelled by a steady-state model, we will not consider it further in this chapter.

However, there are numerous other control structures, for example, control structures where flow ratios are used as manipulators. Because it gets very tedious and complicated to treat such control structures as explicitly as we have done in the previous sections, and the list of such control structures is endless, we will in this section in a unified manner treat all possible control structures that can be described by steady-state models. Basically, we continue to consider simple two-product distillation, but as we show in Section 10-5-4, the results apply to distillation columns of any complexity.

In order to derive and state the general results we have to use a compact matrix notation. The matrix operations used are very elementary, however; it is mainly the compact *notation* we need.

10-5-1 Compact Description of Control Structures

Consider the model for the (L, V) structure, Equations 10-1. If we define $\mathbf{y} = [y \quad x]^{\mathrm{T}}$, $\mathbf{u} = [L \quad V]^{\mathrm{T}}$, $\mathbf{v} = [D \quad B]^{\mathrm{T}}$, and $\mathbf{w} = [F \quad z]^{\mathrm{T}}$, the model can be compactly expressed as

$$\Delta\mathbf{y} = \mathbf{K}_{\mathbf{yu}}\,\Delta\mathbf{u} + \mathbf{K}_{\mathbf{yw}}\,\Delta\mathbf{w} \tag{10-17a}$$

$$\Delta\mathbf{v} = \mathbf{K}_{\mathbf{vu}}\,\Delta\mathbf{u} + \mathbf{K}_{\mathbf{vw}}\,\Delta\mathbf{w} \tag{10-17b}$$

where $\mathbf{K}_{yu}$, $\mathbf{K}_{vu}$, $\mathbf{K}_{yw}$, and $\mathbf{K}_{vw}$ denote the gain matrices explicitly shown in Equations 10-1.

With suitable definitions of **y**, **u**, **v**, and **w**, this model could be used to describe any control structure, but in the following we will assume that **y**, **u**, **v**, and **w** are defined as before, that is, they are vectors of the basic "physical" variables of the process. We will often in general terms refer to this control structure as the "base" structure and to the variables as follows: **y** is a vector of primary outputs, **u** is a vector of primary manipulators, **v** is a vector of dependent (or secondary) manipulators, and **w** is a vector of disturbance variables.

Consider now a control structure where $\boldsymbol{\psi}$ is the vector of primary manipulators and $\boldsymbol{\nu}$ the vector of dependent manipulators (due to inventory control). This control structure can be described by the model

$$\Delta \mathbf{y} = \mathbf{K}_{y\psi}\, \Delta\boldsymbol{\psi} + \mathbf{K}_{y\omega}\, \Delta \mathbf{w} \quad (10\text{-}18a)$$

$$\Delta \boldsymbol{\nu} = \mathbf{K}_{\nu\psi}\, \Delta\boldsymbol{\psi} + \mathbf{K}_{\nu\omega}\, \Delta \mathbf{w} \quad (10\text{-}18b)$$

If the control structure is the (D, V) structure, for example, $\boldsymbol{\psi} = [D \quad V]^{\mathrm{T}}$ and $\boldsymbol{\nu} = [L \quad B]^{\mathrm{T}}$; if it is the (L, B) structure, $\boldsymbol{\psi} = [L \quad B]^{\mathrm{T}}$ and $\boldsymbol{\nu} = [D \quad V]^{\mathrm{T}}$. The gain matrices are defined accordingly (cf. Equations 10-2 and 10-3). Note that the disturbance gains $\mathbf{K}_{y\omega}$ and $\mathbf{K}_{\nu\omega}$ are different from $\mathbf{K}_{yw}$ and $\mathbf{K}_{vw}$.

In the general case, $\boldsymbol{\psi}$ and $\boldsymbol{\nu}$ are some functions of **u**, **v**, and **w**, that is

$$\boldsymbol{\psi} = \boldsymbol{\psi}(\mathbf{u}, \mathbf{v}, \mathbf{w}) \quad (10\text{-}19a)$$

$$\boldsymbol{\nu} = \boldsymbol{\nu}(\mathbf{u}, \mathbf{v}, \mathbf{w}) \quad (10\text{-}19b)$$

The variables $\boldsymbol{\psi}$ and $\boldsymbol{\nu}$ are only allowed to be functions of variables included in the base structure. If the functions contain disturbance variables, the latter have to be measured (or estimated) disturbances. Linearization of Equations 10-19 and introduction of deviation variables give the relationships

$$\Delta\boldsymbol{\psi} = \mathbf{H}_{\psi u}\, \Delta \mathbf{u} + \mathbf{H}_{\psi v}\, \Delta \mathbf{v} + \mathbf{H}_{\psi w}\, \Delta \mathbf{w} \quad (10\text{-}20a)$$

$$\Delta\boldsymbol{\nu} = \mathbf{H}_{\nu u}\, \Delta \mathbf{u} + \mathbf{H}_{\nu v}\, \Delta \mathbf{v} + \mathbf{H}_{\nu w}\, \Delta \mathbf{w} \quad (10\text{-}20b)$$

where the **H** matrices contain partial derivatives of the new variables with respect to the base variables. The columns of $\mathbf{H}_{\psi w}$ and $\mathbf{H}_{\nu w}$ that correspond to unmeasured disturbances in **w** should contain only zeros.

Because **u** and **v** (and **w**) are the variables that physically affect the process, it must be possible to determine **u** and **v** from $\boldsymbol{\psi}$, $\boldsymbol{\nu}$, and measurable disturbances in **w**. This means that also the relationships

$$\Delta \mathbf{u} = \mathbf{M}_{u\psi}\, \Delta\boldsymbol{\psi} + \mathbf{M}_{u\nu}\, \Delta\boldsymbol{\nu} + \mathbf{M}_{u\omega}\, \Delta \mathbf{w} \quad (10\text{-}21a)$$

$$\Delta \mathbf{v} = \mathbf{M}_{v\psi}\, \Delta\boldsymbol{\psi} + \mathbf{M}_{v\nu}\, \Delta\boldsymbol{\nu} + \mathbf{M}_{v\omega}\, \Delta \mathbf{w} \quad (10\text{-}21b)$$

have to exist. These relationships are obtained when $\Delta\mathbf{u}$ and $\Delta\mathbf{v}$ are solved from Equations 10-20. (Note, however, that **u** and **v** should be solved from Equations 10-19 when the control structure is implemented.)

10-5-2 Consistency Relations

The models in Equations 10-17 and 10-18 have to satisfy the consistency relations that can be derived from the external steady-state material balances. Let these material balances be expressed as

$$\mathbf{c}(\mathbf{y}, \mathbf{v}, \mathbf{u}, \mathbf{w}) = \mathbf{0} \quad (10\text{-}22)$$

Linearization and introduction of deviation variables give the relationships

$$\mathbf{F}_{cy}\, \Delta \mathbf{y} + \mathbf{F}_{cv}\, \Delta \mathbf{v} + \mathbf{F}_{cu}\, \Delta \mathbf{u} + \mathbf{F}_{cw}\, \Delta \mathbf{w} = \mathbf{0} \quad (10\text{-}23)$$

From Equations 10-5 it follows that

$$\mathbf{F}_{cy} = \begin{bmatrix} 0 & 0 \\ \overline{D} & \overline{B} \end{bmatrix}, \quad \mathbf{F}_{cv} = \begin{bmatrix} 1 & 1 \\ \bar{y} & \bar{x} \end{bmatrix},$$
$$\mathbf{F}_{cu} = \begin{bmatrix} 0 & 0 \\ 0 & 0 \end{bmatrix}, \quad \mathbf{F}_{cw} = \begin{bmatrix} -1 & 0 \\ -\bar{z} & -\overline{F} \end{bmatrix} \tag{10-24}$$

for a simple two-product distillation column.

In the base structure, $\Delta\mathbf{u}$ and $\Delta\mathbf{w}$ are independent variables. Elimination of $\Delta\mathbf{y}$ and $\Delta\mathbf{v}$ from Equation 10-23 by Equations 10-17 then gives the consistency relations

$$\mathbf{F}_{cy}\mathbf{K}_{yu} + \mathbf{F}_{cv}\mathbf{K}_{vu} + \mathbf{F}_{cu} = \mathbf{0} \tag{10-25a}$$
$$\mathbf{F}_{cy}\mathbf{K}_{yw} + \mathbf{F}_{cv}\mathbf{K}_{vw} + \mathbf{F}_{cw} = \mathbf{0} \tag{10-25b}$$

for the base structure. These consistency relations are equivalent to those given in Equations 10-6 and 10-7. (Equation 10-25a gives Equations 10-6a, 10-6b, 10-7a, and 10-7b.)

Consider now the control structure described by Equations 10-18. In order to derive consistency relations for this control structure, we have to express the material balance equations in terms of $\boldsymbol{\psi}$ and $\boldsymbol{\nu}$ instead of $\mathbf{u}$ and $\mathbf{v}$. If we start from the linearized equations, $\Delta\mathbf{u}$ and $\Delta\mathbf{v}$ can be eliminated from Equation 10-23 by means of Equations 10-21, giving

$$\mathbf{F}_{cy}\,\Delta\mathbf{y} + \mathbf{F}_{c\nu}\,\Delta\boldsymbol{\nu} + \mathbf{F}_{c\psi}\,\Delta\boldsymbol{\psi} + \mathbf{F}_{c\omega}\,\Delta\mathbf{w} = \mathbf{0} \tag{10-26}$$

where

$$\mathbf{F}_{c\nu} = \mathbf{F}_{cv}\mathbf{M}_{v\nu} + \mathbf{F}_{cu}\mathbf{M}_{u\nu} \tag{10-27a}$$
$$\mathbf{F}_{c\psi} = \mathbf{F}_{cv}\mathbf{M}_{v\psi} + \mathbf{F}_{cu}\mathbf{M}_{u\psi} \tag{10-27b}$$
$$\mathbf{F}_{c\omega} = \mathbf{F}_{cv}\mathbf{M}_{v\omega} + \mathbf{F}_{cu}\mathbf{M}_{u\omega} + \mathbf{F}_{cw} \tag{10-27c}$$

Elimination of $\Delta\mathbf{y}$ and $\Delta\boldsymbol{\nu}$ from Equation 10-26 by Equations 10-18 gives the consistency relations

$$\mathbf{F}_{cy}\mathbf{K}_{y\psi} + \mathbf{F}_{c\nu}\mathbf{K}_{\nu\psi} + \mathbf{F}_{c\psi} = \mathbf{0} \tag{10-28a}$$
$$\mathbf{F}_{cy}\mathbf{K}_{y\omega} + \mathbf{F}_{c\nu}\mathbf{K}_{\nu\omega} + \mathbf{F}_{c\omega} = \mathbf{0} \tag{10-28b}$$

If Equations 10-18 is a model for the (D, V) structure, it follows from Equations 10-21 that

$$\mathbf{M}_{u\psi} = \begin{bmatrix} 0 & 0 \\ 0 & 1 \end{bmatrix}, \quad \mathbf{M}_{u\nu} = \begin{bmatrix} 1 & 0 \\ 0 & 0 \end{bmatrix},$$
$$\mathbf{M}_{u\omega} = \mathbf{0}$$
$$\mathbf{M}_{v\psi} = \begin{bmatrix} 1 & 0 \\ 0 & 0 \end{bmatrix}, \quad \mathbf{M}_{v\nu} = \begin{bmatrix} 0 & 0 \\ 0 & 1 \end{bmatrix}, \tag{10-29}$$
$$\mathbf{M}_{v\omega} = \mathbf{0}$$

Equations 10-27 then give ($\mathbf{F}_{cy}$ is given in Equation 10-24)

$$\mathbf{F}_{cy} = \begin{bmatrix} 0 & 0 \\ \overline{D} & \overline{B} \end{bmatrix}, \quad \mathbf{F}_{c\nu} = \begin{bmatrix} 0 & 1 \\ 0 & \bar{x} \end{bmatrix},$$
$$\mathbf{F}_{c\psi} = \begin{bmatrix} 1 & 0 \\ \bar{y} & 0 \end{bmatrix}, \quad \mathbf{F}_{c\omega} = \begin{bmatrix} -1 & 0 \\ -\bar{z} & -\overline{F} \end{bmatrix} \tag{10-30}$$

The consistency relations defined by Equations 10-28 and 10-30 are equivalent to the consistency relations given in Equations 10-9 and 10-10. (Equation 10-28a gives to the first two expressions in Equation 10-9 as well as Equations 10-10a and 10-10b.)

10-5-3 Transformations between Arbitrary Structures

In Section 10-4 we saw how a model for the (L, V) structure can be transformed into models for the (D, V) and (L, B) structures. In a similar way transformations between arbitrary control structure models can be accomplished.

Consider the model for the base structure, given in Equations 10-17, and the model for some other structure, given in Equations 10-18. The linearized relationships between the variables in the two control structures are given both in Equations 10-20 and 10-21. Because it is usually more straightforward to define the variables of the "new" control structure in terms of the variables of the base structure, and not vice versa, we will use Equations 10-20. The steady-state model for the new control structure can then be derived as follows.

Elimination of $\Delta \mathbf{v}$ from Equations 10-20 by Equation 10-17b gives

$$\Delta \boldsymbol{\psi} = (\mathbf{H}_{\psi u} + \mathbf{H}_{\psi v}\mathbf{K}_{vu})\,\Delta \mathbf{u} + (\mathbf{H}_{\psi w} + \mathbf{H}_{\psi v}\mathbf{K}_{vw})\,\Delta \mathbf{w} \quad (10\text{-}31a)$$

$$\Delta \boldsymbol{\nu} = (\mathbf{H}_{\nu u} + \mathbf{H}_{\nu v}\mathbf{K}_{vu})\,\Delta \mathbf{u} + (\mathbf{H}_{\nu w} + \mathbf{H}_{\nu v}\mathbf{K}_{vw})\,\Delta \mathbf{w} \quad (10\text{-}31b)$$

The primary manipulators $\boldsymbol{\psi}$ have to be independent of each other even when $\Delta \mathbf{w} = \mathbf{0}$. This means that the transformations have to be such that the matrix $\mathbf{H}_{\psi u} + \mathbf{H}_{\psi v}\mathbf{K}_{vu}$ is nonsingular. Also, if the manipulators of the base structure are the variables that in reality affect the column, it must be possible to determine $\mathbf{u}$ (and $\mathbf{v}$) from $\boldsymbol{\psi}$, $\boldsymbol{\nu}$, and measurable disturbances in $\mathbf{w}$. Because $\boldsymbol{\nu}$ is a dependent variable, $\mathbf{u}$ must in principle be determinable from $\boldsymbol{\psi}$ and $\mathbf{w}$. This implies that $\mathbf{u}$ can be solved from Equation 10-31a, giving

$$\Delta \mathbf{u} = (\mathbf{H}_{\psi u} + \mathbf{H}_{\psi v}\mathbf{K}_{vu})^{-1} \times \left(\Delta \boldsymbol{\psi} - (\mathbf{H}_{\psi w} + \mathbf{H}_{\psi v}\mathbf{K}_{vw})\,\Delta \mathbf{w}\right) \quad (10\text{-}32)$$

Elimination of $\Delta \mathbf{u}$ from Equations 10-17a and 10-31b by Equation 10-32 then gives the model in Equations 10-18, where

$$\mathbf{K}_{y\psi} = \mathbf{K}_{yu}(\mathbf{H}_{\psi u} + \mathbf{H}_{\psi v}\mathbf{K}_{vu})^{-1} \quad (10\text{-}33a)$$

$$\mathbf{K}_{y\omega} = \mathbf{K}_{yw} - \mathbf{K}_{y\psi}(\mathbf{H}_{\psi w} + \mathbf{H}_{\psi v}\mathbf{K}_{vw}) \quad (10\text{-}33b)$$

$$\mathbf{K}_{\nu\psi} = (\mathbf{H}_{\nu u} + \mathbf{H}_{\nu v}\mathbf{K}_{vu})(\mathbf{H}_{\psi u} + \mathbf{H}_{\psi v}\mathbf{K}_{vu})^{-1} \quad (10\text{-}33c)$$

$$\mathbf{K}_{\nu\omega} = (\mathbf{H}_{\nu w} + \mathbf{H}_{\nu v}\mathbf{K}_{vw}) - \mathbf{K}_{\nu\psi}(\mathbf{H}_{\psi w} + \mathbf{H}_{\psi v}\mathbf{K}_{vw}) \quad (10\text{-}33d)$$

Equations 10-33 provide the relationships between the steady-state models of two arbitrary control structures, Equations 10-17 and 10-18, where the manipulators are related according to Equations 10-20.

Note that when the primary gains $\mathbf{K}_{y\psi}$ and $\mathbf{K}_{y\omega}$ are calculated, the secondary gains $\mathbf{K}_{vu}$ and $\mathbf{K}_{vw}$ are needed if $\mathbf{H}_{\psi v} \neq \mathbf{0}$, that is, if the primary manipulators of the new control structure are functions of the secondary manipulators in the base structure. As we have seen, in some cases the secondary gains can be determined by means of consistency relations; in other cases they have to be provided. Note also that it is unnecessary to specify the secondary manipulators by defining $\mathbf{H}_{\nu u}$, $\mathbf{H}_{\nu v}$, and $\mathbf{H}_{\nu w}$ if only the primary gains are to be determined. The primary gains are, in other words, unaffected by the choice of secondary manipulators.

10-5-4 Complex Distillation Columns

The expressions derived in the preceding text are based on the model given in Equations 10-17. The results thus apply to any process that can be described by a model of the same general form as the model in Equations 10-17. We will show that a distillation column with any number of feed streams, sidestreams, and chemical components can be described by such a model.

Consider a distillation column having n_F feed streams with the flow rates F_i ($i = 1, \ldots, n_F$), n_S sidestreams with the flow rates S_i ($i = 1, \ldots, n_S$), and a total of $n_c + 1$ components in the different streams. Let us define

$$\mathbf{x}_D = \begin{bmatrix} x_{D,1} \\ \vdots \\ x_{D,n_c} \end{bmatrix}, \quad \mathbf{x}_B = \begin{bmatrix} x_{B,1} \\ \vdots \\ x_{B,n_c} \end{bmatrix},$$
$$\mathbf{x}_{S_i} = \begin{bmatrix} x_{S_i,1} \\ \vdots \\ x_{S_i,n_c} \end{bmatrix}, \quad \mathbf{x}_{F_i} = \begin{bmatrix} x_{F_i,1} \\ \vdots \\ x_{F_i,n_c} \end{bmatrix}$$
$$\mathbf{S} = \begin{bmatrix} S_1 \\ \vdots \\ S_{n_S} \end{bmatrix}, \quad \mathbf{F} = \begin{bmatrix} F_1 \\ \vdots \\ F_{nF} \end{bmatrix}, \quad (10\text{-}34)$$
$$\mathbf{x}_\mathbf{S} = \begin{bmatrix} \mathbf{x}_{S_1} \\ \vdots \\ \mathbf{x}_{S_{n_S}} \end{bmatrix}, \quad \mathbf{x}_\mathbf{F} = \begin{bmatrix} \mathbf{x}_{F_1} \\ \vdots \\ \mathbf{x}_{F_{n_F}} \end{bmatrix}$$

where $x_{D,j}$, $x_{B,j}$, $x_{F_i,j}$, and $x_{S_i,j}$ are mole or weight fractions of component j ($j = 1, \ldots, n_c$) in the distillate stream D, the bottoms stream B, the feed stream F_i, and the sidestream S_i, respectively.

Note that the composition of a stream is fully determined by n_c component fractions when the total number of components is $n_c + 1$. In spite of this, the fractions of any number of components, even $n_c + 1$, could be included in the composition vectors of the product streams. However, the composition vectors of the feed streams should contain at most n_c component fractions because we want the input variables to be independent of each other (mathematically). Because a change of any of the n_c component fractions will affect the outputs, the number of component fractions should generally be exactly n_c. The component excluded can, of course, be arbitrarily chosen.

Let us further define

$$\mathbf{y} = \begin{bmatrix} \mathbf{x}_D \\ \mathbf{x}_B \\ \mathbf{x}_S \end{bmatrix}, \quad \mathbf{u} = \begin{bmatrix} L \\ V \\ \mathbf{S} \end{bmatrix}, \quad \mathbf{v} = \begin{bmatrix} D \\ B \end{bmatrix},$$

$$\mathbf{w} = \begin{bmatrix} \mathbf{F} \\ \mathbf{x}_\mathbf{F} \\ \mathbf{q} \end{bmatrix} \tag{10-35}$$

where $\mathbf{q}$ is a vector of all disturbance variables not included in a material balance. The base structure for the column can now be described by Equations 10-17. The model for any other control structure whose variables are related to the variables of the base structure according to Equations 10-20 can be obtained by means of Equations 10-33. The models also have to satisfy the consistency relations in Equations 10-25 and 10-28 derived from the appropriate material balances in Equation 10-22 (see Häggblom and Waller, 1991).

A MATLAB program package for advanced control structure modelling was developed in Häggblom (1988b). Arbitrary (steady-state) control structure transformations can be performed with this package. Routines for identification of consistent models from numerical data are also included in the package (see Section 10-9).

10-6 APPLICATION 1: NUMERICAL EXAMPLES OF CONTROL STRUCTURE TRANSFORMATIONS

In this section we will illustrate by means of numerical examples how models for arbitrary control structure can be derived from a model for a given base structure. As base structure we will use the (L, V) structure.

The examples are control structures for a 15-plate pilot-scale distillation column separating a mixture of ethanol and water. The primary outputs are the temperatures on plates 4 and 14, counting from the top of the column. Table 10-3 (see Section 10-9) shows nominal steady-state data for the column. The steady-state model for the (L, V) structure, which was experimentally determined and reconciled with the models for a few other control structures (see Section 10-9), is

$$\begin{bmatrix} \Delta T_4 \\ \Delta T_{14} \end{bmatrix} = \begin{bmatrix} -0.0435 & 0.0458 \\ -0.2260 & 0.5432 \end{bmatrix} \begin{bmatrix} \Delta L \\ \Delta V \end{bmatrix} + \begin{bmatrix} -0.0010 & 0.0042 \\ -0.1606 & -0.6228 \end{bmatrix} \begin{bmatrix} \Delta F \\ \Delta z \end{bmatrix} \tag{10-36a}$$

$$\begin{bmatrix} \Delta D \\ \Delta B \end{bmatrix} = \begin{bmatrix} -0.6154 & 1.3523 \\ 0.6154 & -1.3523 \end{bmatrix} \begin{bmatrix} \Delta L \\ \Delta V \end{bmatrix} + \begin{bmatrix} 0.0591 & 0.9418 \\ 0.9409 & -0.9418 \end{bmatrix} \begin{bmatrix} \Delta F \\ \Delta z \end{bmatrix} \tag{10-36b}$$

where for simplicity the units have been omitted. (They can be obtained from the fact that the temperature is expressed in degrees Celsius, the flow rates in kilograms per hour, and the feed composition in weight percent.)

In terms of the more compact vector and matrix notation, which is used for the general transformations (Equations 10-33), the primary outputs are $\mathbf{y} = [T_4 \quad T_{14}]^T$, the primary manipulators are $\mathbf{u} = [L \quad V]^T$, the secondary manipulators are $\mathbf{v} = [D \quad B]^T$, and the disturbances are $\mathbf{w} = [F \quad z]^T$. The gain matrices $\mathbf{K_{yu}}$, $\mathbf{K_{yw}}$, $\mathbf{K_{vu}}$, and $\mathbf{K_{vw}}$ in Equations 10-33 follow from these definitions and Equations 10-36 in an obvious way.

10-6-1 (D, V) Structure

In the (D, V) structure, the primary output variables are controlled by D and V, and the secondary outputs (the reflux drum and reboiler holdups) usually by L and B. A generalized inventory control scheme could also be used. Assuming that the conventional inventory control system is used, the primary and secondary manipulators in the (D, V) structure are $\boldsymbol{\psi} = [D \quad V]^T$ and $\boldsymbol{\nu} = [L \quad B]^T$.

The manipulators of the (D, V) structure are related to the manipulators of the (L, V) structure according to

$$\begin{bmatrix} \Delta D \\ \Delta V \end{bmatrix} = \begin{bmatrix} 0 & 0 \\ 0 & 1 \end{bmatrix} \begin{bmatrix} \Delta L \\ \Delta V \end{bmatrix} + \begin{bmatrix} 1 & 0 \\ 0 & 0 \end{bmatrix} \begin{bmatrix} \Delta D \\ \Delta B \end{bmatrix} \tag{10-37a}$$

$$\begin{bmatrix} \Delta L \\ \Delta B \end{bmatrix} = \begin{bmatrix} 1 & 0 \\ 0 & 0 \end{bmatrix} \begin{bmatrix} \Delta L \\ \Delta V \end{bmatrix} + \begin{bmatrix} 0 & 0 \\ 0 & 1 \end{bmatrix} \begin{bmatrix} \Delta D \\ \Delta B \end{bmatrix} \tag{10-37b}$$

which give the transformation matrices

$$\mathbf{H}_{\psi \mathbf{u}} = \begin{bmatrix} 0 & 0 \\ 0 & 1 \end{bmatrix}, \quad \mathbf{H}_{\psi \mathbf{v}} = \begin{bmatrix} 1 & 0 \\ 0 & 0 \end{bmatrix}, \quad \mathbf{H}_{\psi \mathbf{w}} = \mathbf{0}$$
$$\mathbf{H}_{\nu \mathbf{u}} = \begin{bmatrix} 1 & 0 \\ 0 & 0 \end{bmatrix}, \quad \mathbf{H}_{\nu \mathbf{v}} = \begin{bmatrix} 0 & 0 \\ 0 & 1 \end{bmatrix}, \quad \mathbf{H}_{\nu \mathbf{w}} = \mathbf{0} \tag{10-38}$$

Use of the transformations in Equations 10-33 results in

$$\begin{bmatrix} \Delta T_4 \\ \Delta T_{14} \end{bmatrix} = \begin{bmatrix} 0.0707 & -0.0499 \\ 0.3673 & 0.0465 \end{bmatrix} \begin{bmatrix} \Delta D \\ \Delta V \end{bmatrix} + \begin{bmatrix} -0.0052 & -0.0624 \\ -0.1823 & -0.9687 \end{bmatrix} \begin{bmatrix} \Delta F \\ \Delta z \end{bmatrix} \tag{10-39a}$$

$$\begin{bmatrix} \Delta L \\ \Delta B \end{bmatrix} = \begin{bmatrix} -1.6250 & 2.1976 \\ -1.0000 & 0.0000 \end{bmatrix} \begin{bmatrix} \Delta D \\ \Delta V \end{bmatrix} + \begin{bmatrix} 0.0961 & 1.5304 \\ 1.0000 & 0.0000 \end{bmatrix} \begin{bmatrix} \Delta F \\ \Delta z \end{bmatrix} \tag{10-39b}$$

10-6-2 $(L/D, V)$ Structure

A control structure suggested by Ryskamp (1980) is the $(D/(L+D), V)$ structure, where the primary outputs are controlled by $D/(L+D)$ and V. Because $\Delta(D/(L+D)) = -(\bar{D}/(\bar{L}+\bar{D}))^2 \, \Delta(L/D)$, this control structure is equivalent to the $(L/D, V)$ structure when linear steady-state models are used; the only difference is a different scaling of the gains associated with the top primary manipulator. For simplicity, we therefore consider the $(L/D, V)$ structure here. In Ryskamp's control structure, the reflux drum holdup is usually controlled by $L + D$ and the reboiler holdup by B, but we will use L and B as inventory control manipulators. The primary and secondary variables are thus $\boldsymbol{\psi} = [L/D \quad V]^T$ and $\boldsymbol{\nu} = [L \quad B]^T$. Note that the primary steady-state gains are not dependent on the choice of inventory control manipulators (see Section 10-5-3).

In the transformation from the (L, V) structure to the (D, V) structure, the variable transformations only resulted in interchanges between L and D. In the $(L/D, V)$ structure, the manipulator L/D is a nonlinear function of two base variables. Linearization of the variable transformations at the nominal steady state $(\bar{L}, \bar{D})$ gives

$$\begin{bmatrix} \Delta(L/D) \\ \Delta V \end{bmatrix} = \begin{bmatrix} 1/\bar{D} & 0 \\ 0 & 1 \end{bmatrix} \begin{bmatrix} \Delta L \\ \Delta V \end{bmatrix} + \begin{bmatrix} -\bar{L}/\bar{D}^2 & 0 \\ 0 & 0 \end{bmatrix} \begin{bmatrix} \Delta D \\ \Delta B \end{bmatrix} \tag{10-40a}$$

$$\begin{bmatrix} \Delta L \\ \Delta B \end{bmatrix} = \begin{bmatrix} 1 & 0 \\ 0 & 0 \end{bmatrix} \begin{bmatrix} \Delta L \\ \Delta V \end{bmatrix} + \begin{bmatrix} 0 & 0 \\ 0 & 1 \end{bmatrix} \begin{bmatrix} \Delta D \\ \Delta B \end{bmatrix} \tag{10-40b}$$

The transformation matrices are thus

$$\mathbf{H}_{\psi \mathbf{u}} = \begin{bmatrix} 1/\bar{D} & 0 \\ 0 & 1 \end{bmatrix}, \quad \mathbf{H}_{\psi \mathbf{v}} = \begin{bmatrix} -\bar{L}/\bar{D}^2 & 0 \\ 0 & 0 \end{bmatrix}, \quad \mathbf{H}_{\psi \mathbf{w}} = \mathbf{0} \tag{10-41}$$

$$\mathbf{H}_{\nu \mathbf{u}} = \begin{bmatrix} 1 & 0 \\ 0 & 0 \end{bmatrix}, \quad \mathbf{H}_{\nu \mathbf{v}} = \begin{bmatrix} 0 & 0 \\ 0 & 1 \end{bmatrix}, \quad \mathbf{H}_{\nu \mathbf{w}} = \mathbf{0}$$

Numerically, the following model is obtained:

$$\begin{bmatrix} \Delta T_4 \\ \Delta T_{14} \end{bmatrix} = \begin{bmatrix} -1.6167 & 0.0093 \\ -8.3967 & 0.3540 \end{bmatrix} \begin{bmatrix} \Delta(L/D) \\ \Delta V \end{bmatrix} + \begin{bmatrix} -0.0026 & -0.0212 \\ -0.1688 & -0.7545 \end{bmatrix} \begin{bmatrix} \Delta F \\ \Delta z \end{bmatrix} \tag{10-42a}$$

$$\begin{bmatrix} \Delta L \\ \Delta B \end{bmatrix} = \begin{bmatrix} 37.1432 & 0.8372 \\ 22.8568 & -0.8372 \end{bmatrix} \begin{bmatrix} \Delta(L/D) \\ \Delta V \end{bmatrix} + \begin{bmatrix} 0.0366 & 0.5830 \\ 0.9634 & -0.5830 \end{bmatrix} \begin{bmatrix} \Delta F \\ \Delta z \end{bmatrix} \tag{10-42b}$$

10-6-3 $(L/D, V/F)$ Structure

The $(L/D, V/F)$ structure is an example of a control structure where a primary manipulator includes a measurable disturbance. The inclusion of F in the manipulator V/F means that there is such a built-in feedforward from measured disturbances in F that V is changed in the same proportion as F. This results in a better rejection of disturbances in F than is the case, for example, in the $(L/D, V)$ structure. Otherwise, the properties of the $(L/D, V/F)$ structure are similar to those of the $(L/D, V)$ structure.

The primary manipulators for this control structure are $\boldsymbol{\psi} = [L/D \quad V/F]^{\mathrm{T}}$. They give the primary transformation matrices

$$\mathbf{H}_{\psi \mathbf{u}} = \begin{bmatrix} 1/\bar{D} & 0 \\ 0 & 1/\bar{F} \end{bmatrix}, \quad \mathbf{H}_{\psi \mathbf{v}} = \begin{bmatrix} -\bar{L}/\bar{D}^2 & 0 \\ 0 & 0 \end{bmatrix}, \quad \mathbf{H}_{\psi \mathbf{w}} = \begin{bmatrix} 0 & 0 \\ -\bar{V}/\bar{F}^2 & 0 \end{bmatrix} \tag{10-43}$$

If the column inventory is controlled by L and B, the secondary transformation matrices $\mathbf{H}_{\nu \mathbf{u}}$, $\mathbf{H}_{\nu \mathbf{v}}$, and $\mathbf{H}_{\nu \mathbf{w}}$ are the same as in the two previous cases. The resulting model is

$$\begin{bmatrix} \Delta T_4 \\ \Delta T_{14} \end{bmatrix} = \begin{bmatrix} -1.6167 & 1.8639 \\ -8.3967 & 70.7935 \end{bmatrix} \begin{bmatrix} \Delta(L/D) \\ \Delta(V/F) \end{bmatrix} + \begin{bmatrix} 0.0008 & -0.0212 \\ -0.0414 & -0.7545 \end{bmatrix} \begin{bmatrix} \Delta F \\ \Delta z \end{bmatrix} \tag{10-44a}$$

$$\begin{bmatrix} \Delta L \\ \Delta B \end{bmatrix} = \begin{bmatrix} 37.1432 & 167.4339 \\ 22.8568 & -167.4339 \end{bmatrix} \begin{bmatrix} \Delta(L/D) \\ \Delta(V/F) \end{bmatrix} + \begin{bmatrix} 0.3380 & 0.5830 \\ 0.6620 & -0.5830 \end{bmatrix} \begin{bmatrix} \Delta F \\ \Delta z \end{bmatrix} \tag{10-44b}$$

10-6-4 $(L/D, V/B)$ Structure

The control structure considered by many to be the best general control structure is the $(L/D, V/B)$ structure (Takamatsu, Hashimoto, and Hashimoto, 1982; Shinskey,

1984; Skogestad, Lundström, and Jacobsen, 1990). Due to the flow ratios used as primary manipulators, this control structure has a certain built-in disturbance rejection capability. The interaction between the primary control loops as measured by relative gains is also considerably lower than, for example, in the (L, V) structure (Shinskey, 1984; Häggblom, 1988a).

The primary transformation matrices for this control structure are

$$\mathbf{H}_{\psi\mathbf{u}} = \begin{bmatrix} 1/\bar{D} & 0 \\ 0 & 1/\bar{B} \end{bmatrix},$$

$$\mathbf{H}_{\psi\mathbf{v}} = \begin{bmatrix} -\bar{L}/\bar{D}^2 & 0 \\ 0 & -\bar{V}/\bar{B}^2 \end{bmatrix}, \qquad \mathbf{H}_{\psi\mathbf{w}} = \mathbf{0} \tag{10-45}$$

If L and B are used as inventory control manipulators, the model obtained by the transformation is

$$\begin{bmatrix} \Delta T_4 \\ \Delta T_{14} \end{bmatrix} = \begin{bmatrix} -1.5402 & 0.9121 \\ -5.4871 & 34.6410 \end{bmatrix} \begin{bmatrix} \Delta(L/D) \\ \Delta(V/B) \end{bmatrix} + \begin{bmatrix} 0.0006 & -0.0231 \\ -0.0463 & -0.8287 \end{bmatrix} \begin{bmatrix} \Delta F \\ \Delta z \end{bmatrix} \tag{10-46a}$$

$$\begin{bmatrix} \Delta L \\ \Delta B \end{bmatrix} = \begin{bmatrix} 44.0223 & 81.9295 \\ 15.9777 & -81.9295 \end{bmatrix} \begin{bmatrix} \Delta(L/D) \\ \Delta(V/B) \end{bmatrix} + \begin{bmatrix} 0.3265 & 0.4075 \\ 0.6735 & -0.4075 \end{bmatrix} \begin{bmatrix} \Delta F \\ \Delta z \end{bmatrix} \tag{10-46b}$$

10-7 APPLICATION 2: USE OF CONSISTENCY RELATIONS IN TRANSFORMATIONS

In this section we will demonstrate the usefulness of consistency relations in control structure transformations. As an example we will use the well known Wood and Berry (1973) column model, which is probably the most studied model in the distillation control literature over the years. The process is a methanol–water distillation in an eight-plate pilot-plant column. Nominal steady-state data are given in Table 10-2.

TABLE 10-2 Steady-State Data for the Wood and Berry Column

Flow	Flow rate (lb/min)	Composition (wt% methanol)
Feed, F	2.45	46.5
Distillate, D	1.18	96.0
Reflux, L	1.95	96.0
Bottoms, B	1.27	0.5
Steam to reboiler, V	1.71	

The control structure previously studied in the literature is the (L, V) structure. In the original paper on the column (Wood and Berry, 1973) only transfer functions between the primary manipulators (L and V) and the primary outputs (product compositions y and x) were given. Later Ogunnaike and Ray (1979) reported transfer functions between a disturbance in the feed flow rate F and the primary outputs. The steady-state part of the model given by Ogunnaike and Ray is

$$\begin{bmatrix} \Delta y \\ \Delta x \end{bmatrix} = \begin{bmatrix} 12.8 & -18.9 \\ 6.6 & -19.4 \end{bmatrix} \begin{bmatrix} \Delta L \\ \Delta V \end{bmatrix} + \begin{bmatrix} 3.8 \\ 4.9 \end{bmatrix} \Delta F \tag{10-47}$$

The process model is given in the usual form, showing the gains associated with the primary outputs, but not the flow gains. As we have already pointed out, the flow gains are needed if a model for another control structure is to be derived from the model for the (L, V) structure.

However, because the outputs in this case are product compositions, and appropriate steady-state data are known, the consistency relations following from the external material balances can be used to calculate the flow gains. Equations 10-8 give the flow gains

$$K_{DL}^{LV} = -K_{BL}^{LV} = -0.25 \quad (10\text{-}48a)$$

$$K_{DV}^{LV} = -K_{BV}^{LV} = 0.49 \quad (10\text{-}48b)$$

$$K_{DF}^{LV} = 1 - K_{BF}^{LV} = 0.37 \quad (10\text{-}48c)$$

These values differ considerably from the values generally used in the literature, that is, $K_{DL}^{LV} = -1$, $K_{DV}^{LV} = 1$, and $K_{DF}^{LV} = 0$.

Now all necessary data for arbitrary control structure transformations are available. For the (D, V) structure, Equations 10-14a and 10-14b give the model

$$\begin{bmatrix} \Delta y \\ \Delta x \end{bmatrix} = \begin{bmatrix} -52 & 6.7 \\ -27 & -6.2 \end{bmatrix} \begin{bmatrix} \Delta D \\ \Delta V \end{bmatrix} + \begin{bmatrix} 23 \\ 15 \end{bmatrix} \Delta F \quad (10\text{-}49)$$

For the (L, B) structure, Equations 10-16a and 10-16b give the model

$$\begin{bmatrix} \Delta y \\ \Delta x \end{bmatrix} = \begin{bmatrix} 3.3 & 38 \\ -3.1 & 39 \end{bmatrix} \begin{bmatrix} \Delta L \\ \Delta B \end{bmatrix} + \begin{bmatrix} -20 \\ -20 \end{bmatrix} \Delta F \quad (10\text{-}50)$$

Models for other control structures can be calculated by means of the general transformations in Equations 10-33. For Ryskamp's (1980) control structure $(D/(L+D), V)$, they give

$$\begin{bmatrix} \Delta y \\ \Delta x \end{bmatrix} = \begin{bmatrix} -76 & -12 \\ -39 & -16 \end{bmatrix} \begin{bmatrix} \Delta R \\ \Delta V \end{bmatrix} + \begin{bmatrix} 9.4 \\ 7.8 \end{bmatrix} \Delta F \quad (10\text{-}51)$$

where $R = D/(L + D)$.

It is seen from Equations 10-47, 10-49, and 10-50 that a disturbance in F has a much stronger steady-state effect on the product compositions in the (D, V) and (L, B) structures than in the (L, V) structure. This means that the (L, V) structure is probably preferable to the (D, V) and (L, B) structures for elimination of disturbances in the feed flow rate in the Wood and Berry column. As shown by Equations 10-47 and 10-51, Ryskamp's control structure has a significantly larger steady-state sensitivity to disturbances in F than the (L, V) structure. However, the difference is smaller than for the other structures.

We would once more like to stress the importance of using the correct flow gains in control structure transformations. For example, if the standard assumptions ($K_{DL}^{LV} = -1$, $K_{DV}^{LV} = 1$, and $K_{DF}^{LV} = 0$) were used in the calculation of the model for the (D, V) structure from the model for the (L, V) structure, the result would be

$$\begin{bmatrix} \Delta y \\ \Delta x \end{bmatrix} = \begin{bmatrix} -12.8 & -6.1 \\ -6.6 & -12.8 \end{bmatrix} \begin{bmatrix} \Delta D \\ \Delta V \end{bmatrix} + \begin{bmatrix} 3.8 \\ 4.9 \end{bmatrix} \Delta F \quad (10\text{-}52)$$

There is a huge difference between this incorrect model and the correctly derived model in Equation 10-49. For example, according to Equation 10-52 the relative gain (Bristol, 1966) for pairing y with D is 1.3, which is generally considered to be an impossible value for the (D, V) structure (Häggblom, 1988a).

10-8 APPLICATION 3: PROCESS DYNAMICS

The control structure transformations treated so far apply to steady-state models. Similar transformations can also be derived for dynamic models, but because such models consist of cause-and-effect relationships that cannot be reversed, the resulting transformations are in general much more complicated than the steady-state transformations. The dynamic transformations have to include, for example, (design of) the inventory controllers (Yang et al., 1990; Häggblom and Waller, 1991).

There are two situations, however, when dynamic control structure models can be directly transformed according to the relatively simple steady-state transformations. One situation is when the reflux drum and reboiler holdups can be assumed to be perfectly controlled (Häggblom, 1986). In reality, this assumption is always an approximation, but because inventory control is a much easier task than composition (or temperature) control, it can usually be made to work in a much faster time scale than the latter. Unless the inventory controllers are strongly detuned, the approximation is thus often good enough for control structure modelling.

The other situation is when the same inventory control system is used in the control structures (Häggblom and Waller, 1991). A simple example of an inventory control system that can be used with practically any control structure is one where the reflux drum holdup is controlled by $L + D$ and the reboiler holdup by $V' + B$, where V' is an estimate of the vapor boilup rate. This kind of "generalized" inventory control system has many advantages (Shinskey, 1990; Häggblom and Waller, 1991). In Ryskamp's control structure, for example, the reflux drum holdup is usually controlled by $L + D$.

By means of an example taken from the literature, we will next illustrate that the transformations derived for steady-state models can be applied to dynamic models when the reflux drum and reboiler holdups (levels) are perfectly controlled.

In a study of noninteracting control system design methods, Bequette and Edgar (1989) used a distillation model, which they obtained using "standard" assumptions, including perfect level control and constant molar flow rates. From the nonlinear state-space model they first developed for the 17-plate benzene–toluene distillation, they determined two transfer function models: one for the (L, V) structure, one for the (D,V) structure. The models are

$$\begin{bmatrix} \Delta x_3 \\ \Delta x_{17} \end{bmatrix} = \begin{bmatrix} G_{yL}^{LV} & G_{yV}^{LV} \\ G_{xL}^{LV} & G_{xV}^{LV} \end{bmatrix} \begin{bmatrix} \Delta L \\ \Delta V \end{bmatrix} \tag{10-53}$$

with

$$\begin{bmatrix} G_{yL}^{LV} & G_{yV}^{LV} \\ G_{xL}^{LV} & G_{xV}^{LV} \end{bmatrix} = \begin{bmatrix} \dfrac{2.454e^{-s}}{20s + 1} & \dfrac{-3.493e^{-s}}{22s + 1} \\ \dfrac{3.270e^{-s}}{22s + 1} & \dfrac{-5.519e^{-s}}{20s + 1} \end{bmatrix} \tag{10-54}$$

and

$$\begin{bmatrix} \Delta x_3 \\ \Delta x_{17} \end{bmatrix} = \begin{bmatrix} G_{yD}^{DV} & G_{yV}^{DV} \\ G_{xD}^{DV} & G_{xV}^{DV} \end{bmatrix} \begin{bmatrix} \Delta D \\ \Delta V \end{bmatrix} \tag{10-55}$$

with

$$\begin{bmatrix} G_{yD}^{DV} & G_{yV}^{DV} \\ G_{xD}^{DV} & G_{xV}^{DV} \end{bmatrix} = \begin{bmatrix} \dfrac{-1.438e^{-s}}{20s + 1} & \dfrac{0.3986e^{-s}}{5s + 1} \\ \dfrac{-1.917e^{-s}}{20s + 1} & \dfrac{-0.3329e^{-s}}{4s + 1} \end{bmatrix} \tag{10-56}$$

The primary outputs are the mole fractions of the more volatile component on plates 3 and 17. Time is expressed in minutes.

Because the outputs are not product compositions, the consistency relations between the primary gains and the flow gains (which were not given by Bequette and Edgar) do not apply. Furthermore, the consistency relations never apply to complete transfer functions, only to their steady-state parts, because they are derived from steady-state material balances. However, the transformations between models for different control structures are valid for transfer functions when the reflux drum and reboiler holdups are perfectly controlled. According to Equation 10-14a, the relationship between the (D, V) structure and the (L, V)

structure is thus

$$\begin{bmatrix} G_{yD}^{DV} & G_{yV}^{DV} \\ G_{xD}^{DV} & G_{xV}^{DV} \end{bmatrix} = \frac{1}{G_{DL}^{LV}} \begin{bmatrix} G_{yL}^{LV} & -\left(G_{yL}^{LV} G_{DV}^{LV} - G_{yV}^{LV} G_{DL}^{LV}\right) \\ G_{xL}^{LV} & -\left(G_{xL}^{LV} G_{DV}^{LV} - G_{xV}^{LV} G_{DL}^{LV}\right) \end{bmatrix} \tag{10-57}$$

In this case G_{DL}^{LV} and G_{DV}^{LV} are not known. However, from Equation 10-57 two expressions for both G_{DL}^{LV} and G_{DV}^{LV} can be obtained. They are

$$G_{DL}^{LV} = \frac{G_{yL}^{LV}}{G_{yD}^{DV}} \tag{10-58a}$$

$$G_{DL}^{LV} = \frac{G_{xL}^{LV}}{G_{xD}^{DV}} \tag{10-58b}$$

and

$$G_{DV}^{LV} = \frac{G_{yV}^{LV} - G_{yV}^{DV}}{G_{yD}^{DV}} \tag{10-59a}$$

$$G_{DV}^{LV} = \frac{G_{xV}^{LV} - G_{xV}^{DV}}{G_{xD}^{DV}} \tag{10-59b}$$

Numerically, we obtain

$$G_{DL}^{LV} = -1.71 \tag{10-60a}$$

$$G_{DL}^{LV} = -1.71 \frac{20s + 1}{22s + 1} \tag{10-60b}$$

and

$$G_{DV}^{LV} = 2.71 \frac{(20s + 1)(6.74s + 1)}{(22s + 1)(4s + 1)} \tag{10-61a}$$

$$G_{DV}^{LV} = 2.71 \frac{(2.97s + 1)}{(4s + 1)} \tag{10-61b}$$

As seen from Equations 10-60 and 10-61, the two expressions for both G_{DL}^{LV} and G_{DV}^{LV} are identical with respect to steady-state gains, and as similar as can be expected, considering that simple transfer function models are used, when the complete transfer functions are considered. This is thus an illustration of the validity of the transformations for dynamic models when the reflux drum and reboiler holdups are perfectly controlled.

Note that although the flow transfer functions for the (L, V) structure cannot be determined from consistency relations based on steady-state material balances, it is possible to determine them from the primary transfer functions for the (L, V) and (D, V) structures by use of the relationships between these transfer functions. Here this makes the model for the (L, V) structure complete enough for transformations to arbitrary structures.

Note also that the gains $K_{DL}^{LV} = -1.71$ and $K_{DV}^{LV} = 2.71$ are not the expected ones, that is, $K_{DL}^{LV} = -1.0$ and $K_{DV}^{LV} = 1.0$, considering that the assumption of constant molar flows was used in the modelling. The explanation for this is that scaled variables are used in the models (Bequette and Edgar, 1989). Thus, it is meaningless to speak of "constant molar flows" if the scaling of the variables is not given. To avoid ambiguity, it is best always to specify the flow gains.

10-9 APPLICATION 4: IDENTIFICATION OF CONSISTENT MODELS

In the previous sections we have shown that the process gains of a control structure have to satisfy certain consistency relations, which can be derived from basic physical laws such as material balances. We have also shown that the process gains of different control structures are related through exact analytical relationships. However, gains that are determined from numerical data by standard methods do not generally satisfy these relationships (Häggblom, 1988b; Häggblom and Waller, 1988a). The reason is that the process usually is nonlinear with respect to the variables used and that experimental data are inaccurate. In order to obtain "consistent" models, one should estimate the gains subject to the known relationships.

Even if a model for a control structure has been determined subject to known consistency relations, it is never an exact description of the process. If models for other control structures are to be determined from this model, the inaccuracy of the model may be magnified in the calculations. In order to get a model that is accurate enough for arbitrary control structure transformations, it may therefore be necessary to estimate models for a few control structures and to reconcile these models subject to the analytical relationships existing between the models.

In this section we will describe a procedure for estimating consistent steady-state control structure models by means of an example. According to this procedure, models for an arbitrary number of control structures are first determined by some standard method. The next step is to reconcile the models subject to the analytical relationships existing between the models and the consistency relations derived from external material balances. The theory and several applications are described in more detail in Häggblom (1988b, 1989). A program package for solving this kind of estimation and reconciliation problem is also described in Häggblom (1988b).

10-9-1 Reconciliation of Control Structure Models

In principle, consistency relations could be directly taken into account when process gains are estimated from steady-state data. However, if the process gains of several control structures are estimated, it is inconvenient simultaneously to estimate all gains subject to all known consistency relations. It is more convenient first to estimate the gains by standard (unconstrained) methods, then to reconcile the estimates subject to the relevant relationships.

The natural way to reconcile the gains is to use the method of least-squares, that is, to modify the gains subject to the known relationships in such a way that a quadratic objective function of the adjustments is minimized. If also the variances of the gain estimates have been estimated, and the choice of weighting matrices used in the objective function is based on these variances, there is no loss of generality in the proposed two-step procedure (Häggblom, 1988b). However, the method can be used irrespectively of how the initial estimates of the process gains have been obtained and whether estimates of their variances are available or not.

More explicitly, the objective function to be minimized subject to the known relationships is a sum of weighted squares of the adjustments of all gains in all control structures considered. Even if only a few models are reconciled, the dimension of this estimation problem is huge. Moreover, the known analytical relationships between the gains in different control structures are nonlinear with respect to the gains. Therefore, a numerical solution according to this formulation is not practical.

It is possible to reduce the dimension of the problem by elimination of most of the gains in the objective function by means of the known relationships, but the number of remaining gains will still be considerable and the estimation problem will be nonlinear. For example, if two disturbance variables are included in the model and the primary outputs are not product compositions, the number of gains to be simultaneously adjusted is 12. In practice, this means that there will be local minima (at least numerically) and that a unique solution is difficult to obtain.

However, it is also possible to formulate the reconciliation problem as an ordinary linear estimation problem, which has an analytical solution. Suppose that initial estimates of the gains for a number of control structures are known. From the set of gains for each control structure, it is possible to calculate the corresponding gains for the base structure [the (L, V) structure] by means of an "inverse" transformation, which can be derived from Equations 10-33. Each control structure thus gives a set of gains for the base structure, which can be regarded as estimates obtained from different experiments. The task is then to estimate one set of gains for the base structure from these "experimental" gains.

If the objective function is expressed as a sum of weighted squares of the adjustments of each of these gains (i.e., of the "errors" of these gains), the resulting estimation problem is linear. The main problem is the choice of weights for the gains calculated from the gains for different control structures. The simplest solution is to choose the same weight for all calculated values of a given gain. The optimal value of the gain is then the average of the calculated values. However, it is also possible to use weights that are directly related to the accuracy of the gains determined for each individual control structure (Häggblom, 1988b).

When the gains for the base structure have been estimated as outlined, the next step is to reconcile them subject to the consistency relationships obtained from the external material balances. This is a straightforward linear estimation problem.

10-9-2 Numerical Example

In a study of two-point control of a 15-plate pilot-scale distillation column, dynamic models for four different control structures were determined experimentally (Waller et al., 1988a). The liquid distilled was a mixture of ethanol and water, and the variables controlled (in addition to the condenser and reboiler holdups) were the temperatures on plates 4 and 14, counting from the top of the column. The control structures studied were the energy balance structure (L, V), the material balance structure (D, V), Ryskamp's structure $(D/(L + D), V)$, and the two-ratio structure $(D/(L + D), V/B)$.

TABLE 10-3 Nominal Steady-State Data for the Åbo Akademi Column

Feed flow rate, F	200 kg/h
Distillate flow rate, D	60 kg/h
Bottoms flow rate, B	140 kg/h
Feed composition, z	30 wt%
Distillate composition	87 wt%
Bottoms composition	5 wt%
Reflux flow rate, L	60 kg/h
Rate of steam flow to reboiler, V	72 kg/h
Feed temperature	65°C
Reflux temperature	62°C

The nominal steady state at which the models were determined is given in Table 10-3. The actual operating point varied somewhat from experiment to experiment, but the given steady state is used in the reconciliation of the models.

Table 10-4 shows the experimentally determined process gains for the four control

TABLE 10-4 Experimentally Determined Process Gains for the Åbo Akademi Column[a]

(L, V) Structure			
−0.0382	0.0458	−0.001	0.0042
−0.231	0.618	−0.18	−0.622
−0.7	1.3	0.07	0.94
N.A.	N.A.	N.A.	N.A.
(D, V) Structure			
0.075	−0.05	−0.004	−0.16
0.35	0.0	−0.18	−2.25
N.A.	N.A.	N.A.	N.A.
N.A.	N.A.	N.A.	N.A.
$(D/(L + D), V)$ Structure			
6.21	0.0	0.0035	−0.034
34.2	0.34	−0.155	−0.992
N.A.	N.A.	N.A.	N.A.
N.A.	N.A.	N.A.	N.A.
$(D/(L + D), V/B)$ Structure			
8.42	3.0	0.0015	−0.033
46.5	78.4	0.01	−0.974
N.A.	N.A.	N.A.	N.A.
N.A.	N.A.	N.A.	N.A.

[a]N.A. = data not available.

structures. The gains in the gain matrices are ordered in the same way as in Equations 10-36, 10-39, etcetera (see Section 10-6). The fact that complete models for the control structures were not determined does not preclude a reconciliation of the models as previously described. The unknown gains are simply given the weight zero. As a result of the reconciliation, consistent estimates of the unknown gains are also obtained.

Before the process gains can be reconciled, the weighting matrices have to be selected. Obviously, the weights should reflect the reliability of the estimates so that more reliable estimates are given higher weights than less reliable ones. In accordance with this, estimates that are very unreliable, or perhaps not even available, can be given the weight zero. According to statistical considerations, the choice of weights as the inverse of the variances of the initial estimates is optimal under certain conditions (Häggblom, 1988b).

In the study previously mentioned, dynamic models were determined from step responses. Because it was desired to approximate the initial responses fairly well with first order transfer functions, the estimates of the steady-state gains, which are determined by the ultimate responses, are not necessarily as accurate as possible. Another practical difficulty in the estimation of steady-state gains from step responses is that it is not easy to determine when a new steady state has been reached. Based on these facts and on a detailed examination of the original experiments, the weights for the reconciliation were chosen according to Table 10-5, where each entry is the weight for the gain in the corresponding position in Table 10-4.

Table 10-6 shows the result of the linear reconciliation procedure using these weights. [Note that the weights for the gains in each control structure had to be transformed to weights for the gains in the (L, V)

TABLE 10-5 Weights for Reconciliation of Gains for the Åbo Akademi Column

(L, V) Structure			
7	$5 \times 10^{+2}$	$1 \times 10^{+4}$	$6 \times 10^{+4}$
$2 \times 10^{+1}$	3	3×10^{-1}	3
2×10^{-2}	6×10^{-1}	2	1
0	0	0	0
(D, V) Structure			
$2 \times 10^{+2}$	4	$6 \times 10^{+2}$	0
8	4	3×10^{-1}	0
0	0	0	0
0	0	0	0
$(D/(L + D), V)$ Structure			
3×10^{-2}	4	$8 \times 10^{+2}$	9
9×10^{-4}	9	4×10^{-1}	1×10^{-2}
0	0	0	0
0	0	0	0
$(D/(L + D), V/B)$ Structure			
0	0	0	9
0	0	0	1
0	0	0	0
0	0	0	0

TABLE 10-6 Reconciled Gains for the Åbo Akademi Column

(L, V) Structure			
−0.0435	0.0458	−0.0010	0.0042
−0.2260	0.5432	−0.1606	−0.6228
−0.6154	1.3523	0.0591	0.9418
0.6154	−1.3523	0.9409	−0.9418
(D, V) Structure			
0.0707	−0.0499	−0.0052	−0.0624
0.3673	0.0465	−0.1823	−0.9687
−1.6250	2.1976	0.0961	1.5304
−1.0000	0.0000	1.0000	0.0000
$(D/(L + D), V)$ Structure			
6.4670	0.0093	−0.0026	−0.0212
33.5828	0.3540	−0.1683	−0.7545
−57.1458	1.6744	0.0732	1.1660
−91.4271	−0.8372	0.9634	−0.5830
$(D/(L + D), V/B)$ Structure			
6.1607	0.9121	0.0006	−0.0231
21.9485	34.6410	−0.0463	−0.8287
−112.1786	163.8590	0.6530	0.8250
−63.9107	−81.9295	0.6735	−0.4075

TABLE 10-7 Consistent Control Structure Models for the Åbo Akademi Column

$(L, V; D, B)$ Structure			
−0.0435	0.0458	−0.0010	0.0042
−0.2260	0.5432	−0.1606	−0.6228
−0.6154	1.3523	0.0591	0.9418
0.6154	−1.3523	0.9409	−0.9418

$(L/F, V/F; D, B)$ Structure			
−8.7056	9.1520	0.0024	0.0042
−45.2072	108.6396	−0.0328	−0.6228
−123.0736	270.4673	0.3613	0.9418
123.0736	−270.4673	0.6387	−0.9418

$(D, V; L, B)$ Structure			
0.0707	−0.0499	−0.0052	−0.0624
0.3673	0.0465	−0.1823	−0.9687
−1.6250	2.1976	0.0961	1.5304
−1.0000	0.0000	1.0000	0.0000

$(D/F, V/F; L, B)$ Structure			
14.1469	−9.9794	−0.0019	−0.0624
73.4636	9.2920	−0.0554	−0.9687
−325.0087	439.5212	0.3997	1.5304
−200.0000	0.0000	0.7000	0.0000

$(L, B; D, V)$ Structure			
−0.0227	−0.0338	0.0308	−0.0277
0.0211	−0.4017	0.2174	−1.0010
0.0000	−1.0000	1.0000	0.0000
0.4550	−0.7395	0.6958	−0.6964

$(L/F, B/F; D, V)$ Structure			
−4.5410	−6.7675	0.0003	−0.0277
4.2282	−80.3347	−0.0575	−1.0010
0.0000	−200.0000	0.3000	0.0000
91.0081	−147.8922	0.3146	−0.6964

$(L/D, V; L, B)$ Structure			
−1.6167	0.0093	−0.0026	−0.0212
−8.3967	0.3540	−0.1688	−0.7545
37.1432	0.8372	0.0366	0.5830
22.8568	−0.8372	0.9634	−0.5830

$(L/D, V/F; L, B)$ Structure			
−1.6167	1.8639	0.0008	−0.0212
−8.3957	70.7935	−0.0414	−0.7545
37.1432	167.4339	0.3380	0.5830
22.8568	−167.4339	0.6620	−0.5830

$(L/D, B; D, V)$ Structure			
−1.3623	−0.0111	0.0081	−0.0277
1.2685	−0.4228	0.2385	−1.0010
0.0000	−1.0000	1.0000	0.0000
27.3024	−1.1945	1.1508	−0.6964

$(L/D, B/F; D, V)$ Structure			
−1.3623	−2.2265	0.0003	−0.0277
1.2685	−84.5629	−0.0575	−1.0010
0.0000	−200.0000	0.3000	0.0000
27.3024	−238.9003	0.3146	−0.6964

$(D, V/B; L, B)$ Structure			
0.0964	−6.9856	−0.0308	−0.0624
0.3434	6.5044	−0.1584	−0.9687
−2.7552	307.6648	1.2263	1.5304
−1.0000	0.0000	1.0000	0.0000

$(D/F, V/B; L, B)$ Structure			
19.2792	−6.9856	−0.0019	−0.0624
68.6849	6.5044	−0.0554	−0.9687
−551.0481	307.6648	0.3997	1.5304
−200.0000	0.0000	0.7000	0.0000

$(L, V/B; D, V)$ Structure			
−0.0350	3.7785	0.0121	−0.0089
−0.1246	44.8530	−0.0055	−0.7779
−0.3629	111.6653	0.4451	0.5554
0.1867	82.5721	0.2854	−0.2857

$(L/F, V/B; D, V)$ Structure			
−6.9973	3.7785	0.0016	−0.0089
−24.9288	44.8530	−0.0429	−0.7779
−72.5889	111.6653	0.3362	0.5554
37.3315	82.5721	0.3414	−0.2857

$(L/D, V/B; L, B)$ Structure			
−1.5402	0.9121	0.0006	−0.0231
−5.4871	34.6410	−0.0463	−0.8287
44.0223	81.9295	0.3265	0.4075
15.9777	−81.9295	0.6735	−0.4075

$(L/D, V/B; D, V)$ Structure			
−1.5402	0.9121	0.0006	−0.0231
−5.4871	34.6410	−0.0463	−0.8287
−15.9777	81.9295	0.3265	0.4075
8.2171	97.8649	0.3463	−0.2096

structure as described in Häggblom (1988b).] Most of the reconciled gains are close to the experimentally determined gains, but for some gains the differences are considerable. These differences were to be expected from the examination of the experiments. For example, the estimate of $|K_{yL}^{LV}|$ was too small in the original work because a good fit with the initial response was considered more important. Further, the gains in the (D, V) structure associated with a disturbance in the feed composition, and all primary gains in the $(D/(L + D), V/B)$ structure except those associated with this disturbance, were determined at an operating point slightly (but significantly enough) different from the nominal steady state given in Table 10-3. Finally, it can be mentioned that a sign error (already corrected in Table 10-4) in the recording of one of the gains was detected through the reconciliation.

In the reconciliation, only the gains of the base structure were determined when the objective function was minimized. The gains of the other control structures given in Table 10-6 were calculated by means of Equations 10-33. In a similar way the gains for any control structure can be calculated once the transformation matrices are determined. As a numerical illustration, Table 10-7 shows the gains for 15 control structures, which have been calculated from the reconciled gains for the (L, V) structure (which also are included in Table 10-7). An extended control structure notation is used in the table; $(L, V; D, B)$, for example, denotes the (L, V) structure with D and B as inventory control manipulators.

10-10 APPLICATION 5: RELATIVE GAIN ANALYSIS

The relative gain concept, introduced by Bristol (1966) and developed by McAvoy (1983) and Shinskey (1984) among others, is a measure of interaction between control loops. Basically, relative gain analysis is used to pair inputs and outputs in a control structure, but based on the calculated interaction indices it is often used for choosing the control structure (i.e., the inputs and outputs to be used) from a number of alternatives.

In its usual form, relative gain analysis is a steady-state analysis and thus based on process gains only. This means that the transformations and consistency relations previously derived are readily applicable. If the process gains of a system are given for one control structure, such as the conventional (L, V) structure, the process gains for any other feasible control structure can be calculated. The relative gains for any distillation control structure can thus be calculated from known process gains for one structure.

When the reflux drum and reboiler holdups can be assumed perfectly controlled, or when the same inventory control system is used in the control structures considered, the transformations also apply to dynamic models. In this case the relative gains for the control structures can be calculated as functions of frequency. Note, however, that the transformations require that the complete model, including the flow transfer functions, be known because the flow transfer functions cannot be obtained from steady-state consistency relations.

In this section we will not deal with the calculation of relative gains from known models for different control structures. Instead, we will present some analytical relations between the relative gains for a number of control structures. These relations can be used, for example, to calculate relative gains without first deriving models for the various control structures. Some of the expressions also apply to frequency dependent relative gains, but in general they have to be modified for the dynamic case because the consistency relations do not apply to transfer functions. The results given in this section are from Häggblom (1988a), where the subject is treated more extensively.

10-10-1 Some Analytical Relations between Relative Gains

In the (L, V) structure the relative gain for the $y - L$ pairing is given by

$$\lambda_{yL}^{LV} = \left(1 - \frac{K_{xL}^{LV} K_{yV}^{LV}}{K_{yL}^{LV} K_{xV}^{LV}}\right)^{-1} \tag{10-62}$$

Analogously, the relative gain for the $y - D$ pairing in the (D, V) structure is given by

$$\lambda_{yD}^{DV} = \left(1 - \frac{K_{xD}^{DV} K_{yV}^{DV}}{K_{yD}^{DV} K_{xV}^{DV}}\right)^{-1} \tag{10-63}$$

and the relative gain for the $y - L$ pairing in the (L, B) structure by

$$\lambda_{yL}^{LB} = \left(1 - \frac{K_{xL}^{LB} K_{yB}^{LB}}{K_{yL}^{LB} K_{xB}^{LB}}\right)^{-1} \tag{10-64}$$

The models for these control structures are given in Equations 10-1, 10-2, and 10-3, respectively.

The transformations from the (L, V) structure to the (D, V) structure and the (L, B) structure are given by

$$\begin{bmatrix} K_{yD}^{DV} & K_{yV}^{DV} \\ K_{xD}^{DV} & K_{xV}^{DV} \end{bmatrix} = \frac{1}{K_{DL}^{LV}} \begin{bmatrix} K_{yL}^{LV} & -\left(K_{yL}^{LV} K_{DV}^{LV} - K_{yV}^{LV} K_{DL}^{LV}\right) \\ K_{xL}^{LV} & -\left(K_{xL}^{LV} K_{DV}^{LV} - K_{xV}^{LV} K_{DL}^{LV}\right) \end{bmatrix} \tag{10-65}$$

$$\begin{bmatrix} K_{yL}^{LB} & K_{yB}^{LB} \\ K_{xL}^{LB} & K_{xB}^{LB} \end{bmatrix} = \frac{1}{K_{DV}^{LV}} \begin{bmatrix} K_{yL}^{LV} K_{DV}^{LV} - K_{yV}^{LV} K_{DL}^{LV} & -K_{yV}^{LV} \\ K_{xL}^{LV} K_{DV}^{LV} - K_{xV}^{LV} K_{DL}^{LV} & -K_{xV}^{LV} \end{bmatrix} \tag{10-66}$$

From these equations, and Equations 10-62, 10-63, and 10-64, it follows that the relative gains λ_{yL}^{LV}, λ_{yD}^{DV}, and λ_{yL}^{LB} are related by the expression

$$\left(1 - \frac{1}{\lambda_{yL}^{LV}}\right) = \left(1 - \frac{1}{\lambda_{yD}^{DV}}\right)\left(1 - \frac{1}{\lambda_{yL}^{LB}}\right) \tag{10-67}$$

Equation 10-67 has previously been derived by Jafarey, McAvoy, and Douglas (1979) with use of the constant flow assumption $\Delta D = \Delta V - \Delta L$. As no limiting assumptions were made in the preceding derivation, the relationship is exact for any separation process, whether the assumption of constant flows holds or not and regardless of, for example, the type of primary output variables (compositions or temperatures).

A control structure often used and studied is Ryskamp's (1980) control structure $(D/(L + D), V)$, which is essentially equivalent to the $(L/D, V)$ structure. This structure can be interpreted as a hybrid between the (L, V) structure and the (D, V) structure. The relative gain λ_{yR}^{RV}, where $R = L/D$, can be expressed as a weighted average of the corresponding relative gains for the (L, V) structure and the (D, V) structure. The expression is

$$\lambda_{yR}^{RV} = \frac{\lambda_{yL}^{LV} + \rho\lambda_{yD}^{DV}}{1 + \rho} \tag{10-68}$$

where

$$\rho = -\frac{\overline{L}}{\overline{D}} K_{DL}^{LV} \tag{10-69}$$

Equations 10-68 and 10-69 are valid not only for the $(L/D, V)$ structure, but for any control structure where the top and bottom manipulators can be reduced to functions of L/D and V, respectively. This is the case with Ryskamp's control structure, for example.

The control structures where the top and bottom manipulators can be reduced to functions of L and V/B, respectively, form another class of control structures. For these control structures, the relative gain λ_{yL}^{LS}, where $S = V/B$, can be expressed as

$$\lambda_{yL}^{LS} = \frac{\lambda_{yL}^{LV} + \sigma\lambda_{yL}^{LB}}{1 + \sigma} \tag{10-70}$$

where

$$\sigma = \frac{\bar{V}}{\bar{B}} K_{DV}^{LV} \tag{10-71}$$

A control structure often recommended in the literature (Takamatsu, Hashimoto, and Hashimoto, 1982; Shinskey, 1984; Skogestad, Lundström, and Jacobsen, 1990) is the $(L/D, V/B)$ structure. The relative gains λ_{yR}^{RS} for this control structure can be expressed as

$$\lambda_{yR}^{RS} = \gamma \frac{\lambda_{yR}^{RV}\lambda_{yL}^{LS}}{\lambda_{yL}^{LV}} \tag{10-72}$$

where

$$\gamma = \frac{(1 + \rho)(1 + \sigma)}{1 + \rho + \sigma} \tag{10-73}$$

10-10-2 Numerical Example

The relative gain expressions are very useful for analytical purposes. It is possible by means of these and other similar relations to derive, for example, exact and general conditions under which different control structures are superior to one another in terms of relative gains (Häggblom, 1988a).

However, the relations are also convenient for calculation of relative gains for different control structures, especially if the calculations are made by hand. (Note that the models for all control structures considered would have to be derived if these relations were not available.) We will here illustrate the usefulness of the relations for this purpose by means of the model for the Åbo Akademi column previously studied in this chapter. It is possible to check the results of the relative gain expressions from the models for different control structures given in Table 10-7.

Table 10-8 shows the data required for the calculations. From the given gains for the (L, V) structure, the gains for the (D, V) structure can be calculated according to Equation 10-65. The relative gains λ_{yL}^{LV} and λ_{yD}^{DV} for the (L, V) and (D, V) structures can then be calculated by means of Equations 10-62 and 10-63. The relative gain λ_{yL}^{LB} for the (L, B) structure is obtained from Equation 10-67. The parameters ρ, σ, and γ are determined from the given steady-state data and the flow gains K_{DL}^{LV} and K_{DV}^{LV} according to Equations 10-69, 10-71, and 10-73. Finally, the relative gains λ_{yR}^{RV}, λ_{yL}^{LS}, and λ_{yR}^{RS} for the control structures $(L/D, V)$, $(L, V/B)$, and $(L/D, V/B)$ can be calculated by means of Equations 10-68, 10-70, and 10-72. The results are given in Table 10-9.

TABLE 10-8 Data for Calculation of Relative Gains for the Åbo Akademi Column

$\bar{L}/\bar{D} = 1.0000$	$\bar{V}/\bar{B} = 0.5143$
$K_{yL}^{LV} = -0.0435$	$K_{yV}^{LV} = 0.0458$
$K_{xL}^{LV} = -0.2260$	$K_{xV}^{LV} = 0.5432$
$K_{DL}^{LV} = -0.6154$	$K_{DV}^{LV} = 1.3523$

TABLE 10-9 Relative Gains for the Åbo Akademi Column

$\rho = 0.6154$	$\sigma = 0.6955$	$\gamma = 1.1852$
$\lambda_{yL}^{LV} = 1.7795$	$\lambda_{yD}^{DV} = 0.1521$	$\lambda_{yL}^{LB} = 0.9271$
$\lambda_{yR}^{RV} = 1.1595$	$\lambda_{yL}^{LS} = 1.4298$	$\lambda_{yR}^{RS} = 1.1042$

10-11 APPLICATION 6: SYNTHESIS OF DECOUPLED CONTROL STRUCTURES BY TRANSFORMATIONS OF OUTPUT VARIABLES

The usual way of affecting the properties of a control structure is to use transformations of the basic input variables. However, it is also possible to use transformations of the basic output variables for the same purpose. An example of this is the use of logarithmic concentrations as a means of "linearizing" the column (Ryskamp, 1982; Skogestad and Morari, 1988). Another example is the use of sums and differences of concentrations or temperatures as a way of reducing the interaction between the control loops (Waller and Finnerman, 1987).

In this section we will present an extension of the approach last mentioned; we will synthesize one- and two-way decoupled control structures by means of arbitrary (linear) combinations of the output variables. For the (D, V) and (L, B) structures a very interesting result is obtained.

10-11-1 Derivation of Output Transformations

Consider two transformations of the output variables y and x:

$$\eta = \eta(y, x) \tag{10-74a}$$

$$\xi = \xi(y, x) \tag{10-74b}$$

Linearization and introduction of deviation variables give the linear transformations

$$\Delta\eta = H_{\eta y}\,\Delta y + H_{\eta x}\,\Delta x \tag{10-75a}$$

$$\Delta\xi = H_{\xi y}\,\Delta y + H_{\xi x}\,\Delta x \tag{10-75b}$$

Consider now the standard (D, V) structure, described by the model (here we only consider the gains between the primary manipulators and the outputs)

$$\begin{bmatrix} \Delta y \\ \Delta x \end{bmatrix} = \begin{bmatrix} K_{yD}^{DV} & K_{yV}^{DV} \\ K_{xD}^{DV} & K_{xV}^{DV} \end{bmatrix} \begin{bmatrix} \Delta D \\ \Delta V \end{bmatrix} \tag{10-76}$$

Elimination of Δy and Δx from Equations 10-75 by means of Equation 10-76 gives the model

$$\begin{bmatrix} \Delta \eta \\ \Delta \xi \end{bmatrix} = \begin{bmatrix} K_{\eta D}^{DV} & K_{\eta V}^{DV} \\ K_{\xi D}^{DV} & K_{\xi V}^{DV} \end{bmatrix} \begin{bmatrix} \Delta D \\ \Delta V \end{bmatrix} \tag{10-77}$$

where

$$\begin{bmatrix} K_{\eta D}^{DV} & K_{\eta V}^{DV} \\ K_{\xi D}^{DV} & K_{\xi V}^{DV} \end{bmatrix} = \begin{bmatrix} H_{\eta y}K_{yD}^{DV} + H_{\eta x}K_{xD}^{DV} & H_{\eta y}K_{yV}^{DV} + H_{\eta x}K_{xV}^{DV} \\ H_{\xi y}K_{yD}^{DV} + H_{\xi x}K_{xD}^{DV} & H_{\xi y}K_{yV}^{DV} + H_{\xi x}K_{xV}^{DV} \end{bmatrix} \tag{10-78}$$

The condition for decoupling is that the off-diagonal elements should be zero, that is,

$$H_{\eta y}K_{yV}^{DV} + H_{\eta x}K_{xV}^{DV} = 0 \tag{10-79a}$$

$$H_{\xi y}K_{yD}^{DV} + H_{\xi x}K_{xD}^{DV} = 0 \tag{10-79b}$$

Because we have only two equations, two of the parameters $H_{\eta y}$, $H_{\eta x}$, $H_{\xi y}$, and $H_{\xi x}$ can be (almost) arbitrarily chosen. A natural choice is $H_{\eta y} = H_{\xi x} = 1$, because $H_{\eta x} = H_{\xi y} = 0$ then gives the original model. With this choice

$$H_{\eta x} = -\frac{K_{yV}^{DV}}{K_{xV}^{DV}} \tag{10-80a}$$

$$H_{\xi y} = -\frac{K_{xD}^{DV}}{K_{yD}^{DV}} \tag{10-80b}$$

and

$$\Delta\eta = \Delta y - \frac{K_{yV}^{DV}}{K_{xV}^{DV}}\,\Delta x \tag{10-81a}$$

$$\Delta\xi = -\frac{K_{xD}^{DV}}{K_{yD}^{DV}}\,\Delta y + \Delta x \tag{10-81b}$$

will result in a decoupled control structure. If both Equations 10-81a and 10-81b (or 10-80a and 10-80b) apply, there will be two-way steady-state decoupling; if only one of the expressions applies, there will be one-way steady-state decoupling.

Consider now the consistency relations for the standard (D, V) structure, and more precisely, Equation 10-10b, which applies if y and x are product compositions. Elimination of the gains from Equation 10-81a by Equation 10-10b shows that the transformation

$$\Delta\eta = \Delta y + \frac{\overline{B}}{\overline{D}}\Delta x \tag{10-82}$$

decouples η from V in the modified (D, V) structure. From the external material balances, Equations 10-4, it follows that

$$\Delta\eta = \Delta y + \frac{\bar{y} - \bar{z}}{\bar{z} - \bar{x}}\Delta x \tag{10-83}$$

is an equivalent expression.

It is thus possible to achieve perfect one-way steady-state decoupling in the (D, V) structure without knowledge of the model; it is sufficient to know some (nominal) steady-state data. Note that the consistency relations for the (D, V) structure (Equations 10-10) are such that ξ defined according to Equation 10-81b cannot be decoupled from D without knowledge of the model.

By analogy, it follows that one-way decoupling is achieved in the (L, B) structure if

$$\Delta\xi = \Delta x + \frac{\overline{D}}{\overline{B}}\Delta y = \Delta x + \frac{\bar{z} - \bar{x}}{\bar{y} - \bar{z}}\Delta y \tag{10-84}$$

is controlled by B.

It can further be shown that one-way steady-state decoupling is obtained in any control structure where η or ξ defined according to Equations 10-82, 10-83, or 10-84 is controlled by D, B, D/F, or B/F.

Equations 10-80 are necessary and sufficient conditions for decoupling of any control structure when the gains in Equation 10-80 are replaced by the corresponding gains in the control structure considered.

10-11-2 Numerical Examples

Here we will use Equations 10-80 to calculate transformations of the basic output variables y and x so that one- or two-way decoupling is achieved in a number of given control structures.

For the Wood and Berry column, Equations 10-80 give the following results, which are based on the models given in Section 10-7.

(L, V) structure:

$$\Delta\eta = \Delta y - 0.97\Delta x \tag{10-85a}$$
$$\Delta\xi = -0.52\,\Delta y + \Delta x \tag{10-85b}$$

(D, V) structure:

$$\Delta\eta = \Delta y + 1.1\,\Delta x \tag{10-86a}$$
$$\Delta\xi = -0.52\,\Delta y + \Delta x \tag{10-86b}$$

(L, B) structure:

$$\Delta\eta = \Delta y - 0.97\Delta x \tag{10-87a}$$
$$\Delta\xi = 0.93\,\Delta y + \Delta x \tag{10-87b}$$

$(D/(L + D), V)$ structure:

$$\Delta\eta = \Delta y - 0.75\,\Delta x \tag{10-88a}$$
$$\Delta\xi = -0.51\,\Delta y + \Delta x \tag{10-88b}$$

The corresponding calculations for the Åbo Akademi column give the following results (the models are given in Tables 10-6 and 10-7):

(L, V) structure:

$$\Delta\eta = \Delta y - 0.084\,\Delta x \tag{10-89a}$$
$$\Delta\xi = -5.2\,\Delta y + \Delta x \tag{10-89b}$$

(D, V) structure:

$$\Delta\eta = \Delta y + 1.1\,\Delta x \tag{10-90a}$$
$$\Delta\xi = -5.2\,\Delta y + \Delta x \tag{10-90b}$$

(L, B) structure:

$$\Delta\eta = \Delta y - 0.084\,\Delta x \tag{10-91a}$$
$$\Delta\xi = 0.93\,\Delta y + \Delta x \tag{10-91b}$$

$(D/(L + D), V)$ structure:

$$\Delta\eta = \Delta y - 0.026\,\Delta x \tag{10-92a}$$
$$\Delta\xi = -5.2\,\Delta y + \Delta x \tag{10-92b}$$

Note that primary output variables are temperatures in the Åbo Akademi column, whereas they are product compositions in the Wood and Berry column.

10-11-3 Discussion of Output Decoupling Structures Suggested in the Literature

A number of output transformations and noninteracting control structures have been suggested in the literature. These can be examined with respect to noninteraction at steady state in the light of the results and numerical examples in the two previous sections.

As a way to eliminate the often harmful interaction between the primary control loops in dual-composition control of distillation, Rosenbrock (1962) as early as 1962 suggested a scheme, where "the 'top loop' now manipulates the flow of top product to control the sum of the two measured compositions." The bottom loop manipulates V to control the difference between the measured compositions. Rosenbrock stated that the scheme often can be shown to be noninteracting.

As shown in Section 10-11-1, a one-way decoupled structure is obtained if D controls the sum of the product compositions when $\bar{D} = \bar{B}$. The two numerical examples in the previous section illustrate the goodness of the approximation for the (D, V) structure when these conditions are not fulfilled.

Ryskamp (1982) suggested a structure where the reflux is set by the difference between two temperatures in the column, one above and one below the feed plate. He suggested heat input to be set by the sum of the temperatures. For the (L, V) structure, no general results based on consistency relations, like those for the (D, V) and the (L, B) structure, can be derived. Actually, the properties of the (L, V) structure are such that decoupling (at steady state) is obtained when weighted differences of the primary outputs are fed *both* to L and V, as illustrated by the two numerical examples.

More recently, Bequette and Edgar (1989) used a number of methods to calculate noninteracting control schemes for a specific column using D and V as manipulators. They ended up with Rosenbrock's scheme, connecting the composition sum to D and the difference to V.

Weber and Gaitonde (1982) used the manipulators L and V and connected the variable $k(1 - y) + x$ to L and $(1 - y)/x$ to V. The constant k was calculated from the nominal operating point as $\bar{x}/(1 - \bar{y})$.

McAvoy (1983) discussed Weber's and Gaitonde's scheme and stated that a "better," that is, less interacting, choice for k would be $\bar{z}/(1 - \bar{z})$. Further, he suggested a scheme where

$$\tilde{\eta} = x + \left(\frac{\bar{z}}{1 - \bar{z}}\right) y \tag{10-93}$$

is connected to D, stating that the scheme is almost one-way steady-state decoupled (McDonald and McAvoy, 1983). The other output was the separation factor S connected to the boilup V. The separation factor is defined as

$$S = \left(\frac{y}{1 - y}\right)\left(\frac{1 - x}{x}\right) \tag{10-94}$$

In Section 10-11-1 we showed that a one-way steady-state decoupled scheme is obtained when the variable $\Delta\eta = \Delta y + (\bar{B}/\bar{D})\,\Delta x$ is connected to D (or B). The same result is obviously obtained when

$$\eta' = x + \left(\frac{\bar{D}}{\bar{B}}\right) y = x + \left(\frac{\bar{z} - \bar{x}}{\bar{y} - \bar{z}}\right) y \tag{10-95}$$

is connected to D.

A comparison of Equation 10-95 with Equation 10-93 shows that McAvoy's suggestion results in perfect one-way steady-state decoupling when $(\bar{D}/\bar{B}) = \bar{z}/(1 - \bar{z})$, which is equivalent to $\bar{D} = \bar{F}\bar{z}$. This condition is fulfilled, for example, in a "symmetric" separation, where $\bar{z} = 0.5$ and $\bar{D} = \bar{B}$. The comparison also shows that McAvoy's suggestion results in approximate one-way steady-state decoupling for high-purity distillation, where $\bar{y} \approx 1$, $\bar{x} \approx 0$.

It can further be noted that ΔS is equivalent to $\Delta y - \Delta x$ when $\bar{y} = 1 - \bar{x}$. If also $\bar{D} = \bar{B}$, the linearized form of the control scheme suggested by McAvoy is equivalent to a control scheme where $y + x$ is connected to D and $y - x$ to V.

According to Smart (1985), Hobbs (1984) has suggested a scheme where the reflux is manipulated "to control the sum of overhead and bottoms impurities," and bottoms flow is manipulated "to control the difference in impurities." If the overhead impurity is $1 - y$ and the bottoms impurity x, controlling the sum of impurities actually means controlling $x - y$, and controlling the difference means controlling $y + x$. Because $y + x$ is connected to B and $x - y$ to L, the scheme is one-way steady-state decoupled when $\bar{D} = \bar{B}$, provided that y and x denote product compositions.

10-12 APPLICATION 7: A CONTROL STRUCTURE FOR DISTURBANCE REJECTION AND DECOUPLING (DRD)

In the last section we showed how decoupled control structures can be synthesized by transformations of the primary output variables. Another approach, which perhaps is more natural, is to synthesize control structures by transformations of the basic input variables. We have already seen many examples on transformations between models for different control structures with specified input and output variables. However, these control structures are not in any specific way related with the properties of the process. This means that the performance characteristics of the control structures can vary considerably between different columns.

By means of the structural transformation theory presented in Section 10-5, it is possible to synthesize control structures with specified properties. An example of such a control structure is the DRD structure (Häggblom and Waller, 1990), which has the property that, in steady state, the primary control loops are completely decoupled and the primary outputs are insensitive to certain disturbances even in open-loop operation. Next we will show how such a control structure can be synthesized.

Ideally, we would like a control structure with the primary manipulators $\boldsymbol{\psi}$ such that the primary outputs $\mathbf{y}$ depend on the manipulators $\boldsymbol{\psi}$ and the disturbances $\mathbf{w}$ according to

$$\Delta \mathbf{y} = \mathbf{I}\,\Delta\boldsymbol{\psi} + \mathbf{0}\,\Delta\mathbf{w} \tag{10-96}$$

where $\mathbf{I}$ is an identity matrix and $\mathbf{0}$ a zero matrix. The gain matrices of the control structure would then be

$$\mathbf{K}_{\mathbf{y}\boldsymbol{\psi}} = \mathbf{I}, \qquad \mathbf{K}_{\mathbf{y}\boldsymbol{\omega}} = \mathbf{0} \tag{10-97}$$

From the transformations in Equations 10-33 it follows that such a control structure is obtained with the variable transformation

$$\Delta\boldsymbol{\psi} = \mathbf{H}_{\boldsymbol{\psi}\mathbf{u}}\,\Delta\mathbf{u} + \mathbf{H}_{\boldsymbol{\psi}\mathbf{v}}\,\Delta\mathbf{v} \tag{10-98}$$

where

$$\mathbf{H}_{\boldsymbol{\psi}\mathbf{u}} = \mathbf{K}_{\mathbf{yu}} - \mathbf{K}_{\mathbf{yw}}\mathbf{K}_{\mathbf{vw}}^{-1}\mathbf{K}_{\mathbf{vu}} \tag{10-99a}$$

$$\mathbf{H}_{\boldsymbol{\psi}\mathbf{v}} = \mathbf{K}_{\mathbf{yw}}\mathbf{K}_{\mathbf{vw}}^{-1} \tag{10-99b}$$

The same inventory control manipulators are used as in the base structure.

According to this derivation, perfect steady-state disturbance rejection of unmeasured disturbances ($\mathbf{H}_{\boldsymbol{\psi}\mathbf{w}} = \mathbf{0}$) can be obtained even in open-loop operation ($\Delta\boldsymbol{\psi} = \mathbf{0}$) when $\mathbf{v}$ contains such information about the disturbances that $\mathbf{K}_{\mathbf{vw}}$ is nonsingular. The DRD structure can be viewed as containing a disturbance observer, where changes in D and B are used for estimation of the disturbances. Note, however, that this only applies to disturbances that have been taken into account in the design.

Assuming that the (L, V) structure is used as base structure, $\mathbf{u}$ are the variables that physically affect the primary outputs. Therefore, these variables have to be calculated when the control structure is implemented. From Equation 10-98 it follows that they can be determined from

$$\Delta\mathbf{u} = \mathbf{H}_{\boldsymbol{\psi}\mathbf{u}}^{-1}(\Delta\boldsymbol{\psi} - \mathbf{H}_{\boldsymbol{\psi}\mathbf{v}}\,\Delta\mathbf{v}) \quad (10\text{-}100)$$

where the values of $\Delta\boldsymbol{\psi}$ are given by the primary controllers and the values of $\Delta\mathbf{v}$ are measured or given by the inventory controllers.

For the Åbo Akademi column, which is modelled by Equations 10-36, Equations 10-98 and 10-99 yield the manipulators

$$\begin{bmatrix} \Delta\psi_y \\ \Delta\psi_x \end{bmatrix} = \begin{bmatrix} -0.0408 & 0.0398 \\ -0.6330 & 1.4375 \end{bmatrix} \begin{bmatrix} \Delta L \\ \Delta V \end{bmatrix} + \begin{bmatrix} 0.0032 & -0.0013 \\ -0.7828 & -0.1215 \end{bmatrix} \begin{bmatrix} \Delta D \\ \Delta B \end{bmatrix} \quad (10\text{-}101)$$

When the control structure is implemented, L and V are calculated from

$$\begin{bmatrix} \Delta L \\ \Delta V \end{bmatrix} = \begin{bmatrix} -43 & 1.2 \\ -19 & 1.2 \end{bmatrix} \begin{bmatrix} \Delta\psi_y \\ \Delta\psi_x \end{bmatrix} + \begin{bmatrix} 1.1 & 0.09 \\ 1.0 & 0.12 \end{bmatrix} \begin{bmatrix} \Delta D \\ \Delta B \end{bmatrix} \quad (10\text{-}102)$$

where ψ_y and ψ_x are the outputs from the primary controllers in the DRD structure. Experimental results for this control structure are given in Chapter 15. [Note that the numerical values given here differ somewhat from those given in Chapter 15 and in the papers cited in the following text because slightly different models were used for the (L, V) structure.]

It can be noted that both primary manipulators ψ_y and ψ_x are functions of L, V, D, and B, thus in a way combining all structures previously discussed (except those where output variable transformations were used). We note a certain resemblance with a recently suggested "combined balance structure" (Yang, Seborg, and Mellichamp, 1991), where the energy balance structure (L, V) is combined with the two material balance structures (D, V) and (L, B). In the approach of Yang, Seborg, and Mellichamp, the manipulators for the primary loops are chosen as

$$\begin{bmatrix} \Delta\psi_y \\ \Delta\psi_x \end{bmatrix} = \begin{bmatrix} a & 0 \\ 0 & b \end{bmatrix} \begin{bmatrix} \Delta L \\ \Delta V \end{bmatrix} + \begin{bmatrix} 1-a & 0 \\ 0 & 1-b \end{bmatrix} \begin{bmatrix} \Delta D \\ \Delta B \end{bmatrix} \quad (10\text{-}103)$$

Even the recently discovered "super material balance" structure (D, B) (Finco, Luyben, and Polleck, 1989) is covered by Equation 10-103 ($a = b = 0$). The combined balance manipulators can also be viewed as a special case of the DRD manipulators.

The static DRD structure was first studied both experimentally and through simulations in Häggblom and Waller (1990). Its self-regulating properties have further been studied through one-point control in Sandelin, Häggblom, and Waller (1991), where the structure was compared with a number of other control structures for the Åbo Akademi column. The incorporation of "dynamics" into the structure is studied in Waller and Waller (1991) and Häggblom (1992).

ACKNOWLEDGMENT

The results reported have been obtained during a long-range project on multivariable process control supported by the Neste Foundation, the Academy of Finland, Nordisk Industrifond, and Tekes. This support is gratefully acknowledged.

References

Bequette, B. W. and Edgar, T. F. (1989). Noninteracting control system design methods in distillation. *Comput. Chem. Eng.* 13, 641–650.

Bristol, E. H. (1966). On a new measure of interaction for multivariable process control. *IEEE Trans. Autom. Control* AC-11, 133–134.

Finco, M. V., Luyben, W. L., and Polleck, R. E. (1989). Control of distillation columns with low relative volatility. *Ind. Eng. Chem. Res.* 28, 75–83.

Grosdidier, P., Morari, M., and Holt, B. R. (1985). Closed-loop properties from steady-state gain information. *Ind. Eng. Chem. Fundam.* 24, 221–235.

Häggblom, K. E. (1986). Consistent modelling of partially controlled linear systems with application to continuous distillation (in Swedish). Lic. Tech. thesis, Åbo Akademi, Åbo.

Häggblom, K. E. (1988a). Analytical relative gain expressions for distillation control structures. Dr. Tech. thesis, Part V, Åbo Akademi, Åbo.

Häggblom, K. E. (1988b). Estimation of consistent control structure models. Dr. Tech. thesis, Part VI, Åbo Akademi, Åbo.

Häggblom, K. E. (1989). Reconciliation of process gains for distillation control structures. In *Dynamics and Control of Chemical Reactors, Distillation Columns and Batch Processes* (J. E. Rijnsdorp et al., eds.). Oxford: Pergamon Press, pp. 259–266.

Häggblom, K. E. (1991). Modeling of flow dynamics for control of distillation columns. *Proceedings of the American Control Conference*, Boston, pp. 785–790.

Häggblom, K. E. (1992). A combined internal model and inferential control structure for distillation. Report 92-3, Process Control Laboratory, Åbo Akademi, Åbo.

Häggblom, K. E. and Waller, K. V. (1986). Relations between steady-state properties of distillation control system structures. In *Dynamics and Control of Chemical Reactors and Distillation Columns* (C. McGreavy, ed.) Oxford: Pergamon Press, pp. 243–247.

Häggblom, K. E. and Waller, K. V. (1988a). Transformations and consistency relations of distillation control structures. *AIChE J.* 34, 1634–1648.

Häggblom, K. E. and Waller, K. V. (1988b). Transformations between distillation control structures. Report 88-1, Process Control Laboratory, Åbo Akademi, Åbo.

Häggblom, K. E. and Waller, K. V. (1989). Predicting properties of distillation control structures. *Proceedings of the American Control Conference*, Pittsburgh, pp. 114–119.

Häggblom, K. E. and Waller, K. V. (1990). Control structures for disturbance rejection and decoupling of distillation. *AIChE J.* 36, 1107–1113.

Häggblom, K. E. and Waller, K. V. (1991). Modeling of distillation control structures. Report 91-4, Process Control Laboratory, Åbo Akademi, Åbo.

Hobbs, J. W. (1984). Control of a fractional distillation process. U.S. Patent 4,473,443, Sept. 25, 1984.

Jafarey, A., McAvoy, T. J., and Douglas, J. M. (1979). Analytical relationships for the relative gain for distillation control. *Ind. Eng. Chem. Fundam.* 18, 181–187.

Koppel, L. B. (1985). Conditions imposed by process statics on multivariable process dynamics. *AIChE J.* 31, 70–75.

Koung, C. W. and MacGregor, J. F. (1991). Geometric analysis of the global stability of linear inverse-based controllers for bivariate nonlinear processes. *Ind. Eng. Chem. Res.* 30, 1171–1181.

Kridiotis, A. C. and Georgakis, C. (1986). Independent single-input–single-output control schemes for dual composition control of binary distillation columns. International Federation on Automatic Control Symposium, DYCORD, Bournemouth, p. 249–253.

McAvoy, T. J. (1983). *Interaction Analysis*. Research Tringle Park: Instrument Society of America.

McAvoy, T. J. and Weischedel, K. (1981). A dynamic comparison of material balance versus conventional control of distillation columns. Proceedings of the 8th International Federation of Automatic Control, World Congress, Kyoto, pp. 2773–2778.

McDonald, K. A. and McAvoy, T. J. (1983). Decoupling dual composition controllers. 1. Steady state results. Proceedings of the American Control Conference, San Francisco, pp. 176–184.

Ogunnaike, B. A. and Ray, W. H. (1979). Multivariable controller design for linear systems having multiple time delays. *AIChE J.* 25, 1043–1056.

Rademaker, O., Rijnsdorp, J. E., and Maarleveld, A. (1975). *Dynamics and Control of Continuous Distillation Units*. Amsterdam: Elsevier.

Rosenbrock, H. H. (1962). The control of distillation columns. *Trans. Inst. Chem. Eng.* 40, 35–53.

Ryskamp, C. J. (1980). New strategy improves dual composition column control. *Hydrocarbon Processing* 59(6), 51–59.

Ryskamp, C. J. (1982). Explicit versus implicit decoupling in distillation control. In *Chemical Process Control 2* (D. E. Seborg, and T. F. Edgar, eds.). New York: Engineering Foundation/AIChE, pp. 361–375.

Sandelin, P. M., Häggblom, K. E., and Waller, K. V. (1991). Disturbance rejection properties of control structures at one-point control of a two-product distillation column. *Ind. Eng. Chem. Res.* 30, 1187–1193.

Shinskey, F. G. (1984). *Distillation Control*, 2nd ed. New York: McGraw-Hill.

Shinskey, F. G. (1990). Personal communication.

Skogestad, S., Lundström, P., and Jacobsen, E. W. (1990). Selecting the best distillation control configuration. *AIChE J.* 36, 753–764.

Skogestad, S. and Morari, M. (1987a). Control configuration selection for distillation columns. *AIChE J.* 33, 1620–1635.

Skogestad, S. and Morari, M. (1987b). A systematic approach to distillation column control. *Inst. Chem. Eng. Symp. Ser.* 104, A71–A86.

Skogestad, S. and Morari, M. (1988). Understanding the dynamic behavior of distillation columns. *Ind. Eng. Chem. Res.* 27, 1848–1862.

Smart, A. M. (1985). Recent advances in distillation control. *Adv. Instrum.* 40, 493-501.

Takamatsu, T., Hashimoto, I., and Hashimoto, Y. (1982). Multivariable control system design of distillation columns system. Proceedings of the PSE, Kyoto, pp. 243–252.

Takamatsu, T., Hashimoto, I., and Hashimoto, Y. (1987). Selection of manipulated variables to minimize interaction in multivariable control of distillation columns. *Int. Chem. Eng.* 27, 669–677.

Tsogas, A. and McAvoy, T. J. (1981). Dynamic simulation of a non-linear dual composition control scheme. Proceedings of the 2nd World Congress of Chemical Engineering, Montreal, pp. 365–369.

Van Kampen, J. A. (1965). Automatic control by chromatographs of the product quality of a distillation column. Convention in Advances in Automatic Control, Nottingham, England.

Waller, K. V. and Finnerman, D. H. (1987). On using sums and differences to control distillation. *Chem. Eng. Commun.* 56, 253–258.

Waller, K. V., Finnerman, D. H., Sandelin, P. M., Häggblom, K. E., and Gustafsson, S. E. (1988a). An experimental comparison of four control structures for two-point control of distillation. *Ind. Eng. Chem. Res.* 27, 624–630.

Waller, K. V., Häggblom, K. E., Sandelin, P. M., and Finnerman, D. H. (1988b). Disturbance sensitivity of distillation control structures. *AIChE J.* 34, 853–858.

Waller, J. B. and Waller, K. V. (1991). Parametrization of a disturbance rejecting and decoupling control structure. Report 91-13, Process Control Laboratory, Åbo Akademi, Åbo.

Weber, R. and Gaitonde, N. Y. (1982). Noninteractive distillation tower control. Proceedings of the American Control Conference, Arlington, VA, pp. 87–90.

Wood, R. K. and Berry, M. W. (1973). Terminal composition control of a binary distillation column. *Chem. Eng. Sci.* 28, 1707–1717.

Yang, D. R., Waller, K. V., Seborg, D. E., and Mellichamp, D. A. (1990). Dynamic structural transformations for distillation control configurations. *AIChE J.* 36, 1391–1402.

Yang, D. R., Seborg, D. E., and Mellichamp, D. A. (1991). Combined balance control structure for distillation columns. *Ind. Eng. Chem. Res.* 30, 2159–2168.

11

Diagonal Controller Tuning

William L. Luyben
Lehigh University

11-1 INTRODUCTION

Distillation columns are inherently multivariable. As discussed in Chapter 1, even the most simple single-feed, two-product column is a 5×5 system: five manipulated variables and five controlled variables. The column can normally be reduced to a 2×2 system because the control loops for the two liquid levels and the pressure do not interact significantly in many columns. Thus we have in the simplest of distillation columns a system with two compositions or two temperatures that can be controlled and two manipulated variables to achieve this control.

However, there are many more complex distillation systems that occur in high-performance, high-efficiency plants that are designed to compete effectively in a world economy. Examples include heat integration, combined reactor–column processes, vapor recompression, and complex columns (multiple feeds and sidestreams). In such systems the number of controlled and manipulated variables increases, resulting in multivariable processes that are often 3×3, 4×4, or even 5×5.

If we are to operate these multivariable processes safely and efficiently, we must be able to develop control systems for them. There are two basic approaches: use a multivariable controller (where all controlled variables affect all manipulated variables) or use multiloop single-input–single-output (SISO) controllers (where each controlled variable affects only one manipulated variable). The latter is called a diagonal or decentralized controller.

The multivariable-controller approach requires a more complex design procedure and more complex computer control software. Included in this type of controller are dynamic matrix control (DMC; see Chapter 12), internal model control (IMC), linear quadratic (LQ) optimal control, extended horizon control, and several others. The use of multivariable controllers in industry has increased in recent years as more software packages have been developed. The ability of the multivariable controller to handle constraints in a coordinated way is an advantage of these controllers.

However, the majority of chemical processes can be and are handled quite adequately with the more simple diagonal control structure, as long as the interaction among the loops is explicitly taken into account and not ignored. Constraints can be handled by using override controls to supplement the conventional control structure.

The purpose of this chapter is to present a fairly simple approach to designing an effective diagonal control system for a multivariable plant.

11-1-1 The Problem

When the process control engineer is faced with a multivariable process, there are a number of questions that must be answered:

1. What variables should be controlled?
2. What manipulated variables should be selected to control these controlled variables? This selection is more complex than just choosing what control valve to use because we include the possibility of using ratios, sums, differences, and so forth, of various manipulated variable flow rates.
3. How should the controlled and manipulated variables be paired?
4. How do we tune the SISO controllers in this multivariable environment?

We will address each of these in the following sections.

11-1-2 Alternatives

It is important to keep firmly in mind that a good engineer never tries to find a complex solution to a simple problem. This obvious piece of wisdom is, somewhat remarkably, often forgotten in such high-tech areas as multivariable control, computer control, expert systems, and the like. There is an old

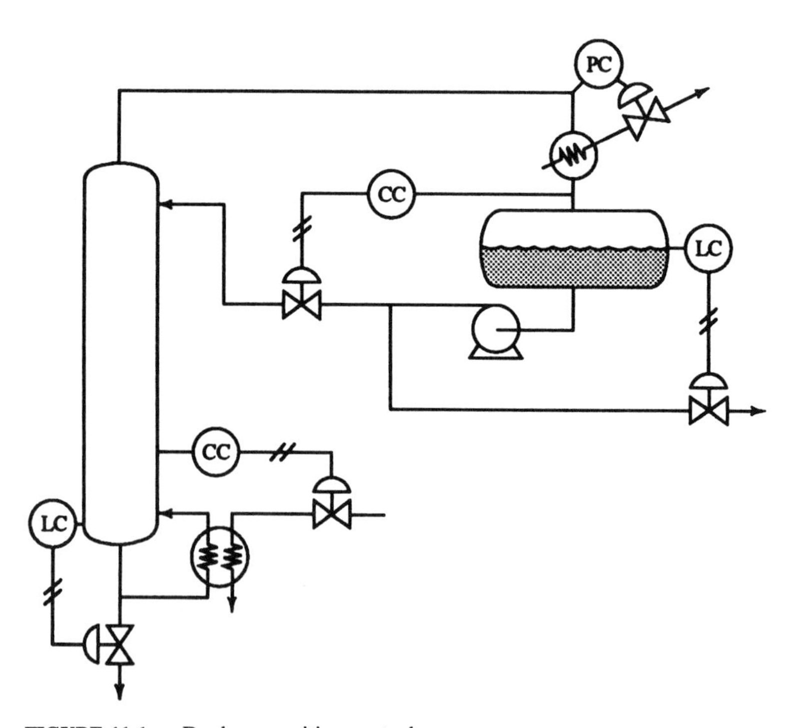

FIGURE 11-1. Dual composition control.

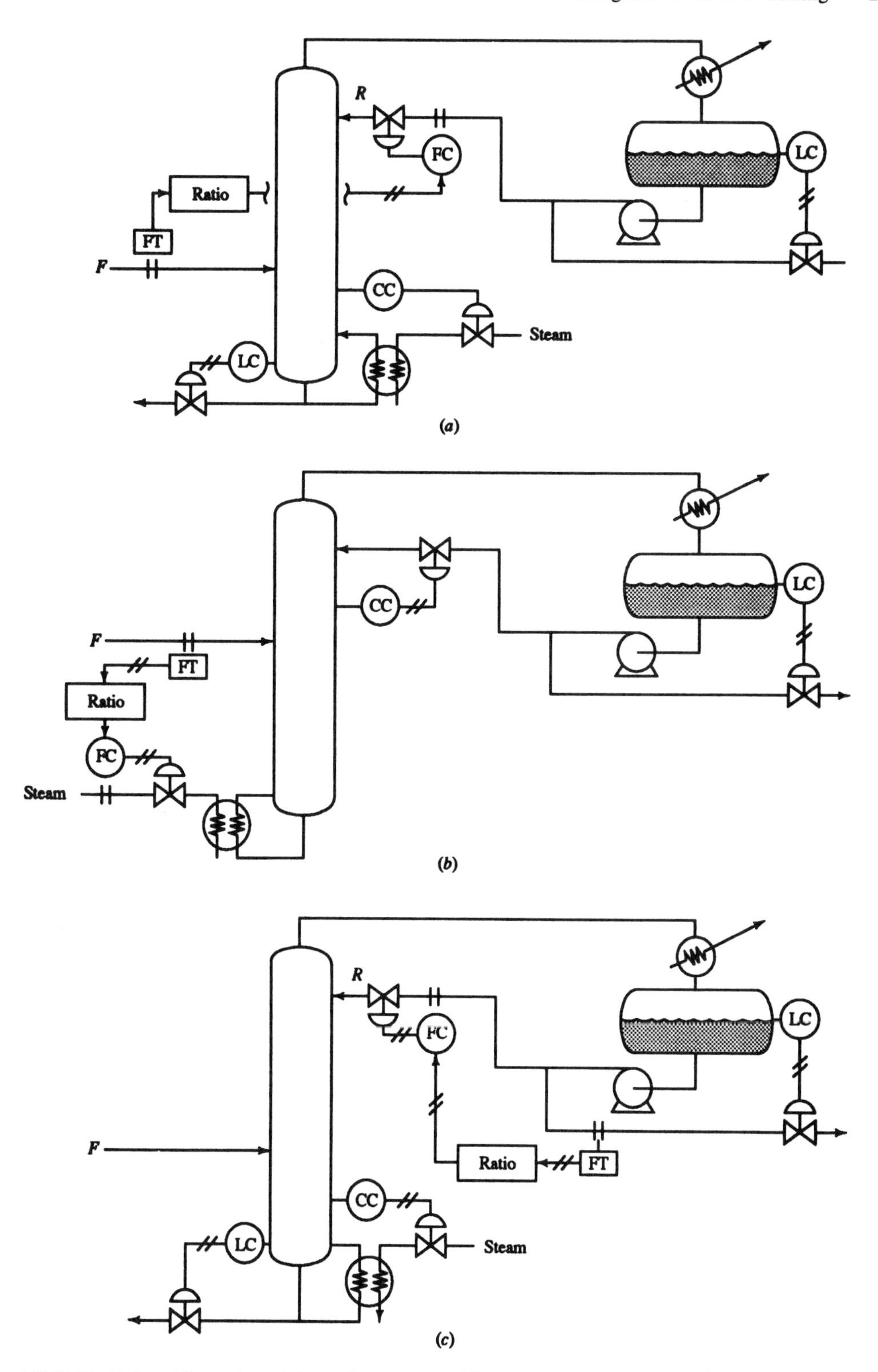

FIGURE 11-2. Alternatives: (a) Constant reflux; (b) constant vapor boilup; (c) constant reflux ratio.

barnyard saying that wisdom is like gold: You usually have to dig through a lot of dirt and break a lot of rock to get to it.

There are many processes that must be dealt with as bona fide multivariable systems. However, there are also many processes that look like they are multivariable but really are not. So before you plunge into the complexities of multivariable control, make sure that you have a problem that really is multivariable.

Let us consider a specific distillation example. Suppose we have a binary separation to make in a simple two-product column. We could immediately say (unwisely) that we want to control both distillate composition x_D and bottoms composition x_B. This "dual composition" control strategy will achieve the minimum consumption of energy when operating the given column at its optimum pressure and specified product purities. See Figure 11-1.

However, the question that you should ask is "How much energy will be saved and can these savings justify the additional expenses and risks associated with designing and operating a more complex control system?" In some columns there will be a substantial economic incentive. In some there may be very little real savings realized by controlling both compositions compared to only controlling one composition and overpurifying the other product.

Figure 11-2 shows three commonly used alternatives to dual composition control: constant reflux, constant vapor boilup, and constant reflux ratio. The single composition (or temperature) control loop is a simple SISO system, so it is easily tuned.

One procedure for quantitatively evaluating the incentive (or lack of it) for dual composition control is to use a steady-state rating program to generate curves that show how reflux, heat input, and reflux ratio must be changed as feed composition changes such that both distillate and bottoms product compositions are held constant. These curves are illustrated in Figure 11-3. In the example shown, the changes in the reflux flow rate are quite small (the curve is fairly flat). This suggests that a fixed reflux flow rate will waste very little energy compared to dual composition control.

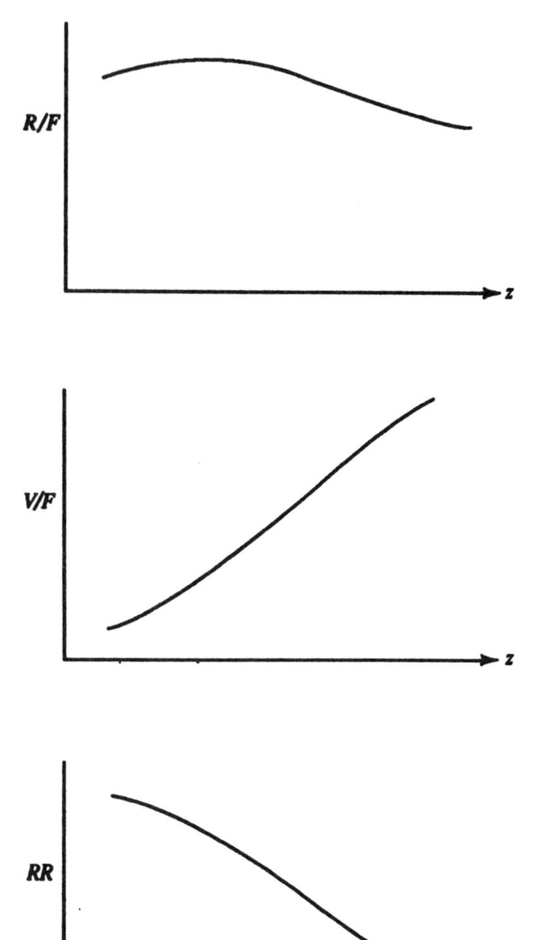

FIGURE 11-3. Steady-state changes required in manipulated variables to hold x_D and x_B constant as feed composition changes.

Note that the disturbance considered is only feed composition and not feed flow rate. Rate changes can usually be effectively handled by using only the appropriate ratio scheme: reflux-to-feed or steam-to-feed. These feedforward control strategies usually take care of throughput changes.

11-1-3 LACEY Procedure

After you have assured yourself that your problem is indeed a multivariable one, the next step is to design a conventional diagonal control system. One approach to this problem is summarized in the following list. Each step is discussed in more detail in later sections of this chapter.

1. *Select controlled variables.* Use primarily engineering judgment based on process understanding.
2. *Select manipulated variables.* Find the set of manipulated variables that gives the largest Morari resiliency index (MRI): the smallest singular value of the steady-state open-loop plant transfer function matrix (the steady-state gain matrix).
3. *Eliminate unworkable variable pairings.* Eliminate pairings with negative Niederlinski indexes.
4. *Find the best pairing from the remaining sets.*
   (a) Tune each combination using the biggest-log modulus tuning procedure (BLT).
   (b) Select the pairing that gives the lowest magnitude closed-loop load transfer function: Tyreus load-rejection criterion (TLC).

Each of these steps is discussed in the following sections.

11-1-4 Nomenclature

It would be useful at this point to define some terms and specify the nomenclature that will be used in this chapter. Multivariable systems are described by sets of equations, either differential equations in the time domain or algebraic equations in the Laplace domain (transfer functions). It is convenient to use matrix notation in order to write these equations in a compact form.

So that we can clearly distinguish between scalar and matrix variables, we will denote column matrices (vectors) with one underscore and matrices with two underscores. For example, the vector $\underline{\mathbf{x}}$ of N controlled output variables $x_1, x_2, x_3, \ldots, x_N$ will be written as

$$\underline{\mathbf{x}} = \begin{bmatrix} x_1 \\ x_2 \\ x_3 \\ \cdots \\ x_N \end{bmatrix} \tag{11-1}$$

The square matrix $\underline{\underline{\mathbf{A}}}$ with N rows and N columns will be written as

$$\underline{\underline{\mathbf{A}}} = \begin{bmatrix} a_{11} & a_{12} & a_{13} & \cdots & a_{1N} \\ a_{21} & a_{22} & a_{23} & \cdots & \\ a_{31} & & & \ddots & \\ \vdots & & & & \vdots \\ a_{N1} & & & & a_{NN} \end{bmatrix} \tag{11-2}$$

The set of equations that describes the open-loop system with N controlled variables x_i, N manipulated variables m_j, and one load input L is

$$\begin{aligned} x_1 &= G_{M_{11}} m_1 + G_{M_{12}} m_2 + \cdots \\ &\quad + G_{M_{1N}} m_N + G_{L_1} L \\ x_2 &= G_{M_{21}} m_1 + G_{M_{22}} m_2 + \cdots \\ &\quad + G_{M_{2N}} m_N + G_{L_2} L \\ &\vdots \\ x_N &= G_{M_{N1}} m_1 + G_{M_{N2}} m_2 + \cdots \\ &\quad + G_{M_{NN}} m_N + G_{L_N} L \end{aligned} \tag{11-3}$$

All the variables are in the Laplace domain, as are all of the transfer functions. This set of N equations is very conveniently represented by one matrix equation:

$$\underline{\mathbf{x}} = \underline{\underline{\mathbf{G}}}_{M_{(s)}} \underline{\mathbf{m}}_{(s)} + \underline{\mathbf{G}}_{L_{(s)}} L_{(s)} \tag{11-4}$$

where $\underline{x}$ = vector of N controlled variables

$\underline{\underline{G}}_M$ = $N \times N$ matrix of process open-loop transfer functions relating the controlled variables and the manipulated variables

$\underline{m}$ = vector of N manipulated variables

$\underline{G}_L$ = vector of process open-loop transfer functions relating the controlled variables and the load disturbance

$L_{(s)}$ = load disturbance

We use only one load variable in this development in order to keep things as simple as possible. Clearly, there could be several load disturbances, and we would just add additional terms to Equations 11-3. Then the $L_{(s)}$ in Equation 11-4 becomes a vector and $\underline{G}_L$ becomes a matrix with N rows and as many columns as there are load disturbances. Because the effects of each of the load disturbances can be considered one at a time, we will do it that way to simplify the mathematics. Note that the effects of each of the manipulated variables could also be considered one at a time if we were looking only at the *open-loop* system or if we were considering controlling only one variable. However, when we go to a multivariable *closed-loop* system, the effects of all manipulated variables must be considered simultaneously.

Figure 11-4 gives the matrix block diagram description of the open-loop system with a feedback control system added. The $\underline{\underline{I}}$ matrix is the identity matrix. The $\underline{\underline{B}}_{(s)}$ matrix contains the feedback controllers. With the multiloop SISO diagonal controller structure the $\underline{\underline{B}}_{(s)}$ matrix has only diagonal elements; all the off-diagonal elements are zero:

$$\underline{\underline{B}}_{(s)} = \begin{bmatrix} B_1 & 0 & \cdots & 0 \\ 0 & B_2 & 0 & \\ & & \ddots & \\ 0 & \cdots & 0 & B_N \end{bmatrix} \qquad (11\text{-}5)$$

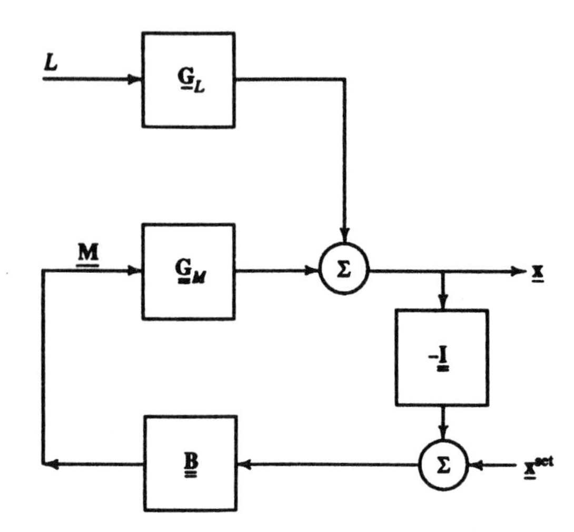

FIGURE 11-4. Block diagram of closed-loop multivariable system.

where $B_1, B_2, \ldots, B_N$ are the individual controllers in each of the N loops.

In the closed-loop system the manipulated variables are set by the controllers:

$$\underline{m} = \underline{\underline{B}}_{(s)}\underline{\underline{E}} = \underline{\underline{B}}_{(s)}\left[\underline{x}^{set} - \underline{x}\right] \qquad (11\text{-}6)$$

Substituting Equation 11-6 into Equation 11-4 and solving for $\underline{x}$ gives

$$\underline{x} = \left[\left[\underline{\underline{I}} + \underline{\underline{G}}_{M_{(s)}}\underline{\underline{B}}_{(s)}\right]^{-1}\underline{\underline{G}}_{M_{(s)}}\underline{\underline{B}}_{(s)}\right]\underline{x}^{set} + \left[\left[\underline{\underline{I}} + \underline{\underline{G}}_{M_{(s)}}\underline{\underline{B}}_{(s)}\right]^{-1}\underline{G}_{L_{(s)}}\right]L_{(s)} \qquad (11\text{-}7)$$

Equation 11-7 gives the effects of setpoint and load changes on the controlled variables in the closed-loop multivariable environment. The matrix (of order $N \times N$) multiplying the vector of setpoints is the closed-loop servo transfer function matrix. The matrix ($N \times 1$) multiplying the load disturbance is the closed-loop regulator transfer function vector.

Remember that the inverse of a matrix has the determinant of the matrix in the denominator of each element. Therefore the denominators of all of the transfer functions in Equation 11-7 will contain $\text{Det}[\underline{\underline{I}} + \underline{\underline{G}}_{M_{(s)}}\underline{\underline{B}}_{(s)}]$.

Now we know that the characteristic equation of any system is the denominator of its transfer function set equal to zero. Therefore the closed-loop characteristic equation of the multivariable system with feedback controllers is the simple *scalar* equation

$$\text{Det}\left[\underline{\underline{\mathbf{I}}} + \underline{\underline{\mathbf{G}}}_{M(s)}\underline{\underline{\mathbf{B}}}_{(s)}\right] = 0 \tag{11-8}$$

We can use this closed-loop characteristic equation to tune controllers in multivariable processes.

11-2 SELECTION OF CONTROLLED VARIABLES

Engineering judgment is the principal tool for deciding what variables to control. A good understanding of the process leads in most cases to a logical choice of what needs to be controlled. Considerations of economics, safety, constraints, availability and reliability of sensors, and so forth must be factored into this decision.

It should be remembered that controlled variables need not be simple directly measured variables. They can also be computed from a number of sensor inputs. Common distillation examples are reflux ratio, reflux-to-feed ratio, reboiler heat input calculated from the flow rate and temperature change of a hot oil stream, and so forth.

Sometimes we are unable to measure directly the variable that we would really like to control. This is frequently true in distillation because it is often difficult and expensive to measure product compositions directly with sensors such as gas chromatographs. Instead, temperatures on various trays are controlled. The selection of the best control tray to use requires a considerable amount of knowledge about the column, its operation, and its performance. Varying amounts of nonkey components in the feed can affect very importantly the best choice of control trays. Some aspects of this inferential control problem were discussed in Chapter 1, and additional aspects are treated in several other chapters.

11-3 SELECTION OF MANIPULATED VARIABLES

Once the controlled variables have been specified, the next step is to choose the manipulated variables. We do not worry at this stage about how to *pair* the variables in a diagonal multiloop SISO controller structure but only about what variables to manipulate. Should we manipulate reflux and vapor boilup (R-V) to control distillate and bottoms compositions, or should we manipulate distillate and vapor boilup (D-V)?

11-3-1 Morari Resiliency Index

The Morari resiliency index (MRI) can be used to give some guidance in choosing the best set of manipulated variables. The set of manipulated variables that gives the *largest* minimum singular value over the frequency range of interest is the best.

The MRI gives an indication of the inherent controllability of a process. It depends on the controlled and manipulated variables, but it does not depend on the pairing of these variables or on the tuning of the controllers. Thus it is a useful tool for comparing alternative processes and alternative choices of manipulated variables.

The MRI is the minimum singular value of the process open-loop transfer function matrix $\underline{\underline{\mathbf{G}}}_{M(i\omega)}$. It can be evaluated over a range of frequencies ω or just at zero frequency. If the latter case, only the steady-state gain matrix is needed:

$$\text{MRI} = \sigma^{\min}_{[\underline{\underline{\mathbf{G}}}_{M(i\omega)}]} \tag{11-9}$$

The singular values of a matrix are a measure of how close the matrix is to being "singular," that is, having a determinant that is zero. A matrix that is $N \times N$ has N singular values. We use the symbol σ_i for singular value. The σ_i that is the biggest in magnitude is called the maximum singular value, and the notation σ^{max} is used. The σ_i that is the smallest in magnitude is called the minimum singular value and the notation σ^{min} is used.

The N singular values of a *real* $N \times N$ matrix are defined as the square root of the eigenvalues of the matrix formed by multiplying the transpose of the original matrix by itself.

$$\sigma_{i[\underline{\underline{A}}]} = \sqrt{\lambda_{i[\underline{\underline{A}}^T\underline{\underline{A}}]}}, \qquad i = 1, 2, \ldots, N \quad (11\text{-}10)$$

The eigenvalues of a square $N \times N$ matrix are the N roots of the scalar equation

$$\text{Det}\left[\lambda \underline{\underline{I}} - \underline{\underline{A}}\right] = 0 \quad (11\text{-}11)$$

where λ is a scalar quantity. Because there are N eigenvalues, it is convenient to define the vector $\underline{\lambda}$, of length N, that consists of the eigenvalues: $\lambda_1, \lambda_2, \lambda_3, \ldots, \lambda_N$:

$$\underline{\lambda} = \begin{bmatrix} \lambda_1 \\ \lambda_2 \\ \lambda_3 \\ \vdots \\ \lambda_N \end{bmatrix} \quad (11\text{-}12)$$

We will use the following notation. The expression $\underline{\lambda}_{[\underline{\underline{A}}]}$ means the vector of eigenvalues of the matrix $\underline{\underline{A}}$.

For example, the MRI for the Wood and Berry column at zero frequency would be calculated

$$\text{MRI} = \sigma^{min}_{[\underline{\underline{G}}_{M(0)}]} = \min \sqrt{\lambda^T_{i[\underline{\underline{G}}_{M(0)}\underline{\underline{G}}_{M(0)}]}} \quad (11\text{-}13)$$

$$\underline{\underline{G}}^T_{M(0)}\underline{\underline{G}}_{M(0)} = \begin{bmatrix} 12.8 & 6.6 \\ -18.9 & -19.4 \end{bmatrix} \begin{bmatrix} 12.8 & -18.9 \\ 6.6 & -19.4 \end{bmatrix}$$

$$= \begin{bmatrix} 207.4 & -369.96 \\ -369.96 & 733.57 \end{bmatrix}$$

The eigenvalues of this matrix are found by solving the equation

$$\text{Det}\begin{bmatrix} (\lambda - 207.4) & 369.96 \\ 369.96 & (\lambda - 733.57) \end{bmatrix}$$

$$= \lambda^2 - 940.97\lambda + 15{,}272 = 0$$

$$\lambda = 916.19, 16.52$$

Therefore the minimum singular value is $\sqrt{16.52} = 4.06 = \text{MRI}$.

As another example, consider the following steady-state gain matrices for three alternative choices of manipulated variables: reflux and vapor boilup (R-V), distillate and vapor boilup (D-V), and reflux ratio and vapor boilup (RR-V).

	R-V	D-V	RR-V
$G_{M_{11}}$	16.3	−2.3	1.5
$G_{M_{12}}$	−18.0	1.04	−1.12
$G_{M_{21}}$	−26.2	3.1	−2.06
$G_{M_{22}}$	28.63	1.8	4.73
MRI	0.11	1.9	0.89

Based on these results, the D-V system should be easier to control (more resilient) because it has the *largest* minimum singular value.

One important aspect of the MRI calculations should be emphasized at this point. The singular values depend on the scaling use in the steady-state gains of the transfer functions. If different engineering units are used for the gains, different singular values will be calculated. Thus the comparison of alternative control structures and processes can be obscured by this effect. The practical solution to the problem is to always use dimensionless gains in the transfer functions. The gains with engineering units should be divided by the appropriate transmitter span and multiplied by the appropriate valve gain. This gives the gains that the control system will see.

The differences in MRI values should be fairly large to be meaningful. If one set of manipulated variables gives an MRI of 10 and another set gives an MRI of 1, you can conclude that the first set is better. However, if the two sets have MRIs that are 10 and 8, there is probably little difference from a resiliency standpoint and either set of manipulated variables could be equally effective.

11-3-2 Condition Number

Another popular index for choosing the best set of manipulated variables is the condition number of the steady-state gain matrix. The condition number is the ratio of the maximum singular value to the minimum singular value,

$$CN = \sigma^{\max}/\sigma^{\min} \tag{11-14}$$

A large condition number is undesirable because it indicates that the system will be sensitive to changes in process parameters (lack robustness). Therefore, sets of manipulated variables that give small condition numbers are favored.

11-4 TUNING DIAGONAL CONTROLLERS IN A MULTIVARIABLE ENVIRONMENT

Tuning is a vital component of any controller design procedure once the structure has been determined. For multivariable systems, we need a procedure that simultaneously tunes all of the multiloop SISO controllers in the diagonal structure, taking into account the interaction that exists among all the loops.

The *biggest log-modulus tuning* (BLT) procedure outlined in the following text is one way to accomplish this job. It provides settings that work reasonably well in many processes. These settings may not be optimum because they tend to be somewhat conservative. However, they guarantee stability and serve as a benchmark for comparisons with other types of controllers and other tuning procedures.

The procedure is easy to use and easy to program, and it is easy to understand because it is based on a simple extension of the SISO Nyquist stability criterion.

11-4-1 Review of Nyquist Stability Criterion for SISO Systems

As you should recall from your undergraduate control course, the Nyquist stability criterion is a procedure for determining if the closed-loop characteristic equation has any roots (zeros) in the right half of the s plane. If it does, the system is closed-loop unstable.

The method is based on a complex variable theorem, which says that the difference between the number of zeros and poles that a function has inside a closed contour can be found by plotting the function and looking at the number of times it encircles the origin. The number of encirclements of the origin (N) is equal to the difference between the number of zeros (Z) and the number of poles (P) that the function has inside the closed contour.

We use this theorem in SISO systems to find out if the closed-loop characteristic equation has any roots or zeros in the right half of the s plane. The s variable follows a closed contour that completely surrounds the entire right half of the s plane as shown in Figure 11-5*a*.

The closed-loop characteristic equation for an SISO system is

$$1 + G_{M_{(s)}}B_{(s)} = 0 = F_{(s)} \tag{11-15}$$

If the process is open-loop stable, the function $F_{(s)}$ will have no poles in the right half of the s plane. Therefore, $P = 0$ for open-loop stable systems and the number of encirclements of the origin (N) is equal to the number of zeros (Z).

Now it is convenient to plot the total open-loop transfer function $G_{M_{(s)}}B_{(s)}$ instead of plotting $F_{(s)}$. This displaces everything to the left by one unit; so instead of looking at encirclements of the origin, we look at encirclements of the $(-1, 0)$ point in the $G_M B$-plane. These curves are illustrated in Figure 11-5*b* and *c*.

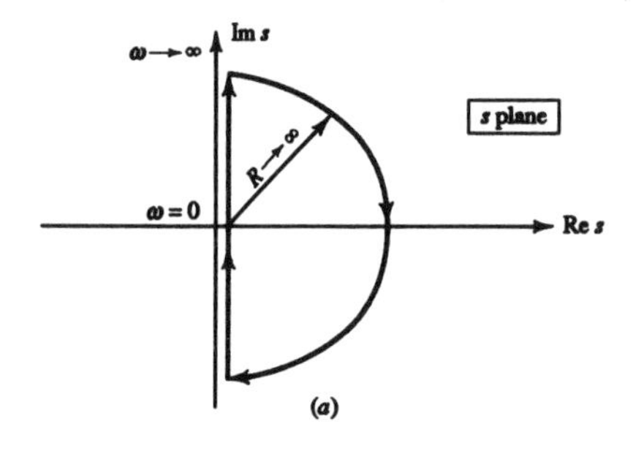

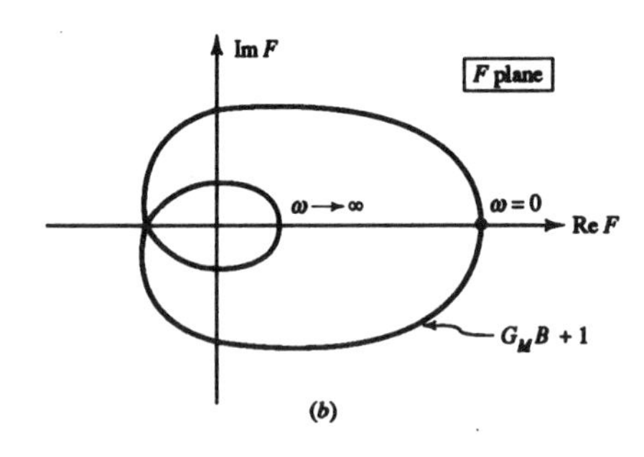

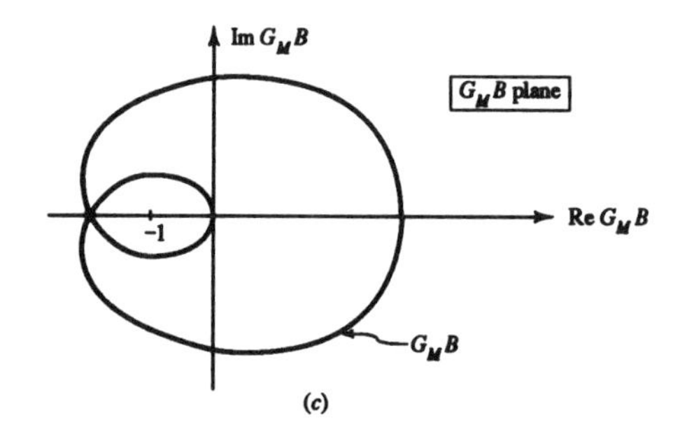

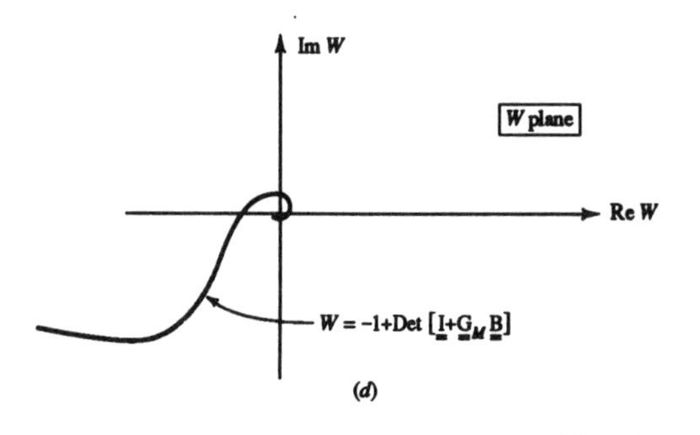

FIGURE 11-5. Nyquist stability criterion. (a) s-plane contour; (b) F-plane contour for SISO system; (c) $G_M B$-plane contour for SISO system; (d) W-plane contour for MIMO system.

11-4-2 Extension to MIMO Systems

The procedure can be extended directly to multivariable processes by simply using the closed-loop characteristic equation for such systems. So instead of using the function $F_{(s)}$ given in Equation 11-15, we use the function given in Equation 11-8 that is the closed-loop characteristic equation of a multivariable process:

$$F_{(s)} = \text{Det}\left[\underline{\mathbf{I}} + \underline{\underline{\mathbf{G}}}_{M_{(s)}} \underline{\underline{\mathbf{B}}}_{(s)}\right] \qquad (11\text{-}16)$$

The contour of this $F_{(s)}$ can be plotted in the F plane. The number of encirclements of the origin made by this plot is equal to the difference between the number of zeros and the number of poles of $F_{(s)}$ in the right half of the s plane.

If the process is open-loop stable, none of the transfer functions in $\underline{\underline{\mathbf{G}}}_{M_{(s)}}$ will have any poles in the right half of the s plane. Additionally, the feedback controllers in $\underline{\underline{\mathbf{B}}}_{(s)}$ are always chosen to be open-loop stable (P, PI, or PID action), so $\underline{\underline{\mathbf{B}}}_{(s)}$ has no poles in the right half of the s plane. Thus if the process is open-loop stable, the $F_{(s)}$ function has *no* poles in the right half of the s plane. So the number of encirclements of the origin made by the $F_{(s)}$ function is equal to the number of zeros in the right half of the s plane.

Thus the Nyquist stability criterion for a multivariable open loop-stable process is

> If a plot of Det$[\underline{\mathbf{I}} + \underline{\underline{\mathbf{G}}}_{M_{(s)}} \underline{\underline{\mathbf{B}}}_{(s)}]$ encircles the origin, the system is closed-loop unstable!

Because the usual way to use the Nyquist stability criterion in SISO systems is to look at encirclements of the $(-1, 0)$ point instead of the origin, we define the function $W_{(i\omega)}$ so that the multivariable plots are similar to the SISO plots:

$$W_{(i\omega)} = -1 + \text{Det}\left[\underline{\mathbf{I}} + \underline{\underline{\mathbf{G}}}_{M_{(i\omega)}} \underline{\underline{\mathbf{B}}}_{(i\omega)}\right] \qquad (11\text{-}17)$$

Then the number of encirclements of the $(-1, 0)$ point made by $W_{(i\omega)}$ as ω varies from 0 to ∞ gives the number of zeros of the closed-loop characteristic equation in the right half of the s plane.

11-4-3 BLT Tuning Procedure

BLT tuning involves the following five steps:

1. *Calculate the Ziegler–Nichols settings for each individual loop.* The ultimate gain and ultimate frequency ω_u of each diagonal transfer function $G_{jj(s)}$ are calculated in the classical SISO way. To do this numerically, a value of frequency ω is guessed. The phase angle is calculated, and the frequency is varied to find the point where the Nyquist plot of $G_{jj(i\omega)}$ crosses the negative real axis (phase angle is $-180°$). The frequency where this occurs is ω_u. The reciprocal of the real part of $G_{jj(i\omega)}$ is the ultimate gain.
2. *A detuning factor F is assumed.* F should always be greater than 1. Typical values are between 1.5 and 4. The gains of *all* feedback controllers K_{ci} are calculated by *dividing* the Ziegler–Nichols gains K_{ZNi} by the factor F,
$$K_{ci} = \frac{K_{ZNi}}{F} \qquad (11\text{-}18)$$
where $K_{ZNi} = K_{ui}/2.2$. Then all feedback controller reset times τ_{Ii} are calculated by multiplying the Ziegler–Nichols reset times τ_{ZNi} by the same factor F,
$$\tau_{Ii} = \tau_{ZNi} F \qquad (11\text{-}19)$$
where
$$\tau_{ZNi} = \frac{2\pi}{1.2\omega_{ui}} \qquad (11\text{-}20)$$
The F factor can be considered as a detuning factor that is applied to all loops. The larger the value of F, the more stable the system will be but the more sluggish will be the setpoint and load responses. The method yields settings that give a reasonable compromise between stability and performance in multivariable systems.
3. *Make a plot of $W_{(i\omega)}$.* Using the guessed value of F and the resulting controller settings, a multivariable Nyquist plot of the scalar function $W_{(i\omega)} = -1 + \text{Det}[\underline{\underline{\mathbf{I}}} + \underline{\underline{\mathbf{G}}}_{M(i\omega)} \underline{\underline{\mathbf{B}}}_{(i\omega)}]$ is made.
4. *Make a plot of L_{cm}.* The closer the W contour is to the $(-1, 0)$ point, the closer the system is to instability. Therefore the quantity $W/(1 + W)$ will be similar to the closed-loop servo transfer function for a SISO loop $G_M B/(1 + G_M B)$. We define a multivariable closed-loop log modulus
$$L_{cm} = 20 \log_{10} \left| \frac{W}{1 + W} \right| \qquad (11\text{-}21)$$
The peak in the plot of L_{cm} over the entire frequency range is the biggest log modulus $L_{cm}^{\max}$.
5. *The F factor is varied until $L_{cm}^{\max}$ is equal to $2N$, where N is the order of the system.* For $N = 1$, the SISO case, we get the familiar $+2$ dB maximum closed-loop log modulus criterion. For a 2×2 system, $a + 4$ dB value of $L_{cm}^{\max}$ is used; for a 3×3, $+6$ dB; and so forth. This empirically determined criterion has been tested on a large number of cases and gives reasonable performance, which is a little on the conservative side.

The method weighs each loop equally, that is, each loop is equally detuned. If it is important to keep tighter control of some variables than others, the method can be easily modified by using different weighting factors for different controlled variables. The less-important loop could be detuned more than the more-important loop.

11-4-4 Examples

Two examples are given in the following text. The first is a 2×2 system and the second is a 3×3 system.

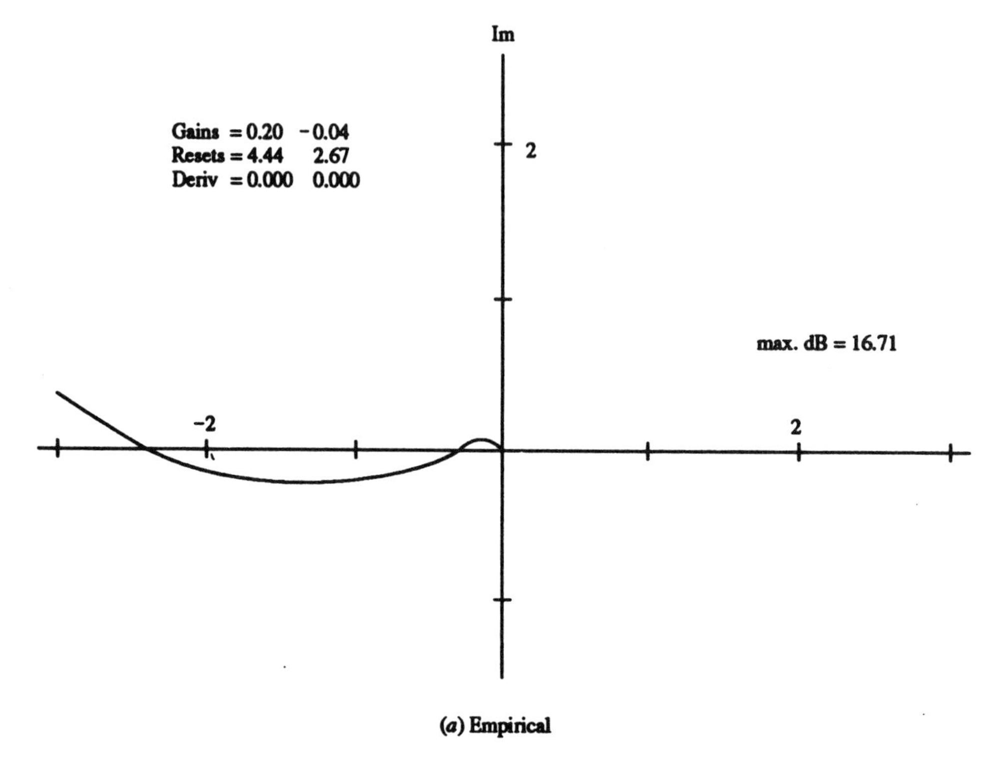

FIGURE 11-6. *W* plots for Wood and Berry column. (a) Empirical settings; (b) Ziegler–Nichols settings; (c) BLT settings.

Wood and Berry Column

The open-loop process transfer functions are

$$\underline{\underline{\mathbf{G}}}_{M(s)} = \begin{bmatrix} G_{M_{11}} & G_{M_{12}} \\ G_{M_{21}} & G_{M_{22}} \end{bmatrix} = \begin{bmatrix} \dfrac{12.8e^{-s}}{16.7s+1} & \dfrac{-18.9e^{-3s}}{21s+1} \\ \dfrac{6.6e^{-7s}}{10.9s+1} & \dfrac{-19.4e^{-3s}}{14.4s+1} \end{bmatrix} \tag{11-22}$$

The process is open-loop stable with no poles in the right half of the s plane. The authors used a diagonal controller structure with PI controllers and found, by empirical tuning, the following settings: $K_{c1} = 0.20$, $K_{c2} = -0.04$, $\tau_{I1} = 4.44$, and $\tau_{I2} = 2.67$. The feedback controller matrix was

$$\underline{\underline{\mathbf{B}}}_{(s)} = \begin{bmatrix} \dfrac{K_{c1}(\tau_{I1}s+1)}{\tau_{I1}s} & 0 \\ 0 & \dfrac{K_{c2}(\tau_{I2}s+1)}{\tau_{I2}s} \end{bmatrix} \tag{11-23}$$

Figures 11-6a and b give W plots for the Wood and Berry column when the empirical settings are used and when the Ziegler–Nichols (ZN) settings for each individual controller are used ($K_{c1} = 0.960$, $K_{c2} = -0.19$, $\tau_{I1} = 3.25$, and $\tau_{I2} = 9.2$). The curve with the empirical settings does not encircle the $(-1, 0)$ point, and therefore the system is closed-loop stable. Figure 11-7 gives the response of the system to a unit

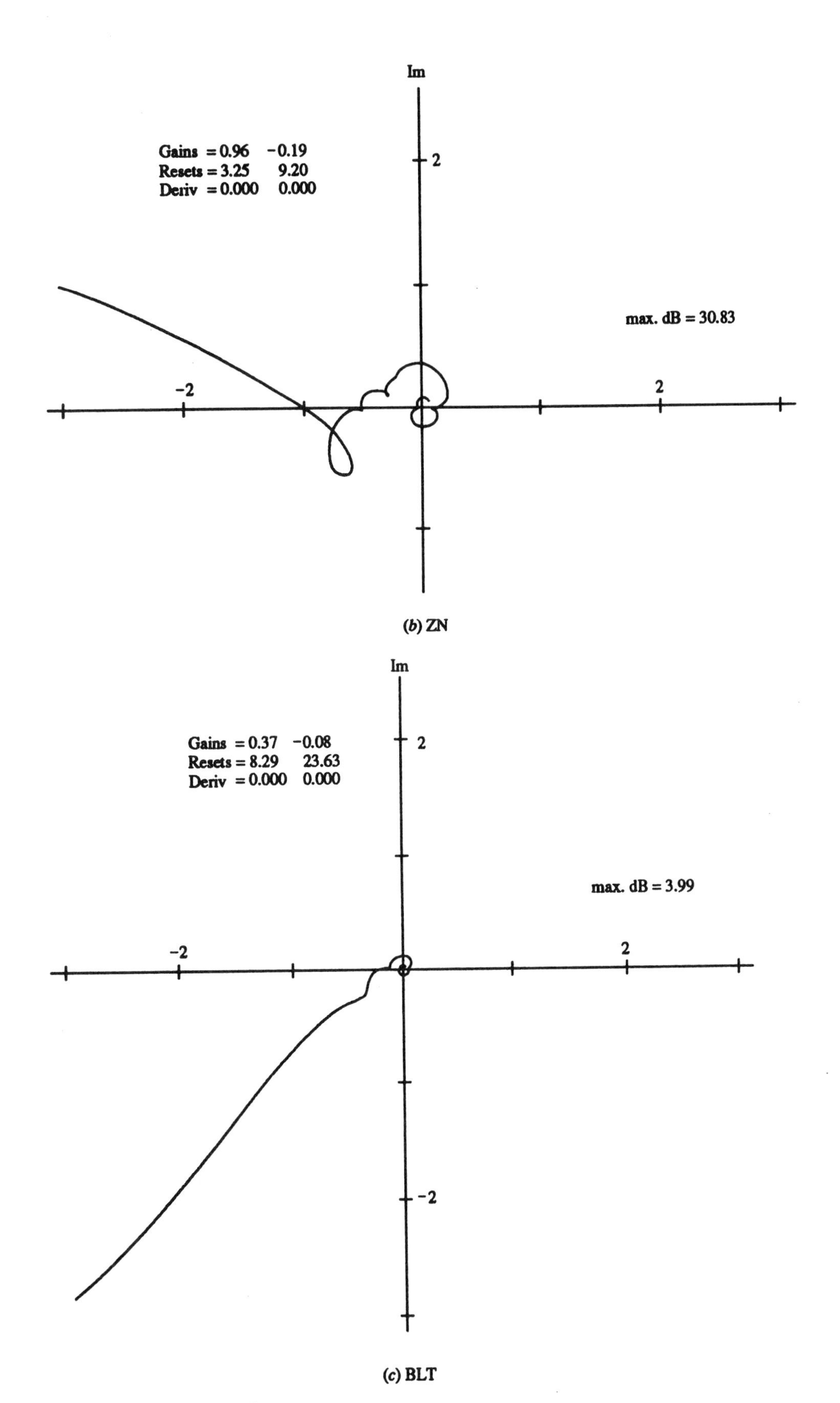

(*b*) ZN

(*c*) BLT

FIGURE 11-6. *Continued.*

step change in x_1^{set}, verifying that the multivariable system is indeed closed-loop stable.

Note that the W plane curve using the ZN settings does encircle the $(-1, 0)$ point, showing that the system is closed-loop unstable with these settings.

Figure 11-6c gives the W plots for the system when the BLT settings are used ($K_{c1} = 0.375$, $K_{c2} = -0.075$, $\tau_{I1} = 8.29$, and $\tau_{I2} = 23.6$ with a detuning factor $F = 2.55$). Time responses using the BLT controller tuning parameters are compared with those using the empirical settings in Figure 11-7.

Ogunnaike and Ray Column

W plots and time responses for this 3×3 system are given in Figures 11-8 and 11-9 for two sets of controller tuning parameters: empirical and BLT. The open-loop transfer function matrix is

$$\underline{\underline{\mathbf{G}}}_{M(s)} = \begin{vmatrix} \dfrac{0.66e^{-2.6s}}{6.7s+1} & \dfrac{-0.61e^{-3.5s}}{8.64s+1} & \dfrac{-0.0049e^{-s}}{9.06s+1} \\ \dfrac{1.11e^{-6.5s}}{3.25s+1} & \dfrac{-2.36e^{-3s}}{5s+1} & \dfrac{-0.012e^{-1.2s}}{7.09s+1} \\ \dfrac{-34.68e^{-9.2s}}{8.15s+1} & \dfrac{46.2e^{-9.4s}}{10.9s+1} & \dfrac{0.87(11.61s+1)e^{-s}}{(3.89s+1)(18.8s+1)} \end{vmatrix} \tag{11-24}$$

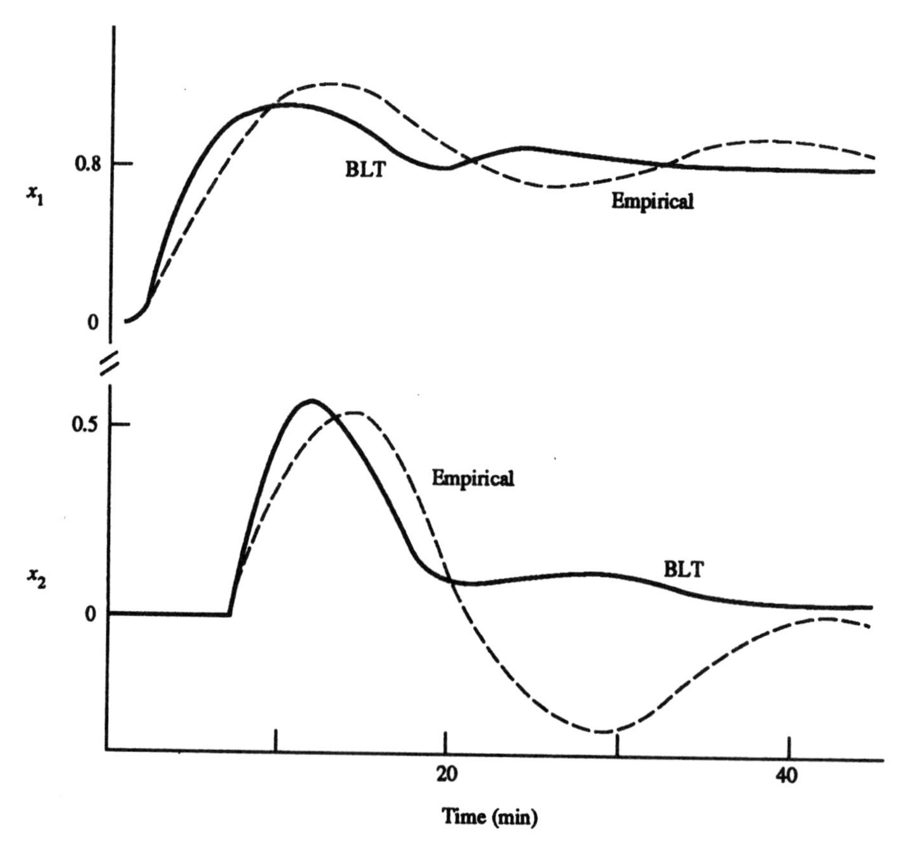

FIGURE 11-7. Time responses for Wood and Berry column for unit step change in x_1 setpoint: empirical and BLT settings.

Empirical controller settings are $K_{c1} = 1.2$, $K_{c2} = -0.15$, $K_{c3} = 0.6$, $\tau_{I1} = 5$, $\tau_{I2} = 10$, and $\tau_{I3} = 4$. BLT controller settings are $K_{c1} = 1.51$, $K_{c2} = -0.295$, $K_{c3} = 2.63$, $\tau_{I1} = 16.4$, $\tau_{I2} = 18$, and $\tau_{I3} = 6.61$ with a detuning factor $F = 2.15$.

Both of these cases illustrate that the BLT procedure gives reasonable controller settings.

If the process transfer functions have greatly differing time constants, the BLT procedure does not work well. It tends to give a response that is too oscillatory. The problem can be handled by breaking up the system into fast and slow sections and applying BLT to each smaller subsection.

The BLT procedure was applied with PI controllers. The method can be extended to include derivative action (PID controllers) by using two detuning factors: F detunes the ZN reset and gain values and F_D detunes the ZN derivative value. The optimum value of F_D is that which gives the minimum value of F and still satisfies the $+2N$ maximum closed-loop log modulus criterion.

11-5 PAIRING

Now that the controlled and manipulated variables have been chosen and we have a procedure to tune controllers once the pairing has been specified, the final question that must be answered is "How do we pair the controlled and manipulated variables?"

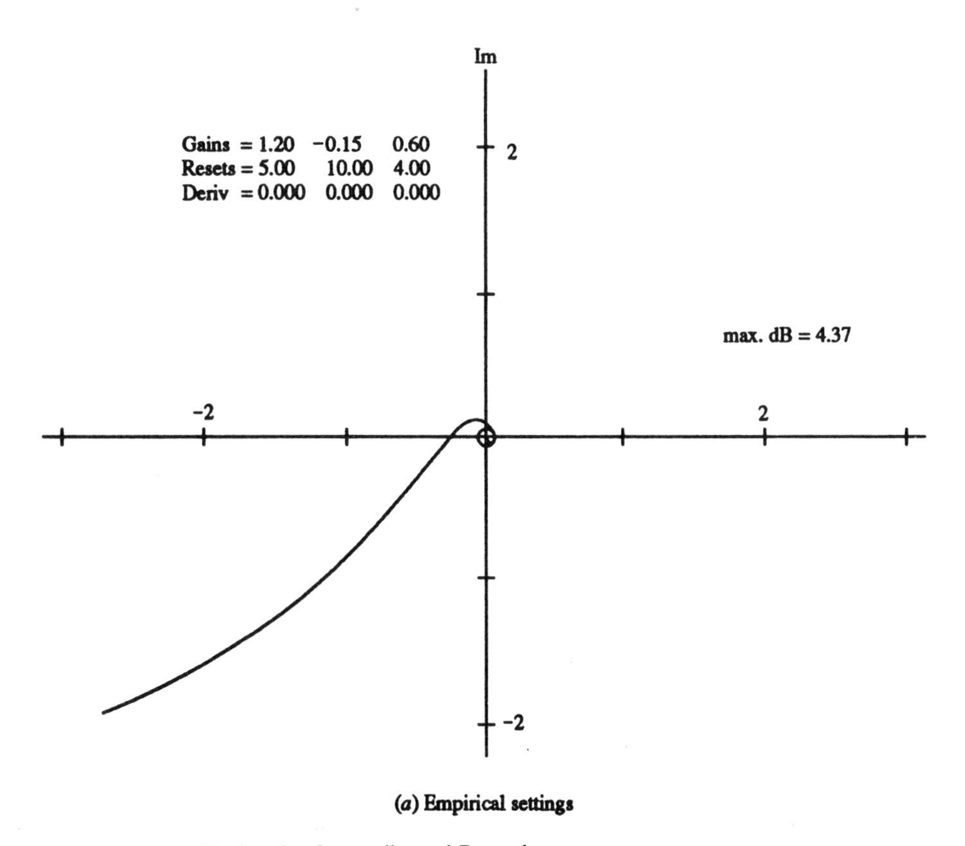

(a) Empirical settings

FIGURE 11-8. *W* plots for Ogunnaike and Ray column.

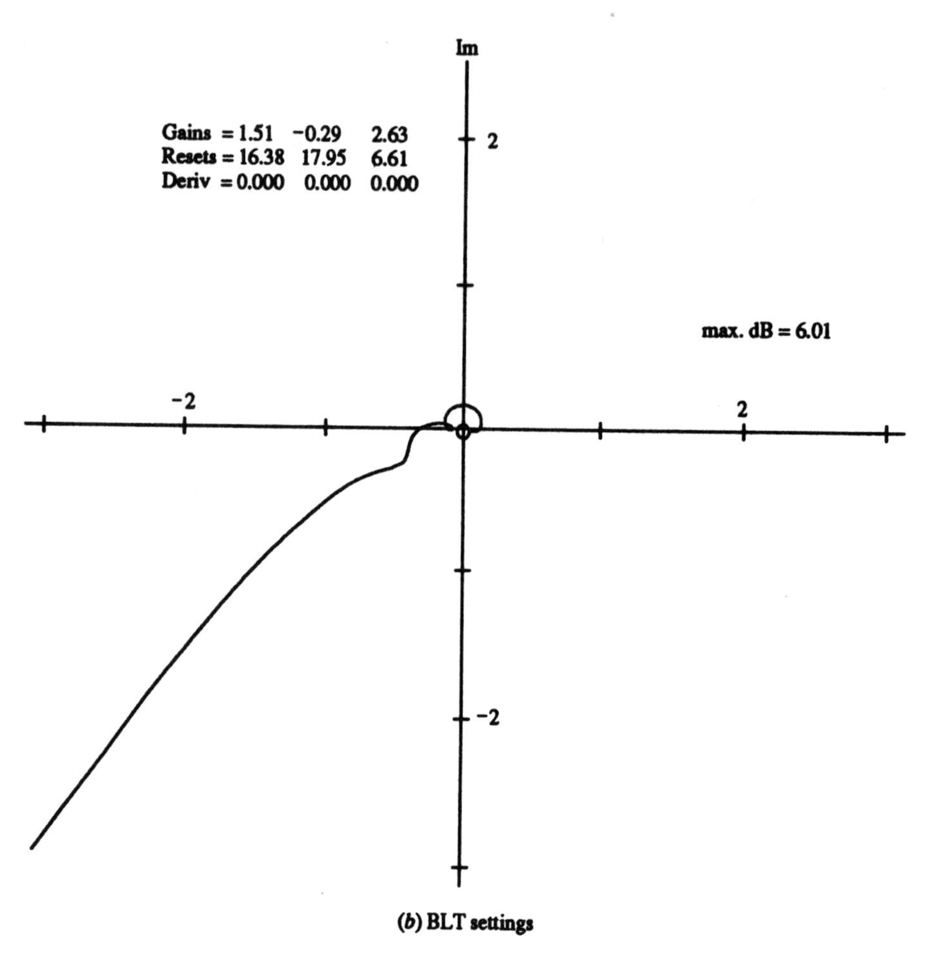

(b) BLT settings

FIGURE 11-8. *Continued.*

11-5-1 Elimination of Unworkable Pairings

The Niederlinski index can be used to eliminate unworkable pairings of variables at an early stage in the design. The settings of the controllers do not have to be known, but it applies only when integral action is used in all loops. It uses only the steady-state gains of the process transfer function matrix.

The method is a "necessary but not sufficient condition" for stability of a closed-loop system with integral action. If the index is negative, the system *will be* unstable for any controller settings (this is called *integral instability*). If the index is positive, the system may or may not be stable. Further analysis is necessary:

$$\text{Niederlinski index} = \text{NI} = \frac{\text{Det}\left[\underline{\underline{\mathbf{K}}}_P\right]}{\prod_{j=1}^{N} K_{P_{jj}}} \tag{11-25}$$

where $\underline{\underline{\mathbf{K}}}_P = \underline{\underline{\mathbf{G}}}_{M_{(0)}}$ = matrix of steady-state gains from the process open-loop $\underline{\underline{\mathbf{G}}}_M$ transfer function

$K_{P_{jj}}$ = diagonal elements in steady-state gain matrix

As an example, the Niederlinski index for

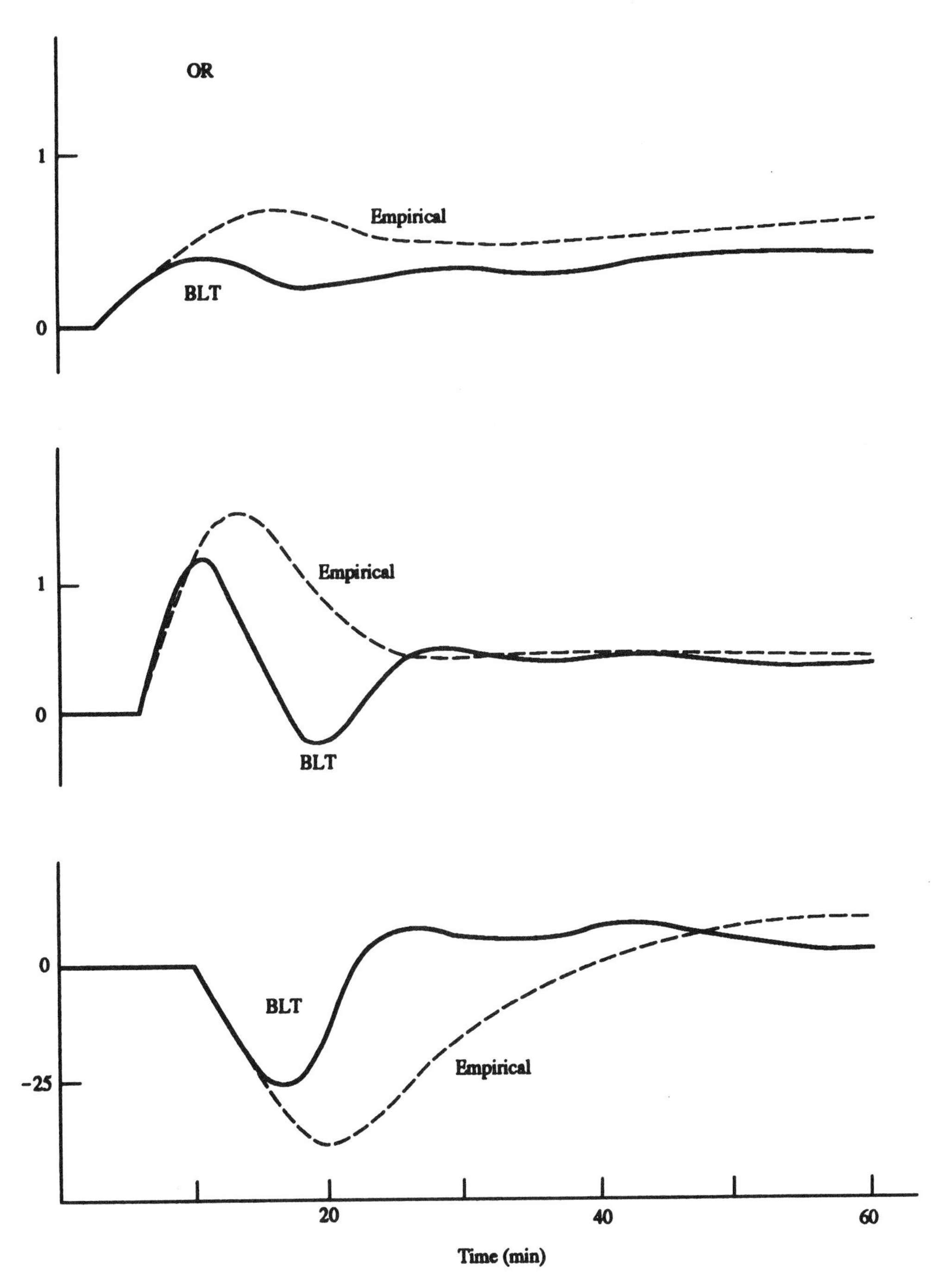

FIGURE 11-9. Time responses for Ogunnaike and Ray column for unit step change in x_1 setpoint: empirical and BLT settings.

the Wood and Berry column is

$$\underline{\underline{K}}_P = \underline{\underline{G}}_{M(0)} = \begin{bmatrix} 12.8 & -18.9 \\ 6.6 & -19.4 \end{bmatrix} \quad (11\text{-}26)$$

$$\begin{aligned} \mathrm{NI} &= \frac{\mathrm{Det}\left[\underline{\underline{K}}_P\right]}{\Pi_{j=1}^{N} K_{P_{jj}}} \\ &= \frac{(12.8)(-19.4) - (-18.9)(6.6)}{(12.8)(-19.4)} \\ &= 0.498 \end{aligned} \quad (11\text{-}27)$$

Because the NI is positive, the closed-loop system with the specified pairing *may* be stable.

Notice that pairing assumes distillate composition x_D is controlled by reflux R and bottoms composition x_B is controlled by vapor boilup V:

$$\begin{bmatrix} x_D \\ x_B \end{bmatrix} = \begin{bmatrix} 12.8 & -18.9 \\ 6.6 & -19.4 \end{bmatrix} \begin{bmatrix} R \\ V \end{bmatrix} \quad (11\text{-}28)$$

If the pairing had been reversed, the steady-state gain matrix would be

$$\begin{bmatrix} x_D \\ x_B \end{bmatrix} = \begin{bmatrix} -18.9 & 12.8 \\ -19.4 & 6.6 \end{bmatrix} \begin{bmatrix} V \\ R \end{bmatrix} \quad (11\text{-}29)$$

and the NI for this pairing would be

$$\begin{aligned} \mathrm{NI} &= \frac{\mathrm{Det}\left[\underline{\underline{K}}_P\right]}{\Pi_{j=1}^{N} K_{P_{jj}}} \\ &= \frac{(-18.9)(6.6) - (12.8)(-19.4)}{(-18.9)(6.6)} \\ &= -0.991 \end{aligned}$$

Therefore, the pairing of x_D with V and x_B with R gives a closed-loop system that is "integrally unstable" for any controller tuning.

11-5-2 Tyreus Load Rejection Criterion

In most chemical processes the principal control problem is load rejection. We want a control system that can keep the controlled variables at or near their setpoints in the face of load disturbances. Thus the closed-loop regulator transfer function is the most important.

The ideal closed-loop relationship between the controlled variable and the load is zero. Of course this can never be achieved, but the smaller the magnitude of the closed-loop regulator transfer function, the better the control. Thus a rational criterion for selecting the best pairing of variables is to choose the pairing that gives the smallest peaks in a plot of the elements of the closed-loop regulator transfer function matrix. This criterion is called the Tyreus load-rejection criterion (TLC).

The closed-loop relationships for a multivariable process are (Equation 11-7)

$$\begin{aligned} \underline{x} &= \left[\left[\underline{\underline{I}} + \underline{\underline{G}}_{M(s)}\underline{\underline{B}}_{(s)}\right]^{-1} \underline{\underline{G}}_{M(s)}\underline{\underline{B}}_{(s)}\right]\underline{x}^{\text{set}} \\ &\quad + \left[\left[\underline{\underline{I}} + \underline{\underline{G}}_{M(s)}\underline{\underline{B}}_{(s)}\right]^{-1} \underline{G}_{L(s)}\right] L_{(s)} \end{aligned} \quad (11\text{-}30)$$

For a 3×3 system there are three elements in the $\underline{x}$ vector, so three curves are plotted of each of the three elements in the vector $[[\underline{\underline{I}} + \underline{\underline{G}}_{M(i\omega)}\underline{\underline{B}}_{(i\omega)}]^{-1}\underline{G}_{L(i\omega)}]$ for the specified load variable L. Figure 11-10 gives the TLC curves for the 3×3 Ogunnaike and Ray column.

11-6 CONCLUSION

This chapter has presented an easy-to-use method for designing simple but effective control systems for multivariable processes.

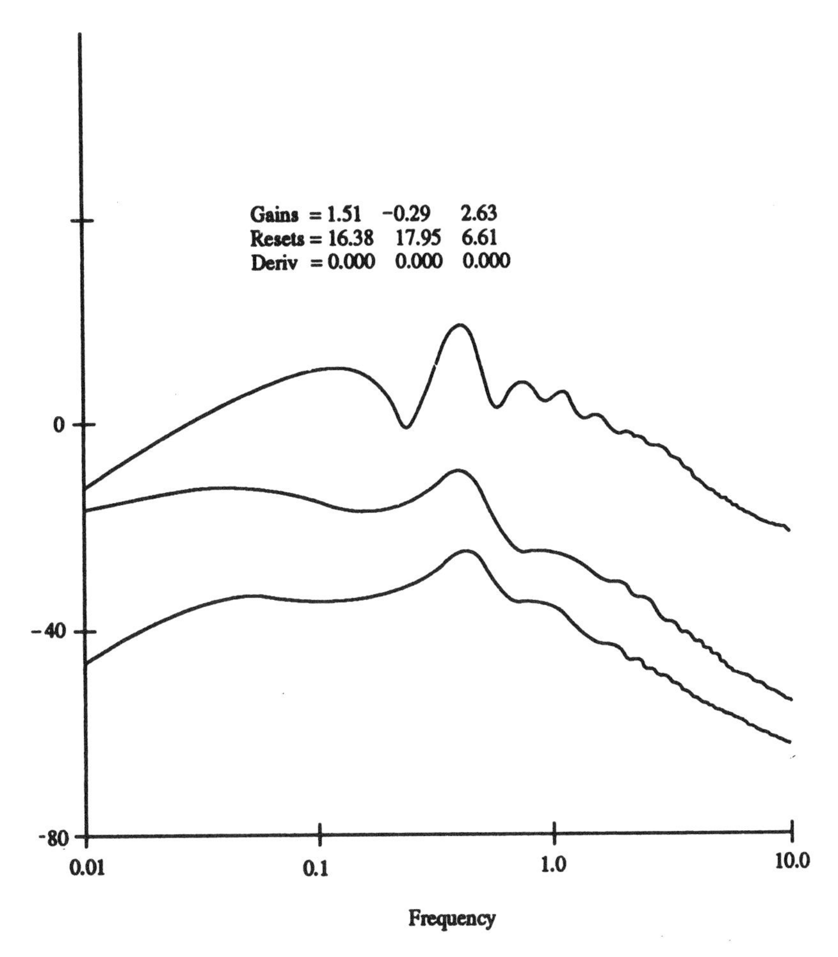

FIGURE 11-10. TLC plots for Ogunnaike and Ray column.

Rational procedures for selecting manipulate variables and pairing variables have been discussed.

One of the key concepts is how one determines stability in a multivariable system. The closed-loop characteristic equation of a MIMO system has been developed.

Another key concept is a procedure for tuning diagonal multiloop SISO controllers in a multivariable environment. The tuning is done considering all controllers simultaneously. The application of these methods will be illustrated in later chapters (Chapters 18, 22, 23, and 24).

12

Dynamic Matrix Control Multivariable Controllers

David A. Hokanson

Exxon Chemical Holland

James G. Gerstle

Exxon Chemical Company

12-1 INTRODUCTION

During the last 10 years, the use of model-based control in industry has increased dramatically. Distillation towers are prime targets for such controllers because of their long dynamics, multivariable nature, and requirement for tight composition control (motivated by efforts to improve product quality through decreased variation). In addition, distillation towers are commonly the unit or plant constraint to higher production. This is due primarily to the high cost of debottlenecking or replacing distillation towers.

The establishment of model-based control in industry has been aided by a number of developments in technology:

- The increasing power of computers used for industrial process control.
- The maturing of model identification techniques (covered in Chapter 7)
- The maturing of model-based control theory and techniques.

This chapter covers the use of a particular model-based control algorithm—dynamic matrix control (DMC)—as it is applied to distillation towers. Two other chapters in this book cover generic model-based control as it is applied to distillation towers. Chapter 14 discusses distillation tower model-based control and its robustness. Chapter 21 discusses the application of a generic model-based controller to a distillation tower.

DMC was originally developed by Charles R. Cutler and others (Cutler and Ramaker, 1979; Prett and Gillette, 1979). A similar algorithm, known commonly as IDCOM, was developed by Richalet and others at approximately the same time (Richalet et al. 1978).

Although DMC can be used for single-input–single-output (SISO) control, most benefit comes from its use as a multiple-input–multiple-output (MIMO) controller. In its original form, a MIMO DMC controller cannot handle constraints and prioritize control objectives effectively. It is this constraint and priority handling that allows DMC to yield significant returns for distillation control. It is also constraint handling and objective prioritization that has given rise to various forms of MIMO DMC.

There are two common forms of constrained DMC control in use within industry. Quadratic dynamic matrix control

(QDMC) (Cutler, Morshedi, and Haydel, 1983; García and Morshedi, 1986) uses quadratic programming to handle constraints. A hybrid linear program–least-squares approach was developed by DMC Corporation and has been documented in a number of recent papers (Cutler and Hawkins, 1987, 1988).

Before pressing forward, a caveat: There are many excellent approaches in applying composition control to distillation towers, and this book describes many of them. In the authors' experience, many of these techniques will equal or exceed the performance of DMC for *unconstrained, dual composition* control. DMC has proven superior in distillation control when faced either with constraints (e.g., tower flooding, heat input limitations, condenser limitations, etc.) or for controlling more than two compositions (e.g., true sidestream towers producing three or more different product streams).

This chapter will cover the basics of DMC as applied to distillation control. It is at most an overview—a "taste" of DMC. The reader is encouraged to study the references for more detailed information on DMC. This chapter starts with a brief mathematical review of the DMC technique, then follows with a very short review of model identification, particularly as it is applied to a multivariable distillation problem. After a discussion on the design aspects of a DMC multivariable controller, the steps of implementing a DMC controller on a distillation tower are covered. Finally, the chapter ends with a review of various industrial DMC distillation controllers that have been published thus far.

12-2 BASICS OF DMC MATHEMATICS

The purpose of this section is to give the reader a "flavor" of the mathematics behind DMC. For a more detailed mathematical description, especially for the multivariable case, see Cutler (1983).

12-2-1 Convolution Models

We start our brief examination of the DMC mathematics with the convolution integral. Given any continuous linear, time-variant, causal, relaxed system (see Chen, 1970), the output $y(t)$ can be always represented in terms of the past input sequence $u(t)$ by the convolution integral

$$y(t) = \int_0^\infty v(\tau) u(t-\tau)\, d\tau \qquad (12\text{-}1)$$

where $v(\tau)$ is a function that weights the contribution over time that the past input sequence $u(t-\tau)$ has on the output $y(t)$. Inputs and outputs are both considered to be deviations from steady-state values.

In discrete or sampled data form (see Oppenheim and Schafer, 1975), the convolution integral becomes

$$y_t = \sum_{j=0}^{\infty} v_j u_{t-j} \qquad (12\text{-}2)$$

Using a unit pulse input given by

$$\begin{aligned} u_t &= 0 \quad \text{at } t \neq 0 \\ u_t &= 1 \quad \text{at } t = 0 \end{aligned} \qquad (12\text{-}3)$$

the process output y_t is

$$y_t = v_t \quad \text{for } t > 0 \qquad (12\text{-}4)$$

and it is shown in Figure 12-1.

In Figure 12-1, the sequence v_j ($j = 0, 1, 2, 3 \ldots$) is referred to as the discrete impulse response function of the process, or the impulse response convolution model. This is the starting point for IDCOM.

Let us now move to the step response world. The step response a_t for a unit step can be obtained from the sum of the impulse response function as

$$a_t = \sum_{j=0}^{t} v_j \qquad (12\text{-}5)$$

This is shown in graphic form in Figure 12-2.

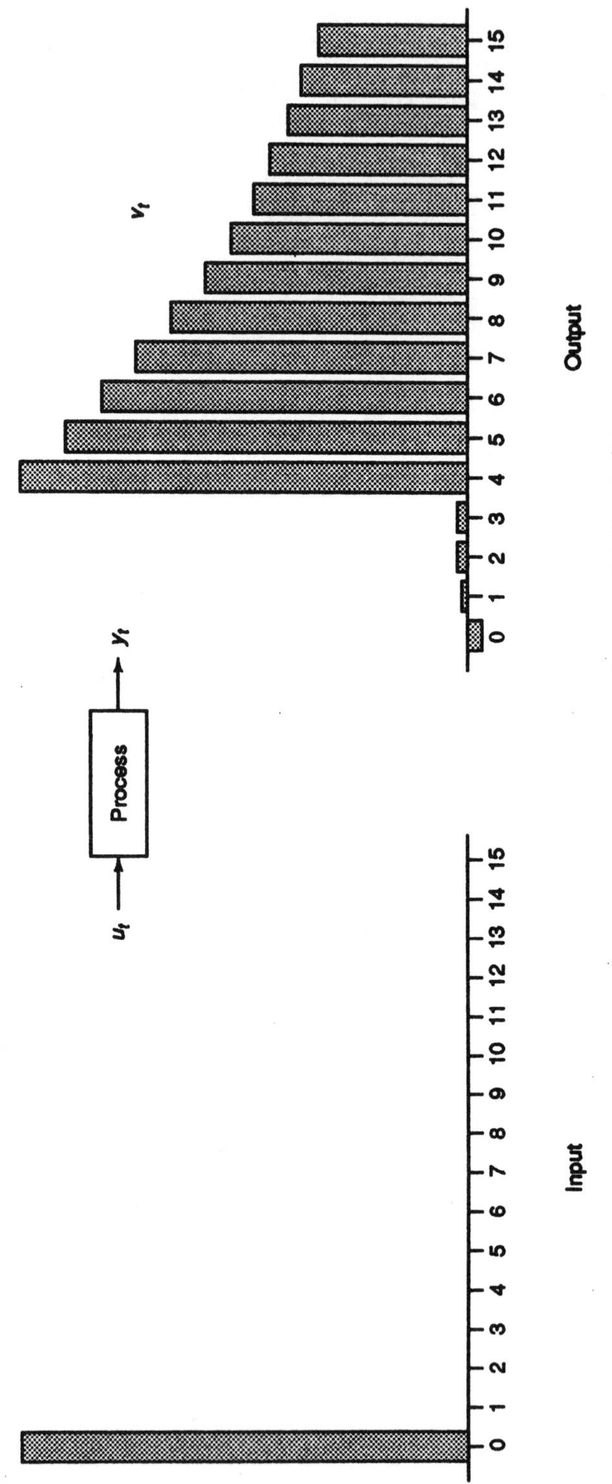

FIGURE 12-1. Impulse response convolution model: discrete form.

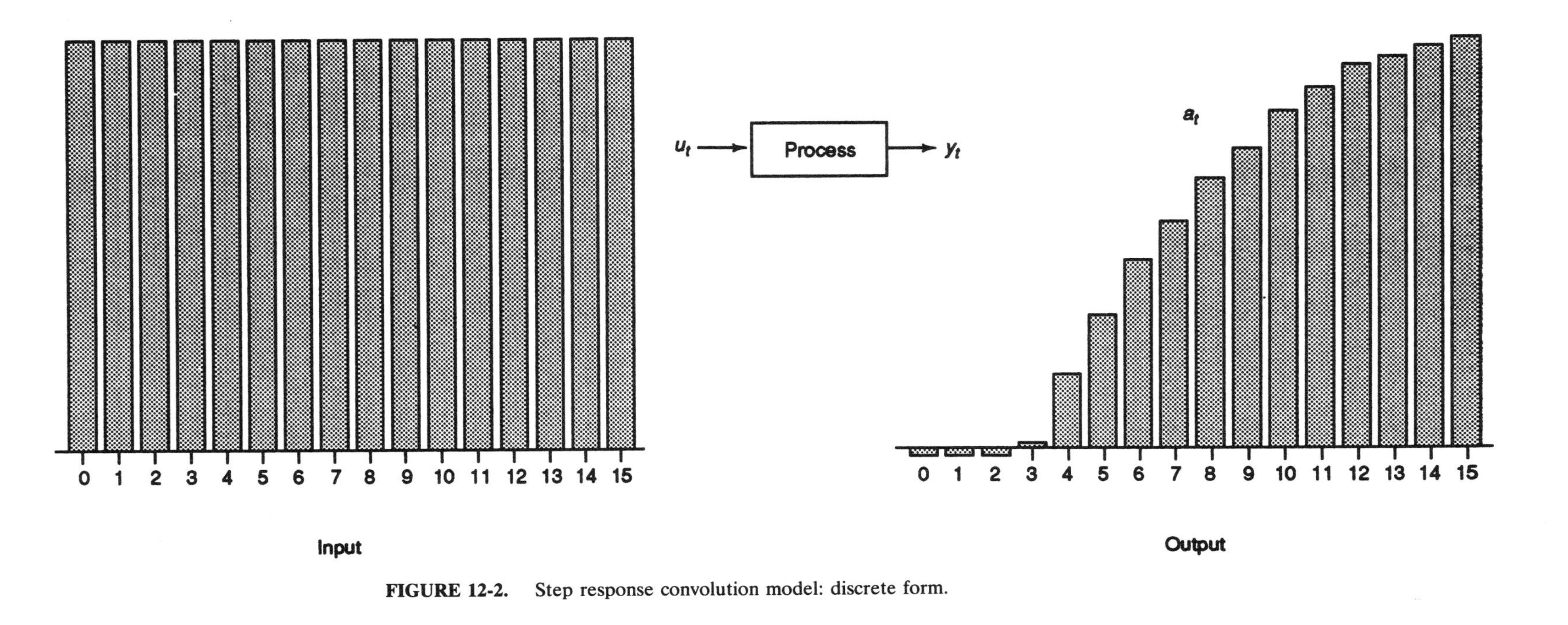

FIGURE 12-2. Step response convolution model: discrete form.

The sequence a_j ($j = 0, 1, 2, 3\ldots$) is the step response convolution model, which is the basis for DMC. Remembering that everything so far is in deviation variables, a_t is no more than a series of numbers based on a set length (time to steady state) and a set sampling interval. There is no set form of the model, other than that steady state is reached by the last interval. In fact, the last value a_t is the process steady-state gain.

Because DMC is based on the step response convolution model, let us look at it in more detail. To keep things relatively simple, we will examine a SISO problem with independent variable u and dependent variable y. Furthermore, we will use a model length of 30 (a popular model length) to minimize the number of indefinite parameters (that is, the n's and m's).

For a lined out process with a single step in u at time 0,

$$\Delta u_0 = u_0 - u_{-1} \tag{12-6}$$

then

$$\begin{aligned} y_1 - y_0 &= a_1\,\Delta u_0 \\ y_2 - y_0 &= a_2\,\Delta u_0 \\ y_3 - y_0 &= a_3\,\Delta u_0 \\ &\vdots \\ y_{30} - y_0 &= a_{30}\,\Delta u_0 \\ y_{31} - y_0 &= a_{30}\,\Delta u_0 \\ &\vdots \end{aligned} \tag{12-7}$$

This is essentially the definition of the step response convolution model described before. What is interesting about this is that we can calculate what y will be at each interval for successive moves in u. For example purposes only, let us examine a series of three moves "into the future"—one now, one at the next time interval, and one two time intervals into the future:

$$\begin{aligned} \Delta u_0 &= u_0 - u_{-1} \\ \Delta u_1 &= u_1 - u_0 \\ \Delta u_2 &= u_2 - u_1 \end{aligned} \tag{12-8}$$

then

$$\begin{aligned} y_1 - y_0 &= a_1\,\Delta u_0 \\ y_2 - y_0 &= a_2\,\Delta u_0 + a_1\,\Delta u_1 \\ y_3 - y_0 &= a_3\,\Delta u_0 + a_2\,\Delta u_1 + a_1\,\Delta u_2 \\ &\vdots \\ y_{30} - y_0 &= a_{30}\,\Delta u_0 + a_{29}\,\Delta u_1 + a_{28}\,\Delta u_2 \\ y_{31} - y_0 &= a_{30}\,\Delta u_0 + a_{30}\,\Delta u_1 + a_{29}\,\Delta u_2 \\ y_{32} - y_0 &= a_{30}\,\Delta u_0 + a_{30}\,\Delta u_1 + a_{30}\,\Delta u_2 \end{aligned} \tag{12-9}$$

Unfortunately, the process is rarely at steady state when control is initiated. If we somehow knew the past 30 moves in u, we could get a series of 30 future predictions of y formulated by extending the preceding logic and going back in time. Let us call this the vector y^p. Note that vector y^p is a full-value prediction and is calculated by

$$\begin{bmatrix} y_0^p \\ y_1^p \\ y_2^p \\ \vdots \\ y_{27}^p \\ y_{28}^p \\ y_{29}^p \end{bmatrix} = \begin{bmatrix} a_{30} & a_{29} & a_{28} & a_{27} & \cdots & a_3 & a_2 & a_1 \\ a_{30} & a_{30} & a_{29} & a_{28} & \cdots & a_4 & a_3 & a_2 \\ a_{30} & a_{30} & a_{30} & a_{29} & \cdots & a_5 & a_4 & a_3 \\ \vdots & & \vdots & & \ddots & & \vdots & \\ a_{30} & a_{30} & a_{30} & a_{30} & \cdots & a_{30} & a_{29} & a_{28} \\ a_{30} & a_{30} & a_{30} & a_{30} & \cdots & a_{30} & a_{30} & a_{29} \\ a_{30} & a_{30} & a_{30} & a_{30} & \cdots & a_{30} & a_{30} & a_{30} \end{bmatrix} \begin{bmatrix} \Delta u_{-30} \\ \Delta u_{-29} \\ \Delta u_{-28} \\ \vdots \\ \Delta u_{-3} \\ \Delta u_{-2} \\ \Delta u_{-1} \end{bmatrix} + \begin{bmatrix} y_{-31} \\ y_{-31} \\ y_{-31} \\ \vdots \\ y_{-31} \\ y_{-31} \\ y_{-31} \end{bmatrix} \tag{12-10}$$

12-2-2 Prediction Errors

A very familiar quotation, "All models are wrong; some are useful," now comes into play. The preceding convolution model will be wrong due to unmodelled load disturbances, nonlinearity in the models, and modelling errors. One way to correct for these is by calculating a prediction error b at the current time and then adjusting the predictions. In equation form,

$$b = y_0^a - y_0^p \tag{12-11}$$

where y^a is the actual, measured value of y. Then the vector y^p (from Equation 12-10) is corrected by b for predictions 1 to 29 as

$$y_i^{\text{pc}} = y_i^p + b \tag{12-12}$$

where y^{pc} is the corrected prediction.

Now, with a set of corrected future predictions, control can be performed. Let us plan for three moves as we did before—one now and two in the future. In Equation 12-9, 32 future values of y could be predicted with three moves into the future. However, this assumed that y was at steady state at time 0. Because we instead have the set of corrected predictions, vector y^{pc}, these equations become

$$\begin{aligned}
y_1 - y_0 &= a_1\,\Delta u_0 + y_1^{\text{pc}} \\
y_2 - y_0 &= a_2\,\Delta u_0 + a_1\,\Delta u_1 + y_2^{\text{pc}} \\
y_3 - y_0 &= a_3\,\Delta u_0 + a_2\,\Delta u_1 + a_1\,\Delta u_2 + y_3^{\text{pc}} \\
y_4 - y_0 &= a_4\,\Delta u_0 + a_3\,\Delta u_1 + a_2\,\Delta u_2 + y_4^{\text{pc}} \\
&\vdots \\
y_{29} - y_0 &= a_{29}\,\Delta u_0 + a_{28}\,\Delta u_1 + a_{27}\,\Delta u_2 + y_{29}^{\text{pc}} \\
y_{30} - y_0 &= a_{30}\,\Delta u_0 + a_{29}\,\Delta u_1 + a_{28}\,\Delta u_2 + y_{29}^{\text{pc}} \\
y_{31} - y_0 &= a_{30}\,\Delta u_0 + a_{30}\,\Delta u_1 + a_{29}\,\Delta u_2 + y_{29}^{\text{pc}} \\
y_{32} - y_0 &= a_{30}\,\Delta u_0 + a_{30}\,\Delta u_1 + a_{30}\,\Delta u_2 + y_{29}^{\text{pc}}
\end{aligned} \tag{12-13}$$

In matrix form, this becomes

$$\begin{bmatrix} y_1 - y_0 \\ y_2 - y_0 \\ y_3 - y_0 \\ y_4 - y_0 \\ \vdots \\ y_{29} - y_0 \\ y_{30} - y_0 \\ y_{31} - y_0 \\ y_{32} - y_0 \end{bmatrix} = \begin{bmatrix} a_1 & 0 & 0 \\ a_2 & a_1 & 0 \\ a_3 & a_2 & a_1 \\ a_4 & a_3 & a_2 \\ \vdots & \vdots & \vdots \\ a_{29} & a_{28} & a_{27} \\ a_{30} & a_{29} & a_{28} \\ a_{30} & a_{30} & a_{29} \\ a_{30} & a_{30} & a_{30} \end{bmatrix} \begin{bmatrix} \Delta u_0 \\ \Delta u_1 \\ \Delta u_2 \end{bmatrix} + \begin{bmatrix} y_1^{\text{pc}} \\ y_2^{\text{pc}} \\ y_3^{\text{pc}} \\ y_4^{\text{pc}} \\ \vdots \\ y_{29}^{\text{pc}} \\ y_{29}^{\text{pc}} \\ y_{29}^{\text{pc}} \\ y_{29}^{\text{pc}} \end{bmatrix} \tag{12-14}$$

or

$$\underline{\mathbf{y}} = \underline{\mathbf{y}}^{\text{pc}} + \underline{\underline{\mathbf{A}}}\,\Delta\underline{\mathbf{u}} \tag{12-15}$$

where $\underline{\underline{\mathbf{A}}}$ is the "dynamic matrix." [Vectors will be denoted by a single underscore; matrices will be denoted by a double underscore.]

12-2-3 Control Solution

From the preceding text, we have a set of future predictions of y based on past moves and a set of future values of y with a set of three moves. Let us now assume that we have a desired target for y called y^{sp}. (We, of course, need this for any control action!) One criterion we can use is to minimize the error squared over the time between now and the future (that is, to use the least-squares criterion). In vector equation form,

$$\text{minimize } \left(\underline{\mathbf{y}}^{\text{sp}} - \underline{\mathbf{y}}\right)^2 \tag{12-16}$$

Substituting into Equation 12-15 yields

$$\text{minimize } \left(\underline{\mathbf{y}}^{sp} - \underline{\mathbf{y}}^{pc} - \underline{\underline{\mathbf{A}}}\,\Delta\underline{\mathbf{u}}\right)^2 \quad (12\text{-}17)$$

If we define an error vector $\underline{\mathbf{e}}$ as

$$\underline{\mathbf{e}} = \underline{\mathbf{y}}^{sp} - \underline{\mathbf{y}}^{pc} \quad (12\text{-}18)$$

then the objective becomes

$$\text{minimize } \left(\underline{\mathbf{e}} - \underline{\underline{\mathbf{A}}}\,\Delta\underline{\mathbf{u}}\right)^2 \quad (12\text{-}19)$$

Solving the least-squares problem, we can find the sets of future moves necessary to minimize the error:

$$\Delta\underline{\mathbf{u}} = \left(\underline{\underline{\mathbf{A}}}^{T}\underline{\underline{\mathbf{A}}}\right)^{-1}\underline{\underline{\mathbf{A}}}^{T}\underline{\mathbf{e}} \quad (12\text{-}20)$$

In the preceding example, the number of moves calculated was set to three for example purposes. Cutler (1983) states that the time covered by the calculated moves should cover the time it takes for the variable to reach 60% of steady state. In actual practice, this number is usually fixed to one-half of the time horizon covered by the overall controller.

12-2-4 Move Suppression

The preceding least-squares control solution has two difficulties:

- It tends to be ill-conditioned or very near singular.
- It results in a very aggressive controller. Without somehow suppressing its outputs, this solution will try to move the plant as fast as mathematically, but not realistically, possible to minimize error.

The solution to both is least-squares smoothing, which in this case is called *move suppression*. To add move suppression, a least-squares penalty is placed on each Δu,

$$\text{minimize } \left(\begin{bmatrix} \underline{\mathbf{e}} \\ \cdots \\ \underline{\mathbf{0}} \end{bmatrix} + \begin{bmatrix} \underline{\underline{\mathbf{A}}} \\ \cdots \\ \underline{\underline{\mathbf{M}}} \end{bmatrix} \Delta\underline{\mathbf{u}}\right)^2 \quad (12\text{-}21)$$

where $\underline{\underline{\mathbf{M}}}$ is a diagonal matrix of move suppression factors and the vector $\underline{\mathbf{0}}$ is a vector of zeros. These factors make the matrix well conditioned and reduce the aggressive moves the controller can impart upon the plant. Figures 12-3 and 12-4 show an example of a setpoint change and its effect on a simulated controlled and manipulated variable with varying move suppression.

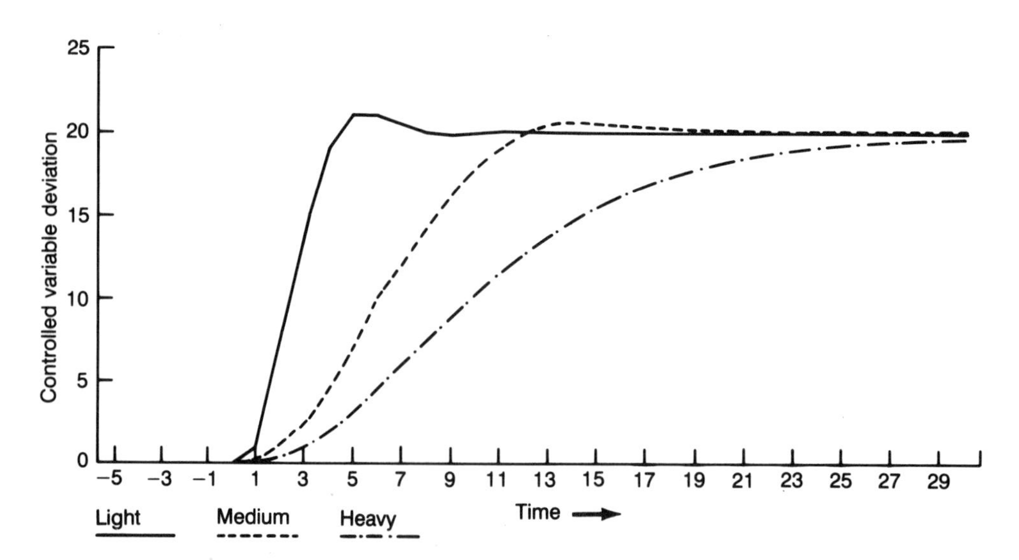

FIGURE 12-3. Move suppression: controlled variable.

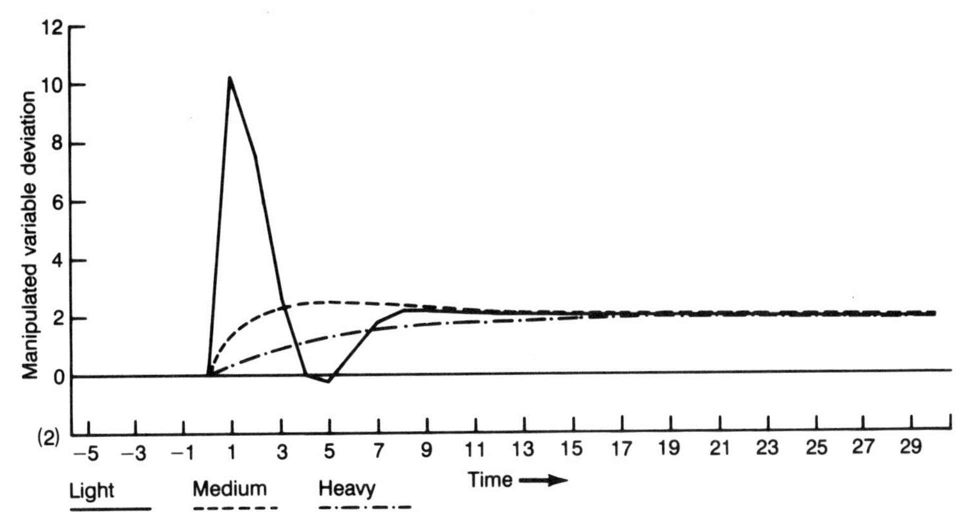

FIGURE 12-4. Move suppression: manipulated variable.

12-3 REVIEW OF MODEL IDENTIFICATION

In this section, we will look at particular topics in the area of model identification that are of interest to the application of DMC. Chapter 7 of this book describes in detail the current state of model identification and touches somewhat on multivariable model identification. Other excellent references are MacGregor (1989) and Box and Jenkins (1976).

12-3-1 DMC Model Identification Background

Basically, there are two ways to derive the DMC convolution model. One way, which is often discussed in DMC research (for example, Georgiou, Georgakis, and Luyben, 1988 and Deshpande, 1985) is to first find a parametric model (usually first or second order plus deadtime). With that in hand, the DMC convolution model can be developed. Although this works quite well in most cases, it does pose the additional task of fitting the plant output to a particular parametric model.

A second approach is to find the convolution model directly. A very simple way of doing this (and the way some of the original DMC controllers were developed) is to induce a step in the manipulated variable when the process is at some steady operating point. The response to this input is then recorded numerically.

For an example, let us look at the response in Figure 12-2. This is actually a first order plus deadtime and noise simulation. Picking off the numbers directly yields the model

$$\underline{\mathbf{a}} = \begin{bmatrix} -0.01 \\ -0.02 \\ 0.00 \\ 0.18 \\ 0.32 \\ 0.47 \\ 0.61 \\ 0.72 \\ \vdots \end{bmatrix} \tag{12-22}$$

One problem this illustrates is that process noise is moved from the plant test directly into the model. Although this example had little noise, a more noisy process could pose a problem. Fitting to a parametric model eliminates this problem.

Another way of deriving a convolution model is to take plant test data and calculate convolution models by using the inverse of the DMC control equations. In other words, with the moves in the process known, find the dynamic matrix $\underline{\underline{A}}$ using least-squares criterion. Least-squares smoothing can again be used to reduce the resulting ill-conditioned matrix and reduce the problems with a noisy signal. This technique has been developed into a commercially available package and is described by Cutler (1991).

12-3-2 Integrating Process Model Identification

All of the preceding DMC control and model identification discussions have dealt with processes that are self-regulating and can be modelled by a linear, steady-state dynamic equation. This does not apply to some processes such as levels and many pressures. These processes are instead integrating processes or "ramps."

The typical integrating process is the level, which can be described mathematically as

$$\frac{dV}{dt} = F_i - F_o \tag{12-23}$$

where F_i = inlet flow
F_o = outlet flow (12-24)
V = drum volume

If we let L be the fractional drum level,

$$L = \frac{V}{V_{\text{tot}}} \tag{12-25}$$

then

$$\frac{dL}{dt} = \frac{F_i - F_o}{V_{\text{tot}}} \tag{12-26}$$

This results in the typical integrating function when a step is entered into the system. Most integrators can be modelled easily by either calculating Equation 12-26 directly from known process flows and volumes or by doing a step test and watching the change in the level.

Often, however, levels and integrating pressures have other dynamics included in the basic integrating model. This is illustrated by the ramp function in Figure 12-5. Transport delays (which result in deadtimes) and other underlying dynamics make it useful to model these processes with more powerful methods. In this case, the output must be differenced before fitting it to the input signal:

$$\begin{matrix} I_1 & I_2 & I_3 & I_4 & \cdots & I_{n-1} \\ \nabla O_1 & \nabla O_2 & \nabla O_3 & \nabla O_4 & \cdots & \nabla O_{n-1} \end{matrix} \tag{12-27}$$

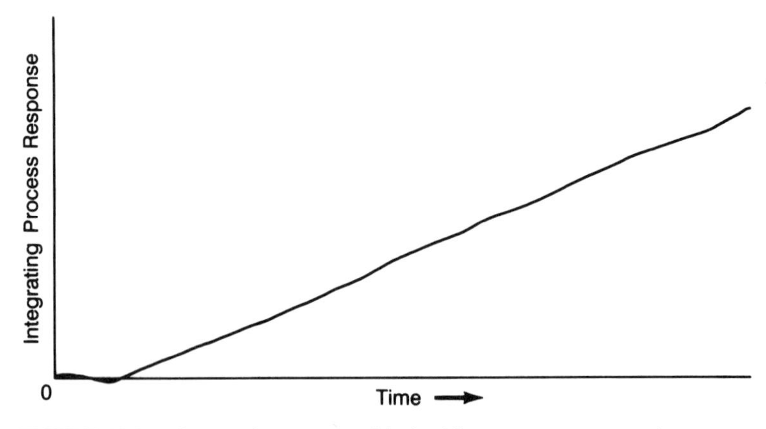

FIGURE 12-5. Integrating process with deadtime: response to step input.

where

$$\nabla O_1 = O_2 - O_1 \tag{12-28}$$

With this, the level (or actually the change in the level) can be fitted either with a parametric model or a convolution model and used in a DMC controller.

12-3-3 Multivariable Model Identification

A designer of a multivariable DMC controller is faced with a set of independent variables that affects a set of dependent variables. In many cases, it is next to impossible to hold everything constant and move only one independent variable at a time. In this case (quite common in the authors' experience) and in general to speed the application of a multivariable DMC controller, multivariable identification is performed.

When moving to the multivariable world from SISO model identification, things are more complicated:

- The independent variables must be truly independent. Independent variables that have some correlation to another independent variable cannot be used.
- When performing multivariable identification, each of the independent variables must have an effect on the dependent variable. If a particular independent variable in a set of independent variables has no effect on a dependent variable, that particular transfer function should be forced to 0 and the identification calculations rerun. Various unwanted correlations could interfere with the resulting multivariable model if this is not done.
- The test length is a multiple of the number of independent variables, assuming each model has the same effective time constant. This is because the number of parameters in the total identification is a multiple of the number of independent variables that are input to the regression. For example, if four independent variables affect the overhead composition of a tower, the test needs to be at least four

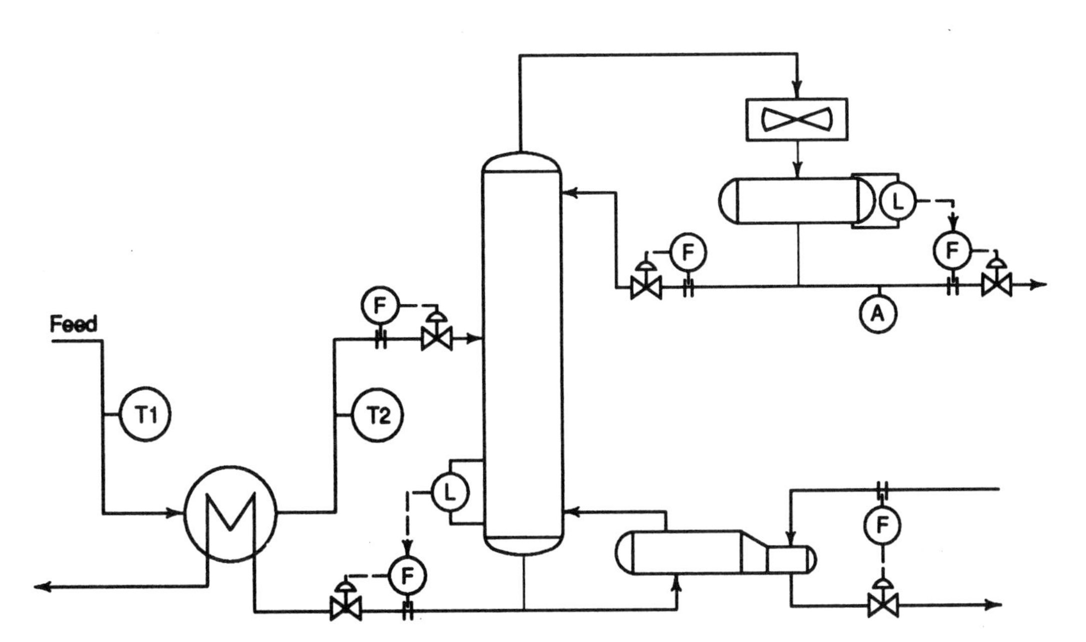

FIGURE 12-6. Energy balance tower with bottoms–feed preheater.

times as long as if simply one variable was moved while holding the others constant.

- The independent variable moves must not be correlated. For example, the independent variables reflux and reboiler heat should not be moved at the same exact time when doing a test for an overhead composition model.

To illustrate the preceding points, let us examine the test of a rather simple energy balance (LV) controlled distillation tower shown in Figure 12-6.

Assuming a relatively constant feed composition, there are four variables that affect the overhead composition:

1. feed rate
2. feed temperature
3. reflux
4. heat input

After identifying these independent variables, the next problem is to select the right feed temperature. Although one could imagine that T_2 would give better control, selecting this can lead both to an incorrect overhead composition model on all four variables and very poor control. The reason is that T_2 is not independent of the other variables. Moving feed rate, reflux, or heat input will affect T_2. Instead, the true independent variable is the temperature of the feed upstream of the bottoms–feed preheater, T_1.

With the four true independent variables selected, the next problem becomes to design a test. Let us assume for this example that a rough "time-to-steady-state" is 4 hours. The test length and sampling interval are determined by the number of parameters that are needed by the model.

If we are doing an analysis that will lead directly to a convolution model (e.g., a 30 parameter DMC model), let us assume that a plant test would require a 1 min sampling time and a 24 hour test (six times the time-to-steady-state) for SISO identification. Manual pseudo-random binary sequence (PRBS) inputs will be used based on a switching time of 0.5 hour (see MacGregor, 1989, for discussion of PRBS signals and switching times). For our four independent variable case, we now need to conduct the test for over 96 hours (because the number of parameters for each independent variable is 4×30, or 120).

Before executing the test directly, we must again make sure that the PRBS moves are randomized. Specifically, correlated moves must be avoided. For example, reflux and reboiler heat should not be moved at the same time even though it might be tempting to do this. Instead, the moves should be made at significant intervals apart (i.e., at least 30 minutes in the preceding example).

One other caution on multivariable tests: Because multivariable tests tend to be long, they have the tendency to include slow rolling effects (i.e., nonstationary load disturbances) that are unmodelled such as day–night shifts in ambient temperature. These disturbances can cause problems with model identification techniques. To minimize the influence of these disturbances, data differencing should be used. (Again see MacGregor, 1989, for details.)

12-3-4 Nonlinear Transformations

Because distillation control problems tend to be nonlinear, there are two ways of dealing with the nonlinearities with a linear controller such as DMC:

- Operate in a region that can be assumed to be linear.
- Transform the variables so that they become more linear.

Three examples of linear transformations that can be used in DMC distillation control follow.

Logarithms for High Purity Distillation Control

The compositions of high purity distillation towers are definitely nonlinear. In many cases, using the logarithm (natural or base 10) will linearize these variables over their normal operating range so that they can be fitted with a linear identification package.

Shinskey (1988) describes the use of logarithms for PID tower composition control, and others (Georgiou, Georgakis, and Luyben, 1988; see others in Section 12-6) have applied it to DMC control. A word of warning, however: The logarithm method does not always work for every high purity tower. The use of logarithms must be evaluated on a case by case basis. Methods of doing this include using a detailed steady-state simulation to find regions where the logarithm transformation can be used, comparing the logarithm and linear results from a model identification package, or closely studying the prediction errors from an on-line DMC controller.

Valve Transformations

Valves that are highly nonlinear should be linearized unless they can be assumed linear over a small operating range. Here, knowledge of the actual valve characteristics can be a big help. Otherwise, doing a fit with one of the standard valve transformations from Perry's Handbook (Perry and Chilton, 1973) can lead to a good model. The linearization equation that takes into account pipe resistances in series with the control valve is

$$Q = \frac{L}{\sqrt{\alpha + (1 - \alpha)L^2}} \tag{12-29}$$

where Q = fraction maximum flow (linear)

L = fraction valve position

α = valve characteristic

Different α's yield the curves shown in Figure 12-7.

Differential Pressure Transformation

As long as a tower remains below the flooding point, a differential pressure measurement can be assumed linear over the normal operating range of a tower. However, if a tower floods easily, the differential measurement will become very nonlinear. A transformation using hyperbolic functions to linearize differential pressure in a case where tower flooding is a problem was developed and used successfully in one application (Hokanson, Houk, and Johnston, 1989).

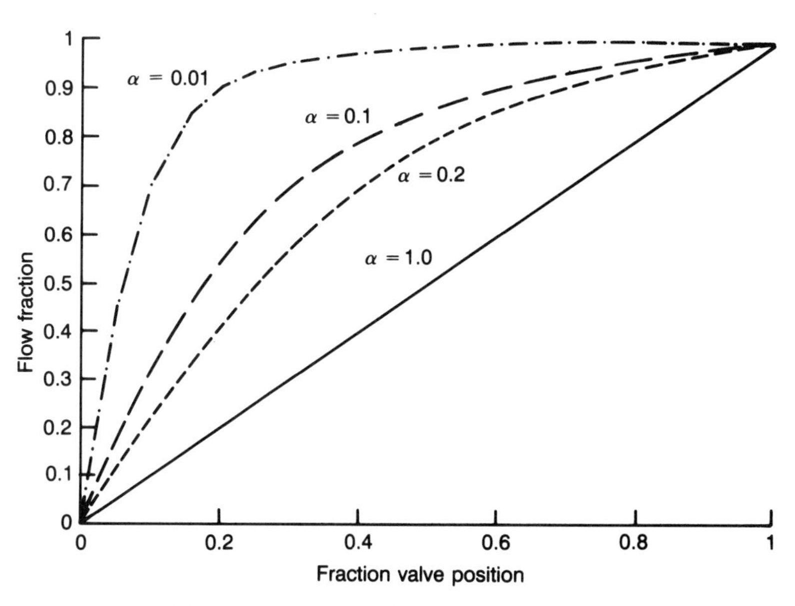

FIGURE 12-7. Linear valve transformation.

12-4 DESIGN ASPECTS OF A MULTIVARIABLE DMC CONTROLLER

Before discussing how to directly apply a multivariable DMC controller to a distillation tower, let us look at the general design aspects of such a controller. In contrast to SISO, two problems arise when moving to the multivariable world: weighting the priority of control and dealing with constraints. To discuss these, let us use the example energy balance tower depicted in Figure 12-6 to pose an example problem for DMC: install two product composition control and include the tower differential pressure in the controller. If the differential pressure gets too high, the tower should violate the bottoms composition target in order to keep the tower from flooding.

In Figure 12-8, the example tower of Figure 12-6 is redrawn with manipulated, feedforward, and controlled variables designated. Also, the desired targets for the controlled variables are indicated in the figure.

12-4-1 Weights

Up to now, all of the DMC mathematics have been done with values in engineering units. Although this causes no problem in the SISO world, in MIMO control with a least-squares solution this could pose a problem. For example, the bottoms target of 2500 ppm most likely will beat the overhead target of 300 ppm and will ignore the differential pressure target of 0.75 bar.

To eliminate this problem, the commercially available DMC Corporation package uses "equal concern error" factors. The equal concern error factors equate errors above and below targets for each controlled variable. These factors both normalize the engineering unit values and prioritize them at the same time. In mathematical terms, the equal concern error factors transform the problem from an ordinary least-squares problem to a weighted least-squares problem (Hsia, 1975).

In the problem in Figure 12-8, one might be equally worried about a 10 ppm error in

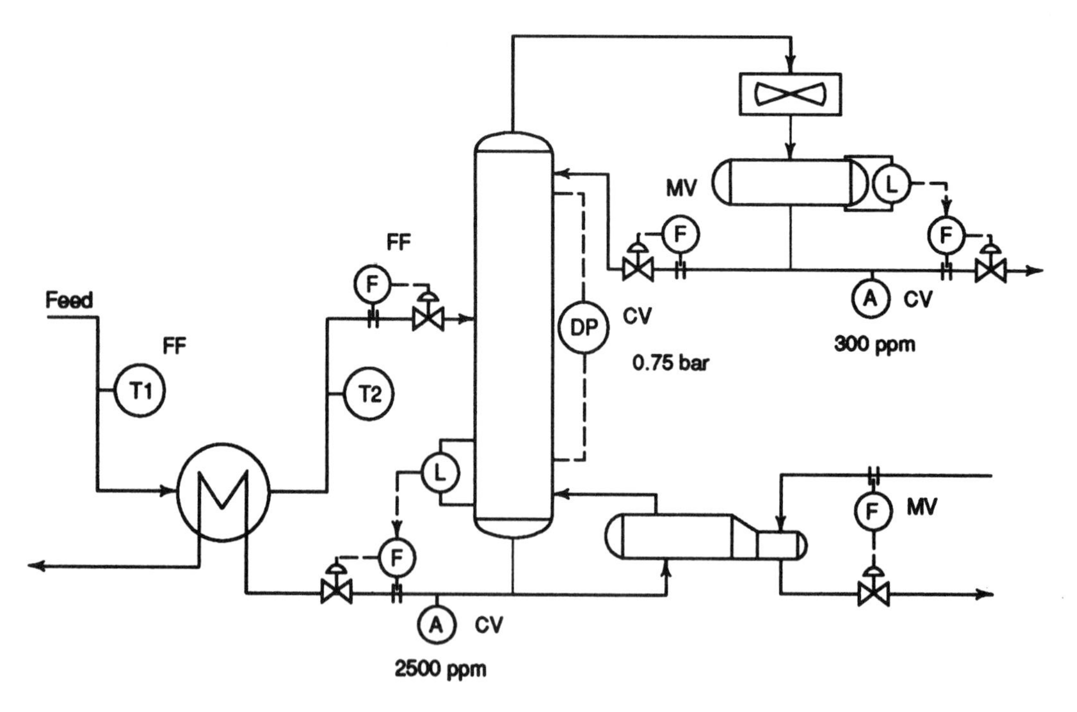

FIGURE 12-8. Energy balance tower with bottoms–feed preheater: DMC example. CV = controlled variable; FF = feedforward variable; MV = manipulated variable.

the top as a 200 ppm error in the bottoms and a 0.02 bar error in the differential pressure. These values are then used by the package to normalize each variable before entering the least-squares control solution.

12-4-2 Constraints

A DMC controller must deal with two different types of constraints:

- Manipulated variable constraints.
- Controlled variable constraints.

Manipulated variable constraints can be a hard limit (e.g., a valve at 100% open), an operator entered limit (e.g., a reboiler maximum heat input), or a maximum desired rate of change to a manipulated variable. All are essentially equivalent, and must be dealt within a DMC control package.

Controlled variable constraints arise when the set of controlled variables targets cannot be reached. For example, if the differential pressure in the preceding example becomes close to the target of 0.75 bar, somehow information must reach the least-squares solution that the bottoms composition specification should be violated. Likewise, if the differential pressure is well below 0.75 bar, the least-squares DMC solution should not be solved to hold the 0.75 bar target.

The handling of these constraints in QDMC is described in Prett and García (1988). Here, the entire DMC controller with manipulated variable constraints is converted into quadratic program (QP). The controlled variable constraints are placed as inequalities that are part of the quadratic objective function. If no constraints exist, QDMC results in the same solution as least-squares DMC.

In the DMC Corporation controller, both manipulated and controlled variable constraints are handled by a linear program (LP), which "directs" the least-squares solution. The LP runs prior to the DMC controller and calculates the optimum steady-state solution. Then it passes its results via the setpoints of the *controlled variables* to the least-squares solution. Prior to implementation, the least-squares controller is checked for any manipulated variable constraint violation in any of the moves. If there are, the moves that are in violation are set at the constraint limits and the least-squares solution is re-solved.

Here is an example of how the LP works within the DMC Corporation controller. Suppose everything in the example tower is fine, and the controller is operating within the range of all variables. Then, the LP directs the DMC controller to target the overhead composition setpoint, bottoms composition setpoint, and passes on the resulting steady-state differential pressure as the differential pressure setpoint to the least-squares solver. The least-squares solver then plans the moves of the reboiler heat and reflux accordingly, taking into account what effects the feed rate has on the controller.

Suppose, however, the feed increases dramatically. Then, the LP sees that the differential pressure may have to be violated and redirects the bottoms composition to a new steady-state value (violating the setpoint) while at the same time directing the differential pressure to hold at its high setpoint. Similar events occur if the heat or reflux is moved to its maximum limit.

12-5 IMPLEMENTATION STEPS FOR A DMC CONTROLLER

The actual implementation of a DMC controller consists of a number of steps:

1. Initial design
2. Pretest
3. Plant test
4. Model identification
5. Off-line simulation
6. Operator interface and on-line controller building
7. Controller commissioning
8. Measuring results

In the following discussion, it is assumed

that the models are generated from plant test data.

12-5-1 Initial Design

The initial controller design phase of the DMC controller involves the selection of the manipulated, feedforward, and controlled variables. Although the design may change over the course of the controller development and implementation, it will always be strongly affected by the decisions taken in the initial design. Thus, the initial design will have a strong impact on the success of the application.

The selection of independent variables (manipulated and feedforward variables) for the DMC application is very important at this stage. Of prime importance is that each independent variable be truly independent. In other words, independent variables must not have an effect on one another. Furthermore, manipulated variables (which can be regulatory controllers or control valves) must have a direct and independent path to each valve. They must not be impeded or aided by overrides, decouplers, and so forth.

Selection of the controlled variables is fairly straightforward at this stage because they can usually be modified once the test is concluded. Tower operating targets and tower constraints are usually the main controlled variables. However, all possible variables are collected during the test so that they can be added as required during the model identification and controller building steps.

To examine this phase, let us look again at the energy balance example in Figure 12-8. The objectives of the controller are to control both the top and bottoms compositions tightly but to give up on the bottoms composition if the tower reaches its flooding limit as determined by the tower differential pressure. These three—overhead composition, bottoms composition, and tower differential pressure—become our prime controlled variables.

There are two manipulated variables that directly affect these controlled variables: reflux flow and reboiler heating medium flow. Depending on the circumstances, these may be modified to internal reflux (especially in this case because the airfin condenser can induce significant changes in subcooling) and reboiler heat input.

The feedforward variables for this case could be the feed rate and T_1. However, in many cases feed temperature changes with feed rate. If there is a significant correlation, T_1 may not be a good feedforward variable either. Any correlation will normally show up in the test data, and if it is found to be significant, the feed temperature can be dropped as a feedforward variable during the model identification and controller building step.

All of this logic results in an initial DMC controller design:

Manipulated variables
- Internal reflux
- Reboiler heat input

Feedforward variables
- Feed rate
- Feed temperature

Controlled variables
- Overhead composition
- Bottoms composition
- Tower differential pressure

However, we are not finished. The bottoms level may cause some problems that need to be examined during the pretest. The existing bottoms level controller impacts the tower through the heat it provides in the bottoms to feed the preheater. If it has been a problem controller and/or has a big impact on tower performance, the bottoms level should be included as a controlled variable. This will also add bottoms flow as a manipulated variable to the controller.

If the bottoms level performs acceptably and does not swing the tower around (perhaps because of the small duty from the preheat exchanger), it can be left out. In this case, it still must be checked for any nonlinear or gap action. Either will ruin the overall quality of the resulting DMC model.

Finally, one more important note about this level. The decision to include the level in the DMC controller must be done prior to the test. Once the test commences, it will be almost impossible to change that decision without a second test.

The decision on whether to use tray temperatures as controlled variables in this tower can be done after the test is completed. Normally, all tray temperatures are collected during the test. If the analyzer response is found to be sluggish when analyzing the test data, a tray temperature can be used in conjunction with the analyzer as a DMC controlled variable. The decision on this is usually made after the test is completed.

12-5-2 Pretest

With the rough design set, the next step is to perform a "pretest." Here, all the independent variables are checked to make sure that each can be moved. For feedforward variables, changes to the process may make enough moves in the variables. If not, a method must be devised to move them.

Again, in the tower depicted in Figure 12-8, there are four independent variables: T_1, feed rate, reboiler heating, and reflux. During the pretest and actual test, the reboiler and reflux setpoints will be collected and there should be no problems moving them. However, during the pretest, T_1 and the feed rate are checked for the type of "natural" process moves that occur and, if possible, how they can be moved.

Also during the pretest, three other things must be accomplished:

1. The data collection package must be tested. In the full plant test the package must operate flawlessly.
2. Move sizes for each independent variable should be roughly determined.
3. An estimation of the time-to-steady-state (i.e., the time horizon that the DMC model will cover) should be made. This will give all involved a rough idea of how long the full plant test must last.

To conduct this test, open loop moves are made to the reflux, reboiler, feed rate, and feed temperature (if possible). The pretest is conducted over one or two shifts, and moves are heavily influenced (of course) by the console operator.

12-5-3 Plant Test

With the pretest complete, the actual test can now be conducted. As previously discussed, the length of the test is set by the number of independent variables and the time-to-steady-state. Using the a rough rule of six times the time-to-steady-state, and assuming the pretest yielded a 4 h time-to-steady-state, the test should then be conducted for four full days.

During the test, each variable should be moved in a PRBS (pseudo-random binary sequence) fashion based on the moves determined during the pretest. Although an automatic PRBS signal can be generated for moving the controller, the authors recommend against using an automatic PRBS signal for plants with limited experience in these types of tests. Controlling the test manually will begin the process of console operator involvement and acceptance and will provide the operators with an excellent learning opportunity. Operators are used to keeping the controlled variables at a constant value. The opportunity to observe the responses to independent variable changes will expand their existing knowledge and dispel any untrue legacies that exist.

In theory, a PRBS test implies that the decision to make a move or stay at the current value is entirely random. In practice, this is not always possible due to the operating objectives of the column. For example, again consider the initial controller design for Figure 12-8. A switching frequency of 0.5 hour has been selected and the reboiler heat input is currently being moved. At the switch time, the process is first examined with the console operator to see if a move "has to be made" in order to keep the compositions within specification, the column out of flood, or any other process event that requires a move.

If no move needs to be made, the process should then be examined to see if a PRBS move should be avoided. This can occur if there are no methods of forcing a change in one of the feedforward variables (e.g., feed temperature) and that variable has had a substantial change. In this case, making no move will allow an uncorrelated response of that variable. Finally, if none of these cases exist, then the PRBS move is made by a random decision (i.e., a coin flip).

From a test management standpoint, it is usually easier to concentrate on moving one independent variable at a time. Concentrating on one variable also minimizes "correlated" independent variable moves. During the test, around 20 moves should be made in each independent variable. These moves are not made consecutively, but rather split between two more different time periods. Further, the order in which the independent variables are moved is changed after a cycle through the variables. For example, in Figure 12-8, the first cycle in the plant test could be to move the reboiler first, reflux second, then feed third. (There is no "easy" way of moving the feed temperature so we must wait for normal process changes.) The second cycle would be slightly different: first reflux, then feed rate, and finally reboiler.

Figures 12-9 and 12-10 illustrate the moves made during a plant test on a tower similar to that depicted in Figure 12-8

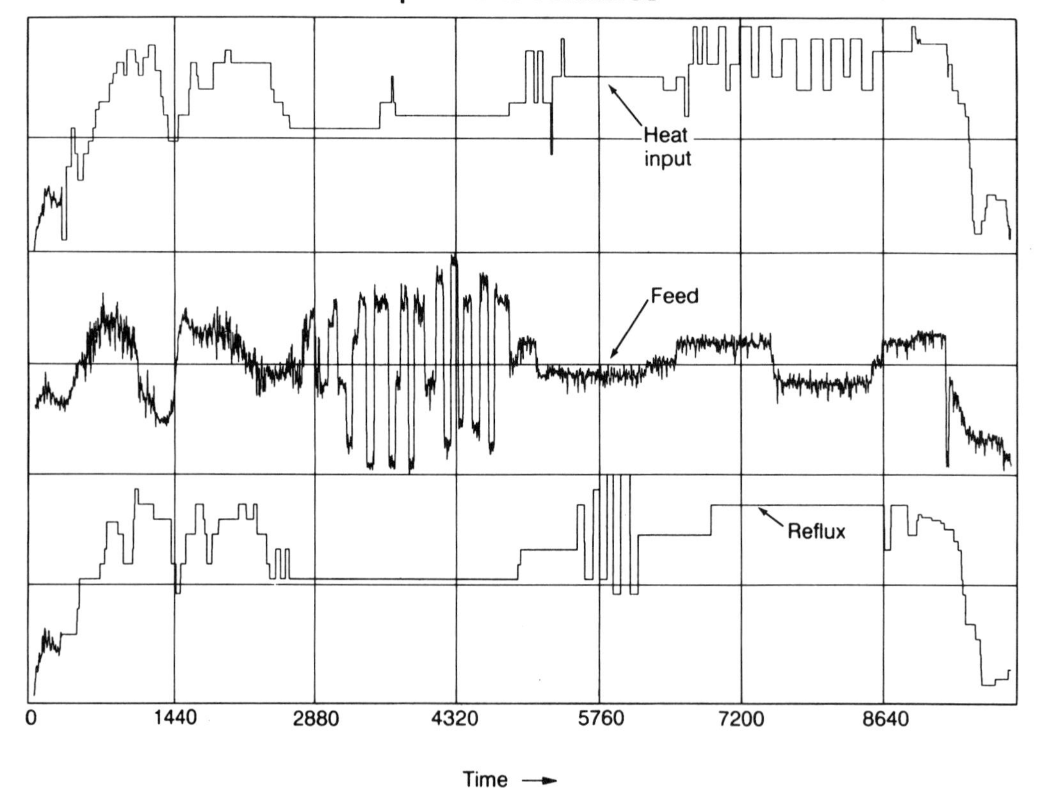

FIGURE 12-9. Independent variable test data.

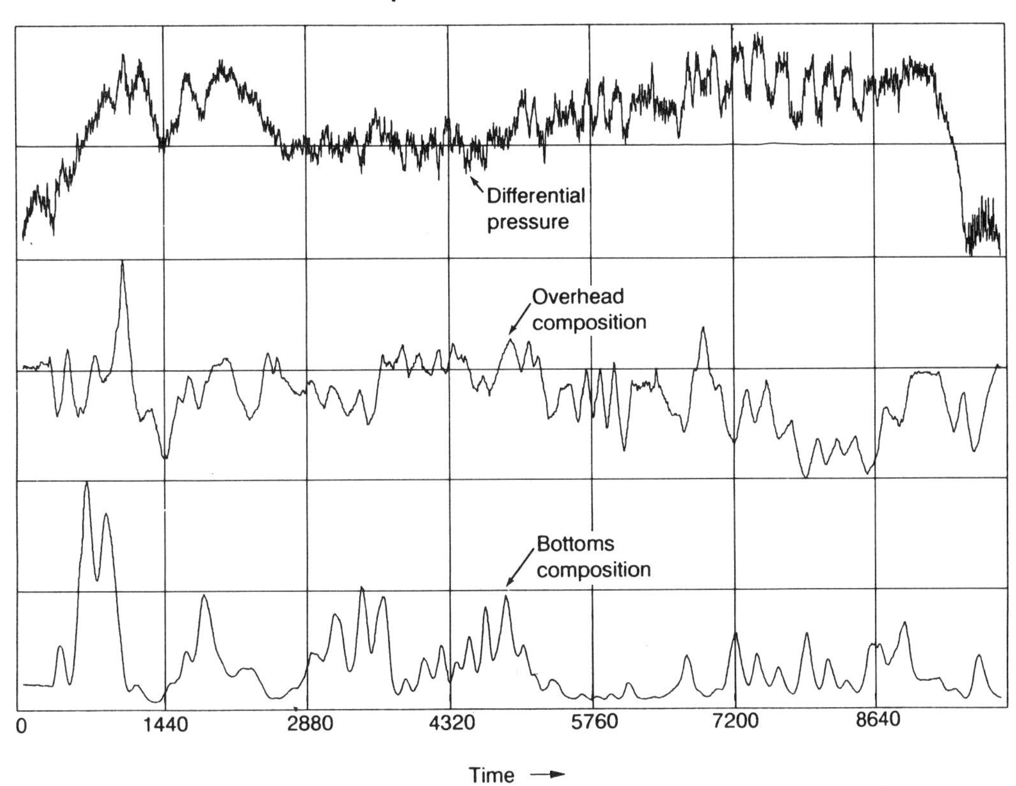

FIGURE 12-10. Dependent variable test data.

but without the feed temperature. It is actually the tower composition controller described by Gerstle, Hokanson, and Andersson (1990).

12-5-4 Model Identification

In this stage, knowledge of both the model identification package and the process is very important. The decision on which models to use and what models to discard depends on both a thorough knowledge of the process plus the goodness of fit information from the identification package. In the example, suppose the feed temperature moved quite frequently during the test, but each move was a function of feed rate. This can be verified by running a cross-correlation check. If such a correlation exists, the process flow should be thoroughly examined to make sure that the correlation between feed rate and feed temperature makes process sense. Then, and only then, should the feed temperature be "thrown out" and the identification run with only three independent variables.

Another example: Suppose the feed rate versus differential pressure showed very little correlation. Before totally dropping this correlation, the fact that there is very little correlation must be verified through process knowledge. (For example, the feed is a saturated liquid and enters below the lower differential pressure tap.) Otherwise, the data must be further analyzed for a better correlation or the test must be run again.

Besides keeping models in and out of the final problem, additional controlled

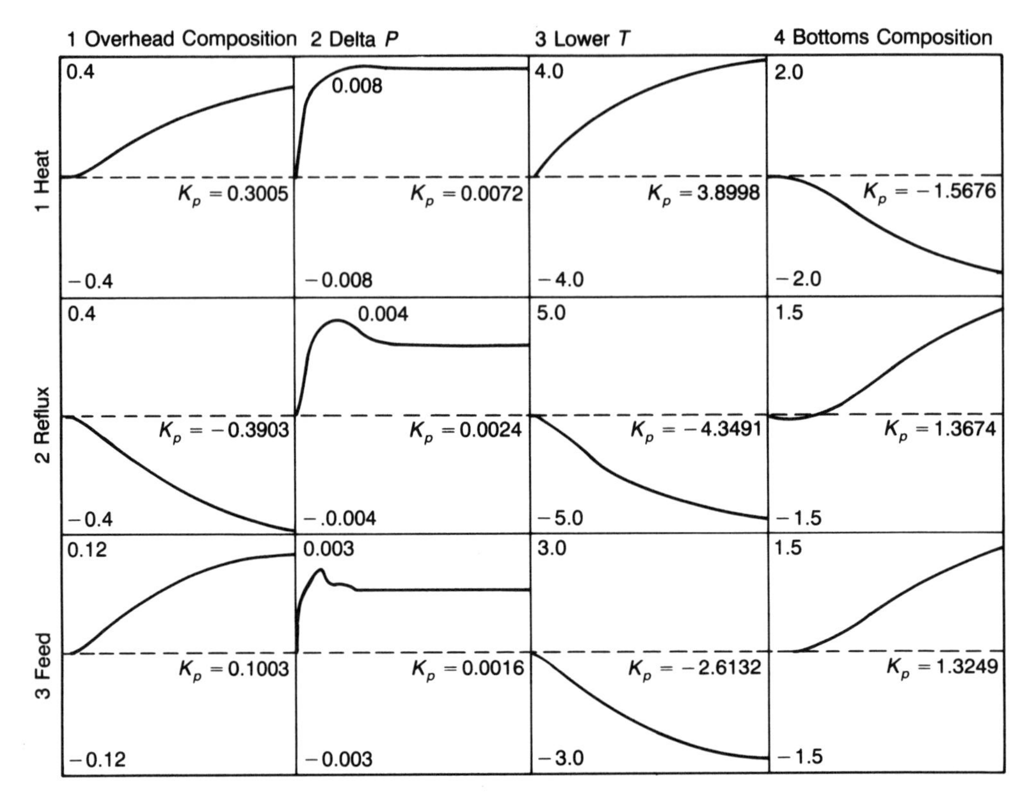

FIGURE 12-11. Overall tower model. Each model represents the response over 240 min.

variables can be added (with care). In the example Figure 12-8, the tray temperatures should be examined to see if they might improve overall response by including them in the model. In the controller discussed in Gerstle, Hokanson, and Andersson (1990), such a tray temperature was included.

Also during the model identification phase, the issue of nonlinearity is also examined. Like the decision on which variables to use in the final model, both process knowledge and identification package goodness of fit are used to determine where linearization should be used. In the example problem, the compositions may be better modelled by using logarithms instead of using the actual data.

At the end of this stage, a multivariable DMC model is produced. Figure 12-11 shows a model developed from the data shown in Figures 12-9 and 12-10.

12-5-5 Controller Building and Simulation

After picking the models, the controller frequency and DMC model size are selected. These parameters are determined by the DMC package used, the dynamics of the process, and a comparison between the longest and shortest dynamics. For example, the DMC Corporation package allows a model size in multiples of 30. If the longest time to steady state is found to be 240 min, the fastest a 30 parameter controller can run is once every 8 min. In the example problem, this may not be fast enough to control the differential pressure, so a larger model (e.g., a 60 parameter model) could be used.

With these parameters selected, the final controller model is built and then simulated off-line. Here the two "tuning constants" —the equal concern error and move suppression factors—are tuned.

12-5-6 Controller and Operator Interface Installation

With the simulation complete, the controller is moved to the on-line control system and supporting software is developed. A key item during this phase is building the operator interface for the controller. If this is the first multivariable controller in the plant, the DMC controller operation will be substantially different than other single loop controllers. Turning the DMC controller on and off alone will be different, so the supporting interface software must be written so that the console operators can easily operate the controller. Besides turning the controller on and off, the interface must aid the console operators in understanding the basics of the DMC multivariable controller.

Besides potentially being faced with the new concepts of multivariable control, the console operators may also be faced with another new concept—that of a "setpoint range." With the controlled variable constraint handling capability comes the ability to set a high and low setpoint for each controlled variable. Again, the interface software should be developed to adequately illustrate this concept.

All these concepts mean both operator training and considerable work on the operator interface. Some of the DMC application papers referenced (e.g., Cutler and Hawkins, 1987; Hokanson, Houk, and Johnston, 1989) contain example displays and discuss this aspect in more detail.

12-5-7 Controller Commissioning

After building all the necessary software, the controller is turned on in "prediction mode." During this time, controller prediction errors (the b's from Equation 12-11) are checked to make sure that the DMC models are reasonable. Also, calculated manipulated variable moves are checked to ensure that they are reasonable.

Finally, the controller is turned on and observed. Move suppression factors, equal concern error weightings, and occasionally the models themselves are modified to improve overall controller effectiveness. Although this phase should be straightforward, the controller commissioning can take as long as the plant test phase.

12-5-8 Measuring Results

The control performance of a tower DMC controller is usually hard to measure with typical control performance measurement tools [e.g., integrated absolute error (IAE)]. This is because DMC controllers are usually installed to handle tower constraints that sometimes keep the tower away from all of its composition setpoints. In the face of no constraints, the authors have found that a well tuned DMC controller on a conventional tower (for example, the one depicted in Figure 12-8) usually equals and sometimes exceeds the tightness of control found from a well designed, conventional dual-composition controller. Of course, if little control is applied on a tower before the implementation of DMC, the improvements in control around a given composition setpoint will be very noticeable.

However, in the case where DMC replaces a well designed, conventional controller, reduced composition variation is usually not the driving force behind its installation. Instead, constraint handling is what is sought from DMC. Here, the results of DMC are usually measured in improvements in plant performance. As the tower is held more tightly against its constraints by the DMC controller, higher unit throughput or less energy consumption are usually easy to measure.

12-6 DMC APPLICATIONS ON INDUSTRIAL TOWERS

This section contains brief reviews of several documented industrial applications of DMC controllers applied to distillation towers. This is in no way a complete list of DMC applications on towers (which may number by now in the hundreds). However, these documented cases do give the reader examples of successful DMC tower controllers.

12-6-1 Hydrocracker C_3 / C_4 Splitter

(Cutler and Finlayson, 1988a)

In this application, DMC was installed on a hydrocracker C_3/C_4 splitter. The feed is a mixture of ethane through pentanes, and the tower makes a separation between propane and butane. The tower has an overhead composition analyzer measuring C_4's. The bottom C_3's are measured in the overhead stream of a following tower.

The motivation for this controller was to reduce tower and product composition variations that, in turn, reduce tower energy consumption. Here is how the DMC controller is arranged:

Manipulated variables

- Lower tray temperature setpoint (manipulates reboiler steam)
- Reflux flow setpoint

Feedforward variables

- Feed flow
- Feed temperature

Controlled variables

- Log of overhead butane (within limits—optimization variable)
- Log of deisobutanizer overhead propane (setpoint)
- Reboiler steam (within limits—a tower constraint)

12-6-2 Hydrocracker Preflash Column

(Cutler and Finlayson, 1988b)

In this application, a DMC controller was installed on a hydrocracker reactor preflash column. This column takes feed directly from the hydrocracker reactors and separates butanes and lighter from a feed stream containing methane through heavy recycle oil. Overhead product is taken as a vapor, and overhead temperature is set by manipulating an airfin blade pitch.

The motivation for the controller was to stabilize tower operation, which had caused many problems for the console operators. This application, when combined with the hydrocracker reactor DMC controllers (Cutler and Hawkins, 1987, 1988), allowed the unit to be operated at higher throughputs. A particular problem in this tower was the control of the overhead reflux drum level, which previously manipulated reflux and was subject to many major feed and heat input disturbances.

The resulting DMC controller has the following arrangement:

Manipulated variables

- Airfin fan pitch and louver position (moved in parallel)
- Reflux flow setpoint
- Valve position on bottoms circulation through the reactor effluent–reboiler exchanger
- Pressure control setpoint on fuel gas to reboiler furnace

Feedforward variables

- Feed flow rate
- Reactor weighted average bed temperature

Controlled variables

- Overhead fan pitch–louver position (within high and low limits)
- Reflux drum level (within high and low limits)
- Upper tray temperature (setpoint)
- Bottom temperature (within limits)

12-6-3 Benzene and Toluene Towers

(Tran and Cutler, 1989)

In this application, two DMC controllers were installed to control first a benzene tower, then a toluene tower, in an aromatics fractionation train. The benzene tower takes a mixture of aromatics from an aromatics extraction unit. The overhead product is a small purge stream, and the actual benzene product is taken as a sidestream a few trays from the top. The bottoms of the benzene tower feed the toluene tower, which produces toluene as its overhead product.

The motivation for these controllers was to minimize energy consumption while holding the tower products tightly at specification. In both applications, tower pressure setpoints are included to minimize tower pressure against various constraints.

The benzene tower controller has the following design:

Manipulated variables
- Internal reflux setpoint
- Tray temperature setpoint (outputting to heat input)
- Pressure setpoint
- Overhead condenser fan pitch and louver position

Feedforward variables
- Feed flow
- Ambient temperature

Controlled variables
- Log of overhead toluene in benzene product
- Log of benzene in toluene
- Compensated temperature
- Valve position of accumulator vent
- Valve position of steam reboiler
- Tower differential pressure
- Overhead accumulator temperature

The toluene tower DMC controller is designed as follows:

Manipulated variables
- Reboiler steam setpoint
- Tray temperature controller setpoint (output to reflux)
- Column pressure setpoint

Feedforward variables
- Feed rate

Controlled variables
- Log of xylenes in toluene product
- Log of toluene in xylenes product
- Log of lower tray temperature difference
- Pressure control valve position
- Reflux flow rate
- Tower differential pressure

12-6-4 Olefins Plant Demethanizer

(Hokanson, Houk, and Johnston, 1989)

In this application, DMC was applied to the overhead of an olefins plant demethanizer. This tower is fed a stream of hydrogen through olefinic C_5's and separates the hydrogen and methane from the heavier components. The overhead of this tower is especially difficult, as the tower has an overhead vapor and liquid that both provide precooling to the tower feed. Supplemental reflux from another tower is fed to this tower to keep liquid rates at reasonable levels in the overhead section of the tower.

The motivation of this controller was to reduce the loss of ethylene in the overhead, which eventually is sent to fuel gas. Bringing some control to the reflux drum level was an additional motivation.

The controller was constructed as follows:

Manipulated variables
- Tower reflux setpoint
- Supplemental reflux setpoint
- Overhead liquid distillate flow setpoint
- Overhead condenser valve position

Feedforward variables
- Overhead pressure setpoint
- Heat input
- Total feed

Controlled variables
- Reflux drum level (within limits)
- Final chill train temperature (within limits)
- Overhead ethylene composition (high limit)

• Tower differential pressure (linearized —high limit)
• Overhead condenser valve target (setpoint)

12-6-5 Olefins Plant C_2 Splitter (Gerstle, Hokanson, and Andersson, 1990)

In this application, DMC was applied to an olefins plant C_2 splitter. This tower receives a liquid feed containing ethane and ethylene. It separates this feed into ethylene product and recycled ethane streams. In this application, tower flooding as determined by tower differential pressure is often the main plant constraint.

The motivation for the controller is to control both compositions and, when necessary, tightly ride the tower differential pressure constraint. Also, tower levels, which had proved difficult to control in the past, were included as a secondary motivation for this controller.

The resulting controller was actually divided into two controllers due to process control computer constraints. The composition controller is designed as follows:

Manipulated variables
• Heat input
• Reflux
Feedforward variable
• Feed flow
Controlled variables
• Log of overhead ethane (setpoint)
• Differential pressure (high limit)
• Bottoms ethylene (within limits)

The level controller DMC controller is designed as follows:

Manipulated variables
• Total overhead product flow (vapor and liquid)
• Bottoms flow
Feedforward variables
• Feed flow
• Heat input
• Reflux flow
Controlled variables
• Reflux drum level (within limits)
• Bottom level (within limits)

12-7 SUMMARY

In this chapter, the application of multivariable DMC on distillation towers has been briefly covered. The chapter started with a short review of DMC mathematics. Although not simple, DMC has a strong mathematical background. Various papers and texts are referenced for the interested reader to further mathematically examine DMC.

Next, various topics on model identification, from the difficulties of multivariable model identification through linear transformations, were covered. Model identification is the key to a successful DMC controller, so the importance of good model identification practice and techniques (discussed in this chapter, Chapter 7, and referenced elsewhere) cannot be over-emphasized.

After a discussion on the design aspects of a multivariable DMC controller, the actual steps of installing a DMC controller on a distillation tower were discussed. Although all steps are important, the initial design, plant test, and identification portions of a DMC application stand out as being crucial to a successful DMC application.

Finally, in the last section, several published DMC applications on industrial distillation towers were summarized. This summary should give the reader an idea for where DMC has improved control.

In summary, DMC is not a panacea for distillation tower control. When simpler techniques are possible, DMC is at best "overkill" and at worst a poor controller when compared to these techniques. However, for towers subject to constraints (even simple towers) and for towers with complex configurations, DMC has the potential of dramatically improving control.

References

Box, G. E. P. and Jenkins, G. M. (1976). *Time Series Analysis*. Oakland, CA: Holden-Day.

Chen, C. T. (1970). *Introduction to Linear Systems Theory*. New York: Holt, Rinehart, and Winston, Inc.

Cutler, C. R. (1983). Dynamic matrix control: A computer control algorithm with constraints. Ph.D. Thesis, University of Houston.

Cutler, C. R. (1991). Experience with DMC inverse for model identification. *Proceedings of the Fourth International Conference on Computer Process Control February 1991, South Padre Island, TX*. Austin, TX: CACHE; New York: AIChE.

Cutler, G. R. and Finlayson, S. G. (1988a). Multivariable control of a C_3/C_4 splitter column. Paper presented at National AIChE Meeting, April 1988, New Orleans, LA.

Cutler, C. R. and Finlayson, S. G. (1988b). Design considerations for a hydrocracker preflash column multivariable constraint controller. Paper presented at IFAC Conference, June 1988, Atlanta, GA.

Cutler, C. R. and Hawkins, R. B. (1987). Constrained control of a hydrocracker reactor. Proceedings of the 1987 American Control Conference, June 1987, Minneapolis, MN.

Cutler, C. R. and Hawkins, R. B. (1988). Application of a large predictive multivariable controller to a hydrocracker second stage reactor. Proceedings of 1988 American Control Conference, June 1988, Atlanta, GA.

Cutler, C. R., Morshedi, A. M., and Haydel, J. J. (1983). An industrial perspective on advanced control. Paper presented at AIChE National Meeting, October 1983, Washington, DC.

Cutler, C. R. and Ramaker, B. L. (1979). Dynamic matrix control—A computer control algorithm. Proceedings of the Joint Automatic Control Conference, June 1980, San Francisco, CA.

Deshpande, P. B. (1985). *Distillation Dynamics and Control*. Research Triangle Park, NC: Instrument Society of America.

García, C. E. and Morshedi, A. M. (1986). Quadratic programming solution of dynamic matrix control (QDMC). *Chemical Engineering Communications* 46, 73–87.

Gerstle, J. G., Hokanson, D. A., and Andersson, B. O. (1990). Multivariable control of a C2 splitter. Presented at AIChE 1990 Annual Meeting, November 1990, Chicago, IL.

Georgiou, A., Georgakis, C., and Luyben, W. L. (1988). Nonlinear dynamic matrix control for high-purity distillation columns. *AIChE Journal* 34(8), 1287–1298.

Hokanson, D. A., Houk, B. G., and Johnston, C. R. (1989). DMC control of a complex refrigerated fractionator. Paper presented at Advances in Instrumentation and Control, ISA 1989, October 1989, Philadelphia, PA.

Hsia, T. C. (1975). *System Identification*. Lexington, MA: Lexington Books.

MacGregor, J. F. (1989). Discrete time process identification: Theory and applications. Text from "An Intensive Short Course on Digital Computing Techniques for Process Identification and Control" given in May 1989 at McMaster University, Hamilton, Ontario.

Oppenheim, A. V. and Schafer, R. W. (1975). *Digital Signal Processing*. Englewood Cliffs, NJ: Prentice-Hall.

Perry, R. H. and Chilton, C. H. (1973). *Chemical Engineers' Handbook*, 5th ed. New York: McGraw-Hill.

Prett, D. M. and Gillette, R. D. (1979). Optimization and constrained multivariable control of a catalytic cracking unit. Proceedings of the Joint Automatic Control Conference, June 1980, San Francisco, CA.

Prett, D. M. and García, C. E. (1988). *Fundamental Process Control*. Stoneham, MA: Butterworth.

Richalet, J. A., Rault, A., Testud, J. D., and Papon, J. (1978). Model predictive heuristic control: Applications to industrial processes. *Automatica* 14, 413–428.

Shinskey, F. G. (1988). *Process Control Systems*, 3rd ed. New York: McGraw-Hill.

Tran, D. and Cutler, C. R. (1989). Dynamic matrix control on benzene and toluene towers. Paper presented at Advances in Instrumentation and Control, ISA 1989, October 1989, Philadelphia, PA.

13

Distillation Expert System

F. G. Shinskey

The Foxboro Company

13-1 INTRODUCTION

An expert system is a computer program that uses an expert's knowledge and experience in a particular domain to solve a narrowly focused, complex problem.

This chapter describes the application of an expert system to the design of regulatory controls for distillation columns. The problem is complex in that the most effective control-system configuration must be found from a set of possible configurations that numbers in the tens or hundreds of thousands. The problem is difficult because of the wide variety of columns, objectives, constraints, and product-quality specifications likely to be encountered. To achieve some measure of success, the scope of the expert system has to be focused only on two-product (simple) and three-product (single-sidestream) columns.

13-1-1 On-line versus Off-line Systems

An on-line expert system would be receiving inputs from measuring devices attached to the process, observing its behavior, and drawing conclusions about whether that behavior was normal, abnormal, or in some need of adjustment. On-line systems are used for process diagnosis, alarm handling, fault detection, and production scheduling (Babb, 1990; Shirley, 1987). Some are also used to assist in process control (Ayral, 1989). The most common application, however, is the tuning of process controllers, both initially, and periodically or continuously as process parameters change (Kraus and Myron, 1984).

By contrast, off-line expert systems receive information entered manually, and produce a decision or message describing the expert's judgment on the conditions described by that information. Off-line expert systems are being used for diagnosing illness, for configuring computer systems, for evaluating insurance risks, and for guiding investors.

The expert system described in this chapter, the Distillation Consultant*, is an off-line system, in that it acts on manually entered information to produce recommendations for configuring control systems. It is essentially a design guide.

*The Distillation Consultant is a product of The Foxboro Company.

13-1-2 Expertise in a Knowledge Domain

The core of any expert system is the expertise of some person or persons in the domain of the problem to be solved, in this case, distillation control. Although distillation is a reasonably exact science, as is control, their combination in configuration of closed loops is capable of producing surprises, due to the variety of conditions previously mentioned. Control engineers approach a distillation column not unlike the blind men approaching an elephant: one stumbles against a leg and decides that it must be a tree; another encounters its trunk and concludes that it is a snake, and so on. Similarly, different experts in distillation control have different experiences, different backgrounds and education, different sets of tools, and possibly different industrial orientations. As a result, their expert systems would be different—perhaps vastly different.

Whether an expert system that includes the experience of two or more experts can be effective is questionable. Often, their conclusions about a specific problem may be in opposition, and both may be incorrect or incomplete. Although it may be possible to blend their expertise to eliminate conflicts, the result may look as ungainly as a camel —the horse designed by a committee.

An expert system may embody not only the experience and understanding, but also the way of thinking of a particular expert. In essence, an expert system carries with it the authority of the expert. Those who would use it must have confidence in that particular expert because the results obtained will reflect the expert's judgment. There can be no "unbiased, impartial, objective" expert system because experts are asked to make judgments based on their own experience and point of view; to do otherwise is to reject some of their expertise and diminish the value of their conclusions.

The Distillation Consultant was designed by the author, and therefore reflects his knowledge and experience in distillation control. The expert was also the programmer, to eliminate the communication gap that can exist between expert and programmer. Although this requires expertise in two domains, it gives full control over the development of the expert system to the expert, an important factor in its success, and essential to prolonged maintenance of the knowledge base.

13-1-3 Logical Rule Base

The central feature of any expert system is the rule base—a set of logical rules that are to be followed to produce the correct results. Some expert systems contain nothing but logical rules, but they are not as likely to treat as complex a problem or to produce as accurate a result as one that also includes calculations and mathematical modelling, as described in the next section. Logical rules follow the conditional branching of the IF... THEN... ELSE statements common to most computer languages. Logical rules such as these can only classify results into distinct categories. Some gradation is possible by accepting multiple-choice inputs. Confidence limits may also be included on the inputs to reflect the degree of certainty of an input, and the results can also include a degree of certainty, and even multiple results with associated degrees of certainty. One means of treating uncertainty is *fuzzy logic*, which has its own mathematical methods for combining inputs (Leung and Lam 1988). However, no fuzzy logic was used in developing the Distillation Consultant.

The two principal methods for manipulating rules are *forward chaining* and *backward chaining*. In forward chaining, the logic flows from a series of statements with their input data to a conclusion. This is the method usually followed in diagnostics and in design, as it rapidly converges on the target while invoking a minimum number of rules. Backward chaining begins with a target or goal, and tries various rules until it finds a set that alone would produce that result. More time is required because more rules must be invoked and tests made. Backward chaining is applied by detectives who solve crimes beginning with a *corpus delecti*, and sort through the suspects by a process of elimination. Actually, the human mind may use both methods in approaching a given problem, depending on the information available and the nature of the problem. The Distillation Consultant uses forward chaining exclusively.

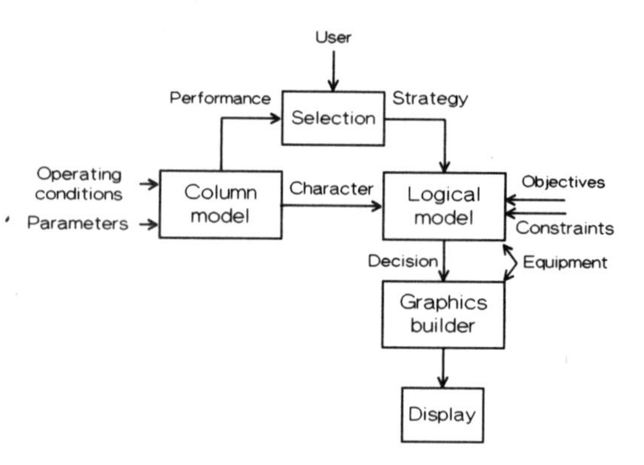

FIGURE 13-1. A block diagram of the Distillation Consultant expert system.

13-1-4 First Principles and Mathematical Modelling

Physical processes such as distillation follow rigorous laws, which if invoked, simplify the logical-rule base and improve the accuracy of the result. For example, the material balance for a two-product column requires that the sum of the two product flows must equal the feed flow in the steady state. For any given value of feed flow, increasing the flow of one product must result in an equal decrease in the other: only one of the product flows can then be manipulated independently to control quality. This principle is used to limit the selection of manipulated variables for composition control in this expert system. Although a system manipulating both product flows for composition control has been used successfully (Finco, Luyben, and Polleck, 1989), the loss of either composition loop also results in the loss of material-balance closure; this system is therefore considered unreliable.

Other applications of first principles include the understanding that an increase in feed rate requires all other flows to increase proportionately if compositions are to remain constant. Similarly, a material balance must exist for each component in the feed, such that an increase in its light-key component will require an increase in the distillate-to-feed flow ratio to maintain constant product compositions.

All of these relationships can be combined into a mathematical model of the distillation process. Such a model should be able to predict the effects of all manipulable variables on product compositions, and the changes in their values required to maintain constant product compositions following certain disturbances. A steady-state model is sufficient to provide this information, but gives no information on the dynamic response of the closed loops that will regulate product quality. However, dynamic models are far more complex than static models, require more process details to complete (and are therefore more process-specific), and require much more time to converge. For these reasons, the Distillation Consultant uses a static column model, and dynamic behavior is inferred from it based on a general understanding of column dynamic response.

Figure 13-1 is a block diagram showing the place of the column model in the expert system. Operating conditions such as product specifications and reflux ratio are entered into the model, along with parameters such as the number of theoretical trays. From this information, the model estimates the interaction between the composition control loops and their sensitivity to disturbances, for each of several candidate strategies. These performance figures are presented to the user and also sorted by the system; the user may accept the system's selection of control strategy or choose otherwise. The selection is then sent to the logical model. The column model also characterizes the column on the basis of its various flow ratios, sending this information to the logical model to assist in configuring the control loops.

The logical model also requires information on the economic objective of the distillation and any constraints placed on it such as the inability to condense all of the overhead vapor. Equipment characteristics also enter into the logic because they limit choices for some control loops. Finally a decision is made on the best control system for this particular column and a drawing is created illustrating the recommended configuration.

13-2 CONFIGURING DISTILLATION CONTROL SYSTEMS

This section is a description of the problem to be solved. Controlling distillation columns is difficult for several reasons: the process is both nonlinear and multivariable, its character varies dramatically from one application to another, and its objectives also vary. A control system that is quite effective on one column may be quite ineffective on another. These facets of the problem are examined in the following text.

13-2-1 Nonlinear Multivariable System

The relationships between product compositions and the flows of streams entering and leaving a distillation column can be determined by combining the internal and external material balances of the components with the vapor–liquid equilibria. However, the modelling effort required is far from trivial, and there are some unknowns, such as tray efficiency. Even when these relationships can be modelled quite accurately, they are complex, nonlinear, and multivariable. As a result, it is difficult to predict, in simple terms, the results of any given control action upon product quality. Models that have been developed for designing columns are of little use for control studies because their output is the recommended number of equilibrium stages. Models used for control purposes must start with a fixed set of stages and report the results of flow changes. They need not be as accurate as models used for column design because their results are principally used to compare one control system against another for purposes of selecting the best. Additionally, they can be fitted to an accurately designed column, or one already operating.

Modelling begins with the overall material balance

$$F = D + B \tag{13-1}$$

where F is the feed flow in moles per hour and D and B are the distillate and bottom product flows in the same units. Similar material balances may be made for each component i:

$$Fz_i = Dy_i + Bx_i \tag{13-2}$$

where z, y, and x are the mole fractions of component i in the feed, distillate, and bottom streams, respectively. Combining the preceding equations produces linear expressions relating product compositions and flow ratios:

$$\frac{D}{F} = \frac{z_i - x_i}{y_i - x_i} \qquad \frac{B}{F} = \frac{y_i - z_i}{y_i - x_i} \tag{13-3}$$

The last two equations are not independent, so that there is essentially one equation having two unknowns: the product compositions. Another equation is required; it relates the compositions of the products on the basis of vapor–liquid equilibrium:

$$\frac{y_l/y_h}{x_l/x_h} = S \tag{13-4}$$

where components l and h represent the light and heavy key, respectively, and S is the separation factor between them.

If the mixture is binary, then $y_h = 1 - y_l$ and a solution can be found. If that is not the case, a solution can still be found for four components in the feed, providing that the lightest does not leave with the bottom product (lighter-than-light key) and the heaviest does not leave with the distillate (heavier-than-heavy key). The complete derivation for this four-component model is given by Shinskey (1988, pages 411–414).

Figure 13-2 plots the preceding relationships as y_h against x_l for a four-component separation. It was generated by the Distillation Consultant for a column having a feed consisting of 1, 20, 40, and 39 mol% lighter-than-light, light, heavy, and heavier-than-heavy keys, respectively. Distillate contained 2 mol% heavy key and the bottom product contained 0.5 mol% light key. The reflux ratio was 3.0 and there were 25 theoretical stages in the column.

The line labelled D/F represents the solution of material-balance Equation 13-3 for a constant distillate-to-feed flow ratio. The hyperbolic curve labelled S is the solution of Equation 13-4 for a constant separation factor. The point where they cross is the current operating point for the column.

The separation factor is a function of reflux-to-distillate flow ratio L/D, relative volatility, the number of theoretical equilibrium stages in the column, and the enthalpy of the feed (Shinskey, 1984, page 79). If D/F is increased, its line will move upward, pivoting about the point where feed composition falls on each axis. But if reflux ratio is held constant as D/F is changed (by changing reflux L proportionately), then both product compositions will change along the S curve, one becoming less pure while the other becomes more pure. If instead, reflux ratio is increased by increasing L while D is held constant, then compositions will move downward along the D/F line, both products becoming purer. This scenario gives a qualitative indication of the interactions that can be observed between product compositions when one flow rate or flow ratio is adjusted.

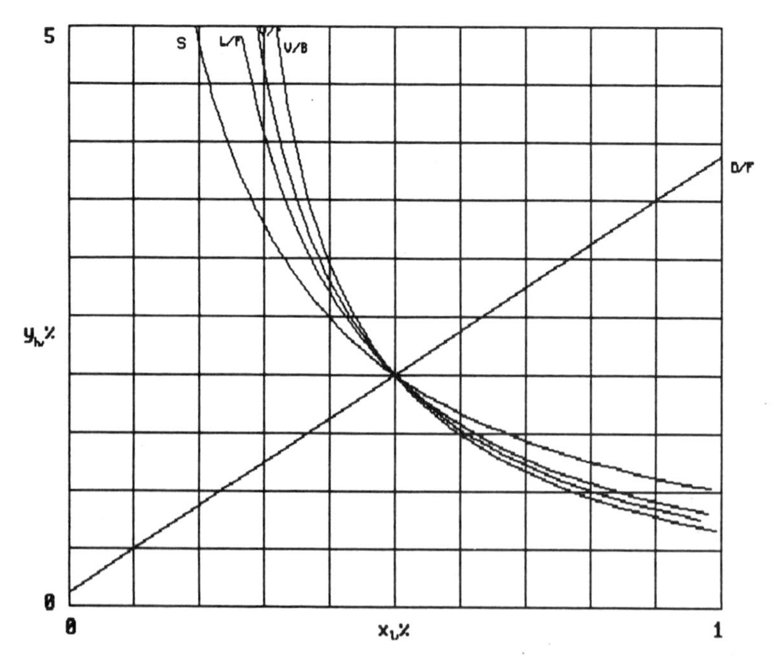

FIGURE 13-2. Operating curves for a typical column plotted by the Distillation Consultant.

These two relationships are sufficient to define a column, but there are other operating curves that may be more meaningful in a given situation. The remaining curves in Figure 13-2 are all derived from the first two. Consider, for example, the case where the reflux-to-feed flow ratio L/F is held constant. Increasing D/F under this condition will cause L/D to fall and thereby, separation will decrease; hence the curve representing constant L/F lies to the right of the S curve above the operating point, and underneath it (higher separation) below the operating point. If overhead vapor flow V is held constant while D/F is increased, reflux falls, causing L/D and hence separation to change more than in the case of constant reflux flow. As a result, the curve for constant V/F is still steeper than the L/F curve. The last curve in Figure 13-2 represents a condition of constant vapor-to-bottom flow ratio V/B, analogous to reflux ratio but at the other end of the column. Notice that it is the steepest of all the curves.

There is one other significant operating curve, corresponding to a constant ratio L/B. It falls between the V/F and V/B curves, but it represents a combination of flows not easily coordinated, being at opposite ends of the column, and so is not given further consideration.

In general, columns having relatively impure products, low reflux ratios, and few trays, tend to have their operating curves well spread as in Fig. 13-2. But those making high-purity cuts from low-volatility mixtures have curves grouped tightly together; in the extreme they are almost superimposed. Yet the curves always fall in the same order. The material-balance line, on the other hand, is quite independent of the curves. Its slope is dependent on feed and product compositions because it is the bottom-to-distillate flow ratio B/D.

The slopes of the various curves at the operating point reveal a tendency for self-regulation of the compositions. Consider the possibility that the V/B curve may be almost vertical at the operating point. As long as the V/B ratio is then held constant, compositions must remain on its curve regardless of variations in the other flow ratios. The verticality of the curve would then make it possible to adjust y_h over a wide range by adjusting reflux, for example, with little influence on x_l; in effect, x_l would be regulated simply by holding V/B constant. Then if bottom composition were to be controlled by manipulating V/B, that controller would have little work to do—a favorable situation.

The material-balance line may be nearly vertical or nearly horizontal, or anywhere in between, depending on the B/D ratio. If it is nearly vertical, the material balance will easily regulate bottom composition; if it is nearly horizontal, top composition will be easily regulated.

13-2-2 Relative Gain Analysis

Relative gain analysis is a simplified method for determining the interaction among control loops in a multivariable process such as distillation (Shinskey, 1984, pages 131–154). Each viable configuration for the two composition loops can be evaluated, and the combination having the least interaction can be selected as a good candidate for controlling the products. The Distillation Consultant uses relative gain as one measure of control-loop performance, and as a basis for estimating another (disturbance rejection) as well.

Relative gain is defined as the ratio of the open-loop gain for a selected loop when all loops on the process are open, to its open-loop gain when all the other loops are closed. In the first case, all other manipulated variables m are held constant, whereas in the second, all other controlled variables c are held constant by their controllers:

$$\lambda_{ij} = \frac{(\partial c_i/\partial m_j)_m}{(\partial c_i/\partial m_j)_c} \qquad (13\text{-}5)$$

Because the same dimensions appear in both numerator and denominator of the relative

TABLE 13-1 Properties of Relative Gains

λ	Property
< 0	Conditional stability—do not close loop
0	Control depends on other loops
0–1	Interaction increases gain and period of oscillation
1.0	Interaction is incomplete, if any
> 1	Interaction reduces control effectiveness
∞	Loops are completely dependent

gain expression, it is a dimensionless number and is unaffected by the choice of scales for the variables. Relative gains are displayed in an $n \times n$ array with the manipulated variables listed across the top and the controlled variables down the side. A relative gain array describing the interactions between the two composition loops in a distillation column would appear as

$$\Lambda_{12} = \begin{bmatrix} \lambda_{y1} & \lambda_{y2} \\ \lambda_{x1} & \lambda_{x2} \end{bmatrix} \tag{13-6}$$

Top and bottom compositions are designated y and x, and the manipulated variables m_1 and m_2 could be any selected flow rates or flow ratios. An important feature of a relative gain array is that the numbers in each column and each row sum to 1.0. In Equation 13-6 then, λ_{y2} is $1 - \lambda_{y1}$, and so is λ_{x1}; as a result, $\lambda_{x2} = \lambda_{y1}$.

Table 13-1 summarizes the properties associated with certain groups of relative gain numbers. A relative gain of 1.0 indicates that m_j will form only one closed loop with c_i, because the same open-loop gain is obtained whether other loops are open or not. There could be one-way interaction—relative gain will not reveal this.

In a 2×2 array, a value of 1.0 must be accompanied by a value of 0 in the same row and column. A relative gain of 0 is caused by the numerator in Equation 13-5 being 0. Control could still be possible in this case (if partial interaction exists), but would depend on both loops being closed all the time. The opposite situation, the denominator of 0, indicates that the loops cannot all be closed at the same time because the variables are not all independent. There are fewer degrees of freedom than there are control loops: A relative gain of infinity is indicated.

Negative relative gains reveal conditional stability because numerator and denominator have opposite signs. A controller given the correct action for negative feedback when other loops are open will produce positive feedback when they are closed. If the action is reversed to handle the multiloop situation, however, it will be incorrect when other controllers are placed in manual or encounter constraints. The dynamic behavior of the loop is also poor, tending to exhibit inverse response and a very long period of oscillation. A relative gain greater than 1 is always accompanied by a negative number in the same row and column.

Interaction between loops having relative gains in the 0–1 range forms only negative feedback loops, but loop gain increases and so does the period of oscillation as λ decreases from 1 to 0. Interaction in the case of relative gains greater than 1 forms weak positive feedback loops having little effect on loop gain or period, but forming a very stiff system. The controllers tend to share disturbances, but recovery is very slow, as is setpoint response, as λ increases.

Controllers in interacting loops may require detuning when all are placed in automatic at once, depending on the relative gain values and on individual loop dynamics. In general, fast loops can upset slow loops, but not vice versa. The slow controller in an interaction between a slow and a fast loop effectively moves both manipulated variables because of the response of the fast controller to its output. As a consequence, its gain must be reduced in direct proportion to its relative gain (for relative gains less than 1). For loops having similar dynamics, the controller time constants also have to be increased as well; the detuning may be shared by the interacting controllers, either equally or unequally as desired. For cases of relative gains greater than 1, less detuning is required; a halving of the controller gain is the largest adjustment necessary, with little or no changes to the time constants.

Relative gains for the two composition loops on a distillation column can be calculated from the slopes of the operating curves:

$$\lambda_{y1} = 1 \Bigg/ \left(1 - \frac{(\partial y/\partial x)_1}{(\partial y/\partial x)_2}\right) \quad (13\text{-}7)$$

The calculated relative gain would appear in the upper left and lower right positions in the 2×2 array, with its complement appearing in the other two positions. If the two slopes have the same sign, then the resulting relative gain will be greater than 1 or less than 0. This would be the case for any combination of the curves in Fig. 13-2. Additionally, the closer together the slopes are, the higher is the degree of interaction. However, the combination of the material-balance line with any of the curves will cancel the negative sign in Equation 13-7 and produce a set of relative gains in the 0–1 region. Formulas for the slopes of all the curves are derived in Shinskey (1984, pages 147–153) for a saturated-liquid feed; the Distillation Consultant includes a correction for feed enthalpy in its calculations, as well.

As mentioned previously, columns making high-purity products will have their operating curves tightly grouped together. Equation 13-7 indicates that curves with very similar slopes will produce very high relative gains. This will limit the viable control-loop configurations for high-purity columns to those manipulating the material balance because its line is independent of the grouped curves. Relatively low-purity separations will have their operating curves spread apart, as those in Figure 13-2. The relative gains for those curves were calculated by the Distillation Consultant and are presented in Table 13-2.

Each of the nine relative gains reported in Table 13-2 represents a 2×2 relative gain array, being the number in the upper-left and lower-right corners of each array.

TABLE 13-2 Relative Gains for the Column of Figure 13-2

	m_1		
m_2	D	L	S
B	∞	0.412	0.487
V	0.613	10.09	2.982
V/B	0.640	5.098	2.457

The two remaining numbers in each array are of no interest because they are either negative or represent dynamically unfavorable pairings, and are therefore not reported by the Distillation Consultant. The pairings appearing in the table therefore are selected as being the most viable candidates for controlling top and bottom compositions simultaneously.

Those combinations using one of the product flows (B or D) as a manipulated variable have relative gains falling in the 0–1 range, whereas the other combinations give numbers greater than 1. Although the highest of the low numbers (0.640) is closer to 1.0 than the lowest of the high numbers (2.457), the latter relative gain is more favorable. Recognize that the range of the lower numbers is severely restricted in comparison to that of the others, and that their control loops are also slower. As a consequence, the best combinations for this column would have top composition controlled by reflux ratio (S) and bottom composition controlled by either boilup (V) or boilup ratio (V/B). (Although the curves in Figure 13-2 are labelled L/F, V/F, etc., the F can be dropped from the relative gain table because as an independent variable it has no effect on relative gain.)

13-2-3 Establishing a Performance Index

Relative gain only reveals one aspect of control-loop performance—that of loop interaction and detuning requirements. Of perhaps more importance, is the response of the product-quality loops to changes in load: feed rate and composition, and heat input and removal. Consider a control loop in a steady state with controlled variable c at setpoint r, at the moment a step change in load q is imposed. This would correspond to the condition at time 0 in Figure 13-3. The change in load then drives c away from r, causing the controller to move manipulated variable m in an effort to recover. Eventually, a new steady state is reached with m settling at a new value corresponding to the current load. The output of an ideal PID feedback controller would respond to the deviation e between c and r as

$$m_n = \frac{100}{P}\left(e + \frac{1}{\tau_I}\int_0^{t_n} e\,dt + \tau_D\frac{de}{dt}\right) + m_0 \tag{13-8}$$

where P = the proportional band expressed in percent
τ_I = integral time constant
τ_D = derivative time constant
t = time

Subscript n represents the conditions at the new steady state, at which time e and its derivative are both 0, so that only the integral term is meaningful. The change in m between the steady states is due entirely to the integration of e between time 0 and time n:

$$\Delta m = \frac{100}{P\tau_I}\int_0^{t_n} e\,dt \tag{13-9}$$

Solving for integrated error, we have

$$\int e\,dt = \Delta m\frac{P\tau_I}{100} \tag{13-10}$$

Integrated error should be minimized because it represents excess operating cost. Suppose the controlled variable in Figure 13-3 were to represent octane number of a gasoline blend, with the setpoint corresponding to the specification limit. The load change raises octane number above specifications, so that temporarily the system is

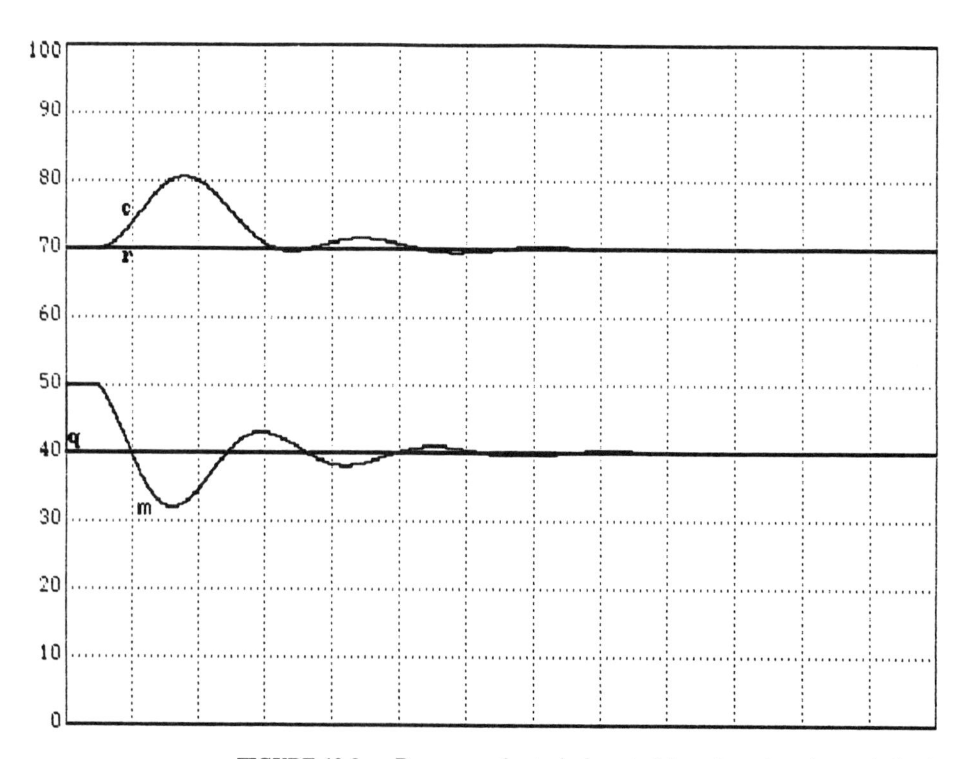

FIGURE 13-3. Response of a typical control loop to a step change in load.

giving away excess octane. This higher-octane product costs more to make, but its selling price remains the same because it is based on the specification. Meanwhile, blending continues at some liter per minute rate. Integrating the octane deviation against time and multiplying the result by the flow rate gives the number of octane-liters given away. This can be converted into a loss in revenue by multiplying by the incremental price difference of an octane number per liter of gasoline. This is the cost of poor control and the incentive to minimize integrated error.

Observe that there are three terms affecting integrated error: Δm, P, and τ_I. Of these, Δm is the only term that can be driven to 0, essentially by uncoupling the manipulated variable from the load. If a particular manipulated variable does not have to change to keep its controlled variable at setpoint in response to a particular load variable, the integrated error will be 0 following changes in that load. Should column feed rate change, all other flow rates must change proportional to it to keep compositions steady, but their flow *ratios* need not change. Consequently, any controller manipulating a flow rate such as reflux or distillate will have to change its output proportional to feed rate and integrate some error. However, if it is manipulating a flow ratio such as D/F or L/D, it would not have to change its output and therefore would not have to integrate any error.

The *load factor* μ for a manipulated variable is here defined as the fractional change in its value required to return product compositions to setpoint following a fractional load change:

$$\mu = \left.\frac{\partial m/m}{\partial q/q}\right|_{y,x} \tag{13-11}$$

Note that μ is dimensionless (like relative gain) and therefore independent of the scales of the variables. It can be calculated by differentiating the equations making up the process model. For example, the load factor for manipulating D in the face of a change in feed composition is found by differentiating Equation 13-3:

$$\left.\frac{\partial D/D}{\partial z_i/z_i}\right|_{y,x} = \frac{z_i}{z_i - x_i} = \frac{1}{1 - x_i/z_i} \tag{13-12}$$

TABLE 13-3 Load Factors for Composition Control

	Load		
m	Q_i	Q_o	z_i
D/F	0	0	$1/(1 - x_i/z_i)$
B/F	0	0	$1/(1 - y_i/z_i)$
L/F	0	-1	$\mu_{Dz} - 1$
V/F	-1	0	$\mu_{Dz} - 1/(1 + D/L)$
L/D	0	-1	-1
$D/(L + D)$	0	-1	$1/(1 + D/L)$
V/B	-1	0	$\mu_{Dz} - \mu_{Bz} - 1/(1 + D/L)$

Table 13-3 lists load factors for many selected manipulated variables in response to variations in heat input Q_i, heat removal Q_o, and feed composition z_i. All manipulated variables are given as flow ratios so that their load factor for feed rate changes is 0. Heat-input disturbances affect vapor rate directly and must be opposed by the controller manipulating V/F or V/B, hence the -1 in that column for those manipulated variables. Similarly, heat-removal upsets affect reflux subcooling and therefore internal-reflux flow, requiring the composition controller to adjust external-reflux flow in the opposite direction.

The term μ_{Dz} in the table is the load factor for D/F in response to z_i listed in the first row. This load factor and μ_{Bz} below it lie in the vicinity of 1.0 for many separations. As a result, that for reflux, $\mu_{Dz} - 1$, is often very small; in other words, reflux typically does not have to change much with feed composition, so that control loops manipulating reflux are not easily upset by feed-composition changes. For columns having high L/D ratios, the load factor for V/F in response to z_i is similarly small. Feed-composition load factors for the last three manipulated variables tend to be in the vicinity of 1.0. Conclusions to be drawn from Table 13-3 are that product-flow manipulation is insensitive to heat-balance disturbances, and reflux and boilup manipulation are relatively insensitive to feed-composition disturbances.

The remaining factors affecting integrated error are the proportional and integral settings of the controller. They should be minimized to minimize integrated error. However, a point is reached where further reduction in the value of either will produce undamped oscillations: This is the limit of stability. Tuning rules are available for controllers (Shinskey, 1988, pages 141–150) that provide stability while minimizing integrated absolute error (IAE). The latter is a preferable criterion to integrated error because minimizing it ensures stability as well as a minimal value of integrated error. The controller producing the response curve in Figure 13-3 was tuned for minimum IAE, and the controlled variable scarcely crosses setpoint at all, so that IAE is only slightly higher than integrated error.

Although the derivative time constant does not appear in the integrated-error expression, it contributes by allowing a reduction in the integral time constant, typically to half the value that produced minimum IAE without derivative action. Integral time should be set relative to the delay or deadtime in response of the controlled variable to the manipulated variable. It is a function of loop interaction only insofar as it should be increased for relative gains less than 1 when the interacting loops have similar dynamics.

The proportional band setting varies directly with the ratio of deadtime to dominant time constant, and with the steady-state process gain. The dominant composition time constant for a column is usually similar for all manipulated variables, but process gain is not. Process gain is included in the relative gain expression Equation 13-5, and therefore can be inferred from calculated values of relative gain. For example, consider the relative gain from Table 13-2 for compositions controlled by reflux and boilup:

$$\lambda_{xV} = \lambda_{yL} = \frac{(\partial x/\partial V)_L}{(\partial x/\partial V)_x} = 10.09 \quad (13\text{-}13)$$

If the top composition loop were then reconfigured to manipulate reflux ratio, a new relative gain would result:

$$\lambda_{xV} = \lambda_{yS} = \frac{(\partial x/\partial V)_S}{(\partial x/\partial V)_x} = 2.98 \quad (13\text{-}14)$$

Interestingly, reconfiguring the top loop results in a reduction in process gain for the bottom loop:

$$\frac{(\partial x/\partial V)_S}{(\partial x/\partial V)_L} = \frac{2.98}{10.09} = 0.295 \quad (13\text{-}15)$$

Similarly, reconfiguring the bottom loop can change the gain of the top loop. Consider changing the manipulated variable for bottom-composition control in the previous example from boilup V to boilup ratio V/B. This changes the process gain of the top loop as follows:

$$\frac{(\partial y/\partial L)_{V/B}}{(\partial y/\partial L)_V} = \frac{5.098}{10.09} = 0.505 \quad (13\text{-}16)$$

Because the proportional band needed by a controller for stability varies directly with the process gain, it follows that selecting the manipulated variables that present the lowest relative gain (of the numbers above 1.0) will allow the lowest setting. This has been demonstrated both in simulation and on real columns. Therefore, the relative gain has as much impact on integrated error as does the load factor.

The case of relative gains less than 1 is roughly the reciprocal of those greater than 1. Controllers in this situation must be detuned for multiloop stability, both in proportional and integral settings, inversely with relative gain. As a result, loops having relative gains less than 0.8 tend to show poor recovery from feed-composition disturbances, and are not recommended unless there are no selections having relative gains less than 5.

The Distillation Consultant estimates integrated errors for each of the combinations in Table 13-2, relative to that of the reflux–boilup system at 100 in response to heat-balance disturbances. For simplicity, the integrated errors for the two composition loops are combined into a single number, and the effects of heat-input and heat-removal disturbances are similarly combined. The results appear in Table 13-4 for the column whose relative gains are given in Table 13-2.

Observe that the reflux–boilup system gives the highest integrated error for heat-balance disturbances and the lowest error for feed-composition disturbances, whereas material-balance configurations (manipulating either B or D) have the opposite behavior. In this situation, the preferred configuration is usually a compromise, manipulating reflux ratio (S) at the top and boilup or boilup ratio at the bottom. These systems combine good response to both disturbances with favorable relative gains. A high-purity column like an ethylene fractionator or butane splitter, however, may have no relative gains between 0.7 and 20, so that the best choice will be a material-balance system. These columns, because of high reflux ratios, also tend to be very sensitive to heat-balance upsets for all but material-balance systems.

The Distillation Consultant selects that combination that gives the best overall performance estimate (in this example the reflux-ratio–boilup-ratio configuration) with second place given to the reflux-ratio–boilup system. The user is, of course, free to override the selection based on preference or conditions not considered by the Consultant.

TABLE 13-4 Integrated Error Estimates

	m_1		
m_2	D	L	S
Heat-Balance Disturbances			
B	...	12	10
V	8	100	15
V/B	8	51	12
Feed-Composition Disturbances			
B	...	51	72
V	42	15	15
V/B	49	15	16

13-2-4 Applications and Objectives

One aspect of distillation that makes control-system selection so difficult is the wide variety of separations that can be encountered from industry to industry, and even within a single industry. Petroleum refining is the most extensive user of distillation, followed closely by the chemical industry. But distillation is also used in beverage production, tall-oil separation from wood-pulping wastes, fatty-acid separation for soap manufacture, and even in the separation of the isotopes of boron for the nuclear industry.

In petroleum refining, we encounter the multiproduct distillation of crude oil, lubricants, and cracked oils, each of which is a specialized operation. Light-hydrocarbon separation is the most common distillation operation, where a mixture of gases or liquids will pass through a series of columns with progressively heavier fractions distilled in each. These columns are named for the light-key component removed: demethanizer, deethanizer, depropanizer, debutanizer, deisobutanizer, and so forth.

Even identically named columns may not have identical functions, however. For example, depropanizers appear in many different parts of a refinery, from crude-oil fractionation to the fluid-catalytic cracking unit to the alkylation unit and the ethylene plant, and each may have a different objective. Feedstocks differ, and so do product values, depending on the presence of olefins and isomers, for example. As a result, a control system that performs satisfactorily on one depropanizer will not necessarily be best for another.

Sources of heating and cooling impact the control system as much as feedstock does. Hot-oil heating should be manipulated differently from steam and direct-firing. Similarly, air-cooled condensers control differently from water-cooled and refrigerated condensers. Heat-pumped operation is still different, and heat integration with other columns imposes further restrictions on operation.

The separation of azeotropic (constant-boiling-point) mixtures involves so much additional technology in terms of solubility limits and nonideal behavior that no attempt has been made to include it in the Distillation Consultant. Similarly, multiple-sidestream columns are too specialized, and heat-pumped columns too uncommon to be included.

Economic objectives offer another variation among otherwise similar columns. In some cases, economic performance will be maximized by maximizing the recovery of the most valuable product from the column, which could be either the top, sidestream, or bottom product. This will usually be the product having the largest flow rate, but not always, because some columns are located within recycle loops around reactors. In situations where the products have similar values, minimizing energy consumption may be the greatest economic incentive. In still other cases, energy could be free, as waste heat from some other operation, which will affect the incentives and therefore must be considered in configuring the controls.

Sidestream columns bring another dimension to the picture: the sidestream could be liquid or vapor, above or below the feed, and a major or minor product. Constraints are also important considerations. The reboiler, condenser, or column itself could be the constraining element to traffic, and the controls must accommodate this to hold product on specification while maximizing throughput. Constraints can also shift with weather conditions and feed composition. Condenser and reboiler design also must be considered because some control schemes will not work properly with some equipment configurations. The Distillation Consultant does take most of these variations into account in choosing a control scheme.

13-3 RULE BASE FOR SIMPLE COLUMNS

The selection made from relative-gain and integrated-error estimates must be converted into a block diagram suitable for indicating functions and connections. Manipulation of reflux ratio, for example, may be implemented in several different ways, each of which may have advantages and disadvantages from the standpoint of accuracy, dynamic responsiveness, ease of operation, and constraint handling. The Distillation Consultant has a logical rule base that satisfies the requirement for manipulating reflux ratio with the dynamic, operational, and other requirements of a particular column. These incorporate the experience and judgment of the author and others to provide a control system that will satisfy all the stated objectives and constraints, and perform with a high degree of confidence.

Because sidestream columns have an added dimension, their rule base is much more complex than that for simple columns. Therefore, simple (two-product) columns are examined first, with additional rules for sidestream columns considered later.

13-3-1 Economic Objective

The purpose of any industrial operation is to maximize revenue. Given fixed selling prices, which are a function of product specifications, this goal is achieved by minimizing operating costs. For a distillation column, operating costs include product devaluation, consumption of utilities, and labor. Product devaluation consists in allowing quantities of a high-value product to go unrecovered, leaving the process with a lower-value stream, assuming its selling price. To increase recovery requires an increase in utilities consumption, principally heating and cooling. Therefore, each separation must be evaluated economically to establish priorities for the control system.

In some cases, an optimum will exist where an incremental decrease in product devaluation is offset by an increase in utilities cost. Typically this optimum is sensitive to feed composition as well as product values and utility costs, and requires frequent recalculation. The optimum operating point is identified as specific compositions of the two products, which therefore requires them both to be controlled. The more valuable product must be held at its specification limits, while the less valuable product will be purer than specification limits to improve the recovery of the other consistent with additional consumption of energy. If there is a large value difference between them, the optimum point may be beyond the upper constraint for heating or cooling, in which case the column must operate at the nearest constraint. If there is little value difference between the products, then minimum cost will be reached by minimizing utility consumption, in which case both product compositions should be controlled at specification limits.

In the case of a large product-value difference then, or of free heating and cooling, only the quality of the more valuable product is controlled. Any other case requires control of both. This is an important distinction in configuring the control system, because in the latter case, composition-loop interaction must be considered, but not in the former. These two situations are now presented separately.

13-3-2 Maximizing Recovery

To maximize recovery of a valuable product requires the column's separation factor to be maximized. This is achieved by operating at the limit of allowable reflux or boilup, depending on which constraint is encountered first. The constraint could be at the condenser, the reboiler, or a flooding limit in the column itself, and it could change from one to another with the weather or other factors.

If vapor or reflux flow is fixed, then changing boilup ratio or reflux ratio amounts to changing the product flow itself. As a consequence, the Distillation Consultant recommends manipulating one of the product flows directly from the composition controller, with either reflux or boilup (whichever is unconstrained) being manipulated indirectly in the opposite direction. This latter link improves the speed of the composition loop, whereas the direct manipulation of the product flow provides accuracy and immunity from energy-balance upsets.

Because the smaller of the two product flow rates can be measured with higher absolute accuracy, it should be the one manipulated by the composition controller. The other product is then placed under liquid-level control to close the external material balance. The internal balance is closed by the other level controller manipulating the unconstrained internal flow. If distillate flow is selected for composition control, reflux should be under accumulator-level control and boilup should be used to control column differential pressure at its constraint. Figure 13-4 was prepared by the Distillation Consultant to illustrate this configuration. Note the feedforward input from feed rate, passing through a dynamic compensator (t) and a nonlinear function block (f). The latter compensates for variations in separation with feed rate. The D/F ratio changes with feed composition; therefore multiplication of the output of the composition controller by feed rate is needed to apply the correct change to distillate flow under all conditions of feed composition. Reflux flow is the difference between the output of the level controller and distillate flow. Thus,

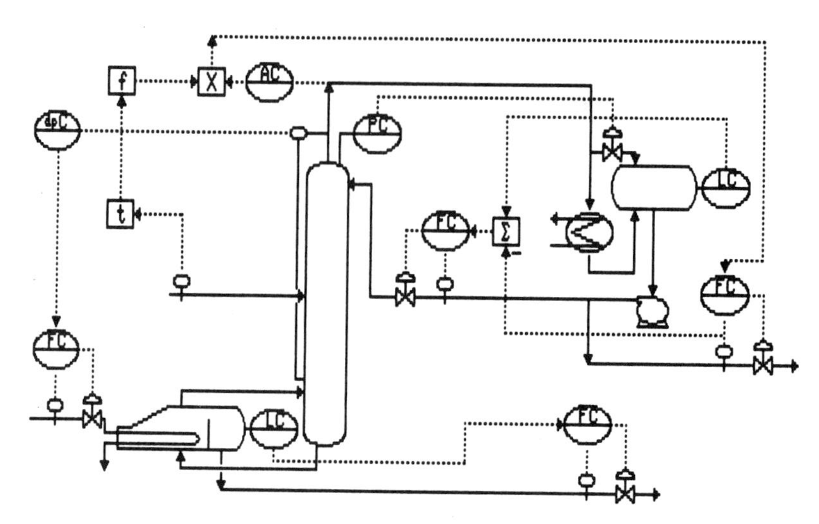

FIGURE 13-4. A column whose controls are configured to maximize recovery of the distillate product.

changes in distillate flow are imposed on reflux to avoid the lag that would otherwise result if the level controller had to do it alone.

Other features of the figure include a submerged water-cooled condenser with pressure controlled by a hot-vapor bypass valve, and a kettle reboiler whose heat input controls column differential pressure. Note that the analyzer samples the overhead vapor rather than the liquid product because its dynamic response avoids the lag of the accumulator; with a total condenser, there will be no steady-state difference between the two compositions.

If bottom composition is to be controlled by manipulating its own flow, base level must be controlled by manipulating heat input. The latter loop may not be successful owing to the possibility of inverse response of level to heat input; this is particularly common when the reboiler is fired. Then an alternative structure must be found. Bottom composition may be controlled by manipulating distillate flow, if heat input can be used to control accumulator level; this loop is successful where there is a narrow temperature gradient across the column, so that most of the heat goes into changing vapor flow rather than raising temperature. Distillate composition can also be controlled by manipulating bottom-product flow, if base level can be controlled by heat input and the temperature gradient is low.

13-3-3 Controlling Both Product Compositions

When economics dictate that both compositions must be controlled, the Distillation Consultant selects the best combination of manipulated variables as described in Section 13-2-3. The selection is then converted to control loops consistent with configurations proven in the field. For example, Ryskamp developed a reflux-ratio configuration (Ryskamp, 1980) that proved to be responsive over a wide range of operating conditions and is included in the recommendations of the Distillation Consultant. The author later applied the same concept to boilup-ratio manipulation and included it in the Consultant. A system illustrating both of those configurations is shown in Figure 13-5.

The reflux ratio is actually manipulated as $D/(L + D)$ rather than L/D, and boilup ratio as $B/(V + B)$ rather than V/B. Again, these have been found to give improved

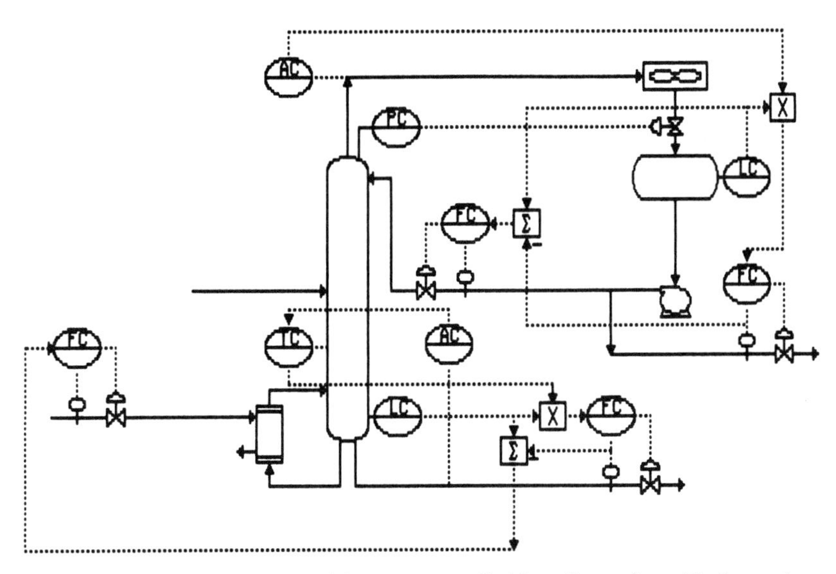

FIGURE 13-5. Product compositions are controlled by reflux ratio and boilup ratio to minimize interaction.

dynamic response and also decouple the level and composition loops. Notice that the lower composition loop sets column temperature in cascade. This configuration improves response to feed composition changes. No feedforward is required to this system because both composition controllers already manipulate flow ratios. (Feedforward is only indicated when a product flow, reflux, or boilup is selected as a manipulated variable.) An overhead air-cooled condenser is shown, with its drain valve manipulated for pressure control.

The Distillation Consultant can also accommodate a partial condenser, a flooded accumulator, several reboiler types as well as direct steam injection, and so forth. With all these possibilities, it can produce more than 30,000 different diagrams, each representing a solution to a potential problem.

13-3-4 Floating Pressure Control

The manipulated variable for column pressure control is selected as a function of the type and location of the condenser because it is the means for adjusting heat removal. Two different examples were previously given. Having controlled pressure, it is often desirable to minimize its value because in most separations relative volatility is improved with falling pressure. This enables energy to be saved per unit of product during periods when plenty of cooling is available. Capacity increases are also possible for light-hydrocarbon separations.

Floating-pressure control applies to those separations that are conducted above atmospheric pressure and ambient temperature. Whether energy savings will be significant depends primarily on the variation likely to be experienced in the cooling medium. Air-cooled condensers experience more temperature variation than those that are water-cooled, and temperate installations show more diurnal, weather-related, and seasonal variations than tropical installations.

To implement a floating-pressure system, a controller sensing the condenser constraint needs to adjust the pressure setpoint. In the case of the hot-vapor bypass valve in Figure 13-4, the constraint is the closed position of the valve. A valve-position controller (VPC) is connected to the output of the pressure controller and set to hold the bypass valve 10% open in the steady state by adjusting the pressure setpoint. The VPC should have integral-only action and be tuned to be slow enough to avoid upsetting composition loops when cooling changes take place, as in a storm. Speed of response may be improved by a feedforward loop, which adjusts the ratio of heat input to feed as a function of pressure (Shinskey, 1988, pages 431–433).

In the case where the valve position does not indicate condenser loading, as when reflux is manipulated to control pressure, another measure must be used, for example, liquid level in the condenser. Then the condenser level controller sets the pressure setpoint. If column temperature is used to control composition, pressure compensation is needed for that temperature measurement to continue to infer and control composition. The Distillation Consultant includes several means for implementing floating-pressure control, including the placement of feedforward loops and application of compensation to temperature measurements.

13-4 RULE BASE FOR SIDESTREAM COLUMNS

To apply the methods developed for simple columns to sidestream columns requires their classification into functional groups. A sidestream column will have most of its stages dedicated to separating its light- and heavy-key components; this much of it resembles a simple column. The third stream is intended to remove a troubling third component: either a lighter-than-light or heavier-than-heavy or a middle key. That third stream is usually the smallest of the three products, and should be treated as a special purge stream apart from the two major products.

13-4-1 Classifications

A Class I column is defined here as one whose smallest product is the distillate. The sidestream is normally removed as a liquid above the feed point, and its light impurity is controlled by the flow of the distillate product. A typical example of a Class I column is the ethylene fractionator, whose sidestream is a high-purity ethylene product, separated from the ethane bottom product at high recovery; methane is removed from the condenser to limit its content in the ethylene. Controls over the key components in the sidestream and bottoms should be selected on the basis of relative gains because this tends to be a very interacting system. However, control over methane in the ethylene product is achieved by manipulating the flow of distillate.

Class II columns have the opposite arrangement: bottom product is the smallest stream and the sidestream is taken as a vapor below the feed. Again, control over the light and heavy keys should be assigned as for a simple column, with bottom flow manipulated to limit the heavier-than-heavy key in the sidestream.

Class III columns have the sidestream as their smallest flow, which may be withdrawn either above or below the feed, depending on the boiling point of the middle key. If not withdrawn, this middle-boiling component can accumulate until it interferes with separation of the major products or, as in the case of fusel oil, reaches a solubility limit and disrupts vapor pressure by creating a second liquid phase. Again, the smallest stream should be manipulated to control the content of its principal constituent in one or the other product.

13-4-2 Configuration Rules

The configuration rules begin with the same consideration as for simple columns—the economics of the separation. They turn out to be essentially the same as for simple columns, the principal distinction being that the smallest stream has no significant economic contribution. However, where the Distillation Consultant posed but one economic decision for the simple column, it must formulate three for the sidestream column—one for each class.

The smallest flow is then arbitrarily assigned to control composition, either of itself or the adjacent major product as desired. Relative gain analysis is used to select the other two composition loops, if economics dictate that they both require control; otherwise, material-balance control is recommended, with feedforward from feed rate.

Sidestream columns present special problems relative to inventory controls. Assignment of a sidestream to control its own composition usually works quite well, but it cannot be assigned to level control without some help. Consider, for example, a Class II column whose largest flow is its vapor sidestream. The largest flow in any process should always be under level control, but the flow of vapor has no direct influence on liquid level. The assignment of base level controlled by sidestream vapor will work, however, if column differential pressure is controlled by heat input (Shinskey, 1984, page 252).

A similar problem appears with a Class I column whose bottom flow is relatively small. Base level control is assigned to heat input and accumulator level control is assigned to reflux. There must be a tie, however, between sidestream and reflux flow to close the material balance, or control of both levels will be lost. Figure 13-6 shows an ethylene fractionator having bottom-product composition controlled by bottom flow, which fits this case. Reflux ratio is manipulated to control the ethane content of the ethylene sidestream. The output of the dynamic compensator acting on the reflux flow signal represents what that flow would be approaching the withdrawal point several trays from the top of the column. Note that a single analyzer on the sidestream reports impurity levels of both the heavy (H) impurity ethane and the light (L) impurity methane, which are both controlled. Methane is withdrawn overhead as a vapor product, so that pressure must be controlled by manipulating the flow of refrigerant to the condenser.

The rule base for sidestream-column configuration is about three times as complex as that for simple columns. The configuration program is capable of drawing about 300,000 different diagrams for

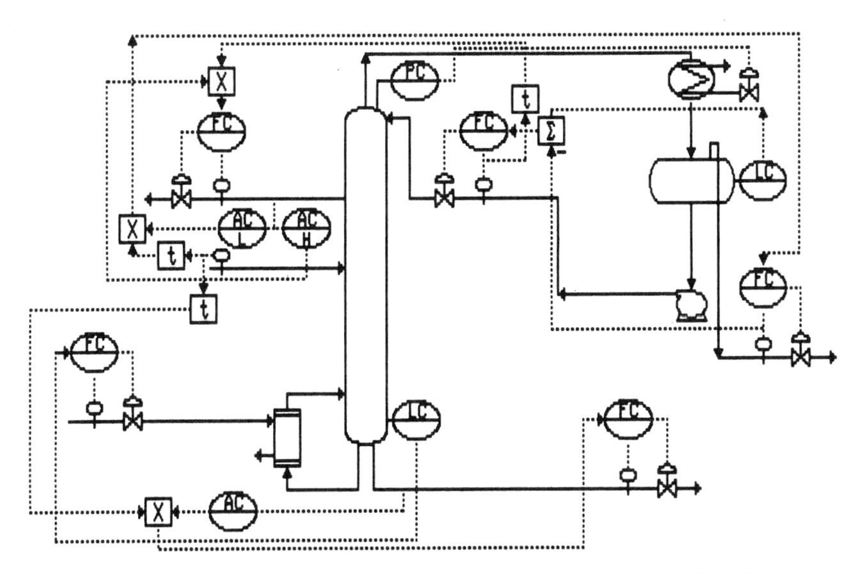

FIGURE 13-6. Both the ethane and methane content are controlled in the sidestream from this ethylene fractionator. (*Note:* Crossing lines do not intersect.)

sidestream columns with their controls. This is about as complex a problem as can be solved using this method, and no consideration has been given to extending it to more streams or less-ideal systems.

13-5 NOMENCLATURE

B	Bottom-product flow
c	Controlled variable
D	Distillate-product flow
d	Derivative
e	Deviation from setpoint
F	Feed flow
L	Reflux flow
m	Manipulated variable
P	Proportional band
Q_i	Heat inflow
Q_o	Heat outflow
q	Load
r	Setpoint
S	Separation
t	Time
V	Vapor boilup
x	Bottom composition
y	Distillate composition
z	Feed composition
∂	Partial derivative
Δ	Difference
λ	Relative gain
Λ	Relative gain array
μ	Load factor
τ_D	Derivative time constant
τ_I	Integral time constant

References

Ayral, T. E. (1989). On-line expert system for process control. *Hydrocarbon Proc.* 68, 61–63.

Babb, M. (1990). New expert system guides control operators. *Control Eng.* 37, 80–81.

Finco, M. V., Luyben, W. L., and Polleck, R. E. (1989). Control of distillation columns with low relative volatilities. *I & EC Res.* 28, 75–83.

Kraus, T. W. and Myron, T. J. (1984). Self-tuning PID controller uses pattern recognition approach. *Control Eng.* 31.

Leung, K. S. and Lam, W. (1988). Fuzzy concepts in expert systems. *Computer* 21, 43–56.

Ryskamp, C. J. (1980). New strategy improves dual-composition control. *Hydrocarbon Processing* 59.

Shinskey, F. G. (1984). *Distillation Control*, 2nd ed. New York: McGraw-Hill.

Shinskey, F. G. (1988). *Process Control Systems*, 3rd ed. New York: McGraw-Hill.

Shirley, R. S. (1987). Status report 3 with lessons: An expert system to aid process control. Paper given at IEEE Annual Pulp and Paper Industrial Technical Conference, 1–5 June 1987.

14

Robust Control

Sigurd Skogestad

University of Trondheim

The objective of this chapter is to give the reader a basic knowledge of how robustness problems arise and what tools are available to identify and avoid them.

We first discuss possible sources of model uncertainty and look at the traditional methods for obtaining robust designs, such as gain margin, phase margin, and maximum peak criterion (M circles). However, these measures are difficult to generalize to multivariable systems.

As an introductory example to robustness problems in multivariable systems we then discuss two-point control of distillation columns using the LV configuration. Because of strong interactions in the plant, a decoupler is extremely sensitive to input gain uncertainty (caused by actuator uncertainty). These interactions are analyzed using singular value decomposition (SVD) and RGA analysis. We show that plants with large RGA elements are fundamentally difficult to control and that decouplers should not be used for such plants. It is shown that other configurations may be less sensitive to model uncertainty.

At the end of the chapter we look at uncertainty modelling in terms of norm bounded perturbations (Δs). It is shown that the structured singular value μ is a very powerful tool to analyze the robust stability and performance of multivariable control systems.

14-1 ROBUSTNESS AND UNCERTAINTY

A control system is robust if it is insensitive to differences between the actual system and the model of the system that was used to design the controller. Robustness problems are usually attributed to differences between the plant model and the actual plant (usually called model–plant mismatch or simply model uncertainty). Uncertainty in the plant model may have several origins:

1. There are always parameters in the linear model that are only known approximately or are simply in error.
2. Measurement devices have imperfections. This may give rise to uncertainty of the manipulated *inputs* in a distillation column because they are usually measured and adjusted in a cascade manner. In other cases, limited valve resolution may cause input uncertainty.
3. At high frequencies even the structure and the model order is unknown, and the

uncertainty will exceed 100% at some frequency.
4. The parameters in the linear model may vary due to nonlinearities or changes in the operating conditions.

Other considerations for robustness include measurement and actuator failures, constraints, changes in control objectives, opening or closing other loops, and so on. Furthermore, if a control design is based on an optimization, then robustness problems may also be caused by the mathematical objective function, that is, how well this function describes the real control problem.

In the somewhat narrow use of the term used in this chapter, we shall consider robustness with respect to model uncertainty and shall assume that a fixed (linear) controller is used. Intuitively, to be able to cope with large changes in the process, this controller has to be detuned with respect to the best response we might achieve when the process model is exact.

To consider the effect of model uncertainty, it needs to be quantified. There are several ways of doing this. One powerful method is the frequency domain (so-called H-infinity uncertainty description) in terms of norm-bounded perturbations (Δs). With this approach one also can take into account unknown or neglected high-frequency dynamics. This approach is discussed toward the end of this chapter. Readers who want to learn more about these methods than we can cover may consult the books by Maciejowski (1989) and Morari and Zafiriou (1989).

The following terms will be useful:

Nominal stability (NS). The system is stable with no model uncertainty.

Nominal performance (NP). The system satisfies the performance specifications with no model uncertainty.

Robust stability (RS). The system also is stable for the worst-case model uncertainty.

Robust performance (RP). The system also satisfies the performance specifications for the worst-case model uncertainty.

14-2 TRADITIONAL METHODS FOR DEALING WITH MODEL UNCERTAINTY

14-2-1 Single-Input–Single-Output Systems

For single-input–single-output (SISO) systems one has traditionally used gain margin (GM) and phase margin (PM) to avoid problems with model uncertainty. Consider a system with open-loop transfer function $g(s)c(s)$, and let $gc(j\omega)$ denote the frequency response. The GM tells by what factor the loop gain $|gc(j\omega)|$ may be increased before the system becomes unstable. The GM is thus a direct safeguard against steady-state gain uncertainty (error). Typically we require GM > 1.5. The phase margin tells how much negative phase we can add to $gc(s)$ before the system becomes unstable. The PM is a direct safeguard against time delay uncertainty: If the system has a crossover frequency equal to ω_c [defined as $|gc(j\omega_c)| = 1$], then the system becomes unstable if we add a time delay of $\theta = \text{PM}/\omega_c$. For example, if PM = 30° and $\omega_c = 1$ rad/min, then the allowed time delay error is $\theta = (30/57.3)$ rad/1 rad/min = 0.52 min. It is important to note that decreasing the value of ω_c (lower closed-loop bandwidth, slower response) means that we can tolerate larger deadtime errors. For example, if we design the controller such that PM = 30° and expect a deadtime error up to 2 min, then we must design the control system such that $\omega_c < \text{PM}/\theta = (30/57.3)/2 = 0.26$ rad/min, that is, the closed-loop time constant should be larger than 1/0.26 = 3.8 min.

Maximum Peak Criterion

In practice, we do not have pure gain and phase errors. For example, in a distillation column the time constant will usually increase when the steady-state gain increases. A more general way to specify stability margins is to require the Nyquist locus of $gc(j\omega)$ to stay outside some region of the -1 point (the "critical point") in the complex plane. Usually this is done by considering the maximum peak M_t of the closed-loop transfer function T:

$$M_t = \max_{\omega} |T(j\omega)|, \qquad T = gc(1 + gc)^{-1} \tag{14-1}$$

The reader may be familiar with M circles drawn in the Nyquist plot or in the Nichols chart. Typically, we require $M_t = 2$. There is a close relationship between M_t and PM and GM. Specifically, for a given M_t we are guaranteed

$$\mathrm{GM} \geq 1 + \frac{1}{M_t},$$
$$\mathrm{PM} \geq 2\arcsin\left(\frac{1}{2M_t}\right) \geq \frac{1}{M_t} \text{ rad} \tag{14-2}$$

For example, with $M_t = 2$ we have GM > 1.5 and PM > 29.0° > $1/M_t$ rad = 0.5 rad.

Comment The peak value M_s of the sensitivity function $S = (1 + gc)^{-1}$ may be used as an alternative robustness measure. $1/M_s$ is simply the minimum distance between $gc(j\omega)$ and the -1 point. In most cases the values of M_t and M_s are closely related, but for some "strange" systems it may be safer to specify M_t rather than M_s. For a given value of M_s we are guaranteed GM > $M_s/(M_s - 1)$ and PM > $2\arcsin(1/2M_s) > 1/M_s$.

14-2-2 Multi-input–Multi-output Systems

The traditional method of dealing with robustness for multi-input–multi-output (MIMO) systems (e.g., within the framework of "optimal control," linear quadratic Gaussian (LQG), etc.) has been to introduce uncertain signals (noise and disturbances). One particular approach is the loop transfer recovery (LTR) method where unrealistic noise is added specifically to obtain a robust controller design. One may say that model uncertainty generates some sort of disturbance. However, this disturbance depends on the other signals in the systems, and thus introduces an element of feedback. Therefore, there is a fundamental difference between these sources of uncertainty (at least for linear systems): Model uncertainty may introduce instability, whereas signal uncertainty may not.

For SISO systems the main tool for robustness analysis has been GM and PM, and as previously noted, these measures are related to specific sources of model uncertainty. However, it is difficult to generalize GM and PM to MIMO systems. On the other hand, the maximum peak criterion may be generalized easily. The most common generalization is to replace the absolute value by the maximum singular value, for example, by considering

$$M_t = \max_{\omega} \bar{\sigma}(T(j\omega)),$$
$$T = GC(I + GC)^{-1} \tag{14-3}$$

The largest singular value is a scalar positive number, which at each frequency measures the magnitude of the matrix T. As shown later, this approach has a direct relationship to important model uncertainty descriptions and is used in this chapter.

Comment In Chapter 11, a different generalization of M_t to multivariable systems is used: First introduce the scalar function $W(j\omega) = \det(I + GC(j\omega)) - 1$ [for SISO systems $W(j\omega) = GC(j\omega)$] and then define $L_c = |W/(1 + W)|$. The maximum peak of $|L_c|$ (in decibels) is denoted $L_c^{\max}$ and is used as part of the biggest log-modulus tuning (BLT) method. For SISO systems $L_c^{\max} = M_t$.

Even though we may easily generalize the maximum peak criterion (e.g., $M_t \leq 2$) to multivariable systems, it is often not useful for the following three reasons:

1. In contrast to the SISO case, it may be insufficient to look at only the transfer function T. Specifically, for SISO systems $GC = CG$, but this does not hold for MIMO systems. This means that although the peak of T [in terms of $\bar{\sigma}(T(j\omega))$] is low, the peak of $T_I = CG(I + CG)^{-1}$ may be large. (We will see later that the transfer function T is related to relative uncertainty at the output of the plant, and T_I at the input of the plant.)
2. The singular value may be a poor generalization of the absolute value. There may be cases where the maximum peak criterion, for example, in terms of $\bar{\sigma}(T)$, is *not* satisfied, but in reality the system may be robustly stable. The reason is that the uncertainty generally has "structure," whereas the use of the singular value assumes unstructured uncertainty. As we will show, one should rather use the structured singular value, that is, $\mu(T)$.
3. In contrast to the SISO case, the response with model error may be poor (RP not satisfied), even though the stability margins are good (RS is satisfied) and the response without model error is good (NP satisfied).

In the next section we give a multivariable example where the maximum peak criterion is easily satisfied using a decoupling controller [in fact, we have $GC(s) = CG(s) = 0.7/sI$, and the values of M_t and M_s are both 1]. Yet, the response with only 20% gain error in each input channel is extremely poor. To handle such effects, in general, one has to define the model uncertainty and compute the structured singular value for RP.

The conclusion of this section is that most of the tools developed for SISO systems, and also their direct generalizations such as the peak criterion, are not sufficient for MIMO systems.

14-3 A MULTIVARIABLE SIMULATION EXAMPLE

This idealized distillation column example will introduce the reader to the deteriorating effect of model uncertainty, in particular for multivariable plants. The example is taken mainly from Skogestad, Morari, and Doyle (1988).

14-3-1 Analysis of the Model

We consider two-point (dual) composition control with the LV configuration as shown in Figure 14-1a. The overhead composition of a distillation column is to be controlled at $y_D = 0.99$ and the bottom composition at $x_B = 0.01$, with reflux L and boilup V as manipulated inputs for composition control, that is,

$$y = \begin{pmatrix} \Delta y_D \\ \Delta x_B \end{pmatrix}, \qquad u = \begin{pmatrix} \Delta L \\ \Delta V \end{pmatrix}$$

This choice is often made because L and V have an immediate effect on the product compositions. By linearizing the steady-state model and assuming that the dynamics may be approximated by first-order response with time constant $\tau = 75$ min, we derive the linear model in terms of deviation variables

$$\begin{pmatrix} \Delta y_D \\ \Delta x_B \end{pmatrix} = G^{LV}\begin{pmatrix} \Delta L \\ \Delta V \end{pmatrix},$$

$$G^{LV}(s) = \frac{1}{\tau s + 1}\begin{pmatrix} 0.878 & -0.864 \\ 1.082 & -1.096 \end{pmatrix} \tag{14-4}$$

Here we have normalized the flows such that the feed rate $F = 1$. This is admittedly a poor model of a distillation column. Specifically, (a) the same time constant τ is used both for external and internal flow changes, (b) there should be a high-order lag in the transfer function from L to x_B to represent the liquid flow down to the column, and (c) higher-order composition dynamics should also be included. However,

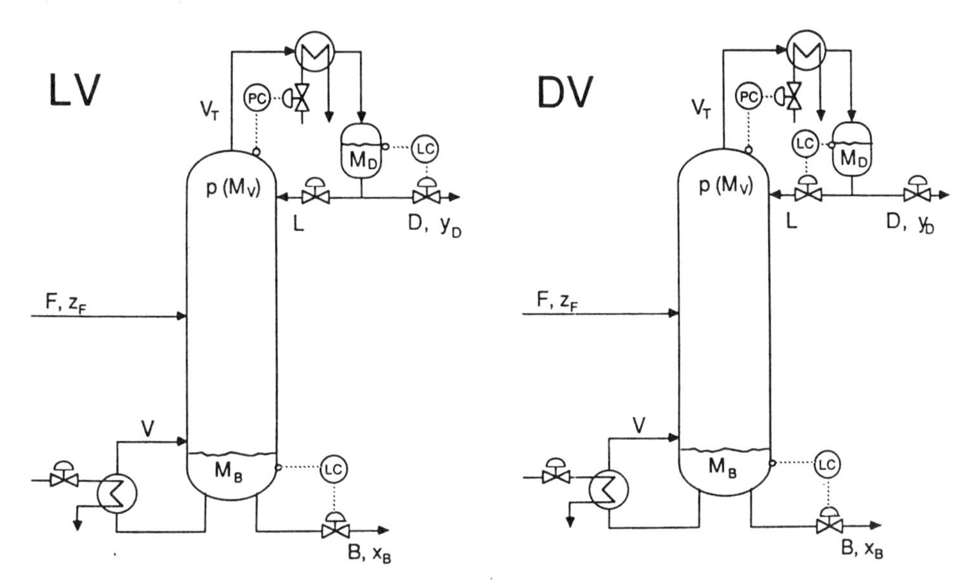

FIGURE 14-1. Control of distillation column with LV and DV configurations.

the model is simple and displays important features of the distillation column behavior. The RGA matrix for this model is at all frequencies:

$$\text{RGA}(G^{LV}) = \begin{pmatrix} 35.1 & -36.1 \\ -36.1 & 35.1 \end{pmatrix} \quad (14\text{-}5)$$

The large elements in this matrix indicate that this process is fundamentally difficult to control.

Interactions and Ill-Conditionedness

Consider the case with no composition control. The effect on top composition of a small change in reflux L with V constant is

$$\Delta y_D(s) = \frac{0.878}{75s + 1} \Delta L(s)$$

If we increase L by only 0.01 (that is, L/F is increased 0.4% from 2.7 to 2.701), then we see that the steady-state increase in y_D predicted from this linear model is 0.00878 (that is, y_D increases from 0.99 to 0.99878). This is a rather drastic change, and the reason is that the column operation is very dependent on keeping the correct product split D/F (with V constant the increase in L yields a corresponding decrease in D); that is, the column is very sensitive to changes in the *external flows* D and B.

Similarly, if we increase V by only 0.01 (with L constant), we see that the predicted steady-state change in y_D is -0.00864. Again, this is a very large change, but in the direction opposite that for the increase in L.

We therefore see that changes in L and V counteract each other, and if we increase L and V simultaneously by 0.01, then the overall steady-state change in y_D is only $0.00878 - 0.00864 = 0.00014$. The reason for this small change is that the compositions in the column are only weakly dependent on changes in the *internal flows* (i.e., changes in the internal flows L and V with the external flows D and B constant).

Summary Because both L and V affect both compositions y_D and x_B, we say that the process is *interactive*. Furthermore, the process is "ill-conditioned," that is, some combinations of ΔL and ΔV (corresponding to changing external flows) have a strong effect on the compositions, whereas other combinations of ΔL and ΔV (corresponding to changing internal flows) have a weak effect on the compositions. The condition number, which is the ratio between the gains in the strong and weak directions, is therefore large for this process (as seen in the following text, it is 141.7).

Singular Value Analysis of the Model
The preceding discussion shows that this column is an ill-conditioned plant, where the effect (the gain) of the inputs on the outputs depends strongly on the *direction* of the inputs. To see this better, consider the SVD of the steady-state gain matrix

$$G = U \Sigma V^T \tag{14-6}$$

or equivalently, because $V^T = V^{-1}$,

$$G\bar{v} = \bar{\sigma}(G)\bar{u}, \qquad G\underline{v} = \underline{\sigma}(G)\underline{u}$$

where $\bar{\sigma}$ denotes the maximum singular value and $\underline{\sigma}$ the minimum singular value. The singular values are

$$\Sigma = \text{diag}\{\bar{\sigma}, \underline{\sigma}\} = \text{diag}\{1.972, 0.0139\}$$

The output singular vectors are

$$V = (\bar{v}\ \underline{v}) = \begin{pmatrix} 0.707 & 0.708 \\ -0.708 & 0.707 \end{pmatrix}$$

The input singular vectors are

$$U = (\bar{u}\ \underline{u}) = \begin{pmatrix} 0.625 & 0.781 \\ 0.781 & -0.625 \end{pmatrix}$$

The large plant gain, $\bar{\sigma}(G) = 1.972$, is obtained when the inputs are in the direction $\begin{pmatrix} \Delta L \\ \Delta V \end{pmatrix} = \bar{v} = \begin{pmatrix} 0.707 \\ -0.708 \end{pmatrix}$, i.e., an increase in ΔL and a simutaneous decrease in ΔV. Because $\Delta B = -\Delta D = \Delta L - \Delta V$ (assuming constant molar flows and constant feed rate), this physically corresponds to the largest possible change in the *external* flows D and B. From the direction of the output vector $\bar{u} = \begin{pmatrix} 0.625 \\ 0.781 \end{pmatrix}$, we see that this change causes the outputs to move in the same direction, that is, it mainly affects the average composition $(y_D + x_B)/2$. All columns with both products of high purity are sensitive to changes in the external flows because the product rate D has to be about equal to the amount of light component in the feed. Any imbalance leads to large changes in product compositions (Shinskey, 1984).

The low plant gain, $\underline{\sigma}(G) = 0.0139$, is obtained for inputs in the direction $\begin{pmatrix} \Delta L \\ \Delta V \end{pmatrix} = \underline{v} = \begin{pmatrix} 0.708 \\ 0.707 \end{pmatrix}$ which corresponds to changing the *internal* flows only ($\Delta B = \Delta L - \Delta V \approx 0$). From the output vector $\underline{u} = \begin{pmatrix} 0.781 \\ -0.625 \end{pmatrix}$ we see that the effect is to move the outputs in different directions, that is, to change $y_D - x_B$. Thus, it takes a large control action to move the compositions in different directions, that is, to make both products purer simultaneously. The condition number of the plant, which is the ratio of the high and low plant gain, is then

$$\gamma(G) = \bar{\sigma}(G)/\underline{\sigma}(G) = 141.7 \tag{14-7}$$

The RGA is another indicator of ill-conditionedness, which is generally better than the condition number because it is scaling independent. The sum of the absolute value of the elements in the RGA (denoted $\|\text{RGA}\|_1 = \Sigma|\text{RGA}_{ij}|$) is approximately equal to the minimized (with respect to input and output scaling) condition number $\gamma^*(G)$. In our case we have $\|\text{RGA}\|_1 = 138.275$ and $\gamma^*(G) = 138.268$. We note that the minimized condition number is quite similar to the condition number in this case, but this does not hold in general.

14-3-2 Use of Decoupler

For "tight control" of ill-conditioned plants the controller should compensate for the strong directions by applying large input signals in the directions where the plant gain is low, that is, a "decoupling" controller similar to G^{-1} in directionality is desired. However, because of uncertainty,

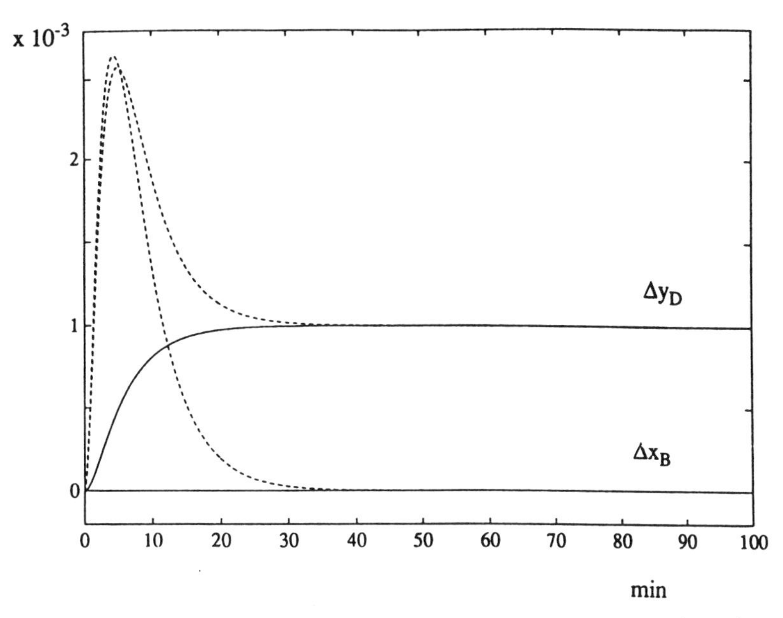

FIGURE 14-2. Response for decoupling controller using LV configuration. Setpoint change in $y_D = 10^{-3}/(5s+1)$. Solid line: No uncertainty; dotted line: 20% input gain uncertainty as defined in Equation 14-9.

the direction of the large inputs may not correspond exactly to the low plant gain direction, and the amplification of these large input signals may be much larger than expected. As shown in succeeding simulations, this will result in large values of the controlled variables y, leading to poor performance or even instability. Consider the following decoupling controller (a steady-state decoupler combined with a PI controller):

$$C_1(s) = \frac{k_1}{s} G^{LV-1}(s) = \frac{k_1(1+75s)}{s} \begin{pmatrix} 39.942 & -31.487 \\ 39.432 & -31.997 \end{pmatrix}, \quad k_1 = 0.7\ \text{min}^{-1} \tag{14-8}$$

We have $GC_1 = 0.7/sI$. In theory, this controller should counteract all the directions of the plant and give rise to two decoupled first-order responses with time constant $1/0.7 = 1.43$ min. This is indeed confirmed by the solid line in Figure 14-2, which shows the simulated response to a setpoint change in top composition.

14-3-3 Use of Decoupler When There is Model Uncertainty

In practice, the plant is different from the model, and the dotted lines in Figure 14-2 show the response when there is 20% error (uncertainty) in the gain in each input channel ("diagonal input uncertainty"):

$$\Delta L = 1.2\,\Delta L_c, \qquad \Delta V = 0.8\,\Delta V_c \tag{14-9}$$

ΔL and ΔV are the actual changes in the manipulated flow rates, whereas ΔL_c and ΔV_c are the desired values (what we believe the inputs are) as specified by the controller. It is important to stress that this diagonal input uncertainty, which stems from our inability to know the exact values of the manipulated inputs, is *always* present. Note that the uncertainty is on the *change* in the inputs (flow rates), not on their absolute values. A 20% error is reasonable for process control applications (some reduction may be possible, for example, by use of cascade control using flow measurements, but there will still be uncertainty because of measurement errors). Regardless, the main objective of this chapter is to demonstrate the effect of uncertainty, and its exact magnitude is of less importance.

The dotted lines in Figure 14-2 show the response with this model uncertainty. It differs drastically from the one predicted by the model, and the response is clearly not acceptable; the response is no longer decoupled, and Δy_D and Δx_B reach a value of about 2.5 before settling at their desired values of 1 and 0. In practice, for example, with a small time delay added at the outputs, this controller would give an unstable response.

There is a simple physical reason for the observed poor response to the setpoint change in y_D. To accomplish this change, which occurs mostly in the "bad" direction corresponding to the low plant gains, the inverse-based controller generates a *large* change in the internal flows (ΔL and ΔV), while trying to keep the changes in the external flows ($\Delta B = -\Delta D = \Delta L - \Delta V$) very *small*. However, uncertainty with respect to the values of ΔL and ΔV makes it impossible to make them both large while at the same time keeping their difference small. The result is a undesired *large* change in the external flows, which subsequently results in large changes in the product compositions because of the large plant gain in this direction. As we shall discuss in the following text, this sensitivity to input uncertainty may be avoided by controlling D or B directly, for example, using the DV configuration.

14-3-4 Alternative Controllers: Single-Loop PID

Unless special care is taken, most multivariable design methods (MPC, DMC, QDMC, LQG, LQG/LTR, DNA/INA, IMC, etc.) yield similar inverse-based controllers, and do generally not yield acceptable designs for ill-conditioned plants. This follows because they do not explicitly take uncertainty into account, and the optimal solution is then to use a controller that tries to remove the interactions by inverting the plant model.

The simplest way to make the closed-loop system insensitive to input uncertainty is to

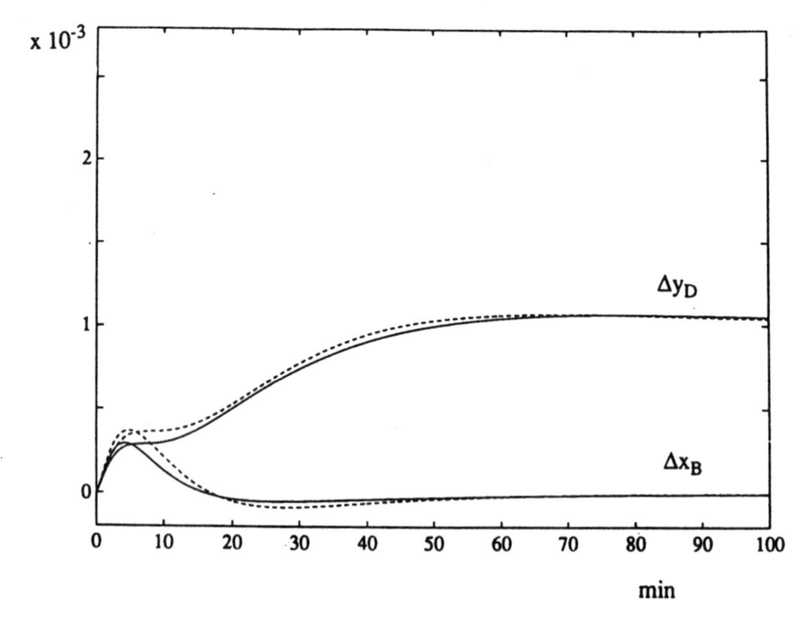

FIGURE 14-3. Response for PID controller using LV configuration.

use a *simple* controller (for example, two single-loop PID controllers) that does not try to make use of the details of the directions in the plant model. The problem with such a controller is that little or no correction is made for the strong interactions in the plant, and then even the nominal response (with no uncertainty) is relatively poor. This is shown in Figure 14-3 where we have used the following PID controllers (Lundström, Skogestad, and Wang, 1991):

$y_D - L$:

$$K_c = 162, \qquad \tau_I = 41 \text{ min}, \qquad \tau_D = 0.38 \text{ min} \tag{14-10}$$

$x_B - V$:

$$K_c = -39, \qquad \tau_I = 0.83 \text{ min}, \qquad \tau_D = 0.29 \text{ min} \tag{14-11}$$

The controller tunings yield a relatively fast response for x_B and a slower response for y_D. As seen from the dotted line in Figure 14-3, the response is not very much changed by introducing the model error in Equation 14-9.

In Figure 14-4 we show response for a so-called mu-optimal controller (see Lundström, Skogestad, and Wang, 1991), which is designed to optimize the worst-case response (robust performance) as discussed toward the end of this chapter. Although this is a multivariable controller, we note that the response is not too different from that with the simple PID controllers, although the response settles faster to the new steady state.

14-3-5 Alternative Configurations: *DV* Control

The process model considered in the preceding text, which uses the *LV* configuration, is fundamentally difficult to control irrespective of the controller. In such cases one should consider design changes that make the process simpler to control. One such change is to consider the *DV* configuration where L rather than D is used for condenser level control (Figure 14-1*b*). The independent variables left for composition control are then D and V:

$$\begin{pmatrix} \Delta y_D(s) \\ \Delta x_B(s) \end{pmatrix} = G^{DV}(s) \begin{pmatrix} \Delta D(s) \\ \Delta V(s) \end{pmatrix} \tag{14-12}$$

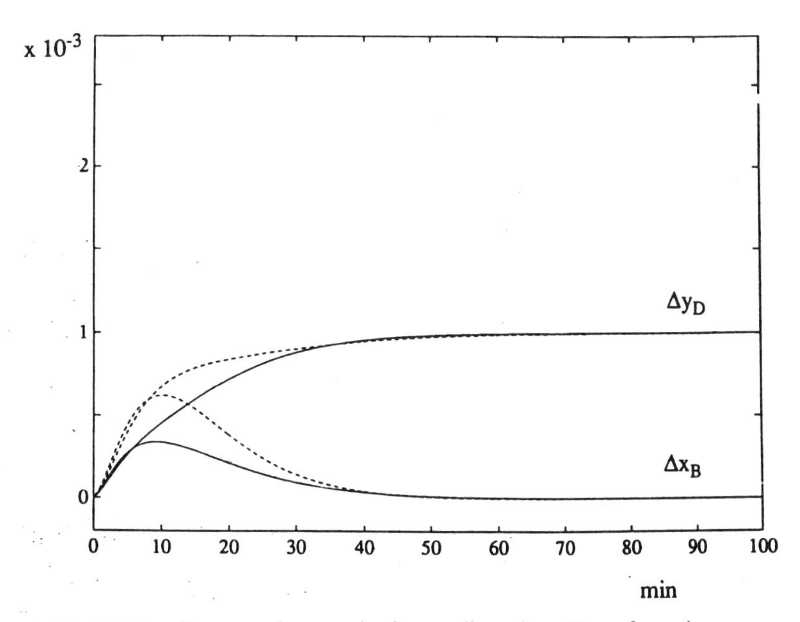

FIGURE 14-4. Response for μ-optimal controller using *LV* configuration.

Comment It is somewhat misleading to consider this a design change because the change from LV to DV configuration is accomplished by a change in the level *control* system. However, many engineers consider the level control system to be such an integral part of the process as to consider it part of the design, although this is, of course, not strictly true.

To derive a model for the DV configuration, assume constant molar flows and perfect control of level and pressure (these assumptions may easily be relaxed). Then $\Delta L = \Delta V - \Delta D$ and we have the following transformation between the two configurations:

$$\begin{pmatrix} \Delta L \\ \Delta V \end{pmatrix} = \begin{pmatrix} -1 & 1 \\ 0 & 1 \end{pmatrix} \begin{pmatrix} \Delta D \\ \Delta V \end{pmatrix} \tag{14-13}$$

and the following linear model is derived from Equation 14-4:

$$G^{DV}(s) = G^{LV}(s)\begin{pmatrix} -1 & 1 \\ 0 & 1 \end{pmatrix} = \frac{1}{75s+1}\begin{pmatrix} 0.878 & -0.864 \\ 0.014 & -0.014 \end{pmatrix} \tag{14-14}$$

This process is also ill-conditioned as $\gamma(G^{DV}) = 70.8$. However, the RGA matrix is

$$\text{RGA}(G^{DV}) = \begin{pmatrix} 0.45 & 0.55 \\ 0.55 & 0.45 \end{pmatrix} \tag{14-15}$$

The diagonal elements are about 0.5, which indicates a strongly interactive system. However, in this case the RGA elements are not large and we may use a decoupler to counteract the interactions.

Simulations using a decoupler are shown in Figure 14-5. As expected the nominal response is perfectly decoupled. Furthermore, as illustrated by the dotted line in Figure 14-5, the decoupler also works well when there is model error. The reason why the model error does not cause problems in this case is that we have one manipulated variable (ΔD) that acts directly in the high-gain direction for external flows and another (ΔV) that acts in the low-gain direction for internal flows. We may then make large changes in the internal flows V without changing the external flows D. This was not possible with the LV configuration,

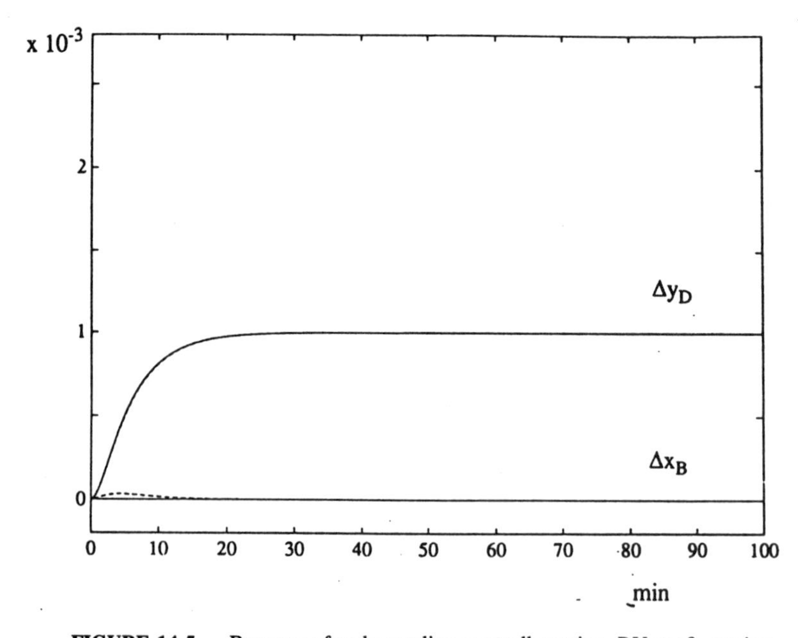

FIGURE 14-5. Response for decoupling controller using *DV* configuration.

where we had to increase both L and V in order to increase the internal flows.

The RGA behavior for various other configurations is treated in detail by Shinskey (1984) for the static case and by Skogestad, Lundström, and Jacobsen (1990) for the dynamic case.

14-3-6 Limitations with the Example: Real Columns

It should be stressed again that the column model used in the preceding text is not representative of a real column. In a real column the liquid lag θ_L(min), from the top to the bottom, makes the initial response for the LV configuration less interactive and the column is easier to control than found here. It turns out that the important parameter to consider for controllability is *not* the RGA at steady state (with exception of the sign), but rather the RGA at frequencies corresponding to the closed-loop bandwidth. For the LV configuration the RGA is large at low frequencies (steady state), but it drops at high frequencies and the RGA matrix becomes close to the identity matrix at frequencies greater than $1/\theta_L$.

Thus, control is simple, even with single-loop PI or PID controllers, if we are able to achieve very tight control of the column. However, if there are significant measurement delays (these are typically 5 min or larger for GC analysis), then we are forced to operate at low bandwidths and the responses in Figures 14-2 through 14-4 are more representative. Furthermore, it holds in general that one should *not* use a steady-state decoupler if the steady-state RGA elements are large (typically larger than 5).

In a real column one must pay attention to the level control for the DV configuration. This is because D does not directly affect composition; only indirectly, through its effect on reflux L through the level loop, does it have influence. In practice, it may be a good idea to let the condenser level controller set $L + D$ rather than L. In this case a change in D from the composition controller will immediately change L without having to wait for the level loop.

14-4 RGA AS A SIMPLE TOOL TO DETECT ROBUSTNESS PROBLEMS

14-4-1 RGA and Input Uncertainty

We have seen that a decoupler performed very poorly for the LV model. To understand this better consider the loop gain GC. The loop gain is an important quantity because it determines the feedback properties of the system. For example, the transfer function from setpoints y_s to control error $e = y_s - y$ is given by $e = Sy_s = (I + GC)^{-1}y_s$. We therefore see that large changes in GC due to model uncertainty will lead to large changes in the feedback response. Consider the case with diagonal input uncertainty Δ_I. Let Δ_1 and Δ_2 represent the relative uncertainty on the gain in each input channel. Then the actual ("perturbed") plant is

$$G_p(s) = G(s)(I + \Delta_I),$$

$$\Delta_I = \begin{pmatrix} \Delta_1 & 0 \\ 0 & \Delta_2 \end{pmatrix} \tag{14-16}$$

In the simulation example we had $\Delta_1 = 0.2$ and $\Delta_2 = -0.2$. The perturbed loop gain with model uncertainty becomes

$$G_pC = G(I + \Delta_I)C = GC + G\Delta_I C \tag{14-17}$$

If a diagonal controller $C(s)$ (e.g., two PIs) is used, then we simply get (because Δ_I is also diagonal) $G_pC = GC(I + \Delta_I)$ and there is no particular sensitivity to this uncertainty. On the other hand, with a perfect decoupler (inverse-based controller) we have

$$C(s) = k(s)G^{-1}(s) \tag{14-18}$$

where $k(s)$ is a scalar transfer function, for example, $k(s) = 0.7/s$, and we have $GC =$

$k(s)I$, where I is the identity matrix, and the perturbed loop gain becomes

$$G_pC = G_p(I + \Delta_I)C = k(s)\left(I + G\Delta_I G^{-1}\right) \quad (14\text{-}19)$$

For the aforementioned LV configuration, the error term becomes

$$G^{LV}\Delta_I(G^{LV})^{-1} = \begin{pmatrix} 35.1\Delta_1 - 34.1\Delta_2 & -27.7\Delta_1 + 27.7\Delta_2 \\ 43.2\Delta_1 - 43.2\Delta_2 & -34.1\Delta_1 + 35.1\Delta_2 \end{pmatrix} \quad (14\text{-}20)$$

This error term is worse (largest) when Δ_1 and Δ_2 have opposite signs. With $\Delta_1 = 0.2$ and $\Delta_2 = -0.2$ as used in the simulations (Equation 14-9), we find

$$G^{LV}\Delta_I(G^{LV})^{-1} = \begin{pmatrix} 13.8 & -11.1 \\ 17.2 & -13.8 \end{pmatrix} \quad (14\text{-}21)$$

The elements in this matrix are much larger than 1, and the observed poor response with uncertainty is not surprising. Similarly, for the DV configuration we get

$$G^{DV}\Delta_I(G^{DV})^{-1} = \begin{pmatrix} -0.02 & 0.18 \\ 0.22 & 0.02 \end{pmatrix} \quad (14\text{-}22)$$

The elements in this matrix are much less than 1, and good performance is maintained even in the presence of uncertainty on each input.

The observant reader may have noted that the RGA elements appear on the diagonal in the matrix $G^{LV}\Delta_I(G^{LV})^{-1}$ in Equation 14-20. This turns out to be true in general because diagonal elements of the error term prove to be a direct function of the RGA (Skogestad and Morari, 1987):

$$(G\Delta G^{-1})_{ii} = \sum_{j=1}^{n} \lambda_{ij}(G)\Delta_j \quad (14\text{-}23)$$

Thus, if the plant has large RGA elements and an inverse-based controller is used, the overall system will be extremely sensitive to input uncertainty.

Control Implications

Consider a plant with large RGA elements in the frequency range corresponding to the closed-loop time constant. A diagonal controller (e.g., single-loop PIs) is robust (insensitive) with respect to input uncertainty, but will be unable to compensate for the strong couplings (as expressed by the large RGA elements) and will yield poor performance (even nominally). On the other hand, an inverse-based controller that corrects for the interactions may yield excellent nominal performance, but will be very sensitive to input uncertainty and will not yield robust performance. In summary, plants with large RGA elements around the crossover frequency are fundamentally difficult to control, and decouplers or other inverse-based controllers should never be used for such plants (the rule is never to use a *controller* with large RGA elements). However, one-way decouplers may work satisfactorily.

14-4-2 RGA and Element Uncertainty / Identification

Previously we introduced the RGA as a sensitivity measure with respect to input gain uncertainty. In fact, the RGA is an even better sensitivity measure with respect to element-by-element uncertainty in the matrix.

Consider any complex matrix G and let λ_{ij} denote the ijth element in its RGA matrix. The following result holds (Yu and Luyben, 1987):

> The (complex) matrix G becomes singular if we make a relative change $-1/\lambda_{ij}$ in its ijth element, that is, if a single element in G is perturbed from g_{ij} to $g_{pij} = g_{ij}(1 - 1/\lambda_{ij})$.

Thus, the RGA matrix is a direct measure of sensitivity to element-by-element uncertainty and matrices with large RGA values become singular for small relative errors in the elements.

Example 14-1

The matrix G^{LV} in Equation 14-4 is nonsingular. The 1, 2 element of the RGA is $\lambda_{12}(G) = -36.1$. Thus the matrix G

becomes singular if $g_{12} = -0.864$ is perturbed to $g_{p12} = -0.864(1 - 1/(-36.1)) = -0.840$.

The preceding result is primarily an important algebraic property of the RGA, but it also has some important control implications:

Consider a plant with transfer matrix $G(s)$. If the relative uncertainty in an element at a given frequency is larger than $|1/\lambda_{ij}(j\omega)|$, then the plant may be singular at this frequency. This is of course detrimental for control performance. However, the assumption of element-by-element uncertainty is often poor from a physical point of view because the elements are usually always *coupled* in some way. In particular, this is the case for distillation columns. We know that the column will not become singular and impossible to control due to small individual changes in the elements. The importance of the previous result as a "proof" of why large RGA elements imply control problems is therefore not as obvious as it may first seem.

However, for process identification the result is definitely useful. Models of multivariable plants $G(s)$ are often obtained by identifying one element at the time, for example, by using step or impulse responses. From the preceding result it is clear this method will most likely give meaningless results (e.g., the wrong sign of the steady-state RGA) if there are large RGA elements within the bandwidth where the model is intended to be used. Consequently, identification must be combined with first principles modelling if a good multivariable model is desired in such cases.

Example 14-2

Assume the true plant model is

$$G = \begin{pmatrix} 0.878 & -0.864 \\ 1.082 & -1.096 \end{pmatrix}$$

By extremely careful identification we obtain the model

$$G_p = \begin{pmatrix} 0.87 & -0.88 \\ 1.09 & -1.08 \end{pmatrix}$$

This model seems to be very good, but is actually useless for control purposes because the RGA elements have the wrong sign (the 1,1 element in the RGA is -47.9 instead of $+35.1$). A controller with integral action based on G_p would yield an unstable system.

Comment The statement that identification is very difficult and critical for plants with large RGA elements may *not* be true if we use decentralized control (single-loop PI or PID controllers). In this case we usually do not use the multivariable model, but rather tune the controllers based on the diagonal elements of G only, or by trial-and-error under closed loop. However, if we decide on pairings for decentralized control based on the identified model, then pairing on the wrong elements (e.g., corresponding to negative RGA) may give instability.

The implication for distillation columns is that one must be careful about using the LV configuration when identifying a model for the column. Rather, one may perform test runs with another configuration, for example, the DV configuration, at least for obtaining the steady-state gains. The gains for the LV configuration may subsequently be derived using consistency relationships between various configurations (recall Equation 14-14). Alternatively, the steady-state gains for the LV model could be obtained from simulations, and test runs for changes in L and V are used only to determine the initial dynamic response (in this case one has the additional advantage that it is not necessary to wait for the responses to settle in order to obtain the gain).

14-5 ADVANCED TOOLS FOR ROBUST CONTROL: μ ANALYSIS

So far in this chapter we have pointed out the special robustness problems encountered for MIMO plants and we have used the RGA as our main tool to detect these robustness problems. We found that plants with *large* RGA elements are (a) fundamentally difficult to control because of sensitivity to input gain uncertainty (and, therefore, decouplers should not be used) and (b) are very difficult to identify because of element-by-element uncertainty.

We have not yet addressed the problem of analyzing the robustness of a given system with plant $G(s)$ and controller $C(s)$. In the beginning of this chapter we mentioned that the peak criterion in terms of M were useful for robustness analysis for SISO systems both in terms of stability (RS) and performance (RP). However, for MIMO systems things are not as simple. We shall first consider uncertainty descriptions and robust stability and then move on to performance. The calculations and plots in the remainder of this chapter refer to the simplified LV model of the distillation column, using the controller with steady-state decoupler plus PI control.

14-5-1 Uncertainty Descriptions

To illustrate that most sources of uncertainty may be represented as norm-bounded perturbations with frequency-dependent magnitudes ("weights"), we shall consider a SISO plant with nominal transfer function

$$g(s) = k\frac{e^{-\theta s}}{1 + \tau s} \tag{14-24}$$

The parameters k, θ, and τ are uncertain and/or may vary with operating conditions. Assume that the relative uncertainty in these three parameters is given by, r_1, r_2, and r_3, respectively. A general way to represent model uncertainty is in terms of norm-bounded perturbations Δ_i. Then the set of possible (or "perturbed") values of the parameters is given by

$$k_p = k(1 + r_1\Delta_1), \qquad |\Delta_1| \le 1 \tag{14-25}$$

$$\theta_p = \theta(1 + r_2\Delta_2), \qquad |\Delta_2| \le 1 \tag{14-26}$$

$$\tau_p = \tau(1 + r_3\Delta_3), \qquad |\Delta_3| \le 1 \tag{14-27}$$

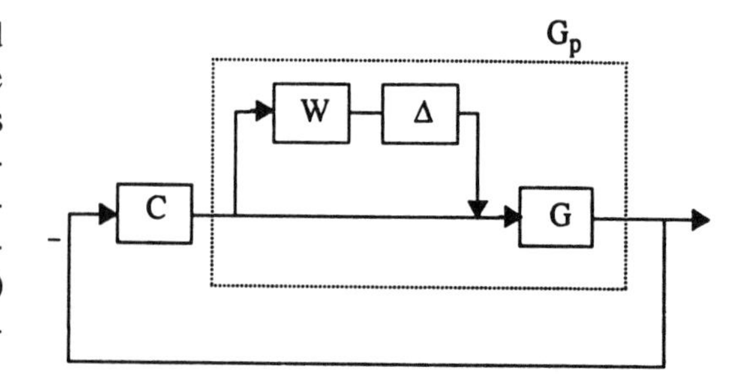

FIGURE 14-6. Multiplicative input uncertainty.

Note that the Δ_i's in the remainder of this chapter are normalized to be less than 1 in magnitude.

First consider the gain uncertainty. For example, assume that k may vary $\pm 20\%$, so that $r_1 = 0.2$. Note that the perturbation on k given by Equation 14-25 may be represented as a relative (or multiplicative) uncertainty as shown in Figure 14-6 with the weight $w = r_1 = 0.2$. In general, the magnitude of the weight varies with frequency, but in this case it is constant only with gain uncertainty.

Now, consider the time delay uncertainty in Equation 14-26. We also want to represent this uncertainty as a relative perturbation. To this purpose use the approximation $e^{-x} \approx 1 - x$ (which is good for small x) and derive

$$e^{-\theta_p s} = e^{-\theta s} e^{-\theta r_2 \Delta_2 s} \approx e^{-\theta s}(1 - r_2 \theta s \Delta_2)$$

or, because the sign of Δ_2 may be both positive or negative,

$$e^{-\theta_p s} \approx e^{-\theta s}(1 + w_2\Delta_2), \qquad w_2(s) = r_2\theta s \tag{14-28}$$

w_2 is the weight for the relative error generated by the time delay uncertainty. With this approximation, $w_2 = 0$ at steady state, reaches 1 (100%) at the frequency $1/(\theta r_2)$ (which is the inverse of the time delay uncertainty), and goes to infinity at high frequencies.

Comment It is also possible to make other approximations for the time delay uncertainty. Skogestad, Morari, and Doyle (1988) and Lundström, Skogestad, and Wang (1991) use an approach where one considers numerically the relative uncertainty generated by the time delay. This results in a complex perturbation Δ_2, but otherwise in the same w_2, except that it levels off at 2 at high frequencies. This approach is used in the computations that follow. (However, for other reasons we would probably have preferred to *not* let w_2 level off at 2 if we had redone this work today.)

We now have two sources of relative uncertainty. Combining them gives an overall 2×2 (real) perturbation block Δ, with Δ_1 and Δ_2 on its diagonal. To simplify, we may include the combined effect of the gain and time delay uncertainty using a single (complex) perturbation by adding their magnitudes together, that is, $w = |r_1| + |w_2|$, or approximately

$$w(s) = r_1 + r_2 \theta s \tag{14-29}$$

For example, with a 20% gain uncertainty and a time delay uncertainty of ± 0.9 min ($\theta r_2 = 0.9$ min), we obtain $w(s) = 0.2 + 0.9s$.

We may also model the time constant uncertainty with norm-bounded perturbations, but we should preferably use an "inverse" perturbation. For example, we may write

$$\frac{1}{1 + \tau_p s} = \frac{1}{1 + \tau s}(1 + w_3 \Delta_3)^{-1},$$

$$w_3(s) = \frac{r_3 \tau s}{1 + \tau s} \tag{14-30}$$

For distillation columns, the time constant τ often varies considerably, but we shall not include this uncertainty here. The reason is that the time constant uncertainty (variations) is generally strongly coupled to the gain uncertainty such that the ratio k_p/τ_p stays relatively constant. Note that k_p/τ_p yields the slope of the initial response and is therefore of primary interest for feedback control.

14-5-2 Conditions for Robust Stability

By robust stability (RS) we mean that the system is stable for all possible plants as defined by the uncertainty set (using the Δ_i's as previously discussed). This is a "worst case" approach, and for this reason one must be careful to not include unrealistic or impossible parameter variations. This is why it is recommended not to include large individual variations in the gain k and the time constant τ for a distillation column model.

Now, consider the distillation column example with combined gain and time delay uncertainty. For multivariable plants it makes a difference whether the uncertainty is at the input or the output of the plant. We will here consider input uncertainty, and the weight w_I then represents, for example, variations in the input gain and neglected valve dynamics. We assume the same magnitude of the uncertainty for each input. The set of possible plants is given by

$$G_p(s) = G(I + w_I \Delta_I),$$

$$\Delta_I = \begin{pmatrix} \Delta_1 & 0 \\ 0 & \Delta_2 \end{pmatrix} \tag{14-31}$$

where Δ_i represents the independent uncertainty in each input channel. This is identical to Equation 14-16, except that w_I yields the magnitude because Δ_i is now normalized to be less than 1. Note that Δ_I is a diagonal matrix (it has "structure"). We assume that the system without uncertainty is stable (we have NS). Instability may then only be caused by the "new" feedback paths caused by the Δ_I block. Therefore, to test for RS we rearrange Figure 14-6 into the standard form in Figure 14-7 where Δ in our case is the matrix Δ_I and $M = -w_I C(I + GC)^{-1} G = -w_I T_I$. M is the transfer function from the output to the input of the Δ_I block. To test for stability we make use of the "small gain theorem". Because the Δ block is normalized to be less than 1 at all frequencies, this theorem says that the system is stable if the M block

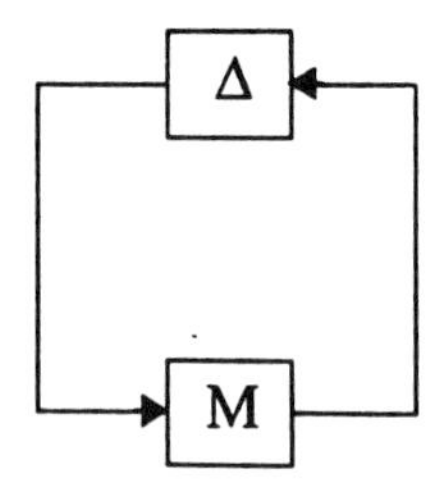

FIGURE 14-7. General block diagram for studying robust stability.

is less than 1 at all frequencies. We use the singular value (also called spectral norm) to compute the magnitude (norm) of M. Robust stability is then satisfied if at all frequencies ω,

$$\bar{\sigma}(M) = \bar{\sigma}(w_I T_I(j\omega)) < 1 \quad (14\text{-}32)$$

However, Equation 14-32 is generally conservative for the following reasons:

1. It allows for Δ to be complex.
2. It allows for Δ to be a full matrix.

It is actually the second point that is the main problem in most cases.

The structured singular value $\mu(M)$ of Doyle (e.g., see Skogestad, Morari, and Doyle, 1988) is defined to overcome these difficulties, and we have that RS is satisfied *if and only if* at all ω,

$$\mu_\Delta(M) = \mu_\Delta(w_I T_I) < 1 \quad (14\text{-}33)$$

This is a tight condition provided the uncertainty description is tight. Note that for computing μ we have to specify the block *structure* of Δ and also if Δ is real or complex. Today there exists very good software for computing μ when Δ is complex. The most common method is to approximate μ by a "scaled" singular value:

$$\mu_\Delta(M) \leq \min_D \bar{\sigma}(DMD^{-1}) \quad (14\text{-}34)$$

where D is a real matrix with a block-diagonal structure such that $D\Delta = \Delta D$. This upper bound is exact when Δ has three or fewer "blocks" (in our example, Δ_I has two blocks).

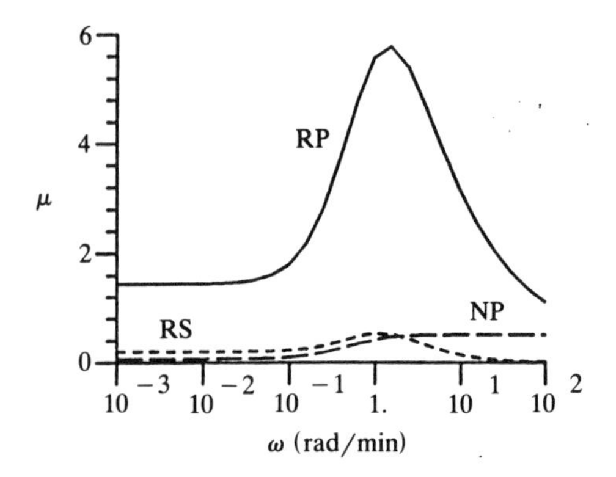

FIGURE 14-8. μ plots for decoupling controller using LV configuration.

As an example, consider the following input uncertainty in each of the two input channels

$$w_I(s) = 0.2 + \frac{0.9s}{0.5s + 1} = 0.2\frac{5s + 1}{0.5s + 1} \quad (14\text{-}35)$$

This corresponds to 20% gain error and a neglected time delay of about 0.9 min. The weight levels off at 2 (200% uncertainty) at high frequency. Figure 14-8 shows $\mu(M) = \mu(w_I T_I)$ for RS with this uncertainty using the decoupling controller. The μ plot for RS shows the inverse of the margin we have with respect to our stability requirement. For example, the peak value of $\mu_{\Delta_I}(M)$ as a function of frequency is about 0.53. This means that we may increase the uncertainty by a factor $1/\mu = 1.89$ before the worst-case model yields instability. This means that we tolerate about 38% gain uncertainty and a time delay of about 1.7 min before we get instability.

Comment For this decoupling controller we have $GC = (0.7/s)I$ and $T_I = T = 1/(1.43s + 1)I$. For this particular case the structure of Δ does not matter and we get a simple analytic expression for μ for robust stability:

$$\mu_\Delta(M) = \bar{\sigma}(w_I T_I) = \left| 0.2 \frac{5s + 1}{(0.5s + 1)(1.43s + 1)} \right|.$$

14-5-3 Definition of Performance

To define performance we shall use the frequency domain and define an upper bound on the sensitivity function S. The sensitivity function gives the change in the response caused by feedback and is probably the best and simplest function to consider when defining performance in the frequency domain. At each frequency we require

$$|S(j\omega)| = |(1 + GC(j\omega))^{-1}| < |w_P^{-1}(j\omega)| \quad (14\text{-}36)$$

or equivalently that the weighted sensitivity is less than 1:

$$\text{NP: } |w_P S(j\omega)| < 1 \quad \text{at all } \omega \quad (14\text{-}37)$$

The peak value (with respect to frequency) of the weighted sensitivity $w_P S$ is also called the H_∞ norm. $w_P(s)$ is the performance weight. Typically, we use the weight

$$w_P(s) = \frac{1}{M_s} \frac{\tau_{cl} s + M_s}{\tau_{cl} s} \quad (14\text{-}38)$$

This requires (a) integral action, (b) that the peak value of $|S|$ should be less than M_s (typically $M_s = 2$), and (c) that the closed-loop response time should be less than τ_{cl} (i.e., the bandwidth should at least be $\omega_B = 1/\tau_{cl}$).

For multivariable systems, the largest singular value of S, $\bar{\sigma}(S)$, is used instead of the absolute value $|S|$. In the introduction we mentioned that the maximum peak on S may be used as a robustness criterion. However, here we are restricting the peak of S at high frequencies primarily to get a good response (without too much oscillation and overshoot). The robustness issues are taken care of much more directly by specifying the allowed uncertainty: see the RS and RP conditions.

NP Specification for our Example

At each frequency the value of $\bar{\sigma}(w_P S)$ should be less than 1. We have selected

$$w_P(s) = \frac{1}{2} \frac{20s + 2}{20s} \quad (14\text{-}39)$$

This requires integral action, a maximum closed-loop time constant of approximately $\tau_{cl} = 20$ min (which of course is relatively slow when the allowed time delay is only about 0.9 min) and a maximum peak for $\bar{\sigma}(S)$ of $M_s = 2$.

As expected, we see from the plot in Figure 14-8 that the NP condition is easily satisfied with the decoupling controller. $\bar{\sigma}(w_P S)$ approaches $1/M_s = 0.5$ at high frequency because of the maximum peak requirement on $\bar{\sigma}(S)$.

14-5-4 Conditions for Robust Performance

Robust performance (RP) means that the performance specification is satisfied for the worst-case uncertainty. The most efficient way to test for RP is to compute μ for RP. If this μ value is less than 1 at all frequencies, then the performance objective is satisfied for the *worst case*. Although our system has good robustness margins and excellent nominal performance, we know from the simulations in Figure 14-2 that the performance with uncertainty (RP) may be extremely poor. This is indeed confirmed by the μ curve for RP in Figure 14-8, which has a peak value of about 6. This means that even with 6 times less uncertainty, the performance will be about 6 times poorer than what we require. Because of a property of μ we may therefore define μ for NP as $\mu_{\Delta_P}(w_P S) = \bar{\sigma}(w_P S)$ where Δ_P is a "fake" uncertainty matrix. Δ_P is a "full" matrix, that is, the off-diagonal elements may be nonzero. μ for robust performance is computed as $\mu_\Delta(N)$ where the matrix Δ in this case has a block-diagonal structure with Δ_I (the true uncertainty) and Δ_P (the fake uncertainty stemming from the performance specification) along the main diagonal and

$$N = \begin{pmatrix} -w_I T_I & -w_I CS \\ w_P SG & w_P S \end{pmatrix} \tag{14-40}$$

The derivation of N is given in, for example, Skogestad, Morari, and Doyle (1988).

The μ-optimal controller is the controller that minimizes μ for RP. For our example we are able to press the peak of μ down to about 0.978 (Lundström, Skogestad, and Wang, 1991). The simulation in Figure 14-4 shows that the response even with this controller is relatively poor. The reason is that the combined effect of large interactions (as seen from the large RGA values) and input uncertainty makes this plant fundamentally difficult to control.

Comment In the time domain our RP-problem specification may be formulated *approximately* as follows: Let the plant be

$$G_p^{LV}(s) = G^{LV}(s)\begin{pmatrix} k_1 e^{-\theta_1 s} & 0 \\ 0 & k_2 e^{-\theta_2 s} \end{pmatrix} \tag{14-41}$$

where $G^{LV}(s)$ is given in Equation 14-4. Let $0.8 \le k_1 \le 1.2, 0.8 \le k_2 \le 1.2, 0 \le \theta_1 \le 0.9$ min, and $0 \le \theta_2 \le 0.9$ min. The response to a step change in setpoint should have a closed-loop time constant less than about 20 min. Specifically, the error of each output to a unit setpoint change should be less than 0.37 after 20 min, less than 0.13 after 40 min, and less than 0.02 after 80 min, and with no large overshoot or oscillations in the response.

14-6 NOMENCLATURE

B	Bottom product flow (kmol/min)
D	Distillate product flow (kmol/min)
G	Nominal plant model
L	Reflux flow (kmol/min)
M	Matrix used to test for robust stability
M_t	Maximum peak of T
M_s	Maximum peak of S
RGA	Matrix of relative gains
s	Laplace variable ($s = j\omega$ yields the frequency response)
$S = (I + GC)^{-1}$	Sensitivity function
$T = GC(I + GC)^{-1}$	Closed-loop transfer function
$T_I = CG(I + CG)^{-1}$	Closed-loop transfer function at the input
U	Unitary matrix of output singular vectors
V	Unitary matrix of input singular vectors
V	Boilup (kmol/min)
x_B	Bottom composition (mole fraction)
y_D	Distillate composition (mole fraction)
w	Frequency-dependent weight function

Greek letters

Δ	Overall perturbation block used to represent uncertainty
Δ_I	Overall perturbation block for input uncertainty
$\Delta_i, \Delta_1, \Delta_2, \Delta_3$	Individual scalar perturbations
$\Delta L, \Delta y_D$, etc.	Deviation variables for reflux, top composition, etc.
$\gamma(A) = \bar{\sigma}(A)/\underline{\sigma}(A)$	Condition number of matrix A
$\mu(A)$	Structured singular value of matrix A
ω	Frequency (rad/min)
$\bar{\sigma}(A)$	Maximum singular value of matrix A
$\underline{\sigma}(A)$	Minimum singular value of matrix A

Subscripts

p	Perturbed (with model uncertainty)
P	Performance

References

Lundström, P., Skogestad, S., and Wang, Z-Q. (1991). Performance weight selection for *H*-infinity and μ-control methods. *Trans. Instrum. Measurement Control* 13, 241–252. London, UK: The Institute of Measurement and Control.

Maciejowski, J. M. (1989). *Multivariable Feedback Design*. Reading, MA: Addison-Wesley.

Morari, M. and Zafiriou, E. (1989). *Robust Process Control*. Englewood Cliffs, NJ: Prentice-Hall.

Shinskey, F. G. (1984). *Distillation Control*, 2nd ed. New York: McGraw-Hill.

Skogestad, S. and Morari, M. (1987). Implication of large RGA-elements on control performance. *Ind. Eng. Chem. Res.* 26, 2323–2330.

Skogestad, S., Morari, M., and Doyle, J. C. (1988). Robust control of ill conditioned plants: High-purity distillation. *IEEE Trans. Autom. Control* 33, 1092–1105.

Skogestad, S., Lundström, P., and Jacobsen, E. W. (1990). Selecting the best distillation control structure. *AIChE J.* 36, 753–764.

Yu, C. C. and Luyben, W. L. (1987). Robustness with respect to integral controllability. *Ind. Eng. Chem. Res.* 26, 1043–1045.

3

Case Studies

15

Experimental Comparison of Control Structures

Kurt V. Waller
Åbo Akademi

15-1 INTRODUCTION

In process control applications multiloop single-input–single-output (SISO) control is the most common approach to solve multi-input–multi-output (MIMO) control problems. These so-called decentralized control systems are usually easier to design and tune than a full MIMO control system. Also tolerance against failure of individual sensors or actuators is generally easier to achieve with properly designed decentralized control systems.

It has been common for a very long time in two-product control of distillation to combine the approach of decentralized control with a suitable transformation of inputs and outputs. The motivation has previously been to reduce the interaction between the main control loops. An example of an output transformation aimed at reducing control loop interaction is the many times suggested approach to control sums and differences between compositions or temperatures. A review and a simulation study of these suggestions is found in Waller and Finnerman (1987). Examples of input transformations are the use of various flow ratios as manipulators, such as L/D for the primary top loop and V/B for the primary bottom loop. ("Primary" is here used to indicate composition or temperature control; "secondary" is used for the inventory control loops.)

However, using such combinations of inputs and outputs actually means that a MIMO control system is used, where part of the MIMO system may consist of multiplications and summations (Waller, 1986, 1990).

The variable transformations discussed in the preceding text and resulting in what, in this chapter, is called different control structures, may have several aims. Early work was motivated by a desire to reduce the interaction between the primary loops. More recently, several researchers (Shinskey, 1985; Waller et al., 1986; Skogestad and Morari, 1987; Skogestad, 1988; Waller et al., 1988a) independently found that the disturbance rejection properties may be very different for different structures.

A third property that may be a function of the control structure chosen is the linearity of the system, where especially logarithmic transformations of the outputs have been suggested as a means of linearizing the system (Skogestad and Morari, 1988).

A fourth property affected by the choice of control structure is the robustness of the system.

This chapter discusses an experimental comparison of different control structures (mainly) for two-point control of a binary separation in a 15-plate pilot plant column separating ethanol and water.

15-2 MANIPULATOR CHOICE FOR DECENTRALIZED CONTROL

In multiloop SISO control of distillation, the manipulator choice for the top composition loop has traditionally been reflux flow rate L or distillate flow rate D or a ratio of the two. For the bottoms composition loop, vapor boilup V' (manipulated through the heat input to the reboiler, here denoted V) or bottoms flow B or a ratio of the two has analogously been used. The structure that uses L and V has been called *energy balance control*; structures using D or B have been called *material balance control*. (It could be mentioned that the manipulators affecting the compositions in the column are L and V. Using D or B is actually only an indirect way of manipulating L or V through the level control loops.)

In addition to the use of simple flows as composition manipulators, as in the (L, V) and (D, V) structures, ratios of flows often have been suggested for use. An early example is Rijnsdorp's suggestion (Rijnsdorp, 1965) to use the ratio of reflux flow and overhead vapor flow $L/(L+D)$ as a manipulator for the top composition control loop. The suggestion was experimentally studied by Wood and Berry (1973), who compared the $((L+D)/L, V)$ structure with the conventional (L, V) structure. The book by Rademaker, Rijnsdorp, and Maarleveld (1975) lists a large number of flow ratio manipulators discussed in the literature.

Renewed interest in ratio control awoke in the 1980s. Ryskamp (1980) proposed the scheme $(D/(L+D), V)$. An extension of this scheme, $(D/(L+D), V/B)$ was suggested and studied by Takamatsu, Hashimoto, and Hashimoto (1982, 1984) and by Shinskey (1984). Increased interest in control structures was also sparked by the findings mentioned in the preceding text that in two-point control, different structures do not only show different loop interactions but they also reject disturbances to various degrees.

An unusual choice of manipulators was recently proposed by Finco, Luyben, and Polleck (1989), who investigated, both through simulation and industrial experiments, the (D, B) structure previously rejected by researchers in the field as representing an infeasible structure.

A manipulator choice like L/D means that both L and D are simultaneously used as manipulators. In linearized form the manipulator can be expressed as $\Delta\psi = a\,\Delta L + b\,\Delta D$. Indeed, Yang, Seborg, and Mellichamp (1991) suggest something they call "combined balance control," where the manipulators are

$$\Delta\psi_1 = a\,\Delta L + (1-a)(-\Delta D) \quad (15\text{-}1)$$

$$\Delta\psi_2 = b\,\Delta V + (1-b)(-\Delta B) \quad (15\text{-}2)$$

Equations 15-1 and 15-2 contain the energy balance control ($a = b = 1$), the two traditional material balance schemes (D, V) and (L, B), and the new super material balance structure (D, B) (for $a = b = 0$), as well as many of the ratio control structures (in linearized form). By choosing a and $b \neq 0$ and $\neq 1$ various combinations of these control structures are obtained, which explains the name "combined balance control."

The combination can be further developed. Actually, by using transformation theory (Häggblom and Waller, 1988) based on consistency relations like total and partial material balances, Häggblom and Waller (1990) derived manipulators that, in the steady state, not only decoupled the two composition control loops, but also eliminated disturbances in feed composition and feed flow rate. The manipulators obtained can be written

$$\Delta\psi_1 = a\,\Delta D + b\,\Delta L + c\,\Delta B + d\,\Delta V \quad (15\text{-}3)$$

$$\Delta\psi_2 = e\,\Delta D + f\,\Delta L + g\,\Delta B + h\,\Delta V \quad (15\text{-}4)$$

The acronym DRD (disturbance rejection and decoupling) has been used for the structure.

15-3 EXPERIMENTAL APPARATUS

A flowsheet for the experimental equipment is shown in Figure 15-1, which also shows the instrumentation. The column is a 15-tray bubble-cap column. From the two feed tanks, each with a volume of 1000 l, feed is pumped to the column through a feed preheater. The reboiler is of a thermosyphon type, which (like the feed preheater) uses fresh steam of 3 bar pressure. The condenser is a water-cooled total condenser. The reflux drum is vented to the atmosphere. Technical data for the system are given in Table 15-1.

TABLE 15-1 Technical Data for the Experimental Distillation System

Column	
Diameter	0.30 m
Plate spacing	0.25 m
Number of plates	15
Number of bubble-caps/plate	18
Holdup area/plate	455 cm^2
Weir height	2.54 cm
Holdup of bottom section	6.2 dm^3
Thermosyphon	
Heat transfer area	6.32 m^2
Holdup (including piping)	9.2 dm^3
Condenser	
Cooling capacity	300 MJ/h
Reflux drum	
Cross area	212 cm^2
Volume	14 dm^3

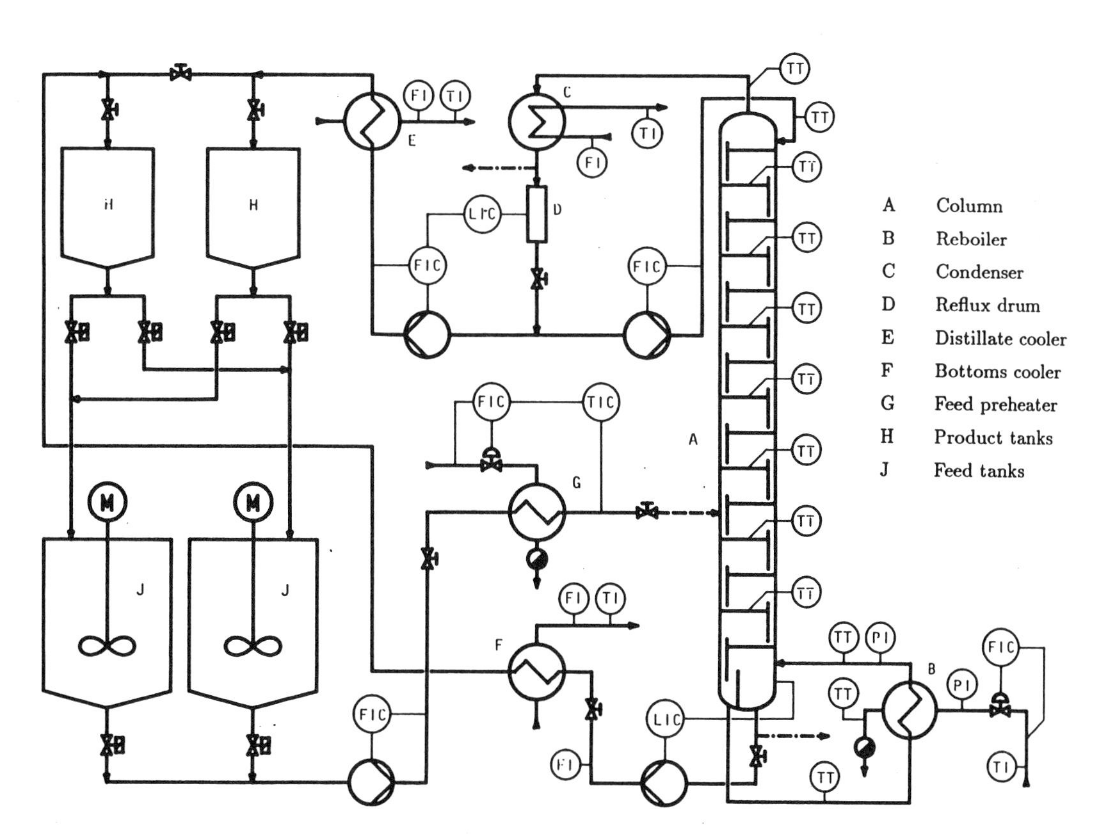

FIGURE 15-1. Flowsheet and instrumentation of experimental pilot plant. Secondary loops shown for (L, V) structure.

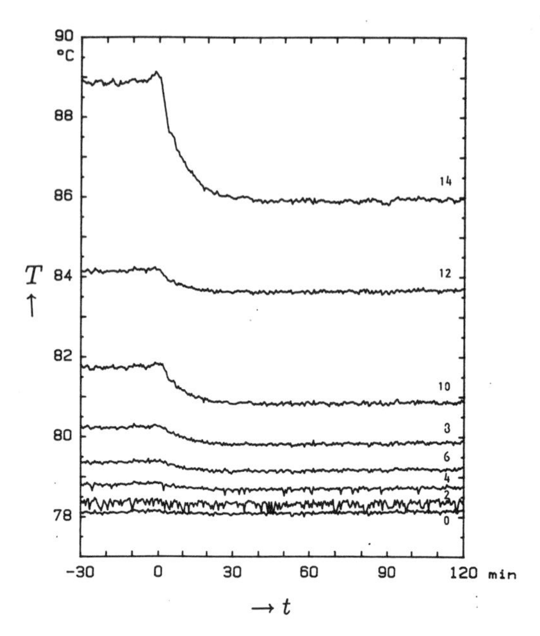

FIGURE 15-2. Illustration of recorded column temperatures after step change of heat input to reboiler.

TABLE 15-2 Nominal Steady-State Data

Feed flow rate F	200 kg/h
Distillate flow rate D	60 kg/h
Bottoms flow rate B	140 kg/h
Feed composition z	30 wt%
Distillate composition x_D	87 wt%
Bottoms composition x_B	5 wt%
Reflux flow rate L	60 kg/h
Steam flow to reboiler V	72 kg/h
Feed temperature	65° C
Reflux temperature	62° C

In the column temperatures are measured on every second plate. An illustration of a recorded temperature profile is shown in Figure 15-2.

15-4 MIXTURE DISTILLED

Ethanol–water is the separation studied. The system is treated on a mass basis, which means that the flows are far from being constant in the column, which they often are assumed to be in simulations. This is thus more in agreement with how columns are operated in practice. Actually, keeping a flow like reflux constant means keeping the flow constant on a volumetric or mass basis or a mixture of the two. This is extensively discussed in Jacobsen and Skogestad (1991). Flows cannot usually be manipulated on a molar basis, which is why it generally seems unsound to make such assumptions in simulations.

This study of a mixture that has such properties as to make the internal flows vary significantly in the column sections has highlighted several interesting properties of the system that would have passed unnoticed if a separation with constant (molar) flows had been investigated. One example is the generally incorrect (steady-state) relation, often used in the literature, $\Delta D = -\Delta L$ (for F and V constant); that is, that a change in reflux flow L eventually brings about an equally large (negative) change in distillate flow D. How incorrect the relation may be is discussed, for example, by Häggblom and Waller (1988, 1989). For the experimental ethanol–water system discussed here, $\Delta D = -0.6\,\Delta L$ at the steady state considered. This large steady-state difference between ΔD and $-\Delta L$ has, in turn, revealed the large difference in disturbance sensitivity that may occur between different control structures.

Data for the nominal steady state studied in this chapter are given in Table 15-2.

15-5 CONTROL STRUCTURES STUDIED

Six control structures are experimentally studied. In all structures the controlled outputs are the same, that is, the temperatures on plates 4 and 14 (counting from the top). It was not possible to move the enriching section measurement closer to the top because the temperature is quite insensitive to composition changes at high ethanol concentrations. Even at plate 4 a specially de-

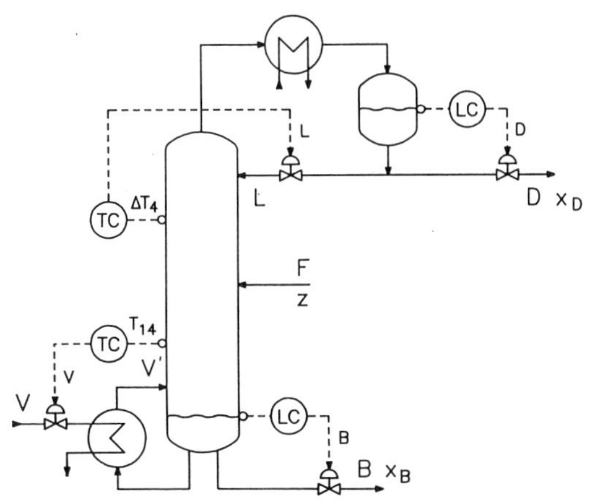

FIGURE 15-3. Implementation of the (L, V) structure.

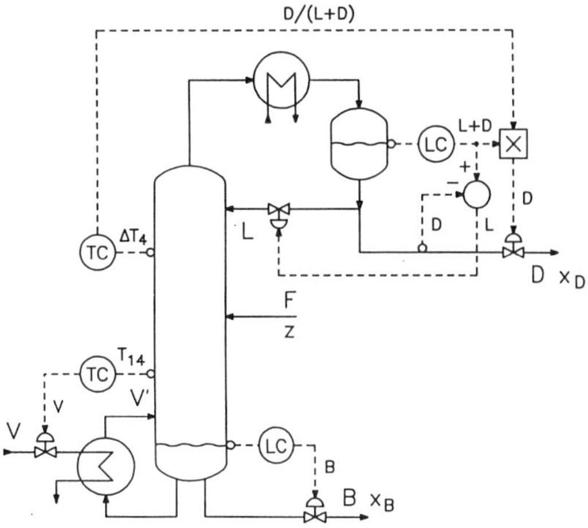

FIGURE 15-5. Implementation of the $(D/(L + D), V)$ structure.

signed temperature measurement system (Gustafsson, 1984) had to be used. The resolution of the system is 0.002° C. It should further be noted that this instrument does not measure temperature directly, but rather deviations from a chosen zero level. This is indicated by the notation ΔT_4 in the figures.

The structures studied are multiloop SISO structures, which differ with the choice of manipulators. The feedback controllers used are all PI controllers.

Five of the six structures are conventional structures, all used in industrial practice. They are the "conventional" or "energy balance" scheme (L, V), two "material balance" schemes, namely, (D, V) and (L, B), Ryskamp's scheme $(D/(L + D), V)$, and the two-ratio scheme $(D/(L + D), V/B)$. The sixth scheme is the DRD scheme, previously discussed. The implementations of four of the structures are shown in Fig-

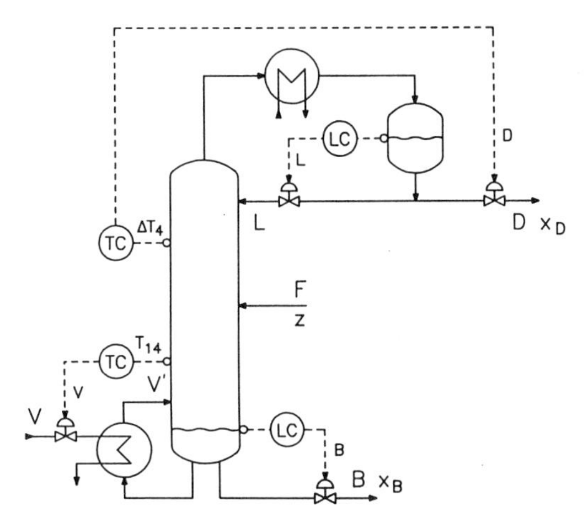

FIGURE 15-4. Implementation of the (D, V) structure.

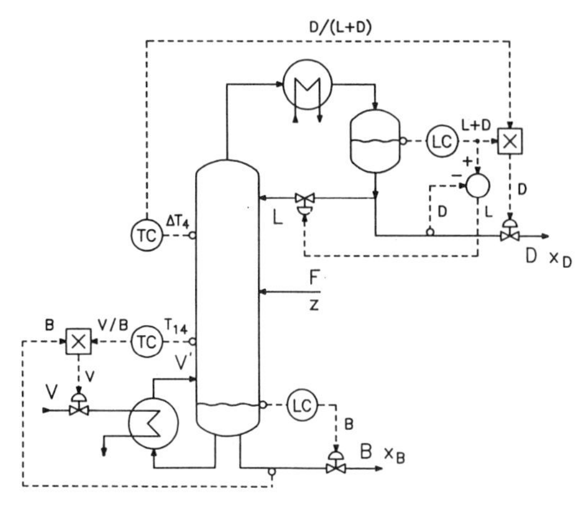

FIGURE 15-6. Implementation of the $(D/(L + D), V/B)$ structure.

ures 15-3 to 15-6, which also show how the level control loops are implemented.

15-6 GENERAL COMMENTS ON THE EXPERIMENTS

The system transfer functions, defined in Figure 15-7, were determined by simple step testing as first-order deadtime transfer functions for four of the structures. The disturbance w can be taken as changes in feed composition or in feed flow rate. The transfer functions obtained are given in Table 15-3 (detailed comments on the experiments are found in Finnerman and Sandelin, 1986, and in Waller et al., 1988b).

It should be noted that the transfer functions are given as experimentally obtained. Because of measurement inaccuracies and slightly varying operating conditions, the transfer functions do not, therefore, necessarily satisfy certain mathematical relationships that can be shown to exist between the transfer functions of different control structures for the same process (Häggblom, 1986; Häggblom and Waller, 1986, 1988). These relationships can be used to reconcile the

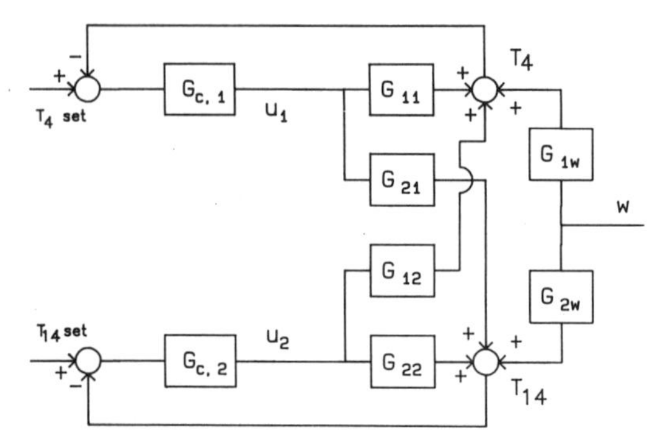

FIGURE 15-7. Block diagram and notation for studied system. [Reprinted by permission from Waller et al. (1988b). Copyright 1988 American Chemical Society.]

TABLE 15-3 Experimentally Determined Transfer Functions for Studied Configurations[a]

Structure	G_{11} / G_{21}	G_{12} / G_{22}	G_{1F} / G_{2F}	G_{1z} / G_{2z}
(L,V)	$\frac{-0.038e^{-0.5s}}{8s+1}$	$\frac{0.046e^{-0.5s}}{11s+1}$	$\frac{-0.001e^{-1.0s}}{10s+1}$	$\frac{0.0042e^{-1.0s}}{8.5s+1}$
	$\frac{-0.23e^{-1.5s}}{8s+1}$	$\frac{0.62e^{-0.5s}}{10s+1}$	$\frac{-0.18e^{-1.0s}}{5.5s+1}$	$\frac{-0.62e^{-1.0s}}{9s+1}$
(D,V)	$\frac{0.075e^{-0.5s}}{14s+1}$	$\frac{-0.054e^{-1.5s}}{14.5s+1}$	$\frac{-0.004e^{-1.0s}}{23s+1}$	$\frac{-0.16e^{-3.0s}}{20.5s+1}$
	$\frac{0.35e^{-1.5s}}{15.5s+1}$	0	$\frac{-0.18e^{-1.0s}}{7.5s+1}$	$\frac{-2.3e^{-1.0s}}{15s+1}$
$\left(\frac{D}{L+D}, V\right)$	$\frac{6.2e^{-0.5s}}{11s+1}$	0	$\frac{-0.0035e^{-1.0s}}{28s+1}$	$\frac{-0.034e^{-2.5s}}{23s+1}$
	$\frac{34e^{-1.25s}}{12.5s+1}$	$\frac{0.34e^{-0.75s}}{10.5s+1}$	$\frac{-0.16e^{-1.0s}}{4.5s+1}$	$\frac{-0.99e^{-1.0s}}{13s+1}$
$\left(\frac{D}{L+D}, \frac{V}{B}\right)$	$\frac{8.4e^{-0.5s}}{15s+1}$	$\frac{3.0e^{-0.5s}}{25.5s+1}$	$\frac{0.0015e^{-1.0s}}{6s+1}$	$\frac{-0.033e^{-7.5s}}{18.5s+1}$
	$\frac{46e^{-1.25s}}{26s+1}$	$\frac{78e^{-0.5s}}{15.5s+1}$	$\frac{0.01e^{-1.0s}}{7.5s+1}$	$\frac{-0.97e^{-1.0s}}{7.5s+1}$

[a]See Figure 15-7 for transfer function notation. Time constants and deadtimes in minutes; gains are in degrees Celsius per kilogram per hour, degrees Celsius per weight percent, and degrees Celsius. Reprinted by permission from Waller et al. (1988b). Copyright 1988 American Chemical Society.

TABLE 15-4 Controller Settings for the Used PI Controllers[a]

Structure	Top Loop K_c	Top Loop T_i	Bottom Loop K_c	Bottom Loop T_i
(L, V)	−180	2.0	12	4.0
(D, V)	130	3.4	9.0	4.5
(V, D)	−120	5.6	18	5.6
$(D/(L+D), V)$	0.68	5.0	17	4.1
$(D/(L+D), V/B)$	1.0	2.4	0.07	3.4

[a]Realized as sampled controllers with sampling time = 0.5 min. Gains in kilograms per hour per degree Celsius (direct flow manipulators) or 1 per degree Celsius (ratio manipulators); integration times in minutes. Reprinted by permission from Waller et al. (1988b). Copyright 1988 American Chemical Society.

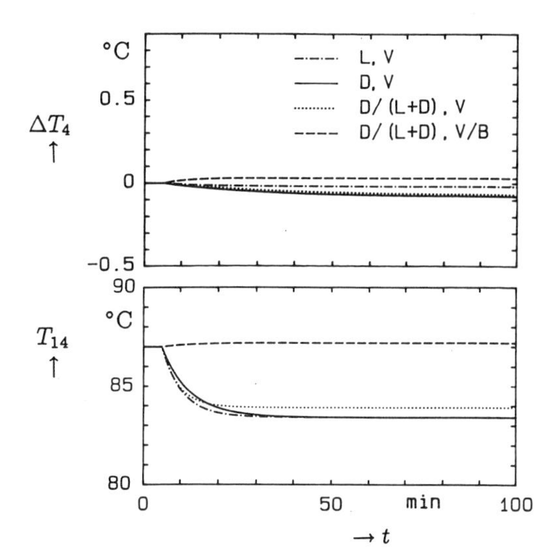

FIGURE 15-9. Open-loop responses to a step change in feed flow rate from 180 to 200 kg/h for four of the strategies studied. Simulations with the experimentally obtained transfer functions. [Reprinted by permission from Waller et al. (1988b). Copyright 1988 American Chemical Society.]

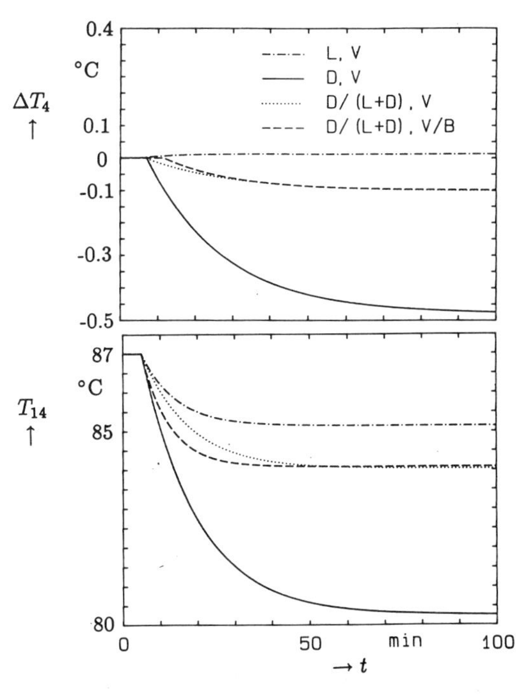

FIGURE 15-8. Open-loop responses to a step increase in feed composition of +3 wt% for four of the strategies studied. Simulations with the experimentally obtained transfer functions. [Reprinted by permission from Waller et al. (1988b). Copyright 1988 American Chemical Society.]

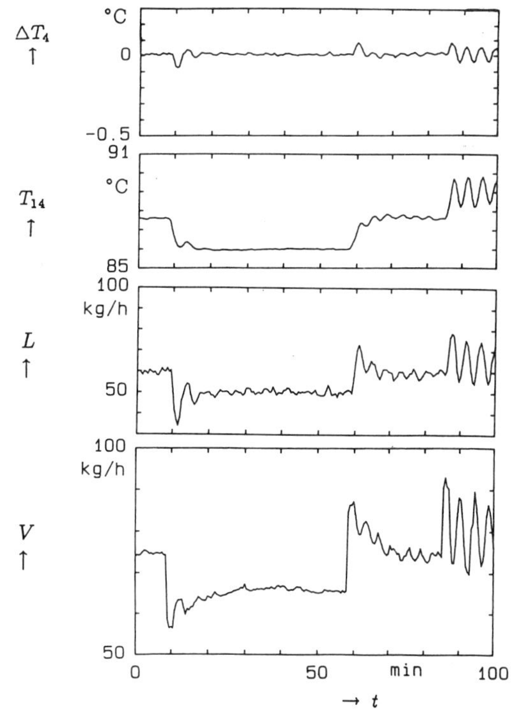

FIGURE 15-10. System response to setpoint changes in T_{14}. Two-point control with (L, V) structure. [Reprinted by permission from Waller et al. (1988b). Copyright 1988 American Chemical Society.]

transfer functions (Häggblom, 1989; Waller et al., 1988a).

The feedback controllers for the temperature loops were tuned on-line to give reasonable responses. The controller settings for two-point control are given in Table 15-4.

Traditionally, when discussing differences between control structures, the interaction between the primary control loops has been the main concern. Table 15-3 reveals another important difference: the (open-loop) disturbance sensitivity may be very different. (Here "open-loop" means that the temperature loops are open; only the level loops are closed.) Figures 15-8 and 15-9 also show that this degree of self-regulation may be very different for different disturbances.

Many times it has been stated in the literature that high-purity separations may exhibit a strongly nonlinear behavior. In such cases it has been suggested that a suitable logarithmic transformation of the outputs would make the system significantly more linear (Skogestad and Morari, 1988).

Although the separation studied in this chapter does not result in very pure products, the system is quite nonlinear at the operating point studied, as illustrated in Figure 15-10, where the system response to

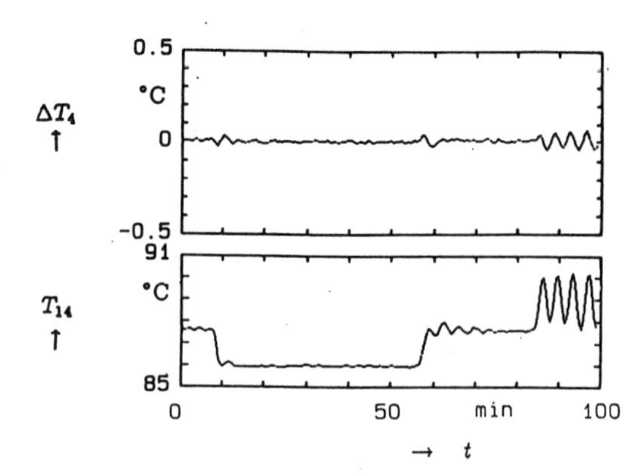

FIGURE 15-11. System response to setpoint changes in T_{14}. Two-point control with $(D/(L+D), V)$ structure.

setpoint changes in T_{14} is shown. The control system is (L, V) with fixed controller parameters. As seen, the system stability considerably decreases when the setpoint for T_{14} is increased.

In addition to having different disturbance sensitivity and control loop interaction, different structures could also be expected to show different degrees of nonlinearity. This aspect is highlighted in Figures 15-11 and 15-10. In both cases fixed controller settings were used (Table 15-4). No difference concerning nonlinearity between (L, V) and $(D/(L+D), V)$ can be observed.

15-7 (D,V), (V,D), AND (L,B) STRUCTURES

In the experimental identification of the dynamics for the (D, V) structure, $G_{22} = 0$ was obtained (Table 15-3). This means that a change in V results in such a change in L that the effects on T_{14} cancel. A low value of K_{22} in the (D, V) scheme was also obtained by McAvoy and Weischedel (1981) and by Tsogas and McAvoy (1981) for columns with high product purities and high reflux ratio. In the study reported in this paper the result was obtained for a low-purity column with a low reflux ratio.

$G_{22} = 0$ means that T_{14} is not controllable by V if the $(D \leftarrow T_4)$ loop is not in operation (a backward arrow denotes feedback). The relative gain for the scheme is zero, which means that "control depends on other loop(s)" (Shinskey, 1984).

Because (D, V) thus lacks integrity, it would be natural to use another scheme. One appealing alternative is the other "material balance" scheme (L, B), which has a relative gain of 1 when (D, V) has a relative gain of 0 (Häggblom and Waller, 1988).

The (L, B) scheme was experimentally tried on the column. However, satisfactory control of the temperatures was not obtained during the experiments. One prob-

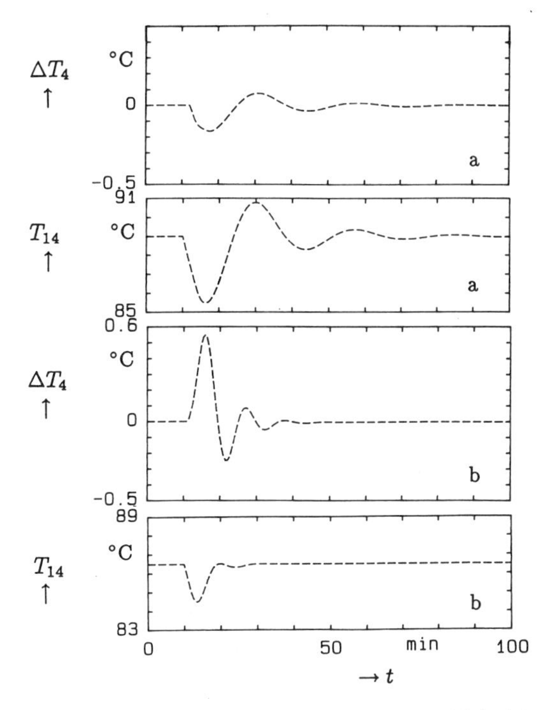

FIGURE 15-12. Simulated two-point control (a) with (D, V) structure and (b) with (V, D) structure. [Reprinted by permission from Waller et al. (1988b). Copyright 1988 American Chemical Society.]

lem was the inverse response between boilup and reboiler level, which made it very difficult to make this level control loop function properly, even for very small disturbances. Experiments made with disturbances in feed flow rate indicate that the (L, B) structure is extremely sensitive to these disturbances.

Another possibility to improve the (D, V) scheme is to reverse the loops, that is, to control T_4 by V and T_{14} by D, here denoted a (V, D) scheme. In this case, the scheme has a relative gain of 1 and both loops would work also in case of failure of the other loop.

Figure 15-12a shows the simulated response of the two controlled temperatures to a step disturbance in feed composition for the (D, V) scheme, whereas Figure 15-12b shows the corresponding responses for the (V, D) scheme. The simulations indicate that one loop is better controlled by scheme number one, and the other by scheme number two. The (V, D) scheme was not, however, experimentally evaluated.

15-8 $(D/(L+D), V)$ STRUCTURE

This scheme was originally suggested by Ryskamp (1980), who motivated it in the following way.

The scheme "holds reflux ratio constant if the top composition controller output is constant. An increase of heat input from the bottom composition controller does not

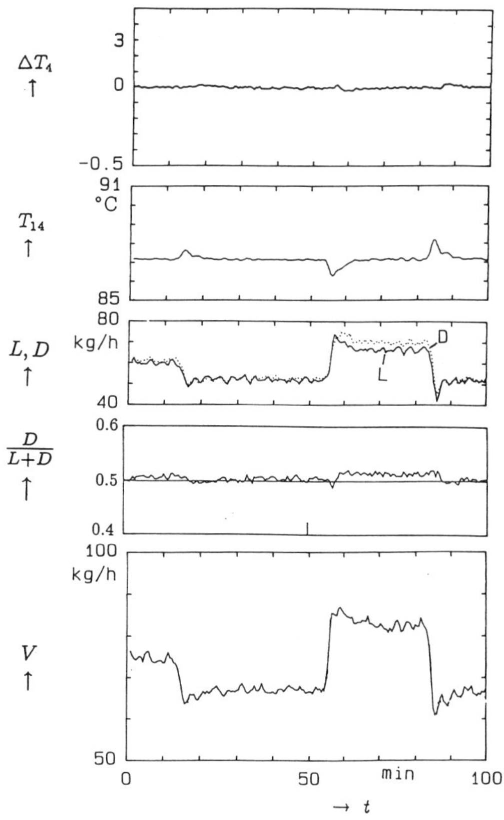

FIGURE 15-13. Experimental illustration of control by $(D/(L+D), V)$. Step changes in feed composition have occurred at $t \approx 12$ min (z changes from 30 to 27 wt%), at $t \approx 55$ min (from 27 to 33 wt%), and at $t \approx 80$ min (from 33 to 27 wt%). [Reprinted by permission from Waller et al. (1988b). Copyright 1988 American Chemical Society.]

make top product as impure as would occur with reflux constant (conventional control) nor as overpure as would occur with distillate flow constant (material balance control)." Thus, this property of the scheme results in a certain decoupling effect. The scheme is sometimes said to result in "implicit decoupling," in contrast to "explicit decoupling" accomplished by external decoupling elements.

For this scheme, $G_{12} = 0$ in our column, which means that the relative gain is 1 and that changes in V do not affect T_4 or, more precisely, that a change in V causes such a change in L that the net effect on T_4 is 0.

The implementation of the scheme caused no practical problems. The two loops were separately tuned and worked well together. Figure 15-13 is an experimental illustration of how the scheme $(D/(L + D), V)$ works.

15-9 $(D/(L+D), V/B)$ STRUCTURE

This scheme is an extension of Ryskamp's scheme, suggested and studied by Takamatsu, Hashimoto, and Hashimoto (1982, 1984) and by Shinskey (1984), who have shown that the steady-state interaction between the two primary loops often is small.

Shinskey (1984, page 156) has calculated eight separation cases, and in all of them he obtained a lower relative gain (less interaction at steady state) for the scheme $(D/(L + D), V/B)$ than for Ryskamp's scheme. In fact, it can be theoretically shown (Häggblom, 1988) that the relative gain for the $(D/(L + D), V/B)$ scheme is always lower than the relative gain for Ryskamp's scheme when the latter has a relative gain larger than 1. If the relative gain for Ryskamp's scheme is 1, it is 1 also for the two-ratio scheme. In this investigation, the operating conditions for the two-ratio scheme were slightly different from those for Ryskamp's scheme. This is one explanation why the relative gain calculated from the process gains given in Table 15-3 (see Table 15-5) do not satisfy the preceding conditions.

The response of T_{14} to changes in the manipulated variables $D/(L + D)$ and V/B did not follow a simple first-order deadtime response, as is illustrated in Figure 15-14. In spite of this, the transfer functions have been identified as first-order deadtime transfer functions (see Table 15-3). For this scheme, a change in feed flow rate did not affect the uncontrolled temperatures T_4 and T_{14} much (Figure 15-15). At the same time, however, this change quite strongly affected the flows in the column, as shown in the figure. It was not possible to eliminate the oscillations shown in the figure by detuning the bottom level controller. Actually, it turned out to be very difficult to make the $(V/B \leftarrow T_{14})$ loop operate satisfactorily because of stability problems. An explanation

TABLE 15-5 Relative Gains of Studied Structures

Structure	Relative Gain
(L, V)	1.8
(D, V)	0.0
(L, B)	1.0
$(D/(L + D), V)$	1.0
$(D/(L + D), V/B)$	1.3
DRD	1.0

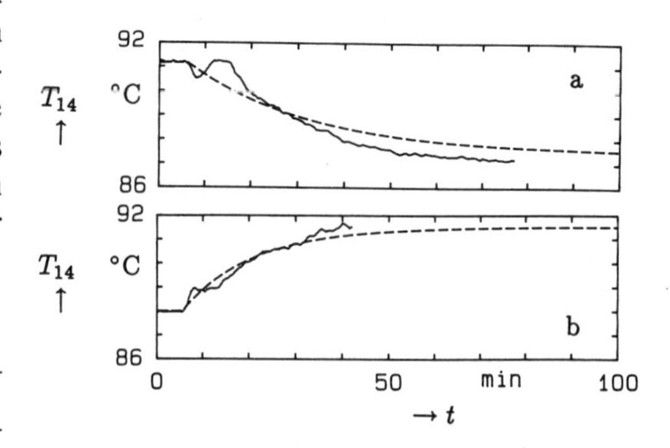

FIGURE 15-14. Experimental (—) response of T_{14} to step change in (*a*) manipulator $(D/(L + D)$ and (*b*) manipulator (V/B). Also responses of the transfer functions given in Table 15-3 are plotted (- - -). [Reprinted by permission from Waller et al. (1988b). Copyright 1988 American Chemical Society.]

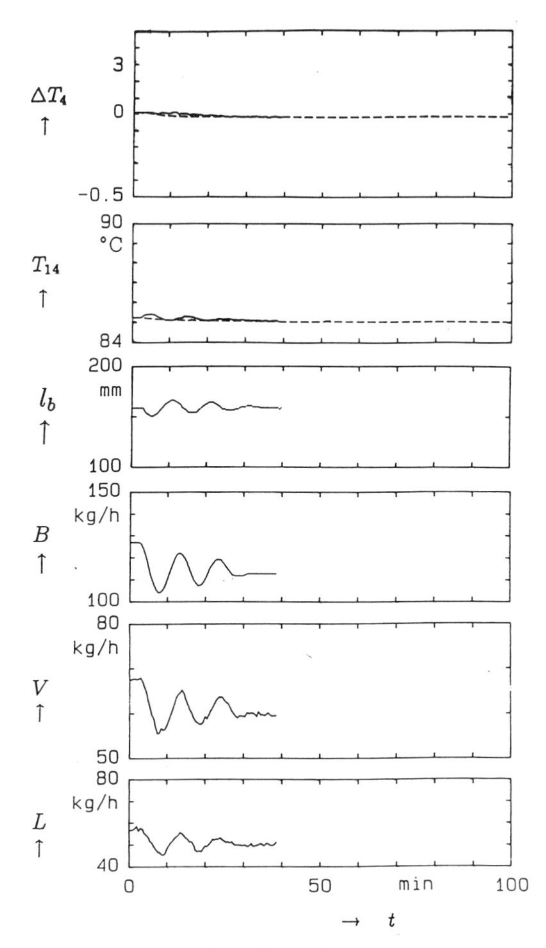

FIGURE 15-15. Experimental response to a step change in feed flow rate from 200 to 180 kg/h at $t \approx 1$ min for the $(D/(L + D), V/B)$ strategy when both the temperature loops are open. The reboiler level is denoted by l_b. [Reprinted by permission from Waller et al. (1988b). Copyright 1988 American Chemical Society.]

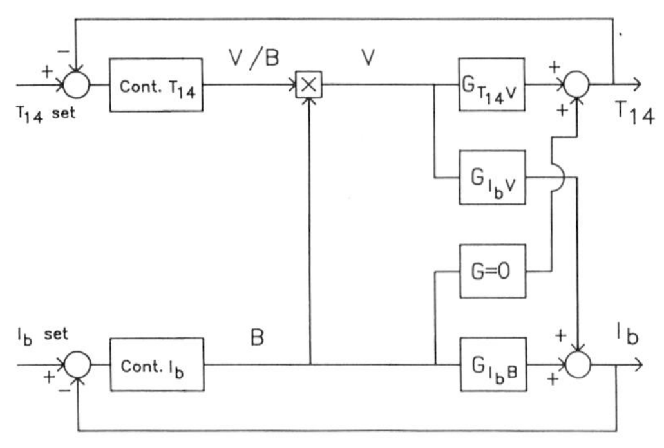

FIGURE 15-16. Illustration of the coupling between the reboiler level control loop and the T_{14} control loop for the scheme $(D/(L + D), V/B)$. [Reprinted by permission from Waller et al. (1988b). Copyright 1988 American Chemical Society.]

is that there is coupling, as illustrated in Figure 15-16, between the level and the temperature loops, introduced by use of V/B as the manipulator. On the other hand, except for the level control loop manipulator (see following text) the same basic structure is valid for the top part of the column, and no difficulties were encountered there.

Also in the investigation of the other conventional configurations studied, there were problems with the bottom level control and the bottom level measurement. There was not, however, much incentive to improve these configurations because the reboiler level does not affect the T_{14} control (the block $G = 0$ in Figure 15-16 is the same in all conventional control structures studied). The stability problems are accentuated by an inverse response of the reboiler level and the bottoms flow B to changes in V.

Attempts to improve the T_{14} control by improving the properties of the level control loop by way of adjusting the feedback controller parameters in the level loop were not successful. Instead, before being fed to the T_{14} loop, the measurement of B was filtered by a first-order, low-pass filter with a time constant of the order of 100 s. This improved the T_{14} control, but it did not eliminate all problems: the level loop could still turn unstable quite unpredictedly. Of the conventional schemes studied, this $(D/(L + D), V/B)$ structure was by far the most time-consuming to implement.

A consequence of having a more complicated control scheme is that the system becomes more difficult to supervise. During operation of the scheme $(D/(L + D), V/B)$, it was sometimes hard to get a picture of what was happening in the column. An illustration of how the scheme operates is provided by Figure 15-17.

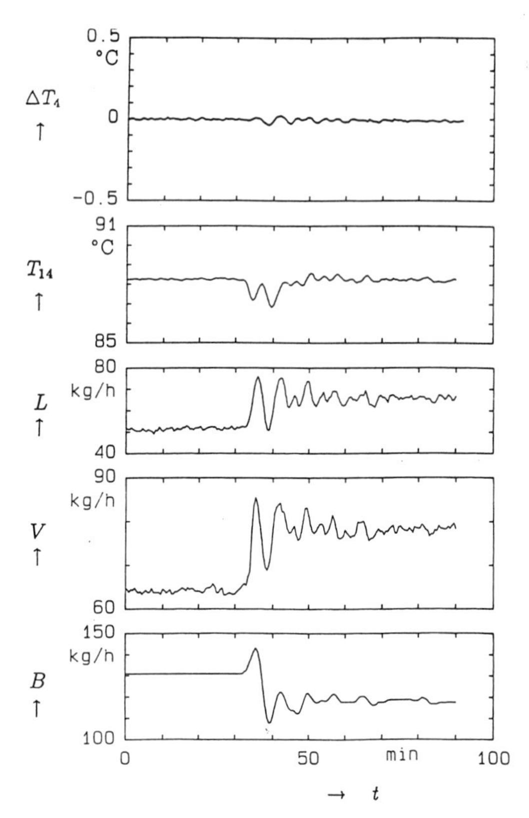

FIGURE 15-17. Experimental illustration of two-point control with the $(D/(L+D), V/B)$ strategy for a step change in feed composition from 27 to 33 wt% at $t \approx 32$ min. [Reprinted by permission from Waller et al. (1988b). Copyright 1988 American Chemical Society.]

Similar difficulties to get the bottom level control loop to function properly have been experienced on industrial columns (Shinskey, 1987). As a remedy Shinskey (1990) suggests using $V' + B$ as the level control manipulator. Because V' is not normally measurable, the manipulator is chosen to be $aV + B$ (see Figure 15-3 for notation), where a is estimated so that $aV \approx V'$.

15-10 COMPARISON OF FOUR CONVENTIONAL CONTROL STRUCTURES

In this section the four conventional control structures (L, V), (D, V), $(D/(L+D), V)$, and $(D/(L+D), V/B)$ are compared, as implemented in Figures 15-3 to 15-6.

15-10-1 One-Point Control

A rough comparison of the control qualities for the controlled variable in one-point control using different control structures for a distillation column is obtained by the parameter $|K_w/T_w|$, that is, the absolute value of the ratio between the gain and the time constant of the disturbance transfer function (when expressed by a first-order transfer function as in Table 15-3). The ratio $|K_w/T_w|$ has been found to be roughly proportional to the integrated absolute error of the controlled variable after a step disturbance (Sandelin, Häggblom, and Waller, 1991a).

The parameter $|K_w/T_w|$ is shown in Table 15-6 for the two disturbances, the two loops, and the four structures. The values for the bottoms loop in the (D, V) structure are in parentheses because of the zero process transfer function for this loop (which means that this loop is not useful for one-point control). In the $(D/(L+D), V)$ structure the prediction for the bottom loop is inaccurate because of the longer deadtime in the process transfer function for this loop in Table 15-3.

Neglecting differences in implementation difficulties for one-point control, Table 15-6 indicates that for disturbances in feed flow rate the best control of the top loop is obtained by (L, V) and $(D/(L+D), V)$.

TABLE 15-6 Disturbance Sensitivity Parameter $|K_w/T_w| \times 10^3$

Structure	F		z	
	Top Loop	Bottom Loop	Top Loop	Bottom Loop
(L, V)	0.10	33	0.49	69
(D, V)	0.17	(24)	7.8	(150)
$(D/(L+D), V)$	0.13	(36)	1.5	(76)
$(D/(L+D), V/B)$	0.25	1.3	1.8	130

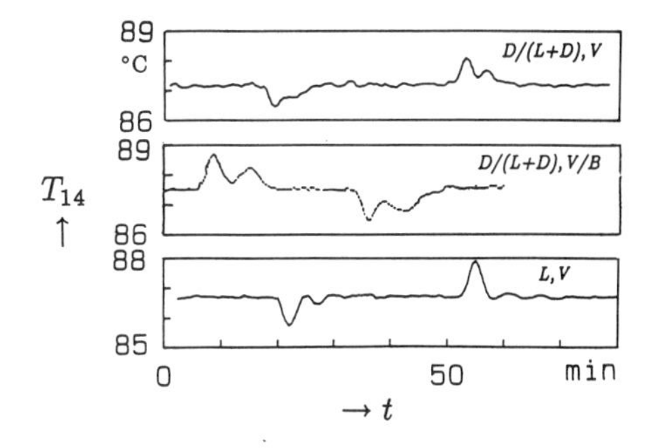

FIGURE 15-18. Experimental comparison of one-point control of T_{14} for three structures. The disturbances are step changes in feed composition of the following size: $(D/(L + D), V)$: 27.1 wt% → 33.0 wt% → 27.1 wt%; $(D/(L + D), V/B)$: 32.1 wt% → 27.1 wt% → 32.1 wt%; (L, V): 27.0 wt% → 33.0 wt% → 27.0 wt%.

For the bottoms loop the best by far is obtained by $(D/(L + D), V/B)$.

For disturbances in feed composition, the best top loop control correspondingly is achieved by (L, V). This structure is also the best structure for the bottom loop, closely followed by $(D/(L + D), V)$. (It should be remembered, however, that the transfer functions given in Table 15-3 are quite inaccurate, which is also why the results in Table 15-6 are quite inaccurate.)

A rough experimental illustration is given in Figure 15-18. Considering that the disturbance to the $(D/(L + D), V/B)$ structure was only 5 wt%, compared to 6 wt% for the other two structures shown, the agreement between Figure 15-18 and Table 15-6 is reasonable.

The behavior of the uncontrolled output in one-point control of a two-output distillation process is extensively discussed in Sandelin, Häggblom, and Waller (1991b).

15-10-2 Two-Point Control

An experimental comparison of four conventional strategies is shown in Figure 15-19. Although the levels of operation are somewhat different, this does not, in our experi-

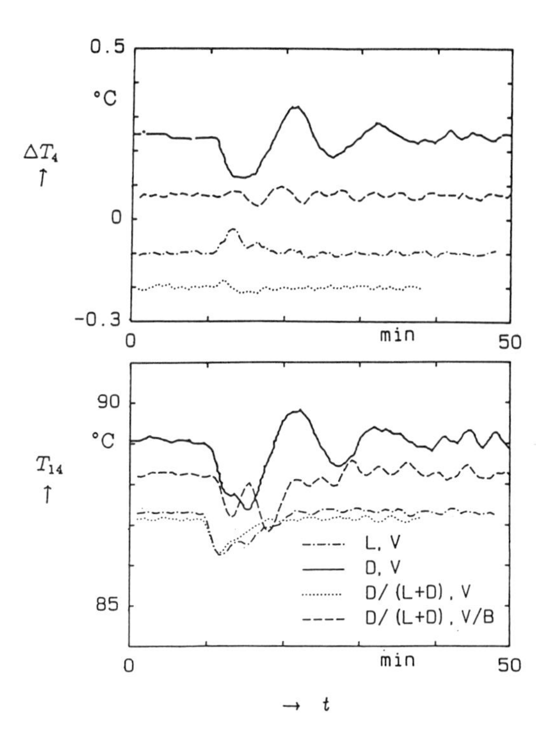

FIGURE 15-19. Experimental comparison of four studied conventional structures. At $t \approx 10$ min, a step change from 27 to 33 wt% in feed composition is introduced. (The measuring device on plate 4 measures temperature differences from a reference temperature; thus no recording of the temperature level is obtained.) [Partly reprinted from Waller et al. (1988b). Copyright 1988 American Chemical Society.]

ence, affect the control qualities to such a degree that the conclusions would be affected. The relative gains as calculated from the experimentally obtained process gains for the studied structures are given in Table 15-5. Of the four strategies, $(D/(L + D), V)$ and (L, V) are the ones most suitable to use when coping with disturbances in feed composition at the studied steady state in our column. The first one was also found quite insensitive to controller tuning for the temperature loops.

For disturbances in feed flow rate, on the other hand, the scheme $(D/(L + D), V/B)$ is probably the one resulting in the best control quality, as indicated by Figure 15-9. The conventional scheme (L, V), Ryskamp's scheme $(D/(L + D), V)$, and the two-ratio scheme $(D/(L + D), V/B)$ resulted in roughly the same control quality for step disturbances in feed composition. The material balance scheme (D, V) resulted in the worst control in addition to being only conditionally operable [the $(V \leftarrow T_{14})$ loop works only when the $(D \leftarrow T_4)$ loop is closed].

It was difficult to implement and operate the scheme $(D/(L + D), V/B)$. One explanation is that the bottoms level control loop (measurement and control) is incorporated into the temperature control system and that difficulties in the level loop were transmitted to the temperature loops (see Figure 15-16).

15-11 STRUCTURE FOR DISTURBANCE REJECTION AND DECOUPLING

If a (dynamic or static) process model for one control structure is known, the corresponding process models for other control structures can be calculated through transformations (Häggblom and Waller, 1988; Yang et al., 1990; Häggblom and Waller, 1991). Not only can a model for a given structure be calculated from another structure, such as (D, V) from (L, V), but models for structures with specified properties can be calculated, too. One such structure is called DRD and it denotes a structure that both rejects certain disturbances and decouples the two primary loops. The manipulators in DRD are functions of D, L, B, and V, as in Equations 15-3 and 15-4. DRD is treated in detail by Häggblom and Waller (1990).

As an illustration, consider a static DRD, calculated on the basis of reconciled gains of the (L, V) structure and designed so as to (at steady state) eliminate disturbances in feed composition and feed flow rate. If the equations are solved with respect to L and V, which are the manipulators with which the controlled temperatures are affected, the following expressions are obtained (Häggblom and Waller, 1990):

$$\Delta L = -42\,\Delta\psi_1 + 1.3\,\Delta\psi_2 + 1.1\,\Delta D + 0.12\,\Delta B \quad (15\text{-}5)$$

$$\Delta V = -18\,\Delta\psi_1 + 1.3\,\Delta\psi_2 + 1.0\,\Delta D + 0.14\,\Delta B \quad (15\text{-}6)$$

Equations 15-5 and 15-6 were calculated on a process model experimentally determined in 1986. Because simulations indicate (Häggblom and Waller, 1990) that the control properties of DRD may be quite sensitive to the controller parameters, new identification of a process model for the (L, V) structure was made in 1990 (Nurmi, 1990). Based on this identification, the following DRD controller was calculated:

$$\Delta L = -41.0\,\Delta\psi_1 + 2.4\,\Delta\psi_2 + 0.89\,\Delta D + 0.22\,\Delta B \quad (15\text{-}7)$$

$$\Delta V = -16.2\,\Delta\psi_1 + 1.4\,\Delta\psi_2 + 0.92\,\Delta D + 0.16\,\Delta B \quad (15\text{-}8)$$

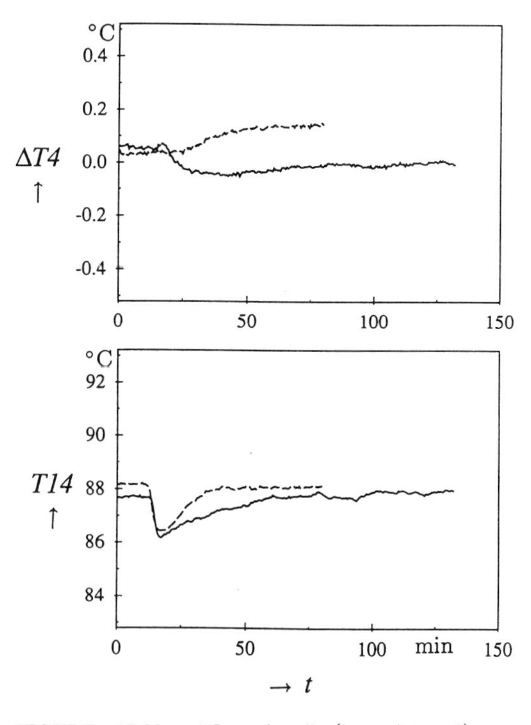

FIGURE 15-20. "Open-loop" (i.e., $\Delta\psi = 0$) responses of DRD structure to a +6 wt% step change in feed composition. Structure given by Equations 15-5 and 15-6 (—) and by Equations 15-7 and 15-8 (- - -).

Figure 15-20 shows the experimental "open-loop" responses (i.e., $\Delta\psi = 0$) with the DRD structures for both controllers given by Equations 15-5, 15-6, and 15-7, 15-8 (Nurmi, 1990). The disturbance in Figure 15-20 is a +6 wt% change in feed composition, whereas it is only a +3 wt% change in Figure 15-8 for four conventional structures. Especially for the bottom loop, the DRD structure very well eliminates the disturbance in the steady state. The deviation from the desired steady state during the first 20 min following the step disturbance was expected. It is caused by the process dynamics neglected in the static DRD controller used (see Häggblom and Waller, 1990, for a more detailed illustration).

In two-point feedback control there seems to be no direct advantage in using DRD compared to standard structures. The actual advantages are more indirect: a certain control is obtained over both products (for the disturbances considered in the design at the DRD structure) even if one or both feedback controllers fail or are turned to manual. Illustration of how indirect two-point control is obtained through one-point control using DRD is given in Figures 15-21 and 15-22, where also the corresponding experimental results with the (L, V) structure are shown (Sandelin, Häggblom, and Waller, 1991b).

It should be noted that closing one of the feedback loops may destroy the self-regulating properties of the other output, as illustrated in Figures 15-21 and 15-22 for the (L, V) structure (compare with Figure 15-8 where both feedback loops are open). This property is extensively discussed and illustrated in Sandelin, Häggblom, and Waller (1991b).

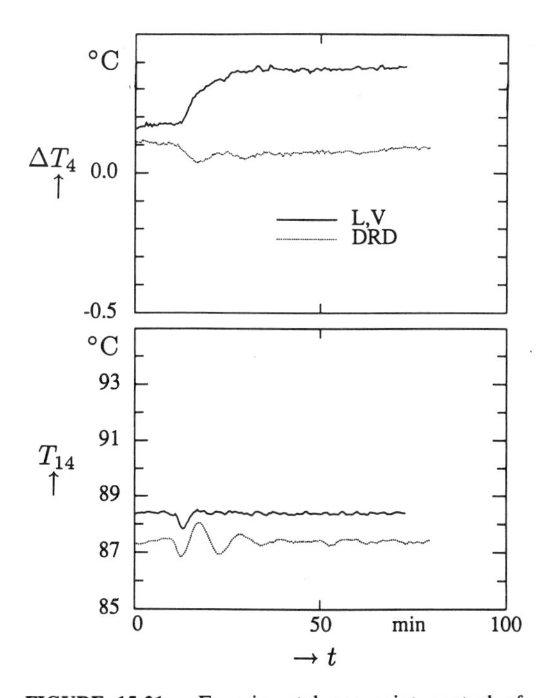

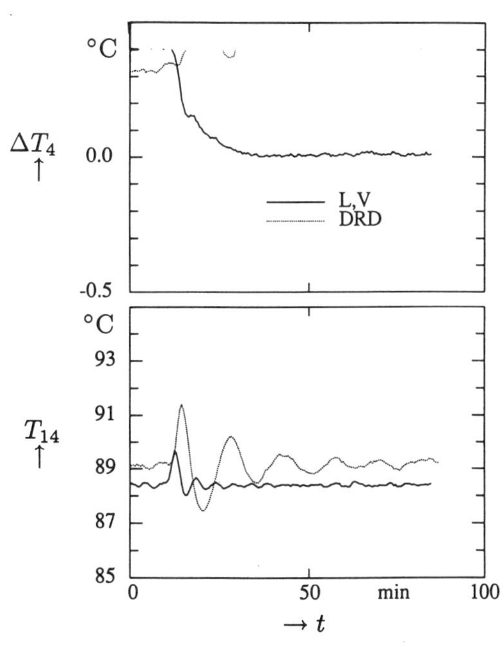

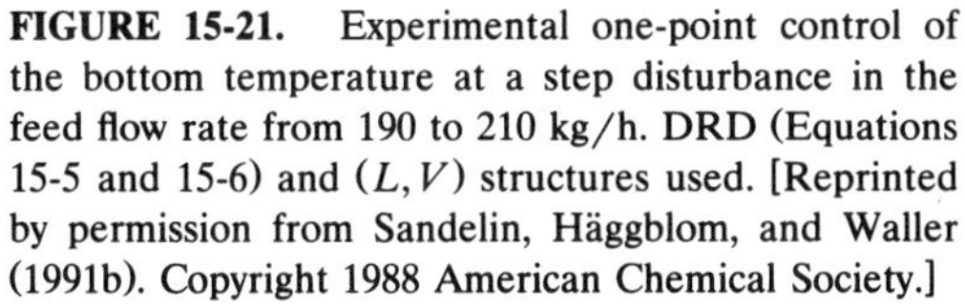

FIGURE 15-21. Experimental one-point control of the bottom temperature at a step disturbance in the feed flow rate from 190 to 210 kg/h. DRD (Equations 15-5 and 15-6) and (L, V) structures used. [Reprinted by permission from Sandelin, Häggblom, and Waller (1991b). Copyright 1988 American Chemical Society.]

FIGURE 15-22. Experimental one-point control of the bottom temperature at a step disturbance in the feed composition from 33 to 27 wt%. DRD (Equations 15-5 and 15-6) and (L, V) structures used. [Reprinted by permission from Sandelin, Häggblom, and Waller (1991b). Copyright 1988 American Chemical Society.]

It should further be emphasized that DRD in general is efficient only for the disturbances considered in the calculation of the parameters of the DRD structure, which in the preceding examples are feed composition and feed flow rate.

15-12 CONTROLLER TUNING FOR ROBUSTNESS AGAINST NONLINEARITIES

It has often been stated that high-purity distillation shows nonlinear behavior. As illustrated by Figures 15-10 and 15-11, the low-purity system studied in this chapter shows significant nonlinearity. If one wants to operate at several operating points like the ones shown in Figures 15-10 and 15-11, there are several different ways to design the control system. One is to use adaptive control. If one wants to use a controller with fixed settings over the whole operating range, there are several approaches to tune the controller. One technique is demonstrated by Sandelin, et al. (1991). The approach is a multi-objective, constrained min-max technique, using a number of models for the process over the operating range. An experimental illustration of the resulting control system is shown in Figure 15-23 (cf. Figures 15-10 and 15-11). The controllers used in Figures 15-23, 15-10, and 15-11 are two SISO PI controllers with fixed parameters in all cases. More details are given in Sandelin, et al. (1991).

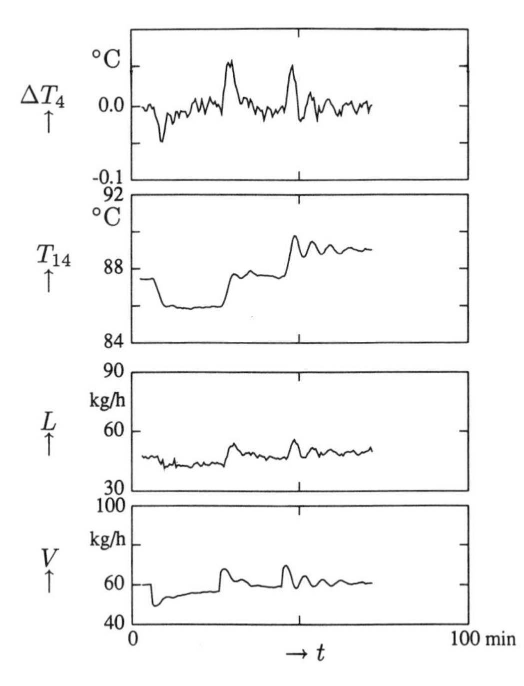

FIGURE 15-23. Experimental responses of the outputs and the inputs to setpoint changes in T_{14}. The control structure is (L, V) and two PI controllers are used for feedback. The controller parameters have been calculated by a min-max approach to achieve robustness against nonlinearities over operating range considered. Note the different scaling for ΔT_4 compared to previous figures. The controller settings are (cf. Table 15-4): top loop: $K_c = -125$ kg h^{-1}/°C, $T_i = 5.8$ min; bottom loop: $K_c = 5.4$ kg h^{-1}/°C, $T_i = 2.7$ min. [Reprinted with permission from Chemical Engineering Science, Sandelin, et al. (1991). Copyright 1991 by Pergamon Press.]

15-13 SUMMARY AND CONCLUSIONS

Various ways to control distillation have been investigated on a 15-plate pilot-plant column separating ethanol and water. Emphasis has been put on two-point feedback control using multiloop SISO PI control. Various structures have been studied through different manipulated variables for the two primary loops. Five conventional structures have been investigated both with respect to control quality and difficulties of implementation: (L, V), (D, V), (L, B), $(D/(L + D), V)$, and $(D/(L + D), V/B)$.

By far the most difficult structures to implement were the two structures (L, B) and $(D/(L + D), V/B)$, and by far the worst control quality for the disturbances considered (feed composition and feed flow rate) was obtained by (D, V), a structure that for the studied system lacks integrity because it has a relative gain of 0. No results concerning control quality were obtained for (L, B).

Some properties of a special disturbance rejection and decoupling strategy, called DRD, are also illustrated. One feature of this strategy is that it approximately gives indirect two-point control through one-point control for the two disturbances considered.

Finally, an illustration is given of a multiloop PI controller with fixed parameters, tuned by a min-max linear quadratic theory, which copes with the nonlinearities of the system studied when operated over a larger operating range.

ACKNOWLEDGMENT

The results reported have been obtained during a long-range project on multivariable process control supported by the Neste Foundation, the Academy of Finland, Nordisk Industrifond, and Tekes. A number of people have been involved in this project over the years and the results reported stem from their endeavors: Dan Finnerman, Sten Gustafsson, Kurt-Erik Häggblom, Päivi Nurmi, Peter Sandelin, and Kai Wikman. Their contributions are gratefully acknowledged. Some of the results presented have been published previously by the author and co-workers in *American Institute of Chemical Engineers Journal*, *Industrial and Engineering Chemistry Research* and *Chemical Engineering Science*. They are reprinted here by permission.

References

Finco, M. V., Luyben, W. L., and Polleck, R. E. (1989). Control of distillation columns with low relative volatility. *Ind. Eng. Chem. Res.* 28, 75–83.

Finnerman, D. and Sandelin, P. (1986). M.Sc. Theses, Åbo Akademi, Åbo, Finland (in Swedish).

Gustafsson, S. (1984). Unpublished work, Åbo Akademi.

Häggblom, K. E. (1986). Tech. Lic. Thesis, Åbo Akademi (in Swedish).

Häggblom, K. E. (1988). Consistent control structure modeling with application to distillation control. Ph.D. Thesis, Åbo Akademi.

Häggblom, K. E. (1989). Reconciliation of process gains for distillation control structures. In *Dynamics and Control of Chemical Reactors, Distillation Columns and Batch Processes*, J. E. Rijnsdorp et al., eds. Oxford: Pergamon Press, pp. 259–266.

Häggblom, K. E. and Waller, K. V. (1986). Relations between steady-state properties of distillation control system structures. In *Dynamics and Control of Chemical Reactors and Distillation Columns*, C. McGreavy, ed. Oxford: Pergamon Press, pp. 243–247.

Häggblom, K. E. and Waller, K. V. (1988). Transformations and consistency relations of distillation control structures. *AIChE J.* 34, 1634–1648.

Häggblom, K. E. and Waller, K. V. (1989). Predicting properties of distillation control structures. Proceedings of the American Control Conference, Pittsburg, pp. 114–119.

Häggblom, K. E. and Waller, K. V. (1990). Control structures for disturbance rejection and decoupling of distillation. *AIChE J.* 36, 1107–1113.

Häggblom, K. E. and Waller, K. V. (1991). Modeling of distillation control structures. Report 91-4, Process Control Lab., Åbo Akademi.

Jacobsen, E. W. and Skogestad, S. (1991). Multiple steady states in ideal two-product distillation. *AIChE J.* 37, 499–511.

McAvoy, T. J. and Weischedel, K. (1981). A dynamic comparison of material balance versus conventional control of distillation columns. Proceedings IFAC World Congress, Kyoto, Japan, pp. 2773–2778.

Nurmi, P. (1990). M.Sc. Thesis, Åbo Akademi (in Swedish).

Rademaker, O., Rijnsdorp, J. E., and Maarleveld, A. (1975). *Dynamics and Control of Continuous Distillation Units*. Amsterdam: Elsevier.

Rijnsdorp, J. E. (1965). Interaction in two-variable control systems for distillation control: II. Application of theory. *Automatica* 3, 29–52.

Ryskamp, C. J. (1980). New strategy improves dual composition column control. *Hydrocarbon Process.* 59, (6) 51–59.

Sandelin, P. M., Häggblom, K. E., and Waller, K. V. (1991a). A disturbance sensitivity parameter and its application to distillation control. *Ind. Eng. Chem. Res.* 29, 1182–1186.

Sandelin, P. M., Häggblom, K. E., and Waller, K. V. (1991b). Disturbance rejection properties of control structures at one-point control of a two-product distillation column. *Ind. Eng. Chem. Res.* 29, 1187–1193.

Sandelin, P. M., Toivonen, H. T., Österås, M. and Waller, K. V. (1991). Robust multiobjective linear-quadratic control of distillation using low-order controllers. *Chem. Eng. Sci.* 46, 2815–2827.

Shinskey, F. G. (1984). *Distillation Control*, 2nd ed. New York: McGraw-Hill.

Shinskey, F. G. (1985). Disturbance-rejection capabilities of distillation control systems. In Proceedings of the American Control Conference Boston, pp. 1072–1077.

Shinskey, F. G. (1987). Personal communication.

Shinskey, F. G. (1990). Personal communication.

Skogestad, S. (1988). Disturbance rejection in distillation columns. Proceedings of Chemdata 88, Gothenberg, Sweden, June 13–15, pp. 365–371.

Skogestad, S. and Morari, M. (1987). Control configuration selection for distillation columns. *AIChE J.* 33, 1620–1635.

Skogestad, S. and Morari, M. (1988). Understanding the behavior of distillation columns. *Ind. Eng. Chem. Res.* 27, 1848–1862.

Takamatsu, T., Hashimoto, I., and Hashimoto, Y. (1982). Multivariable control system design of distillation columns system. Proceedings of PSE, Kyoto, Japan, Technical Session, pp. 243–252.

Takamatsu, T., Hashimoto, I., and Hashimoto, Y. (1984). Dynamic decoupler sensitivity analysis and its application in distillation control. Preprints IFAC World Congress, Budapest, Hungary, Vol. III, pp. 98–103.

Tsogas, A. and McAvoy, T. J. (1981). Dynamic simulation of a non-linear dual composition control scheme. Proceedings of the 2nd World Congress of Chemical Engineering, Vol. 5, pp. 365–369.

Waller, K. V. (1986). Distillation column control structures. In *Dynamics and Control of Chemical Reactors and Distillation Columns*, C. McGreavy, ed. Oxford: Pergamon Press, pp. 1–10.

Waller, K. V. (1990). Distillation: Control structures. In *Systems & Control Encyclopedia, Supplementary Volume 1*, M. G. Singh, ed. Oxford: Pergamon Press, pp. 174–181.

Waller, K. V. and Finnerman, D. H. (1987). On using sums and differences to control distillation. *Chem. Eng. Commun.* 56, 253–258.

Waller, K. V., Finnerman, D. H., Sandelin, P. M., and Häggblom, K. E. (1986). On the difference between distillation column control structures. Report 86-2, Process Control Lab., Åbo Akademi.

Waller, K. V., Häggblom, K. E., Sandelin, P. M., and Finnerman, D. H. (1988a). Disturbance sensitivity of distillation control structures. *AIChE J.* 34, 853–858.

Waller, K. V., Finnerman, D. H., Sandelin, P. M., Häggblom, K. E., and Gustafsson, S. E. (1988b). An experimental comparison of four control structures for two-point control of distillation. *Ind. Eng. Chem. Res.* 27, 624–630.

Wood and Berry (1973). *Chem. Eng. Sci.* 28, 1707–1717.

Yang, D. R., Seborg, D. E., and Mellichamp, D. A. (1991). Combined balance control structure for distillation columns. *Ind. Eng. Chem. Res.* 29, 2159–2168.

Yang, D. R., Waller, K. V., Seborg, D. E. and Mellichamp, D. A. (1990). Dynamic structural transformations for distillation control configurations. *AIChE J.* 36, 1391–1402.

16

Industrial Experience with Double Quality Control

Henk Leegwater
DSM

16-1 INTRODUCTION

In this chapter the background of multivariable control will be discussed. Two real industrial experiences are presented. The control structure of two similar separations, but with different process configurations, will be worked out in more detail. Both applications are ethylene–ethane separations. One has a separate reboiler and condenser, and an intermediate reboiler, whereas the other has heat integration by combining a reboiler–reflux condenser. Both applications have been in operation for more than six years.

16-2 QUALITY CONTROL

In many chemical plants separation of products from nonconverted raw materials is done usually by a train of several distillation columns downstream of the reactor section. Frequently each product has its own column. The product is recovered from either the top or bottom of that column (depending on whether it is a light or heavy product); the rest of the material is then fed to another column for further separation. Typically in industrial practice only the quality of the product is controlled to meet specifications; quality control of the other stream from the column is not addressed. It is quite common to find that even the optimal concentration of the nonproduct end is not known.

16-3 SINGLE QUALITY CONTROL

To meet the product specification, which is mostly a maximum impurity level, a margin of safety is provided. The quality is controlled at a purity greater than called for by the specification to prevent producing off-spec material. To operate at minimum energy consumption, however, the column should be operated as close as possible to the maximum impurity level. The better the control, the smaller is the variability of quality around the setpoint. Additionally, operation closer to specifications is possible and the energy consumption is lower. Constant product purity is as important to the processes downstream that use the product as is absolute purity because it stabilizes their operation. So, constant product quality (smaller variances around setpoints) is important to the economics of the product columns discussed in this chapter and to the downstream users of the product. This more

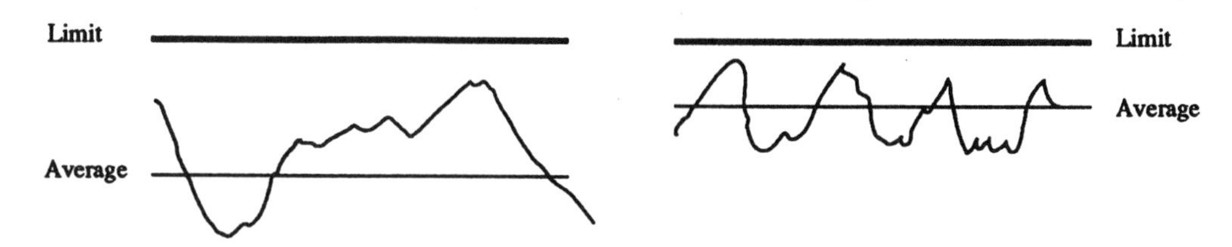

FIGURE 16-1. The better the control, the closer to maximum allowed impurity level the unit can be operated, resulting in lower costs.

constant product (smaller variances) results from good control, which is, however, more erratic; see Figure 16-1.

16-4 WHY DOUBLE QUALITY CONTROL?

Although most of the time quality is specified for only one stream, there are several reasons to control the other stream as well. The justification is usually economics: balancing the energy consumption versus the loss of valuable product, increasing the throughput, or stabilizing downstream units.

16-4-1 Energy Consumption versus Degradation of Valuable Product

Consider the simplified part of a steam cracker involving the ethane and ethylene streams seen in Figure 16-2.

Top Impurity

The top stream of this column, ethylene, is the main product stream of the cracker. Its impurity, primarily ethane, has to be below a certain level: exceeding that level automatically results in dumping the ethylene stream to the flare. However, there are two incentives to have the maximum allowable

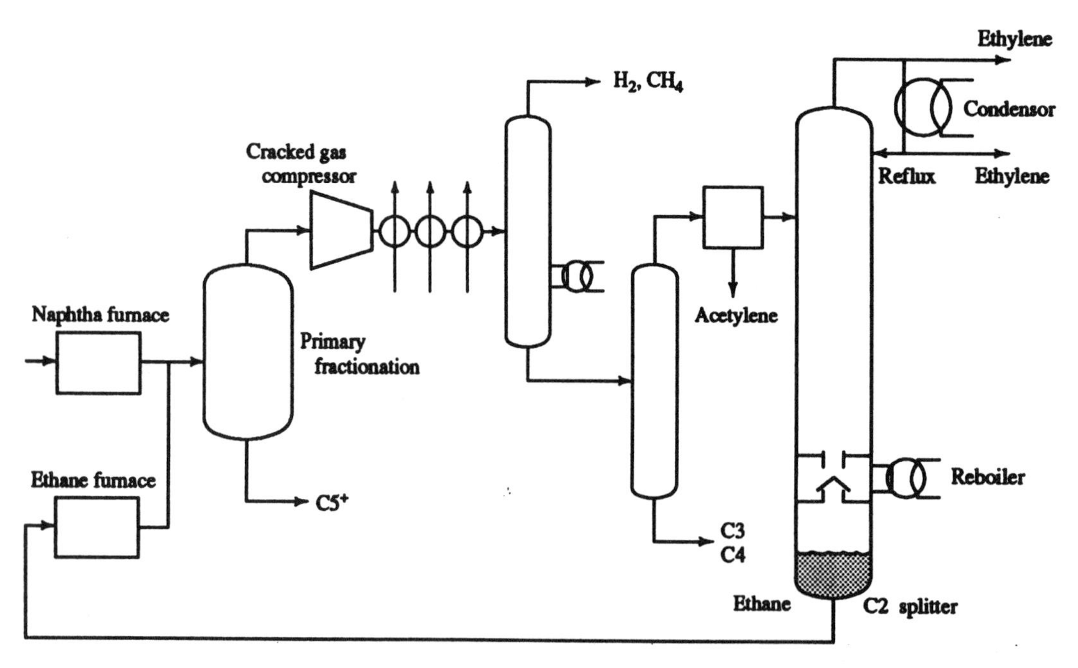

FIGURE 16-2. The C2 splitter with ethane recycle in a naphtha cracker.

amount of ethane in the ethylene product stream. First it reduces the energy consumption of the column. Second the bottom stream will decrease and the more valuable ethylene stream will increase. The value of the top product is much higher: in fact ethane is sold as ethylene.

Bottom Impurity
Although the bottom stream, mainly the ethane, is recycled via a furnace where it is cracked, the amount of impurity is also important for optimizing the column and even the plant operation. Increasing impurities, primarily ethylene, increases recycle costs (compression and separation) and reduces capacity. Most significant perhaps, valuable ethylene is lost because it is partly cracked into less valuable products such as hydrogen, methane, propene, and so forth.

On the other hand, as the ethylene in the bottom stream decreases, separation costs in the C2 splitter go up. So there is an optimum impurity level, which can be calculated if the separation costs, compression costs, and the behavior of ethylene in the ethane furnace are known. See Figure 16-3.

The optimal bottom impurity is not a constant; it varies as a function of energy costs, product values, and the current plant operating environment. If, for instance, the refrigerating compressors are running with open recycle valves to prevent surge, additional cooling energy is free of charge. In that case the optimal bottom impurity will decrease.

16-4-2 Optimizing Throughput

In the last column of a butadiene plant, the product 1,3-butadiene leaves over the top and the C4-raffinate (1,2-butadiene, butenes, etc.) leaves at the bottom. This column is the bottleneck of the whole plant, not only for the butadiene production itself, but also for the C4-raffinate 1, which serves as the feedstock of the methyl-tertiary butyl ether (MTBE) plant on the same site. The column will be operated at the maximum hydraulic load (reflux, vapor load). A lower reflux–feed ratio (R/F) results in higher feed: the lower the purities in both ends, the lower the reflux–feed ratio and the higher the throughput.

Of course the product needs to meet its specifications regarding impurity level: the top needs to be operated as close as possible to the maximum allowed impurity level.

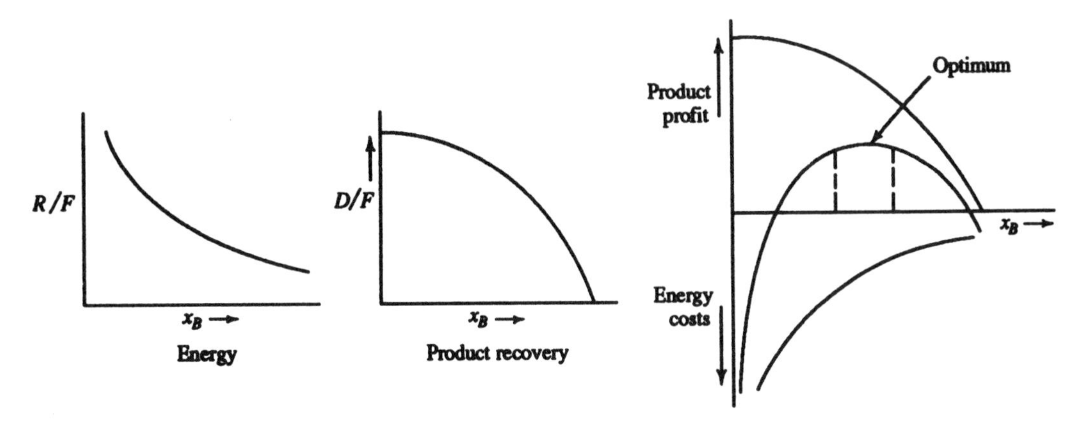

FIGURE 16-3. Given a certain top impurity, putting more energy (reflux and reboiler heat) in a column results in a purer bottom stream and a higher top product recovery. Too much energy, however, is not worth the extra gain of top product; too little allows the loss of valuable top product to the less valuable bottom stream.

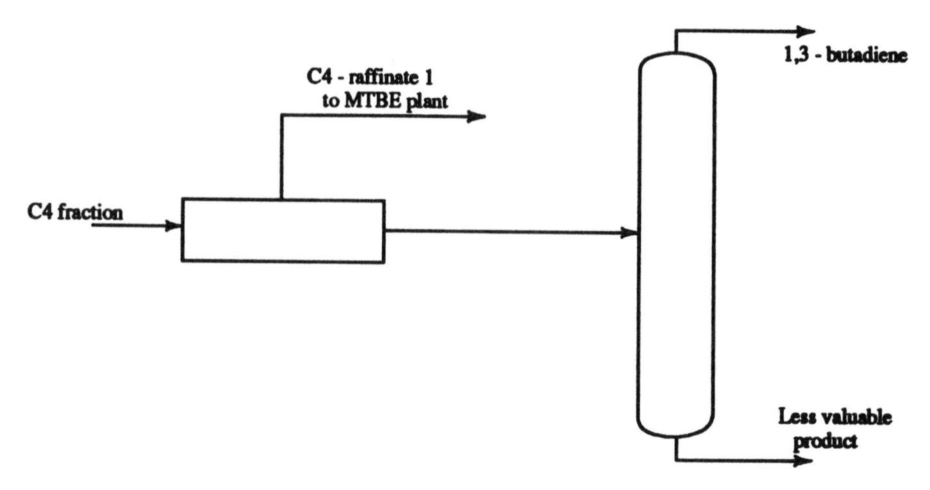

FIGURE 16-4. Optimizing the throughput of a 1,3-butadiene plant looking at butadiene recovery over the top, loss at the bottom, and increase of C4-raffinate 1 production.

So the two targets to optimize the operation of the column are: throughput and bottom impurity.

To calculate the optimal bottom impurity we need to know the values of 1,3-butadiene, the bottom stream, C4-raffinate 1, and the C4 fraction. In the bottom stream the 1,3-butadiene is devalued to a less valuable product. We do not have to consider the costs of the reflux and heat input because the column will be operated at its hydraulic constraints: the amount of reflux and cooling will hardly vary as functions of throughput and bottom impurity. See Figure 16-4. In general, double quality control reduces the internal streams in columns and therefore tends to maximize throughput.

16-5 QUALITY MEASUREMENTS

Normally gas chromatographs are installed to measure top and bottom qualities. Usually the impurities are measured because their concentrations are lower and therefore the accuracy of the measurement is higher. Due to the nature of analyses by gas chromatographs, these measurements are noncontinuous sampled signals: they are afflicted with sample time and deadtime. The *sample time* can vary from a few minutes to half an hour. Because of the costs of installation and operation, one single gas chromatograph is sometimes shared to measure the impurity of several different streams. In that case, the sample time can be an hour or even several hours. The *deadtime* of the measurement consists of two components: the travel time through the sample line from the field to the place where the gas chromatograph is situated (1 to 5 min) plus the analysis time of the gas chromatograph itself; after the sample has been taken, it takes time before the *result of the measurement* is available. See Figure 16-5.

In the case of the manually taken samples, which are hand processed in a laboratory, the deadtime can be an hour or more and the sample time even a day or longer.

Besides gas chromatographs, infrared-based measurement devices are available for quality measurements. These are usually continuous analyzers: however, a deadtime (travel time) is added to the control loop dynamics. Usually they are not sensitive enough for low impurity measurement ranges and are used only as a final measurement in separations that are not high purity.

If the dynamics of a gas chromatograph preclude optimal tuning of a quality control loop, an infrared (IR) analyzer can be used as the measurement to a slave controller, the master being the chromatographic-based quality loop. In such a configuration, the IR quality measurements are made at a point in the column where the impurities are in the range at which the sensitivity of the IR analyzer is satisfactory. The slower gas chromatograph acts, then, as the measurement for the master controller that gives the setpoint to the slave controller.

A cheaper and usually adequate solution is to use temperature as a first indication of change in the concentration profile in place of the IR analyzer. Like the IR measurement, temperature is also an inferential measurement and is not uniquely related to the concentration. In binary separations the relationship between temperature and concentration is influenced only by the pressure at the point of measurement. If the correlation between temperature, pressure, and quality is known, the measurement can compensate for pressure variations. In non-

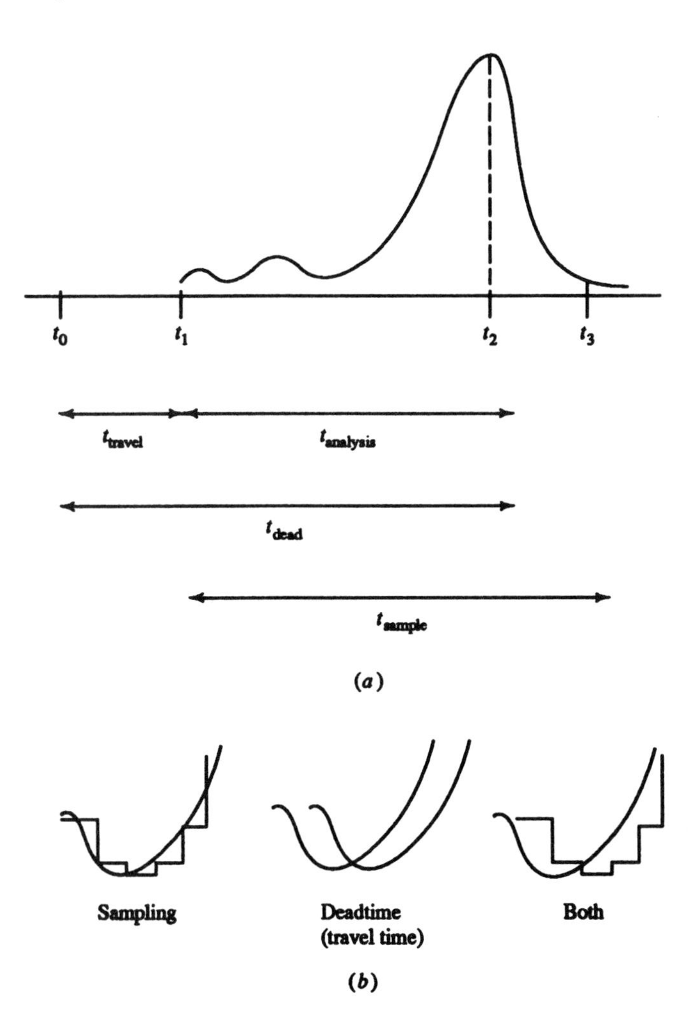

FIGURE 16-5. (*a*) Measurement by gas chromatograph. t_0 sample leaves the column; t_1 sample arrives at the gas chromatograph; t_2 measurement available; t_3 new sample arrives at the gas chromatograph. (*b*) Gas chromatograph measurement: sampling and deadtime are the dynamic consequences.

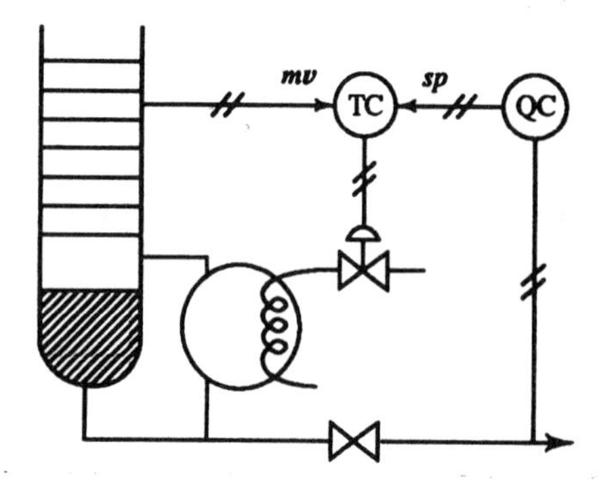

FIGURE 16-6. Master–slave quality control. Master: gas chromatograph; slave: temperature or infrared.

binary separations this correlation depends also on the concentration of the nonkey components. Correlations between concentrations of tertiary systems and temperature measurements are much lower and also much more difficult to compensate. Therefore temperature measurements and controllers act generally as a slave controller in a quality control loop. Frequently the master controller is an operator. See Figure 16-6.

16-6 WHY MULTIVARIABLE CONTROL FOR DOUBLE QUALITY CONTROL?

A column needs basic controls to let it work properly: both an upgoing vapor flow and a downgoing liquid flow are needed for separation. To take care of this, the following basic controls are needed:

- Bottom level control
- Reflux vessel level control
- Pressure control

The feed is mostly determined by upstream equipment or is flow controlled. Looking at double quality control, 2 degrees of freedom are left. Let us assume that these are the reflux and reboiler heat as in Figure 16-7. The quality control can then be considered as a 2 × 2 system as shown in Figure 16-8.

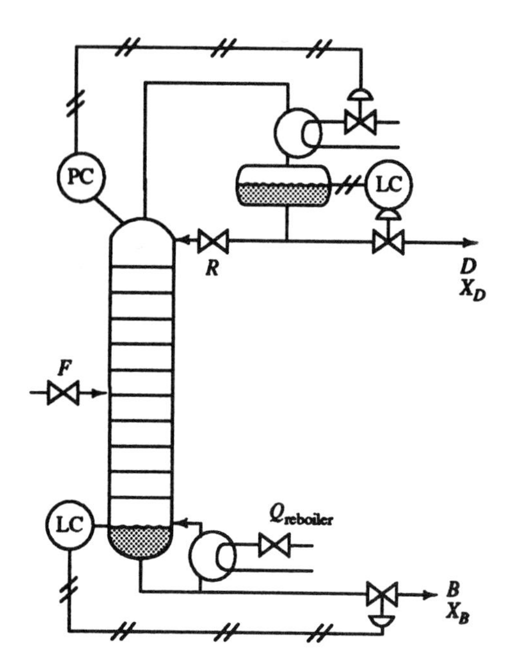

FIGURE 16-7. Arbitrarily chosen basic controls.

Increasing the amount of reflux while holding the heat input constant, increases the top quality (decreases top impurity) and decreases the bottom quality (increases bottom impurity). Increasing the reboiler heat input while holding the reflux constant, increases the bottom quality (decreases bottom impurity) and decreases top quality (increases top impurity). This interaction often leads to instability for the traditional control scheme: bottom quality controlled by the reboiler heat input and top quality controlled by the amount of reflux. The reason is clear. Even if the loops are properly tuned,

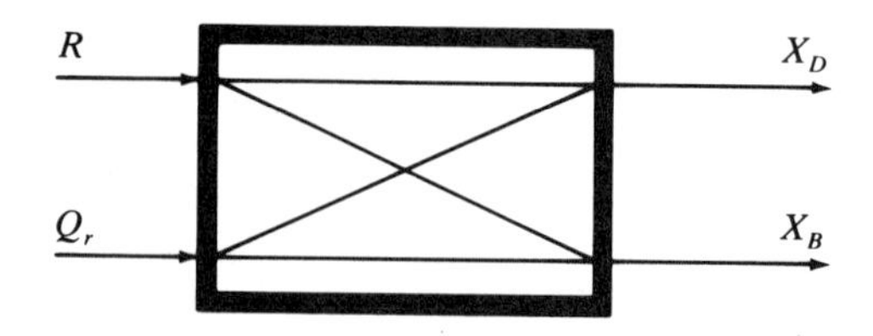

FIGURE 16-8. After basic controls, quality control is a 2 × 2 problem.

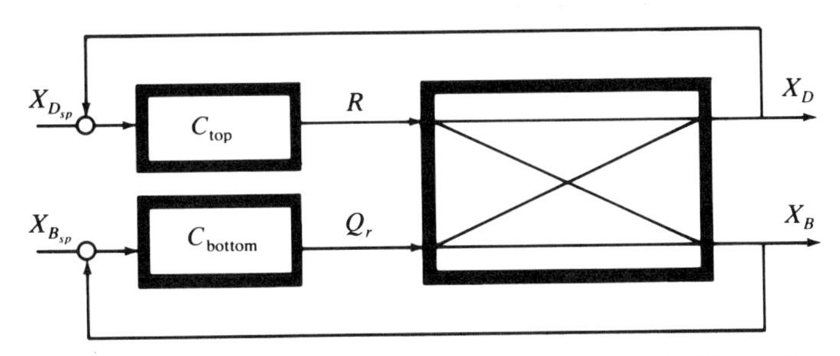

FIGURE 16-9. Traditional but not recommendable double quality scheme.

they tend to fight each other. The purer the product streams, the higher the interaction and the more severe the effects. If, for instance, the top impurity is too high, the top quality controller will increase the amount of reflux. However, more reflux also causes a more impure bottom, resulting in more reboiler heat input via the bottom quality controller, which causes a less pure top, which results again in more reflux, and so on. In many distillation systems this interaction can cause flooding. See Figure 16-9 for the traditional double quality scheme.

Only if one of the two loops is tuned slow compared to the other might this scheme work. Dependent on the disturbances, large deviations for the weak controlled quality can be expected. Often this scheme will need operator assistance if there are disturbances; at best it gives marginal performance.

16-7 HEAT AND MATERIAL BALANCE IN RELATION TO SEPARATION

In fact the separation of a column depends on the ratio between the down-going liquid stream and the up-going vapor stream. Rectification or separation in the top section is

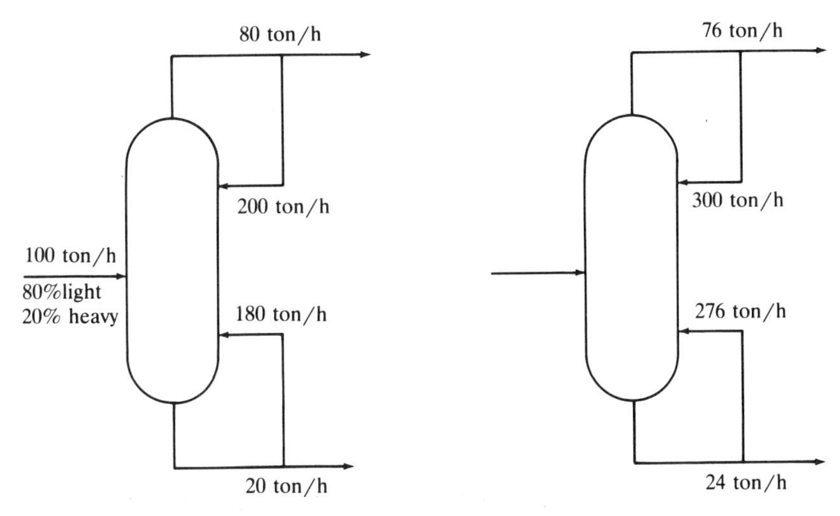

FIGURE 16-10. An improperly balanced reflux and bottom heat causes a high impurity in the bottom.

improved by a higher L/V ratio (more reflux), whereas stripping or separation in the bottom section is improved by a lower L/V ratio (more heat input). Both internal flows act on both flows leaving the column: the heat and mass balance are coupled. Thus increasing the amount of reflux to increase the top purity must also be followed by an increase in heat input to compensate for the higher amount of liquid to the bottom.

From the material balance one can conclude also that an imbalance between these streams causes a unnecessarily high impurity in one of the two ends, as seen in Figure 16-10.

Figure 16-10 shows two situations for a binary distillation column. In both cases the feed, which is at dew point and thus just vaporized, contains 80 ton/h light component and 20 ton/h heavy component. On the left the reflux–feed ratio is $R/F = 2$; on the right $R/F = 3$. In the right-hand case there is an imbalance between the amount of reflux and the heat input: the amount of flow out of the bottom is even more than the heavy component in the feed. The surplus of 4 ton/h must be light component, causing an impurity of at least 20%. The effect will be less pronounced in a real column, but the thoughts behind it remain true.

We can conclude that after a major change in reflux, the bottom heat input also needs to be adapted, and vice versa.

16-8 NET HEAT INPUT

In the case of normal separations the magnitudes of the product streams are broadly the amount of heavy component in the feed for the bottom stream and the amount of light component for the top stream. Therefore the heat input and the amount of reflux need to be in balance with each other: for example, the heat input must be adjusted after a change in the reflux in such a way that the amount of product flow out of the bottom (and thus also over the top) will remain the same. Otherwise, from a mass balance point of view it can be seen that the separation will prejudice one of the ends as pointed out in Section 16-7. In fact, for a given feed, the net heat input needs to be constant:

$$Q_{\mathrm{net}} = R\Delta H_R + F_{\mathrm{reb}}\Delta H_{\mathrm{reb}}$$

where R = reflux (ton/h)
ΔH_R = difference of heat contents between top flow and reflux (kJ/ton)
F_{reb} = reboiler flow (ton/h)
ΔH_{reb} = difference of heat contents between reboiler flow inlet and outlet (kJ/ton)

To compensate for feed flow variations the Q_{net} is normalized by the feed flow itself: Q_{net}/F.

The background of this factor can also be clarified by the following example. Consider a vapor feed at its dew point. Normalizing Q_{net}/F again by dividing it by the heat needed to condense a ton of feed, the factor $Q_{\mathrm{net}}/(F\Delta H_F)$ can be considered as the fraction of the feed that is condensed and leaves the column at the bottom. That fraction is about the same as the fraction of heavy component in the feed. In case of subcooled feed, ΔH_F is the heat needed to vaporize a ton of feed, and the factor can be considered as the amount of feed that needs to be vaporized to form the top product, which is about the same as the fraction of light component in the feed.

An operator gave the following metaphor. Increasing R/F while holding Q_{net}/F constant presses the concentration profile like the bellows of an accordion. Increasing Q_{net}/F while holding R/F constant is moving the concentration profile upward like moving the accordion as a whole without changing the form of the bellow. See Figure 16-11.

16-9 SEPARATION INDICATORS

To understand separation and Q_{net} better, two separation factors are introduced. The

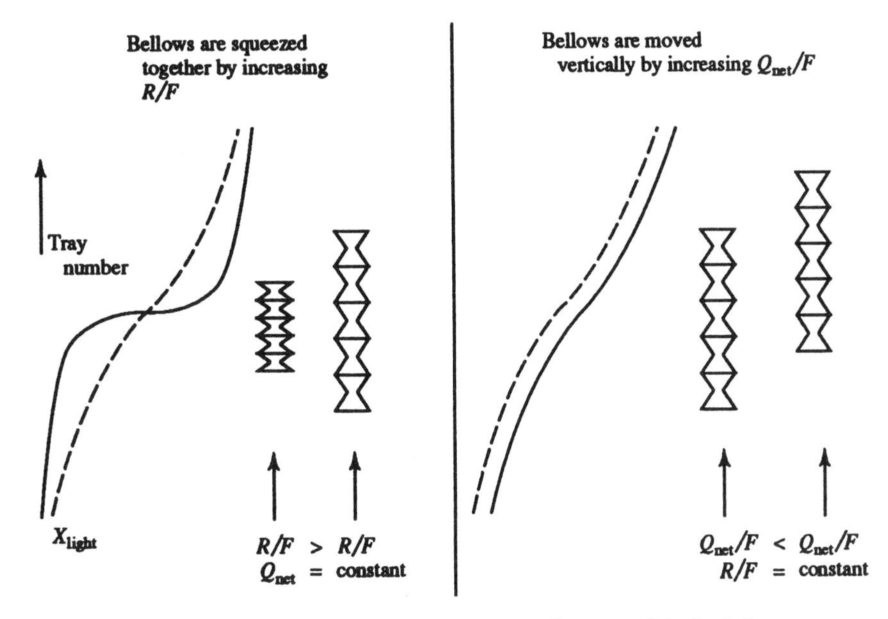

FIGURE 16-11. Metaphor of the bellows of an accordion to explain the influences on the concentration profiles.

philosophy behind the control scheme was derived from the inverse Nyquist array method: combinations of measured variables give new variables; the same is true for manipulated variables.

16-9-1 Separation Performance Indicator

The separation performance indicator (SPI) indicates the degree of total separation of the column: the lower this number is, the lower is the "total" impurity and the better is the separation:

$$\text{SPI} = X_D + cX_B$$

The separation performance is related to the amount of energy put into the column.

16-9-2 Separation Accent Indicator

The separation accent (SAI) indicates where the accent of the separation is placed: bottom, top, or divided equally. The lower this number is, the more the concentration profile is pushed to the lower end (the bottom) of the column. This means that the top is more in favor than the bottom, regarding the impurities. The separation accent indicator can be compared with the cutpoint:

$$\text{SAI} = X_D - cX_B$$

The separation accent is related to Q_{net}/F.

16-9-3 Interpretation

The constant c accounts for different scales of top and bottom measurements and the design of the column, especially the location of the feed tray in relation to the feed composition. By decreasing the reboiler heat input a little, a column has the following steady-state conditions:

$$X_D = 400\ \text{ppm}, \quad X_B = 0.5\%, \quad R/F = 2.5$$

$$X_D = 300\ \text{ppm}, \quad X_B = 0.9\%, \quad R/F = 2.5$$

It can be seen that $c = 250$ ppm/%, and the SPI (separation performance indicator) is for both cases 525 ppm, whereas SAI (separation accent indicator) changed from 275 to 75 ppm. So the total separation of the column is in both case equal, but the separation accent has gone more to the top.

In the case of more reflux, keeping Q_{net} constant by adapting the reboiler heat input, the same column has the following steady-state conditions:

$$X_D = 400 \text{ ppm}, \quad X_B = 0.5\%, \quad R/F = 2.5$$
$$X_D = 350 \text{ ppm}, \quad X_B = 0.3\%, \quad R/F = 2.6$$

The total impurity SPI was reduced from 525 to 425 ppm. This better separation was equally used to decrease the impurities in both top and bottom, which can be seen because the SAI remained the same: 275 ppm.

Theoretically it could be expected that it would have been better to use the logarithms of the impurities, or the reciprocals, instead of the impurities themselves. Examination of experimental results as well as results from simulation showed no significant difference between the different approaches for control purposes.

When using the separation indicators for dynamic control purposes, it might be better to derive them from dynamic responses by dividing the response speeds (slopes), or first derivatives with respect to time of impurity responses after a change in reflux or reboiler heat. See Figure 16-12.

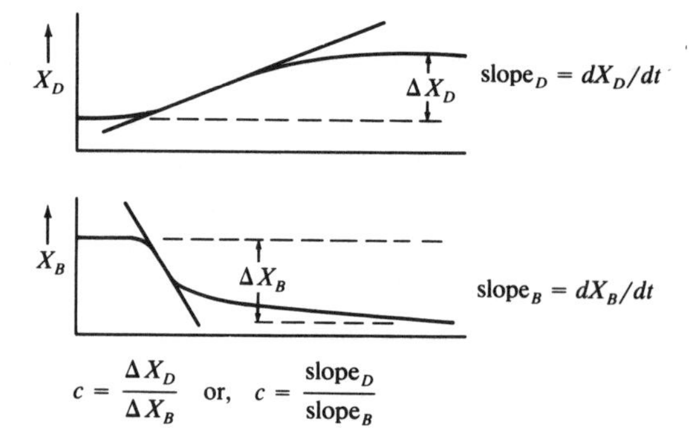

FIGURE 16-12. Two possibilities for the determination of the factor c in the separation indicators, based on concentration response to a change in reflux or reboiler heat.

16-10 DEVELOPMENT OF THE CONTROL SCHEME FOR THE C2 SPLITTERS

16-10-1 Characterization of the Column Operation

Before starting to develop a new control scheme, it is important to have good insight into the way the column is operated, its dynamic behavior, and the kind of disturbances. In the case of the C2 splitter without heat integration, a total of 72 measurements were collected every 10 min via the plant computer for a period of several months. After recalculation to 30 measurements like R/F and Q/F, they were plotted on a graph that measured about 1×6 m. This graph has been studied extensively by the process control engineer and the designer of the control scheme. From this insight the control scheme was developed and tested with real disturbances. We learned from this method of looking at the column that the original infrared measurement in the bottom was not selective for ethylene: it measured the sum of ethylene and propene. Propene can be found in the bottom after an upstream disturbance. The controller that sees this as ethylene will react totally wrong. Instead of leaving this propene in the bottom stream, it tries to strip it out by adding heat. Of course this makes no sense. For this reason this instrument has been replaced by a chromatograph.

16-10-2 Column Simulation

A dynamic mathematical model was especially made for developing a new control strategy for both cases. Due to the high purity separation, the behavior of the column is highly nonlinear in both a static and a dynamic sense. Therefore a rigorous nonlinear model is needed. It consists of dynamic tray-to-tray simulations based on differential equations for conservation of mass, mass of individual components, and energy. Algebraic equations are needed to describe the equilibrium of ethylene and ethane, enthalpies, hold-ups, and so forth. The model was run interactively on a mainframe computer and was time-scaled faster than real time because of the slow responses of the actual column. The reliability of this model was checked by process operators who normally operate this column in the plant. It appeared to be a "hifi" model.

16-11 INTRODUCTION OF THE NEW CONTROL SCHEMES INTO INDUSTRIAL PRACTICE

Much attention has been paid to the introduction of the new control schemes to the operators. A good understanding of a control scheme is vital for its automatic operation for the long run. No control scheme can be in automatic for 100% of the time. Instruments can fail or need to be calibrated. To cope with very severe disturbances, not necessarily related to the column itself, control must sometimes be returned to manual mode. When the operators have a good understanding of the control scheme, they will trust it, know its limitations, and restore it to automatic mode after a major disturbance. If not, controllers will be put in manual and remain in manual.

A lecture of several hours for every shift was presented in the control room during the more quiet evening and night shifts, sometimes together with the dynamic simulation. The operators could then apply disturbances to the column simulator for better understanding of the new control scheme and gain confidence in it.

16-12 C2 SPLITTER WITH HEAT INTEGRATION

The C2 splitter in DSM's naphtha cracker number 3 has 120 trays. The column has a combined reboiler/condenser. The plant has an open-type ethylene refrigerating system, which means that product ethylene itself flows through the refrigerating system

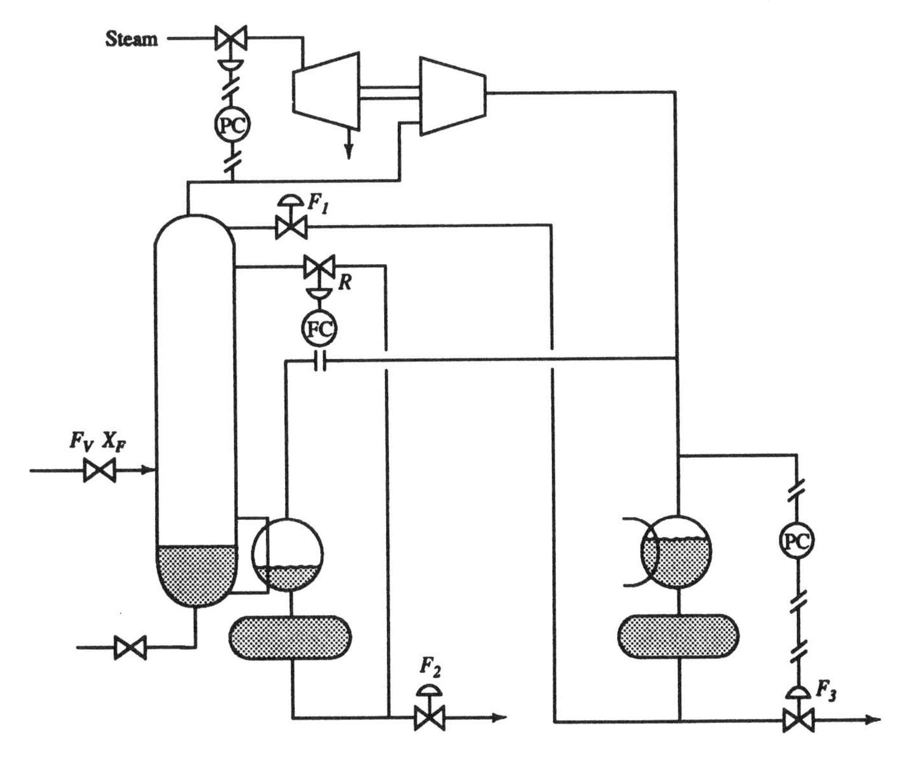

FIGURE 16-13. The C2 splitter with heat integration.

and acts at the same time as low temperature (-110 to $-40°C$) coolant. The ethylene is condensed in the reboiler to provide the bottom heat input and serves then as reflux. One trim stream (F_2) is provided to give more bottom heat than the condensing reflux alone, and one stream (F_1) to provide more reflux than the main reflux itself. See Figure 16-13 for a visual representation. Unfortunately the operators called the two trim streams bottom heat and auxiliary reflux. In fact, the two streams act in the same way, except for sign: 1 ton/h more bottom heat extracts 1 ton/h from the main reflux, while the amount of heat input remains constant. The same is true for 1 ton/h less auxiliary reflux. For better understanding the bottom heat stream was renamed "negative-auxiliary reflux."

16-12-1 Basic Controls

Pressure is controlled by the suction pressure controller of the third stage of the ethylene refrigerating compressor. The reflux/condenser vessel is normally full due to flooded condenser operation. The column bottom level is controlled by regulating the outgoing liquid stream. Two flow controllers are provided: one main reflux flow and one that is switchable to auxiliary reflux and bottom heat (negative-auxiliary reflux).

16-12-2 Quality Measurements

Two gas chromatograph are installed to measure top and bottom quality. The sample time is 12.5 min and the total deadtime is 16 min (2.5 min transport time through the sample line from the field to the gas chromatograph building plus 12.5 min for analyzing). Two temperature measurements provide the operator continuous early warning about changes in concentration.

16-12-3 Previous Control Scheme

According to the instructions the top concentration was controlled by the auxiliary reflux/bottom heat controller and the bottom concentration was controlled by the main reflux. After a change in the amount of feed, which is not measured directly but can be derived from the upstream load, the operators changed both flows manually. There appeared to be no consensus among operators about the influence of the main reflux. Because the main reflux not only acts as reflux, but also as bottom heat, the final influence was not clear to all. Probable reasons for lack of consensus are the large time constant in the order of duration of an operator shift (8 h) and the inverse response: After an increase of the main reflux, the bottom starts to become more pure because the influence of bottom heat dominates first; at the end, the bottom will be less pure. Calculations and plant observations show that an extra 1 ton/h main reflux gives, after flashing, an extra internal reflux of 0.96 ton/h and, due to the differences in heat of condensation of the ethylene reflux and the heat of evaporation of ethane in the bottom, an extra up-going vapor flow of 0.91 ton/h. From the perspective of a mass balance it can be seen that the bottom flow at the end will be 0.05 ton/h higher. If the bottom flow is 10 ton/h this means an increase of at least 0.5% in impurity.

16-12-4 Improved Control Scheme

Compared with other columns, through its nature, this column is relatively stable: A change in main reflux automatically causes more bottom heat, which is necessary for mass balance reasons, as pointed out in a previous section. Although the feed flow is not measured, the top flow immediately shows the feed flow variations because the feed enters the column at its dew point:

$$V_{\text{top}} = F + V_{\text{bottom}}$$

$$V_{\text{bottom}} = aR$$

where

$$a = \frac{\text{heat of vaporization 1 kg ethane}}{\text{heat of condensation 1 kg ethylene}} \approx 0.91$$

The factor a depends on the discharge pressure of the compressor (the higher the pressure, the lower the heat of condensation):

$$\frac{R}{V_{\text{top}}} = 1 \Big/ \left(a + \frac{1}{R/F} \right)$$

By controlling the main reflux–top vapor flow ratio, the main reflux–feed flow is inherently controlled, which is necessary to cope with feed flow variations.

The flow from the lowest tray to the bottom compartment is

$$F_{\text{bottom}} = b(R + r) - aR$$

where

$$b = \frac{\text{internal reflux after flashing}}{\text{external reflux}} \approx 0.96$$

r = auxiliary reflux (positive or negative)

The main reflux determines the separation of the column. The auxiliary reflux can balance the separation between top and bottom impurity: given a certain separation, increase the purity of the top or the bottom at the cost of decreasing the purity of the other product.

The control scheme in Figure 16-14 appeared to be the most simple to understand and was adequate at the same time. Because of the perfect pressure control, a temperature could be used for measuring the position of the concentration profile in the column. The temperature at tray 14, which is controlled by the auxiliary reflux, appeared to be the most suitable. In fact it acts as Q_{net} control. There was no need to make the scheme more complicated by using internal reflux control, feed–auxiliary reflux ratio, separator indicators, and so forth.

16-12-5 Condenser–Reboiler Level Control

The level in the condenser–reboiler is self-regulating as long as the discharge pressure is sufficiently high:

$$R = UA(L)\left[T_{\text{cond. eth.}}(P_{\text{dis}}) - T_{\text{bottom}}(P_{\text{suct}}, X_B)\right]$$

The amount of condensing ethylene depends on the temperature difference between the condensing ethylene, the bottom temperature, and the available area at the condensing side of the reboiler–condenser. The temperature of the condensing ethylene reflux depends on the discharge pressure. If the temperature difference is sufficient, the level in the condenser will be self-regulating: the condensing area is exactly as large as is needed to condense the requested reflux by the reflux flow con-

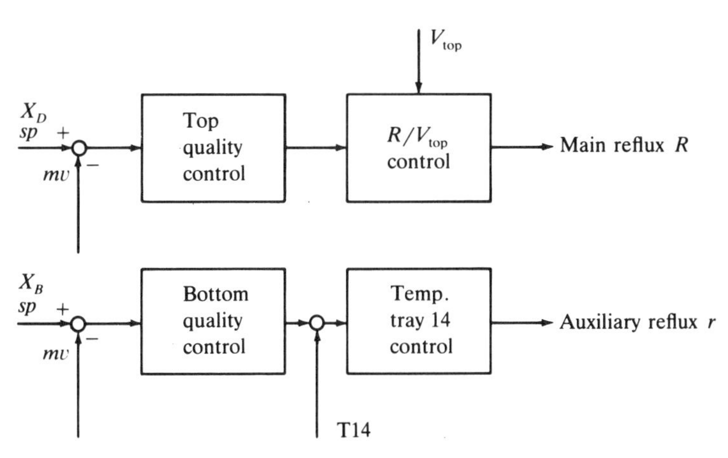

FIGURE 16-14. Improved control scheme.

troller. If, however, the temperature difference is so low that even the total area is not sufficient, less reflux will be condensed than used. This causes the level in the reflux–reboiler vessel to decrease more rapidly than can be expected by the difference between requested reflux and the actual condensed reflux. The reflux transmitter is located in the vapor flow and the valve in the liquid flow; therefore, the controller does not see the actual reflux entering the column. Integral action (reset) of the reflux flow controller causes a continuous increase of liquid flow far above its setpoint, whereas the amount of condensing vapor remains the same. See Figure 16-15.

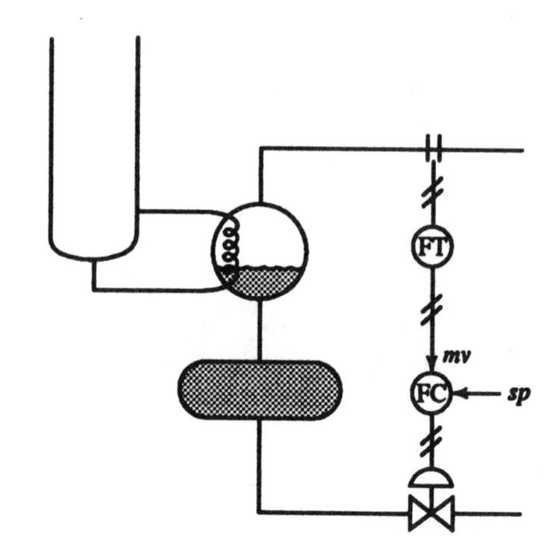

FIGURE 16-15. Reflux flow control.

Finally the stabilizing connection between reboiler heat and reflux vanishes, which results in a upset that ultimately causes a breakthrough of gas to the column, no reflux to the column, increasing pressure, off-spec products, and so forth. To prevent this a level controller is installed that can override the reflux flow controller. It is better to have too little reflux than no reflux at all.

To ensure sufficient pressure, a discharge pressure controller is provided to guaran-

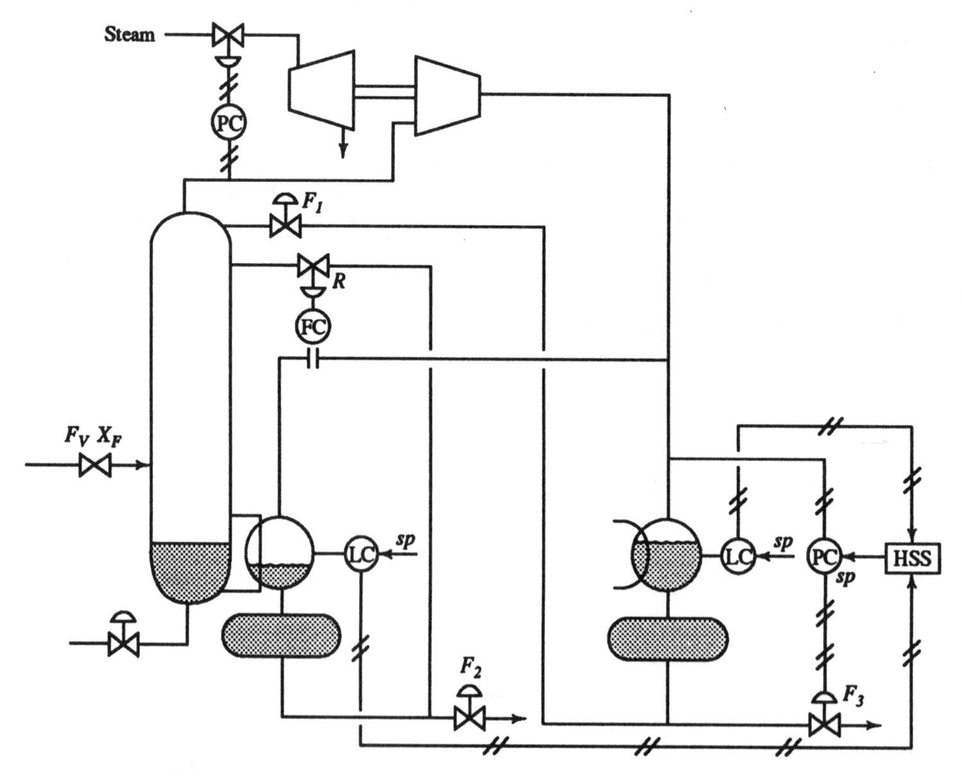

FIGURE 16-16. Discharge pressure control of the third stage of the ethylene compressor, preventing a too low pressure, which results in breakthrough of gas and an excessive pressure, which spoils compressor energy.

tee flooded condenser–reboiler operation. However, if this discharge pressure is too high the energy consumption of the compressor is also higher than necessary. Therefore the setpoint of the discharge pressure controller is determined, via a high selector switch, by the level controllers of the condensers in which the compressed ethylene of the third stage is condensed. See Figure 16-16

16-12-6 Experiences

Influence of Reboiler–Condenser Vessel Level At the beginning the setpoint of the reflux–reboiler vessel level override controller was chosen too low. The background was to use this level change to maximize the time for the discharge pressure controller to adjust its setpoint and real value. However, during a level change there is an imbalance

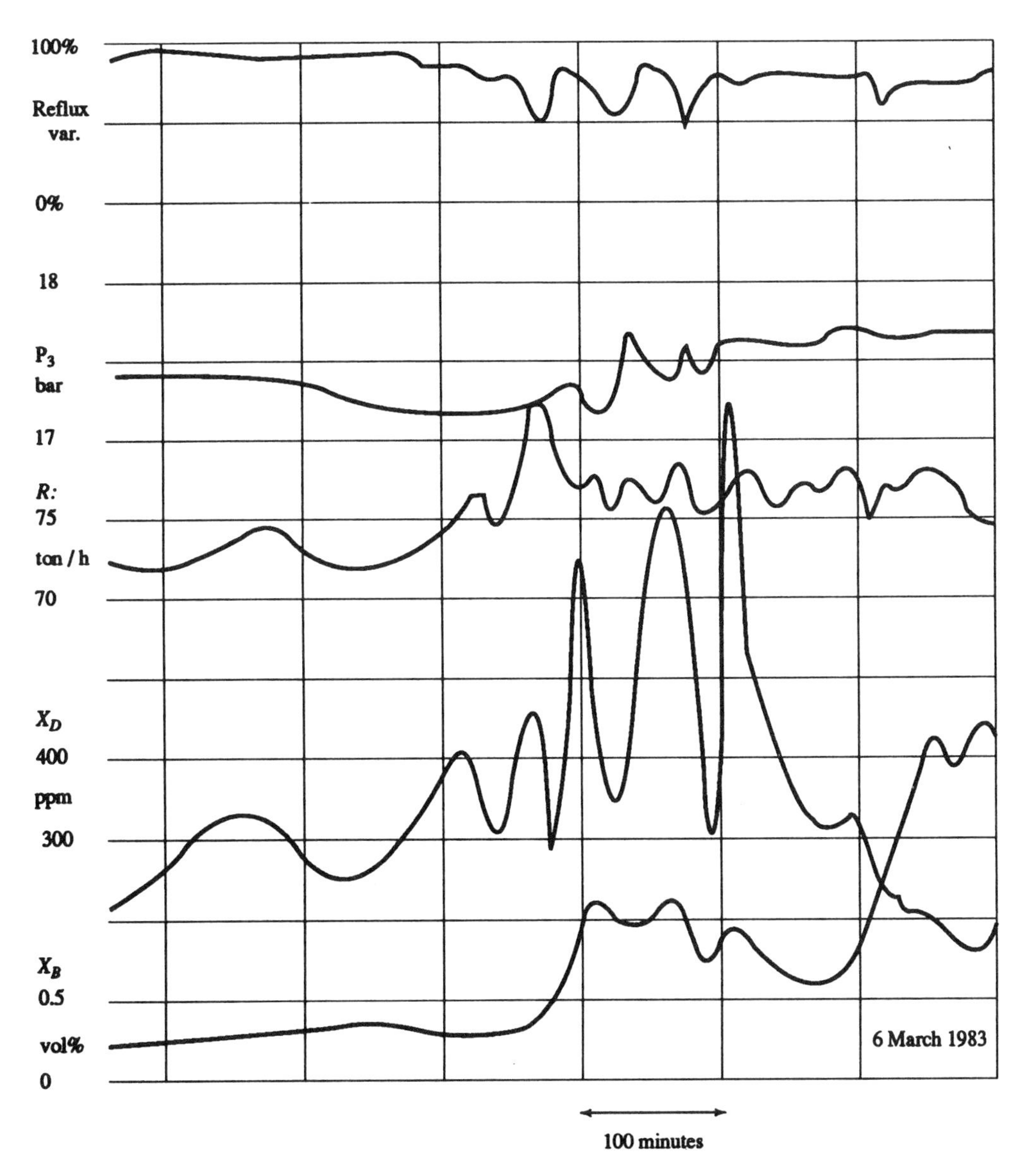

FIGURE 16-17. Influence after an imbalance between heat input and reflux, caused by discharge pressure too low to condense the reflux. The fluctuating level in the reflux–condenser vessel indicates this imbalance.

between reboiler heat and reflux (i.e., in Q_{net}). This causes a major upset in the column. Therefore, this setpoint needs to be as close to 100% as possible. Severe upset occurred when the discharge pressure of the third stage of the compressor became too low to condense the requested reflux and the setpoint of the level controller was 50%. See Figure 16-17.

Reflux Measurement Not Properly Pressure Compensated

After an increase in discharge pressure the reflux flow started to rise, which resulted in a purer top stream. After the controller put the concentration back to its setpoint, the reflux was 1.5 ton/h less than before. See Figure 16-18. What occurred? Because the reflux flow measurement takes place in the vapor phase, this measurement needs to be compensated for pressure variations: a ΔP measurement is in fact a volume flow measurement. If the pressure increases, the same ΔP will be maintained by the reflux flow controller, resulting in a greater mass flow. This increase was not detected by the flow controller, but it is seen by the V_{top} measurement; therefore, the effect was magnified by the R/V_{top} controller. The increase in discharge pressure from 17.5 to 18.2 atm causes a direct increase of $\sqrt{(18.2+1)/(17.5+1)} = 1.02$ or 2%. Thus V_{top} will increase. To keep an R/V_{top} ratio of 0.69, the reflux will be increased by the ratio controller:

$$\frac{R_{new}}{R_{old}} = \frac{1 - 0.69}{1 - 0.69\sqrt{(18.2+1)/(17.5+1)}}$$

$$= 1.05$$

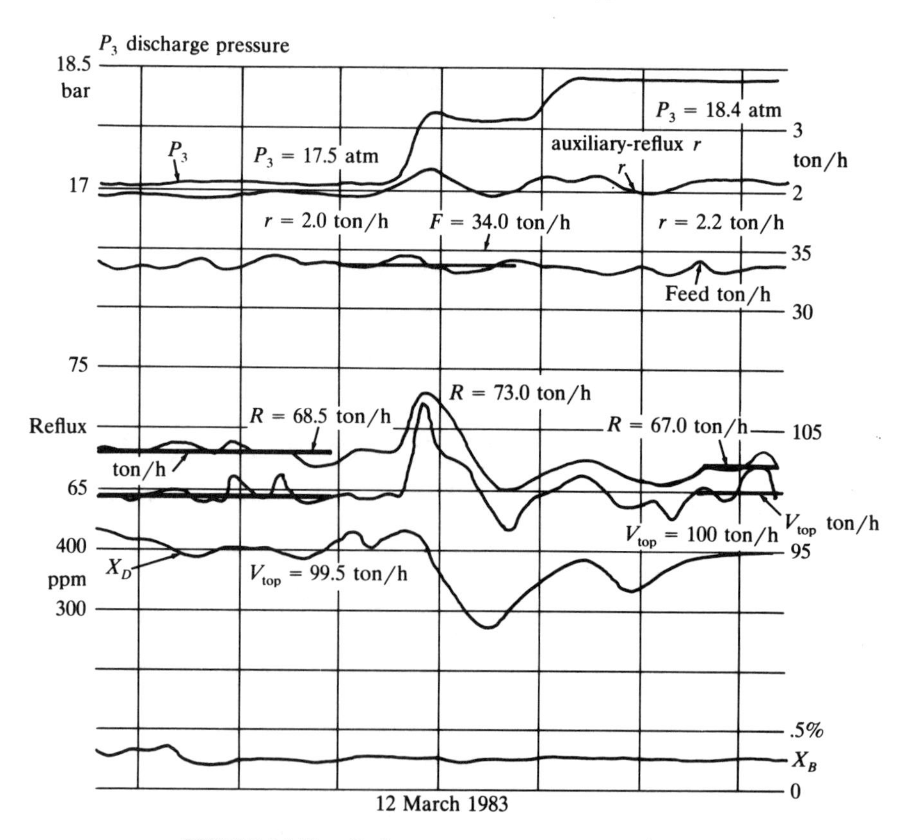

FIGURE 16-18. Reflux measurement not properly pressure compensated. After an increase in discharge pressure the same ΔP of the flow transmitter results in more tons per hour, which will be amplified by the V_{top}/reflux ratio controller. The quality controller ultimately corrects this increase.

This ratio is in accord with the plant data depicted in Figure 16-18: 73/68.5 = 1.065.

Because of excess reflux, the top concentration starts to be too pure, and therefore the concentration controller will reduce the reflux until the top concentration setpoint is met. At the end the discharge pressure was 18.4 atm, which gives an aberration of 2.3%. The concentration controller corrects, in fact, the deviation in flow measurement: the reflux measurement shows 67.0 ton/h and the previous measurement was 68.5 ton/h, 67.0/68.5 = 0.978, a reduction of 2.2%. (Remember the flow started 2% too high after the pressure change). In fact, the amount of reflux is then the same as before the change in discharge pressure.

To prevent such offsets, good pressure compensated flow measurement is necessary: it reacts in a feedforward way instead of via the concentration controller. Note that during the offset the bottom concentration controller did its work perfectly.

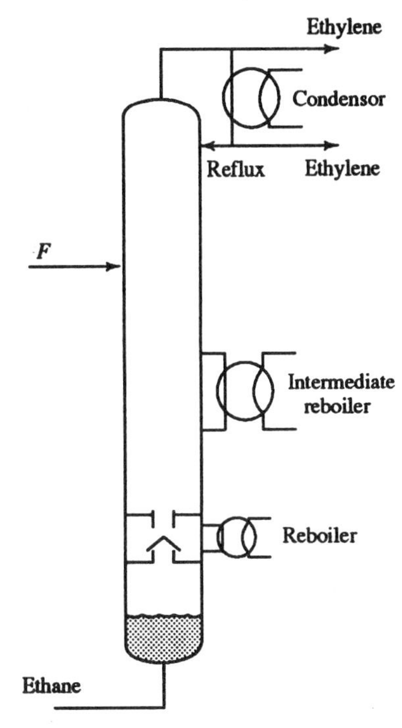

FIGURE 16-19. The C2 splitter without heat integration.

16-13 C2 SPLITTER WITHOUT HEAT INTEGRATION

The C2 splitter in DSM's naphtha cracker number 4 has 160 trays and, in addition to a reboiler and condenser, is equipped with an intermediate reboiler between the bottom and the feed tray that is designed to utilize an energy stream from elsewhere in the plant. See Figure 16-19.

16-13-1 Basic Controls

The basic loops are standard: pressure is controlled by the condenser, and the top and bottom levels are controlled with the product streams.

16-13-2 Quality Measurements

The concentrations are measured with gas chromatographs. The sample time is 20 min and the total deadtime is 23 min, caused by the same 20 min analysis time and 3 min transport lag from the column to the chromatograph. Temperature measurements in both top and bottom sections are available.

16-13-3 Previous Control Scheme

The previous control scheme in Figure 16-20 was a "standard" double quality control: Top quality was controlled with the reflux and bottom quality was controlled with the reboiler.

16-13-4 Interaction

The two available manipulated variables (reflux and bottom heat input) influence both the top and the bottom impurity in such a way that the two independent loop controllers of the original control scheme did not work satisfactorily. Although feed variations are feed forward corrected via ratio control, it does not account directly for changes in the flow over the intermediate

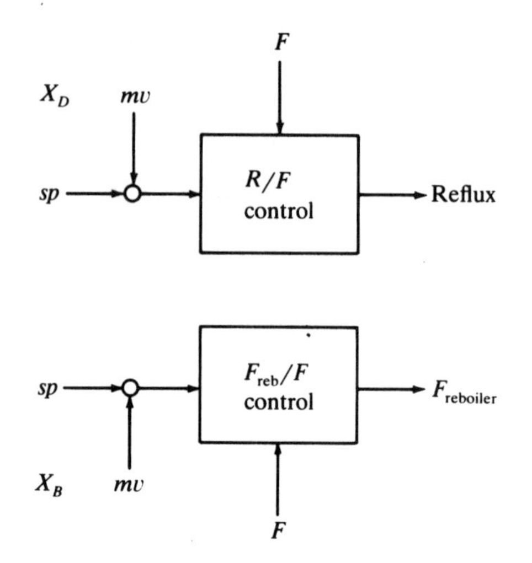

FIGURE 16-20. Previous quality control scheme.

reboiler. In fact the heat input via the intermediate reboiler should also be controlled proportional to the feed, but because this reboiler is part of a complex, tightly balanced refrigerating system it could not be involved in the control scheme.

16-13-5 Improved Control Scheme Using Q_{net} / F and the Separation Factors

Column simulations were used to develop the new control scheme. Because of the presence of the intermediate reboiler Q_{net}/F control contributed a great deal to stabilizing the column. Of course the heat

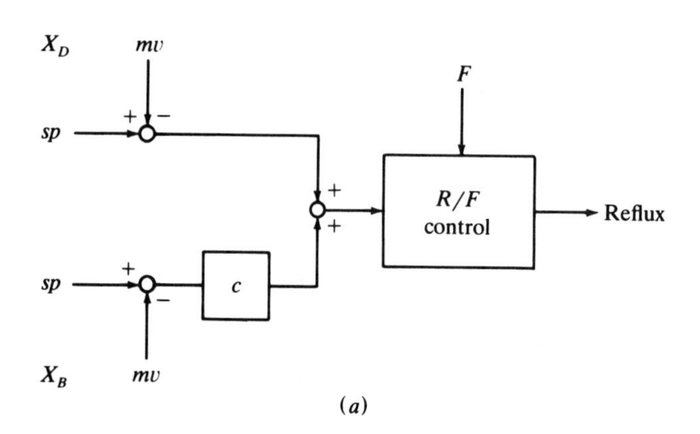

(a)

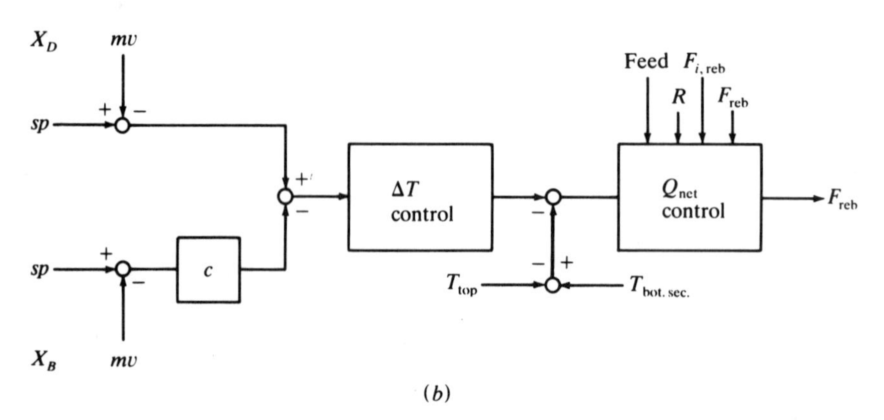

(b)

FIGURE 16-21. (a) Separation performance control. (b) Separation accent control.

input via the intermediate reboiler is taken into account. Because of the interaction, two separation indicators were used to partly decouple the influence of reflux and reboiler heat.

Simulations of the column showed further that much better control could be achieved if information about changing concentrations was available sooner. This was especially true for Q_{net}/F control loop. Due to the high purity of the column, temperatures in the bottom and the top hardly changed with concentration variations. A temperature measurement somewhere in the bottom section appeared to be a good indication for a moving concentration profile. To make this indication independent of pressure variations, the temperature difference with the top temperature was used. So the Q_{net}/F became the master controller and the Δ temperature became the slave controller. See Figure 16-21*b*.

16-13-6 Experiences

The effect of the new control scheme was evaluated by collecting plant data. The standard deviation of the impurities was reduced by a factor of 7 and energy consumption of the column was reduced by 5%. In addition, the intermediate reboiler could be operated for optimizing the refrigerating system without introducing disturbances in the column.

After six years of experience this control approach has proven to be robust and is

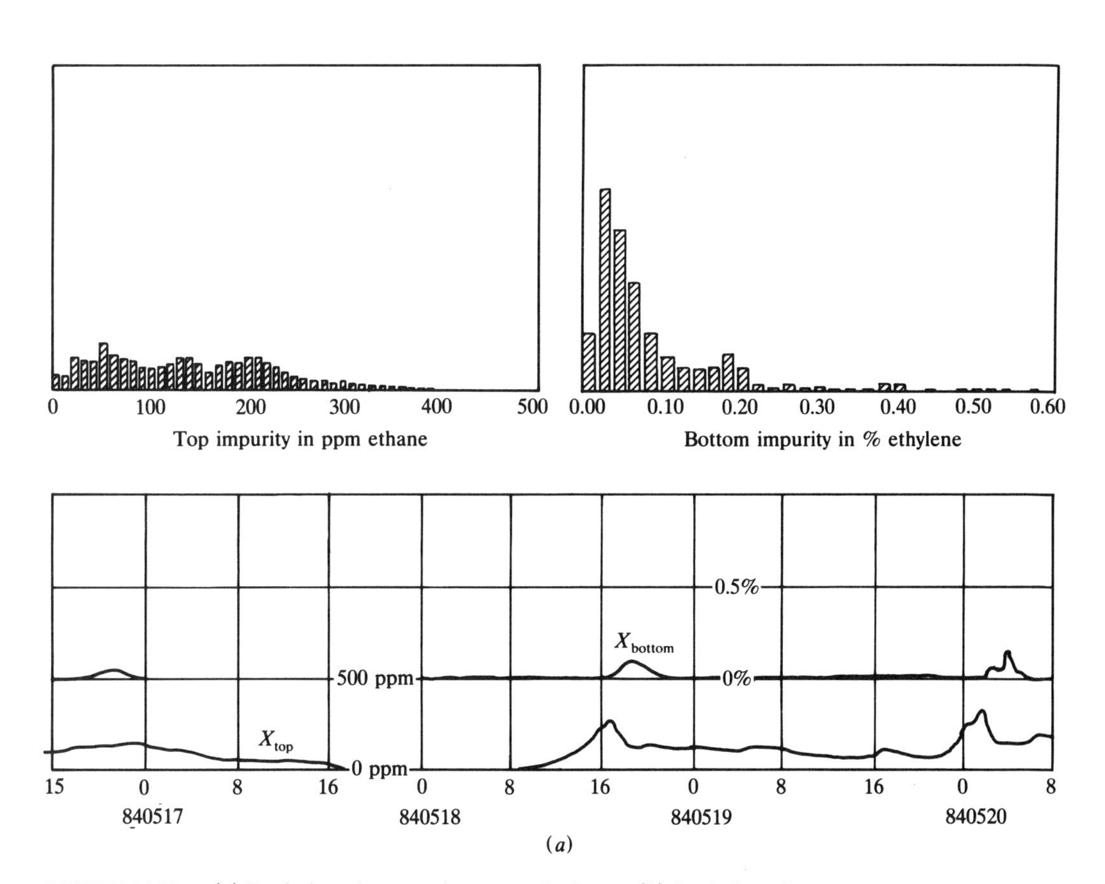

FIGURE 16-22. (*a*) Real plant data: previous control scheme. (*b*) Real plant data: new control scheme.

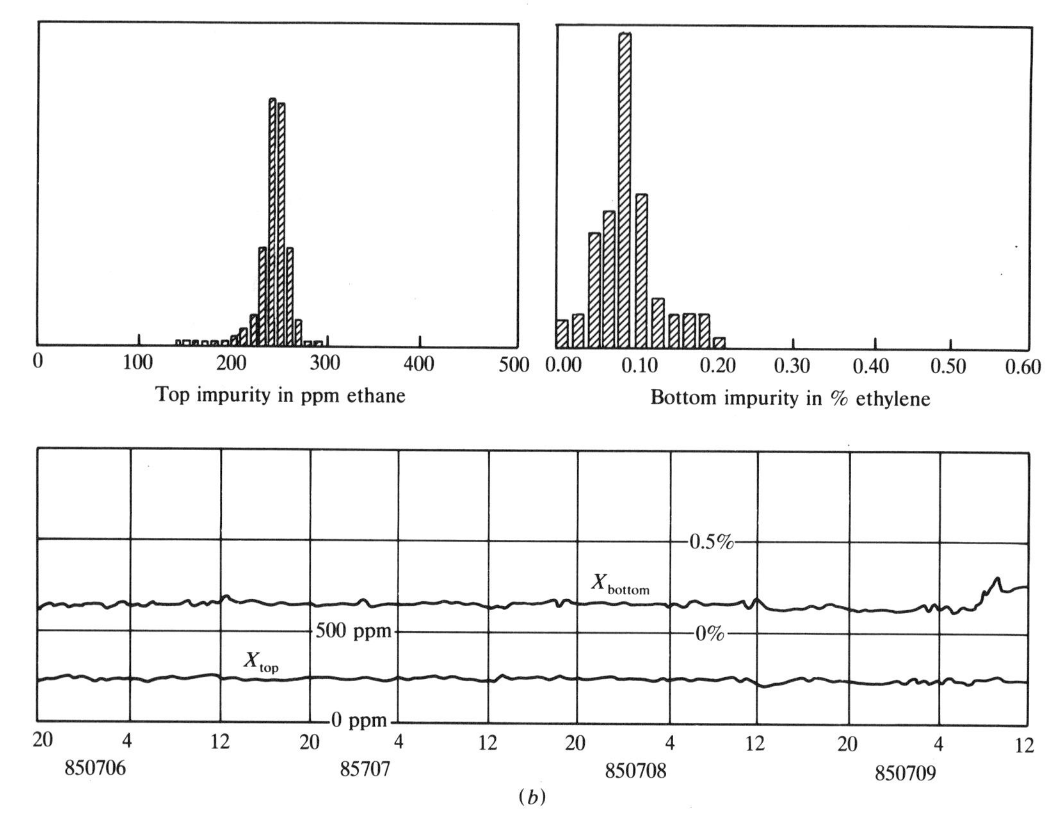

FIGURE 16-22. *Continued*

well accepted by the operators. Outside of periods of instrument maintenance, the control scheme operates on automatic. See Figure 16-22 for the plant data.

16-14 CONCLUSION

A cookbook approach will never work for solving control problems in general, and particularly not for more complicated control problems like double quality control of distillation columns. No two columns are the same, and even when they are said to be the same, the operating conditions, including disturbances affecting the column, are not identical. In addition to the knowledge of the (im)possibilities of control engineering, it is important to have a good insight into the dynamic behavior of the column itself.

Thinking in terms of Q_{net}/F and the indicators for separation performance and separation accent appears to be very useful. Sometimes they can be applied literally and sometimes they serve as a way of approaching and solving the double quality control problem. After our success with these C2 splitters, control systems for other columns were designed using this approach. In some cases the use of the separation indicators was not needed: the top impurity was controlled by R/F and the bottom impurity by Q_{net}/F. For operators the approach appeared to be very well understandable and thus well accepted.

17

Control of Distillation Columns via Distributed Control Systems

T. L. Tolliver

Monsanto Company

17-1 INTRODUCTION

The intent of this chapter is to provide a review of three distillation column control applications as implemented via distributed control systems. The control schemes have been developed by a typical industrial design approach, applying process insight in order to reject disturbances and achieve a hierarchy of process control objectives (Buckley, 1978; Hougen and Brockmeier, 1969; Rademaker, Rijnsdorp, and Maarleveld, 1975; Shinskey, 1984).

Process insight can be obtained for new distillation columns by using steady-state simulation of parametric operating cases, determined by rigorous material and energy balance design programs (Luyben, 1975; Tolliver and McCune, 1978; Thurston 1981a, b). This approach is also effective for analysis of existing distillation columns, particularly if a good base case model has already been developed. The implementation of the column control schemes via distributed control systems is emphasized because this instrumentation has had significant impact on what can be practically implemented and maintained in the manufacturing environment (McMillan, 1991).

17-1-1 Historical Background on Distributed Control Systems

In the latter half of the 1970s, a new form of control instrumentation hardware was introduced into the processing industries. This new instrumentation is referred to as distributed control systems (DCS). What rapidly emerged as the unique characteristic of this DCS hardware is the functional distribution of control over many microprocessor-based controllers with a common, console-based, video display unit (VDU) as the operator interface. The proposed benefits of this hardware approach over the prior usage of analog instrumentation, even when used in conjunction with process minicomputers, were increased visibility, functionality, and reliability (Dallimonti, 1980).

In the late 1970s, DCS systems were being designed and installed only on large grassroot plants. However, throughout the 1980s, DCS hardware became the accepted practice both for new plants and for modernization projects on existing plants (Tolliver et al., 1989). The widespread acceptance of DCS as a replacement for analog instrumentation is a good indication that the proposed benefits are being achieved.

17-1-2 Advantages over Analog Instrumentation

Due to the increased visability, functionality, and reliability, DCS instrumentation has provided an excellent tool for the application of sophisticated control schemes. As an example, consider the prior practice with an analog implementation of a control scheme to maintain the duty on a heat exchanger. There would be three field transmitters, including a differential pressure transmitter for flow rate and two temperature transmitters for the inlet and outlet heat exchanger temperatures.

This prior approach would then require three additional analog modules, one to take the square root of the dp signal to provide a linear flow, one to take the difference between the two temperatures, and one to provide the product of the linear flow and temperature difference. These modules typically would be blind (i.e., no indication) and require special bench calibration methods. Each would be subject to the inaccuracy and nonrepeatability of analog computation. Design would have to take into account proper scaling and the resulting system would be somewhat inflexible to changes in the field transmitter ranges. Indication and recording of the individual transmitter signals, in addition to the computed heat duty, would be expensive, requiring manual retrieval from multipen strip chart recorders.

Another previous approach to this control scheme implementation would have been to use a process minicomputer in conjunction with analog instrumentation. The minicomputer would read the three field transmitters, perform the computations digitally, and provide a supervisory setpoint to an analog flow controller. Supervisory control, as opposed to direct control, would be necessary because the availability of a process minicomputer is not satisfactory for continuous control applications. This was a very good approach for the late 1960s and early 1970s, before DCS instrumentation became available. It can be effectively implemented.

However, this latter approach still has significant shortcomings, which would often result in the loop reverting to simple flow control at the operator's discretion. Primarily, these shortcomings center around the operator interface, which is now divided between a VDU to the minicomputer and the panel mounted analog controllers. Also, the diverse engineering disciplines involved in using and maintaining both the analog and minicomputer system present a problem in support of ongoing process troubleshooting (McMillan, 1989; Hanley, 1990).

This heat exchanger control scheme implemented via DCS instrumentation would have the same three field transmitters wired directly to a microprocessor-based single loop controller, which reads the signals, performs the computation digitally, and provides a heat duty controller with a standard VDU interface to the operator. Availability of the microprocessor-based control is significantly *better* than an analog system. Real-time digital simulation can easily be applied for loop checkout and operator training. The heat duty loop now tends to stay in service, without heroic design and ongoing maintenance efforts. The process engineer can now focus his attention more toward smart control strategies and less on the implementation and system maintenance requirements.

The difference in these approaches is not black and white. There have been, and still are, successful implementations of heat duty control with analog instrumentation. Also, there are certainly some highly successful and sophisticated supervisory control strategies implemented with a process minicomputer. The difference is a matter of degree. Before DCS, these successes were the exception; now with DCS they are widely practiced. In a review paper of distillation column control in the Chemical Processing Industries (CPI), it was lamented that during the 1970s, there were fewer than 150 successful applications of feedforward control (Tolliver and Waggoner, 1980). Now, feedforward control is routinely applied.

17-1-3 Future Trends

In spite of this seemingly glowing endorsement of DCS instrumentation, there remain several areas for improvement, which are being addressed by major instrumentation manufacturers. One trend is the extension of the data highway (digital signals on a coaxial cable) out of the control room and into the field. In addition to reduced wiring costs, perhaps the main benefit of this approach will be easier access to additional control loop information from smart transmitters and smart valves.

Smart transmitters are transmitters that perform self-diagnostics, compensate themselves for changes in ambient conditions, and can be automatically recalibrated with varying ranges. In general, they provide more accurate and reliable data. Likewise, smart valves will also provide self-diagnostics and true valve positioning and indication. Valve positioning problems have been reported as the primary cause of poor loop performance in basic control loops (Ender, 1990). True valve position, rather than the implied position based on controller output, will certainly help in control loop troubleshooting.

Other areas of improvement will address the ever-growing data base of process information and the need for smart alarms, data reconciliation, and troubleshooting diagnostics. All of these areas require improved data management methods as the amount of process data increases. Critical operating information needs to be brought to the operators attention on an exception basis through the use of smart alarms. The extensive process data must be verified by data reconciliation methods before subsequent integrated business systems propagate errors. Maintenance of sophisticated control strategies becomes much more complex with the increased number of field transmitter signals involved. Therefore continuous monitoring is desirable to provide preventive maintenance, and on-line fault tree analysis is also needed to reduce troubleshooting effort and downtime when problems do occur.

17-2 CASE I

This first example of distillation column control, implemented via DCS instrumentation, will be referred to as Column A. An overview of the process material balance and column control strategy is provided by Figure 17-1. The process background and considerations leading to this particular control scheme design, implementation details, and results are discussed in the subsections that follow.

17-2-1 Process Background — Debottlenecking

Column A was an existing column, being operated via DCS instrumentation installed on a prior modernization project. At the time the DCS hardware was installed, it was believed that a subsequent debottlenecking project would soon replace this column; Therefore, the existing analog control strategy, without improvement, was implemented for the short period of interim operation. Control of this column was normally stable, although somewhat oscillatory, and during larger than normal disturbances, the column control required intervention and considerable operator attention.

As the debottlenecking project was being developed, an alternative process design was proposed, which made modifications to the existing column rather than replacing it. It was determined that the stripping section of the existing column was limiting capacity, primarily because a considerable amount of sensible heat was having to be provided by the reboiler. The alternative design therefore provided another source of heat, by partially vaporizing the feed with a new feed preheater, and increased capacity by modifying the trays below the feed for oper-

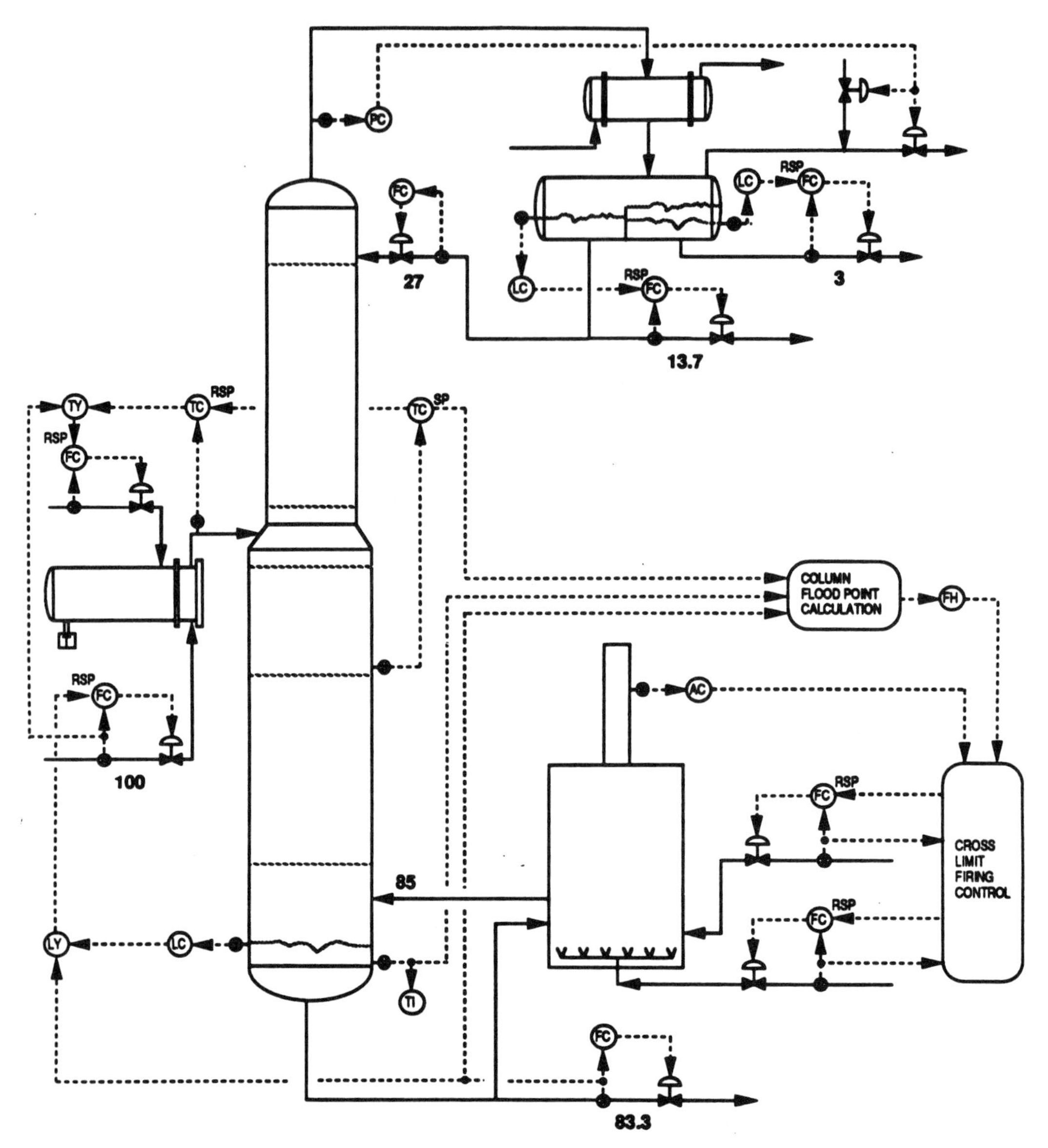

FIGURE 17-1. Column A.

ation at the resulting new liquid and vapor rates. The modified column would then be able to support a 50% capacity increase.

The proposed design required the column to operate close to its flood point almost all the time. There was much skepticism among those with prior experience operating the existing column that this new design would really handle the proposed rates, or that the column could be satisfactorily controlled near its flood point. However, the proposed column design was finally accepted, due to its large reduction in capital requirements, and a detailed control review was recommended as part of the debottlenecking project. It is doubtful that this aggressive design would have been approved without a distributed control system available to handle the difficult control requirements.

17-2-2 Control Scheme Design

Column A represents a truly multicomponent distillation with nearly 25 components present in significant quantities in the feed. However, from a control perspective, the feed can be considered to contain the following four lumped components:

3% component A, lighter than the light key
12% component B, the light key
70% component C, the heavy key
15% component D, heavier than the heavy key

where components A, B, and C are recycled back to the reaction area of the process and component D is the final product. Column A makes the split between the keys, taking components A and B overhead with approximately 12 to 15% component C, whereas components C and D are taken out the bottom with less than 0.05% component B. All compositions and specifications are expressed in weight percent. The overall column material balance is given in Figure 17-1 on the basis of 100 lb/h of feed flow.

The bottom specification is considered to be the primary control objective because increases in component B leaving with the bottom stream represent an unacceptable yield loss. The overhead specification impacts the amount of recycle material, which affects operating costs; however, the operating cost curve is relatively flat near the overhead specification. Although transients in the overhead composition are acceptable, a permanent increase in recycle material is undesirable because it eventually impacts column capacity, returning to the column as additional feed.

The control scheme arrangement of the column overhead is somewhat dictated by the use of the large horizontal reflux drum as a decanter for separating the more dense and immiscible component A from component B. This decanter also represents the only process surge capacity for component B; therefore, loose level control is generally desirable. With these considerations, in addition to the loose overhead composition specification and the relatively low reflux ratio, it is logical to establish a controlled flow of reflux for optimum column separation, and to adjust the distillate flow for decanter level control.

Lumped components C and D, together consist of more than 15 individual components, and product D actually becomes three products with light, medium, and heavy boiler distributions, Hence, this process includes a refining train consisting of three additional distillation columns and other processing equipment fed by the bottom stream of Column A. There are no intermediate storage tanks in this refining train; therefore, it is highly desirable to isolate those operations from transients due to disturbances in the front end of the process. For this reason, the bottom stream of Column A is placed on flow control and its setpoint establishes production rate for the refining process.

Column A could have been considered part of the refining train, and then its feed would have been flow controlled in a more typical column control strategy. However, an analysis of the types of disturbances indicated that the most frequent and severe upset to Column A is a variation in the amount of component B contained in the feed. If the feed to Column A were flow controlled, then any such large variation in component B would cause a significant disturbance to the bottom flow. However, with flow control of the bottom stream, the feed flow to Column A is adjusted as necessary to satisfy the column material balance, as indicated by the sump level, and the refining rate is thus maintained.

The new feed preheater, which partially vaporizes the feed, was originally viewed by manufacturing as a potential operating problem. However, from a control perspective, it gave a new opportunity to react to the most frequent disturbance, feed variations in component B, before they affected the column. By applying temperature control to the feed leaving the preheater, as the amount of component B varies, so does the amount flashed immediately upward. This control strategy maintains a more constant flow of the liquid entering the stripping section.

The key components in Column A have a high relative volatility and therefore separate rather easily. As is often the case with easy separations, the boiling point temperature varies considerably, in this case over 300° F, from top to bottom of Column A. The majority of the temperature increase occurs below the feed, in the stripping section of the column. The feed preheater is able to use steam; however, the reboiler requires a gas-fired heater due to the higher process temperature. Because the majority of the feed leaves with the bottom flow from the column, a significant portion of the reboiler duty is still providing sensible heat.

In order to operate with the stripping section near its flood point, it is necessary to keep the vapor rate constant. Often differential pressure can be used as a measure of vapor rate, and for this reason, a pressure sensor was installed just below the feed tray. However, a reliable measurement could never be demonstrated on Column A. The constant vapor rate could be maintained only if the amount of latent heat provided by the reboiler was held constant. The gas-firing rate of the reboiler is thus determined by the sum of the sensible heat required by the external column energy balance and the latent heat necessary to provide a constant internal boilup.

The sensible heat requirements vary considerably as the distribution of components lumped as components C and D change. For example, the bottom temperature of Column A can vary as much as 50°F with different feed stocks. For this reason, it is necessary to measure and compute the sensible heat requirements under the current operating conditions.

The sensible heat required is computed as

$$Q_s = BC_p(T_c - T_b) \tag{17-1}$$

where Q_s = sensible heat duty (MBtu/h)
B = bottom flow rate (klb/h)
C_p = specific heat constant (MBtu/klb-° F)
T_c = control tray temperature (°F)
T_b = bottom temperature (°F)

At startup, it was found that the measured temperature on the control tray had transients that significantly affected the sensible heat calculation. With the control scheme shown in Figure 17-1, column temperature is controlled by manipulating the feed temperature setpoint, which in turn adjusts the steam flow to the feed preheater. The intent of this arrangement is to allow the vapor rate in the stripping section to remain constant. However, as the temperature on the control tray would change, the sensible heat calculation would also respond to the transient. The solution to this problem was to use the column temperature controller setpoint, rather than the measured temperature.

After the sensible heat requirement is calculated, a latent heat determined in proportion to the bottom flow rate is added, and then the sum is converted to a gas-firing rate. The latent heat used has upper and lower limits that correspond to vapor rates at the flood and weep points of trays in the stripping section. The resulting firing rate can be manually biased before being used as a setpoint to the firing control system.

The manual bias is implemented as a trim to the computed setpoint in terms of percent of the full scale firing rate. At a 100% setting, the computed value is used with a 0% bias. At 101% or 99%, a $\pm 1\%$ bias is used. The bias station indicates both the percentage used and the percentage requested. This allows the operator to manually balance the firing rate before putting the firing system in remote setpoint. Following startup, the flood calculation was checked periodically by raising this bias until incipient flooding was observed.

This latter capability proved quite helpful in gaining the confidence and support of the operators. In spite of extensive training with regard to the sensible and latent heat aspects of the reboiler duty, the operators often felt that the flood point calculation was unduly limiting the firing rate. They observed operation under some conditions where the computed reboiler firing rate was as much as 30% less than at other times. However, every time they attempted to bias up the firing rate, they experienced incipient flooding, usually within 3% of the computed value.

17-2-3 Implementation Details

During the control study, it was determined that the reboiler gas flow had been previously cycling due to almost 8% hysteresis in the gas flow control valve. These cycles were significant enough to have been causing the oscillatory column temperature experienced prior to the debottlenecking project, and which would be unacceptable with post project operation near the flood point of the column. Installation of a valve positioner has since allowed the gas flow to be controlled essentially at setpoint.

The combustion control system on the reboiler was also upgraded with a cross limit firing control scheme using an oxygen controller to trim the air flow setpoint. This latter upgrade was necessary to achieve the maximum heat duty required at the highest rates and temperatures. Prior operating practice using excessive air flow, in addition to being energy inefficient, would have limited the heat duty available to less than that required.

The feed temperature control loop uses feedforward control to set the steam flow in proportion to feed rate with feedback trim from the feed temperature controller. The feed temperature loop was at first tuned as a PID controller with tight setpoint response. Later, in order to get the best response for the column temperature controller, the feed temperature loop was retuned with mostly proportional action, slow reset, and no derivative. Feedforward control based on the feed rate provides most of the action required to handle load disturbances, minimizing the need for reset. Also, offset from setpoint is not critical to overall performance because the feed temperature setpoint is being adjusted by the column temperature controller.

This triple cascade controller arrangement with two temperatures and a flow was an operational concern prior to startup. However, the feed temperature controller proved its value in rejecting disturbances during startup, when the column is typically run in a manual mode prior to putting the column temperature controller on automatic. During startup conditions, the column now runs more stably, and manual adjustments are required much less often than with previous operation.

The column temperature control loop maintains a characteristic column temperature used to infer the column's separation objectives. The prior sensor location was verified using a procedure that computes parametric cases from a steady state material and energy balance design program (Tolliver and McCune, 1980). With steady state temperature profiles at slightly greater and less distillate–feed ratio than the design case, the temperature sensor is located at the stage that exhibits the greatest, most symmetrical, steady state temperature deviation from the base case. Results of these parametric studies are presented in Figures 17-2 and 17-3 for the two alternate feed stocks used by this process.

The existing temperature sensor location, on a column tray equivalent to stage 4, was found acceptable for both feed stocks. However, the resulting process gain in this temperature control loop was significantly different between these two feed cases. Hence, it was found necessary to apply alternative

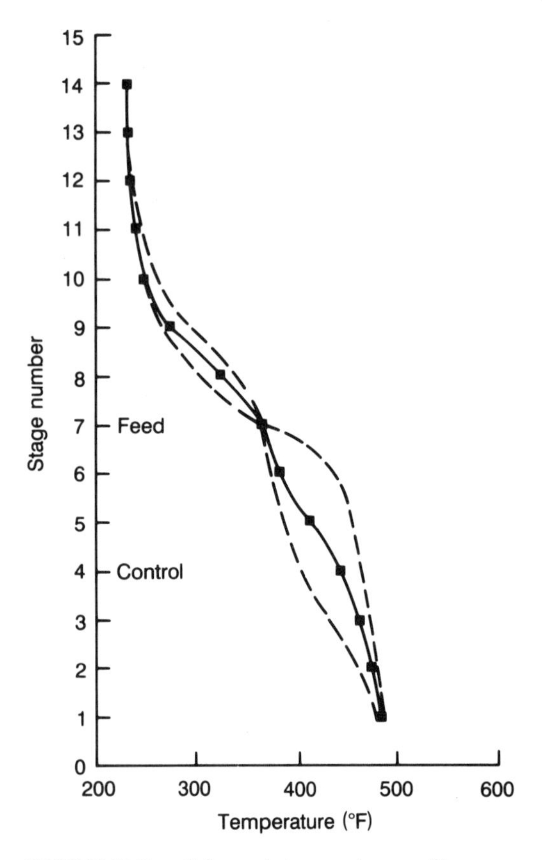

FIGURE 17-2. Column A temperature profile.

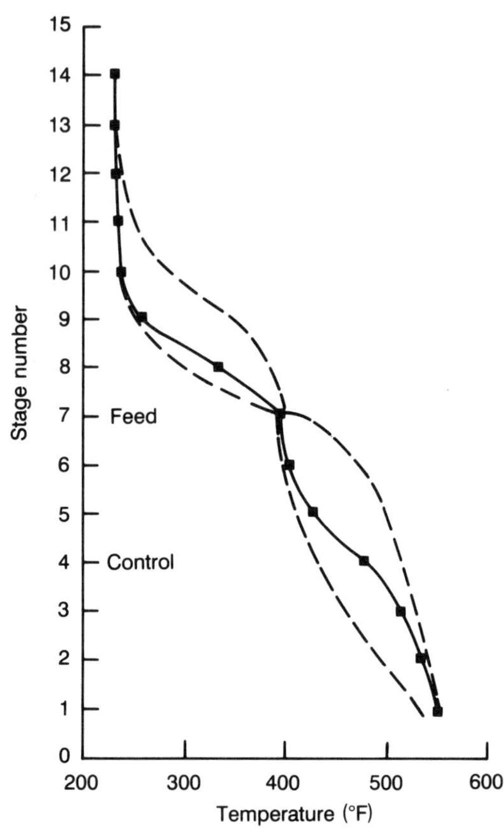

FIGURE 17-3. Column A alternate feed temperature profile.

PID tuning, dependent upon the feed stock being used. With the prior control scheme, this loop was tuned to provide control under the high process gain shown in Figure 17-3; however, the loop then performed poorly under the lower process gain evident with the feed in Figure 17-2. The DCS instrumentation allowed these tuning changes to be automated and thus transparent to the operator.

Although the performance of the column temperature control loop was generally much better than before the project, there were still infrequent disturbances of a magnitude such that the column would enter flooding conditions if additional corrective actions were not immediately taken. However, because the column runs so much better, the operators now pay less attention to it, so when these infrequent but major disturbances occur, the operators are less likely to manually intervene in time.

After observing three of four of these events occurring over a six month period, it became apparent that a major disturbance could allow the column temperature to move far enough above setpoint that the use of the temperature setpoint in the flood calculations becomes a poor assumption. Flooding can occur because the flood calculation predicts more sensible heat than is actually being used during the transient, and hence temporarily allows too much reboiler duty. For a temperature decrease, the assumption presents no problem because it errs in the conservative direction.

The solution to these infrequent events was to implement a high, positive temperature deviation interlock. When this interlock condition occurs, the reboiler duty is immediately decreased by 5%, and as the condition clears, the reboiler duty is slowly ramped back to the computed firing rate. An alarm is used to annunciate that the interlock has occurred, but further operator intervention has not been required.

17-2-4 Results

A direct comparison of column control performance before and after the project is not completely valid because of the process and equipment modifications. However, it is obvious that in addition to a 50% capacity increase, tighter control with less operator attention is also being achieved. It is also reasonable to assume that this degree of complexity could not have been implemented and maintained with an analog system.

Historical trends provided excellent troubleshooting tools as performance was fine tuned and the infrequent mishaps were addressed. The flexibility of the DCS software allowed the implementation to evolve as operating experience was gained. Early success gained operator confidence and support, which in turn, lead to their taking an active role in troubleshooting and suggesting several of the subsequent modifications that were implemented.

17-3 CASE II

This second example of distillation column control, implemented via DCS instrumentation, will be referred to as Column B. An overview of the process material balance and column control strategy is provided by Figure 17-4. The process background and considerations leading to this particular control scheme design, implementation details, and results are discussed in the subsections that follow.

17-3-1 Process Background — Reduced Capital

Column B, as shown in Figure 17-4, is not a simple distillation because the column has multiple feeds and a sidedraw. This column was designed to replace a simple distillation column and two single-effect evaporators in series. Also, a large intermediate storage tank was eliminated between this column and the distillation column that follows it. Column B is now the first column in a three column refining process.

The project to install this column also established inventory management procedures aimed at minimizing in-process inventories. As part of those procedures, feed from the crude area now routinely bypasses the crude storage tank and is fed directly to Column B. The crude tank is still used during start-ups, shutdowns, and to balance operating rates between the crude area and refining, but generally it operates with less than a 10% level.

A second area of inventory management addresses the recovery of material, dilute with water, from the periodic washout of several batch operations throughout the process. The material from many of these sources is collected in a recovered material storage tank. This recovered material is then processed along with the feed to Column B.

A third area of inventory management is the reduction of material processed by a batch recovery column, fed from several sources of high boiling materials. This reduction is accomplished by a recycle feed of high boiling material from one of the more dilute sources to the reboiler of Column B. This recycle feed flashes some of the more volatile product into the reboiler vapor and concentrates the high boiling impurities along with the bottom stream of Column B, which then is fed to the batch recovery column.

17-3-2 Control Scheme Design

Column B operates as a purge column, removing both a low boiling distillate and a high boiling bottom stream from the product, which leaves the column as a liquid sidedraw above the feed. (Luyben, 1966) The overall column material balance, on the basis of 100 lb/h feed flow, is given in Figure 17-4. The column is designed for the following feed composition, given in terms

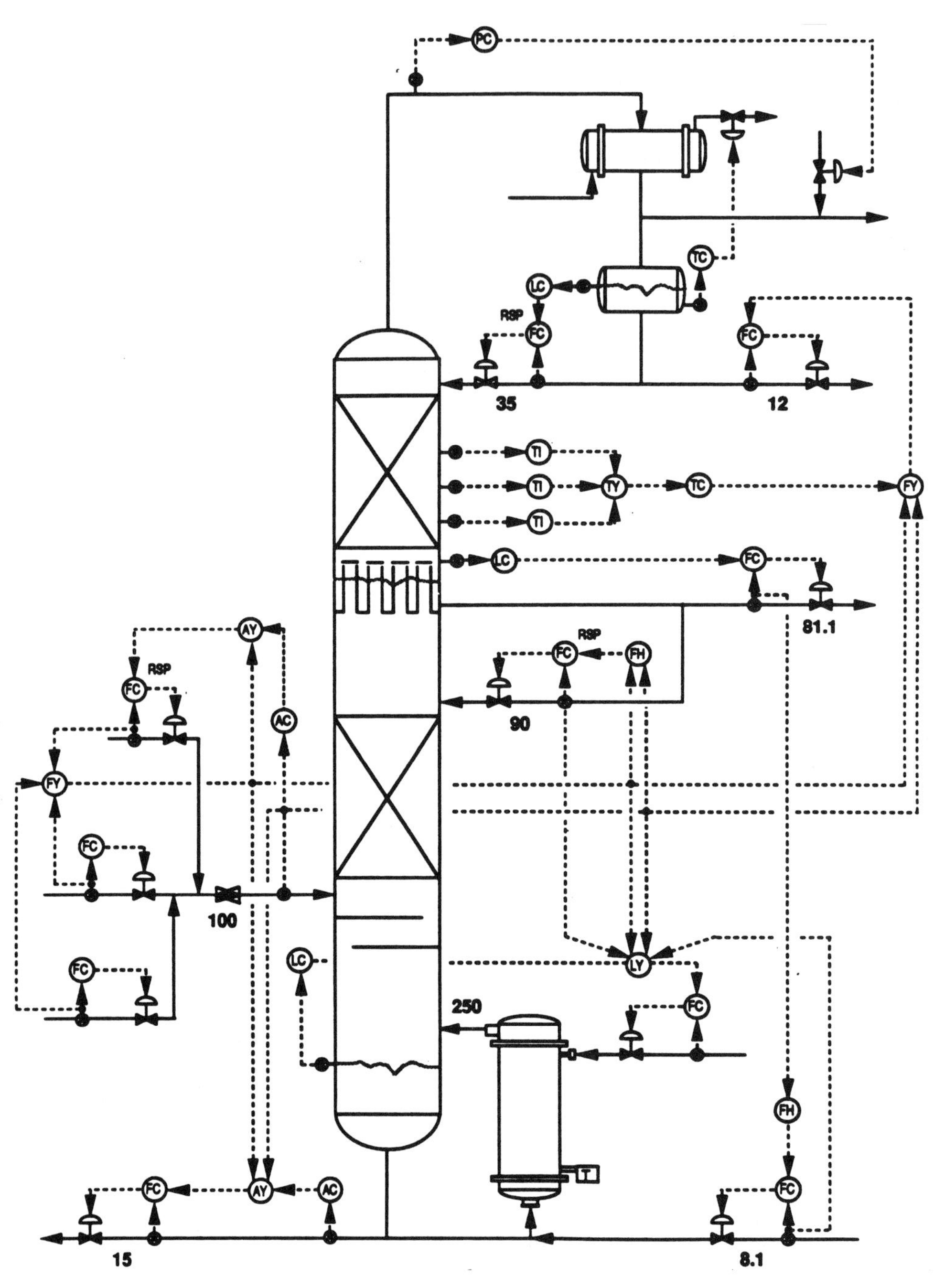

FIGURE 17-4. Column B.

of four lumped components:

12.0% water
87.2% organic product
0.2% intermediate boiling impurities
0.6% high boiling impurities

where all compositions are expressed in weight percent. The top section of Column B is making the separation between the water and organic product, with the recovery objective that less than 1.0% organic product leaves with the distillate. The product sidedraw must contain less than 1.0% water, primarily due to the problems excess water creates in the vacuum system of the next column. Variations in the amount of water leaving with the sidedraw are not important, as long as total amount leaving is acceptable.

In the bottom section of the column, the intermediate and high boiling impurities are separated from the sidedraw product, with a specification of less than 15 ppm total impurities in the product. The recycle feed stream is fed to the reboiler of Column B and typically contains 99.8% organic product and 0.2% intermediate boiling impurities, which is approximately equal to the vapor composition leaving the reboiler.

The only requirement for the bottom stream is to concentrate the intermediate and high boiling impurities, together with the recycle feed, to a minimum flow, with manageable viscosity, for further processing by the batch recovery column. These conditions are met when the bottom stream contains approximately 95% organic product.

Inventory management is achieved by blending fresh crude, crude from the crude storage tank, and dilute recovered material, through an in-line static mixer, to become the feed to Column B. The fresh feed from the crude area is flow controlled at a rate slightly less than the crude being produced. Excess crude flows to the crude storage tank. The flow from the crude storage tank is flow controlled, with the setpoint adjusted manually to maintain 10% inventory, using total feed rates within the capacity limits of Column B.

The flow of the dilute recovered material is manipulated to control the water composition of the total feed to Column B. The level in the dilute recovered material tank is loosely controlled by manually adjusting the setpoint on the total feed water composition controller. If the level is high, then the water concentration setpoint is increased and the dilute material is worked off. This approach compensates for large water disturbances in the fresh crude, which occur periodically during washouts, thereby providing a fairly constant feed rate and water composition to Column B.

A direct material balance approach is taken with the overall column control scheme shown in Figure 17-4, as is often recommended for purge column control (Tolliver, 1986). This approach directly sets the top and bottom purge stream flows to maintain compositions, whereas the corresponding internal flows are manipulated to maintain the reflux drum and sump levels. The sidedraw flow must then be manipulated to maintain the collection tray level and thus satisfy the overall column material balance. The reflux flow below the sidedraw is set to establish the reflux ratio and separation in the lower section of the column, and because of the energy balance, this results in a fairly high reflux ratio and excess separation for the top section of the column.

One concern with the proposed direct material balance control scheme is the sump level controller, which must manipulate the reboiler heat to maintain level. This control scheme, in some instances, can be subject to an "inverse response" resulting in poor level control (Buckley, 1974). Adding to this concern was the process requirement to minimize holdup in the sump. However, three factors indicate that Column B should not experience an inverse response. The column has a forced circulation reboiler, structured packing internals with low liquid holdup, and the liquid in the lower section of the column is nearly all product. Some further considerations to address this concern are discussed with the implementation details.

17-3-3 Implementation Details

At the top of the column, the reflux drum level is tuned with high gain for tight level control, such that excess boilup is immediately returned to the column. The distillate water purge rate is determined by a feedforward and feedback control scheme. Feedforward control, based on the water entering the column, is calculated from the total feed flow and the feed composition. The total feed flow is computed from the sum of the three individual flows that combine at the static mixer. The water composition in the combined feed is determined from a continuous and reliable density measurement. Dynamic compensation in the form of a lead–lag plus deadtime is provided for each feedforward signal and field tuned to match the process dynamics.

Feedback control uses a characteristic column temperature to infer overhead and sidedraw composition. The temperature sensor location is determined from parametric cases of the steady state material and energy balance. Column temperature profiles from the base case design and two parametric cases at slightly greater and less distillate–feed ratio are shown in Figure 17-5. The temperature sensor is located at stage 21, which exhibits the greatest, most symmetrical, steady state temperature deviation from the base case. Because of uncertainty in the height of a theoretical plate (HETP) in the structured packing and in the reflux ratio in the top section of the column, alternate sensor locations are provided one stage above and below stage 21.

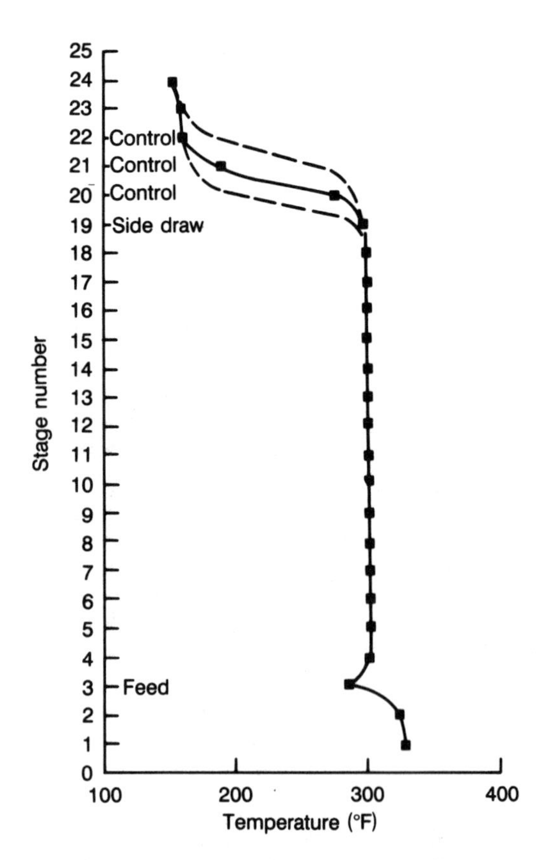

FIGURE 17-5. Column B temperature profile.

In the top section of Column B, which is separating water from organic material, the relative volatility is very large and there is a sharp temperature break across the control stage located with the preceding procedure. For this high gain situation, the feedback temperature control scheme actually utilizes a weighted average temperature signal as a form of profile position control (Luyben, 1972). With this approach, any one, or all of the measured stage temperatures may be utilized to whatever extent is found appropriate.

Also, a notch gain PID control algorithm is used to compensate for the extreme nonlinearity that occurs with this high process gain (Fisher Controls, 1984). The output of the feedback temperature controller is then combined with the feedforward signal described previously to provide a remote setpoint for the distillate flow controller. Because the feedforward signal is based on both feed flow and composition, the feedback signal is combined as a bias, rather than as a multiplier (Tolliver, 1991b).

The internal liquid from the top section of Column B is collected on a chimney tray. Level control of this tray is achieved by manipulating the sidedraw flow rate. This level loop is tuned loosely, with mostly proportional action, to minimize flow disturbances to downstream equipment. A notch gain PID algorithm is again used, in this case to allow the gain to be further reduced while the level is close to setpoint. A notch gain factor of 0.5, with the notch defined as 10% above and below setpoint, has been found to be the most effective tuning for averaging out temporary flow disturbances. The recycle feed flow, which is fed to the column reboiler, is set via an operator adjustable percentage of the sidedraw flow.

Typically, the recycle flow is maintained as 10% of the sidedraw flow.

The reflux below the sidedraw is flow controlled with a setpoint that provides the necessary separation to keep the organic impurities below specification. There is no on-line measurement of these impurities; therefore, feedback control is implemented manually based on periodic off-line laboratory analysis. Feedforward control is also implemented based on a reflux–total organic feed ratio, which is computed from the total feed flow, less the water entering with the feed. Upper and lower reflux limits are enforced to avoid flooding and weeping constraints.

At the bottom of Column B, a similar feedforward signal establishes a tails–total organic feed ratio, and is used to set the bottom purge flow rate. The recycle feed flow is not included as a separate disturbance because its impurities are accounted for, at least in the steady state, with the ratio of the total organic feed taken. There is no temperature sensitivity to infer compositions in the lower section of Column B; rather an on-line viscosity measurement is used to provide feedback trim of this ratio. Viscosity is actually the stated control objective for the bottom stream. The viscosity measurement is continuous, but less reliable than the feed density measurement. When the viscosity measurement is not available, the tails–total organic feed ratio is still maintained and may be manually adjusted with the viscosity controller output.

As mentioned previously, the sump level is controlled by manipulating the column boilup. This is achieved by providing a remote setpoint to the reboiler steam flow controller. In order to address concerns about the performance of this loop, feedforward control is provided to help correct for potential disturbances. The remote setpoint for the reboiler steam flow is determined from the sum of the level controller output plus a feedforward steam flow calculation based on several measured disturbance signals. Individual steam–disturbance ratios are applied to the total organic feed flow, the water flow entering with the feed, the lower section reflux flow, and the recycle feed flow. This feedforward signal provides most of the level control action; therefore, the feedback loop is tuned with moderate gain and little reset.

17-3-4 Results

This control strategy has worked quite well, in spite of many process problems that were encountered during start-up. The ease with which this column control strategy allows total reflux operation and manual composition control before commissioning the automatic temperature and viscosity feedback loops proved invaluable during the start-up. The extensive use of feedforward control has allowed the column to run smoothly even before the feedback loops were put in service.

Loop checkout and operator training via a real-time digital simulation of the process also contributed greatly to the success of this project (McMillan, 1985). Inventory management by adjusting the feed composition was a difficult concept to explain; however, it was easily demonstrated by operating the simulator. The rejection of water disturbances, as a result of this approach, was probably one of the most significant factors contributing to overall successful operation.

The main process problem that was encountered at start-up was poor separation of the organic impurities from the column sidedraw. A modification of column internals to allow a higher reflux ratio was eventually required to solve this problem. This higher reflux ratio used below the sidedraw also causes a higher reflux ratio to result in the top section of the column; however, no control scheme revisions were required. The effective location of the characteristic column temperature was easily modified by adjusting the weighting factors on the three individual temperature sensors.

17-4 CASE III

This third example of distillation column control, implemented via a DCS, actually consists of two columns operating in conjunction because of energy integration. The main column, with the product sidedraw, will be referred to as Column C and the second column will be referred to as Column D. An overview of the process material balance and column control strategy is provided by Figure 17-6. The process background and considerations leading to this particular control scheme design, implementation details, and results are discussed in the subsections that follow.

17-4-1 Process Background — Energy Conservation

These two columns represent an extensive process revision providing additional capacity and energy conservation. The replacement project was justified primarily on reduced operating cost. By replacing two separate distillation columns with these two energy integrated columns and utilizing high efficiency structured packing in Column C, high pressure steam usage is reduced by 30%. Furthermore, 90% of the high pressure steam used in the reboiler is now recovered as low pressure steam from the Column C condenser–steam boiler. An existing distributed control system was expanded to include control of this new refining process as part of the cost reduction project.

The feed to Column C comes from a crude storage tank. The product, taken as a sidedraw from Column C, flows directly to a product storage tank. The impurities in the feed are purged from the bottom of Column C and the top of Column D. A portion of the bottom stream from Column C is eventually recycled after further processing; however, the distillate from Column D becomes a waste stream.

17-4-2 Control Scheme Design

These two columns operate in conjunction to purge both low boiling and high boiling impurities from the product, similar to the single column previously described in Case II. The two column system is designed for the following feed to Column C, given in terms of weight percent:

1.5% low boiling impurities
92.9% organic product
5.6% high boiling impurities

where the primary objective is to produce a sidedraw of 99.5% organic product, with less than 100 ppm high boiling impurities. Secondary objectives are to minimize product losses in the top and bottom purge streams; however, temperature constraints at the reboiler actually limit product recovery more than separation in Column C. This system operates under very high vacuum also to minimize column temperatures and product degradation.

Relative flow rates, on the basis of 100 lb/h feed flow to Column C, are shown for both columns in Figure 17-6. Because of the small impurity purge flows, a direct material balance approach is appropriate for this system. Notice that internal column flows are relatively large. Column D has a reflux ratio (R/D) of 8.0, and Column C has a reflux ratio near 20, if the total vapor feed to Column D is considered as the distillate from Column C.

Energy balance disturbances are a major concern due to these relatively large internal flows. A 5% change in the vapor rate in Column C can become a 100% disturbance to Column D. For this reason, it was decided that the reboiler duty should be fixed, relative to the feed rate, to establish separation. With the bottom flow being adjusted for direct material balance composition control and the reboiler duty establishing separation, the sidedraw flow is then ma-

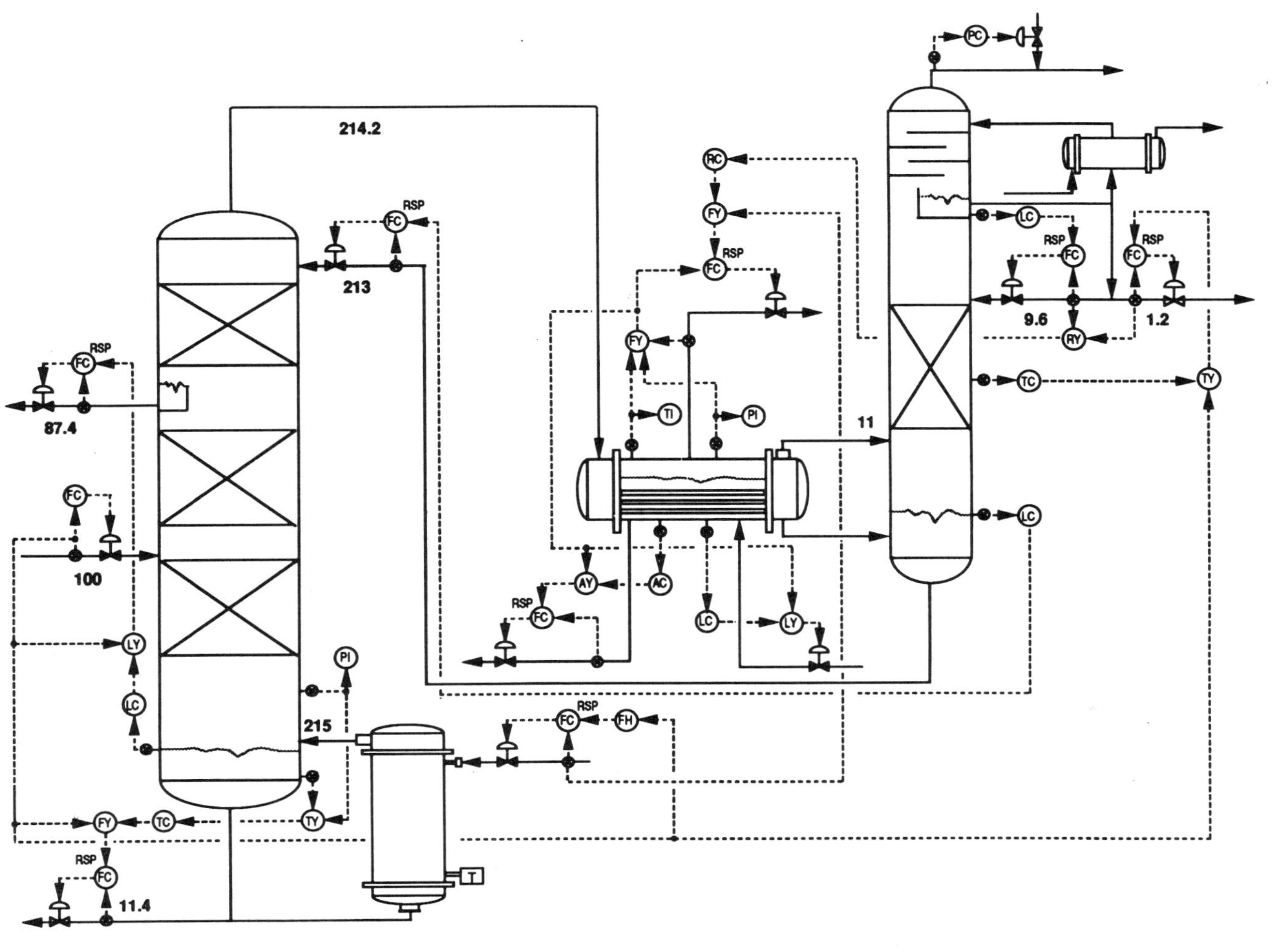

FIGURE 17-6. Columns C and D.

nipulated to maintain the sump level in Column C.

In order to maintain the energy balance split between columns, the Column C condenser–steam boiler duty is set to remove approximately 90% of the reboiler duty. Because this condenser generates steam, the heat removal duty may be set simply by flow controlling the low pressure steam leaving the boiler. The steam pressure inside the boiler is allowed to float, which establishes the temperature difference available for heat transfer.

Thus, the steam flow from the condenser is set in ratio to the steam used at the reboiler. Feedback control of the Column D reflux ratio is then used to trim this steam ratio. The Column D reflux ratio is computed from the reflux flow, which is manipulated for reflux drum level control, and the distillate flow, which is manipulated for direct material balance composition control.

Temperature control is used to infer and maintain composition in both Column C and Column D. Steady state temperature profiles are shown in Figure 17-7 for both columns. Parametric cases with alternate values of D/F and B/F are used to determine the sensitivity and best location for these temperature control sensors. The parametric cases are shown as dashed lines about the solid line for the design case. The deviations toward the top are due entirely to changes in D/F, and the deviations at the bottom are due entirely to changes in B/F. Thus there is little steady-state interaction between these two temperature loops, with sensors located in Column D equivalent to stage 22, and in Column C in the vapor space above the sump.

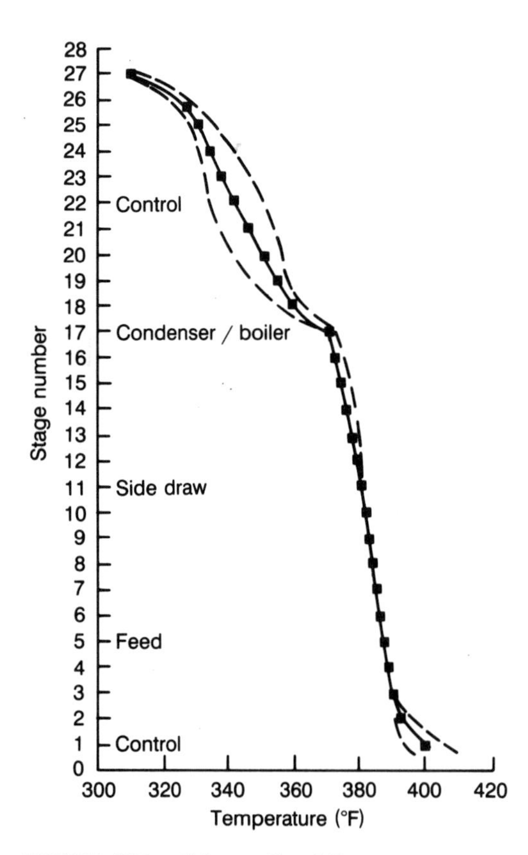

FIGURE 17-7. Columns C and D temperature profile.

17-4-3 Implementation Details

The Column C bottom temperature loop is implemented with both feedforward and feedback manipulation of the bottom flow. The feedforward signal is based on a design B/F fraction of the feed flow. The feedback temperature controller output is used as a bias to this feedforward signal, providing a higher bottom flow if temperature is above setpoint and less flow if the temperature is low.

Temperature sensitivity to composition changes is low. However, pressure variations can contribute significantly to the measured temperature; therefore, pressure compensation is applied to the measured temperature. Pressure effects on the temperature at the bottom of Column C are negated by applying a pressure correction bias to the measurement. If the measured pressure is greater than design, then measured temperature is decreased to approximate the equivalent temperature at the design pressure. Because the inferred composition varies slowly, a significant dynamic lag, with a time constant of several minutes, is used on this pressure correction term in order to filter any rapid pressure transients.

The reboiler steam flow receives its remote setpoint from a feedforward signal designed to provide optimum separation at various feed rates. Upper and lower setpoint limits are used to avoid flood and weep conditions. A manual bias is available to provide more or less heat to the reboiler depending on ambient weather conditions. The Column C reflux–feed ratio is a good indicator for manual feedback to the reboiler steam flow.

The Column C base level control loop, which manipulates the sidedraw flow, is also implemented with a feedforward signal based on a design S/F fraction of the feed flow. The level controller output is a bias to this feedforward signal. The feedback level controller is tuned as tightly as possible; however, the deadtime, due to the trays between the sidedraw and the sump, limits the gain that can be used.

As mentioned previously, condenser–boiler steam flow is set to establish the design energy balance split between Column C and Column D. Because the steam pressure is allowed to float, both pressure and temperature measurements are used to calculate the mass steam flow. The remote setpoint for the condenser steam flow is determined from a ratio times the reboiler steam flow. This ratio is adjusted by integral-only control of a computed Column D reflux ratio, which is a good steady state indicator of the design energy balance split. The integral-only control algorithm maintains the desired steady state reflux ratio while avoiding dynamic interactions with the other control loops (Tolliver, 1991a).

The condenser steam flow is used as a feedforward signal added to the level control feedback signal to position the boiler feedwater valve. This application of feedforward control does not provide a remote setpoint because there is no boiler feedwater flow controller. It is important that when the operator puts the level controller into manual mode, he/she has complete control of the valve, without changes from the feedforward signal. Therefore, in this application, the feedforward signal is combined with the feedback signal internal to the level controller (Fisher Controls, 1984; Tolliver, 1991b).

The boiler blowdown flow is set in ratio to the boiler steam flow, with feedback adjustment of this ratio provided by a conductivity controller. The blowdown flow controller allows a simple flow ratio to be maintained when the conductivity measurement is not available for feedback.

The Column D base level control loop, which manipulates the reflux flow returning to Column C, is tightly tuned. Likewise, the Column D reflux drum level controller, which manipulates reflux flow returning to Column D, is also tightly tuned. For both these level loops, tight tuning is achieved with a high gain and little reset.

The Column D temperature control loop is implemented with both feedforward and feedback manipulation of the distillate flow. The feedforward signal is based on a design D/F fraction of the feed flow. The feedback temperature controller output is used as a bias to this feedforward signal, providing a higher distillate flow if the temperature is low, and less flow if the temperature is high.

17-4-4 Results

This control strategy has worked quite well. The columns are easily started up under total reflux operation, and by the time feed is introduced, the sidedraw product is typically well within specification and remains that way.

Loop checkout and operator training via a real-time digital simulation of the process contributed greatly to the success of this project. The integral-only control of the reflux ratio in Column D, as a means of maintaining the energy balance split, was the most difficult concept to explain. However, the operation and dynamics were easily demonstrated on the simulator.

During the initial start-up, the only operational problem experienced was related to achieving the desired operating pressure. There were many leaks, making the high vacuum unattainable. As a result, the sump temperature was higher than design, which limited the heat input capability.

At first, the condenser–steam generator was not cooling properly, which was allowing too much vapor feed to Column D. Once the steam valve was opened manually, all the inerts that had been trapped vented into the low pressure steam system. Although this caused problems with all the other low pressure steam users, it resolved the immediate problem with the condenser.

Then, the opposite problem of too much cooling became apparent. The steam pressure increased, as expected, in order to reduce cooling; however, a miscalibrated pressure relief valve limited the steam pressure. With higher than normal column temperatures, too much cooling was being achieved and total condensation was preventing any vapor feed from reaching Column D. A shutdown was finally required to fix the vacuum leaks and properly set the pressure setting on the relief valve. The following start-up proceeded without problems.

17-5 SUMMARY

As Buckley states in his paper on the status of distillation control system design, "In designing a process control system we must consider a number of nontechnical factors such as: (a) availability of skills in the design organization, (b) availability of skills in the plant organization, (c) time and money budgeted for the project, and (d) the ultimate competitive position of the plant" (Buckley, 1978).

Distributed control system instrumentation has had a major impact in allowing more sophisticated control strategies to be designed, implemented, operated, and maintained. For each of these examples, the design effort lasted for approximately six person weeks. This time included review of historical plant data, parametric case studies, conceptual design and review, detailed design, software configuration, and implementation.

Real-time simulations were also developed during this six week effort and used for configuration checkout. These models were then used to demonstrate the control strategies to plant supervisory personnel, who reviewed the implementation, and in turn used the models to conduct operator training. Time for start-up coverage was not included, as process and mechanical delays obscured those numbers.

It is interesting to note, that although these three examples included highly instrumented, complex distillation columns, there were no composition specific measurements, such as with gas chromatograph (GC) analyzers. Compositions were inferred by nonspecific property measurements such as temperature, density, and viscosity. This approach is quite typical in the chemical processing industries where analyzers and sampling systems are quite expensive to implement and maintain. The computational capabilities of distributed control systems are frequently utilized to make these inferred measurements.

Feedforward control was the primary "advanced" control technique that was used throughout these examples. Some optimization, adaptive tuning, constraint, and decoupling control considerations were addressed. Two point composition control was performed with both examples having sidedraw products, but was not required to satisfy the control objectives for Column A. The control strategy for each of these examples has withstood the test of time, having now been in continuous service for more than two to three years.

References

Buckley, P. S. (1974). Material balance control in distillation columns. AIChE Workshop on Industrial Process Control, Tampa.

Buckley, P. S. (1978). Status of distillation control system design. ISA Analytical Instrumentation Symposium, Houston, Vol. 16, pp. 11–39.

Dallimonti, R. (1980). Experiences with distributed systems in process control: 1975–1980. Proceedings of the Joint Automatic Control Conference, Vol. 1, paper no. TA-4A.

Ender, D. B. (1990). Use of personal computers in control loop analysis. ISA Instrum. Chem. Petroleum Ind. 21, 49–57.

Fisher Controls (1984). Configuring interactive and computing controllers. *PRoVOX Instrumentation Regulatory Controllers*. Marshalltown, IA: Fisher Controls.

Hanley, J. P. (1990). How to keep control loops in serivce. *InTech* 37(10), 30–32.

Hougen, J. O. and Brockmeier, N. F. (1969). Developing process control strategies I—Eleven basic principles. *Instrum. Tech.* 16(8), 45–49.

Luyben, W. L. (1966). 10 schemes to control distillation columns with sidestream drawoff. *ISA J.* 13(7), 37–42.

Luyben, W. L. (1972). Profile position control of distillation columns with sharp temperature profiles, *AIChE J.* 18(1), 238–240.

Luyben, W. L. (1975). Steady state energy conservation aspects of distillation column control system design. *Ind. Eng. Chem. Fund.* 14(4), 321–325.

McMillan, G. K. (1985). Real time simulations in distributed control systems for training and engineering. Proceedings of the American Control Conference, Vol. 1, pp. 1718–1720.

McMillan, G. K. (1989). Continuous control techniques for distributed control systems. *ISA Independent Learning Module*. Research Triangle Park, NC: ISA.

McMillan, G. K. (1991). Editor's viewpoint on distributed control systems—Selection, implementation and maximization. *ISA Trans.* 30, No. 2, 5–7.

Rademaker, O., Rijnsdorp, J. E., and Maarleveld, A. (1975). *Dynamics and Control of Continuous Distillation Units*. New York: Elsevier.

Shinskey, F. G. (1984). *Distillation Control for Productivity and Energy Conservation*, 2nd ed. New York: McGraw-Hill.

Tolliver, T. L. (1986). Purge column control via a distributed control system. *ISA Trans.* 25(3), 49–54.

Tolliver, T. L. (1991). Continuous control experiences. *ISA Trans.* 30, No. 2, 63–68.

Tolliver, T. L. (1991). Feedforward control implementation on a distributed control system, *ISA Instrum. Chem. Petroleum Ind.* 22, 99–117.

Tolliver, T. L., Dugar, U., Wintjen, D., Ahmad, P., Rominger, M., and Hill, K. (1989). Panel discusses control systems upgrading. *Hydrocarbon Process.* 68(11), 63–68.

Tolliver, T. L. and McCune, L. C. (1978). Distillation control design based on steady state simulation. *ISA Trans.* 17(3), 3–10.

Tolliver, T. L. and McCune, L. C. (1980). Finding the optimum temperature control tray for distillation columns. *Instrum. Tech.* 27(9), 75–80.

Tolliver, T. L. and Waggoner, R. C. (1980). Distillation column control; A review and perspective from the CPI. *ISA Adv. Instrum.* 35(1), 83–106.

Thurston, C. W. (1981a). Computer-aided design of distillation column controls, Part 1. *Hydrocarbon Process.* 60(7), 125–130.

Thurston, C. W. (1981b). Computer-aided design of distillation column controls, Part 2. *Hydrocarbon Process.* 60(8), 135–140.

18

Process Design and Control of Extractive Distillation

Vincent G. Grassi II

Air Products and Chemicals, Inc.

18-1 OVERVIEW

This chapter presents case studies to improve process insight and develop a methodology for the process design and control of extractive distillation systems. This material attempts to bridge the gap between process design and control by introducing dynamic methods into the process design.

A computerized process design procedure will be developed. Graphical residue maps and column operating curves of the extractive distillation column improve process insight into the critical design variables. Dynamic simulation and multivariable control system analysis will lead to robust and very practical single-input–single-output control scheme solutions.

These methods will be illustrated by three industrially significant extractive distillation systems:

System 1—methyl acetate, methanol, and water.

System 2—methyl acetate, methanol, and ethylene glycol.

System 3—ethanol, water, and ethylene glycol.

The case studies will be referenced as Systems 1, 2, and 3 throughout this chapter. These systems illustrate significant variation but enough similarity so a generic understanding of the process is obtained.

Air Products and Chemicals, Inc. practices the methyl acetate, methanol, and water system. The models and methods developed in this chapter have been verified against operating data from this real plant. I will point out the significant findings from a critical evaluation of these cases in this chapter. A more detailed and complete description of these cases studies is contained in Grassi (1991).

Extractive distillation affects the liquid phase activity of the components so the mixture may be efficiently separated into pure products. This is done by adding a third, heavy, component termed the solvent. The solvent has an affinity for one component, causing it to boil intermediate in the ternary mixture. The overhead product from the extractive distillation tower contains the lightest component. The intermediate component leaves with the solvent in the bottoms stream of the extractive distillation tower.

The extractive tower bottoms stream feeds a solvent recovery distillation tower. The solvent recovery tower overhead product contains the intermediate component. The solvent recycles from the bottom of the solvent recovery tower to the extractive distillation tower.

Figure 18-1 presents a simplified flow sheet of the extractive distillation process. This process requires very nonideal vapor–liquid equilibrium and tight process integration within the double column system. The process dynamics are highly nonlinear and multivariable with many interactions. Various methods of design and operation have been published, but industrial experience has shown that the process design and control of this process is not obvious.

18-1-1 Extractive and Azeotropic Distillation

Extractive and azeotropic distillation are industrially important separation processes. These are used when key components form an azeotrope or have a sufficiently low rela-

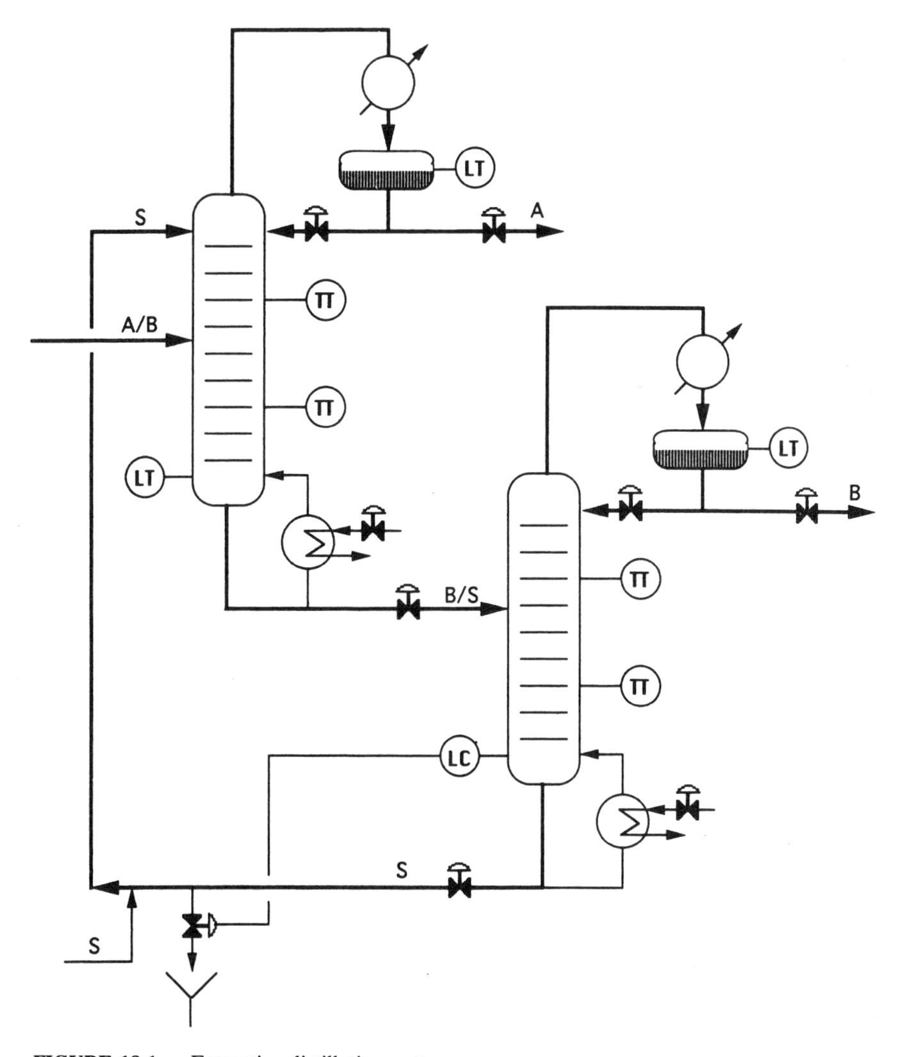

FIGURE 18-1. Extractive distillation system.

tive volatility that conventional distillation becomes impractical.

A component not present in the feed is added to the process. This new component enhances the relative volatility of the key components so the mixture can be separated efficiently by distillation. Figure 18-2 illustrates this effect.

Extractive distillation is characterized by the addition of a heavy component, called the solvent, added toward the top of the distillation column. Examples of industrially significant extractive distillations are:

1. Methyl acetate from methanol using water.
2. Ethanol from water using ethylene glycol.
3. Butene-2 from *n*-butane using acetone.

Addition of a light component, called the entrainer, differentiates *azeotropic* distillation from *extractive* distillation. The entrainer forms a minimum boiling azeotrope with one of the key components. The remaining component is removed from the azeotropic tower as the bottoms product. Overhead product containing the minimum boiling azeotrope is fed to a second distillation tower to recover the second key component.

Extractive Distillation
Vapor Liquid Equilibrium

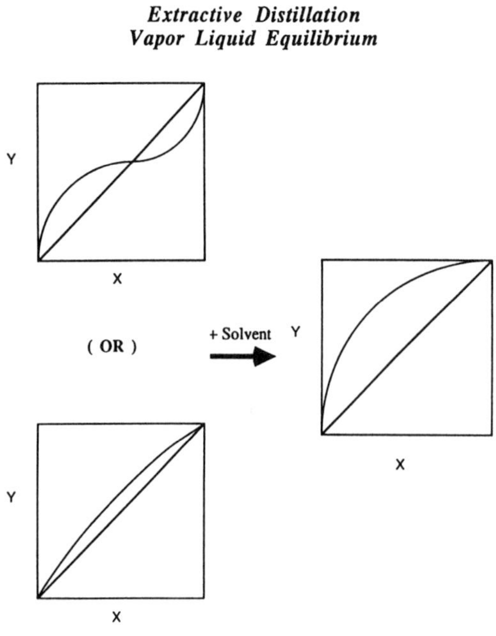

FIGURE 18-2. Vapor–liquid equilibrium effect.

If the overhead product condenses to two liquid phases in the reflux accumulator, the entrainer rich phase serves as reflux to the azeotropic tower. The second phase feeds the recovery tower. This tower recovers the key component as the bottoms product. If the overhead of the azeotropic tower condenses to a single phase we can operate the recovery tower at a different pressure to remove the key component. Examples of industrially significant azeotropic distillations are:

1. Acetic acid from water using ethyl acetate.
2. Ethanol from water using benzene.
3. Acetic acid from water using isopropyl acetate.

Figure 18-3 contains a simplified flow diagram of a generic heterogeneous azeotropic system.

Azeotropic distillation is closely related to extractive distillation. This chapter will be confined to the generics of extractive distillation, but the reader will see many extensions to azeotropic distillation.

Extractive distillation has highly nonideal vapor–liquid equilibrium. It is a dual column, integrated system and it presents many difficulties in process simulation, design, and control. I will present concepts that improve our engineering insight and develop methodologies for the process design and control of these systems. These concepts will be applied to the three extractive distillation case studies. These systems are highly nonlinear and multivariable. It is important to consider the process dynamics during the process design of these systems. This can be done easily in a way that is productive and provides greater process engineering insight into the nature of extractive distillation systems.

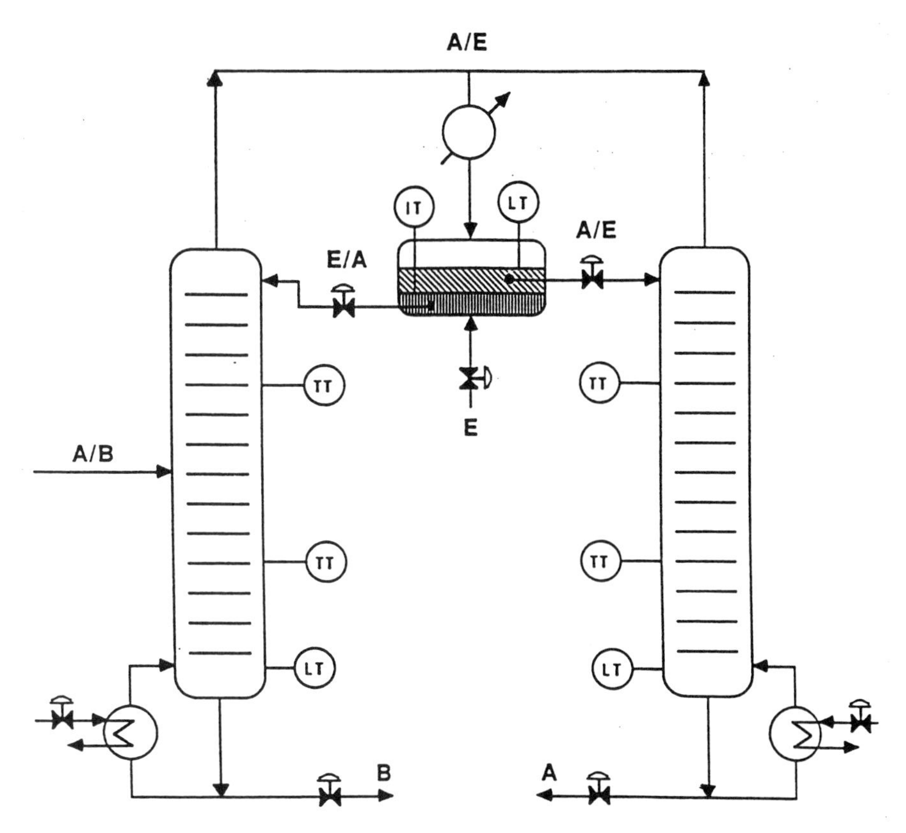

FIGURE 18-3. Azeotropic distillation system.

18-1-2 History

The study of azeotropic and extractive distillation began early in the twentieth century. By 1915 a continuous azeotropic distillation process to produce anhydrous alcohol was known. The need for synthetic rubber in both world wars promoted the need for extractive distillation of butadiene. Worldwide interest in the 1970s for synthetic fuel alternatives catalyzed additional interest in extractive distillation processes.

Benedict and Rubin (1945) wrote a landmark paper on the theoretical and practical aspects of azeotropic and extractive distillation. Othmer, in the 1950s, pioneered laboratory techniques to determine the nonideal equilibrium data needed to design azeotropic systems. During this period many engineers published works on the simulation, design, and application of azeotropic systems (Atkins and Boyer, 1949; Chambers, 1951; Gerster, 1969; Black and Ditsler, 1972).

The 1980s brought dynamic simulations of these systems (Prokopakis and Seider, 1983; Kumar and Taylor, 1986). Doherty developed significant insights into the fundamental topology of the properties and synthesis of azeotropic systems (Doherty and Perkins, 1979; Doherty and Caldarola, 1985). Andersen, Laroche, and Morari (1990) developed an understanding of steady state indices for azeotropic systems. Publications on the behavior of heterogeneous extractive distillation columns have appeared (Lee and Coombs, 1987).

My personal experience with extractive distillation began in 1981. We were experiencing operating instabilities in one extrac-

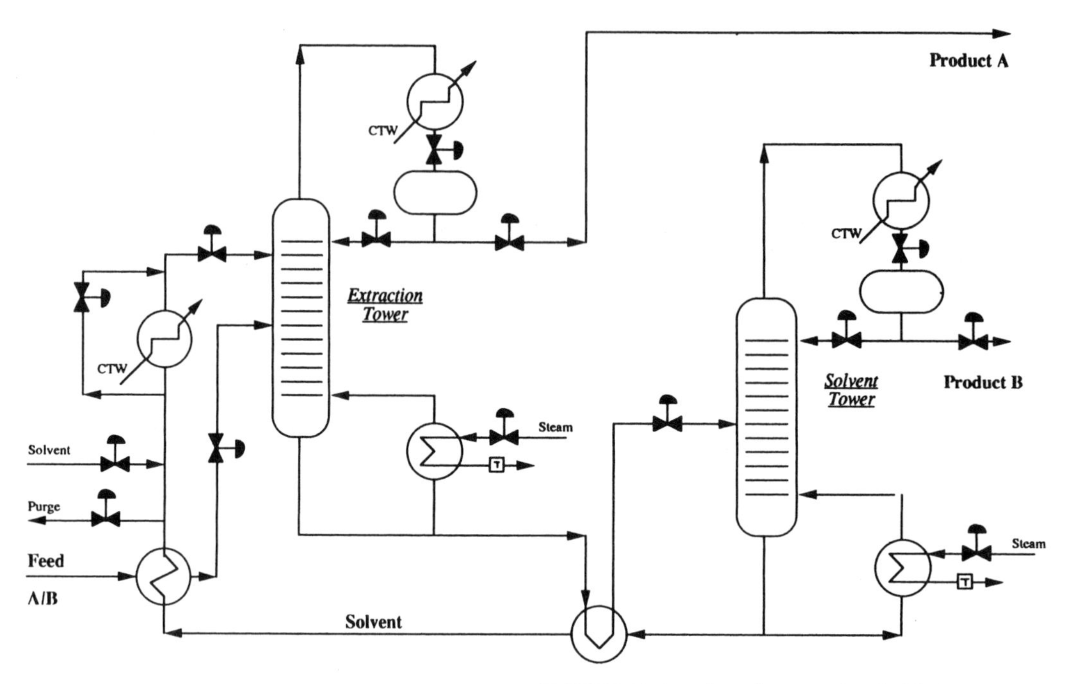

FIGURE 18-4. Generic extractive distillation process.

tive distillation system and needed additional capacity in another. These systems are heterogeneous with multiple azeotropes. When we started working on these columns, we could not even simulate the distillation columns at steady state. Over the years we have successfully expanded their capacity almost threefold through process improvements without increasing the size of the distillation towers. These process improvements came about only by a fundamental understanding of the distillation process that was gained through detailed steady state and dynamic process simulation verified by plant testing.

18-1-3 Process Description

Figure 18-4 is a generic process flow diagram of the extractive distillation process we will consider. This figure shows the key manipulative variables and interstream heat exchangers. Our aim in this chapter is to develop a general procedure that optimally designs the extractive distillation process and finds the best control scheme needed to operate it.

A binary mixture of components A and B feeds the extractive distillation process. These two components either form an azeotrope or have sufficiently low relative volatility so conventional distillation is impractical. The feed is introduced to the middle of the extractive distillation tower. A heavy solvent is chosen that has an affinity for component B. We feed the solvent a few trays from the top of the extractive distillation tower. These trays remove solvent from the overhead product. The distillate contains component A as product. The bottoms contains the solvent and component B. This stream feeds the solvent recovery tower. The distillate contains component B as product. The bottoms contains the solvent that is recycled to the extraction tower.

The solvent is the highest boiling component in the system. We must cool the solvent before returning to the extraction tower to prevent it from vaporizing. Much of its heat can be recovered in the process by using tower feed preheaters. A final trim cooler cools the solvent to its required feed temperature.

The solvent tower sump maintains solvent inventory. Solvent tower sump level controls the fresh solvent to makeup for solvent losses through the product streams. Fresh solvent makeup flow gives us an additional degree of freedom. We can use the solvent tower bottoms flow (solvent recycle) as a control variable for something other than sump level control.

Discounting pressure and inventory controls, there are 6 degrees of freedom in this process. These control five compositions and the temperature of the solvent feed to the extractive tower.

Extractive distillation affects a key component by reducing its volatility and removing it in the bottoms stream with the heavy extractive solvent. As in conventional distillation, both overhead and bottom composition controllers interact. Material recycle of the solvent and heat integration of the intercoolers create dynamic interactions. Rigorous process design and control of this process must consider the entire system as a coupled system of distillation towers.

Figure 18-1 shows the manipulative variables, that is, control valves for this process. As in conventional distillation the reflux, distillate, heat input, and bottom flows can be manipulated. However, solvent flow is also a manipulative variable in extractive distillation. Its control is not obvious. Additionally, there could be an optimum addition rate when considering process dynamics that is different from that by considering only steady state effects.

Extractive distillation can produce a heterogeneous azeotropic overhead. In this case, the overheads will condense into two liquid phases. Reflux is not a manipulative variable if only one phase is used as reflux. Its flow is set by the liquid phase split.

18-2 PHASE EQUILIBRIA

Extractive distillation requires nonideal phase equilibria behavior. The solvent is a new component added to the system that alters the phase behavior of the mixture to be distilled. For example, consider a system where component A is nonpolar and component B is polar. A polar solvent will attract component B greater than component A. This attraction, or intermolecular interaction, will cause the solvent to pull component B down the column into the bottoms stream.

The components of ideal mixtures do not interact. The solvent is chosen so nonideal interaction provides a sufficiently large increase in relative volatility between the two key components.

18-2-1 Relative Volatility

The fugacity of the vapor and liquid phases are equal at equilibrium. Our model for the vapor–liquid equilibrium for these systems is the same as that presented in Chapter 3. We will consider low pressure, less than 50 psia, nonreactive systems. The vapor is essentially ideal so the vapor phase fugacity coefficient is equal to 1. The liquid phase activity coefficient is not unity due to interaction between components in the liquid phase.

The activity coefficient is a function of the liquid composition and temperature. The relative volatility between any two components is a function of the activity coefficient and the pure components vapor pressures. Figure 18-5 describes these relationships.

The NRTL correlation is used to compute the liquid phase activity coefficient for

$$f_{\text{vapor}} = f_{\text{liquid}}$$

$$\phi_i y_i P = \gamma_i x_i P_i^s$$

$$\text{low pressure, ideal vapor} \quad \Rightarrow \quad \phi_i = 1$$

$$K_i = \frac{y_i}{x_i} = \frac{\gamma_i P_i^s}{P}$$

$$\alpha_{1,2} = \frac{K_1}{K_2} = \left(\frac{\gamma_1}{\gamma_2}\right)\left(\frac{P_1^s}{P_2^s}\right)$$

FIGURE 18-5. Relative volatility. The extractive effect is obtained by selectively increasing the relative volatility between the two key components by altering the liquid phase activity.

these case studies. You can choose your own model, or equation of state, for your particular application. Corresponding states models work well for these nonideal chemical systems. Good results may also be obtained using the Wilson and UNIQUAC activity coefficient models.

We must consider the relative volatility for all of the binary pairs in the extractive distillation process. The feed contains components A and B. The solvent component S has an affinity for component B. The relative volatility between components A and B, when solvent is present, determines how effective the extraction tower is. The relative volatility between components A and S determines how well we will be able to keep solvent out of the extraction tower distillate. The relative volatility between B and S determines the effectiveness of the solvent recovery tower.

Other phenomena that can occur from the addition of solvent to the mixture are additional azeotropes and two liquid phases. A ternary system can have up to four azeotropes. Each binary pair can form an azeotrope and the mixture can have one ternary azeotrope. Azeotropes can create distillation composition boundaries over which a single distillation tower cannot cross.

The Gibbs phase rule relates the number of phases and azeotropes in a mixture:

$$F = c + 2 - Ph \qquad (18\text{-}1)$$

where F = number of degrees of freedom
c = number of components in mixture
Ph = number of phases present

Our system contains three components and a homogeneous azeotrope contains two phases. Therefore, $F = 3 + 2 - 2$, or 3 degrees of freedom. The vapor and liquid compositions are equal at an azeotrope. We therefore have two constraint equations, $y_1 = x_1$ and $y_2 = x_2$, at the azeotrope, which take 2 degrees of freedom; only 1 degree of freedom is left. Therefore at constant pressure there are no degrees of freedom left and there must be only one solution that fixes temperature. This solution is the ternary azeotrope.

Similarly, there are only two components for each binary pair in the mixture and one constraint, $y_1 = x_1$. Setting pressure fixes the system at the azeotrope for the binary pair.

The phase rule is also useful to find operating conditions when multiple liquid phases can occur. For example, the total number of phases in a distillation process with two liquid phases is 3. The phase rule reduces to: $F = c + 2 - 3$, or $F = c - 1$. A ternary system has 2 degrees of freedom, that is, temperature and pressure. It is therefore possible to have more than one tray with two liquids on it. However, a binary system has only 1 degree of freedom. Because pressure is set, only one composition results in two liquids. It is very unlikely that any trays in a binary tower will have two liquid phases.

The presence of two liquid phases on some trays in a distillation tower can create operating instabilities. If the two phases have significantly different transport properties they can reduce tray efficiency. Small changes in operating variables can cause trays to switch from single to two liquid phases. Strong operating instabilities, such as wide temperature swings, can result. Systems that exhibit two liquid phases have an increased tendency to foam if large surface tension gradients exist. We need to be circumspect of additional azeotropes and heterogeneous phase behavior during the process design and control of extractive distillation towers.

I recommend that a solvent be chosen so it yields a relative volatility of at least 2 between A and B, 10 between A and S, and 5 between B and S. The solvent should not create two liquid phases or any new azeotropes. The larger the relative volatility is enhanced, the better.

Sometimes a less desirable solvent is more economical than the ideal solvent. A less desirable solvent might be one that is already present in the plant, inexpensive, nontoxic, or plentiful. We must learn how to design a process and control it for these solvents as well.

18-2-2 Residue Curves

Residue curve maps are an insightful way to visualize the phase equilibrium of a ternary system. They graphically depict azeotropes, heterogeneous phase location, separation feasibility, and distillation paths for the extractive distillation tower. These curves are very useful in determining the process design of extractive distillation systems. Doherty presents the theory of residue curve maps quite well (Doherty and Perkins, 1979; Doherty and Caldarola, 1985).

A residue curve map is the time trajectory of the liquid phase composition during a simple single stage batch distillation with no rectification. Figures 18-6*a* and *b* illustrate the residue curve map for the first two of our case studies. These are the residue curve maps for methyl acetate (A) and methanol (B) using either water ($S1$) or ethylene glycol ($S2$) as the extractive solvent.

The residue curve map for our ternary system is drawn as a right triangle. The liquid composition of component A is the ordinate and the liquid composition of component B is the abscissa. The vertices of the triangle locate the pure components. To construct the residue curve map we consider the simple single stage batch distillation process. A mixture is batch charged into a pot. Heat is added to boil the mixture. Vapor is removed, but no additional feed is added to the pot. We plot the liquid composition trajectory on the residue curve map as the boiling process proceeds.

The residue curve map using $S2$ has one binary azeotrope. It is a minimum boiling azeotrope between components A and B. Pure components and azeotropes are stable points on the residue curve map. Residue curves either start or end at these points. The residue curve map for this system using $S2$ is easy to interpret. The edges of the residue curve map are the residue curves for the binary mixtures. The residue curves always proceed from low to higher temperatures. The residue curves end at the pure solvent vertex because it is the highest boiler in the system. Residue curves diverge from the minimum boiling azeotrope. No distilla-

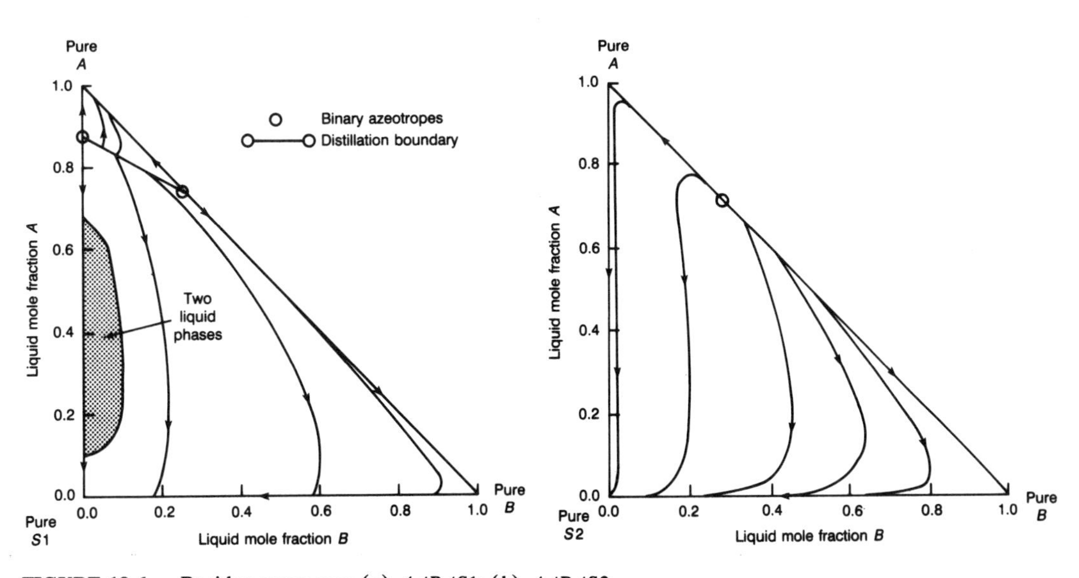

FIGURE 18-6. Residue curve map (a) $A/B/S1$; (b) $A/B/S2$.

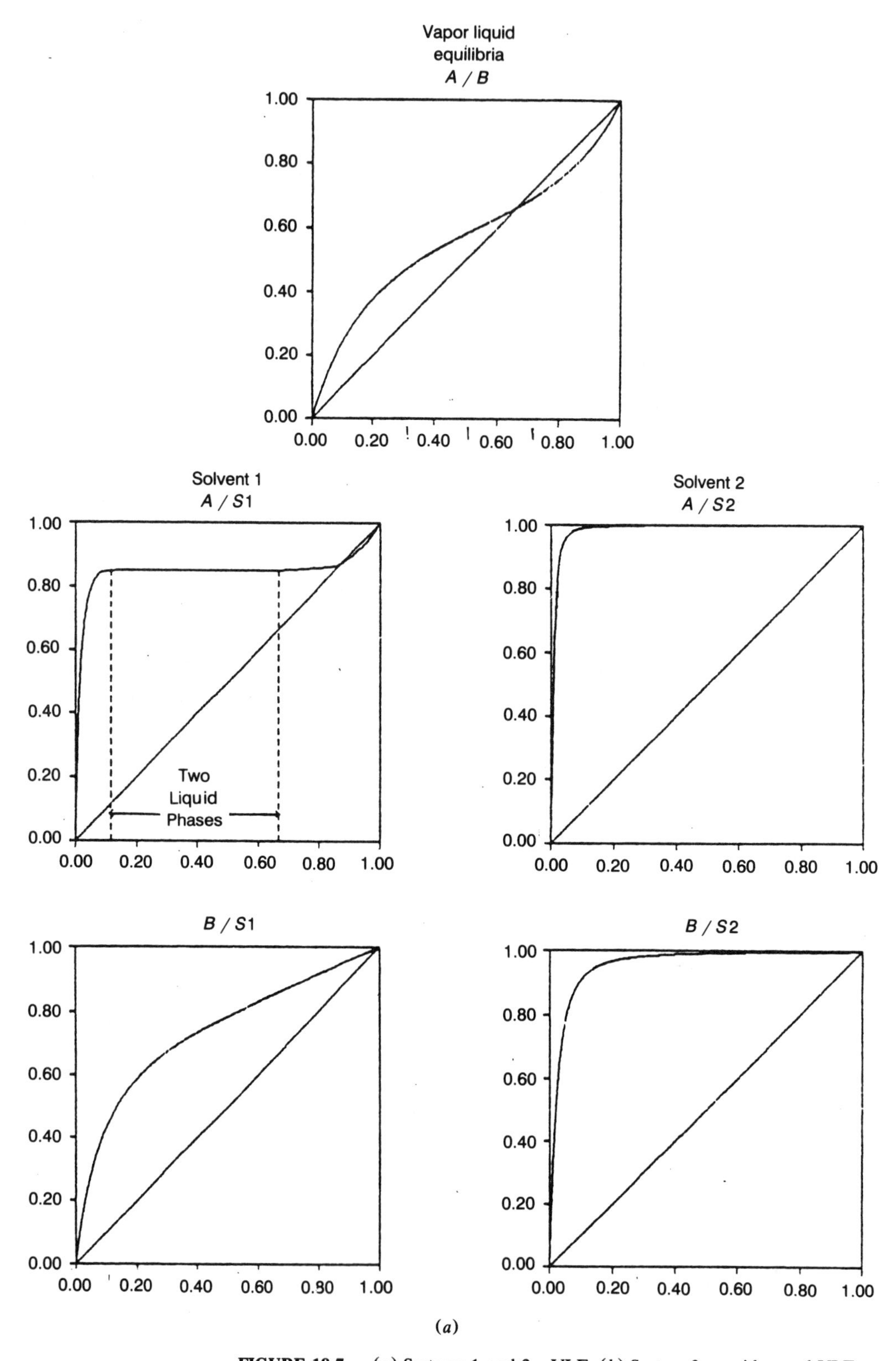

FIGURE 18-7. (*a*) Systems 1 and 2—VLE. (*b*) System 3—residue and VLE.

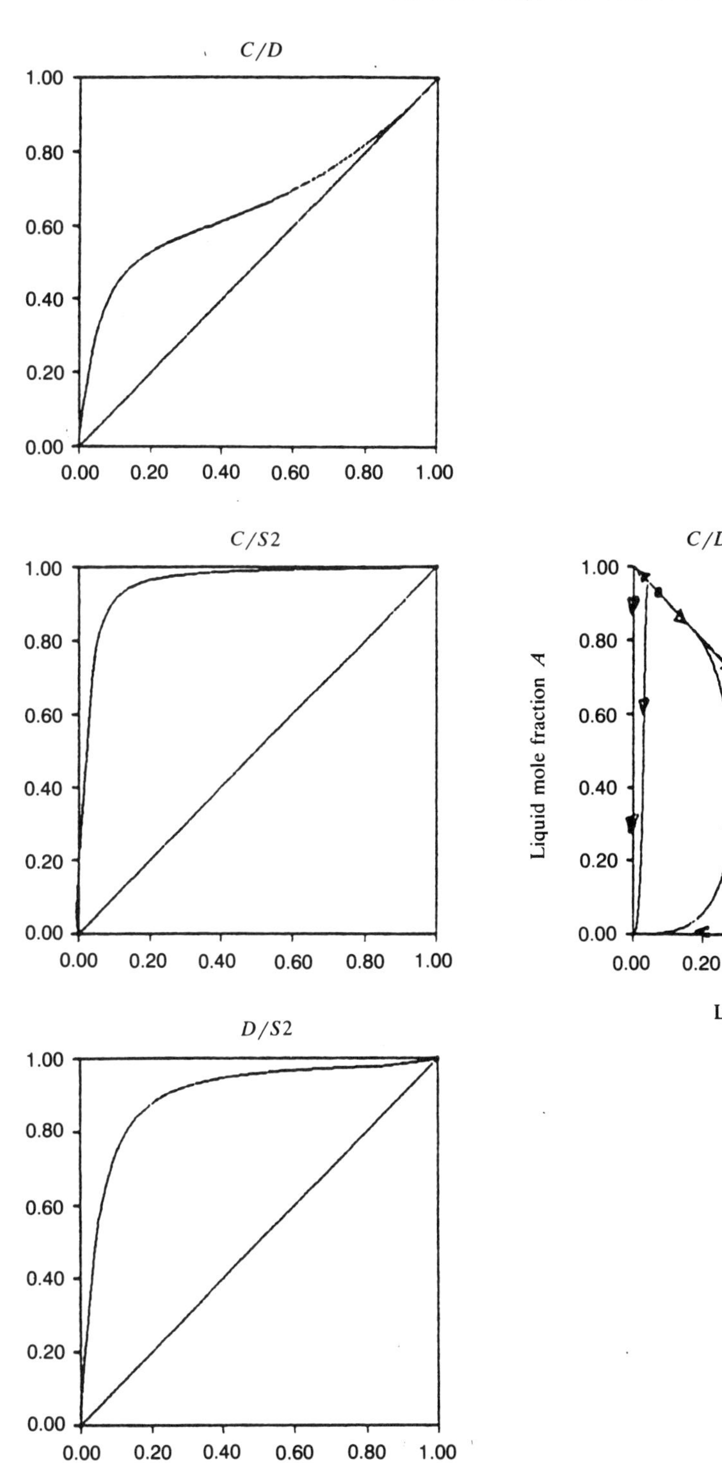

(b)

FIGURE 18-7. *Continued.*

tion boundary or two liquid phase region exist in the residue map for System 2.

Let us now examine a more complex residue curve map that results for the same A and B components, but using $S1$ instead of $S2$. Solvent 1 ($S1$) forms a heterogeneous azeotrope with component A. Open circles label the azeotropes on the residue map. The presence of the second azeotrope creates a distillation boundary. This boundary divides the composition plane into two distinct regions. The distillation process cannot cross the boundary and therefore cannot make products in both. The bottoms composition of the extractive distillation tower must be a binary mixture of B and $S1$. It is therefore impossible to make a pure distillate product consisting of component A. The overhead product cannot lie beyond the line passing between the two binary azeotropes.

We can see the two liquid phase region using this residue map. This is a great aid to help us avoid the two liquid phase region when we design the extractive distillation tower. The residue curve map of this complex system helps our understanding of the process and aids in the design and control of extractive distillation systems.

Figure 18-7*a* and *b* contains the residue curve maps and binary vapor–liquid equilibrium curves for our three case studies. The components used in System 3 are ethanol (C), water (D), and ethylene glycol ($S2$).

18-3 PROCESS DESIGN

Our goal is to design an extractive distillation process at minimum economic cost. An optimum process design minimizes total cost: the fixed capital cost of the equipment plus the variable cost of operation. Capital costs are the total fixed project costs to design, construct, and start-up the process. Variable costs are the operating costs of utilities, raw materials, and workforce. Stability of process control significantly affects variable cost. If the process is difficult to control it will require additional labor to operate and create higher utility and raw material costs due to upsets. To get the optimal economic design, it is very important to consider the controllability of the extractive distillation process during process design.

We will develop a computerized design procedure that can be run on a small personal computer. Computer programs that rate an existing distillation tower are more common than design programs. You can use commercial process simulators to rate an existing extraction distillation tower. These programs are cumbersome to use to design a new extractive distillation system and do not leave the process engineer with as much process insight as a design procedure.

18-3-1 Degrees of Freedom

To begin the design of our process we must first determine which variables we will set and what variables will be calculated by the design procedure. The feed stream is given and the operating pressures of the distillation towers are fixed by the plant cooling medium. We will fix a design limitation on the minimum approach temperatures in our interstage heat exchangers.

Given these conditions we need to find the number of trays, reflux ratio, and condenser and reboiler duties for the towers. If we have a procedure to design the extractive distillation tower, we can find the solvent temperature, solvent recycle rate, and the cross exchanger areas. Conventional methods can design the binary solvent recovery tower. The extractive distillation tower is much more complicated because it is multicomponent, has multiple feeds, and has nonideal phase equilibrium.

Figure 18-8 details the degrees of freedom available for the design of the extractive distillation tower. The degrees of freedom are the total number of variables less the total number of equations describing the problem. The extractive distillation tower has two feeds and three components. We have one total material balance equa-

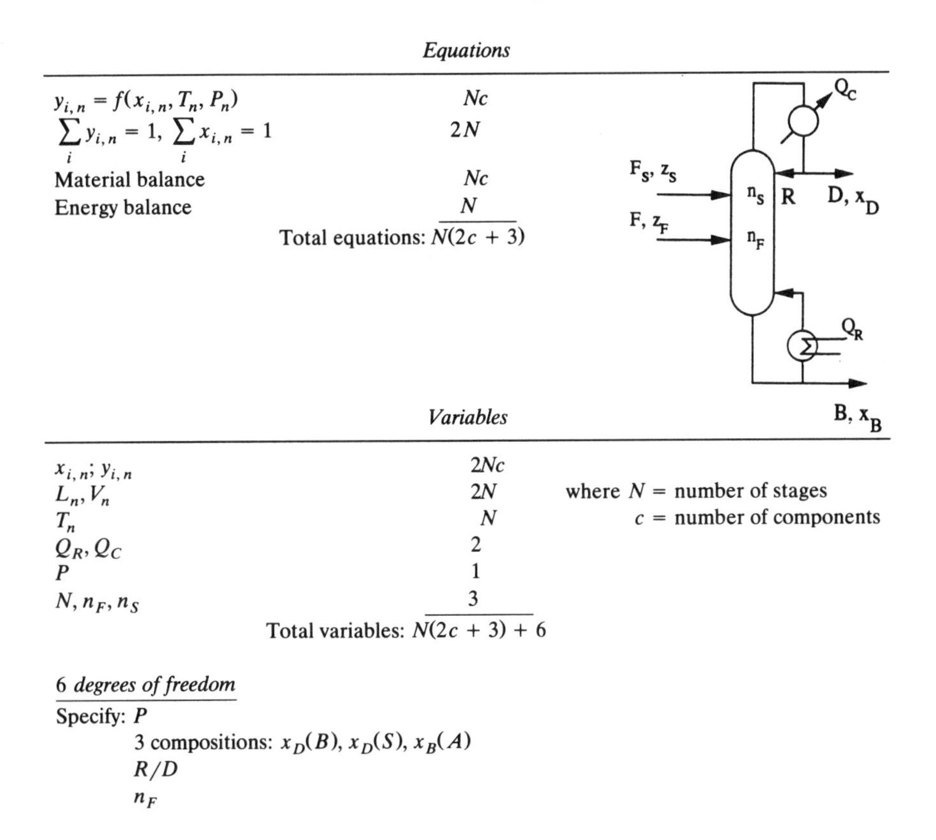

Equations

$y_{i,n} = f(x_{i,n}, T_n, P_n)$	Nc
$\sum_i y_{i,n} = 1, \sum_i x_{i,n} = 1$	$2N$
Material balance	Nc
Energy balance	N
Total equations:	$N(2c + 3)$

Variables

$x_{i,n}; y_{i,n}$	$2Nc$
L_n, V_n	$2N$
T_n	N
Q_R, Q_C	2
P	1
N, n_F, n_S	3
Total variables:	$N(2c + 3) + 6$

where N = number of stages
c = number of components

6 degrees of freedom

Specify: P

3 compositions: $x_D(B), x_D(S), x_B(A)$

R/D

n_F

FIGURE 18-8. Extractive distillation design degrees of freedom.

tion for every stage and one component balance for each component for every stage. The vapor and liquid mole fractions must sum to 1 on every stage. We have one phase equilibrium relationship for each component on every stage and we have one total energy balance on every stage. This produces $N(2c + 3)$ equations, where N is the total number of stages and c is the number of components. I include the condenser and reboiler as stages using this nomenclature. The total number of stages is therefore the total number of trays plus 2.

We now count the number of variables. Each stage contains a liquid and vapor mole fraction for each component. Each stage has one liquid and vapor flow rate leaving the stage. The reboiler and condenser duties and column pressure are variables. Finally the total number of trays and the two feed stage locations are variables. This gives us a total of $N(2c + 3) + 6$ variables.

The design of the extractive distillation tower has 6 degrees of freedom. There are six more variables than equations. We must therefore set six variables to design this tower. The six variables that I will set in my design procedure are:

1. The column pressure.
2. Three product compositions.
3. The column reflux ratio.
4. One feed stage location.

Column pressure will be set by plantwide utility considerations. The three product compositions that we will set are:

1. The overhead B composition.
2. The overhead S composition.
3. The bottoms A composition.

These all represent impurities in the final products. The impurity of A in the extraction tower bottoms affects product specifications because all of it will end up in the overhead product of the solvent tower.

The last two variables to be set are the reflux ratio and one feed stage location. We will initially set these variables to design the distillation tower. An optimization step will follow the design to find what reflux ratio and feed stage location produce the most economical design of the entire extractive distillation system.

18-3-2 Process Design Procedure

We will now develop a procedure to design the extractive distillation process. The procedure involves four steps.

1. Separation screening.
2. Process specification.
3. Equipment design.
4. Optimization.

Separation Screening

The first step is to screen various solvents to find the best one for the application. You can easily screen various solvents by generating the residue curves for your system with each solvent. Possible distillation products and phase equilibrium are evident by inspection of these residue curves. This will allow you to discard solvents that are incapable of achieving the desired separation immediately. We can rigorously test feasible solvents in the next steps to determine which produces the best design.

Process Specification

The second step sets design specifications. These process specifications provide a unique design of the extractive distillation system. The design procedure takes these specifications, determines the minimum reflux ratio for each tower, and determines the optimum solvent flow rate and temperature.

I have found the following specifications to be the most useful in the design of extractive distillation systems:

1. Feed stream: flow, composition, temperature, pressure.
2. Extraction tower: $x_D(B), x_D(S), x_B(A), P$.
3. Solvent tower: $x_D(S)$, $x_B(B)$, P.
4. Tower design: $(R/D)/(R/D)_{\min}$.
5. Heat exchanger minimum temperature approach.
6. Initial value for the extractive distillation tower feed stage location (n_F).
7. Initial values for the solvent recycle temperature and solvent/feed ratio.

where x_D = overhead composition (mole fraction)
x_B = bottoms composition (mole fraction)
A, B, S = components
P = pressure
R/D = actual reflux ratio
$(R/D)_{\min}$ = minimum reflux ratio

The feed stream is fully specified at the onset as are the product streams from each column. Overall process economics set these specifications on a product and solvent recovery basis. The solvent recycle, that is, solvent tower bottoms, composition is specified by assuming that all of its impurity is carried overhead in the extractive distillation tower. This is a good assumption because the rectifying stages above the solvent feed stage do not contain any solvent. Any component B leaving the solvent feed stage as vapor will exit in the distillate.

A design factor in excess of the minimum reflux ratio is chosen based on engineering judgment or individual optimization of the extractive and solvent distillation towers. The distillation design procedure will determine the minimum reflux ratio. We apply this factor to the minimum reflux ratio to set the actual reflux ratio.

Specifications must be made to design the solvent intercoolers. We can set cross exchanger approach temperatures; e.g., 5° F. The duty and heat exchange area can then

Enriching Section

$$y_{i+1} = \frac{x_i r_i + x_D}{r_i + 1}$$

$$r_i = \frac{L_i}{D} = \frac{H_D - Q_C/D - H_{i+1}^V}{H_{i+1}^V - H_i^L}$$

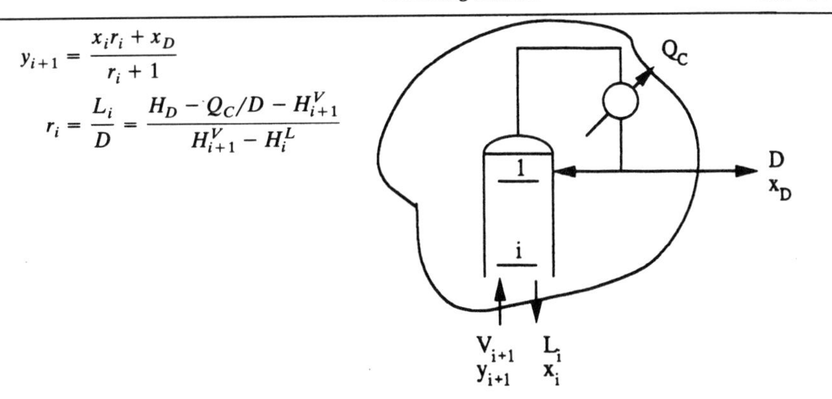

Solvent Section

$$x_{j+1} = \frac{y_j s_j + x_B - z_F F/B}{s_j + 1 - F/B}$$

$$s_j = \frac{H_B - Q_R/B - H_F F/B + H_{j+1}^L(F/B - 1)}{H_{j+1}^L - H_j^V}$$

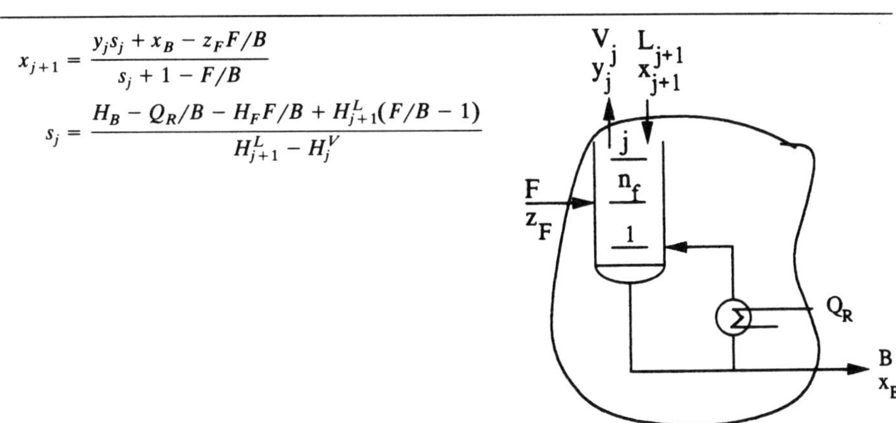

Stripping Section

$$x_{k+1} = \frac{y_k s_k + x_B}{s_k + 1}$$

$$s_k = \frac{V_k}{B} = \frac{H_B - Q_R/B - H_{k+1}^L}{H_{k+1}^L - H_k^V}$$

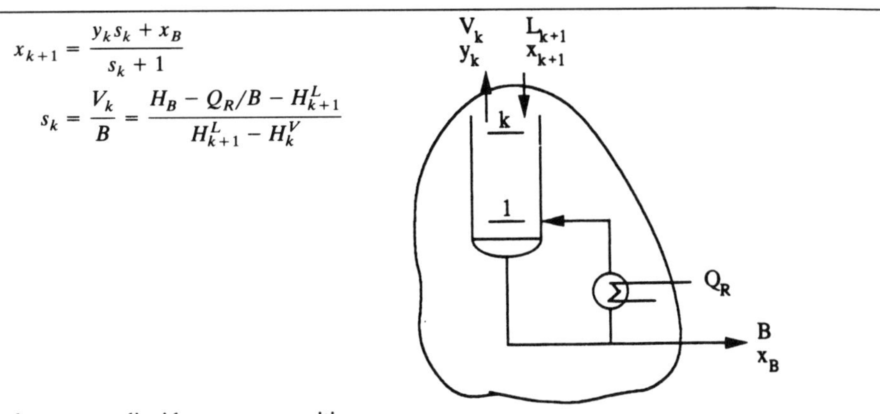

where x, y = liquid, vapor composition
Q_C, Q_R = condenser, reboiler duty
D = distillate molar flow rate (or stream if subscript)
B = bottoms molar flow rate (or stream if subscript)
F = feed molar flow rate
H = stream enthalpy
r = stage liquid–distillate flow ratio
s = stage vapor–bottoms flow ratio
V = vapor flow rate (or stream if superscript)
L = liquid flow rate (or stream if superscript)
i, j, k = stage indices

FIGURE 18-9. Extractive distillation design balances.

be calculated. We may fix the feed stage location at a starting point. The optimum feed stage location will be determined in the next step for the extractive distillation tower by an optimization procedure.

The last set of specifications (solvent recycle temperature and solvent recycle rate) are the two important economic optimization variables. The optimization procedure will find the global optimum for these.

Equipment Design

The next step is to design the process equipment. We need to design two distillation towers, two cross exchangers, a trim cooler, and the column sump holdups. Design of the trim cooler and solvent column base holdup is straightforward. The trim cooler is sized to cool the solvent to its design feed temperature. The solvent column base holdup is sized to have significantly greater solvent holdup than the distillation tower trays. Design of the towers and the cross exchangers is interdependent.

The process variables that we have chosen in the previous section allow us to design the distillation system using a systematic approach. Given product compositions and pressure for each of the distillation towers, we can compute the sump temperatures of the towers by bubble point calculations. Next we design the cross exchangers from which we can calculate the column feed stream temperatures. The solvent recycle stream temperature, flow, and composition are set. We will optimize the solvent recycle stream later. All of the feeds and design conditions for the distillation towers are known and we can complete the design of the towers.

This design method uses an equilibrium stage model and is rigorous for all distillation systems. The method gives the number of ideal stages, feed stage locations, reflux ratio, and boilup required to meet your product specifications. Divide the number of stages by an overall tray efficiency to obtain the actual number of stages.

Extraction Tower

To design the extractive distillation tower we compute the minimum reflux ratio and the remaining feed stage location. We can design the extractive distillation tower by computing the operating composition profiles above and below a feed stage. This is done by starting at each end of the tower and computing the composition profile at each stage moving toward a feed stage. Figure 18-9 gives the material and energy balances over various sections of the distillation tower.

Heat and material balances are written for the enriching, solvent, and stripping sections of the tower. The enriching section is the collection of stages above the solvent feed point. The solvent section is the collection of stages between the solvent feed and the organic feed stage. The stripping section is the collection of stages below the organic feed stage.

Inspecting the balance equations reveals that at the top and bottom stages only the flow rate leaving and the composition of the stream entering are unknown. These two unknowns can be computed by solving the heat and material balance equations simultaneously. The procedure is repeated for the next stage in the section, and so on. If the organic feed stage is known, the stripping section equations switch to the solvent section upon reaching this feed stage. We compute the two profiles until they intersect on the residue map. The location at which the two profiles cross is the solvent feed stage location. The separation is only possible if the profiles cross. This procedure can be used to design the extractive distillation tower. Figure 18-10 depicts this procedure graphically.

To design the tower we must first compute the minimum reflux ratio. The minimum reflux ratio is the reflux ratio at which the two composition profiles just touch with an infinite number of stages. We can implement this method in a simple computer program: The program logic is sequential and does not require sophisticated numeri-

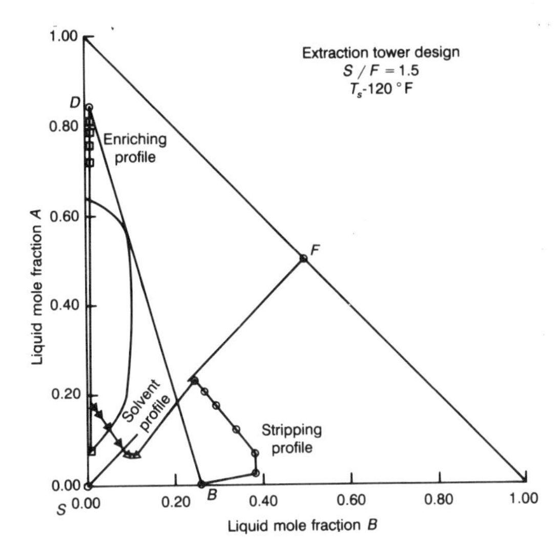

FIGURE 18-10. Extractive distillation design composition profiles.

cal methods. We specify the organic feed stage, choose a sufficiently large maximum number of section stages, and compute the minimum reflux ratio by trial and error. I usually set the organic feed stage and the maximum section stages at 50. Assume a reflux ratio that is reasonable for the system. A reflux ratio of 1.0 is a good starting estimate for the three case studies.*

If the composition profiles cross, a lower reflux ratio is selected and the procedure is repeated. If the composition profiles do not cross, a larger reflux ratio is used. The bisection method can be used to choose the next estimate of the reflux ratio between iterations. The calculations are very fast and can be run in less than 1 minute on a personal computer. Once the minimum reflux ratio is known, you can set the design reflux ratio at a factor larger than the minimum. The economic optimum factor for these case studies is about 1.1 for the extraction and solvent towers. Now that the design reflux ratio is known, we need to compute the optimum location of the feed stage.

The best location for the organic feed is the stage that results in the least number of total stages in the column. The feed location can be found by repeating the design procedure for various feed stage locations until the best one is found. This completes the design of the extractive distillation tower.

Solvent Tower

The solvent tower can be designed by the preceding procedure without the complication of a second feed. The solvent tower is designed as a binary distillation tower because the composition of component A is negligible. The Ponchon–Savarit method can be used.

Optimization

Now that we have a way to design the equipment in the process we must optimize the system considering process economics. You could write all the design equations that we have developed so far and plug them into a simulation package that would produce the economic optimum. I personally do not like that general approach. I prefer to break large problems down into parts and solve each by iteration to reach the optimum. I believe that an engineer learns more about the process by maintaining a physical picture of it during the solution. The engineer learns how design variables affect each piece of equipment, rather than the mathematics of its solution.

The important stream to optimize in this process is the solvent recycle stream. We need to find the best solvent stream composition, temperature, and flow.

Solvent composition is not a strong optimization variable. The impurity in the solvent stream is component B. Most of component B that enters the extractive tower will leave in the distillate as an impurity, so a good assumption is that all of component B that enters the extractive tower leaves in the distillate. Use this criterion to set the composition of the recycle stream.

The two important variables to optimize are the solvent temperature and flow. Solvent flow (solvent recycle ratio) has the strongest effect on system economics and

*Extractive distillation columns also have a *maximum* reflux ratio.

has a unique optimum. Increased solvent flow increases the relative volatility between components A and B in the extractive tower. This reduces the energy requirement in the extractive tower, but increases the energy requirement in the solvent tower because its feed rate increases. These two variables are easily optimized for any extractive distillation system by using the design procedure presented here. The variables were fixed in the design procedure. They can be varied as optimization parameters in this final procedure.

The design procedure runs in less than 1 min on a personal computer. The program designs the entire distillation system and computes its total cost for a given solvent rate and temperature. It is easy to simply run the program for multiple cases of solvent flow and temperature. You can plot total cost as a function of solvent flow for various solvent feed temperatures. This procedure can be added to the design program to compute the optimum system design in one run, which would only take a few minutes on a personal computer.

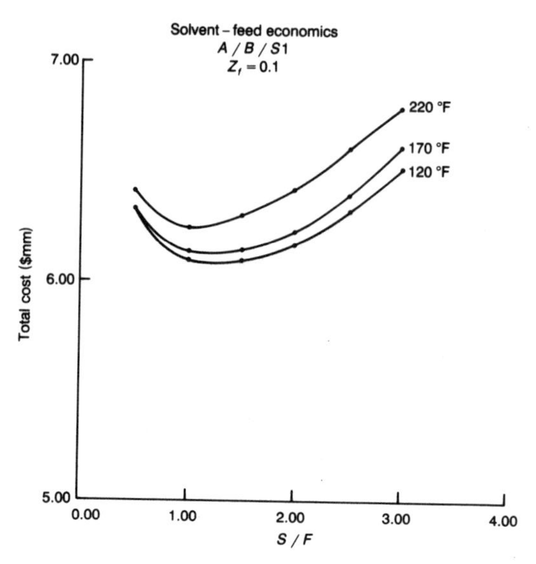

FIGURE 18-11. Cost impact of solvent to feed ratio and temperature.

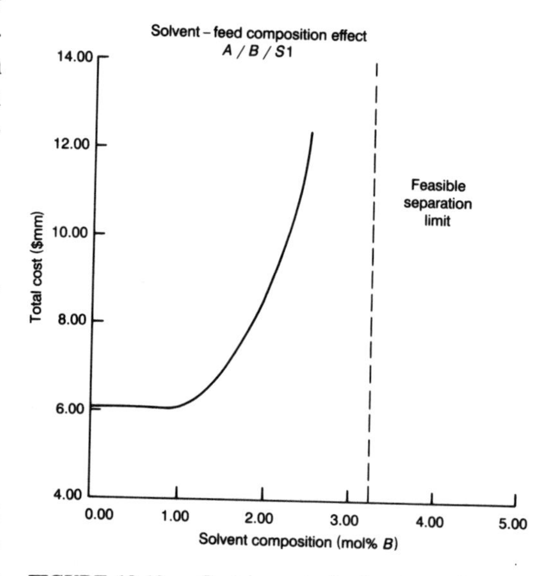

FIGURE 18-12. Cost impact of solvent recycle composition.

18-3-3 Total Cost Relations

The design variables with the most significant effect on cost are:

1. Solvent to feed ratio.
2. Solvent feed temperature.
3. Solvent recycle composition.

Total cost is the sum of installed equipment and three years of operating. I use only the cost of steam for this basis because this is the largest component of operating cost. Figures 18-11 and 18-12 illustrate the effect of these three variables on total cost for one of our case studies using the procedure described in the last section.

Figure 18-11 is typical of extractive distillation processes. Total cost is a function of the solvent recycle rate, temperature, and composition. The design procedure gives us a way to determine the optimum design point. Inspection of Figure 18-11 reveals that in a practical sense the total cost relation is relatively flat. The solvent ratio and temperature can be specified somewhat loosely without a large cost penalty. The same is true for operating the plant. You can move the solvent ratio and solvent temperature without a large operating cost penalty. This can be of great benefit if feed composition is altered and you need to balance the tower loadings.

Figure 18-12 shows that we must be sure that the solvent recycle composition does not exceed a maximum limit. This limit is easily calculated by assuming that all of the impurity entering the extraction tower leaves in the distillate product. A good rule to follow is to design for one-half of this limit. We can also see there is not a strong cost penalty in overpurifying the bottoms product in the solvent tower.

18-4 PROCESS CONTROL

To the process control engineer a two product single distillation tower typically has 5 degrees of freedom. There are usually five control valves:

1. The distillate product flow.
2. Reflux flow.
3. Bottoms product.
4. Heat input.
5. Pressure control.

Our extractive distillation process contains two distillation towers. The extraction tower bottoms feeds the solvent tower. The bottoms stream from the solvent tower is recycled back to feed the extraction tower. Independently both towers have 10 degrees of freedom. When the two are coupled we lose 1 degree of freedom by recycling the solvent tower bottoms to the extraction tower. It is the process control engineer's job to determine what to control with each valve. It is equally important that the process design engineer supplies the process control engineer with a system design that is controllable.

The solvent makeup feed theoretically creates an additional degree of freedom. We can use the solvent makeup to control the solvent inventory in the system. However, the solvent makeup flow is very small relative to the solvent recycle flow. Solvent inventory cannot be controlled precisely during dynamic upsets; it can only be controlled long term. The solvent tower sump must be sized larger than the solvent holdup in the towers to prevent the sump from overfilling or emptying during upsets.

Extractive distillation is a highly integrated process. We have seen that changes in process design variables have a significant effect on product specifications at steady state. The controllability for this process is also sensitive to the choice and pairing of the manipulative and control variables.

Our first step to determine the pairing of the control and manipulative variables of the extractive distillation process is to reduce the number of degrees of freedom.

Our objective is to control the product compositions in this process. Pressure dynamics are very fast compared to composition dynamics in these systems. Therefore we will treat pressure control as a static problem. We can control column pressure by manipulating cooling water flow in the condenser, manipulating the condensate flow leaving the condenser, or manipulating the vapor flow into the condenser. There are many other possibilities, but these are the most common. Selecting a pressure control scheme in each distillation tower eliminates 2 degrees of freedom. See Chapter 1 for a more detailed discussion of pressure control schemes.

Two material inventories that must be controlled in conventional distillation towers are the material in the reflux accumulator and bottoms sump. Therefore two control valves must be used for these two control variables.

The reflux ratios of the three case studies are relatively low. Experience (Shinskey, 1983) shows that it is best to use the product streams to control the column inventories for such systems. The extraction tower distillate and bottoms streams control the extraction tower reflux accumulator level and bottoms sump, respectively.

The solvent tower distillate controls the solvent tower reflux accumulator level. However, the solvent tower bottoms stream cannot control the solvent tower sump level because it is recycled within the process. Solvent makeup flow loosely controls the solvent tower bottoms sump, as discussed previously.

Inventory control eliminates 3 degrees of freedom. This leaves 2 degrees of freedom for each tower with an additional degree of freedom for control of the solvent recycle stream. Extractive distillation is a process with 5 degrees of freedom, therefore five variables to control. These could be five compositions or any combination of five flows, compositions, or tray temperatures.

The solvent column is a binary distillation tower. The distillate stream is a product from our system. The bottoms stream is an intraprocess stream. Composition control of the distillate is necessary to meet customer product specifications, and composition control of the bottoms stream is critical to ensure operability of the system and to meet the extraction tower overhead specification.

At most, we will control two compositions in the solvent tower. The simplest combination of control pairings is to use the reflux or distillate to control the overhead composition. If reflux is chosen to control overhead composition, then the distillate will control the reflux accumulator level. Similarly, if the distillate is chosen to control the overhead composition, then the reflux will control the accumulator level.

Heat input controls the bottoms composition of the solvent tower. The solvent tower bottoms flow should not control the bottoms composition because this stream is recycled within the process.

Using reflux to control the overhead composition and heat input to control the bottoms composition on the solvent tower is common. The solvent is usually much heavier than component B, which leads to a moderate to low reflux ratio in the solvent tower. Reflux is usually the best control variable when the reflux ratio is low to moderate. The solvent makeup rate is usually too small to use for composition control. Therefore, heat input is usually the best variable to control the solvent tower bottoms composition.

We can use multivariable or model based controls if none of these control schemes is satisfactory to control the product specifications. They are, however, rarely needed in industrial practice.

The extraction column contains 3 degrees of freedom. We will therefore consider controlling, at most, three compositions.

The distillate is a product stream from the system. Composition control of this stream is important to meet customer product specifications. The bottoms stream is an intraprocess stream. All the light material (component A) in this stream will leave in the distillate of the solvent tower. The light composition of the bottoms stream directly affects our customer product specifications, so we must control it. The remaining two components in the bottoms stream do not directly affect customer products, but influences the solvent tower and, hence, plant operability.

These considerations lead us to conclude that there are two combinations of three composition control specifications for the extraction tower. We can either control two compositions in the extraction tower overhead and one composition in the bottoms, or one composition in the overhead and two compositions in the bottoms. We can control the compositions of components B and S in the overhead and the composition of component A in the bottoms stream if the extraction tower overhead product specifications are important. If we need to minimize feed composition disturbances to the solvent tower (because it is operating very close to flood), we can control one impurity in the extraction overhead, the compositions of components A and B in the bottoms.

18-4-1 Control System Economics

The engineer must consider overall plant economics to find the best control scheme for the extractive distillation system. We found that we can control five variables in the extractive distillation system. We now need to determine how many of these variables and which ones are needed to control the system. We do not need to control variables that are insensitive to plant operation. My approach is to find the least number of control variables. This has the advantage of reducing project cost, simplifying the operation, and providing greater process integrity by reducing the chance of failure.

The first step in determining control system economics is to define our control objectives for the plant. For our case studies they are to control the product compositions at 99 $\pm$ 1 mol%. We must meet these specifications for feed rate and feed composition disturbances of $\pm$10%. This is a very ambitious goal, so it serves well to test this design methodology.

We will examine the steady state effect of various control configurations by holding manipulated or control variables constant when we change the feed composition. Holding a controlled variable constant as feed composition changes quantifies the value of placing that variable in a control loop. Holding a manipulative variable constant quantifies the value of leaving a loop off control. Figure 18-13*a* contains the possible control schemes for extractive distillation systems. Figures 18-13*b* through *e* give simplified instrument diagrams for the four most common control schemes.

We assume that at least one composition will be controlled in each distillation tower. The solvent tower has three possibilities:

1. Control the bottoms composition.
2. Control the overhead composition.
3. Control both bottoms and overhead compositions.

It is very important to control solvent tower bottoms composition because we need to prevent the extraction tower overhead product from being affected during an upset. We must control the extraction tower bottoms composition of the light component. If this composition is not controlled, any light component in the bottoms stream will end up concentrated in the distillate product of the solvent tower. The extractive distillation tower has four possible schemes:

1. Control the bottoms A composition and the overhead B composition.
2. Control the bottoms A composition and the bottoms B composition.
3. Control the bottoms A composition, the overhead B composition, and the bottoms B composition.

Control schemes that control the amount of solvent in the overhead product of the extractive tower will not be considered. This is because the solvent is a very heavy component in the extractive tower and is in trace composition in the overhead stream. The composition of component B is the important consideration for the overhead stream.

The three systems presented in this chapter all have the same economic control scheme; in each, the best plantwide control scheme is E1-S1. The bottoms composition of both the extraction and solvent towers are controlled. Figure 18-14 contains four possible control schemes for the System 1 extractive distillation process. The total reboiler energy requirements are calculated for each scheme over the range of feed compositions for which we want to design. The heat duties are calculated so the product specifications are met. We accomplished this by fixing the composition if it is a control variable or fixing the manipulative variable if the composition is not controlled. The manipulative variable is at a constant value over the feed composition range so that the desired product composition is always met. This means that the product is overpurified over most of the feed composition range. Overpurification is not as efficient as maintaining the product specification over the entire feed composition range by direct control. However, this analysis quantifies the cost of inefficiency.

Control Scheme	Control Variable(s)	Manipulative Variable(s)
	Extraction Tower Control Schemes	
E1	$x_B(A)$	Q_R
E2d	$x_B(A)$ and $x_D(B)$	Q_R and R
E2b	$x_B(A)$ and $x_B(B)$	Q_R and S
E3	$x_B(A)$, $x_D(B)$, and $x_B(B)$	Q_R, R, and S
	Solvent Tower Control Schemes	
S1	$x_B(B)$	Q_R
S2	$x_B(B)$ and $x_D(B)$	Q_R and R

where $x_B(A)$ = bottoms composition of component A
$x_B(B)$ = bottoms composition of component B
$x_D(B)$ = distillate composition of component B
$x_D(S)$ = distillate composition of component S
Q_R = reboiler duty
R = reflux flow
S = solvent flow

(*a*)

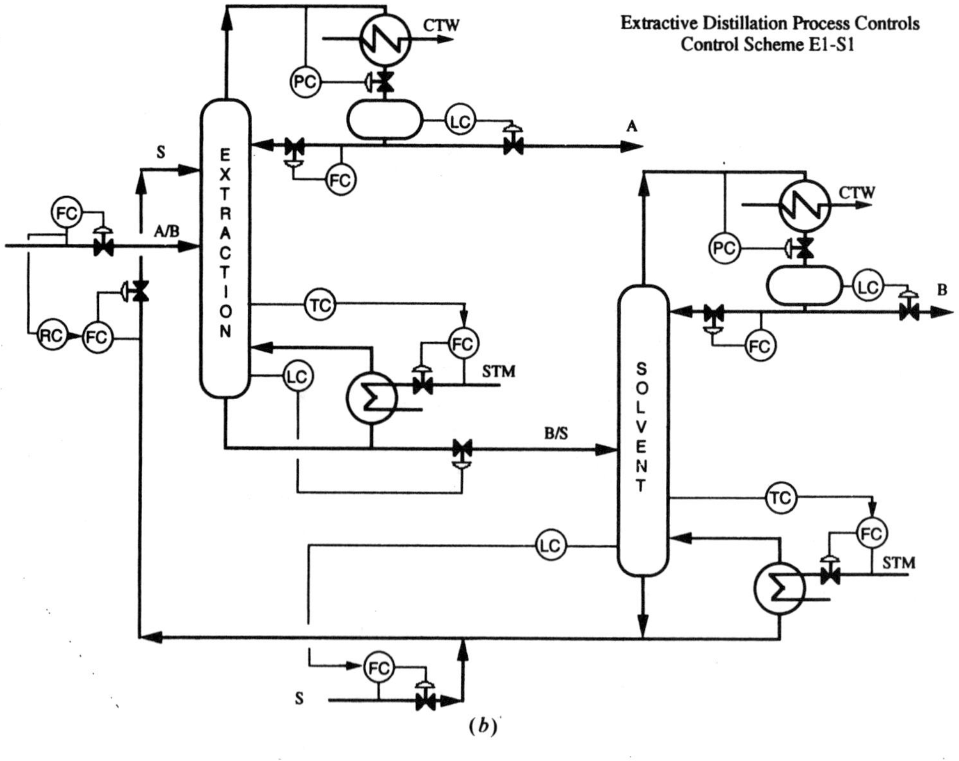

(*b*)

FIGURE 18-13. Extractive distillation control schemes.

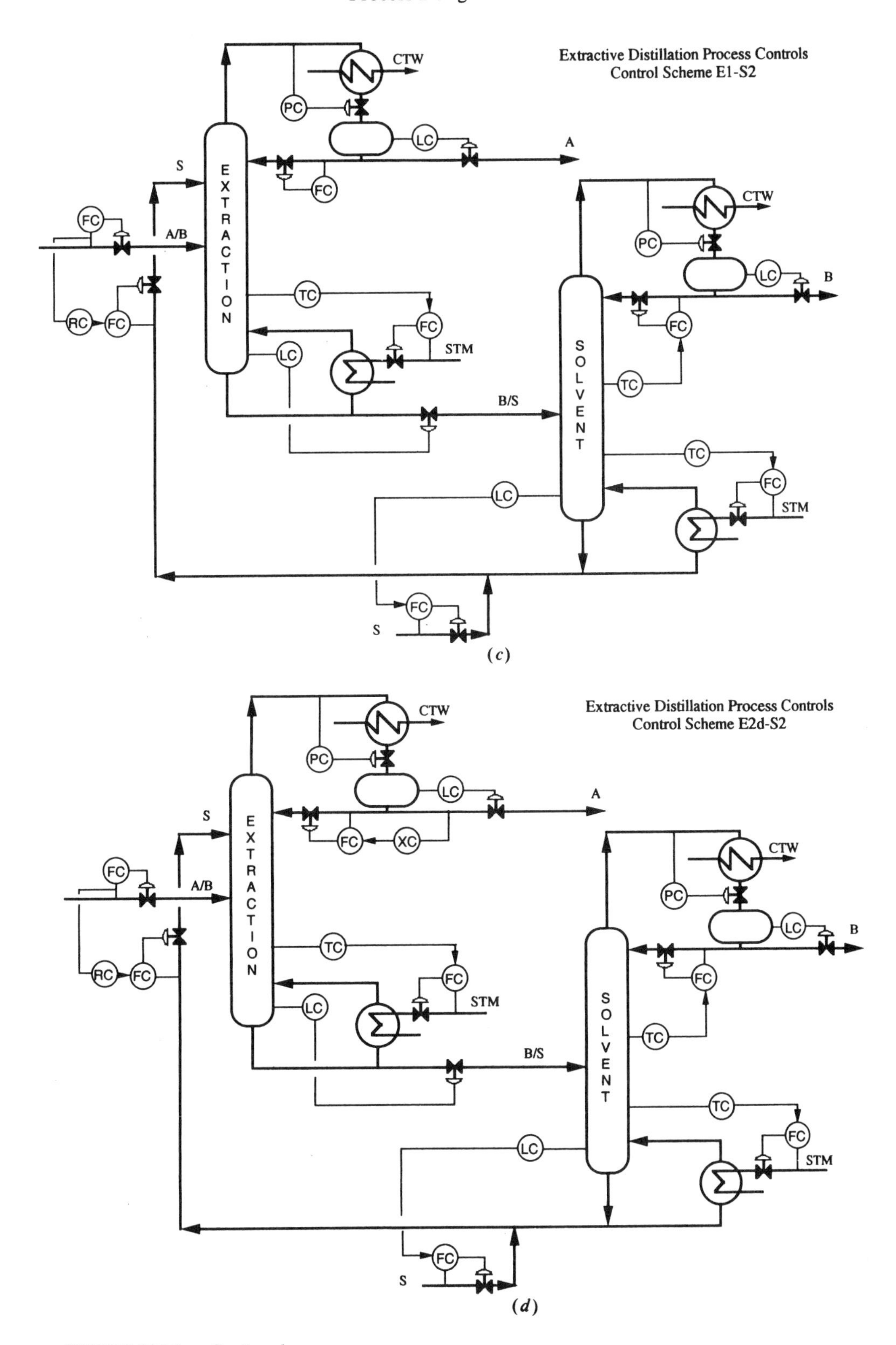

FIGURE 18-13. *Continued.*

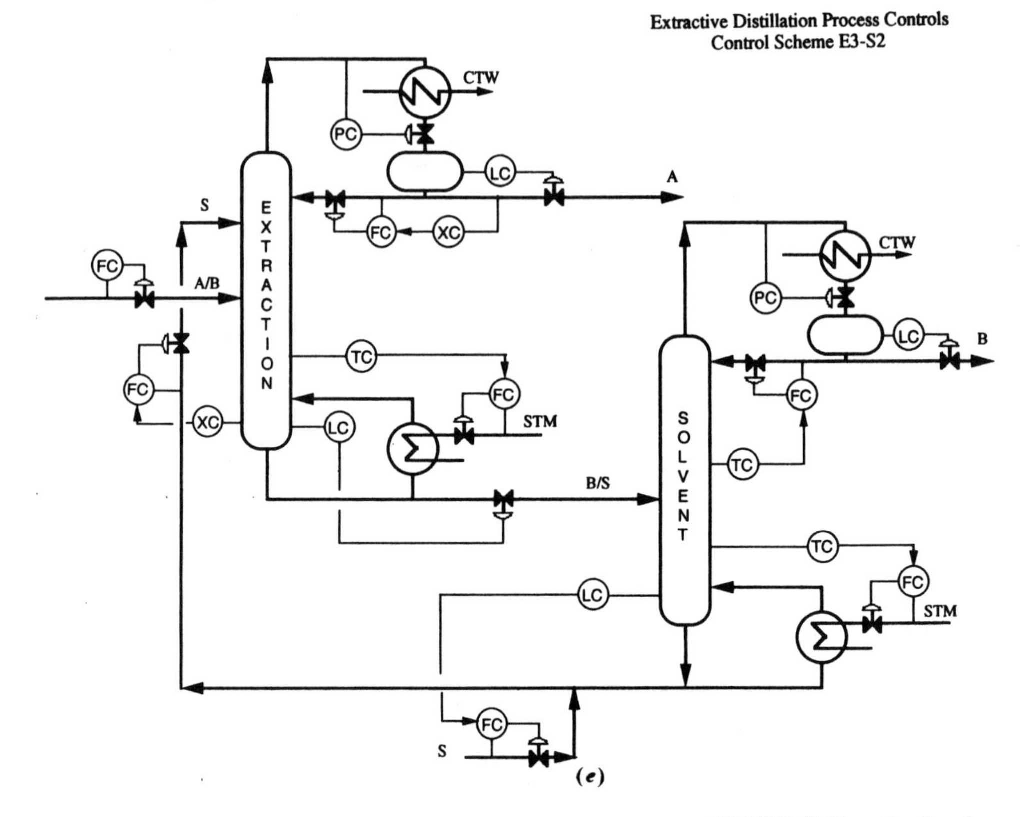

FIGURE 18-13. *Continued.*

Inspection of Figure 18-14 reveals that the total energy requirement does not vary significantly over the feed composition range. We cannot justify the cost of adding higher control beyond E1-S1.

Figure 18-14 also shows that control scheme E3-S2 results in a less flexible process design than the other schemes. The energy is significantly higher using E3-S2 control at a feed composition of 0.6 mole fraction of component A. This is a result of trying to control the extraction tower bottoms composition too tightly. There is not enough solvent flow to separate the feed with this composition, so the reflux has to increase, thereby increasing the reboiler duty. I recommend that you not control the bottoms composition of component B in the extraction tower.

18-4-2 Measured Variables

We must select measured variables that either equal or infer the variables we wish to control. We need to control product stream compositions. Composition can be measured directly by analyzers or can sometimes be inferred by tray temperatures. I prefer to use tray temperatures, which are less expensive to measure than composition and also faster to measure. A temperature measurement usually has a time constant of about 30 s, whereas composition measured by gas chromatography usually has a deadtime of 5 to 10 min. Newer analyzers based on infra-red or near infra-red eliminate the deadtime, but are two orders of magnitude more expensive than simple temperature measurements.

E1-S1 Control Fixed Variables: Extraction Tower: Reflux, $x_B(A)$, S/F; Solvent Tower: Reflux, $x_B(B)$

$x_F(A)$	Extraction Reboiler (MMBtu/h)	Solvent Reboiler (MMBtu/h)	Total Duty (MMBtu/h)
0.4	19.19	26.91	46.10
0.5	21.20	25.31	46.51
0.6	23.23	23.72	46.95

E1-S2 Control Fixed Variables: Extraction Tower: Reflux, $x_B(A)$, S/F; Solvent Tower: $x_D(B)$, $x_B(B)$

$x_F(A)$	Extraction Reboiler (MMBtu/h)	Solvent Reboiler (MMBtu/h)	Total Duty (MMBtu/h)
0.4	19.21	26.68	45.89
0.5	21.20	23.1	44.30
0.6	23.21	19.52	42.73

E2d-S2 Control Fixed Variables: Extraction Tower: $x_B(A)$, $x_D(B)$, S/F; Solvent Tower: $x_D(B)$, $x_B(B)$

$x_F(A)$	Extraction Reboiler (MMBtu/h)	Solvent Reboiler (MMBtu/h)	Total Duty (MMBtu/h)
0.4	18.51	26.62	45.13
0.5	20.97	23.07	44.04
0.6	23.21	19.52	42.73

E3-S2 Control Fixed Variables: Extraction Tower: $x_B(A)$, $x_D(B)$, $x_B(B)$; Solvent Tower: $x_D(B)$, $x_B(B)$

$x_F(A)$	Extraction Reboiler (MMBtu/h)	Solvent Reboiler (MMBtu/h)	Total Duty (MMBtu/h)
0.4	17.56	27.95	45.51
0.5	20.92	23.09	44.01
0.6	48.84	18.04	66.88

FIGURE 18-14. System 1—Plantwide control economics.

We can control most of the product compositions in our extractive distillation process by measuring temperature. The solvent is much less volatile than either key component. The solvent tower product compositions are usually sensitive to reflux and heat input, so we will control it by enriching and stripping tray temperatures.

The stripping section of the extraction tower is rich in solvent and intermediate component, and by specification, the composition of component A is very low in the bottoms stream. Stripping section temperatures are sensitive to component A composition because it is very light relative to the solvent. Higher concentrations of the light component in the bottoms stream lower the stripping section temperature profile considerably. Therefore stripping section temperatures are good measurements for control of the bottoms light component.

The overhead composition of the extraction tower is not well suited to temperature control. The solvent section contains a sig-

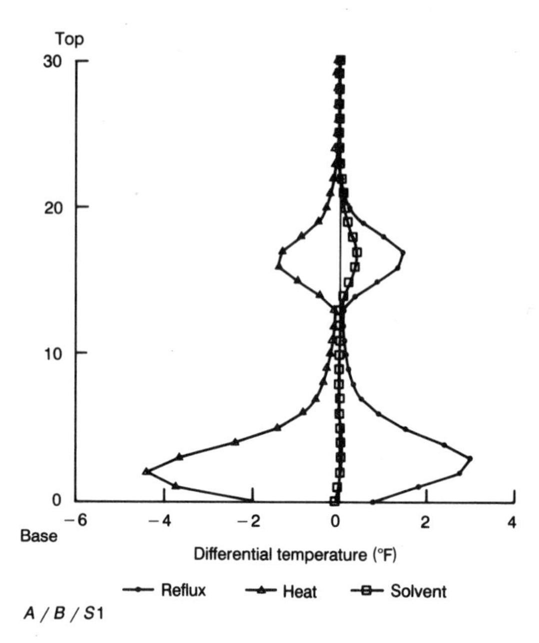

FIGURE 18-15. Extractive tower temperature process gains.

nificant amount of solvent in the liquid phase, and the two key components boil close to each other. Temperature is not responsive to operating changes and cannot be used for overhead composition control. A composition analyzer must be used.

Figure 18-15 illustrates the process gains between manipulative variables and tray temperature. Differential temperature profiles have been plotted, which are the difference in temperature from steady state when the manipulative variables change. The manipulative variables changed are reflux, heat input, and solvent feed rate. I decreased each manipulative variable, while holding the others constant, by 1% from its steady state value. It is evident from this figure that bottoms composition is easily controllable with a stripping section temperature. The top section temperatures are insensitive and therefore not controllable.

Selection of the proper control tray location is made from two criteria. First the tray temperature must be insensitive to changes in feed composition at constant product composition. Second, the tray temperature must be sensitive to changes in the manipulative variable at constant feed composition. Balancing these two criteria will lead to the best tray location for control.

We can infer the bottoms composition of component *B* from bottoms temperature and pressure, if needed. There is a large boiling point difference between the intermediate and the solvent. A correlation between bottoms temperature, pressure, and composition is easy to find. It is usually linear in temperature and pressure over a small range of composition. This correlation is easily implemented in a computer control system.

I recommend that you not control the bottoms component *B* composition under normal circumstances. This composition has to be controlled by manipulating the solvent recycle flow. Manipulating solvent flow increases interaction between the extraction and solvent towers. I have found it unnecessary to hold the bottoms composition constant unless the solvent tower is capacity limited. If the solvent tower must operate at flood, it may be necessary to have tight control over the extraction tower bottoms composition. Although it can be done, do not do it unless it is absolutely necessary. This control scheme will be discussed later in the chapter.

18-4-3 Extraction Tower Nonlinearities

Distillation processes are highly nonlinear. Nonlinearities occur in the governing set of differential equations as shown in Chapter 3. Distillation nonlinearities are commonly seen in practice, and they make it harder to produce higher product purities than lower purity products. A decrease in reflux flow by 5% will yield a larger variation in product purity than a 5% increase in reflux flow.

Conventional distillation processes usually have consistent vapor–liquid equilibrium characteristics. The only operating effect that changes the phase equilibrium is stage efficiency. This effect is usually small until you flood the column. We intentionally design extractive distillation processes to vary the stage equilibrium by solvent flow. In this respect, extractive distillation is more nonlinear than ordinary distillation. Nonlinearities in extractive distillation can lead to counterintuitive behavior of the column. The process control engineer must deal with these nonlinearities carefully in the design and tuning controls for extractive distillation.

Figure 18-16 contains the steady state process gains between the manipulative and control variables for our three case studies. These gains are linear and are calculated for very small changes in the manipulative variable. The results are scaled to instrument ranges so they are dimensionless and are similar to what you would find in the field.

These gains illustrate many of the general cause and effect variables in the extractive distillation column. A first observation is that heat input has a large effect on the bottoms composition of the lightest component. This agrees with our previous conclusion that a stripping section temperature can control the bottoms composition by manipulating heat.

A second important observation is that the overhead solvent composition is insensitive to the manipulative variables. The solvent is very heavy in the extractive distillation tower and moves down the tower quickly. Extraction tower overhead solvent composition control is not important.

Reflux or solvent flow can control the overhead B component. I prefer to use reflux rather than solvent flow to minimize the interactive effects of solvent flow.

Further inspection of Figure 18-16 reveals a good example of a counterintuitive process gain in extractive distillation. The process gain between the extraction tower overhead B composition and solvent flow is negative for Systems 1 and 3. This is what we expect. Increasing solvent flow in the extraction tower will extract more of component B at the bottoms of the column, which will lower the overhead composition of component B and produce a negative gain. However, System 2 has a positive gain! This is not what we would expect; it is counterintuitive.

Further inspection of the nature of System 2 reveals why this process gain is positive. The positive gain results from an impure solvent recycle stream composition and a high purity extraction tower distillate. Systems 1 and 3 use a solvent recycle stream composition of 0.001 mole fraction B. The

	$\partial x_D(A)$	$\partial x_D(B)$	$\partial x_D(S)$	$\partial x_B(A)$	$\partial x_B(B)$	$\partial x_B(S)$
			System 1			
∂R	0.072	−0.13	0.058	< 0.01	−0.45	0.45
∂Q	−0.111	0.192	−0.081	−0.652	0.200	0.452
∂S	0.072	−0.131	0.059	< 0.01	−0.453	0.445
			System 2			
∂R	0.069	0.068	< 0.01	14.02	24.45	−38.47
∂Q	0.196	−0.196	0.0	−65.37	39.75	26.62
∂S	−0.71	0.70	< 0.01	7.36	−17.70	10.34
			System 3			
∂R	25.7	−25.7	0.0	36.8	−7.3	−29.5
∂Q	−15.0	15.0	0.0	−57.9	13.5	44.4
∂S	21.6	−21.6	0.0	26.2	−44.2	18.0

FIGURE 18-16. Extractive distillation column steady state gains.

extraction tower overhead compositions are 0.01 mole fraction *B*. System 2 operates with a solvent recycle composition and extraction tower overhead *B* composition of 0.005 mole fraction *B*. Our design guideline says that the recycle composition should be about 0.001 mole fraction *B*. If we operate the solvent tower bottoms composition at 0.001 mole fraction *B*, the gain is negative. Increasing the bottoms purity from 0.005 to 0.001 mole fraction *B* only increases the reboiler duty 0.5%.

This process gain for System 2 is very nonlinear. It is positive for small changes in solvent flow but becomes negative for large changes in solvent flow. This occurs as a result of two competing effects: An increase in solvent flow rate lowers the volatility of component *B* and, at the same time, increases the amount of impurities fed into the top of the extraction tower. When we design the system too close to the limiting solvent impurity, the later effect becomes very pronounced. Therefore, when we

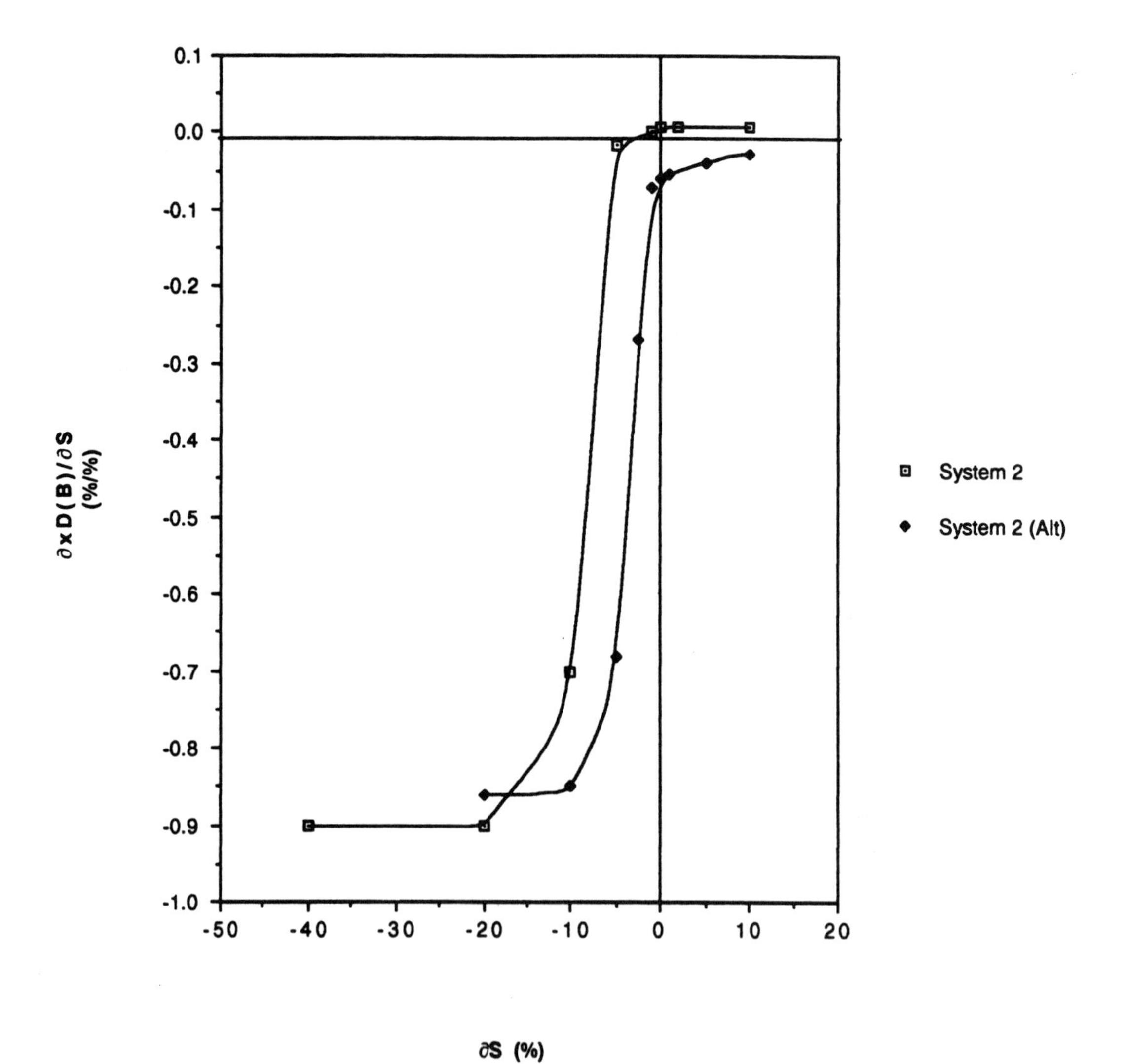

FIGURE 18-17. Nonlinearity of solvent flow.

slightly increase the solvent, the effect of lowering the volatility of component B is less than the effect of feeding more impurities into the extraction tower, which results in an increase in the distillate composition of component B. The effect on volatility is greater than the effect of increased impurities when we make a large change in solvent flow.

Figure 18-17 contains a plot of the process gain between the extraction tower's overhead B composition and solvent flow for two systems. The results for System 2 show that the process gain is positive for small changes in solvent flow and negative for large decreases in solvent flow. Additionally, Figure 18-17 includes a second case, labelled "System 2 (Alt)," which is a redesign of the system that produces negative process gains for all changes in solvent flow. I designed the alternate based on an extraction tower overhead composition and a solvent recycle composition of 0.01 and 0.001 mole fraction B, respectively.

If the recycle composition in System 2 is lowered to 0.001 mole fraction B, or if the extraction tower overhead composition is raised to 0.01 mole fraction B, a curve similar to System 2-Alt results. These effects are qualitatively valid for any extractive distillation system with impure solvent recycle and tight extraction tower overhead composition specifications.

I have indicated many ways to avoid this nonlinearity. The solvent tower can be operated with a lower impurity in the bottoms, or the extraction tower can be operated with a higher overhead specification. In this

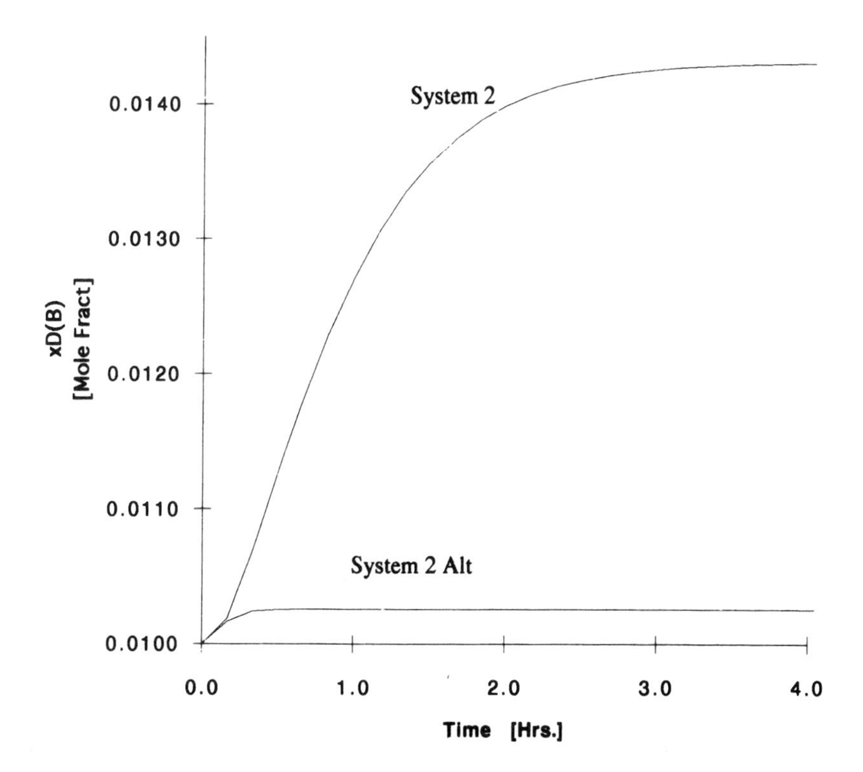

FIGURE 18-18. Extraction tower overhead composition response.

case I decided to redesign the process, not only for the steady state gain effect but also for dynamic considerations.

18-4-4 Extraction Tower Open Loop Dynamics

Distillation towers usually have well behaved open loop characteristics. The open loop eigenvalues are negative and real. This results in open loop dynamics that decay exponentially without oscillation. Extractive distillation columns have similar open loop dynamics. One important exception is a caveat that the process engineer needs to consider during the design of an extractive distillation system: Too many trays in the solvent section of the extraction tower can lead to slow overhead composition response for changes in solvent flow. Figure 18-18 illustrates this effect.

The curves in Figure 18-18 are the open loop responses of the extraction tower overhead B composition for a 1% decrease in solvent flow. The two curves represent different designs for System 2. The design overhead B composition is 0.01 mol%, the solvent impurity is 0.005, and the solvent to feed ratio is 1 in both cases. The extraction tower for System 2 has 52 trays with 33 solvent section trays and the extraction tower for System 2-Alt has 30 trays with only 11 solvent section trays. The alternative design has considerably fewer trays! It is important to note that neither system has the positive steady state gain effect described in the previous section. Table 18-1 contains a summary of the process design specifications for the three case studies. Each system is designed with a feed composition of 50 mol% A and B, and tray Murphree vapor phase efficiencies in all towers are 80%. Both systems are design for the same product specifications. Although the alternative system consumes about 3% additional energy because it has less trays, the total cost of both systems is essentially equal. We save in lower initial capital cost what we pay in additional energy.

TABLE 18-1 Case Study Designs

	System 1	System 2	System 2-Alt	System 3
Extraction Tower				
Enriching trays	8	5	5	7
Solvent trays	9	33	11	48
Stripping trays	13	14	14	14
Diameter (ft)	6	6	5	5.5
Reboiler duty (MMBtu/h)	21	19.6	19	26.8
Reflux ratio	0.9	1.5	1.5	1.4
Solvent/feed	1.0	0.8	1.0	1.2
Weir height (in)	2.5	2.5	2.5	2.5
Sump holdup (lb mol)	150		150	170
Accum. holdup (lb mol)	30		33	25
Avg. tray holdup (mol)	10		7	6
Solvent Tower				
Enriching trays	9	5	5	3
Stripping trays	11	3	3	8
Diameter (ft)	6	4	3.5	3.5
Reboiler duty (MMBtu/h)	23	10.6	15	15.4
Reflux ratio	2.0	0.28	0.1	0.16
Weir height (in)	2.5	2.5	2.5	2.5
Sump holdup (lb mol)	300		250	250
Accum. holdup (lb mol)	40		40	60
Avg. tray holdup (mol)	6		4	4

Figure 18-18 shows that the alternative design has better solvent process dynamics and is the design of choice. System 2 takes almost 3 h to reach steady state from a change in solvent flow. System 2 Alt reaches steady state in 30 min. Even if overhead composition control is not important for this tower, distillate disturbances from this tower will upset a downstream tower.

18-4-5 Control Schemes

E1-S1

We found that the most economical control scheme for our case studies was E1-S1. We control the stripping temperature in both the extraction and solvent towers by manipulating the heat input of the tower's respective reboiler. The solvent recycle and both reflux flows are constant.

We held the solvent and reflux flows constant based on changes in feed composition. We want to also design our control system for feed rate changes. Product compositions are controlled during feed rate disturbances by fixing a constant solvent-to-feed and reflux-to-feed ratio. Figures 18-13*b* through *e* show schematics of these control schemes.

The extraction and solvent towers interact dynamically because the solvent recycle stream couples the distillation towers. The E1-S1 control scheme is a single-input–single-output control scheme in which the two controllers do not interact. The solvent recycle changes slowly relative to the stripping temperatures and the stripping temperatures do not affect each other significantly. This control scheme is treated as a diagonal control structure without any cross terms. This was discussed in detailed in Chapter 11.

We select the desired control trays for these control loops by the procedure described in Section 18-4-2. We can then tune the two temperature control loops individually because interaction is not significant. I have found that the Ziegler–Nichols tuning rules for PI controllers work well for these types of control loops. The ultimate gain and period for the Ziegler–Nichols settings are found by relay testing (Luyben, 1990). The loops are easy to tune, so you can use any other single-input–single-output tuning method with which you are comfortable.

I used a simulation based on model 2A from Chapter 3 to simulate the towers and control system. The simulation includes both

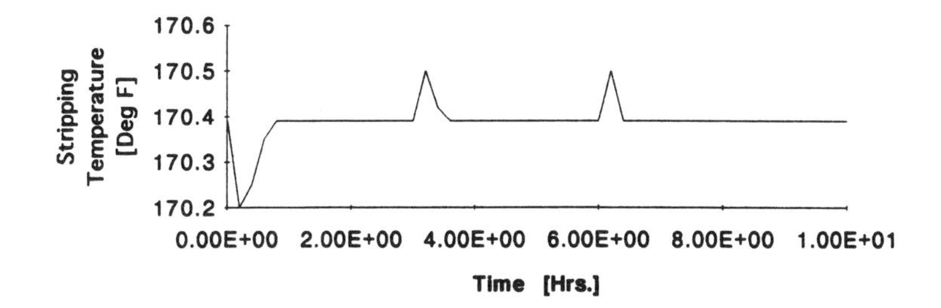

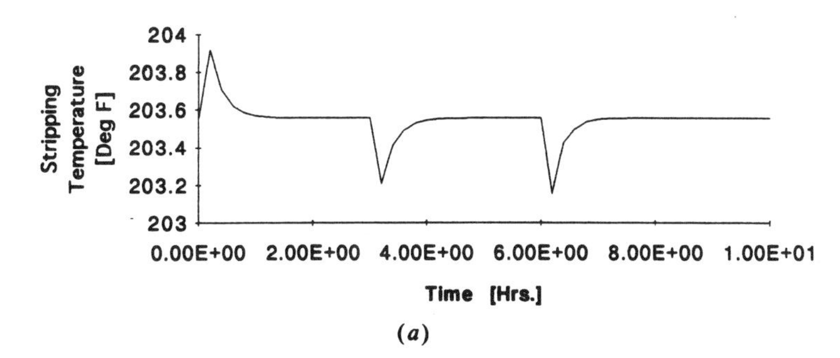

(*a*)

FIGURE 18-19. Extractive distillation E1-S1 temperature response.

the extraction and solvent towers as a coupled system using the E1-S1 control scheme shown in Figure 18-13*b*.

Figures 18-19*a* and *b* contain sample results for System 1 using this control scheme. They illustrate the closed-loop response of the E1-S1 control scheme for extractive distillation systems. Figure 18-19*a* contains the temperature response for the extraction and solvent tower temperature control loops. Figure 18-19*b* contains the composition response of the important streams. These results are generically representative of extractive distillation systems. All three systems presented in this chapter respond similarly.

Three feed composition disturbances were used to test the control system performance. Initially the feed composition to the extraction tower is decreased from 0.5 to 0.4 mole fraction of component *A*. The simulation is run for 3 h. Then the feed composition is changed back to 0.5. The run continues for another 3 h before the feed composition is increased to 0.6.

These are significant disturbances to the column. Disturbances in feed composition are much more difficult to control than disturbances in feed rate. Feed rate disturbance are controlled by the ratio controls discussed before. Feed composition is usually not measured, and cannot be controlled by feedforward schemes. Feed composition, unlike feed rate, also affects the steady state temperature profile.

Inspection of the response curves in Figures 18-19*a* and 18-19*b* reveals that E1-S1 provides excellent control for this system.

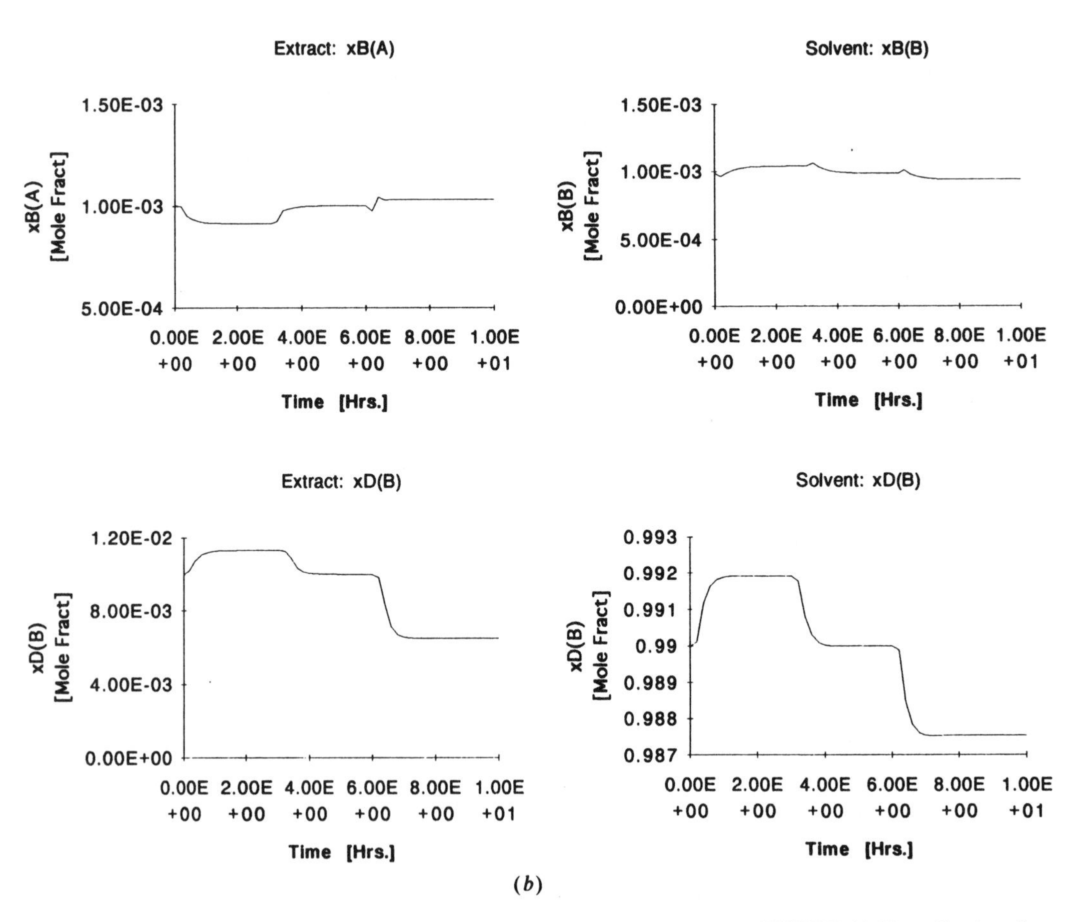

(*b*)

FIGURE 18-19. *Continued.*

There are steady state offsets in the product compositions because product compositions are not controlled directly, but they meet our control system requirements.

Higher Order Control Schemes

We have seen that E1-S1 is the most economical control scheme. It is very easy to implement and provides good control. Our economics are based on minimizing the total cost of operation and control scheme implementation. I have shown that E1-S1 provides good economic control. Current operating costs do not justify higher order control schemes for these case studies. The control objectives used allow the product compositions to vary within 1% of design. If operating costs increase significantly or there is a need for tighter product composition specifications, higher order control schemes will be necessary. The higher order control schemes are E1-S2, E2d-S1, E2b-S1, E2d-S2, E2b-S2, E3-S2. These are diagonal multivariable control schemes.

Figure 18-13 reveals that there are seven higher order control schemes for extractive distillation. All of these schemes are controllable if you follow the guidelines given at the introduction to process control of extractive distillation. It is beyond the scope of this chapter to detail each of these alternatives. A more complete description can

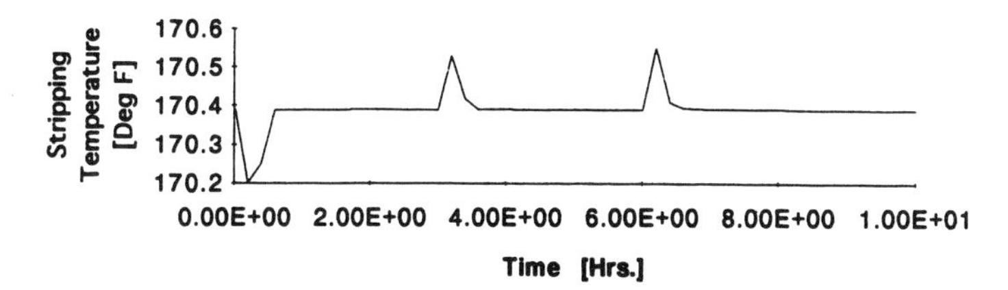

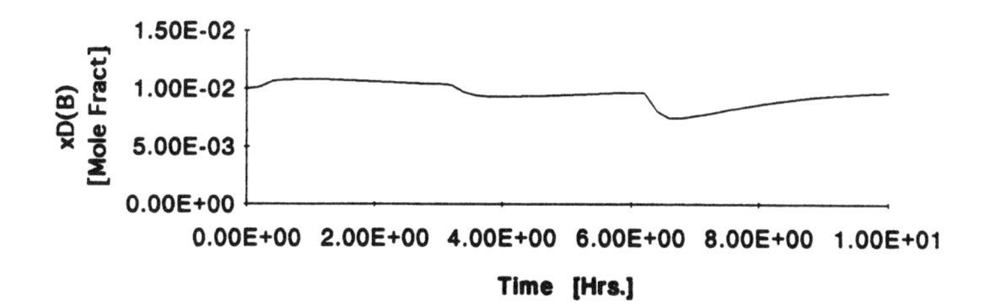

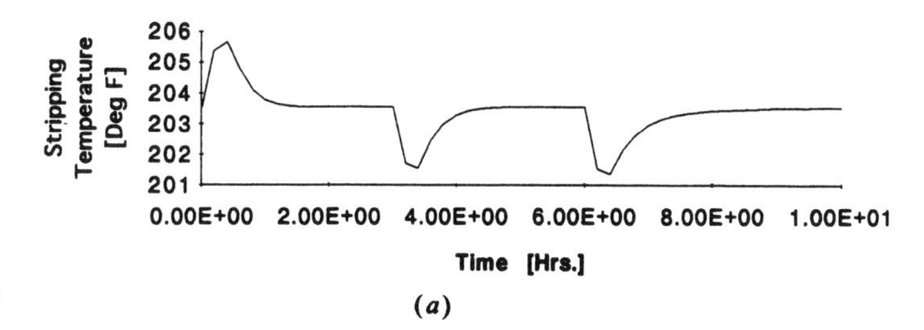

(*a*)

FIGURE 18-20. System 1 E2d-S2 (*a*) temperature response and (*b*) composition response.

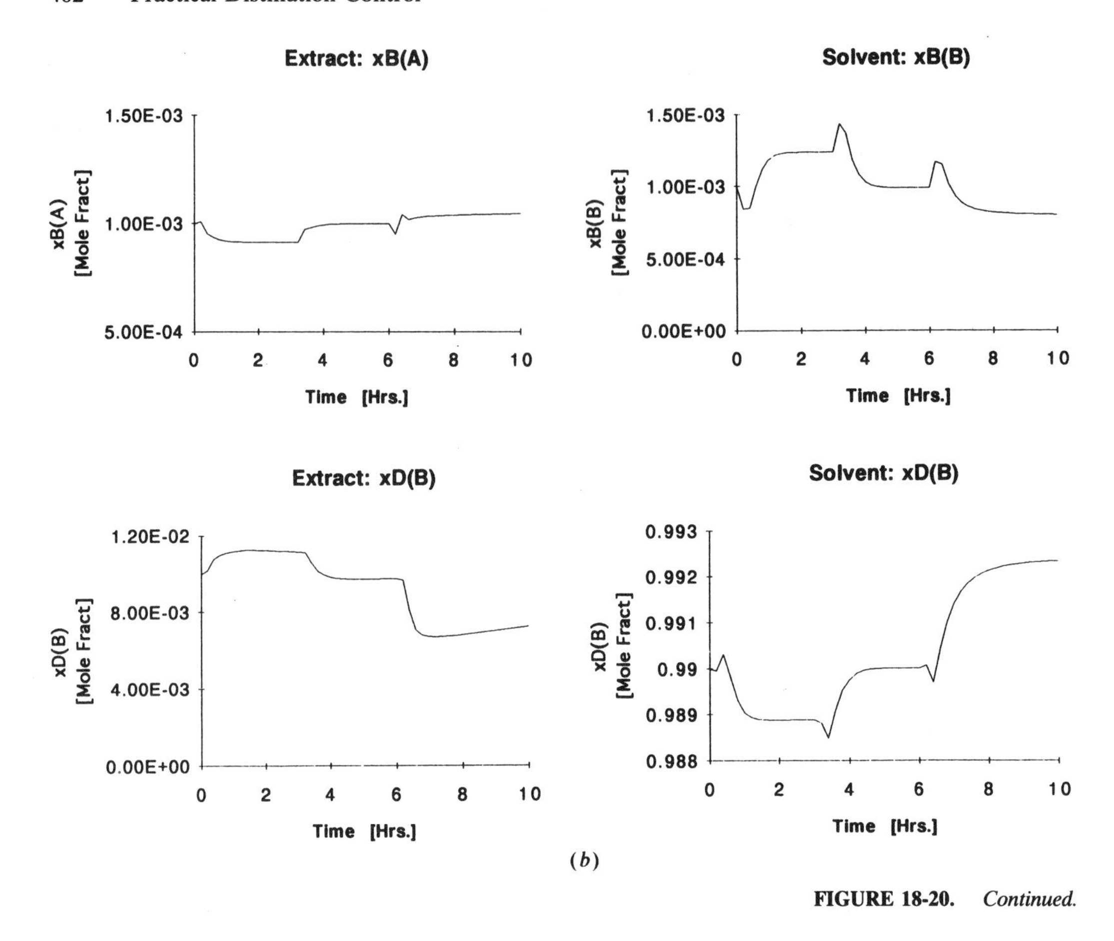

(*b*)

FIGURE 18-20. *Continued.*

be found in Grassi (1991). Figures 18-20*a* and *b* illustrate the performance of the E2d-S2 control scheme for System 1. This system is a 4×4 multivariable control problem. We control two variables on the extraction tower and two on the solvent tower.

The two variables controlled on the extraction tower are a stripping section temperature and the overhead composition of component *B*. Heat input controls the stripping section temperature. Reflux controls the overhead composition. An analyzer with a 5 min deadtime is used to control the overhead composition of component *B*.

One stripping and one enriching section temperature are measured in the solvent tower. Reflux flow and heat input controls the enriching and stripping section temperatures, respectively.

I tuned the loops using the ATV and BLT methods (Luyben, 1990). There is significant interaction between the two control loops within each tower. These must be tuned as a multivariable diagonal controller. A more detailed description of this can be found in Chapter 11.

The temperature control loops respond very well. The extraction tower overhead composition loop is slow, but meets our control objective very well. Control loop interaction between the distillation towers does not significantly affect the tuning of the temperature loops. However, intercolumn interaction must be taken into account to tune the extraction tower overhead composition loop. I took this interaction into account by identifying transfer functions for the extraction tower control loops with both towers coupled. I then used the BLT method to tune the 2 × 2 extraction tower control system. Intercolumn interaction is not significant for the solvent tower. Transfer functions for the solvent tower can be identified with the tower decoupled from the extraction tower.

Intercolumn interaction can be recognized by examining the steady state process gains for the entire extractive distillation process. First compute the steady state process gains for the control system with the two towers coupled. Next compute the steady state gains for each column by itself, that is, uncoupled. If the gains are similar, then the control system can be tuned without considering intercolumn interaction. However, if there are significant differences you will need to consider intercolumn interaction. For the case studies presented in this chapter, System 1 has significant intercolumn interaction; Systems 2 and 3 do not.

The E2d-S2 control scheme is the highest order control scheme needed to control the process. Because all product compositions are under control with E2d-S2 we do not need E3-S2. E3-S2 is more difficult to control than E2d-S2 because it adds a control variable (the solvent flow) to the multivariable control problem. I have found for each of these systems that the E3-S2 control scheme is not needed, and E1-S1 is economically the best.

18-5 CONCLUSIONS

A methodology to design and control extractive distillation systems has been presented. Three general extractive distillation systems illustrate the methods and highlight the important insights. I hope that these insights will prove useful to you in your future work with extractive distillation systems.

The procedures are suitable for application on small personal computer systems. The process design procedure runs in less than 1 min on a PC. A dynamic simulation using model 2A from Chapter 3 is capable of accurately representing these systems without unnecessary model complication. Model 2A is also suitable for a personal computer. The columns presented in these case studies run about 30 times faster than real time on a 80386 based PC.

The results show that extractive distillation processes are controllable. The preferred control scheme is E1-S1. This control scheme is very robust, easy to implement, operate, and maintain. Higher order control schemes are not needed unless you have unique process requirements.

It is very important that recycle composition and flow are given enough conservatism during the process design to ensure that the system is controllable. These considerations are very important when you want to implement control schemes of higher order than E1-S1. Effective design of extractive distillation processes relies on a strong interaction between process design and process control.

References

Andersen, H. W., Laroche, L., and Morari, M. (1991). Dynamics of homogeneous azeotropic distillation columns. *Ind. Eng. Chem. Res.* 30(8), 1846–1855.

Atkins, G. and Boyer, C. M. (1949). Application of the McCabe–Thiele method to extractive distillation columns. *Chem. Eng. Prog.* 49(9), 553–562.

Benedict, M. and Rubin, L. C. (1945). Extractive and azeotropic distillation—Parts I and II. *Trans. AIChE* 41, 353–392.

Black, C. and Ditsler, D. E. (1972). *Azeotropic and Extractive Distillation. Advances in Chemistry Series.* Washington, DC: American Chemical Society, Vol. 115, pp. 1–15.

Chambers, J. (1951). Extractive distillation design and applications. *Chem. Eng. Prog.* 47(11), 555–565.

Doherty, M. F. and Caldarola, G. A. (1985). Design and synthesis of homogeneous azeotropic distillations III. The sequencing of towers for azeotropic and extractive distillations. Topical Report, University of Massachusetts, Amherst, MA.

Doherty, M. F. and Perkins, J. D. On the dynamics of distillation processes—III. The topological structure of ternary residue curve maps. *Chem. Eng. Sci.* 34, 1401–1414.

Gerster, J. A. (1969). Azeotropic and extractive distillation. *Chem. Eng. Prog.* 65(9), 43–46.

Grassi, V. G. (1991). Process design and control of extractive distillation. Ph.D. Dissertation, Lehigh University.

Kumar, S. and Taylor, P. A. (1986). Experimental evaluation of compartmental and bilinear models of an extractive distillation column. Report, Department of Chemical Engineering, McMasters University.

Lee, F. and Coombs, D. M. (1987). Two-liquid-phase extractive distillation for aromatics recovery. *Ind. Eng. Chem. Res.* 26(3), 564–573.

Luyben, W. L. (1990). *Process Modeling, Simulation and Control for Chemical Engineers*, 2nd ed. New York: McGraw-Hill.

Prokopakis, G. J. and Seider, W. E. (1983). Dynamic simulation of azeotropic distillation towers. *AIChE J.* 29(6), 1017–1029.

Shinskey, F. G. (1983). *Distillation Control.* New York: McGraw-Hill.

19

Control by Tray Temperature of Extractive Distillation

John E. Anderson

Hoechst Celanese

19-1 SITUATION

Separation by simple distillation of formic acid (HFo) from acetic acid (HAc) in the presence of water is very difficult because of the azeotropic boiling mixture formed. As a result the separation in this application is done by extractive distillation using heptanoic acid as the high boiling extracting agent or solvent (U.S. Patent No. 4, 692, 219).

Figure 19-1 is a schematic diagram of the extractive distillation system, a main column followed by a solvent recovery column. It also shows the controls on the main column as they were prior to this control study. HAc is extracted and carried out the bottom of the main column with the solvent while HFo and water are distilled overhead. The main column is operated at a fixed ratio of solvent flow to feed flow and at a constant reflux rate. The concentration of HFo in the HAc product is the product specification of most concern and it is measured by an analyzer in the overhead of the solvent recovery column. However, the location of the HFo analyzer introduces too much delay in the HFo measurement to allow the analyzer signal to be used effectively in closed loop control. As a result the plant controlled the temperature at tray 10 of the main column as shown in Figure 19-1. In effect the tray 10 temperature loop was used to control the HFo losses or the concentration of HFo in the HAc product.

For the temperature controller to be effective in controlling the HFo losses there must be a tray 10 temperature corresponding to each value of HFo concentration in the HAc product. However, the relationship between tray 10 temperature and HFo concentration changes when the feed and solvent rates to the main column change. Past operating data showed that temperature control would give fair control of HFo concentration during periods when the feed and solvent rates were steady. However, the feed rate and the solvent rate (ratioed to the feed) changed almost on a daily basis, and the operators would have to hunt for a new temperature setpoint at which to control. The output from the analyzer on the overhead of the solvent recovery column was used to help in this hunting process. Poor control resulted until the feed flow had been steady for quite a while.

19-2 ANALYSIS

Let us define two variables *XD* and *XB* for the main column as follows.

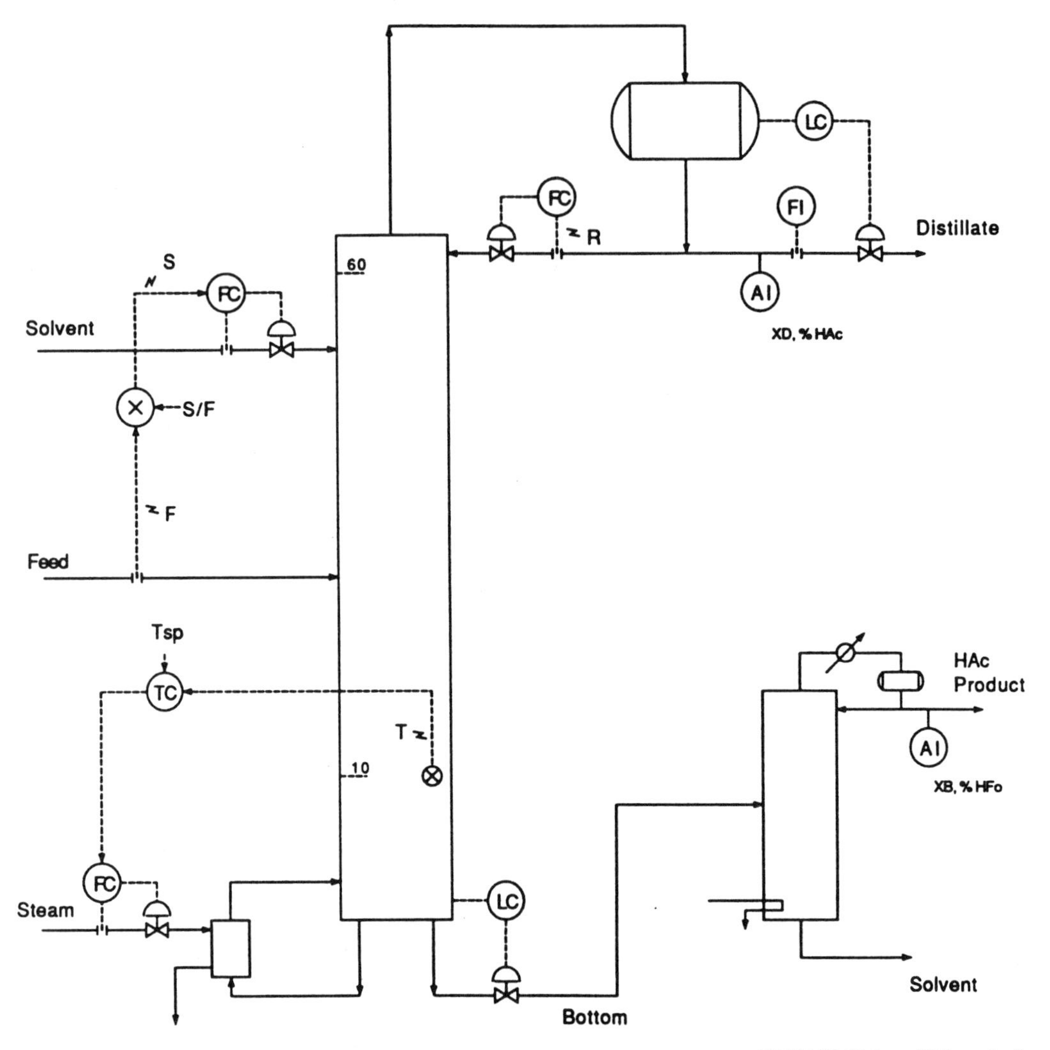

FIGURE 19-1. Old controls.

XD is the weight percent of HAc in the main column distillate.

XB is the weight percent of HFo in the main column bottom product on a solvent-free basis. The HFo analyzer on the distillate of the recovery column gives the same value but is delayed.

To get a better understanding of how *XD* and *XB* change with reflux, solvent, and distillate rates, a number of simulation cases were run using a tray to tray steady state simulation of the main column. A constant feed rate was assumed for all cases. Figures 19-2 and 19-3 show the resulting contour plots of *XD* and *XB* on a grid of the manipulated variables: reflux flow versus distillate flow and solvent flow versus distillate flow. From Figure 19-2 we see that reflux flow has almost no effect on either *XD* or *XB* over the range of reflux rates examined. Reflux flow would be of little to no use as a manipulated variable. Thus the practice of operating with a constant reflux rate is a reasonable one.

The choice of the pairing of control and manipulative variables is not so obvious from Figure 19-3. However, a graphical calcula-

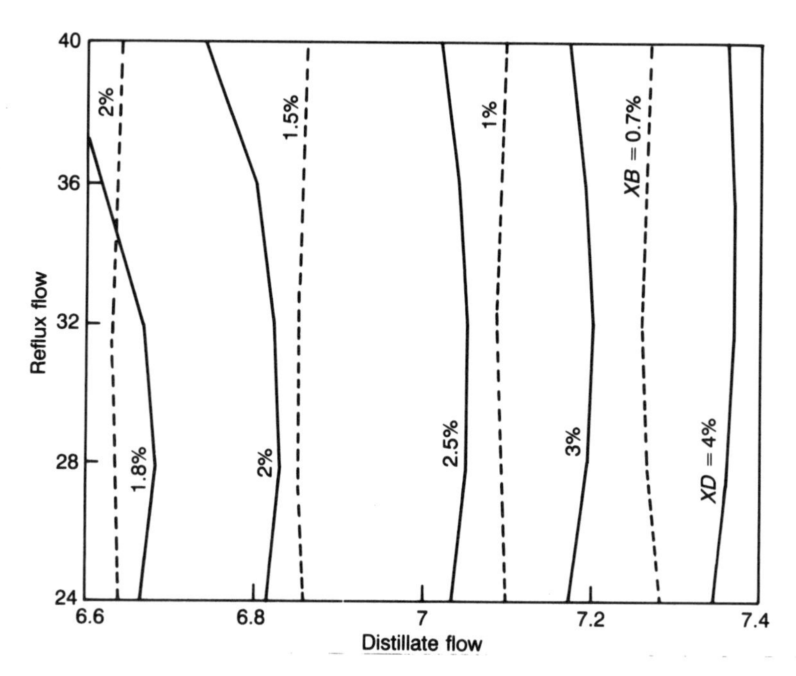

FIGURE 19-2. *XD* and *XB* contours on reflux versus distillate.

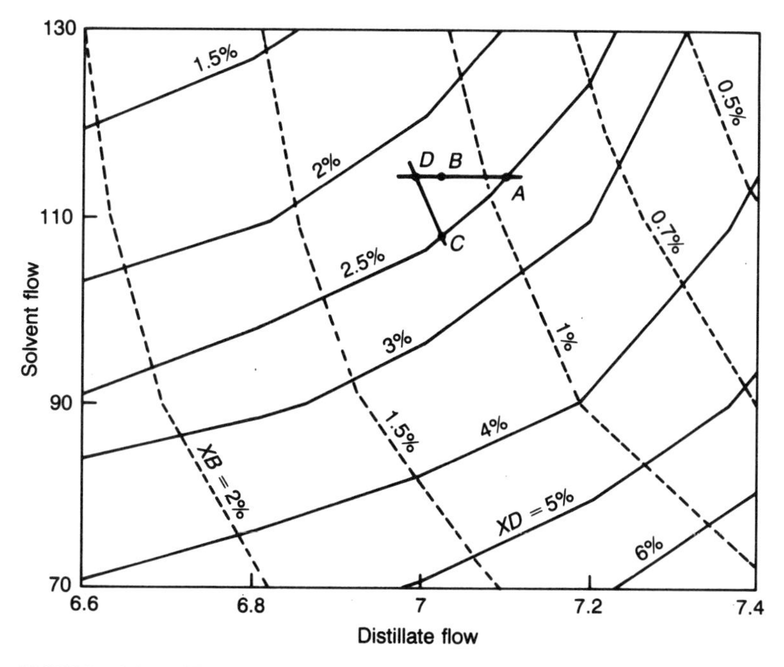

FIGURE 19-3. *XD* and *XB* contours on solvent versus distillate.

tion will help. Step from point A to point B along a horizontal line by changing distillate flow while keeping the solvent flow constant. Step from point A to point C using the same size change in distillate flow but move along an XD contour line. Project point C up to point D on the horizontal line through A by moving along an XB contour line. The line segments AB and AD are the changes in XB for a given change in distillate flow at constant solvent rate and constant XD, respectively. Then the ratio AB/AD is the element of the relative gain array corresponding to the pairing of XB and distillate flow. Because the ratio is greater than 0.5, changing distillate flow is the better way to move between XB contour lines. Stated another way: XB is best controlled by manipulating distillate flow. However, under the constraint of constant reflux flow any change in steam flow is translated into a corresponding change in distillate flow. Thus manipulating steam flow is equivalent to manipulating distillate flow when the reflux flow is held constant.

In order to develop a tray temperature as an inferential measurement of XB, it helps to think of the liquid on a tray as being a mixture of solvent and light components. The light components are the mixture of HFo, water, and HAc in the tray liquid. The relative amount of solvent to the total light components is the major factor in determining the boiling point of the liquid. The relative amounts of the light components to each other has a smaller effect on the boiling point of the liquid. Thus only if the major effect of solvent on tray temperature is compensated, or removed, can the smaller effect be found. A restatement of the problem is: "Is it possible to find and use for control the smaller effect of XB on tray temperature when it is masked by the larger temperature effect due to solvent?"

Equation 19-1 was chosen as the model giving the relation between solvent, the light components, XB, and boiling temperature of the mixture:

$$XB = a + bT + c(R/S) \tag{19-1}$$

where XB = percent HFo in the overhead of the solvent recovery column
T = tray temperature
R/S = reflux flow to solvent flow ratio
a, b, c = constants

The ratio of reflux flow to solvent flow was used as an indicator of the ratio of the light components to solvent on a tray. Other models could be developed from theory. However, the empirical model of Equation 19-1 was quite adequate over the normal range of operations. Regressions of plant operating data to Equation 19-1 gave both good fits of the data and the values of the model constants.

Before doing the regressions on the operating data, it was necessary to shift the XB data forward in time so as to reconcile it in time with the T and R/S data from the main column. Of the existing tray temperatures on the main column it was tray 10 temperature that gave the best correlation. Rearranging Equation 19-1, a better statement of the relation is obtained for our purposes:

$$XB = a + b(T + (c/b) * (R/S)) \tag{19-2}$$

$$TC \equiv T + (c/b)(R/S) \tag{19-3}$$

$$XB = a + bTC \tag{19-4}$$

Equation 19-1 is factored to give Equation 19-2. A compensated temperature TC is then defined by Equation 19-3. Substitution of that definition into Equation 19-2 yields Equation 19-4. Equation 19-4 shows XB to be a function of TC only. Thus a compensated temperature has been created, which can be used as a measure of XB. It is important to note that the compensated temperature TC has a number of other advantages for good control when compared to the analyzer. TC is continuous, fast in response, and reliable.

19-3 SOLUTION

The new controls for the main column are shown in Figure 19-4. They are the same as

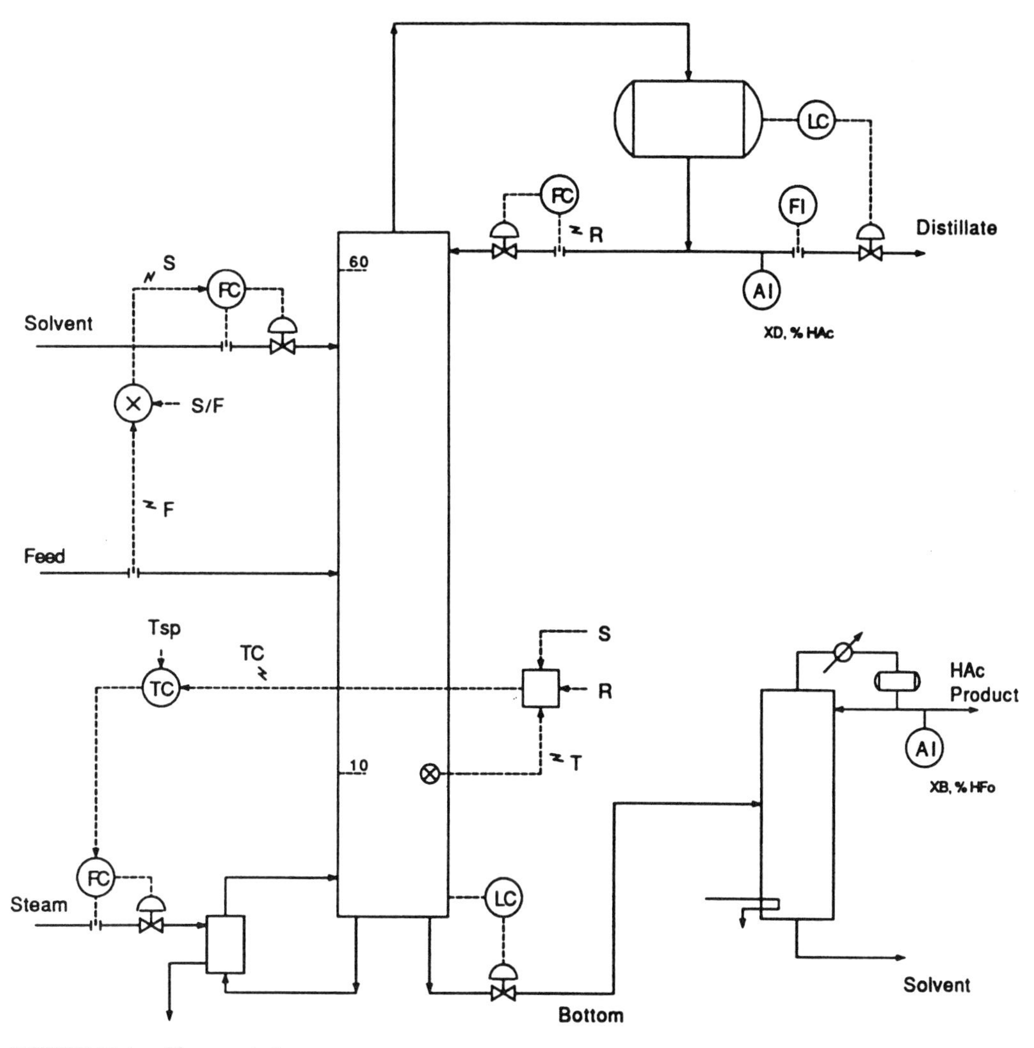

FIGURE 19-4. New controls.

the old controls except that the measurement to the temperature controller has been replaced by a compensated temperature as calculated by Equation 19-3. The operator is able to select a setpoint for the compensated temperature corresponding to the desired *XB*. Then the column will control at that compensated temperature and its corresponding *XB* even though the feed and the solvent rates change. If the compensation Equation 19-3 is not exact, the operator may find it necessary to make small adjustments to the temperature setpoint based on the average output of the HFo analyzer. Another method to correct for model misfit would be to automatically update a bias to the compensation equation based on the HFo analyzer output.

An examination of some operating data shows both the need for and the benefits of using a compensated temperature in the control scheme. Figure 19-5 contains plots of data collected before the new controls were applied. The solid curve in Figure 19-5*b* is the measured temperature on tray 10 and the dotted curve is the calculated compensated temperature. Solvent flow is the solid curve in Figure 19-5*c*. Note that the measured temperature changed during the four days of operation even though the

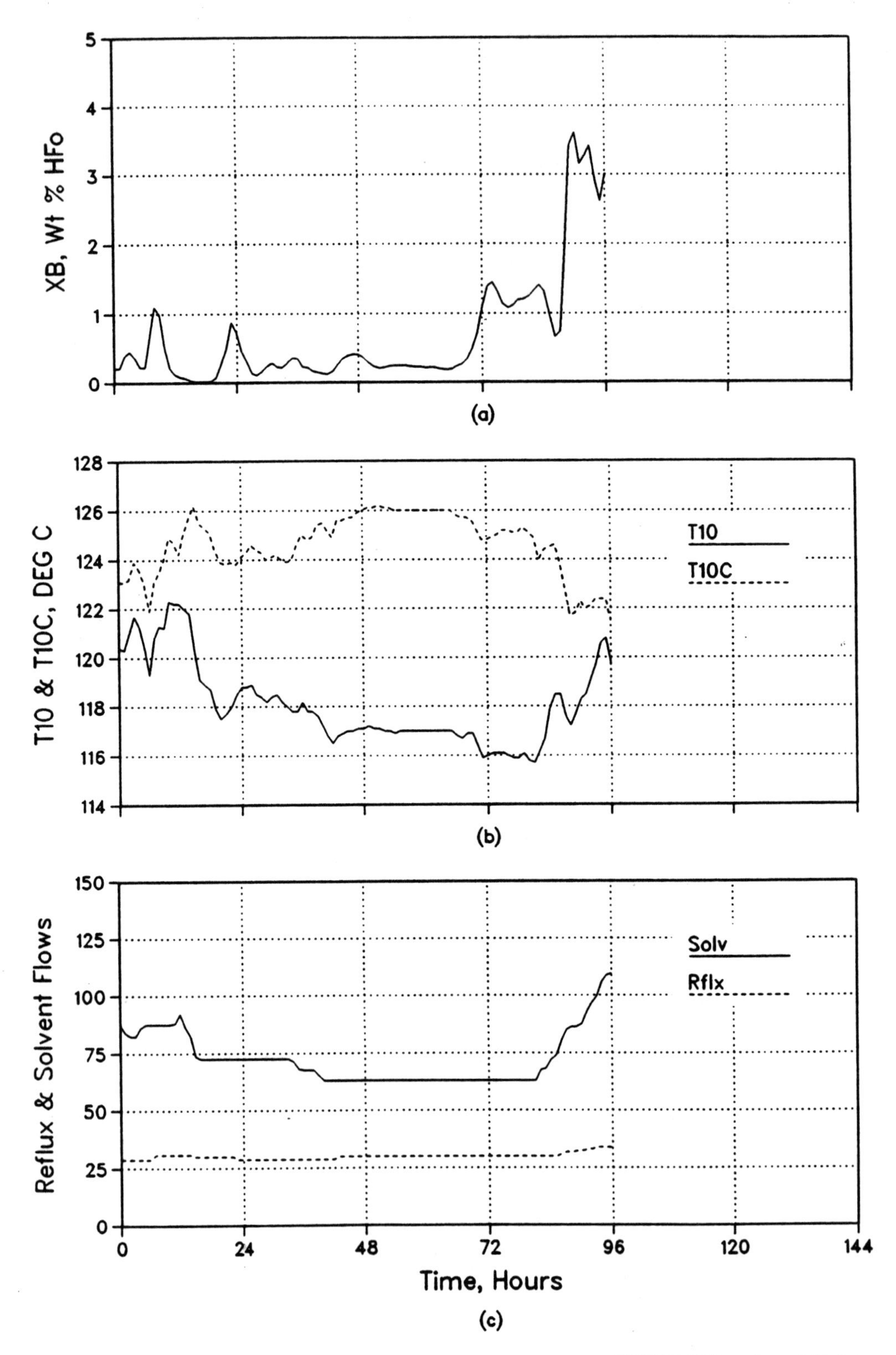

FIGURE 19-5. August 26 data.

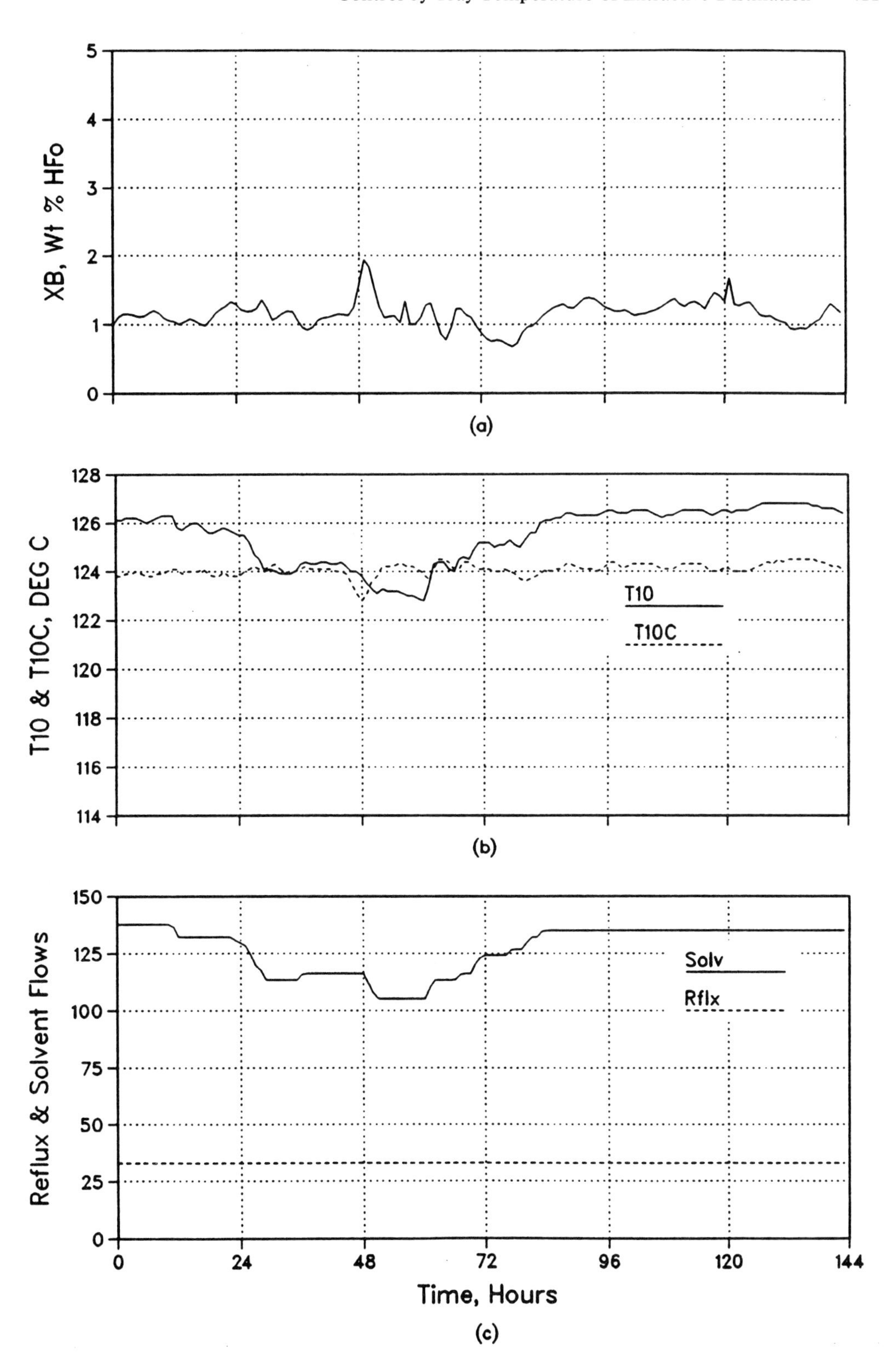

FIGURE 19-6. October 17 data.

measured temperature was on closed loop control. This was because the operator was changing the setpoint of the temperature controller. Every time the solvent flow changed with a feed rate change, the desired control temperature for tray 10 would also change (the desired control temperature being the temperature that corresponds to the specified *XB*). Each change in solvent flow required that the operator hunt for the new desired control temperature. This trial and error hunt took place on a very slow responding system. It was only in the third day when the solvent flow did not change that stable operation was achieved. At the beginning of the fourth day the *XB* specification, or *XB* control point, was raised to 1%. At midday another change in feed and solvent flows were made. This caused *XB* to go off-spec to 3%. Overall the control was fair to poor. The difference between the measured temperature and the compensated temperature in Figure 19-5*b* graphically illustrates the effect of solvent flow on the tray 10 temperature. The difference ranged from 2 to 8°. The temperature effect of solvent flow with a dynamic range that large will completely mask any temperature changes due to *XB*. During the fourth day the two temperatures actually go in the opposite directions.

Figure 19-6 shows operating data after the new controls were applied. During the first four days the solvent rate was changing as shown by the solid curve in Figure 19-6*c*. The changing solvent flow caused the measured temperature to vary (solid curve in Figure 19-6*b*). However, it was the compensated temperature that was being controlled. The operators did not have to adjust the temperature controller setpoint during solvent flow changes as they had in the past. This is seen in the trace of the compensated temperature that was controlled around the same setpoint for six days. By comparing the *XB* curves in Figures 19-5*a* and 19-6*a*, we find that the control of *XB* was significantly improved, and there was much less operator intervention.

19-4 CONCLUSIONS

Finding a method to compensate the tray 10 temperature improved temperature as an indicator or measurement of *XB*. Because the compensated temperature has the same dynamic response characteristics as the tray 10 temperature, using the compensated temperature in place of the HFo analyzer output eliminates the delays in the HFo measurement caused by the location of the analyzer sample point and the analyzer itself. Simple feedback with the faster measurement gives good control.

There are some comments and conclusions about this project that could be of general value to other control problems.

1. Tray temperature has some very desirable qualities as a measurement for a control scheme. Temperature is continuous. It is faster and more mechanically reliable than an analyzer. One drawback to temperature for distillation control is that it is not specific to composition. (It is not a single variable function of composition.) However, the compensated temperature is much more specific to composition and still retains the other desirable qualities of a temperature.
2. Simple linear models can often be used very effectively to incorporate process knowledge into a control scheme over the normal operating range of the unit.
3. The behavior of a tray temperature as a strong function of the solvent content on the tray is typical of extractive distillation. The use of a compensated temperature for control should be applicable in most other extractive distillation columns.
4. This control scheme was easily implemented on a distributed control system. The digital system produced the quality of data that allowed the compensation of a large effect in order to find and use a smaller effect.

20

Distillation Control in a Plantwide Control Environment

James J. Downs

Eastman Chemical Company

20-1 INTRODUCTION

The development of distillation column control strategies involves not only the understanding of how to control columns as isolated unit operations but also the understanding of how its operation influences and is influenced by the other unit operations that comprise a chemical process. The role of distillation as a unit operation that separates components usually makes the operation of the distillation column central to the control of component inventories within a process. This can be contrasted with unit operations that do not change stream compositions such as heat exchangers and heat exchanger networks.

The operating objectives of distillation columns are certainly dictated by the overall process operating objectives. In addition, the local operation of distillation columns is more often than not directed by situations and occurrences external to the column itself. For example, the feed rate to a column used to purge by-products from a process is typically changed to balance the rate of generation to the rate of removal of the by-products. The behavior and sensitivity of equipment downstream of the distillation column is also important in the decisions regarding column control objectives. It is usually the overall operating objectives of the process that will dictate what strategy is best for a column.

Although we usually think about material balance and energy balance equations applying to a unit operation, they also apply to whole processes and to entire chemical complexes. The time it takes to accumulate and deplete inventories may be longer for large processes or chemical complexes, but the laws of accumulation and depletion of material hold nonetheless. Whereas for a process, we think of the rate of accumulation of each component to be zero, the fact that the control system must ensure that to be the case is often overlooked. The manipulation of flows, utilities, and the readjustment of process operating conditions to maintain a balance of material and energy entering and leaving a process is the overriding priority for the control system. The material balance must be maintained not only from an overall viewpoint but also for each component in the system. It is the understanding of how the component material balances are controlled that is the key to linking the distillation column control strategy to the overall plant control strategy.

This chapter contains a discussion of the concepts needed to understand the component material balance issues that must be addressed during the design of plantwide control strategies. The interaction of unit operation control strategy design with the overall plantwide control strategy is illustrated using examples. A case study is then presented to show in detail the relationship between column control strategies and an overall plantwide control strategy.

20-2 PLANTWIDE COMPONENT INVENTORY CONTROL

20-2-1 Component Inventory Control for a Tank

The control of the overall material balance for unit operations is usually taken for granted because of the obviousness of an imbalance of material flows. For the overall control of inventories that are liquids, we typically control liquid levels; for gas inventories we control pressures; for solid inventories we control weights. When the unit operation equipment overfills or goes empty, the underlying cause must be found and corrected for the unit operation to have any chance of operating as intended. However, a much more subtle problem is the control of component inventories. Each component, whether important or insignificant, must have its inventory controlled within each unit operation and within the whole process.

The control of component inventories may occur by process self-regulation or by manipulating process flows or conditions. For example, many processes have innumerable trace components that "recycle to extinction." That is, they are ever present in the process but have no obvious source or sink. The rate of formation and of consumption of these components is set equal by mechanisms such as reaction equilibrium or concentration dependent irreversible reactions. For components that occur in significant quantities, the control strategy for the unit operations must take into account how the inventory of each component will be maintained.

The material balance equation for each component can be written as

$$\text{input} + \text{generation} - \text{output} - \text{consumption} = \text{rate of accumulation}$$

Based on the terms of the material balance equation one can identify the stream flows and the process operating conditions that affect the accumulation or depletion (i.e., inventory) of each component.

Example 20-1

Consider the tank shown in Figure 20-1. Water is the feed to the tank and the water leaving the tank is determined by the height of liquid in the tank, which follows the

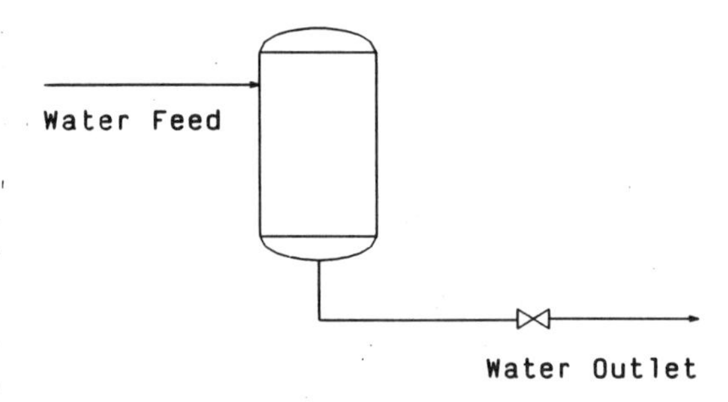

FIGURE 20-1. Water tank with gravity outlet flow.

TABLE 20-1 Component Material Balance for Example 20-1

	Input	+ Generation	− Output	− Consumption	= Accumulation
Component	Enters Through System Boundaries	Produced Within the System	Leaves Through System Boundaries	Consumed Within the System	Inventory Controlled by
Water	Water feed	0	Water outlet	0	Height of water in the tank

equation

$$F_{out} = k\sqrt{(\text{height})}$$

Each term of the material balance equation can be identified as shown in Table 20-1.

For this example the inventory of water in the tank is "self-regulating." For an increase in water feed, the inventory of water increases, the flow out of the tank increases, and, as a result, the balance between the water feed and water outlet is reestablished. Note also that a change in the outlet resistance causes a difference in the inlet and outlet flows, but the water height will again seek a level where the outflow identically equals the inflow. No controller is needed to ensure that the material balance for the water will be satisfied.

Example 20-2

Consider the tank shown in Figure 20-2*a*. Water is the feed to the tank and the water leaving the tank is pumped out. The rate of water entering and leaving the tank is not affected by the amount of water in the tank. Each term of the material balance equation can be identified as shown in Table 20-2.

For this example the inventory of water in the tank is *not* "self-regulating." For an increase in the water feed, the inventory of water increases, the flow out of the tank remains constant, and, as a result, the balance between the water feed and water outlet is not reestablished. This is referred to as an *integrating process*. Integrating variables require a controller to make them self-regulating. Moreover, it is impractical to expect the inventory to remain constant when the flows in and out of the tank are "set equal" in a feedforward fashion as shown in Figure 20-2*b* because of inaccuracies in measurements.

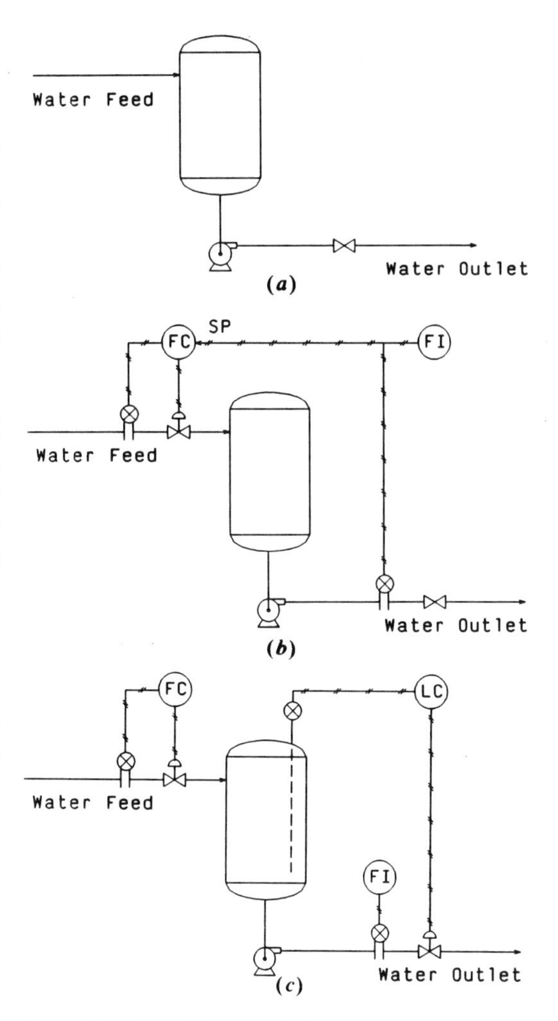

FIGURE 20-2. Water tank with pumped outlet flow.

To control the inventory of water in the tank requires a level controller to change the flow of water entering or leaving the

TABLE 20-2 Component Material Balance for Example 20-2

	Input	+ Generation	− Output	− Consumption	= Accumulation
Component	Enters Through System Boundaries	Produced Within the System	Leaves Through System Boundaries	Consumed Within the System	Inventory Controlled by
Water	Water feed	0	Water outlet	0	Not self-regulating

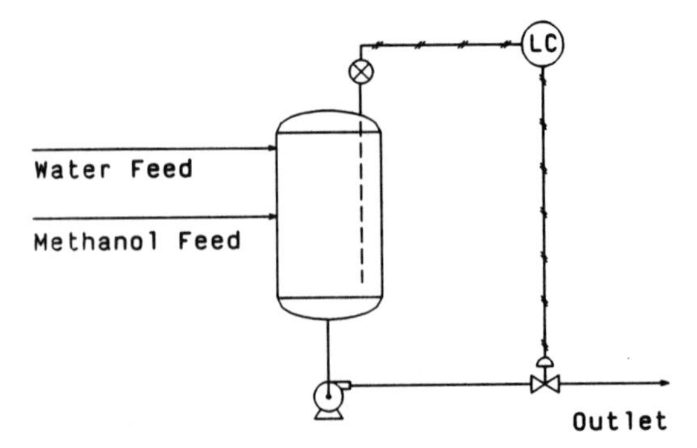

FIGURE 20-3. Mixing tank for methanol and water.

tank. For example, Figure 20-2c illustrates a level controller that manipulates the outlet flow from the tank to control the inventory of water in the tank. With the level controller in automatic, the inventory of water in the tank is again self-regulating. More water fed into the tank will result in more water being removed from the tank. The controller has changed an integrating inventory into a self-regulating inventory.

Example 20-3

Consider the tank shown in Figure 20-3. Water and methanol are the feeds to the tank. The outlet flow is manipulated to control the total inventory in the tank. Each term of the material balance equation for both the methanol and water can be identified as shown in Table 20-3.

Without the level controller, neither the methanol inventory nor the water inventory is self-regulating. However, with the level controller, both of the component inventories are self-regulating. The outlet flow for methanol is given by

$$\begin{aligned}&(\text{flow methanol out})\\&\quad = (\text{total flow out})\\&\quad\quad \times (\text{fraction methanol in the tank})\end{aligned}$$

The fraction methanol in the tank will approach the ratio of the methanol feed to the total feed as time progresses. This causes the flow of methanol leaving the tank to follow the relationship

$$\begin{aligned}&(\text{flow methanol out})\\&\quad \to (\text{total flow out})\\&\quad\quad \times \left(\frac{\text{methanol feed}}{\text{methanol feed} + \text{water feed}}\right)\end{aligned}$$

Without the level controller, there is nothing to ensure that the flow of methanol out of the tank is equal to the flow in at steady state. However, by virtue of the level controller, as time progresses

$$\begin{aligned}&(\text{total flow out})\\&\quad \to (\text{methanol feed} + \text{water feed})\end{aligned}$$

and this relationship substituted into the

TABLE 20-3 Component Material Balance for Example 20-3

	Input	+Generation	−Output	−Consumption	= Accumulation
Component	Enters Through System Boundaries	Produced Within the System	Leaves Through System Boundaries	Consumed Within the System	Inventory Controlled by
Methanol	Methanol feed	0	Outlet	0	Self-regulating via the level controller
Water	Water feed	0	Outlet	0	Self-regulating via the level controller

methanol material balance yields

(flow methanol out) → methanol feed

If the feed rate of methanol increases, the flow out of the tank will be increased by the level controller. Furthermore, as the methanol accumulates in the tank, the exit stream composition has an increased fraction methanol. As a result, the composition of methanol in the tank will increase until a balance between the methanol fed to the tank and methanol leaving the tank is reestablished. The inventory of water in the tank is controlled in a similar manner. To control the inventory of methanol or water in the tank does not require a methanol inventory controller because the level controller has made the component inventories self-regulating.

20-2-2 Component Inventory Control for a Process

Understanding how the inventories of each component in a process are controlled is the key to overall plantwide control. The concept of self-regulating and integrating component inventories is useful in the analysis of plantwide control strategies. The distillation unit operation plays a central role in the control of component inventories as do other unit operations that separate components. The component inventories in an entire process can be imagined as being contained in one large tank as illustrated in Figure 20-4. The choice of how to control each unit operation in a process is linked closely with the overall process control of component inventories. The control strategy for each unit operation must be developed within the framework of the overall component inventory control strategy.

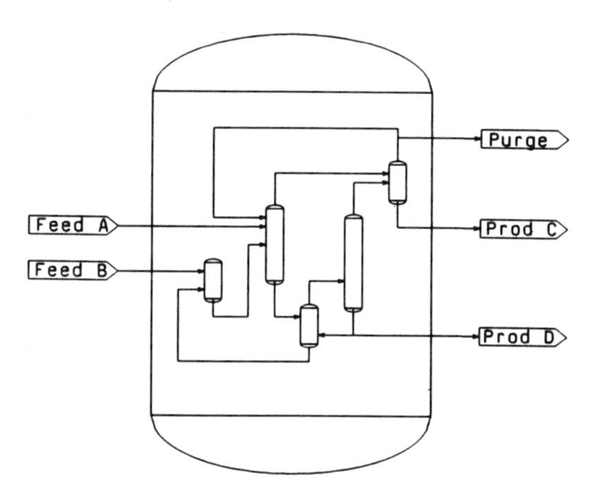

FIGURE 20-4. Plantwide inventory viewed as a tank inventory.

Example 20-4

Consider the process shown in Figure 20-5*a*. Stream 1 is a 10 mol/h water feed used to scrub a 100 mol/h nitrogen offgas stream; stream 2 contains 10 mol% methanol. The scrubbed nitrogen is vented and the water containing the methanol is fed to a distillation column. The column concentrates the methanol to a 90% methanol and 10% water distillate product. The water that leaves the bottom of the column is returned to the scrubber. The total liquid inventory in the base of the scrubber is controlled by manipulating the feed to the distillation column. The distillation column has a control strategy that maintains the distillate composition at setpoint by manipulating the column reflux and has the steam to the reboiler set on flow control. The setpoint on the composition controller is 90% methanol. Each term of the overall process material balance equation for each component can be identified as shown in Table 20-4 developed for the control strategy shown in Figure 20-5*a*.

For this process the component inventory of water is not self-regulating. If the component feed rates of water and methanol are not in a 10/90 ratio (the ratio of water and methanol leaving the process), what happens to the water fed in excess? There is no mechanism for the water entering or leaving the process to change based on how much of it is in the process. The control strategy as drawn in Figure 20-5*a* will not work from an overall point of view, even though the

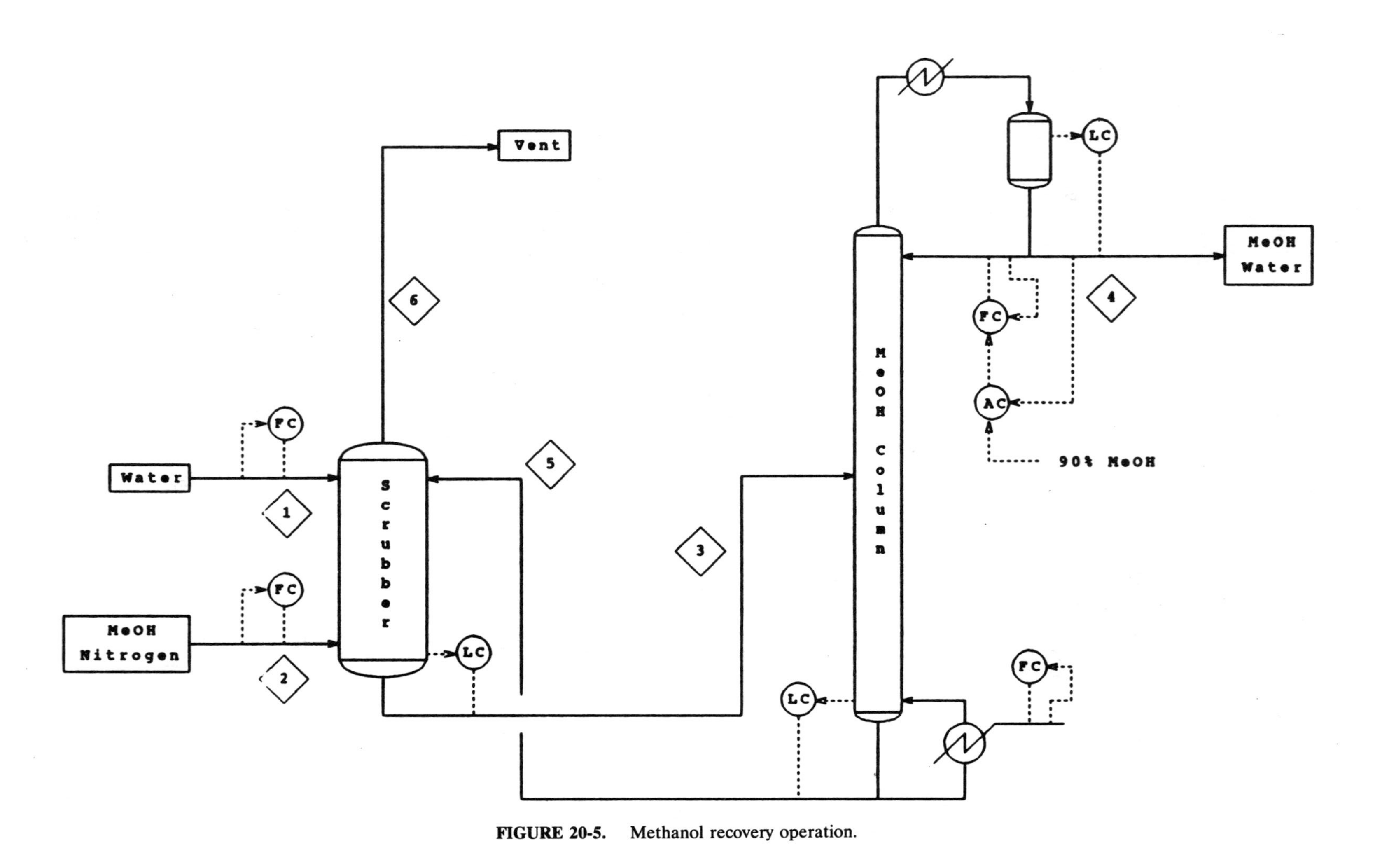

FIGURE 20-5. Methanol recovery operation.

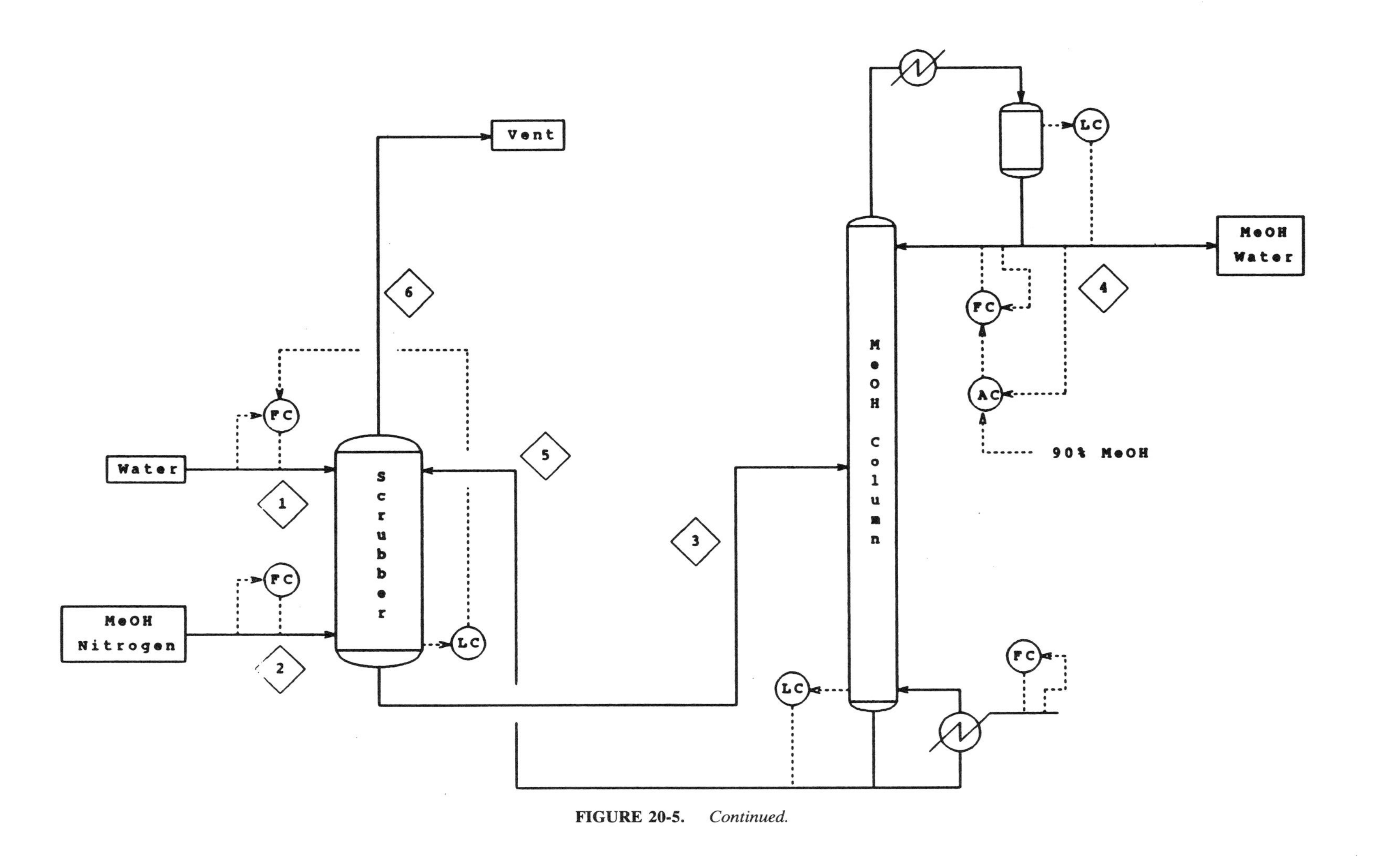

FIGURE 20-5. *Continued.*

TABLE 20-4 Component Material Balance for Example 20-4

	Input	+ Generation	− Output	− Consumption	= Accumulation
Component	Enters Through System Boundaries	Produced Within the System	Leaves Through System Boundaries	Consumed Within the System	Inventory Controlled by
Water	Stream 1	0	Stream 4 and small amount in stream 6	0	Not self-regulating
Methanol	Stream 2	0	Stream 4	0	Self-regulating via the distillate composition controller
Nitrogen	Streams 2	0	Stream 6	0	Self-regulating via the vent to atmosphere

control of the individual unit operations may be satisfactory.

Neglecting the small amount of water that leaves in the vent, the outlet flows for methanol and water are given by

$$\begin{aligned}(\text{flow methanol out}) &= (\text{distillate flow})(90\%)\\ &= 0.9D\\ (\text{flow water out}) &= (\text{distillate flow})(10\%)\\ &= 0.1D\end{aligned}$$

The inlet flows of each component are given by

$$\begin{aligned}&(\text{flow methanol in})\\ &\quad= (\text{flow of stream 2})\\ &\qquad\times(\text{fraction methanol in stream 2})\\ &\quad= F_2 x_{\text{MeOH},2}\\ &(\text{flow water in}) = (\text{flow of stream 1}) = F_1\end{aligned}$$

The overall component balances that result for each component are

Water:

$$F_1 = 0.1D \quad \Rightarrow \quad D = 10F_1$$

Methanol:

$$F_2 x_{\text{MeOH},2} = 0.9D \quad \Rightarrow \quad D = 1.11F_2 x_{\text{MeOH},2}$$

If the flow of stream 1, F_1, and stream 2, F_2, are fixed, then there is no value of D that will satisfy both equations unless $10F_1 = 1.11F_2 x_{\text{MeOH},2}$, that is, the feed rates of methanol and water are in a 90/10 ratio. Because of the fractionating ability of the distillation unit operation, the methanol will concentrate in the top of the distillation column. The column control strategy will ensure that distillate stream flow will satisfy the removal of the methanol from the process. The distillate rate will follow the equation $D = 1.11F_2 x_{\text{MeOH},2}$ (= 11.1 mol/h). For a 10 mol/h feed rate of water and a water removal rate of 1.11 mol/h (= 0.1 D), the water will accumulate in the process at the rate of 8.89 mol/h.

Consider the control strategy illustrated in Figure 20-5*b*. By virtue of the level controller on the base of the scrubber being linked to the feed rate of water to the process, the water inventory is self-regulating. As more methanol enters the process and thus drags with it more water from the process in the column distillate stream, less water is returned to the scrubber from the base of the column. As a result of less water being fed to the scrubber, the scrubber base level falls. The level controller for the scrubber base increases the amount of water being fed to the process. The balance of

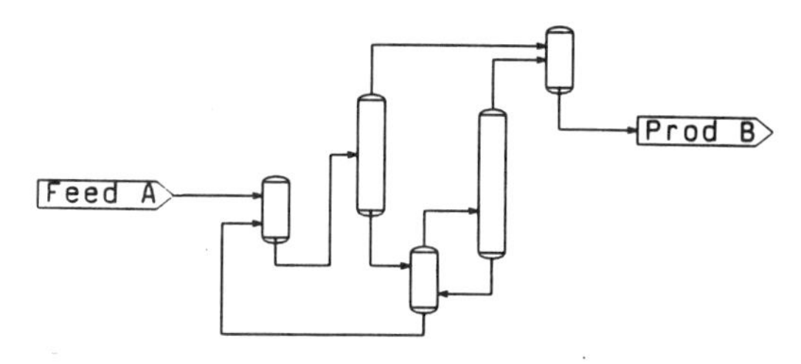

FIGURE 20-6. Isomerization process.

water entering and leaving the process is then restored.

In Figure 20-5*b*, the inventory of water and methanol are controlled by the scrubber base level controller and the distillation column control scheme, respectively. The behavior of the column and its control strategy determines how the individual component material balances are satisfied.

Example 20-5

Consider an isomerization plant shown in Figure 20-6. The reaction $A \rightarrow B$ takes place in a captive carrier liquid and follows first order kinetics with respect to A. The B product is separated from the carrier and from A by distillation and removed from the process as a pure component. A component inventory table is given in Table 20-5.

Provided that the reactor is sized large enough, the inventory of component A is self-regulating. If conditions change to cause A to accumulate or deplete, then the concentration of A in the reactor will also change in the way needed to change the reaction rate of A to reestablish zero accumulation or depletion. For example, if the reactor temperature decreases and less A is consumed by reaction, A will accumulate in the system until its concentration is high enough to again react all the A that is fed. Similarly, if the feed rate of A is decreased, then A will react and deplete until its concentration is low enough to slow down the reaction rate to match the feed rate of A. If the feed rate of A exceeds the capacity of the reactor, then A is no longer self-regulating and A will continue to build in the process because it has no exit except by reaction.

The inventory of the product B does not self-regulate. If more B is produced, there is no mechanism to slow down its production rate as its inventory increases. The removal of B from the process must be explicitly changed by a controller to match the rate B is produced. There are two issues to decide. First, what measurement is indicative of how the inventory of B is changing; second, what manipulated variable(s) should be chosen to control the B inventory.

Component inventories in distillation columns are frequently good indicators of plant component inventories because of the concentration and accumulation properties of the distillation operation. In this example, if the removal rate of B is less than its

TABLE 20-5 Component Material Balance for Example 20-5

	Input	+ Generation	− Output	− Consumption	= Accumulation
Component	Enters Through System Boundaries	Produced Within the System	Leaves Through System Boundaries	Consumed Within the System	Inventory Controlled by
A	A feed stream	0	0	$Vk_0e^{-E/RT}[A]$	Self-regulating via reaction
B	0	$Vk_0e^{-E/RT}[A]$	B product stream	0	Not self-regulating
Carrier	0	0	0	0	Not self-regulating

production rate in the reactor, then B will tend to accumulate in the tops of the distillation columns due to the fractionation of A, B, and carrier. The removal rate of B from the column is typically adjusted to achieve a temperature or composition target in the column. As more B is produced, the column control scheme must result in more B being removed as product to satisfy plantwide component inventory requirements.

The alternative to using the product rate to control the plant inventory of component B is to manipulate the production rate of B by reaction. The reactor conditions of level, temperature, or concentration of A can be changed to alter the formation rate of B to match the product rate. When this choice is made, the process inventory of component A is no longer self-regulating and the feed rate of A must now be manipulated to control the inventory of A in the process.

The carrier component inventory is not self-regulating. Because it is not entering or leaving the process, its inventory in the process should be constant. However, there is no mechanism to prevent small amounts being lost in vents, the product stream, sludging operations, and so forth. In practice, small amounts of captive components are lost over time and the inventory of the carrier may be made up batchwise on a daily or weekly basis.

Example 20-6

Consider the process described in Example 20-5, Figure 20-6. To illustrate the interplay

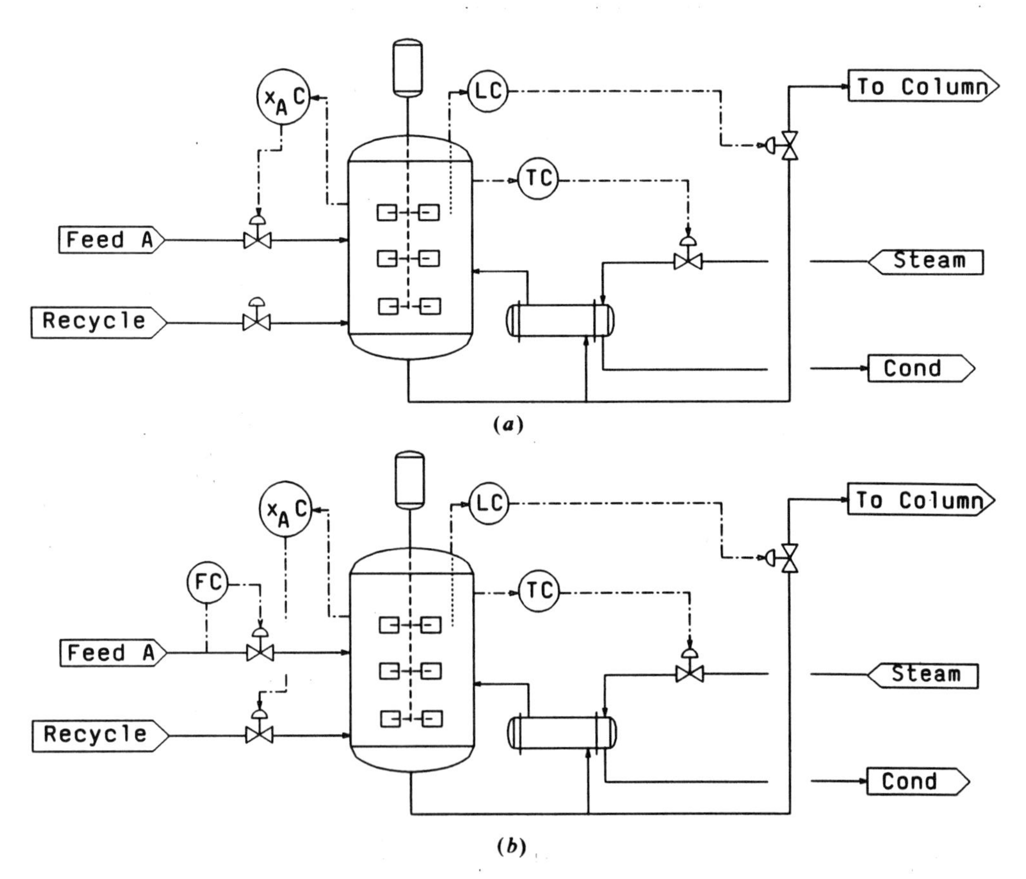

FIGURE 20-7. Isomerization reactor control strategies.

between unit operation control strategy design and the overall component inventory control, two reactor control strategies are sketched in Figure 20-7. Each strategy is designed to control the reactor level, temperature, and composition of component A. When considering only the reactor, the choice of using the recycle or the fresh feed of A to control the composition of A in the reactor is primarily influenced by steady state and dynamic behavior of each manipulated variable and how each manipulated variable may interact with the temperature and level control loops. However, from a plantwide viewpoint the choice is to use the fresh feed of A for composition control.

Refer to Table 20-5. The inventory of component A is determined solely by the feed rate of A and the consumption of A by reaction. The reactor control strategies shown in Figure 20-7 set each term in the reaction expression. The level controller sets the volume V of the liquid in the reactor. The temperature controller sets the reactor temperature and thus the term $e^{-E/RT}$. The composition controller sets the concentration of A. These three controllers together set the consumption of A by reaction. If the reactor control strategy in Figure 20-7*b* is used, the feed rate of A and the consumption rate of A are set independently and the inventory of component A will not self-regulate. This is similar to Example 20-2 where the flows of water into and out of the tank were set independently. However, the control strategy that uses the feed rate of A to control the reactor composition links the inventory of A to the process feed of A and the inventory of A is once again self-regulating by virtue of the composition controller.

If the feed rate of A is used to set the overall process production rate, then a mechanism is necessary for one of the terms in the reaction expression, V, T, or $[A]$, to change for component A inventory control. If an increase in the process inventory of A is detected, then increasing V, T, or $[A]$ can be increased to react more A.

20-3 ACETALDEHYDE OXIDATION PROCESS CASE STUDY

The application of the principles of plantwide component inventory control are illustrated in a case study of a process to produce acetic acid. To examine the linkage between the design of distillation column control strategies and the overall process control design problem, several candidate column control strategies will be proposed. The behavior of the candidate column control strategies and how they fulfill plantwide control objectives will then be discussed.

20-3-1 Description of the Problem

The partial oxidation of acetaldehyde (HAc) is a common route to the production of acetic acid (HOAc). Figure 20-8 is the schematic and Table 20-6 is a material balance for a process to partially oxidize acetaldehyde using atmospheric air. The oxidation is a liquid phase reaction using water as a diluent and heat sink and is given by the reaction

$$\text{HAc (liq)} + \tfrac{1}{2}\text{O}_2\text{ (gas)} \rightarrow \text{HOAc (liq)}$$

The reaction rate is first order with respect to acetaldehyde concentration in the liquid phase and first order with respect to oxygen partial pressure in the gas phase. The kinetics are described by

$$r_A = 2.557 \times 10^6 \exp\left(\frac{-10000}{1.987(T + 273)}\right) \times [p_{O_2}][\text{HAc}]$$

where r_A = formation rate of HOAc (lb-mol/h)

T = temperature (°C)

$[p_{O_2}]$ = partial pressure of oxygen (psia)

[HAc] = concentration of acetaldehyde (lb-mol/ft^3)

The reactor operates adiabatically at 127.1°C

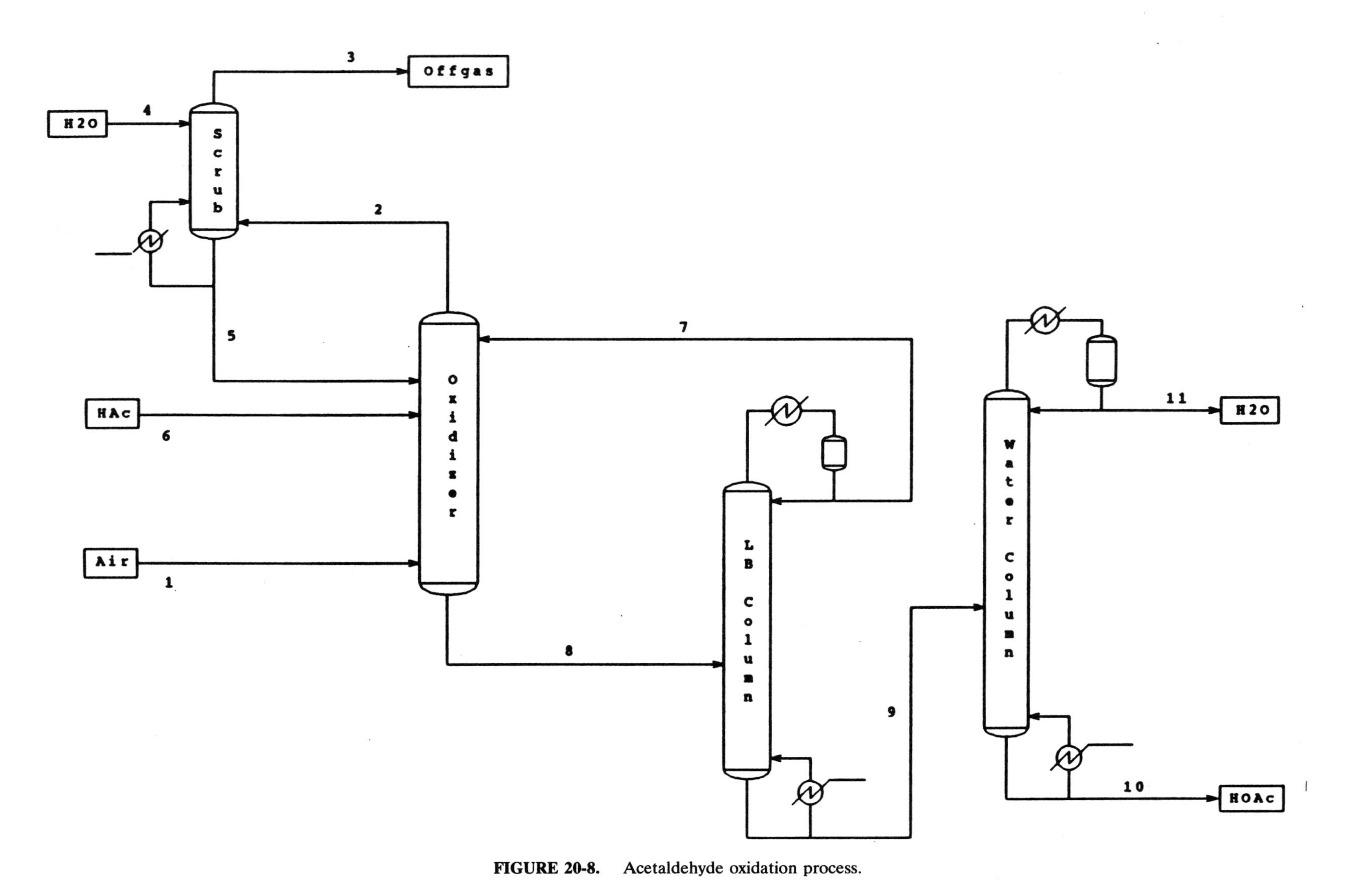

FIGURE 20-8. Acetaldehyde oxidation process.

TABLE 20-6 Material Balance for the Acetaldehyde Oxidation Process

Data for Streams 1 through 5

Stream Number	1	2	3	4	5
Stream Label	Air feed	Ox vapor product	Vent	Water feed to scrub	Scrubber underflow
Flow (lb-mol/h)	120.000	124.947	102.288	2.58855	25.2475
Temp (deg C)	25.0000	127.065	20.0000	25.0000	20.0000
Stream Molar Composition (mole frac)					
1 Nitrogen	0.800000	0.768325	0.938525	0.00000	0.000000
2 Oxygen	0.200000	0.037778	0.046147	0.00000	0.000000
3 Acet ald	0.000000	0.045863	0.014084	0.00000	0.169910
4 Acet aci	0.000000	0.037111	0.000000	0.00000	0.183659
5 Water	0.000000	0.110923	0.001244	1.00000	0.646432

Data for Streams 6 through 10

Stream Number	6	7	8	9	10
Stream Label	HAc feed	LB col distillate	Ox underflow	LB col underflow	Water col underflow
Flow (lb-mol/h)	40.0000	66.3519	107.373	41.0207	38.5594
Temp (deg C)	25.0000	40.0000	127.065	120.239	125.000
Stream Molar Composition (mole frac)					
1 Nitrogen	0.00000	0.000000	0.00000	0.0000	0.0000
2 Oxygen	0.00000	0.000000	0.00000	0.0000	0.0000
3 Acet ald	1.00000	0.048547	0.03000	0.0000	0.0000
4 Acet aci	0.00000	0.017610	0.37000	0.9400	1.0000
5 Water	0.00000	0.933843	0.60000	0.0600	0.0000

Data for Stream 11

Stream Number	11
Stream Label	Water col distillate
Flow (lb-mol/h)	2.46127
Temp (deg C)	100.000
Stream Molar (mole frac) Composition	
1 Nitrogen	0.0000
2 Oxygen	0.0000
3 Acet ald	0.0000
4 Acet aci	0.0000
5 Water	1.0000

and 193.4 psia with a liquid holdup of 20 lb-mol (12.1 ft^3 of liquid).

The vapor stream leaving the reactor flows to a water scrubber to remove acetic acid and some acetaldehyde from the vapor stream before the stream leaves the process boundaries. The scrubber uses water as a scrubbing medium and has a recirculation loop and cooler at the bottom. Most of the recovery is due to the cooling of the vapor product to condense the acid and acetaldehyde.

The liquid product leaving the reactor is the feed to the low boiler (LB) removal column. The primary purpose of the LB column is to recover the unreacted ac-

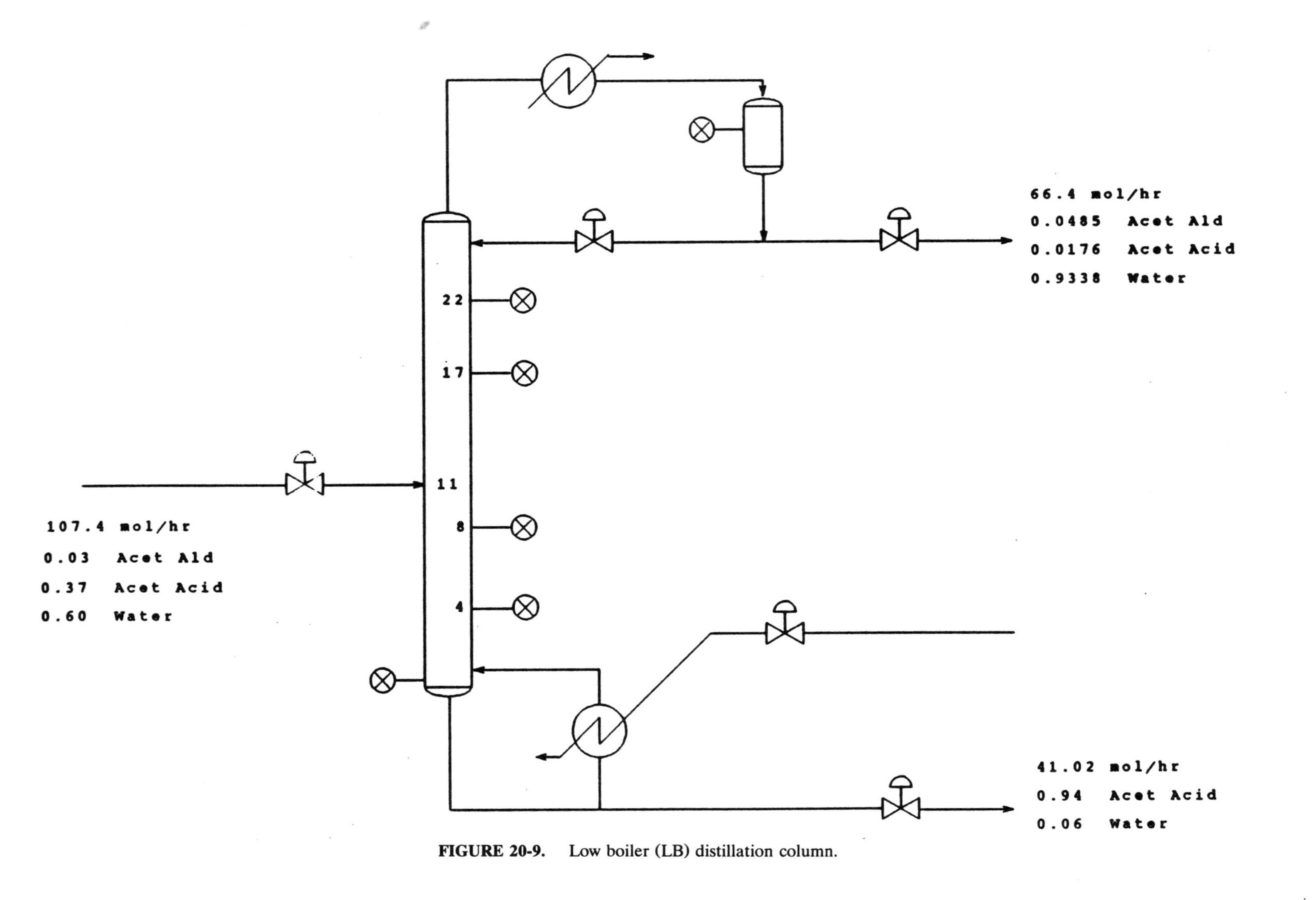

FIGURE 20-9. Low boiler (LB) distillation column.

etaldehyde and recycle it to the oxidizer. In addition, water is recycled to the oxidizer to maintain a water concentration of 0.6 mol fraction in the oxidizer. The acid product and some water leave the bottom of the LB column and feed the water column where the water and acid are separated into relatively pure product streams.

The focus of this case study is to develop an understanding of how the control of the LB distillation column affects the overall plantwide control strategy.

20-3-2 Low Boiler Column Analysis

The low boiler (LB) distillation column sketched in Figure 20-9 is used to separate acetaldehyde (nbp 20°C), water (nbp 100°C), and acetic acid (nbp 115°C). The column has 24 crossflow trays, a thermosyphon reboiler, and an overhead total condenser. The column is operated at atmospheric pressure by venting the primary condenser to a dedicated column vent scrubber and then to the atmosphere. Measurements on the column consist of temperatures on trays 4, 8, 17, and 22. No composition measurements are available except by lab sample once per 8 h shift. Differential pressure type level transmitters provide level measurements for the base of the column and the reflux drum.

The column material balance is listed in Table 20-6. The vapor–liquid equilibrium relationships for this column reveal that the acetaldehyde has a high relative volatility to water and acetic acid and as a result it is removed exclusively in the distillate product. The water has a low relative volatility to acetic acid of about 1.15.

Base case temperature and composition profiles illustrated in Figure 20-10 indicate that the key split is between the water and acetic acid, and that the temperature changes from tray to tray are the greatest in the stripping section of the column.

Without knowing the purpose of this column from a plantwide viewpoint, it is difficult to state an operating objective. Common perception is to assume that the operating objective is to separate the feed components into two streams having compositions equal to the compositions given in the base case or design basis.

The control system design needs to provide disturbance rejection properties appropriate to the type and nature of the disturbances. Typically, feed rate and feed composition are common disturbances that the control system must deal with. In addition, variability in the utility supply, environmental upsets, and slow degradation of column internals are common as well.

Steady State Analysis

The steady state gains relating the column temperatures to the manipulated variables are determined using a rigorous steady state simulation of the column. An example calculation is the determination of the gains relating temperature on tray j to the reflux rate while holding bottom product rate, feed rate, and feed composition constant. Being a steady state calculation implies that the distillate rate and steam rate to the reboiler are used to control reflux drum level and base level, respectively. The gains are found using the equation

$$\left(\frac{\partial T_j}{\partial L}\right)_{B,F,x_F} \approx \frac{T_j(L+\Delta L) - T_j(L-\Delta L)}{2\,\Delta L}$$

where $T_j(L+\Delta L)$ is the temperature on tray j corresponding to a reflux rate of $L+\Delta L$ and similarly for $T_j(L-\Delta L)$. The value of ΔL used for this numerical calculation of the gains is $1 \times 10^{-6}L$. Steady state process gains between the column temperatures and the manipulated variables reflux L and bottom product rate B are plotted in Figure 20-11. Steady state gains for other combinations of manipulated variables are found in a similar manner.

For each pair of manipulated variables a matrix of steady state gains relating the temperatures to the manipulated variables is calculated. Each matrix has 25 rows (tem-

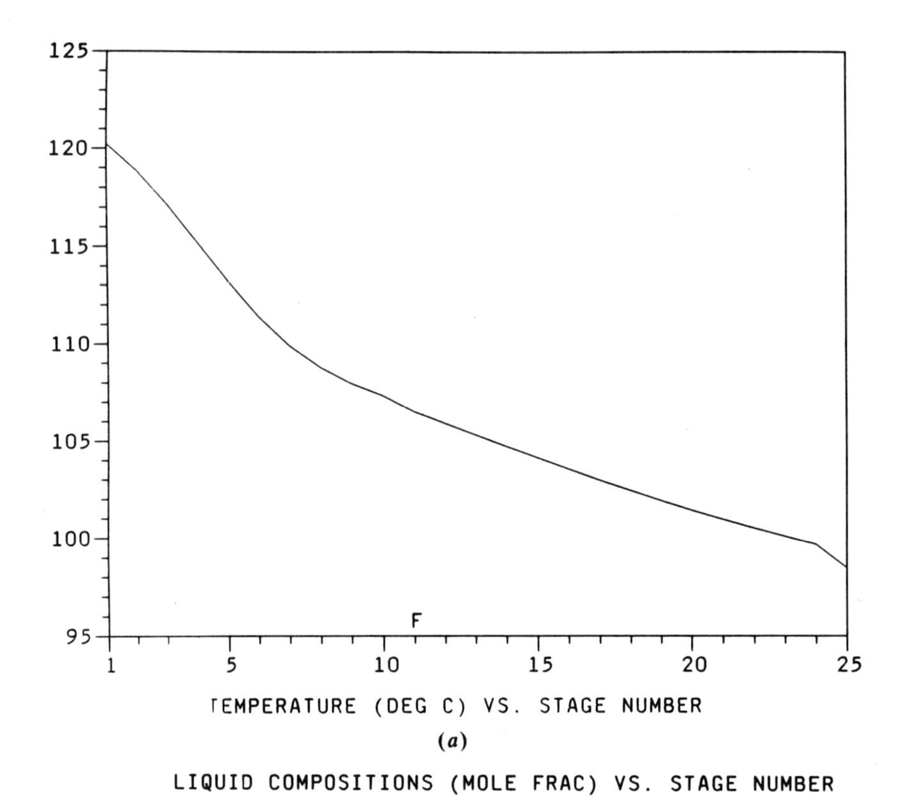

(*a*)

LIQUID COMPOSITIONS (MOLE FRAC) VS. STAGE NUMBER

1.0
0.8
0.6
0.4
0.2
0.0
F
1 5 10 15 20 25

1 = ACET ALD 2 = ACET ACI 3 = WATER

(*b*)

FIGURE 20-10. Low boiler column temperature and composition profiles.

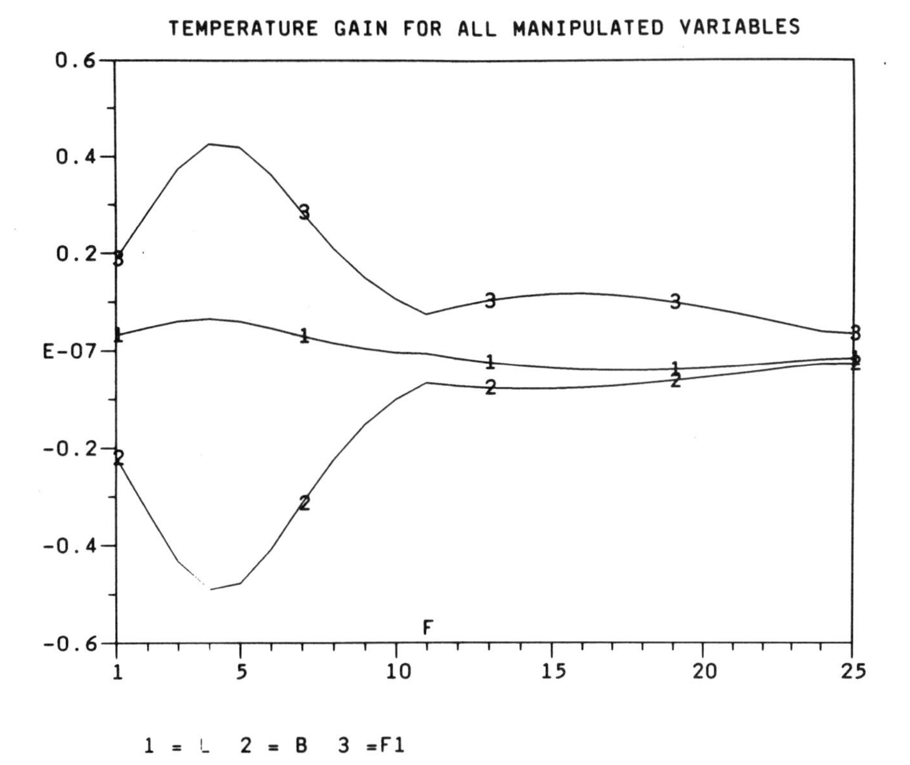

FIGURE 20-11. Low boiler column steady state gains.

peratures) and 2 columns (manipulated variables). Singular value decomposition can be used to estimate good locations for temperature measurements. A detailed description of singular value decomposition and how it is used in the design of distillation column control strategies is presented in Chapter 8.

The singular value decomposition of the gain matrices can be written as

$$\left[\frac{\partial \underline{\mathbf{T}}}{\partial m_1} \; \frac{\partial \underline{\mathbf{T}}}{\partial m_2}\right] = \underline{\underline{\mathbf{U}}}\underline{\underline{\Sigma}}\underline{\underline{\mathbf{V}}}^T$$

$$= [\underline{\mathbf{u}}_1 \quad \underline{\mathbf{u}}_2]\begin{bmatrix} \sigma_1 & 0 \\ 0 & \sigma_2 \end{bmatrix}\begin{bmatrix} \underline{\mathbf{v}}_1^T \\ \underline{\mathbf{v}}_2^T \end{bmatrix}$$

The results for the manipulated variable set (L, B) are shown in Figure 20-12. The values of the elements of $\underline{\mathbf{u}}_1$ and $\underline{\mathbf{u}}_2$ are plotted versus tray number as curves 1 and 2. The peak in the curve for $\underline{\mathbf{u}}_1$ at trays 4 and 5 indicates that the temperature on those trays will be the most sensitive to changes in the manipulated variables. The peak in the curve for $\underline{\mathbf{u}}_2$ occurring on trays 15 to 18 indicates that this zone is the second most sensitive location in the column whose temperature movement is *independent* of trays 4 and 5. The two $\underline{\mathbf{u}}$ vectors define output directions that the column temperature profile will follow when changes are made to the manipulated variables. The $\underline{\mathbf{u}}$ vectors for other sets of manipulated variables $[(L, V)$, (D, V), $(L/D, V/B)$, and $(L/D, B)]$ are the same as those for (L, B).

The SVD results suggest using tray 4 for single point temperature control and trays 4 and 17 for two point temperature control. The choice of trays 4 and 17 for two point control is based on the attempt to use column temperature measurements that contain the most independent information

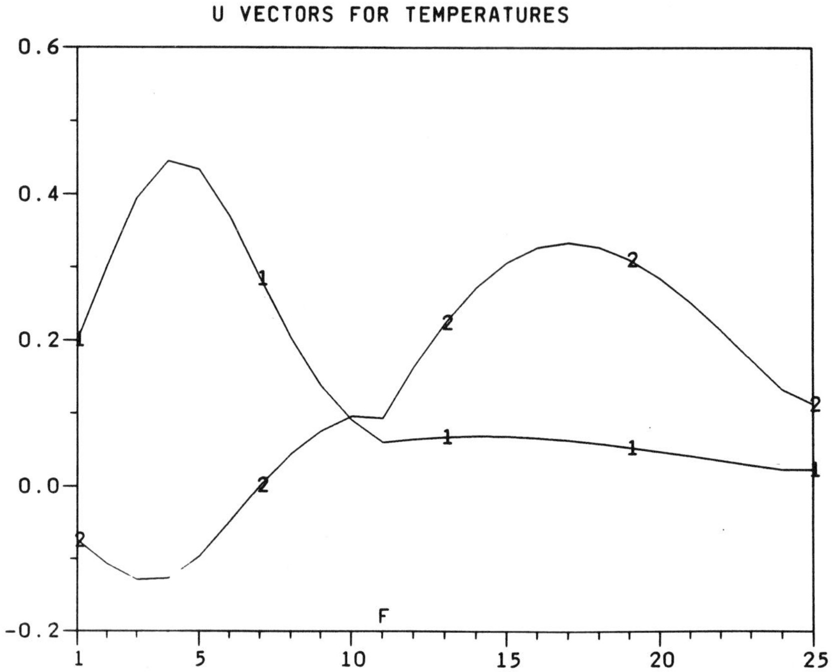

U VECTORS FOR TEMPERATURES

FIGURE 20-12. Low boiler column SVD $\underline{u}$ vectors.

about the column. Selection of temperature measurements based on this criteria does not include the issue of how well these temperatures correlate with end compositions.

After choosing the location of column temperature measurements, the many choices of manipulated variables to be used for temperature control can be screened using various steady state measures. Table 20-7 lists four steady state measures evaluated for five sets of manipulated variables. The RGA is listed for the pairing (T_{17}-m_1) and (T_4-m_2). It is desirable to have the relative gain λ_{11} close to 1, the condition number (CN) small, the second singular value σ_2 large, and the intersivity index σ_2/CN large. Using these guidelines the choice of (L, B), $(L/D, V/B)$, and $(L/D, B)$ appears to be better than (L, V)

TABLE 20-7 Low Boiler Column Interaction Measures for 2 × 2 Temperature Control Using T_4 and T_{17}

m_1	m_2	λ_{11}	CN	σ_2	σ_2/CN ($\times 10^6$)
L	V	6.22	81.2	0.0387	476
D	V	0.193	12.0	0.0671	5592
L	B	0.833	10.8	0.0471	4361
L/D	V/B	1.2	6.3	0.0384	6095
L/D	B	0.891	12.3	0.0443	3602

and (D, V). The manipulated variable screening can be done using other sets of temperatures (or compositions if they are available) for control.

Based on the steady state information, it appears that two point temperature control is feasible. The performance of two point temperature control is compared to single point control to assess the advantages of each strategy. Single point temperature control should be based on T_4 and the column steam rate chosen as the manipulated variable. The column reflux is ratioed to the feed rate in a feedforward fashion. This arrangement is typical for single point temperature control when the control point is in the stripping section and the feed is liquid. For two point temperature control, T_4 and T_{17} are chosen as the controlled variables and the (L, B) set is chosen as the manipulated variables. Other sets can also be evaluated.

Dynamic Results

Figures 20-13 and 20-14 show the dynamic responses for the single and dual temperature control strategies, respectively, for a feed composition disturbance. For this column both strategies have similar dynamic characteristics and both perform satisfactorily. The steady state values for the tray 17 temperature are different and this gives rise to slightly different top compositions.

Decisions about the control strategy for this column are not obvious based on the control analysis. Both strategies perform satisfactorily (as do strategies using other manipulated variables). The decision in this case is dictated more by the underlying operating objectives for the column than control performance. Were the column a final product refining column where the top product was sold to outside customers, then the dual temperature control strategy that holds the top composition more constant

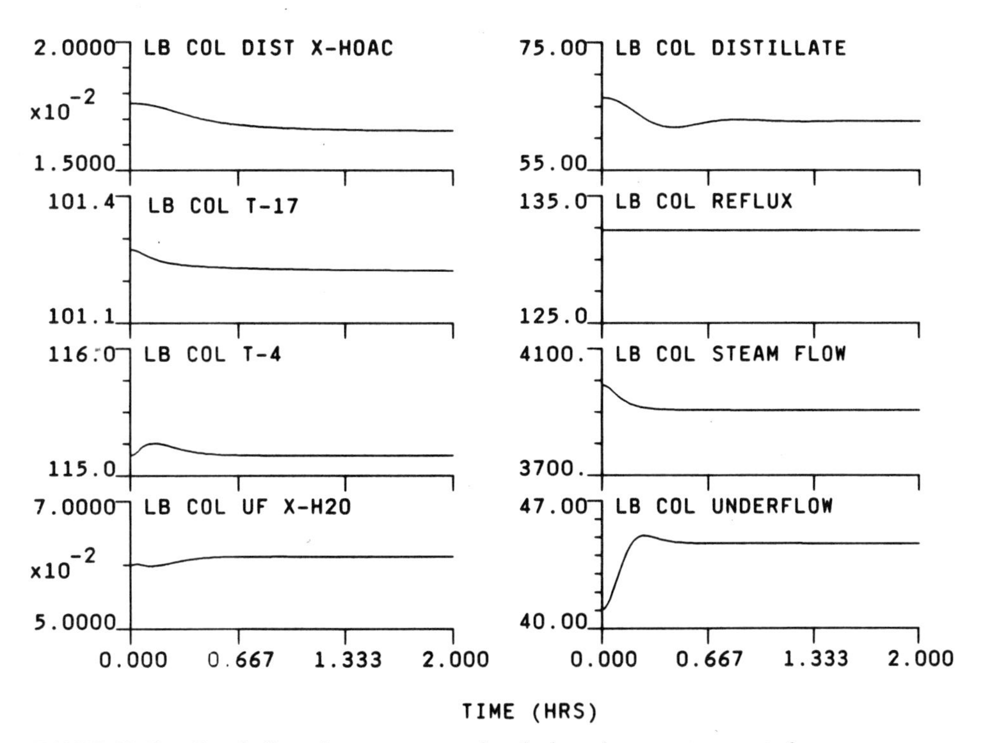

FIGURE 20-13. Low boiler column response using single-end temperature control.

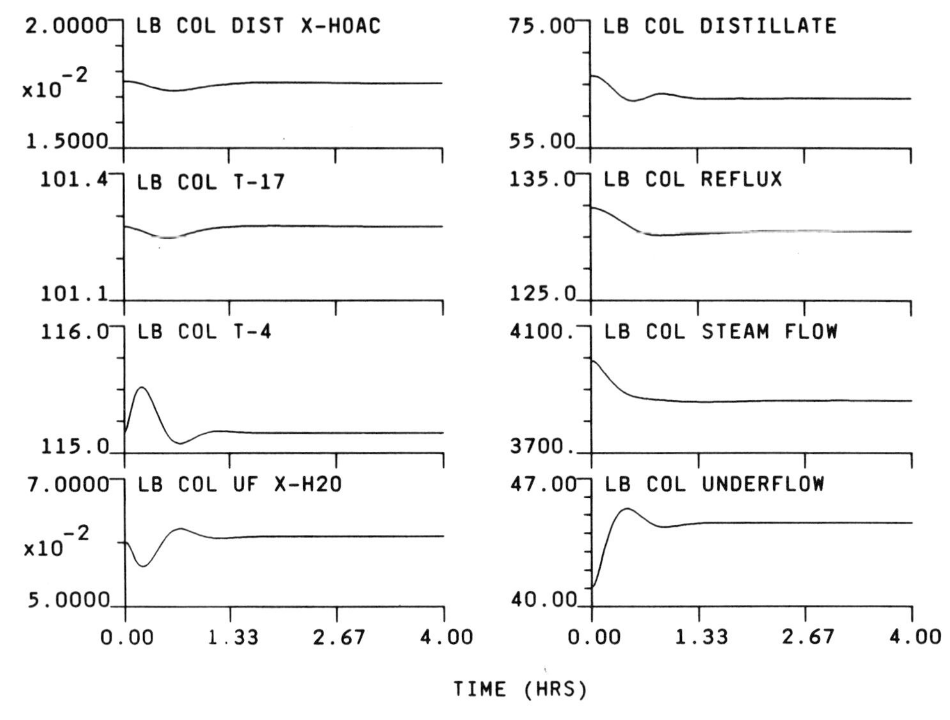

FIGURE 20-14. Low boiler column response using dual-end temperature control.

than the single end control strategy would be preferable. However, if the top composition were not critical, then the simplicity of the single end control strategy may be desired. The key point is that the choice of strategy in this case is more dependent on operating objectives than on control performance. The determination of the overall plant operating objectives and how they influence the objectives of the distillation columns within a process is key to the column strategy design.

From the local analysis of the LB column we have determined that both single- and double-ended temperature control are workable. Because single-end control works about as well as dual-end control and is simpler to implement, we will consider the use of single point temperature control using tray 4 and manipulating the column heat input. If the designer of the column control system is not aware of the overall plant objectives, this strategy is a likely design.

20-3-3 Component Inventory Control Analysis

A candidate overall control strategy for the acetaldehyde oxidation process is illustrated in Figure 20-15 using the single point temperature control on the LB column that resulted from the LB column analysis. The tray 4 temperature controller for the LB column does a good job holding the bottom composition of the LB column constant.

Consider the material balance envelope shown in Figure 20-15. Based on the candidate plantwide control strategy, the material balance for each component can be listed as in Table 20-8 for the envelope shown.

Provided that the reactor is sized large enough, the inventory of acetaldehyde will self-regulate similarly to component A in Example 20-5. The oxygen and the nitrogen will self-regulate similarly to the water and methanol in Example 20-3. The pressure

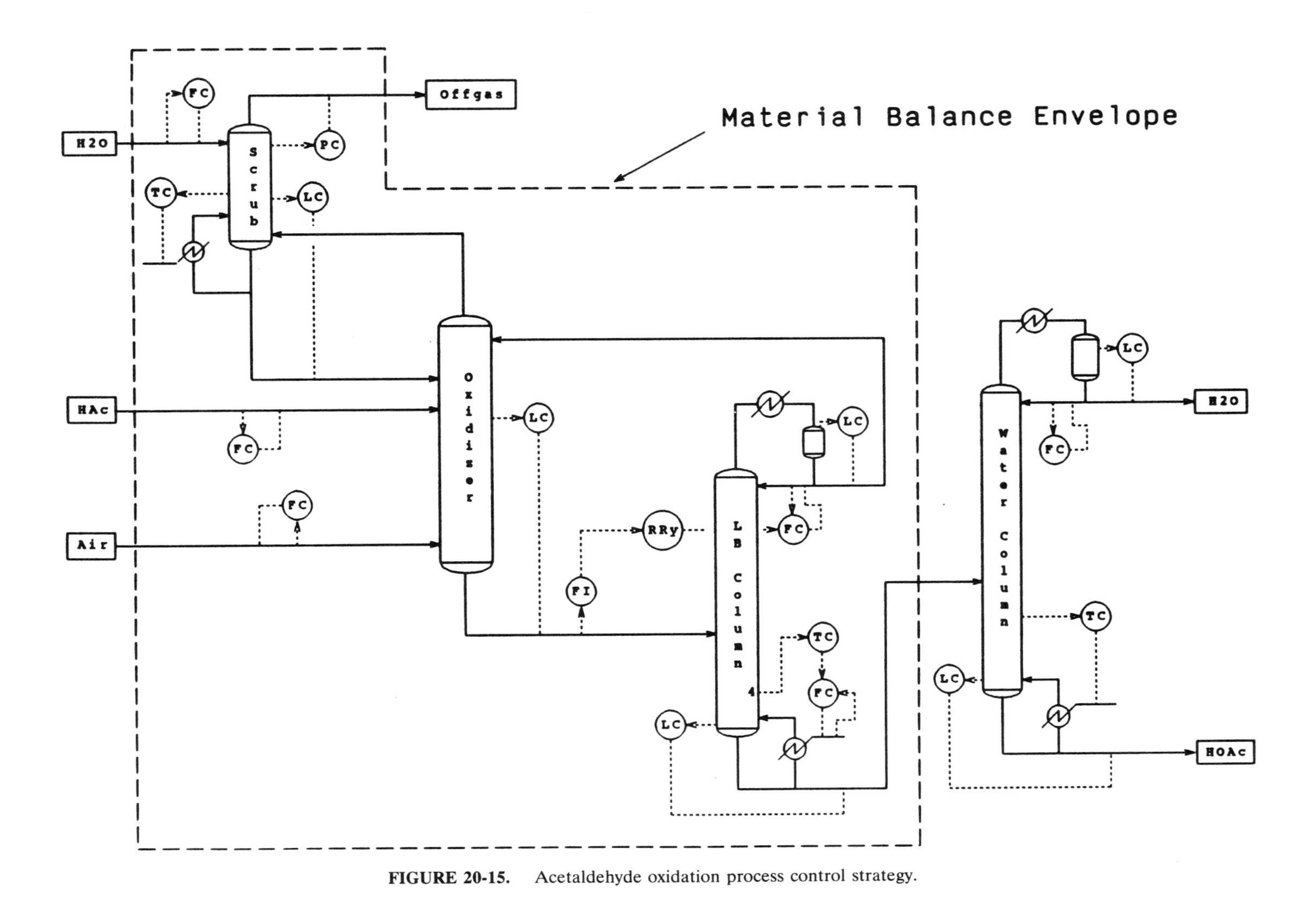

FIGURE 20-15. Acetaldehyde oxidation process control strategy.

TABLE 20-8 Component Material Balance for the Acetaldehyde Oxidation Process

	Input	+ Generation	− Output	− Consumption	= Accumulation
Component	Enters Through System Boundaries	Produced Within the System	Leaves Through System Boundaries	Consumed Within the System	Inventory Controlled by
HAc	HAc feed	0	Small amount in offgas	$Vk_0e^{-E/RT}[\text{HAc}][p_{O_2}]$	Self-regulating via reaction (Ex. 20-5)
O_2	Air feed	0	Offgas	$Vk_0e^{-E/RT}[\text{HAc}][p_{O_2}]$	Self-regulating via reaction and pressure controller (Ex. 20-3 and 20-5)
N_2	Air feed	0	Offgas	0	Self-regulating via pressure controller (Ex. 20-3)
HOAc	0	$Vk_0e^{-E/RT}[\text{HAc}][p_{O_2}]$	Bottom of LB column	0	Not self-regulating (Ex. 20-5)
H_2O	Scrubber water feed	0	Bottom of LB column and small amount in offgas	0	Not self-regulating (Ex. 20-4)

controller is analogous to the level controller in that example. The acetic acid will self-regulate via the temperature controller on the LB column similar to the methanol in Example 20-4. The water is not self-regulating.

The temperature controller at the bottom of the LB column provides good composition control of the bottom product stream. If we assume that the bottoms composition is held constant by the temperature controller at the base case value of 4% water and 96% acid, consider the consequences of changing either the scrubber water feed or the acetaldehyde feed rate. If either rate is changed, then the ratio of water to acid that must exit the process is no longer 4/96. Because the LB column temperature controller maintains this ratio in its bottoms stream (and hence the process exit streams leaving the water column), the inventory of water and acid in the process will change. In particular this change will manifest itself in the composition of the oxidizer.

For example, assume an increase in the scrubber water feed. The inventory of water in the process will begin to increase in the oxidizer. As a result the feed composition to the LB column will become richer in water. However, the tray 4 temperature controller in maintaining the 4/96 water to acid ratio in the bottom stream forces the extra water in the feed out the distillate stream and back to the oxidizer. The oxidizer then continues to become a higher and higher fraction water. In practice the water concentration becomes so high as to stop the oxidation reaction and results in a plant shutdown. A similar scenario occurs following a change in acetaldehyde feed rate.

From a plantwide viewpoint it becomes obvious that the composition of the bottoms stream leaving the LB column must be al-

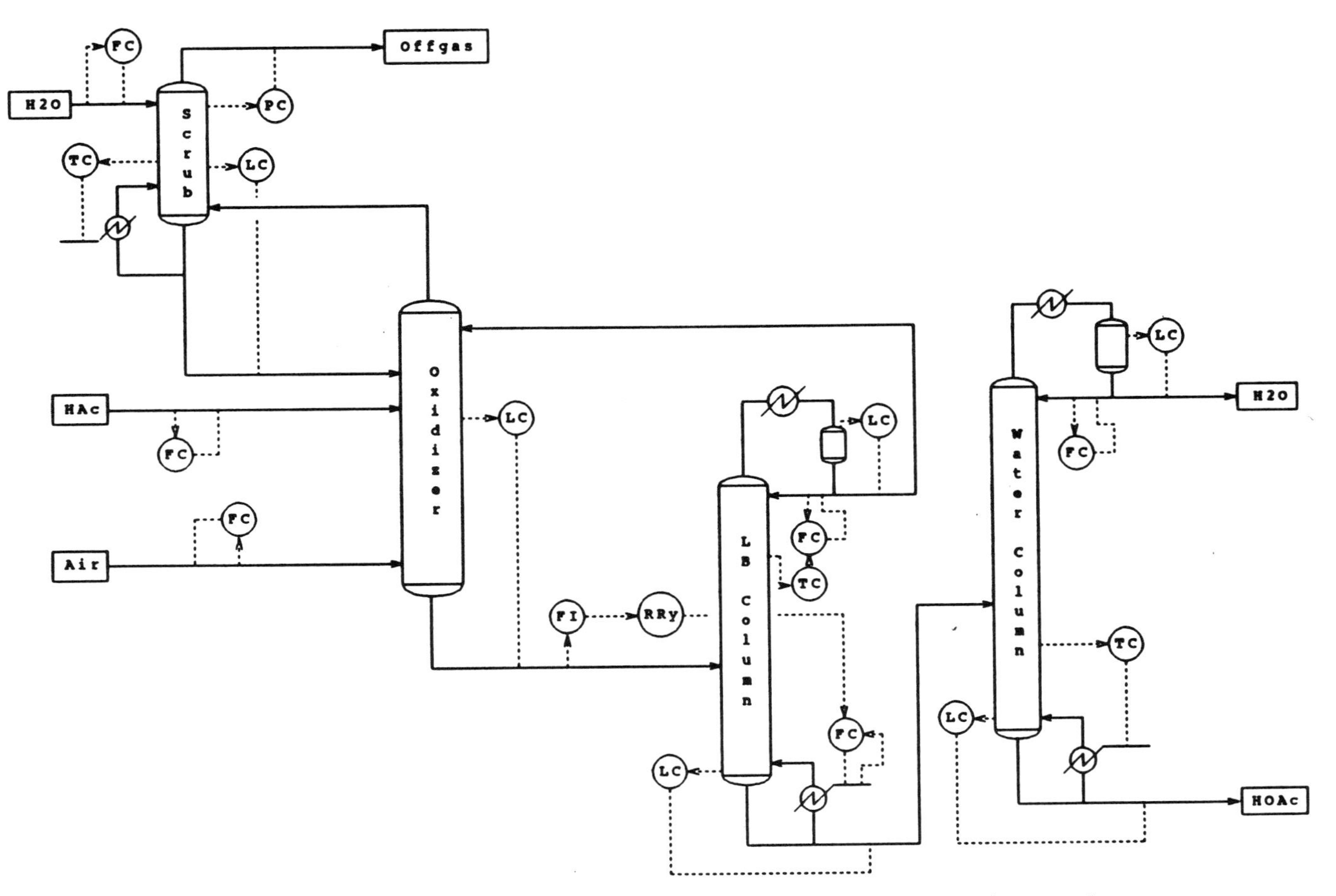

FIGURE 20-16. Acetaldehyde oxidation process control strategies without control of LB column tray 4 temperature.

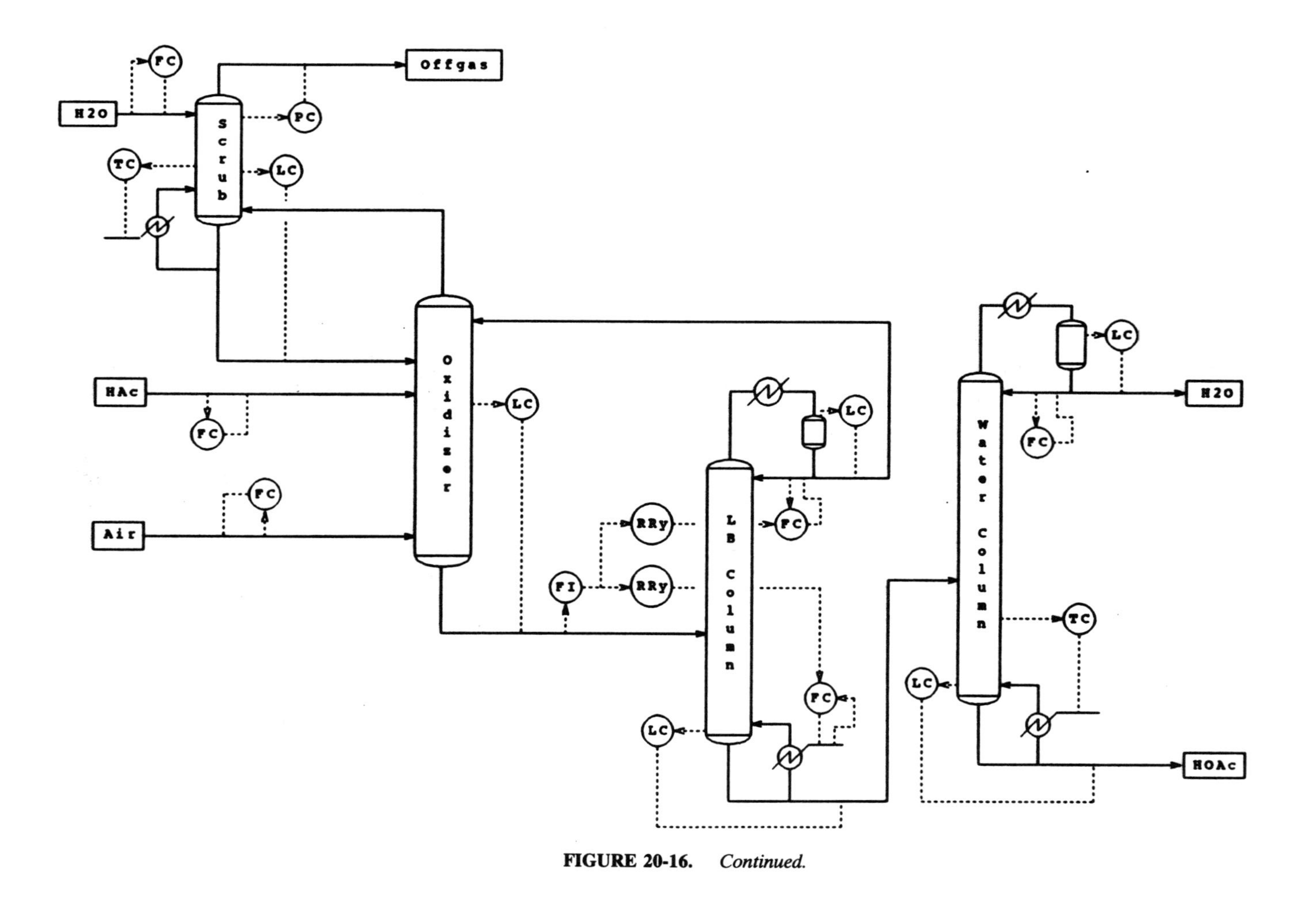

FIGURE 20-16. *Continued.*

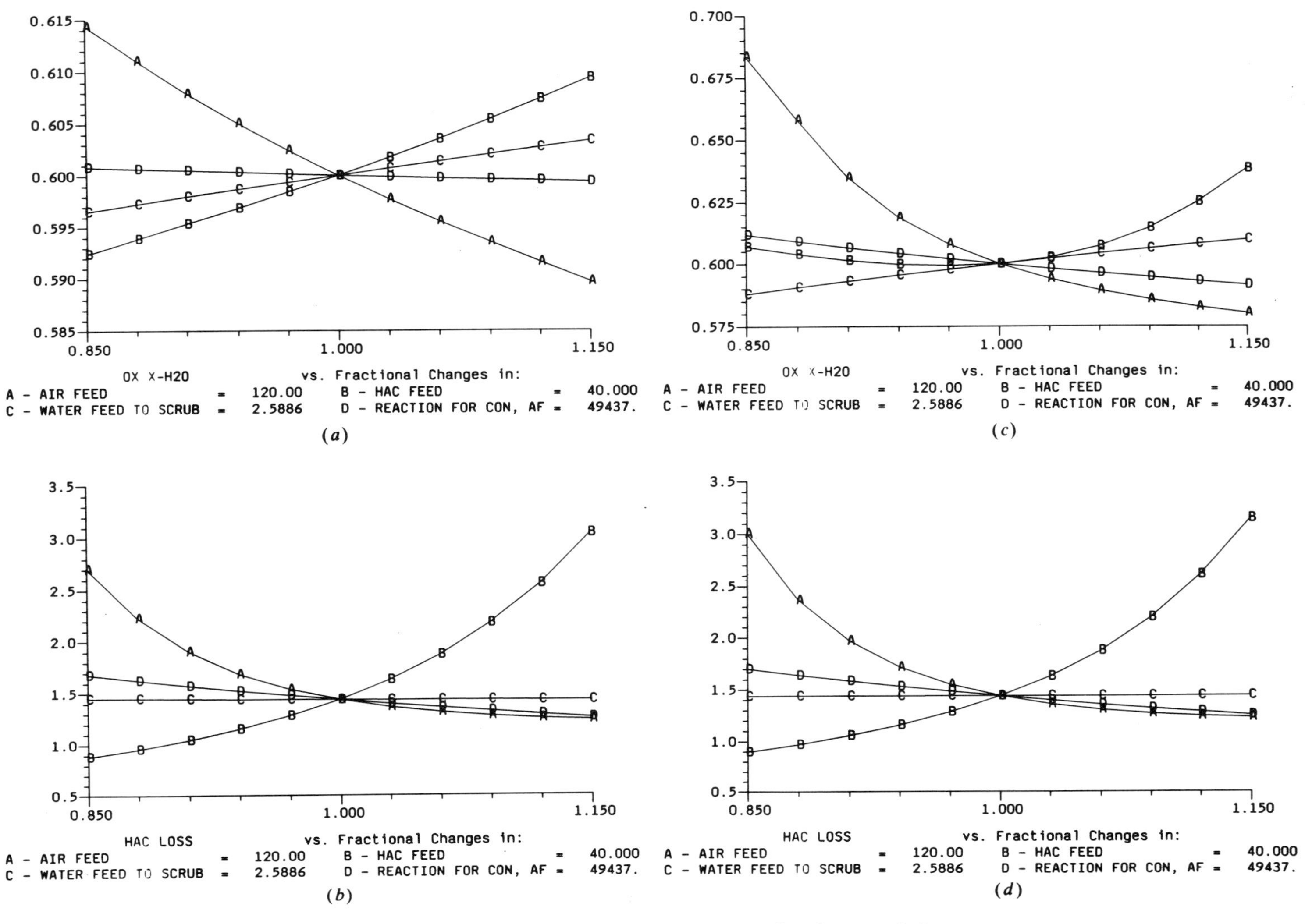

FIGURE 20-17. Acetaldehyde oxidation process disturbance analysis.

lowed to float to match the ratio of water being fed and acid being produced in the process. The control and operation of the LB column is the key to controlling the inventory of the water and acid.

20-3-4 Control of the Acetaldehyde Oxidation Process

To control the inventory of acid and water in the process, the component inventory analysis suggests that the composition of the bottom product of the LB column must be allowed to vary as the feed ratio of water and acetaldehyde changes. To illustrate how the control of the LB column influences overall plantwide control, two control strategies are compared from a steady state disturbance rejection viewpoint.

Figure 20-16 illustrates two plant control strategies that differ only in how the LB column is controlled. In Figure 20-16*a*, the top composition of the LB column is controlled by means of the tray 17 temperature controller, which manipulates the reflux rate to the column. This strategy attempts to hold a relatively constant composition in the distillate stream that recycles back to the oxidizer. The control strategy shown in Figure 20-16*b* has no feedback control on the LB column. The reflux and the reboiler duty are simply ratioed to the feed rate. This strategy accounts for process rate changes but allows the top and bottom compositions to float.

A disturbance analysis as described in Chapter 6 is carried out to determine the effect of four disturbances on the acetaldehyde losses and the oxidizer water composition. In practice other disturbance analyses are done for other disturbances and other process variables as well. Figure 20-17 illustrates the results for the two control strategies. The strategy that controls tray 17 temperature has a greater sensitivity of oxidizer water composition and acetaldehyde loss to the disturbances than the LB column ratio strategy. In particular, for a change in air feed rate to the oxidizer, the ratio strategy results in much less sensitivity of the oxidizer water composition. Other variables such as the sensitivity of acetaldehyde loss to acetaldehyde feed rate are not affected much by the choice of the LB column con-

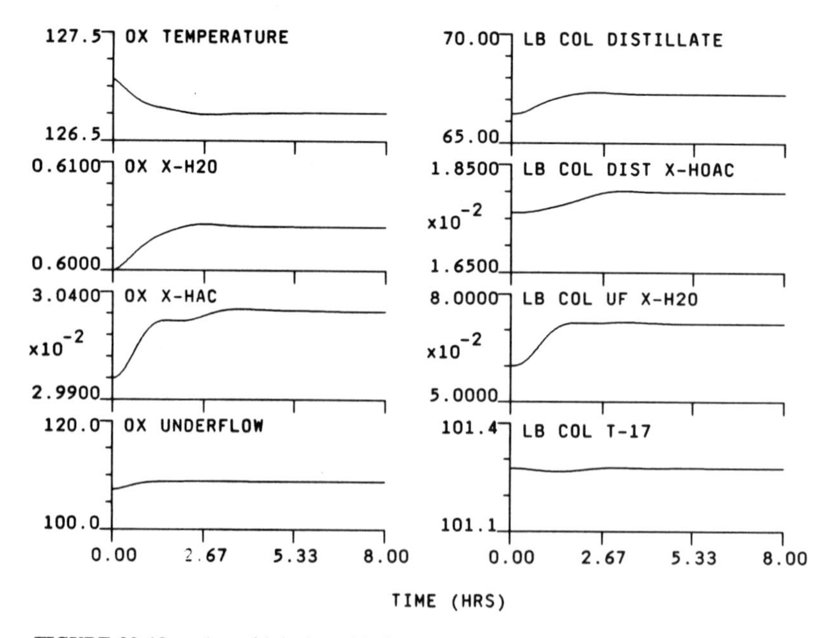

FIGURE 20-18. Acetaldehyde oxidation process response for a change in the water feed rate to the scrubber. Control strategy shown in Figure 20-16*a*.

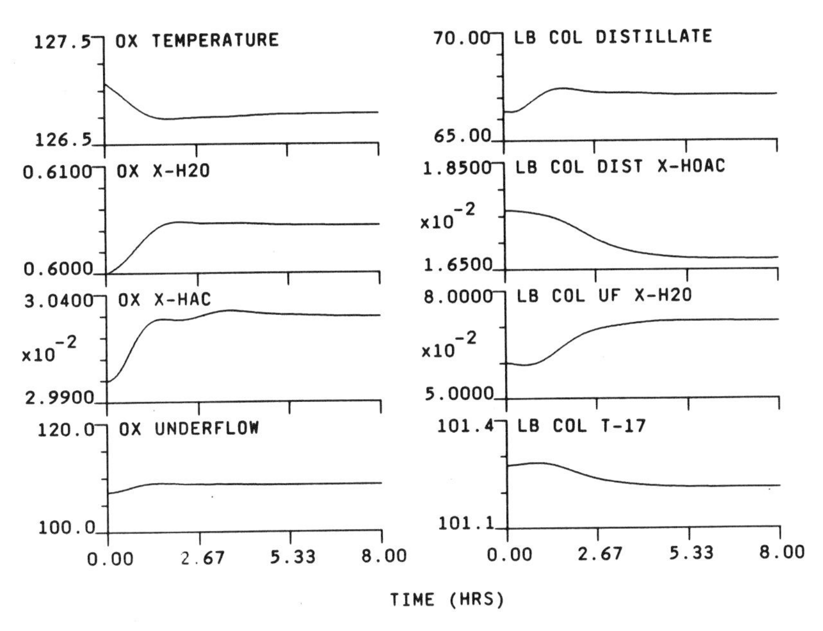

FIGURE 20-19. Acetaldehyde oxidation process response for a change in the water feed rate to the scrubber. Control strategy shown in Figure 20-16*b*.

trol strategy. This limited disturbance analysis suggests that the ratio strategy for the LB column may be the better choice if oxidizer water composition is important.

A dynamic simulation is used to compare the dynamic performance of the alternatives. Figures 20-18 and 20-19 illustrate the response of the process to a water feed rate change to the scrubber for each candidate strategy. For this disturbance, the strategy that has no temperature or composition control on the LB column performs as well as the strategy where tray 17 temperature is controlled. Other dynamic simulations can be performed to compare results for other disturbances. As a result of the limited disturbance analysis and dynamic simulations, the ratio strategy with no temperature feedback looks to be a good choice for the LB column.

For this case study the control of the LB column turned out to be a simple ratio control strategy even though dual-ended control is feasible and performs satisfactorily. The dependence of the LB column control strategy on the overall plantwide strategy is the key point. Different overall strategies will admit different LB column control choices. Each strategy will have strengths and weaknesses when faced with different process disturbances. The job of the control system designer is to ferret out the strengths and weaknesses of each strategy while ensuring that the component inventory control is maintained.

20-4 CONCLUSIONS

The development of control strategies for distillation columns is highly dependent on the overall process inventory control objectives. Distillation column control strategies must be designed using an understanding of how column control affects the overall component inventory control picture. By looking at the overall picture and understanding how each component's inventory is controlled, one can develop criteria for the objectives of the distillation column control strategy. After this perspective is developed, the tools and techniques for the design of particular column control strategies can be used effectively.

21

Model-Based Control

F. F. Rhiel

Bayer AG

21-1 INTRODUCTION

As shown in practice, distillation control design is successful when chemical engineering know-how of column behavior is included in the control concept as a priori knowledge. Incorporation of an observer model represents one such possibility. This is a mathematical process model that operates in the process computer in parallel to the actual process. Using the model, significant process variables, which are difficult or impossible to measure on-line, are estimated from easily measured variables.

As a rule, dynamic column models, which must be reduced for use in the process computer, are used as a basis for the development of observer models. However, for many applications, for example, columns with a large number of plates and multicomponent systems, the development of observer models, suitable for use in a process computer, is extremely difficult or impossible to achieve.

In the present case, static observer models have been developed. Static in the present case means that the observer is based on a steady state model. The observer model has been incorporated into control concepts with an adaptive multivariable controller. These concepts have been put into practice in the operation of two different plant columns. The first plant column separates water from a solvent; the second one separates isomeric components. The results obtained in practice are presented. Only a brief mention is made of the adaptive multivariable controller, which is described in Krahl, Pallaske, and Rhiel (1986) and Krahl, Litz, and Rhiel (1985).

21-2 DESIGN OF THE CONTROL CONCEPT

21-2-1 Process Description

The first column (solvent column) separating water from a solvent is shown in Figure 21-1. The mixture fed into the column contains roughly 85 wt% solvent. The remainder consists of water and a small amount of higher boiling component. At the bottom, solvent and the higher boiling component are taken off. The bottom product, which is recycled to the chemical process preceding the distillation column, is not allowed to exceed 500 ppm of water. The top product, which is passed to the biological water treatment as waste water, should contain about 50 ppm of solvent.

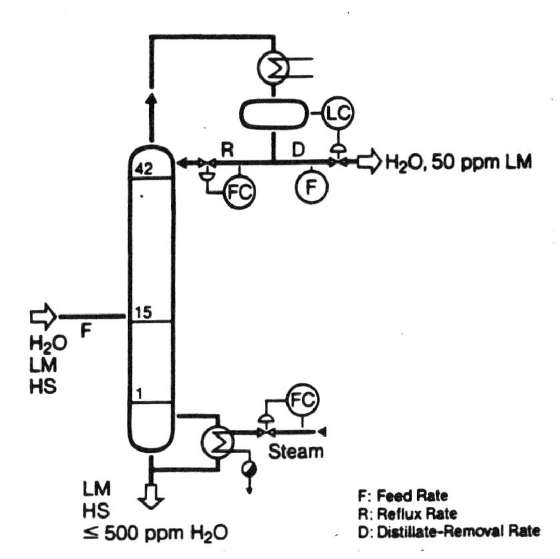

FIGURE 21-1. Solvent column. LM denotes solvent; HS denotes high boilers.

In the conventional control concept the water concentration of the bottom product has been controlled via a temperature in the stripping section of the column. By this concept the water concentration could be kept within the required limits without problems. On the other hand, it was considerably more difficult to control the solvent concentration in the top product. Samples of the top product were taken off at 4 h intervals and analyzed by gas chromatography. The operating personnel obtained the analysis results 1 h after sampling, and could then carry out possible corrections by changing the reflux and steam rates.

The second column (isomeric column) has 40 sieve trays and separates isomeric components with the following product specifications. In the bottom product the concentration of the low boiling key component is not allowed to exceed 1.5 wt%. In the top product the sum of the two high boiling key components should be 300 ppm. The conventional control concept of this column is similar to the concept of the solvent column.

Because avoidance of fluctuations in the top product concentrations is by far the greater problem in the present case, this chapter deals essentially with this problem and its solution.

As already mentioned in the introduction, an adaptive multivariable controller, operating on the basis of a self-tuning principle, is used. This controller takes into account the control circuit coupling between the rectifying and stripping sections of the column, and automatically tunes itself to the column dynamics.

It is relatively simple to control the concentrations in the bottom products. Control of these concentrations is achieved via temperatures and pressures of plates in the stripping section (sixth plate in the case of the solvent column as shown in Figure 21-4). The so-called pressure-compensated temperature, which is correlated with the concentration of the bottom product (water concentration in the case of the solvent column), is formed from the two variables.

In contrast to the control of the bottom product concentrations, it is much more difficult to control the concentrations of the top products because their correlations with the temperatures and pressures in the rectifying section are not sufficient for control. Therefore, in the conventional control concept the concentrations of the top products have to be analyzed by gas chromatography. The concentrations are only measured at fairly long intervals and the results are not immediately available to the operating personnel. This disadvantage is to be eliminated with an observer model that estimates the concentration of the top product from the process variables capable of direct measurements and immediately conveys this information both to the operating personnel and to the controller.

Static Column Behavior

The static behavior of the solvent column is illustrated in Figure 21-2. The curves in this figure are based on measurements recorded at the plant column and on simulation calculations. The temperature curves in the diagram were obtained by variation of the steam and reflux rates. This diagram shows

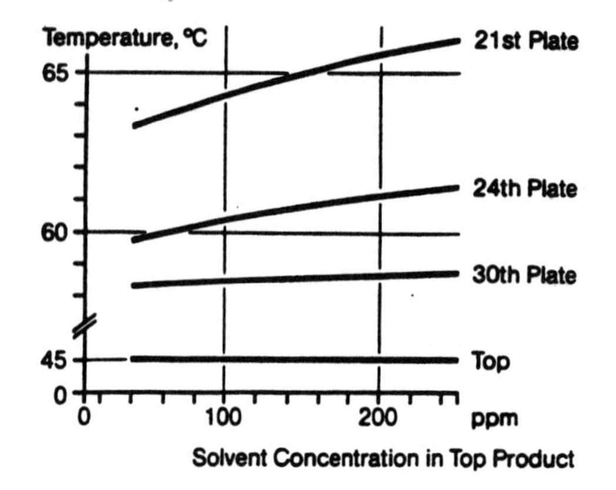

FIGURE 21-2. Relationship between solvent concentration in the top product and column temperatures.

that the top temperature is unsuitable for the control of the solvent concentration in the top product because this temperature remains practically constant over a change in the solvent concentration from 50 to 200 ppm. Only the temperatures of the 21st and 24th plates are sensitive to a change in solvent concentration.

Dynamic Column Behavior

The dynamic column behavior is illustrated in Figure 21-3. The diagram shows the transient response of the temperatures when the steam rate is increased by 1 metric ton/h at time 1 h. The plotted curves confirm that the top temperature is unsuitable for control of solvent concentration. These curves also show that the temperatures of the 21st and 24th plates show the shortest response time, whereas for the 30th plate this time is already 20 min longer. Variation of the feed rate leads to similar findings. The results show that disturbances are detected at a very early stage through the temperature of the 24th plate. Thus, this plate represents an important measurement point for the control of solvent concentration in the top product.

21-2-2 Design of the Observer Model

Principle of the Observer Model

The solvent concentration in the top product of the solvent column cannot be controlled directly by the temperature of the 24th plate, as an example. This would also be true if the effect of pressure on temperature were taken into account. Only the concentration on the 24th plate can be kept constant by this method. The concentration in the top product is additionally determined by the separation efficiency of the column in the region between the 24th plate

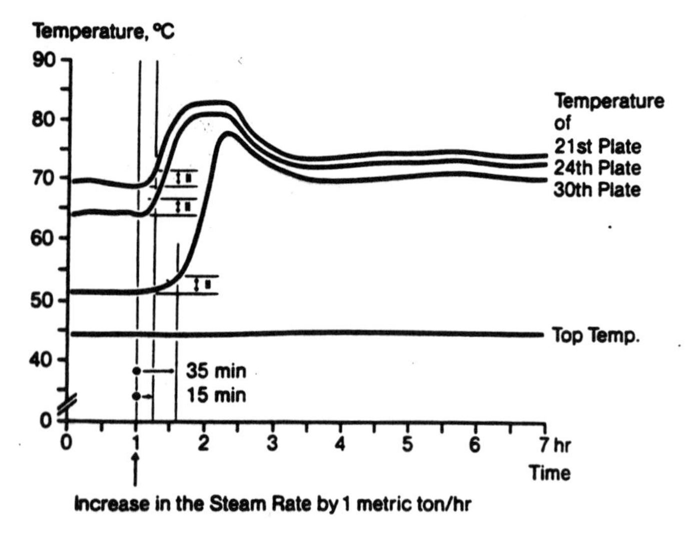

FIGURE 21-3. Transient responses with a change in the steam rate.

and the top. This efficiency varies with the other process variables. An observer model has to include these variables for the estimation of solvent concentration in the top product.

In the present case, a steady state observer model is developed for the upper region of the rectifying section of the column. In the case of the solvent column only the section between the 24th plate and the top, which contains solely water and solvent, is considered. Thus solvent concentration at this plate can be calculated from measured temperature and pressure on this plate. The solvent concentration in the top product can be regarded as a function of the concentration at the 24th plate and the separation efficiency between the plate and the column top. Using a steady state model, this separation can be calculated from the liquid to vapor ratio, the number of plates, and the tray efficiency in this section of the column. The liquid to vapor ratio can be determined by the reflux rate, the distillation removal rate, the reflux temperature, and the top temperature of the column. These values can be measured directly. The tray efficiency does not remain constant, but varies with the hydrodynamic load. The relationship between efficiency and hydrodynamic load must still be determined by empirical tests. The formulas describing the relationship between the solvent concentration of the 24th plate and the top of the column are published in Rhiel and Krahl (1988a). These formulas are simple and can easily be implemented in process computers or process control systems.

In the case of the isomeric column there are three components in the upper region of the rectifying section. Therefore, it is necessary to measure temperature and pressure at two different plates in order to enable the determination of the concentration in the top product. Also, the algorithm for the calculation of the top product concentrations is more complex compared to the formulas for the solvent column.

Properties of the Observer Model

The observer model is based on a static column model. It follows from this that the values estimated by the observer model agree with the actual, that is, analytically determined values, only when the column is at a steady state. If the column is disturbed

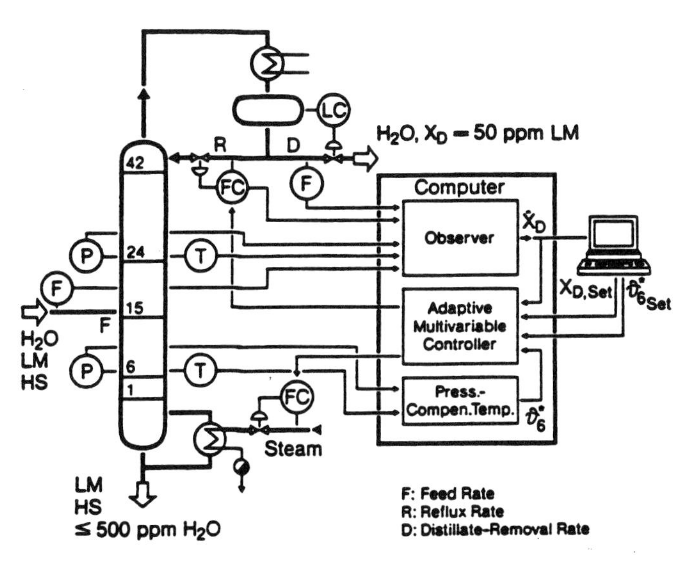

FIGURE 21-4. Solvent column with observer model and adaptive multivariable controller. LM denotes solvent; HS denotes high boilers.

by a change in rate or concentration in feed, the estimated and experimental values no longer agree. In this case, the concentration in the top product, estimated by the observer model, approximately corresponds to the value that would be obtained in the absence of corrective control measures. Thus, as is apparent from the transition functions (Figure 21-3) of the solvent column, the observer model provides predictions of the top concentration 20 min in advance. This property is especially favorable for a controller because the disturbances are recognized at an early stage, thus allowing the controller to carry out a rapid but smooth correction. This is shown clearly by the tests described in the following section.

21-2-3 Control Concept

In the control concept a self-tuning adaptive multivariable controller with a minimum variance control algorithm is used. The concept, including the observer model and the controller for the solvent column, is illustrated in Figure 21-4.

21-3 RESULTS OF MODEL-BASED CONTROL

The diagrams presented in the following text display the recorded operating data of the plant columns together with the analytical findings.

21-3-1 Observer and Controller Behavior

Figures 21-5 and 21-6 show the operating data of the solvent column over a period of 110 h. After 82 h, the feed rate is suddenly increased by 15% and, after 2 h, reduced by the same amount.

The pressure compensated temperature of the sixth plate initially falls as a result of the increase in feed rate. The steam rate undergoes an appropriately large increase in order to prevent an increase in the water concentration of the bottom product. The increase in the steam rate also influences the rectifying section of the column. The increase in the observer value shows that the disturbance is detected promptly. The controller makes corrections just as

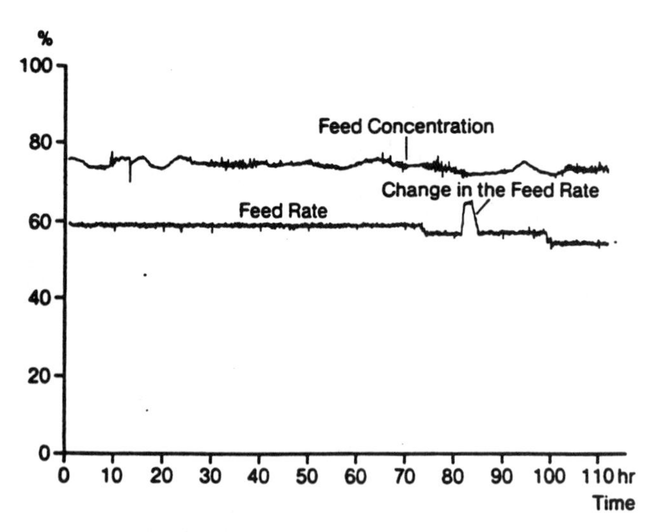

FIGURE 21-5. Rate disturbance: 15% change in feed rate by manual adjustment.

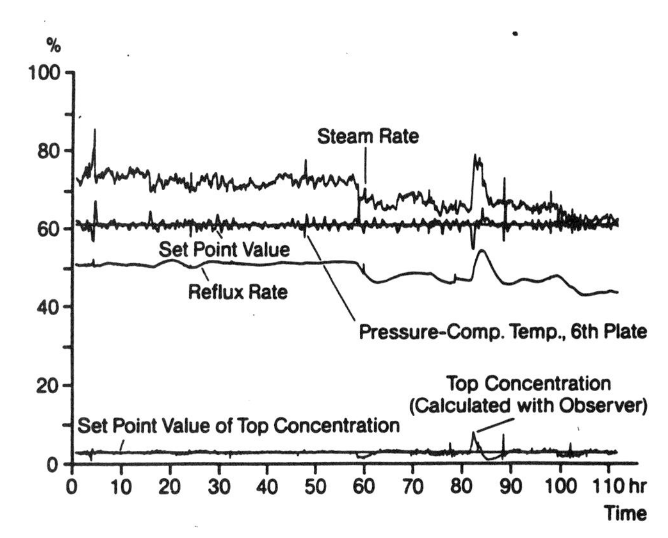

FIGURE 21-6. Reaction of the observer: coordinated manipulation of steam and reflux rates by adaptive multivariable controller.

promptly, as is apparent from the increase in the reflux rate.

Figure 21-7 shows a comparison between the observer and analytical results. It is apparent that the two sets of results agree when the column is at steady state. Only when the disturbance occurs does the observer model produce a higher value, which causes the controller to make prompt corrections. As a result, a very flat curve of solvent concentration in the top product is obtained in spite of this disturbance.

21-3-2 Introducing the New Control Concept in the Production Plant

The implementation of the new control concept in the production plant has been done in three steps. This is advisable in order to avoid any risk as far as possible.

In the first step, only the observer model is installed and the column is operated with the conventional control concept. The observer values are recorded by the computer, but they are not transmitted to the operat-

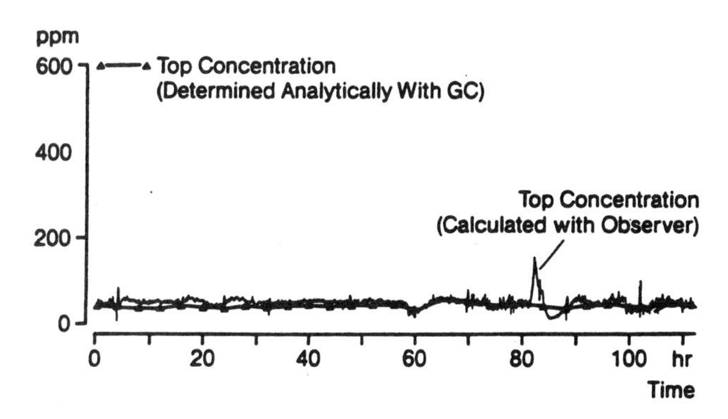

FIGURE 21-7. Results: The top concentration (analytical value) remains constant in spite of large rate disturbance.

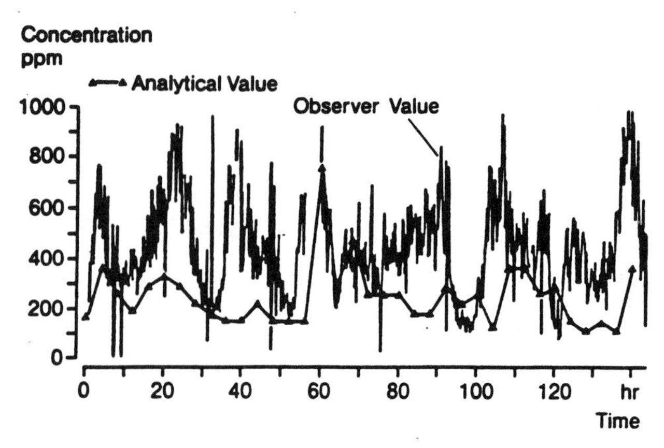

FIGURE 21-8. Conventional operation: Observer values are *not* displayed to operating staff.

ing staff. Figure 21-8 shows the observer and analytical values of the solvent concentration in the top product. In this way, the correlation between the observer and the analytically measured values can be tested. Additionally this diagram illustrates that the fluctuations in column operation are shown even more clearly by the observer values.

In the second step, the column also is operated with the conventional control concept, but, in contrast to the first step, the observer values are continuously displayed on a screen for the operating staff. This leads, as shown in Figure 21-9, to a considerably smoother concentration curve. It is clear from the diagram that the observer values represent satisfactory predictions.

21-4 COMPARISON BETWEEN CONVENTIONAL AND NEW CONTROL CONCEPT

Figure 21-10 reproduces the analytical values of the solvent concentration in the top product for the column operation with the conventional and the new control concept. The comparison between the two curves highlights the much smoother curve obtained with the observer model, combined

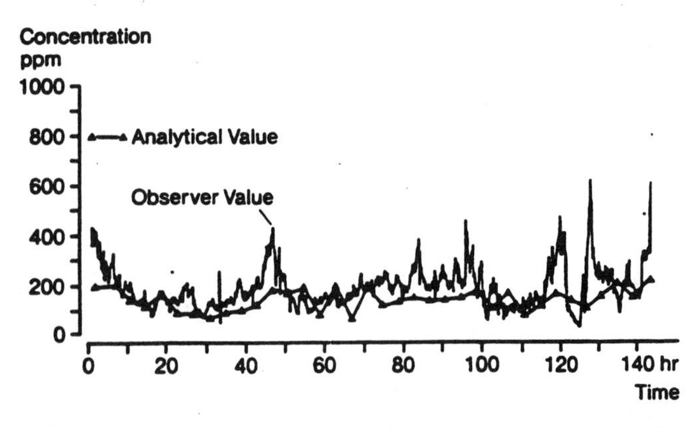

FIGURE 21-9. Conventional operation: Observer values are displayed to operating staff.

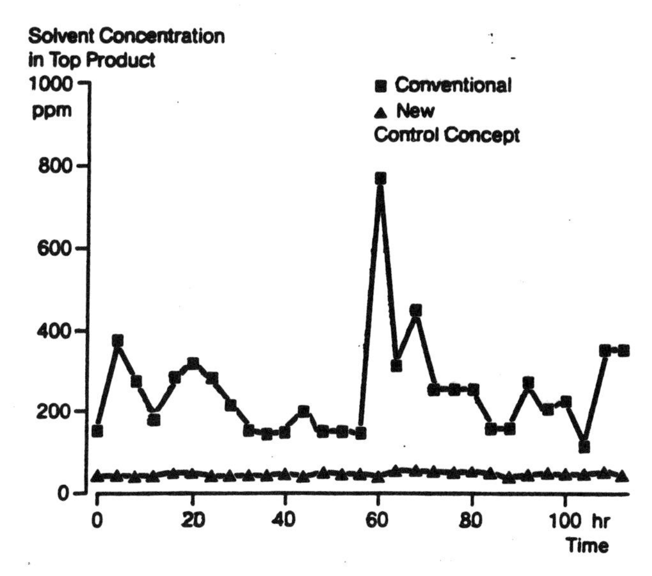

FIGURE 21-10. Comparison between conventional and new control concept (solvent column).

with the adaptive multivariable controller. In this way, the fluctuations could be reduced to a standard deviation of 5 ppm, at an average value of 50 ppm.

The temperatures plotted in Figure 21-11 are averages recorded over a lengthy period of time. It is apparent that the column temperatures, particularly in the stripping section, could be substantially lowered. Lowering of the temperatures would permit a reduction in the reflux ratio and thus in the energy consumption.

The improvement of the top product quality of the isomeric column by the new

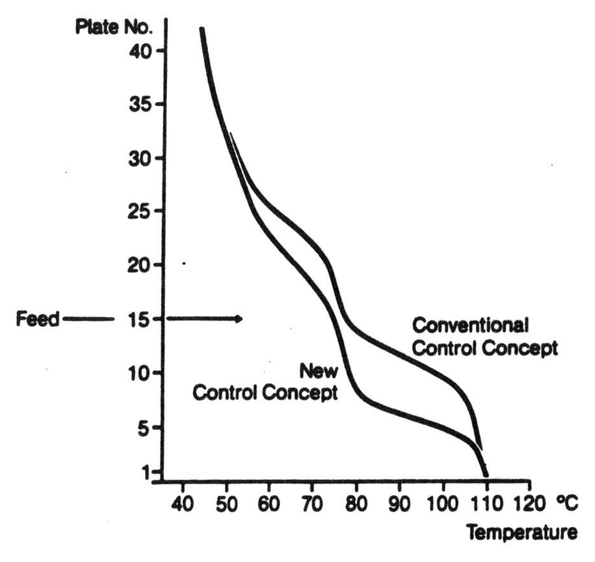

FIGURE 21-11. Temperature profiles: conventional versus new control concept (solvent column).

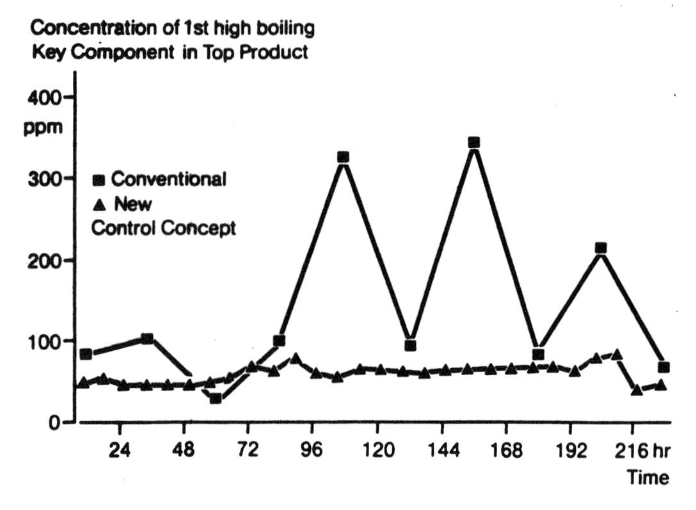

FIGURE 21-12. Comparison between conventional and new control concept (isomeric column): first high boiling key component.

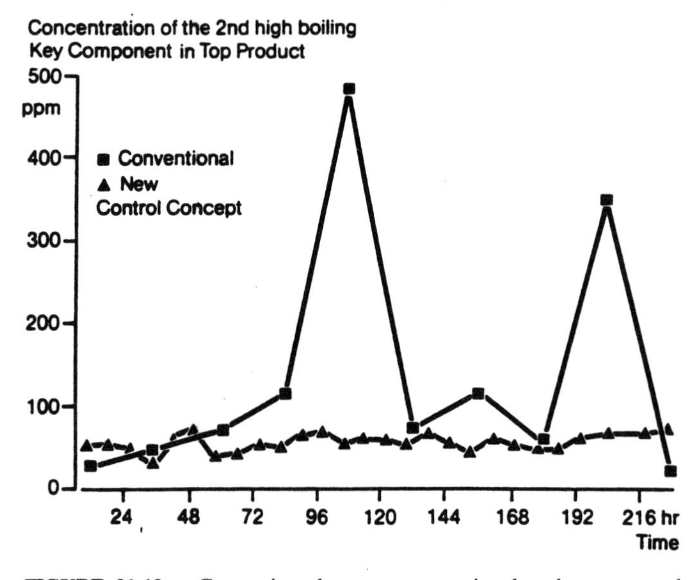

FIGURE 21-13. Comparison between conventional and new control concept (isomeric column): second high boiling key component.

control concept is illustrated in Figures 21-12 and 21-13. The diagrams reproduce the concentrations of the two high boiling key components in the top product. The curves also show that the control of distillation columns containing two high boiling key components in the upper region of the rectifying section can be substantially improved by the new control concept. The differences of the boiling points are 30 K between the top product component and the first high boiling key component and 2.5 K between the two high boiling key components. Therefore, three components have to be taken into account in the observer model.

21-5 CONCLUSIONS

On the basis of operating tests, the advantages of the control concept with observer model and adaptive multivariable controller can be summarized as follows:

- More uniform product quality.
- Higher yields of products (lower solvent losses as an example).
- Lower energy consumption.
- Lower analysis costs.
- Reduced sensitivity to disturbances (control involves only temperatures, pressures, and rates; it does not depend on analytical equipment).
- Robust control with fine adjustment capabilities because disturbances are recognized at an early stage and can be promptly corrected.
- Eases the burden on operating personnel through completely automatic operation.

21-6 APPENDIX

Formulas of the Model for the Solvent Column

Principle of the observer model:

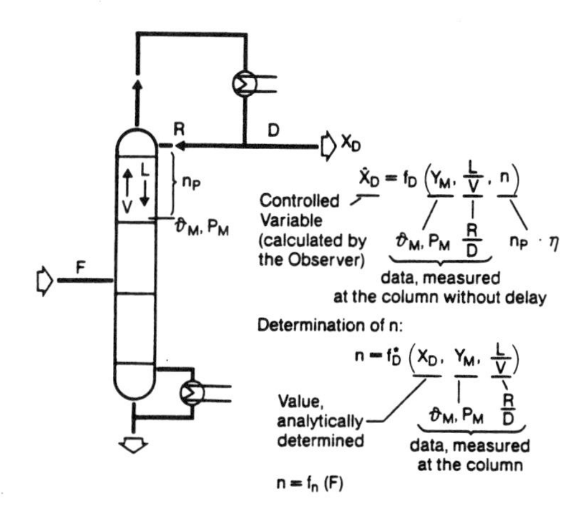

Thermodynamic equilibrium of water–solvent:

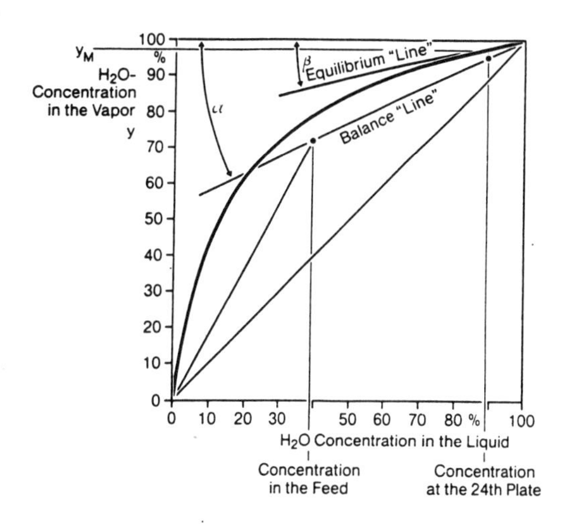

Calculation of the solvent concentration in the top product:

$$x_D = y_M \frac{1}{m + (1-m)\dfrac{\left(\dfrac{a}{m}\right)^{n+1} - 1}{\dfrac{a}{m} - 1}} \tag{21-1}$$

$$n = \frac{\ln\left[\left(1 - \dfrac{m}{a}\right)\dfrac{y_M - mx_D}{X_D - mX_D} + \dfrac{m}{a}\right]}{\ln\left(\dfrac{a}{m}\right)} \tag{21-2}$$

$$m = \tan\beta \tag{21-3}$$

$$a = \frac{L}{V} = \tan\alpha \tag{21-4}$$

$$\frac{L}{V} = \frac{R/D}{1 + R/D} \tag{21-5}$$

$$n = n_P\eta \tag{21-6}$$

$$\eta = \eta(F) \tag{21-7}$$

where x_D = concentration of top product
y_M = concentration of vapor at measure point
m = slope of equilibrium "line"
a = slope of balance "line"
L = liquid rate
V = vapor rate
R = reflux rate
D = distillate removal rate
n = number of theoretical stages
n_P = number of plates
η = tray efficiency
F = feed rate

References

Gilles, E. D. (1985). Von der MSR-Technik zur wissensbasierten Prozeßleittechnik. *Chem.-Anlagen Verfahren*. 34(4), 34–35, 138.

Krahl, F. (1981). Multivariable adaptive control of a stochastically disturbed process. Thesis, Gesamthochschule Wuppertal (in German).

Krahl, F., Litz, W., and Rhiel, F. F. (1985). Experiments with self-tuning controllers in chemical plants. IFAC Workshop on Adaptive Control of Chemical Processes, pp. 127–132.

Krahl, F., Pallaske, U., and Rhiel, F. F. (1986). Einsatz der adaptiven Mehrgrößenregelung an Destillationskolonnen. *Chem.-Ing.-Tech.* 58(7), 608–609.

Krahl, F. and Rhiel, F. F. (1986). Energieoptimale Regelung von Destillationskolonnen mit adaptiven Mehrgrößenreglern. Fachberichte Messen, Steuern, Regeln, 14. Fortschritte in der Meß- und Automatisierungstechnik durch Informationstechnik, INTERKAMA Congress, pp. 385–396.

Rhiel, F. F. and Krahl, F. (1988a). A model-based control system for a distillation column. *Chem. Eng. Technol.* 11, 188–194.

Rhiel, F. F. and Krahl, F. (1988b). Modellgestuetzte Regelung einer Destillationskolonne. *Chem.-Ing.-Tech.* 60, 649.

Rhiel, F. F., Perne, R., Pauen, S., and Steffens, F. (1989). Model-based control systems for distillation columns. DYCORD + '89, Maastricht, Netherlands.

22

Superfractionator Control

William L. Luyben
Lehigh University

22-1 OCCURRENCE AND IMPORTANCE

Distillation columns that perform difficult separations are called superfractionators. By difficult separations we mean that the relative volatility between the chemical components is quite low: less than 1.2 and approaching 1.05. Such low relative volatilities mean that there is very little difference between the composition of the liquid phase and the composition of the vapor phase in equilibrium with it. Hence many equilibrium stages are required to accomplish the separation.

There are a fair number of these types of columns. The separation of heavy water from regular water is probably the most striking example. The relative volatility is only 1.05 and multiple columns are required with a total of 750 trays.

The most common industrial example is the separation of propylene from propane. This separation is performed in almost every petroleum refinery in the world and in many chemical plants. Relative volatility is about 1.1 and columns contain typically 150 to 200 trays. The propylene–propane separation is a major consumer of energy. However, it should be emphasized that this energy is at a very low temperature level (about 135°F). This means that waste heat often can be used. Other industrial examples include many types of isomer separations (e.g., isopentane–normal pentane, ortho-xylene–meta-xylene, etc.).

It is important to note that the separations that use superfractionators are almost always binary, that is, only two components are present. This is because it is more efficient to make these difficult separations with binary systems than when other heavier-than-heavy key or lighter-than-light key components are present. Thus superfractionators usually are positioned at a point in the process where other lighter and heavier components have been removed in previous processing steps. This means that disturbances from upstream units are usually frequent because the superfractionator is at the end of the line. The control system must be able to effectively reject these disturbances.

22-2 FEATURES

Because of the low relative volatilities, superfractionators have the following features: many trays, high reflux ratios (large energy consumptions), flat temperature profiles, and very slow dynamic responses.

22-2-1 Many Trays

The number of trays is over 100. This means that superfractionators are very tall. Often they are constructed in two shells because of physical limitations on heights. This is called a *split-shell* design. The vapor from the top of the first shell is fed into the base of the second shell. The liquid from the base of the second shell is pumped up to the top of the first shell. Feed is introduced on the appropriate tray in either shell.

There is another type of split-shell design that is used in systems where the pressure drop through the many trays would produce base temperatures that would exceed some maximum value as set by polymerization, degradation, or safety limitations. These columns normally operate under vacuum to keep temperatures low, and there are significant changes in pressure through the column due to pressure drop. In this type of system each shell has its own reboiler and condenser, and the pressure at the top of each shell is kept as low as possible (usually limited by cooling water temperatures and the need to subcool the condensed vapor to reduce losses of product to the vacuum system). Each shell contains as many trays as possible without exceeding the maximum temperature limitation in its base. Vapor from the top of each shell is totally condensed and pumped into the base of the next shell in series. Vapor for each shell is produced in its reboiler. Liquid from the base of the shell is pumped to the top of the previous shell in the series. These systems are very energy inefficient because of the multiple vaporizations and condensations.

Distillation columns that are used to produce very high purity products (ca. 10 ppm) can sometimes also contain many trays. However, these columns are usually not considered to be superfractionators because relative volatilities are not low and reflux ratios are not high. This is because the minimum reflux ratio does not change very much as product purity is increased, even though the minimum number of trays does increase.

22-2-2 High Reflux Ratios

The low relative volatility produces a vapor–liquid equilibrium line on a McCabe–Thiele diagram that is quite close to the 45° line. Thus the operating line in the rectifying section must have a slope that approaches 45° where the reflux ratio is infinite. Typical reflux ratios are 10 to 20 or higher.

The high reflux ratio means that large amounts of vapor are required in the reboiler per unit of feed (high energy consumption) and that the diameter of the column must be large to handle the vapor rate. As a result, superfractionators are large in both height and diameter. Both of these physical features mean that a large amount of liquid is held up on the trays in the column.

22-2-3 Flat Temperature Profile

Low relative volatilities mean that the boiling points of the two components are quite close. Therefore the temperature at the top of the column where the concentration of the lighter component is high is not much different than the temperature at the bottom of the column where the heavier component is high. The result is that temperature can seldom be used in superfractionators to infer compositions. Direct online analyzers are usually required in order to effectively control product quality in superfractionators.

Gas chromatographs are frequently used. Their intermittent operation (sampling times from 1 to 20 min) imposes an inherent limitation on the effectiveness of any control system. This is one of the reasons why superfractionator control is more difficult than columns where continuous temperature measurements can be used to infer compositions.

22-2-4 Slow Dynamics

The large amount of liquid in these columns produces dynamic responses that can be very slow (time constants that are many hours). This makes control more difficult.

22-3 ALTERNATIVE CONTROL STRUCTURES

There are a number of alternative control structures for superfractionators. Some of the schemes are sketched in Figure 22-1. Because of the high reflux ratio, control of the reflux drum is almost always achieved by manipulating reflux and not distillate.

1. *D-V*: Figure 22-1*a* shows the *D-V* scheme where distillate flow controls dis-

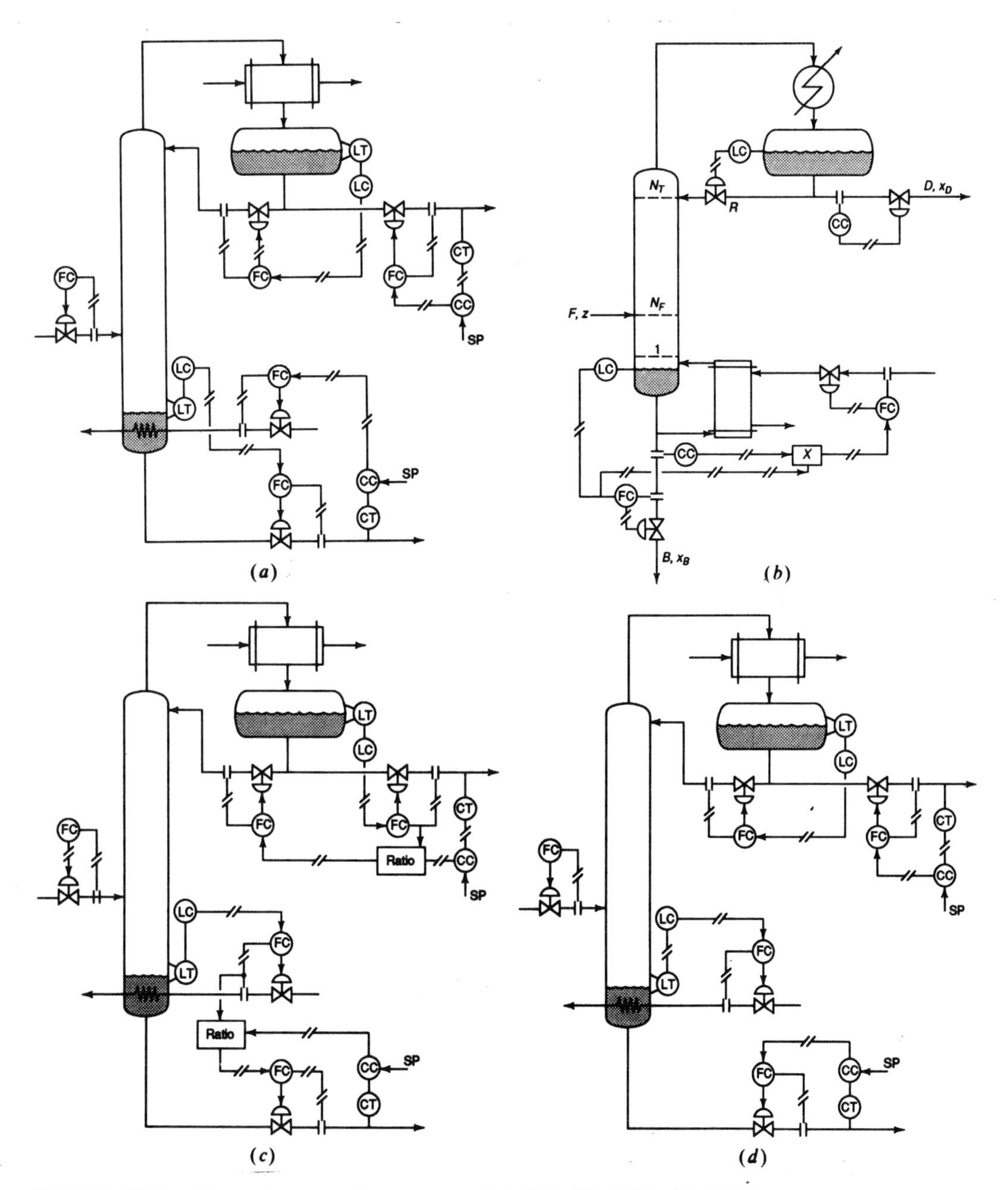

FIGURE 22-1. Alternative control structures: (*a*) *D-V*; (*b*) *D-BR*; (*c*) *RR-BR*; (*d*) *D-B*.

tillate composition and heat input controls bottoms composition.

2. *D-BR*: Figure 22-1*b* shows a scheme where distillate flow rate controls distillate composition and boilup ratio (heat input over bottoms flow rate) controls bottoms composition.
3. *RR-BR*: Figure 22-1*c* shows a scheme where reflux ratio (reflux flow rate over distillate flow rate) controls distillate composition and boilup ratio controls bottoms composition. This scheme requires the measurement of four flows and the calculation of two ratios. Therefore, it is more complex and more susceptible to problems with noisy flow signals.
4. *D-B*: Figure 22-1*d* shows an unusual scheme where distillate composition is controlled by distillate flow rate and bottoms composition is controlled by bottoms flow rate.

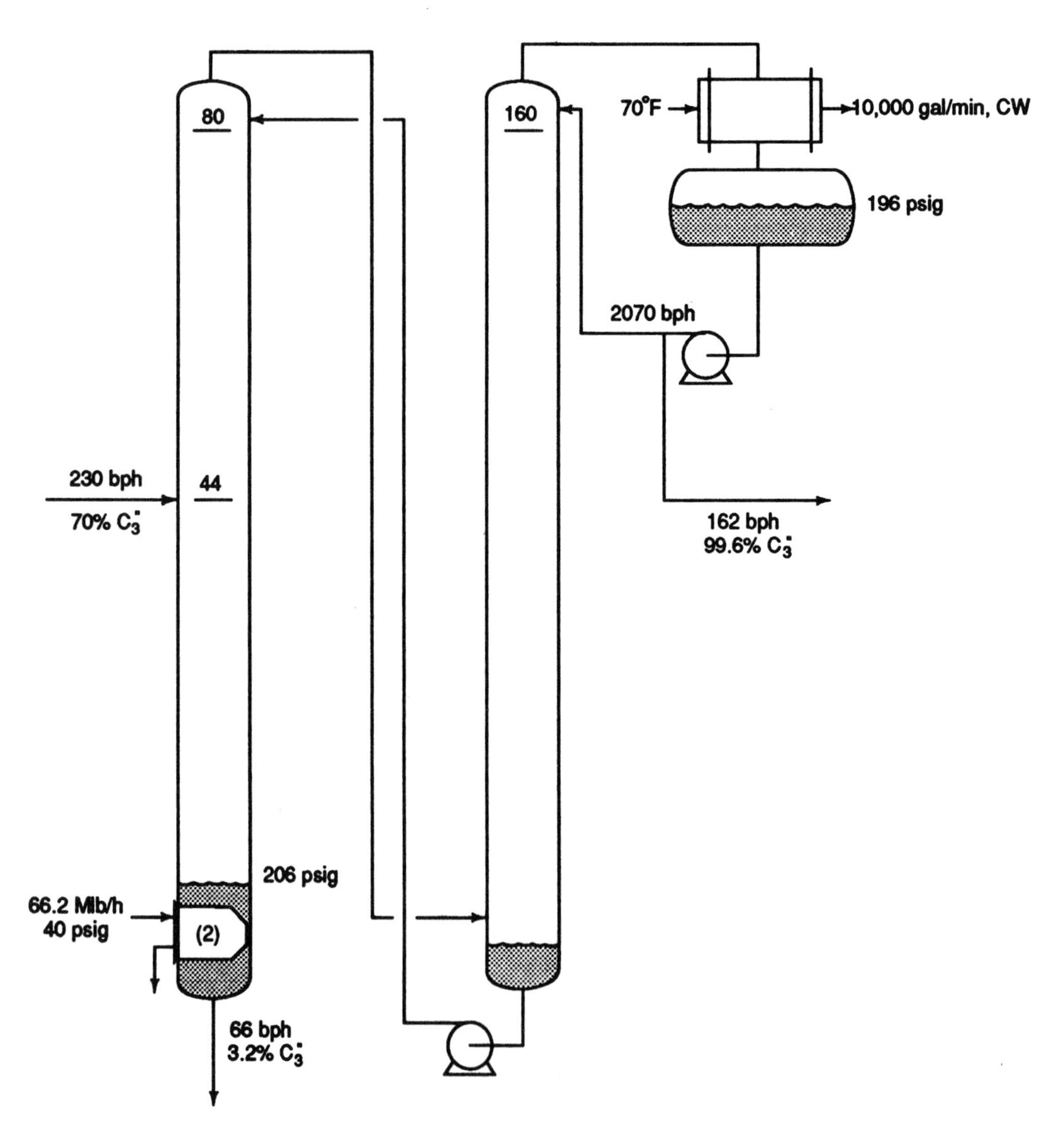

FIGURE 22-2. Sun Oil column.

We will study the performances of these alternatives later in this chapter on a propylene–propane separation column. The methods discussed in Chapter 11 can be used to tune the controllers once transfer function models have been obtained (see Chapter 7).

It is worth noting that obtaining transfer function models for the *D-B* scheme is not a straightforward procedure. This is because the steady-state gains are indeterminant: One cannot find the steady-state gain between distillate flow rate and any composition with bottoms flow rate fixed since *D* plus *B* must add up to *F* at steady state. Dynamically $D + B$ does not have to equal *F*. We will present a method for determining the dynamic *D-B* transfer functions later in this chapter.

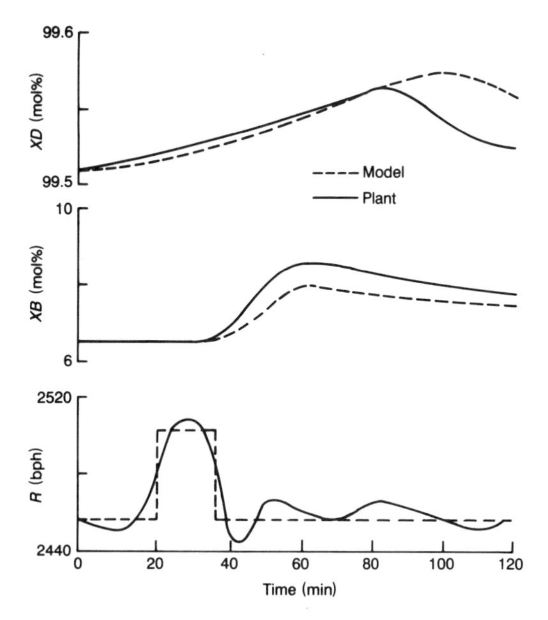

FIGURE 22-3. Plant tests compared to dynamic model predictions (Sun Oil column).

22-4 INDUSTRIAL EXAMPLE

At this point a specific example might be useful.

22-4-1 Process

The column to be considered is a 160 tray, 15 ft diameter, split-shell tower studied by Finco, Luyben, and Polleck (1989). Figure 22-2 summarizes the process flow rates, compositions, pressures, and heat duties. Note that this column uses 66,200 lb_m/h of steam in its two stab-in reboilers and operates with a reflux ratio of about 14. Polymer-grade propylene (99.6 mol%) is produced overhead. The bottoms product is 3.2 mol% propylene and goes to LPG.

22-4-2 Plant Dynamic Tests

Steady-state plant data were used to calculate tray efficiencies (about 100%). Then several dynamic plant tests were conducted to validate a dynamic mathematical model of the process. Figure 22-3 shows one of these tests in which a pulse in reflux was made. The comparison between the model and the plant data was considered good enough to proceed with simulation comparisons of alternative control structures.

22-4-3 Simulation

The dynamic model used was similar to that presented in Chapter 3. Relative volatilities were calculated as functions of composition and pressure. Equimolal overflow was assumed because latent heats of vaporization are essentially equal in the propylene–propane system. The model accurately replicated plant data except for rapid changes in pressure.

Level controllers were assumed to be perfect. Composition controller tuning was done by using an empirical BLT approach (Chapter 11): controller settings were detuned from Ziegler–Nichols settings until the resulting time-domain responses showed closed-loop damping coefficients of about 0.3.

22-4-4 Results

Figure 22-4 gives the responses of several control structures to step changes in feed

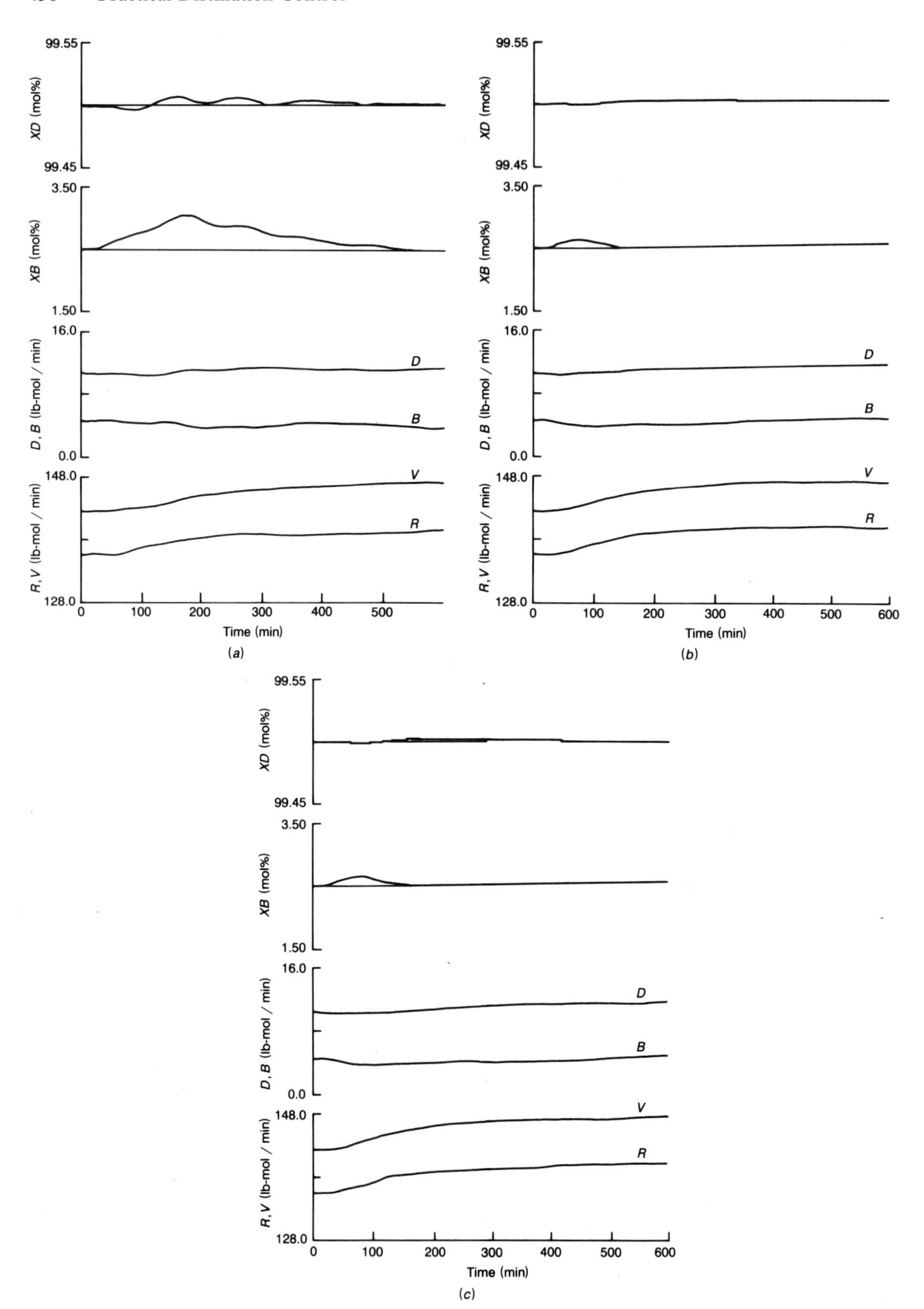

FIGURE 22-4. Simulation results for alternative control structures (Sun Oil column): (*a*) *D-V*; (*b*) *RR-BR*; (*c*) *D-B*.

composition. The *RR-BR* and the *D-B* structures give significantly better responses than the traditional *D-V* structure. Because the *RR-BR* requires more instrumentation complexity, the *D-B* structure looks promising.

22-5 TUNING THE *D-B* STRUCTURE

One of the principle difficulties with using the *D-B* structure is the problem of obtaining the transfer functions necessary for quantitative controller tuning. A procedure for achieving this is presented below (Papastathopoulou and Luyben, 1990). The technique uses ideas originally developed by Skogestad, Jacobsen, and Morari (1990) for transforming from one control structure to another in order to calculate frequency-dependent relative gain arrays.

22-5-1 Transformations

The transfer functions for the *D-B* structure can be found by transforming the transfer function for the *D-V* structure or the *R-V* structure. Either of these can be found using conventional identification methods.

In order to make these transformations, several assumptions are made: equimolal overflow, perfect level control in reflux drum and base, and liquid flow dynamics described by

$$G_{L(s)} = \frac{L_1}{R} \tag{22-1}$$

The assumption of perfect level control in the base gives

$$V = L_1 - B \tag{22-2}$$

and the assumption of perfect level control in the reflux drum gives

$$R = V - D \tag{22-3}$$

Combining Equations 22-1 through 22-3 gives

$$\begin{aligned} V &= G_L R - B \\ V &= G_L(V - D) - B \\ (1 - G_L)V &= G_L D - B \\ V &= \frac{-G_L}{1 - G_L} D - \frac{1}{1 - G_L} B \end{aligned} \tag{22-4}$$

Equation 22-4 in matrix form is

$$\begin{bmatrix} D \\ V \end{bmatrix} = \begin{bmatrix} 1 & 0 \\ \dfrac{-G_L}{1 - G_L} & \dfrac{-1}{1 - G_L} \end{bmatrix} \begin{bmatrix} D \\ B \end{bmatrix} \tag{22-5}$$

Let us assume that the *D-V* transfer functions have been found in the same conventional way:

$$\begin{bmatrix} x_D \\ x_B \end{bmatrix} = \underline{\underline{G}}_M \begin{bmatrix} D \\ V \end{bmatrix} = \begin{bmatrix} G_{11} & G_{12} \\ G_{21} & G_{22} \end{bmatrix} \begin{bmatrix} D \\ V \end{bmatrix} \tag{22-6}$$

Now Equation 22-5 can be substituted into Equation 22-6 to obtain the transfer functions for the *D-B* structure:

$$\begin{bmatrix} x_D \\ x_B \end{bmatrix} = \underline{\underline{G}}_M \begin{bmatrix} D \\ V \end{bmatrix}$$

$$= \underline{\underline{G}}_M \begin{bmatrix} 1 & 0 \\ \dfrac{-G_L}{1 - G_L} & \dfrac{-1}{1 - G_L} \end{bmatrix} \begin{bmatrix} D \\ B \end{bmatrix}$$

$$\begin{bmatrix} x_D \\ x_B \end{bmatrix} = \begin{bmatrix} G_{11} - \dfrac{G_{12}G_L}{1 - G_L} & \dfrac{-G_{12}}{1 - G_L} \\ G_{21} - \dfrac{G_{22}G_L}{1 - G_L} & \dfrac{-G_{22}}{1 - G_L} \end{bmatrix} \begin{bmatrix} D \\ B \end{bmatrix} \tag{22-7}$$

Note that at steady state, $s \to 0$ and $G_L \to 1$. Because each element in the transfer function matrix of Equation 22-7 contains the term $1 - G_L$, these transfer functions are indeterminant at zero frequency, which we know has to be true because we cannot obtain steady-state gains for them.

22-5-2 Example

As a specific example, consider a 150 tray column separating a 75 mol% propylene, 25 mol% propane feed into a 99 mol% distillate and a 2 mol% bottoms product. The *D-V* transfer functions are

$$\begin{bmatrix} x_D \\ x_B \end{bmatrix} = \begin{bmatrix} \dfrac{-37.7e^{-6s}}{(7200s+1)(2s+1)} & \dfrac{0.647e^{-6s}}{(159s+1)(2.86s+1)^2} \\ \dfrac{-43.8e^{-6s}}{(2338s+1)(12.7s+1)^2} & \dfrac{-1.71e^{-6s}}{(26s+1)(3.82s+1)^2} \end{bmatrix} \begin{bmatrix} D \\ V \end{bmatrix} \tag{22-8}$$

The tray liquid hydraulics are assumed to be a series of first order lags, one for each tray:

$$G_L = \frac{1}{(1 + \tau_L s)^{NT}} \tag{22-9}$$

where NT = total number of trays in the column = 150
τ_L = tray hydraulic time constant

The parameter τ_L can be calculated from the Francis weir formula by linearizing the total mass balance for the tray:

$$\tau_L = 1 \bigg/ \left(1200 \frac{w_L}{\pi D^2} \left[\frac{4\, m_n\, MW}{\pi D^2 \rho} - w_h \right]^{0.5} \right) \tag{22-10}$$

where w_L = weir length
D = column diameter
m_N = molar holdup on the nth tray
MW = molecular weight
ρ = density
w_h = weir height

The specific numerical example gives a τ_L of 4.5 s.

Figure 22-5 gives Bode plots for the G_{11} elements of both the *D-V* and the *D-B* transfer function matrices. They are almost identical at high frequencies, but at low frequencies the *D-B* transfer functions look like integrators (approaching infinite magnitude and $-90°$ phase angle).

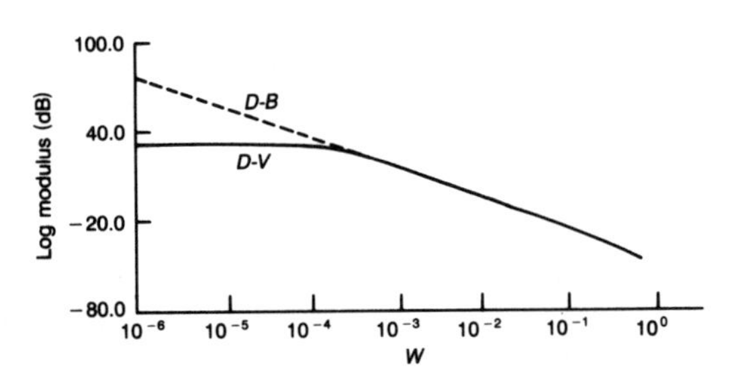

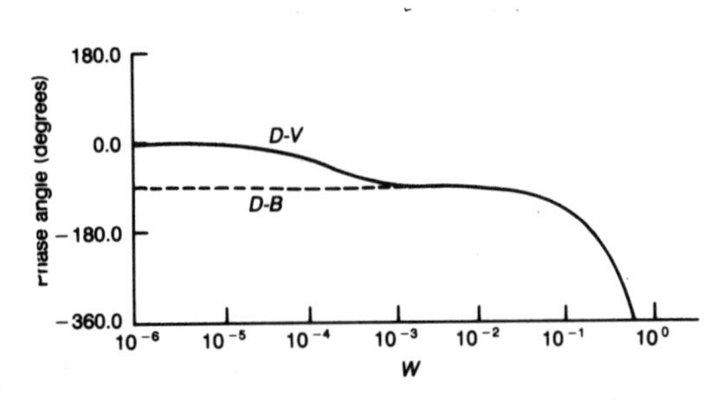

FIGURE 22-5. Bode plot of the (1, 1) element of the *D-V* (solid line) and (dashed line) *D-B* transfer function matrices.

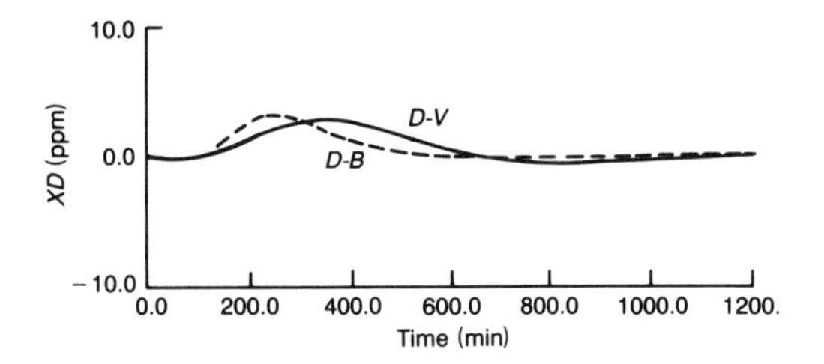

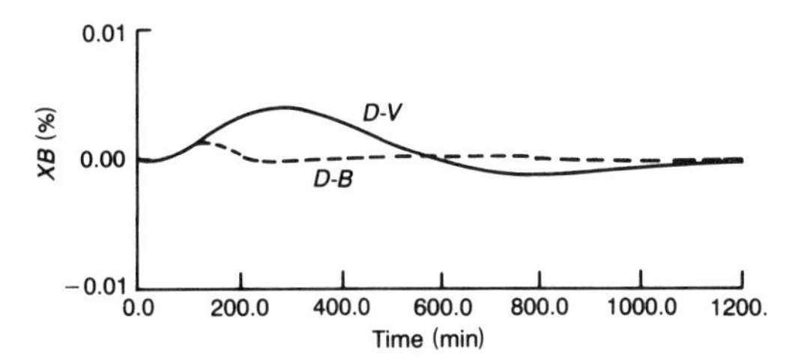

FIGURE 22-6. Responses of *D-V* (solid line) and *D-B* (dashed line) structures for a feed composition disturbance.

The BLT tuning constants for the two structures are:

	K_c	τ_I
D-V	$-7.38, -0.517$	66.3, 93.3
D-B	$-7.52, 13.8$	65.0, 91.6

Figure 22-6 compares responses for a step change in feed composition. These results confirm the results from the Sun Oil Column, where the controller tuning was done empirically. There is not too much difference in the control of x_D, but the control of x_B is much better with the *D-B* structure.

22-5-3 Fragility of *D-B* Structure

One important aspect of the *D-B* structure should be recognized. If any sensor fails or any valve saturates, the *D-B* structure cannot work. This is because distillate or bottoms flow rates cannot be held constant and still control any composition in the column. Thus the *D-B* structure degrades poorly or is "fragile."

This problem can be easily avoided by the use of an appropriately designed override control system that will switch the control structure in the event of an instrument failure or valve saturation. For example, suppose the base level is increasing and the steam valve opens in an attempt to bring the level back down. If the steam valve saturates wide open, the base level would continue to rise. A high-base-level override controller should take control of the bottoms control valve and begin to control base with bottoms flow, giving up bottoms composition control.

22-6 PITFALLS WITH RATIO SCHEMES

The schemes using the boilup ratio gave some unexpected and interesting open-loop dynamic responses (open-loop underdamped behavior). They also required the use of transformations to determine the open-loop plant transfer functions because some of them could not be determined by conventional identification methods. Details of the transformation methods, a theoretical analysis that explains the simulation results, and comparisons of performance are given in Papastathopoulou and Luyben (1990c). Here we will only illustrate some of the interesting dynamics that were observed.

Consider the *D-BR* structure shown in Figure 22-7. Distillate composition is controlled by distillate flow rate, and bottoms composition is controlled by changing the ratio of vapor boilup to bottoms flow rate. Base level changes both bottoms flow rate and steam flow rate through the ratio element.

Unusual open-loop dynamics were discovered when autotune tests were run. Figure 22-8 shows an attempt to perform an autotune on the x_D-*BR* loop. Although oscillation were established initially, they stopped after about 40 min because the distillate product drifted away from the setpoint.

Figure 22-9 gives the open-loop response of the column (level controllers on automatic but composition controllers on man-

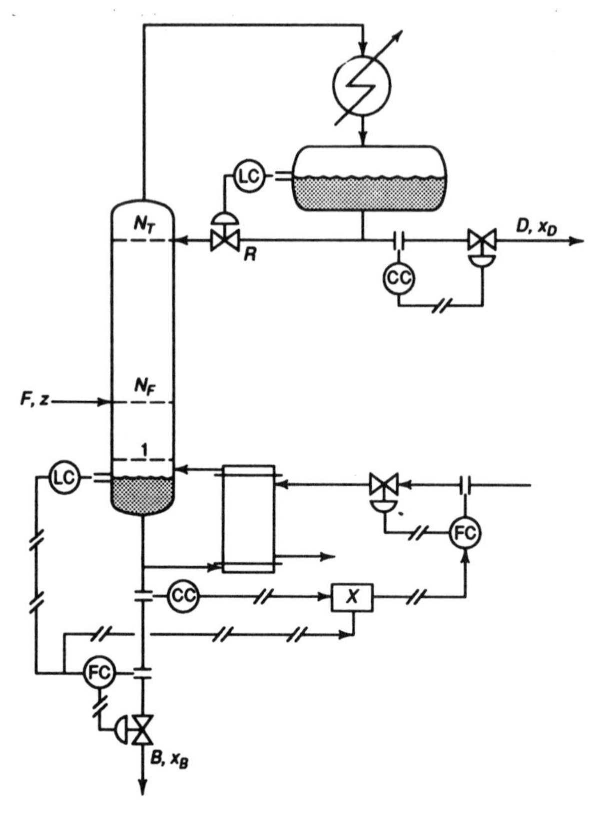

FIGURE 22-7. *D-BR* structure.

ual) for a feed flow rate disturbance. Note the unusual open-loop *underdamped* behavior. The presence of the boilup ratio controller in a tall column that has significant liquid hydraulic lags produces an oscillatory open-loop system. This can be explained by following through the sequence of events that occur following a rise in base level. The level controller increases bottoms flow rate and brings the base level back down.

But the increase in bottoms flow also will cause an increase in heat input. This vapor

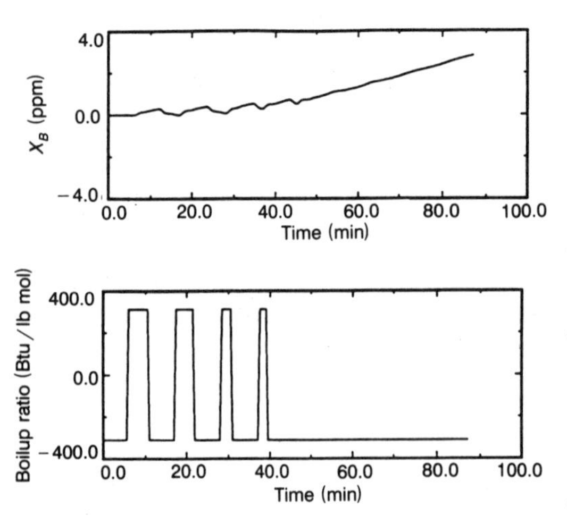

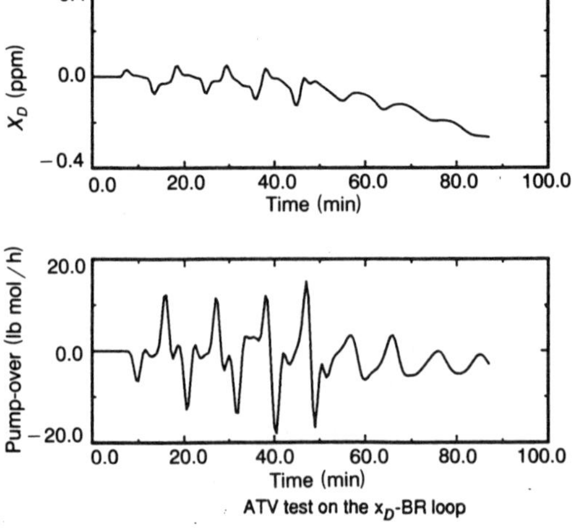

FIGURE 22-8. Autotune of x_D-*BR* loop.

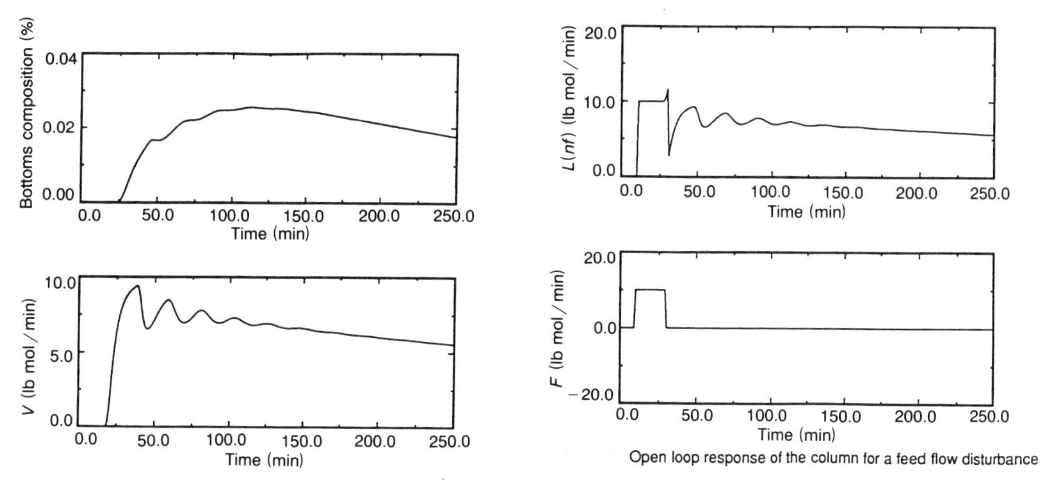

FIGURE 22-9. Open-loop response to feed flow rate disturbance.

quickly goes overhead and begins to increase reflux drum holdup. The reflux drum level controller then increases reflux. Remember, distillate flow rate is fixed because the composition controllers are on manual. After some significant time due to the liquid hydraulic lags, the increase in reflux will cause an increase in the liquid entering the base of the column. This causes another temporary increase in B and V, and the cycle repeats itself.

The open-loop damping coefficient becomes smaller as the number of trays in the column increases and as the tray hydraulic time constant increases.

This unusual dynamic behavior means that it is difficult to obtain the open-loop transfer functions and tune the controllers.

22-7 SUPERFRACTIONATOR WITH SIDESTREAM EXAMPLE

A second and somewhat more complex example is presented in the following text. This column produces a sidestream in addition to distillate and bottoms products. This means that there is one more controlled variable: sidestream composition. There is one more manipulated variable that is obvious: sidestream flow rate (S). There is an additional manipulated variable that is not so obvious: location of the sidestream drawoff tray (NS).

22-7-1 Process

The column studied was based on an actual column at the Exxon Chemical Baytown chemical plant. The column separates a 70 mol% propylene, 30 mol% propane feed into a 99.66 mol% propylene polymer-grade distillate product, a 94 mol% propylene chemical-grade liquid sidestream product, and 0.02 mol% propylene bottoms product that goes to fuel (LPG). The column has 190 trays and operates with a reflux ratio of 13.3. Because a vapor recompression system is used, the operating pressure in the column is 110 psia. This pressure is lower than would be required in a conventional column with a water-cooled condenser. Therefore, the relative volatility is increased, making the separation easier. Table 22-1 summarizes the steady-state design and operating parameters for the sidestream column, Figure 22-10 shows the system, and Table 22-2 gives the properties of the various streams numbered in Figure 22-10.

Rigorous, nonlinear steady-state and dynamic models were used. The tray efficiency in the steady-state model was adjusted until the steady-state model matched the plant

TABLE 22-1 Parameter Values for Sidestream Column

F, lb-mol/h	1000
S, lb-mol/h	196
D, lb-mol/h	535
z, mole fraction propylene	0.70
x_D, mole fraction propylene	0.995
x_S, mole fraction propylene	0.93
x_B, mole fraction propylene	0.0002
NT	200
NS	150
NF	60
P, psia	110
M_B, lb-mol	1060
M_D, lb-mol	780
M_n, lb-mol	29
R, lb-mol/h	7450
RR	13.3
d, ft	10
w_h, in.	1
w_L, ft	6.41

operating data. An efficiency of 95% was found.

22-7-2 Plant Dynamic Tests

The dynamic model was validated by plant tests. Several types of tests were conducted. Figure 22-11*a* compares the plant experimental data with the model predictions for a dual pulse on distillate flow. Figure 22-11*b* shows the same comparison for an autotune test on the x_D-D loop (see Chapter 7). It should be noted that the switching times in both the model and the plant were determined by x_D crossing the setpoint, that is, the input to the model was *not* the sequence of pulses in distillate flow found from the plant data. Thus the autotune test was a particularly sensitive evaluation of model fidelity. These results demonstrated that the dynamic mathematical model was accurate enough to use for comparative studies of alternative control structures.

22-7-3 Steady-State Analysis

The steady-state model was used to calculate steady-state gains for a large number of alternative control structures for the sidestream column. An efficient procedure was developed to determine these gains for any choice of manipulated variables with only three runs of the rating program (Papastathopoulou and Luyben 1990a).

The controlled variables were x_D, x_S, and x_B. A wide variety of manipulated variables were explored in the 3×3 multivariable system. The basic manipulated variables were D, S, and V or D, NS, and V. Ratio schemes were also included such as boilup ratio ($BR = V/B$), reflux ratio ($RR = R/D$), and sidestream-to-vapor ratio ($SV = S/V$).

Figure 22-12 shows the D-NS-V control structure. Three sidestream drawoff trays are used with five trays separating them ($NS_3 - NS_2 = NS_2 - NS_1 = \Delta NS = 5$). Liquid streams from two of the three trays are mixed together. The signal from the composition controller (CO) sets this split at any point in time.

TABLE 22-2 Stream Properties

Stream Number	Flow Rate (lb-mol/h)	Temperature (°F)	Pressure (psia)	Vapor Fraction
1	8913.6	47.8	108.9	1.0
2	7800.3	89.2	190.3	1.0
4	1113.3	105.8	223.9	1.0
6	195.7	100.1	223.9	0.0
7	917.6	100.1	223.9	0.0
8	917.6	47.8	108.9	0.248
10	7800.3	87.2	190.3	0.0
11	7800.3	47.8	108.9	0.179

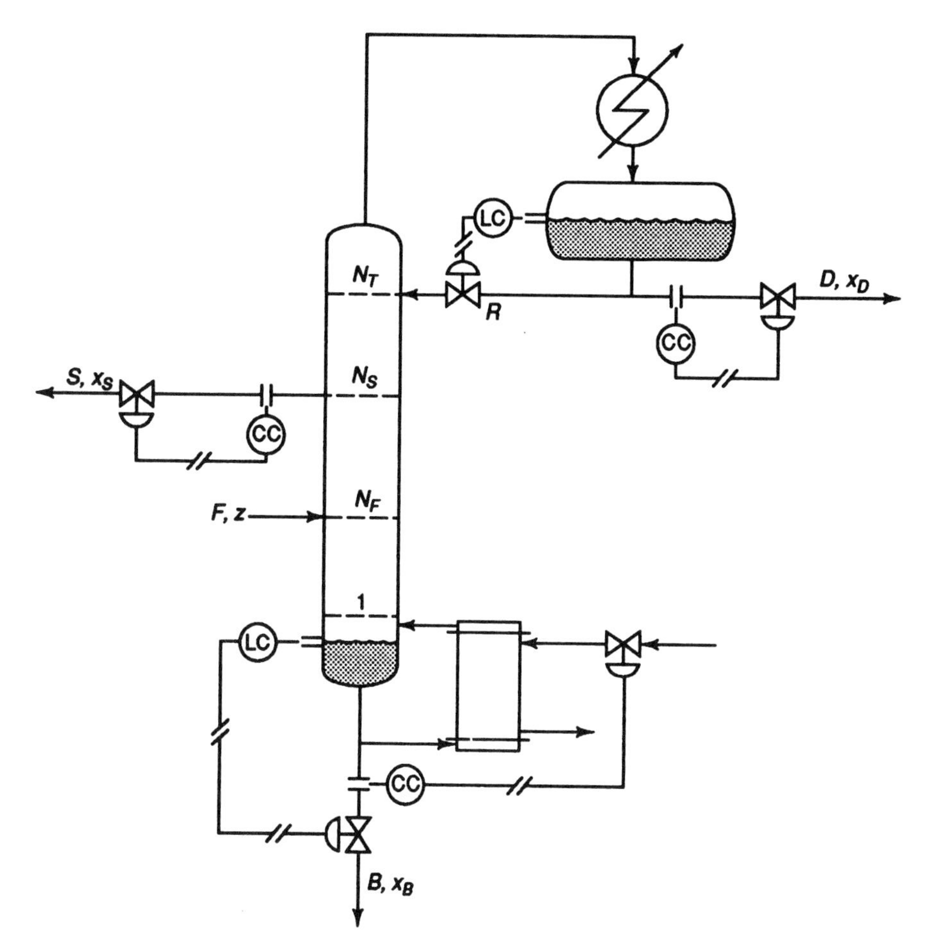

FIGURE 22-10. Sidestream column.

Several steady-state indices were calculated for the alternative control structures when all three compositions were controlled using various pairings and various choices of manipulated variables. The indices used are summarized as follows:

1. MRI: The Morari resiliency index is the minimum singular value of the steady-state gain matrix. The larger its value, the better the control.
2. RGA: The relative gain array is the Hadamard product of the steady-state gain matrix and its inverse. Diagonal elements (λ_{ii}) close to unity indicate little interaction.
3. NI: The Niederlinski index is the determinant of the steady-state gain matrix divided by the product of its diagonal elements. This index must be positive.
4. CN: The condition number is the ratio of the maximum singular value of the steady-state gain matrix to its minimum singular value. The smaller this number, the better the control.
5. JEC: The Jacobi eigenvalue criterion is the largest eigenvalue of the matrix A, where A satisfies the equation

$$K = D(I - A) \tag{22-11}$$

where K = steady-state gain matrix
D = a diagonal matrix whose elements are those of K
I = identity matrix

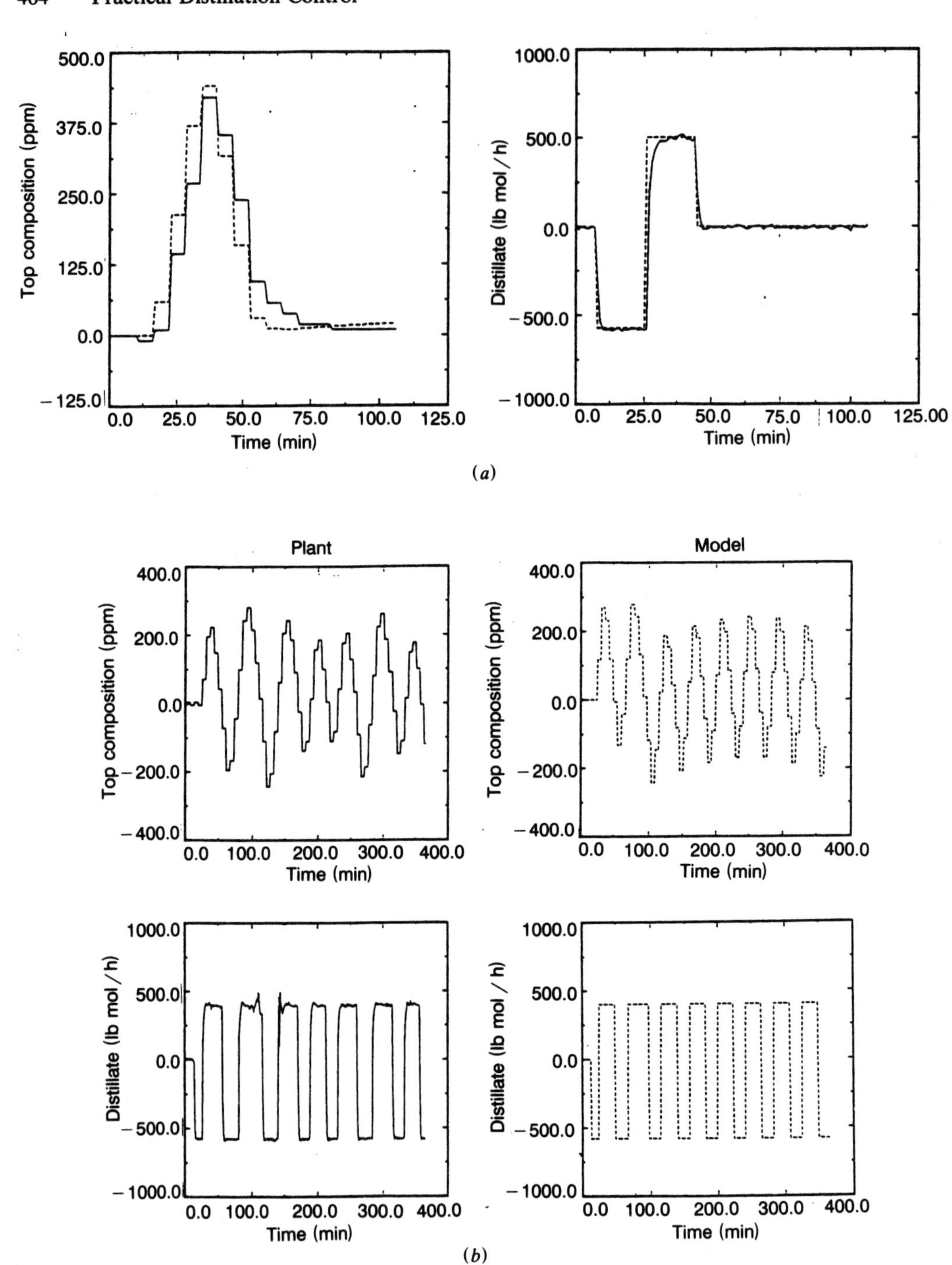

FIGURE 22-11. Plant tests compared to dynamic model predictions (Exxon column): (*a*) Dual pulse in distillate flow rate; (*b*) autotune test on x_D-D loop.

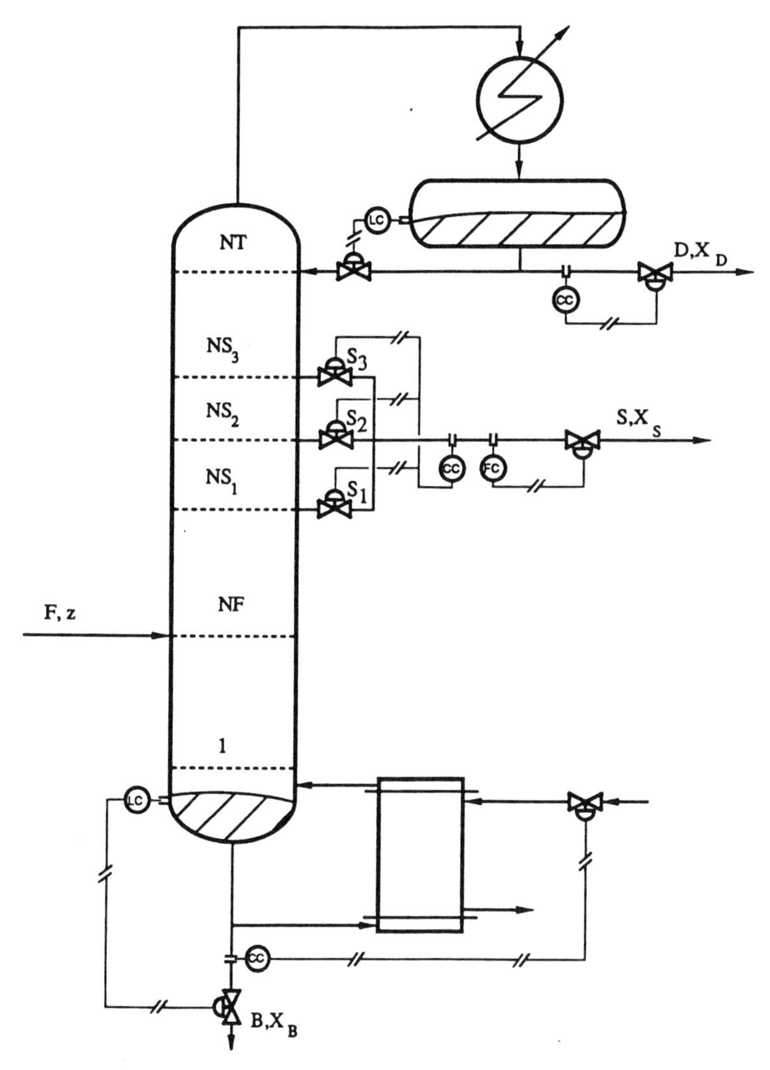

FIGURE 22-12. Manipulation of sidestream location (NS).

The values of JEC should be between -1 and 1, and the best pairing is that with the smallest JEC.

All schemes with negative NIs or RGAs were eliminated. All schemes with very large CNs were eliminated. This still left a large number of possible schemes. Heuristic dynamic considerations were used to reduce the number of cases. All schemes that use B to control sidestream composition or S (or NS) to control overhead composition were eliminated because these manipulated variables have no direct effect on these controlled variables. Of the remaining schemes, the D-S-V and the D-NS-V were selected for further evaluation using dynamic analysis.

One scheme that cannot be analyzed using the steady-state indices is the D-S-B structure because the steady-state gains are indeterminant. However, the transfer functions for this structure can be obtained at higher frequencies by transformations that are similar to those discuss earlier for the D-B structure.

22-7-4 Simulation Results

Tuning the schemes where sidestream drawoff location (NS) was used as a manipulated variable was initially thought to be straightforward because the blending of two streams should be essentially instantaneous except for the 6 min deadtime from the composition analyzer. Therefore we assumed that the transfer function relating x_S to the sidestream composition controller output signal CO would be just the steady-state gain and the 6 min deadtime.

However, when the sidestream composition controller was designed using this assumed transfer function, the closed-loop response of this loop (x_S-CO) was very oscillatory. After considerable investigation, the source of the problem was discovered. The open-loop response of x_S to a step change in CO shows a large instantaneous jump in x_S and a slow decay to the lower steady-state value. This is the type of response associated with a derivative unit. It results from the initial blending of the two streams with different compositions, followed by a slow shift in the compositions on the liquid drawoff trays.

Therefore, the open-loop transfer function relating x_S and CO was not just a deadtime and gain but had to contain a lead–lag element:

$$\frac{x_S}{\text{CO}} = \frac{0.076(325s + 1)e^{-6s}}{81s + 1} \tag{22-12}$$

Because the sidestream composition loop was much faster than the distillate and bottoms composition loops, it was tuned independently. A reset time equal to half the deadtime was selected ($\tau_I = 3$ min), and the controller gain was found that gave a +2 dB maximum closed-loop log modulus for this single-input–single-output loop ($K_c = 0.836$). With these settings held constant in the sidestream composition loop, the other two controllers were tuned using the BLT procedure, considering the entire 3×3 multivariable process and designing for +6 dB maximum closed-loop log modulus $L_c^{\max}$.

Figure 22-13 shows the dependence of the $L_c^{\max}$ on the detuning factor F_{BLT}. Note the very interesting multiple solutions, that is, there are three values of the detuning factor that give a +6 dB maximum closed-loop log modulus. This phenomenon has not been reported before in the literature. Naturally the desired solution is the one that gives the smallest detuning factor.

Figure 22-14 gives the closed-loop response of the nonlinear rigorous model of sidestream column for a 5% step decrease in feed composition using three control schemes: *D-S-V*, *D-S-B*, and *D-NS-V*. Note that the distillate compositions are plotted as parts per million deviations. All three schemes do a fairly good job in controlling the column, but the *D-NS-V* does the best job in holding sidestream composition constant. This structure also has the advantage of holding the sidestream flow rate constant. Therefore, if tight control of the sidestream is important, the *D-NS-V* scheme is recommended.

Other types of disturbances were introduced into the system (including feed rate and sidestream flow rate), and the *D-NS-V* structure was found to work well for all.

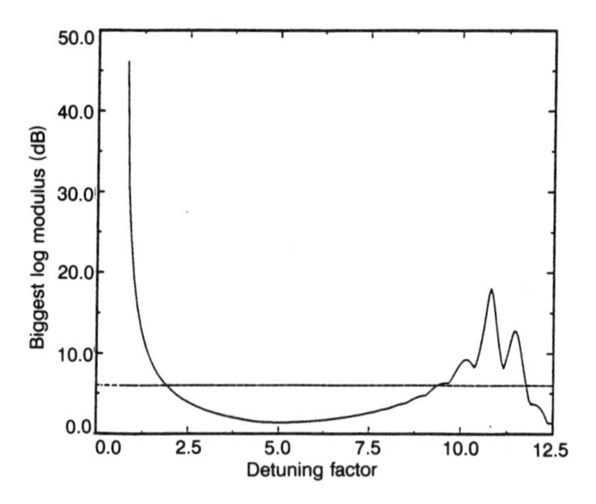

FIGURE 22-13. Maximum closed-loop log modulus versus detuning factor.

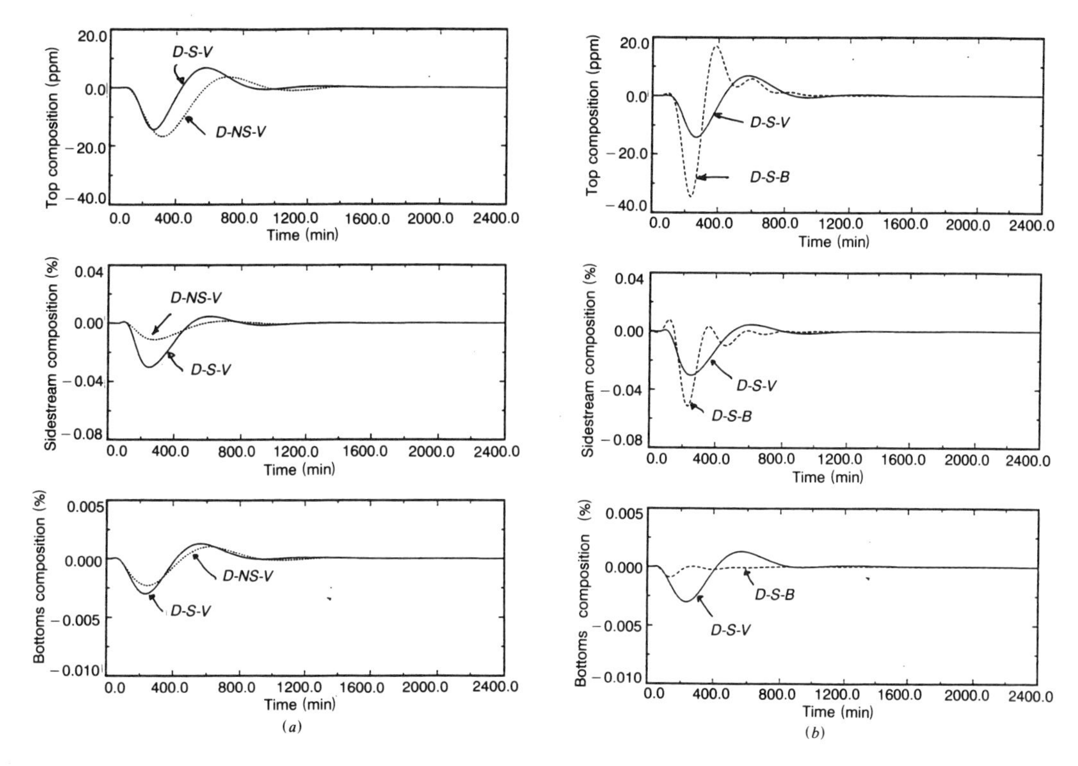

FIGURE 22-14. Responses of sidestream superfractionator with various controller structures.

The structure was also found to be robust for changes in the purity level of x_D.

22-8 CONCLUSION

This chapter has dealt with the problem of controlling distillation columns that have many trays and high reflux ratios. An unorthodox control structure (DB) is shown to provide effective control of this type of column because of the significant liquid hydraulic lags and the very large internal flow rates compared to the external flow rates.

Methods for tuning the DB structure have been reviewed. Some interesting control difficulties have been revealed when certain ratio control systems are used.

References

Finco, M. V., Luyben, W. L., and Polleck, R. E. (1989). Control of distillation columns with low relative volatilities. *Ind. Eng. Chem. Res.* 28, 75–83.

Papastathopoulou, H., and Luyben, W. L. (1990a). A new method for the derivation of steady-state gains for multivariable processes. *Ind. Eng. Chem. Res.* 29, 366–369.

Papastathopoulou, H., and Luyben, W. L. (1990b). Tuning controllers on distillation columns with the distillate-bottoms structure. *Ind. Eng. Chem. Res.* 29, 1859–1868.

Papastathopoulou, H., and Luyben, W. L. (1990c). Potential pitfalls in ratio control schemes. *Ind. Eng. Chem. Res.* 29, 2044–2053.

Skogestad, S., Jacobsen, E. W., and Morari, M. (1990). Inadequacy of steady-state analysis for feedback control: Distillate-bottoms control of distillation columns. *Ind. Eng. Chem. Res.* 29, 2339–2346.

23

Control of Vapor Recompression Distillation Columns

Cristian A. Muhrer

Air Product and Chemicals, Inc.

23-1 INTRODUCTION

This chapter examines the dynamics and control of direct vapor recompression columns. Two specific, industrially important separations are looked at in detail. These are the ethanol–water system and a hydrocarbon system of commercial interest, such as might be found in a petroleum refinery (e.g., a propylene–propane splitter). For both systems there are substantial economic incentives for using vapor recompression.

Distillation consumes a large percentage of the energy used in the chemical process industries. Consequently, there is a significant incentive to improve the energy efficiency of this widely applied separation process. Vapor recompression is one technique that can be applied in some separation systems (Muhrer, Collura, and Luyben, 1990).

Vapor recompression has been discussed in the literature for almost half a century (Robinson and Gilliland, 1950). However, heat pump systems in distillation have been accepted by industry only in recent years (Meili, 1987) because of the sharp rise in energy costs in the 1970s.

The key concept of vapor recompression (VRC) is to raise the temperature level of a fluid by compression. Vapor recompression designs are typically classified by the fluid. Figure 23-1 shows a system that uses an external fluid (refrigerant). When either the top or bottom column fluid is suitable for compression, we can use direct vapor recompression or reboiler liquid flashing, respectively. These configurations are shown in Figures 23-2 and 23-3. Most applications use direct vapor recompression: process vapor from the top of the column is compressed to a high enough pressure so that it can be condensed in the reboiler-condenser in (or somewhere near) the base of the column. From now on, we will limit the discussion to only direct vapor recompression.

Most of the literature on vapor recompression focuses on the steady-state economic aspects (capital costs, operating costs, and optimum steady-state operating conditions): Null (1976), Mostafa (1981), Quadri (1981), Brousse, Claudel, and Jallut (1985), Meili and Stuecheli (1987), Ferre, Castells, and Flores (1985), Collura and Luyben (1988). These steady-state designs have shown that vapor recompression is economical in those separations where the following conditions apply:

- Heat is not available from other process sources (heat integration).

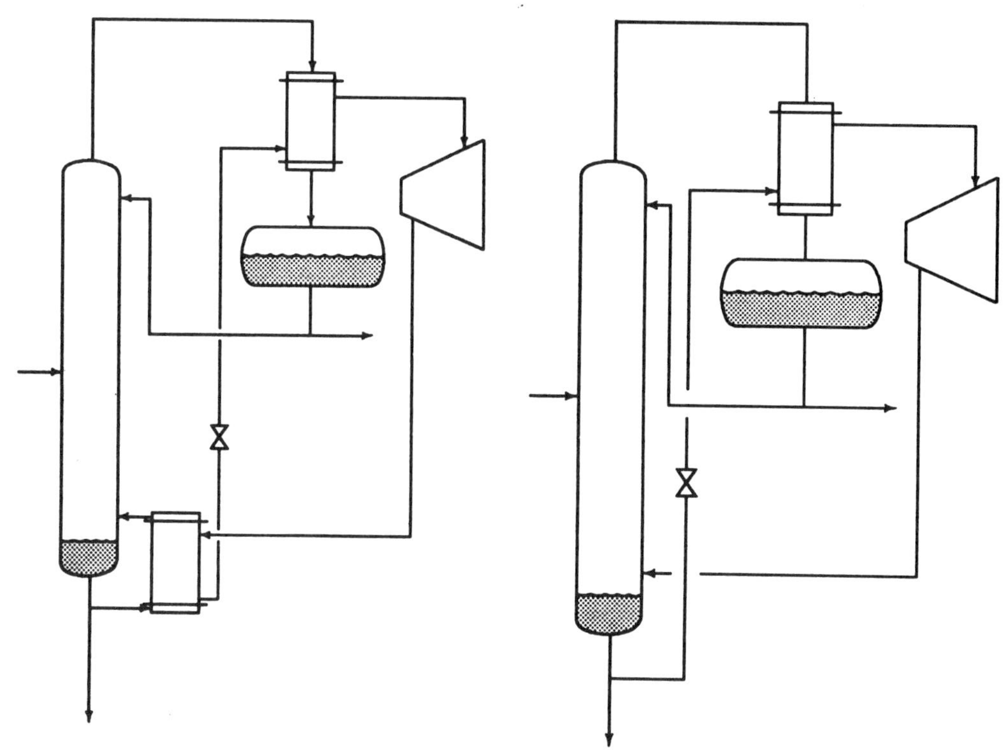

FIGURE 23-1. VRC with external fluid.

FIGURE 23-3. VRC with compression of flashed bottom vapor.

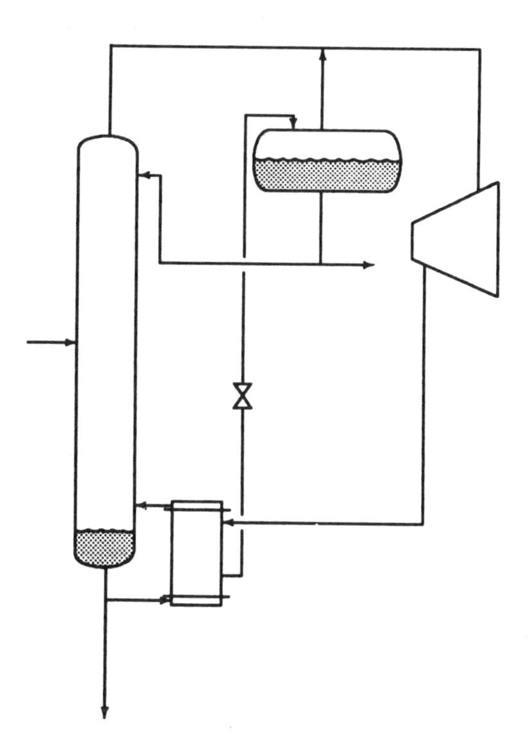

FIGURE 23-2. VRC with compression of overhead vapor.

- Low-temperature operation requires refrigeration.
- Temperatures are not too high (compressor thermal limitations).
- Pressures are not too low (large vapor volumes).
- The temperature difference between the top and the bottom of the column is small (moderate compression ratios).

The last item is probably the most restrictive. It limits vapor recompression to separations that have small boiling point differences between the components. Thus, most practical applications of vapor recompression occur in binary systems where the separation is difficult (low relative volatility).

23-2 DESIGN

The choice of column operating pressure on the design of the vapor recompression col-

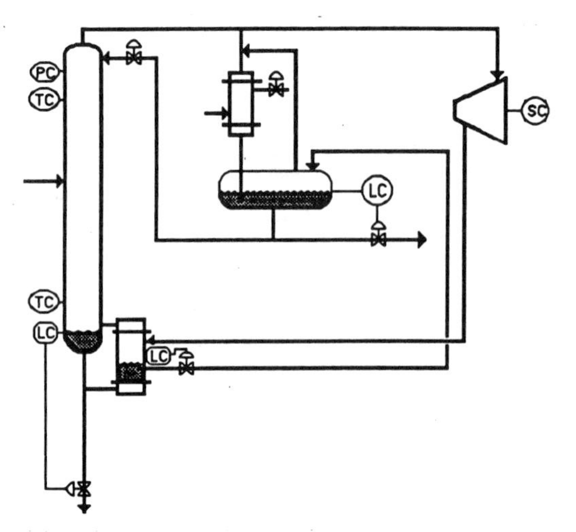

FIGURE 23-4. VRC design with low pressure reflux drum.

umn is significant. Figure 23-4 shows a column in which the reflux drum operates at the same pressure as the column. This scheme is used when the column operates at a high enough top temperature so that cooling water can be used in the trim condenser. This configuration minimizes compressor work because only the amount of vapor needed in the reboiler passes through the compressor.

Lowering the column pressure reduces energy consumption due to the increase in relative volatility. At some point, however, the reflux drum temperature will become too low to allow condensation with water. Figure 23-5 shows a configuration in which the reflux drum operates at the compressor discharge pressure. Now all the vapor from the top of the column passes through the compressor.

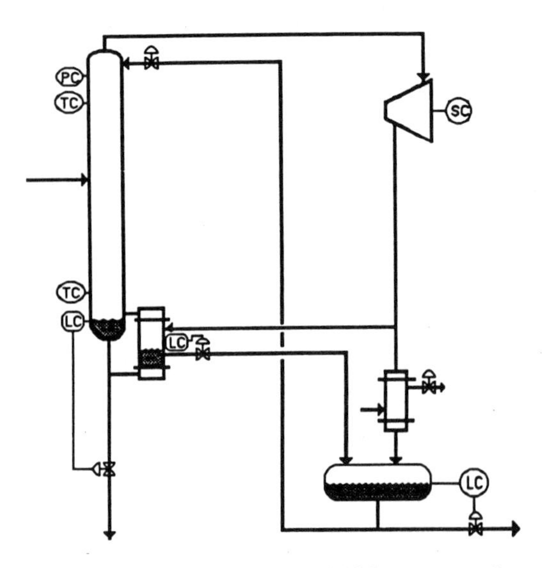

FIGURE 23-5. VRC design with high pressure reflux drum.

23-3 DYNAMICS AND CONTROL

The literature covering the dynamics and control aspects of vapor recompression in distillation is limited. Mosler (1974) and Quadri (1981) give qualitative discussions of control schemes for maintaining column inventories. Nielsen et al. (1988) reported the use of an adaptive multivariable controller in a pilot-plant column with external refrigerant. The system separated methanol–2-propanol. Dual composition control was successfully implemented using reflux and high pressure in the external heat pump circuit as the manipulated variables.

The main difference between vapor recompression and conventional distillation is the way in which energy is added to and removed from the column. In a conventional steam-heated, water-cooled distillation column, the dynamic changes in energy addition (reboiler duty) or removal (condenser duty) can be varied independently. In addition, the dynamics of the reboiler and condenser are typically very fast compared to the dynamics of the column.

In a vapor recompression column, energy addition and removal are linked together in the reboiler-condenser. The trim cooler does allow for some independent energy removal, but the quantity of energy removed is small compared to the total amount of energy flowing through the column.

The behavior of pressure and boilup also appears to be different. In a conventional column, pressure dynamics are normally fast and can be controlled by manipulating condenser duty. Vapor boilup can be directly and independently manipulated by heat input to the reboiler.

In a vapor recompression column, the effect of compressor speed on pressure seems to be complex. For example, one would expect an increase in compressor speed to increase the flow through the compressor. This should pull more vapor from the column and decrease the pressure in the column. However, the same increase in speed will increase the heat added to the column, and thus increase vapor rates. The latter, might produce an increase in column pressure.

Despite the complexity added by the vapor recompression design, the process does not require a complex control structure (Muhrer, 1989). We can use the conventional column control strategy in the corresponding vapor recompression column by simply replacing heat input control with compressor control. The pressure loop is faster than the composition loops, so we can handle it independently. For this to be valid, two conditions are necessary:

1. The composition time constants of the column must be at least 5 times larger than the pressure time constant.
2. The time constants of the reboilers of the conventional and vapor recompression columns should be approximately the same.

In terms of a distillation design parameter, the first condition corresponds to having a temperature difference between the top and the bottom of the column that is no greater than approximately 60°F, which represents the upper limit for vapor recompression under today's economic conditions.

The inherent composition dynamics are primarily determined by the holdup of material in the column. Because almost all systems considered for vapor recompression represent difficult separations (low relative volatility), the column will always have a large number of trays, which results in slow composition dynamics. Composition control loop dynamics will also be slowed by composition sensors (typically chromatographs) because it is often not possible to use temperatures to infer compositions due to the flat temperature profile.

Pressure dynamics are fast in almost all distillation columns because the vapor residence time (time constant) of the column is small. Changes in the rate of vapor boilup or condensation rapidly affect pressure.

The second condition means that if the dynamics of the reboiler are fast in the conventional columns, as is usually the case, the dynamics of the reboiler in the vapor recompression column must also be fairly fast. This is indeed the typical situation in most vapor recompression systems. Vapor holdup in the compressor is relatively small compared to the vapor holdup in the column (ratio 1/20); thus, changes in compressor speed (or other compressor control variables) rapidly affect the vapor rates in the column, emulating the effect of steam in conventional columns.

These conditions normally exist in most vapor recompression distillation systems for reasons discussed in the preceding text. Thus, I believe this is a generic solution to the control of vapor recompression columns.

23-4 ALTERNATIVE COMPRESSOR CONTROL SYSTEMS

The control strategies for vapor recompression towers presented so far have been designed with a variable-speed compressor. If the compressor is driven by a steam turbine, changing speed is straightforward. However, variable-speed electric motor drives are expensive. Compressor control can be accomplished in several other ways: suction throttling, bypassing flow back into the compressor suction, or varying the heat-transfer area in the reboiler-condenser using a

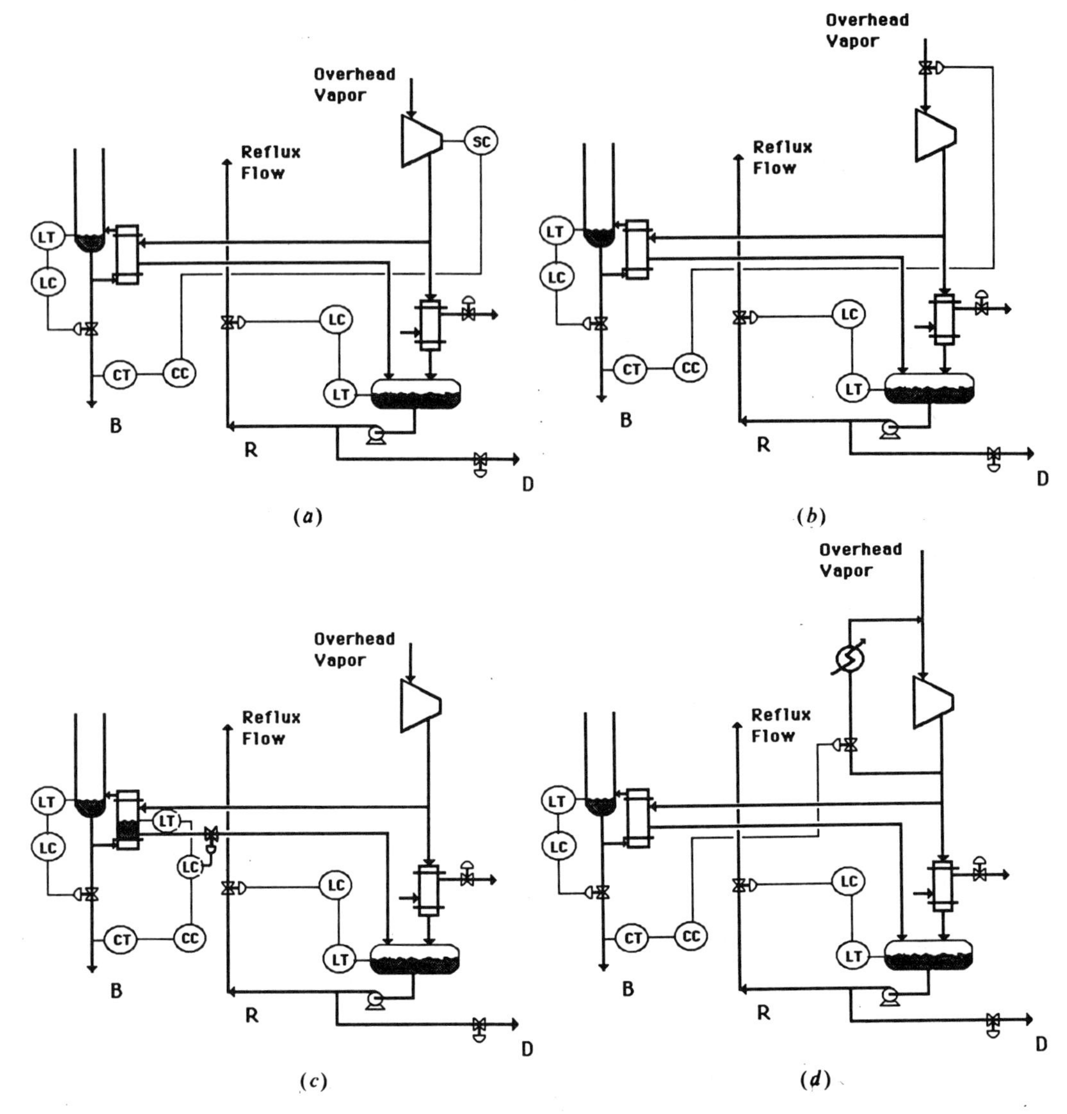

FIGURE 23-6. Compressor control alternatives. (*a*) Variable speed; (*b*) suction throttling; (*c*) variable heat-transfer area; (*d*) bypassing.

flooded reboiler design (condensate throttling).

In this section, these alternatives will be compared in terms of both steady-state design (energy consumption) and dynamic controllability. The compressor control regulates heat input to the column, thus controlling bottoms composition. Figure 23-6 shows control schemes for bottoms composition control by using four different compressor control structures.

23-4-1 Compressor Performance Curves

The performance of a centrifugal compressor cannot be exactly predicted from theoretical analysis. Instead it must be experimentally determined and reported in the form of performance curves. A set of performance curves is shown in Figure 23-7*a*. The stability limit or "surge line" denotes the minimum flow at which the compressor can operate at a given speed. Operation below the surge point could cause violent vibration to the point of permanent damage.

A typical arrangement for a centrifugal compressor with speed control is shown in Figure 23-7*b*. An antisurge control loop (Shinskey, 1979) is designed to adjust the bypass flow in order to keep the flow rate through the compressor above the surge value. The square of the surge flow is approximately linear with the pressure drop across the compressor. This relation sup-

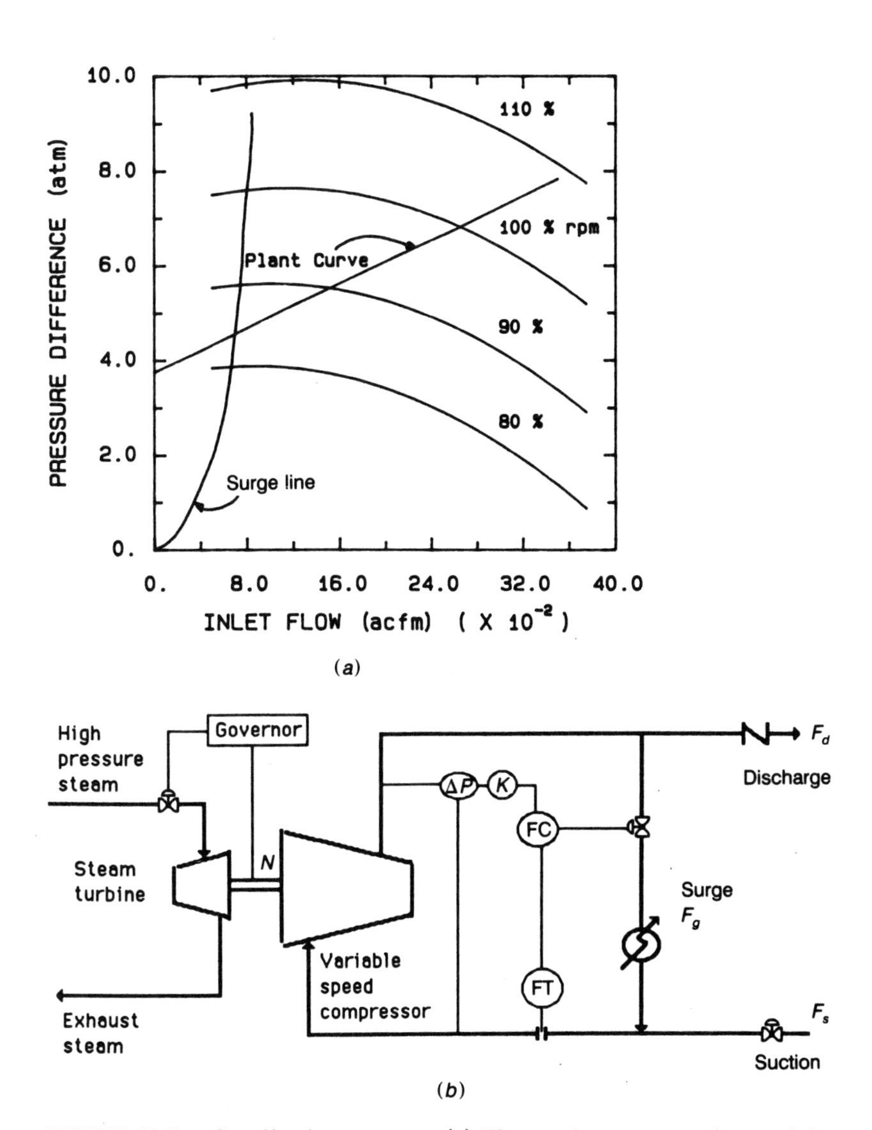

FIGURE 23-7. Centrifugal compressor. (*a*) Plant and compressor characteristic curves. (*b*) Antisurge control scheme.

plies the setpoint of the antisurge controller, which measures compressor inlet flow.

23-4-2 Plant Characteristic Curve

We need the plant characteristic curve to analyze the effect of the compressor control alternatives and the compressor characteristic curves on the process. Figure 23-7*a* gives the compressor characteristic curves as well as the plant characteristic curve for a propylene–propane vapor recompression system. This system is examined in detail in Section 23-5. It is interesting to note that the plant curve displays a unique, almost linear relationship between head and flow. This is in sharp contrast to the typical quadratic plant characteristic curve that is usually found in compressor applications. These normal plant curves reflect the second-order relationship between pressure drop through a system and flow that is typical for compression systems. The usual plant curves start at the origin: The head is zero when the flow is zero.

However, in vapor recompression distillation applications, the plant characteristic curve is strikingly different. Not only is it linear, as opposed to quadratic, but it also does not go to zero pressure drop as the flow goes to zero.

This unique behavior can be explained by the fact that the plant curve is determined by heat addition and removal in the reboiler-condenser, which is given by

$$Q_R = UA(T_R - T_B) \qquad (23\text{-}1)$$

where Q_R = heat-transfer rate in reboiler-condenser (Btu/h)
A = heat-transfer area (ft^2)
U = overall heat-transfer coefficient (Btu/h ft^2 °F)
T_R = saturation temperature (°F) of distillate product (mostly propylene) at discharge pressure
T_B = bottoms temperature (°F), which is the saturation temperature of bottoms product (mostly propane) at base pressure

From the compressor discharge side, the heat is given by

$$Q_R = 60\lambda\rho V \qquad (23\text{-}2)$$

where λ = heat of vaporization of distillate product (Btu/lb_m)
ρ = vapor density (lb_m/ft^3)
V = volumetric flow rate of vapor to reboiler (ft^3/min)

For the small range over which discharge pressure changes, a linear relationship between saturation pressure and temperature for the distillate product can be assumed:

$$T_R = \alpha + \theta P_D \qquad (23\text{-}3)$$

where α and θ are constants. Pressure has units of atmospheres and temperature has units of degrees Fahrenheit. Then the discharge pressure can be calculated from

$$P_D = \frac{60\lambda\rho}{UA\theta}V + \frac{T_B - \alpha}{\theta} \qquad (23\text{-}4)$$

So the pressure differential across the compressor must be

$$\Delta P = P_D - P_s = \frac{60\lambda\rho}{UA\theta}V + \frac{T_B - \alpha}{\theta} - P_s \qquad (23\text{-}5)$$

where P_s is the compressor suction pressure or the column pressure.

Equation 23-5 is the equation of a straight line:

$$\Delta P = mV + b \qquad (23\text{-}6)$$

where m and b are both constants:

$$m = \frac{60\lambda\rho}{UA\theta} \tag{23-7}$$

$$b = \frac{T_B - \alpha}{\theta} - P_s \tag{23-8}$$

This explains the linear shape of the plant characteristic curve. At the zero flow condition, the temperature differential driving force in the reboiler-condenser must be zero. That means that the temperature of the condensing distillate product must be equal to the temperature of the boiling bottoms product. Because the bottoms is less volatile than the distillate, the pressure in the column (determined by bottoms composition and temperature) will be less than the compressor discharge pressure (determined by distillate composition and temperature).

This can be seen mathematically by looking at the intercept of the linear relationship given in Equation 23-6:

$$\lim_{V \to 0} \Delta P = b = \frac{T_B - \alpha}{\theta} - P_s = P_{d(T_B)} - P_s \tag{23-9}$$

where $P_{d(T_B)}$ = pressure that the distillate product (mostly propylene) exerts at the base temperature T_B

P_s = column pressure

If we neglect the pressure drop through the column (which will be small at zero vapor boilup), the column pressure will be the pressure $P_{s(T_B)}$ that the bottoms product (mostly propane) exerts at temperature T_B. Distillate product is more volatile than bottoms product, so it will exert a higher pressure $P_{d(T_B)}$ at the same temperature. Therefore, $P_d > P_s$ and a pressure differential exists at zero flow.

Assuming pure propylene and propane products,

$$\lim_{V \to 0} [\Delta_P] = [P^s_{\text{propylene}}]_{T_B} - [P^s_{\text{propane}}]_{T_B} \tag{23-10}$$

This type of almost linear plant characteristic curve should be found in most vapor recompression distillation systems. For the propylene–propane system, the values for m and b are 0.0014 min atm/ft^3 and 3.72 atm. These values compare very well with the values obtained by fitting data from a rigorous rating program: 0.0012 and 3.75.

23-4-3 Description of Alternative Compressor Controls

Variable Speed

Compressor speed is adjusted to compensate for process disturbances. Figure 23-8*a* shows how the compressor speed must change to maintain compositions when additional heat is required by the column. The discharge pressure must increase to provide the higher temperature differential driving force in the reboiler and the flow must increase. This is achieved by moving the operating point to a higher speed curve.

Suction Throttling

The suction throttling strategy adjusts compressor suction pressure to compensate for process disturbances. Figure 23-8*b* shows the case when additional heat is required by the column. The operating point moves along a constant speed curve to a higher flow by reducing the pressure difference across the compressor. The higher temperature differential driving force in the reboiler-condenser increases the discharge pressure. The increased opening of the suction valve decreases the pressure drop over the valve and increases suction pressure enough to provide the needed increase in both discharge pressure and flow. The pressure differential across the compressor goes down.

Variable Heat-Transfer Area

The heat-transfer area of the reboiler-condenser is adjusted to compensate for process disturbances. As shown in Figure 23-8*c*,

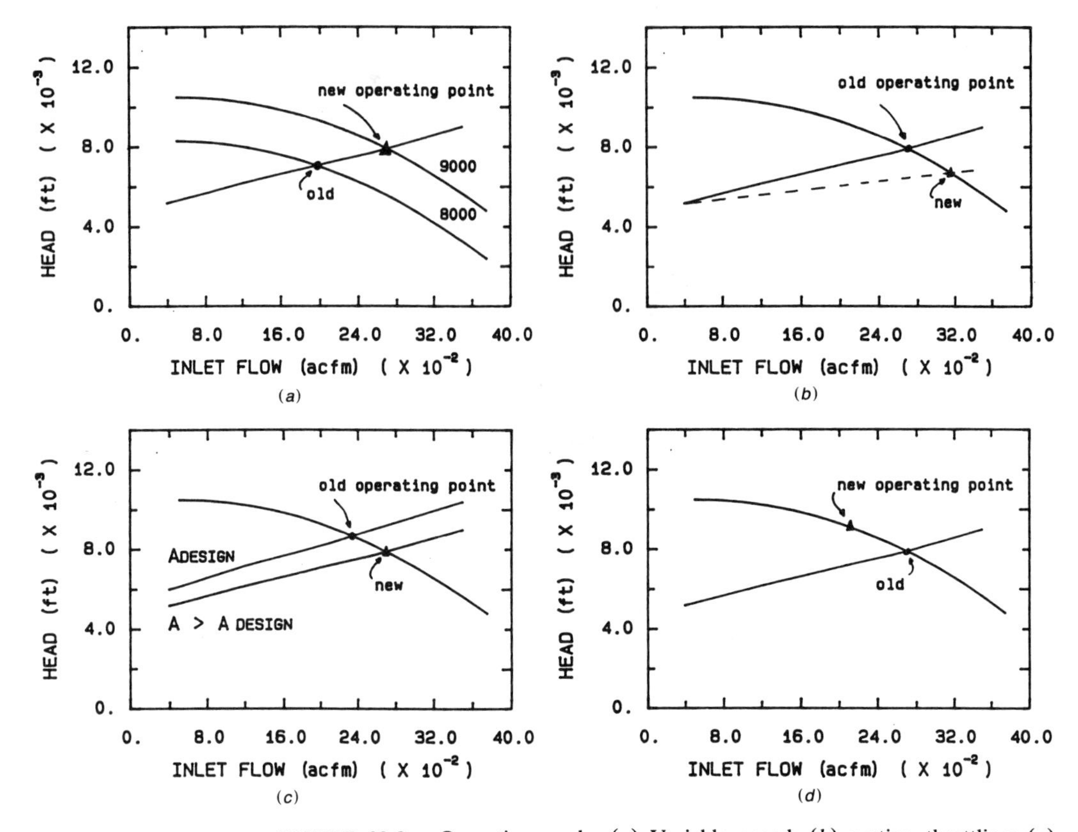

FIGURE 23-8. Operation mode. (*a*) Variable speed; (*b*) suction throttling; (*c*) variable heat-transfer area; (*d*) bypassing.

the heat input to the column is regulated by changing the setpoint of the level controller in the reboiler-condenser, thus varying the heat-transfer area available.

This flooded-condenser design changes the plant characteristic curve. Figure 23-8*c* shows how the operating point of the compressor moves to a higher flow rate by increasing the heat-transfer area when the column requires more heat. The new operating point is farther down the constant speed curve, allowing more flow to go through the compressor. The compressor head is lower because increasing the available heat-transfer area decreases the temperature differential driving force in the reboiler-condenser. This means that the compressor discharge pressure decreases.

At the design operating point, the process conditions are the same as in the variable-speed compressor case, but a larger reboiler-condenser must be used to make it possible to handle disturbances.

Bypassing (Spill Back or Recycle)

The bypass control scheme operates with more flow through the compressor than is necessary to meet the column heat input requirements. The excess flow is recycled to the compressor suction. Figure 23-8*d* shows that to increase the heat input to the column the bypass flow must decrease, causing a corresponding increase in the compressor head. The heat input increases due to the higher temperature differential driving force in the reboiler-condenser. The net effect is increased flow through the reboiler, less flow through the compressor, and much less flow through the bypass.

This configuration requires operation with some bypass flow to provide a reasonable capability to handle process disturbances. This extra flow causes a dramatic increase in energy consumption in comparison with the other alternatives.

23-4-4 Dynamic Performance

For the propylene–propane system, Figure 23-9 shows the closed-loop response of the four different compressor control schemes for a step change in feed composition from 0.6 to 0.7. Both the controlled variable (bottoms composition) and the four manipulated variables corresponding to the control scheme used are presented (speed, suction pressure, liquid height in the reboiler, and bypass flow).

The results show that all of the compressor control alternatives perform well. They differ in control action effort and energy consumption. Variable-speed and suction throttling require smaller adjustments in the manipulated variable than bypassing or variable heat-transfer area. Variable heat-transfer area is inherently slower than the other alternatives because of the liquid dynamics in the heat exchanger. For the propylene–propane system, this difference is not appreciable due to the slow composition responses. This should be true for most vapor recompression systems because of the relatively slow composition dynamics typical of the separations for which vapor recompression is economical.

Figure 23-10 compares the energy consumptions of the compressor control alternatives over a range of feed compositions. Bypass control is the simplest alternative but requires much more energy than the others. Variable heat-exchanger area could be advantageous when feed composition disturbances are usually over the design value because energy consumption is less sensitive than in the variable-speed or suction throttling cases. For a plant that operates over a wide range of feed rates and compositions, it would be advisable to have

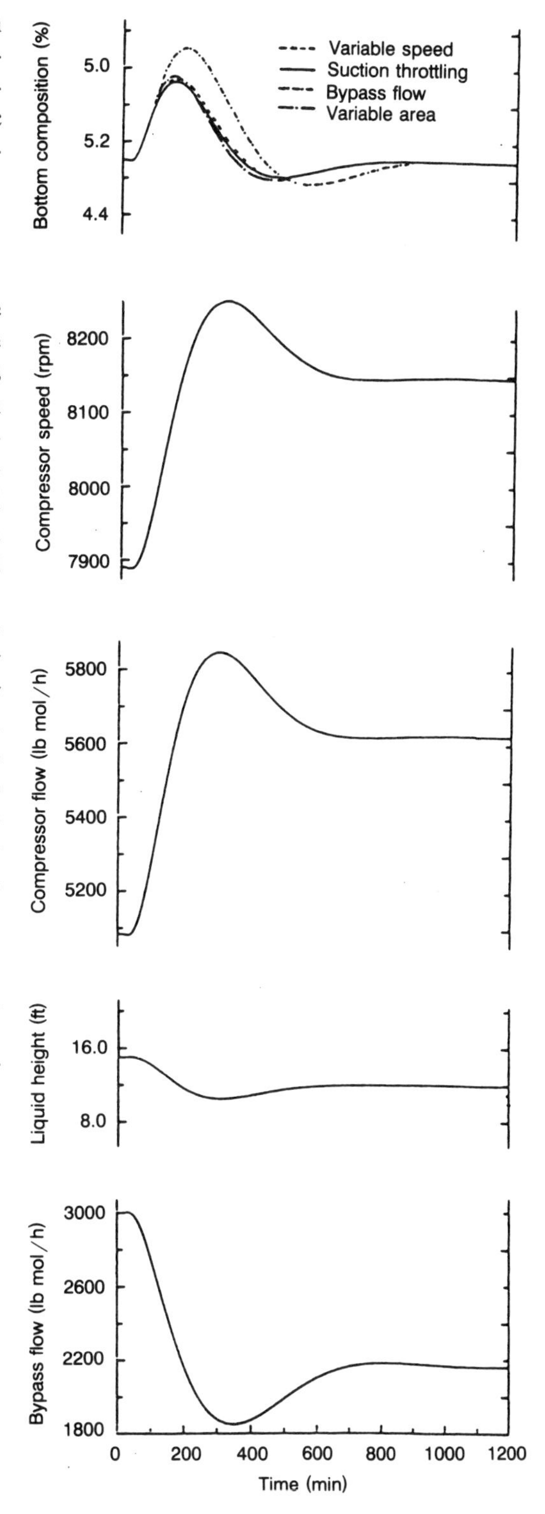

FIGURE 23-9. Comparison of dynamic performances of compressor control alternatives.

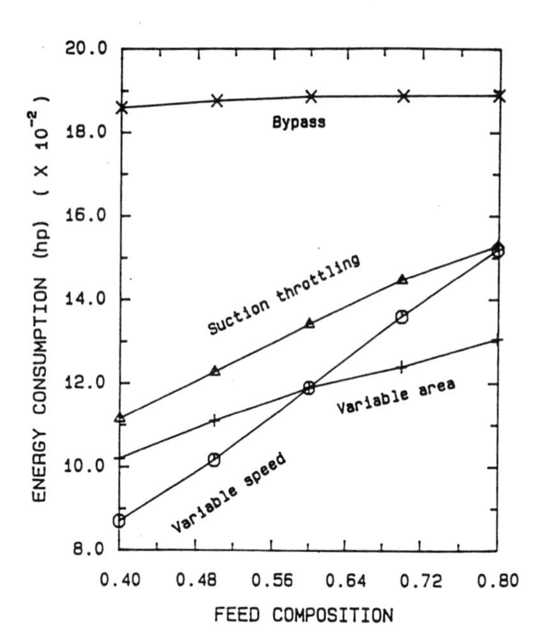

FIGURE 23-10. Comparison of energy consumptions of compressor control alternatives.

a variable-speed compressor. However, in practice, the selection of the compressor control scheme used depends on many factors of the specific application: availability of high-pressure steam for turbine drives, variability of process operating conditions, and so forth.

23-5 CASE STUDIES

In this section, we will examine two specific industrial separations: ethanol–water and propylene–propane. There are substantial economic incentives for using vapor recompression in both of these systems, but particularly in the propylene–propane system, as will be shown later. These two systems differ in many ways and span the region in which most practical applications of vapor recompression will lie.

Pressure The ethanol–water system operates at atmospheric pressure. The propylene–propane system operates at 18 atm for a conventional design and at 10 atm for a vapor recompression design.

Vapor–Liquid Equilibrium The ethanol–water mixture has very nonideal liquid behavior and ideal vapor-phase behavior because of the low operating pressure. The propylene–propane mixture behaves as an ideal liquid, but its vapor is nonideal because of the high pressure.

Vapor Holdup Because of the low pressure, the vapor holdup of the ethanol–water system is negligible. Vapor holdup must be considered in the propylene–propane system.

Reflux Drum Pressure A low-pressure reflux drum can be used with ethanol–water; a high pressure reflux drum must be used with propylene–propane.

Temperature Difference across Column The ΔT between the top of the column and the bottom of the column is 62 and 22°F for the ethanol–water and propylene–propane systems, respectively.

The ΔT has a major impact on the economic attractiveness of vapor recompression systems because a large ΔT implies a high compression ratio and increased compressor power. Vapor recompression is generally considered to be economically viable (late 1980s energy costs) for separations that have ΔTs that are about 60°F or less. In light of this, the propylene–propane and ethanol–water systems can be seen as approximating the boundary conditions of the region where economic incentives exist.

23-5-1 Steady-State Design

Ethanol–Water System

The design and optimum operating parameters for a conventional distillation column at the base case conditions are shown in Figure 23-11*a*. The column contains 78 trays (40% efficiency in the stripping section and 50% in the rectifying section). The reflux ratio (1.75) is 1.1 times the minimum. The energy consumption for this design is 117.4 $\times 10^6$ Btu/h (120,000 lb/h of 50 psia steam). Feed is preheated by overhead vapor and bottoms product.

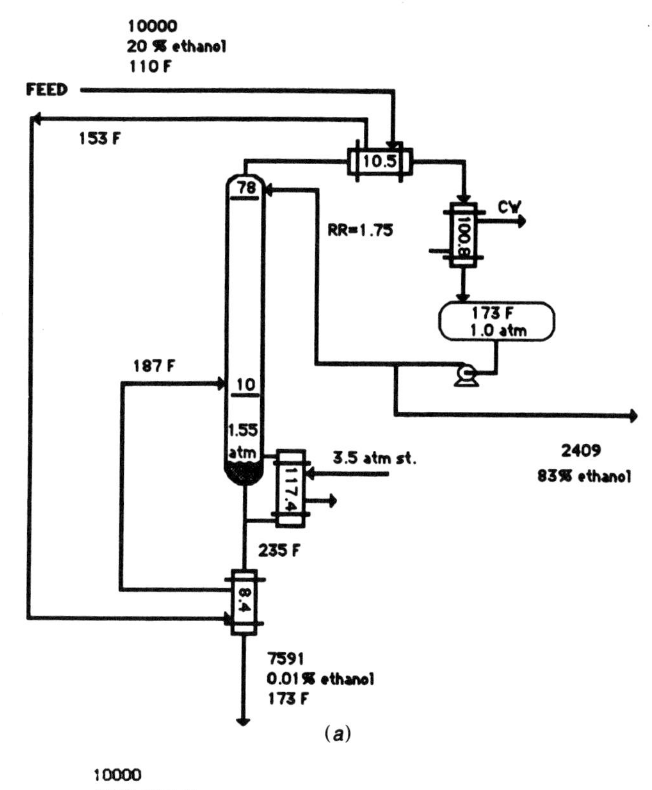

(*a*)

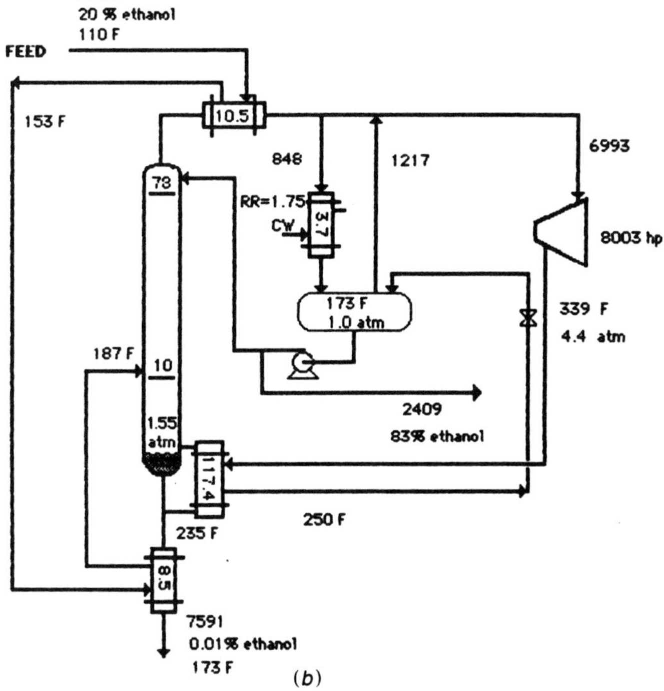

(*b*)

FIGURE 23-11. Ethanol–water column design. (*a*) Conventional column. (*b*) Vapor recompression column (numbers inside heat exchangers are 10^6 Btu/h; numbers without units along lines are flow rates in pound-moles per hour; compositions are in mole percent).

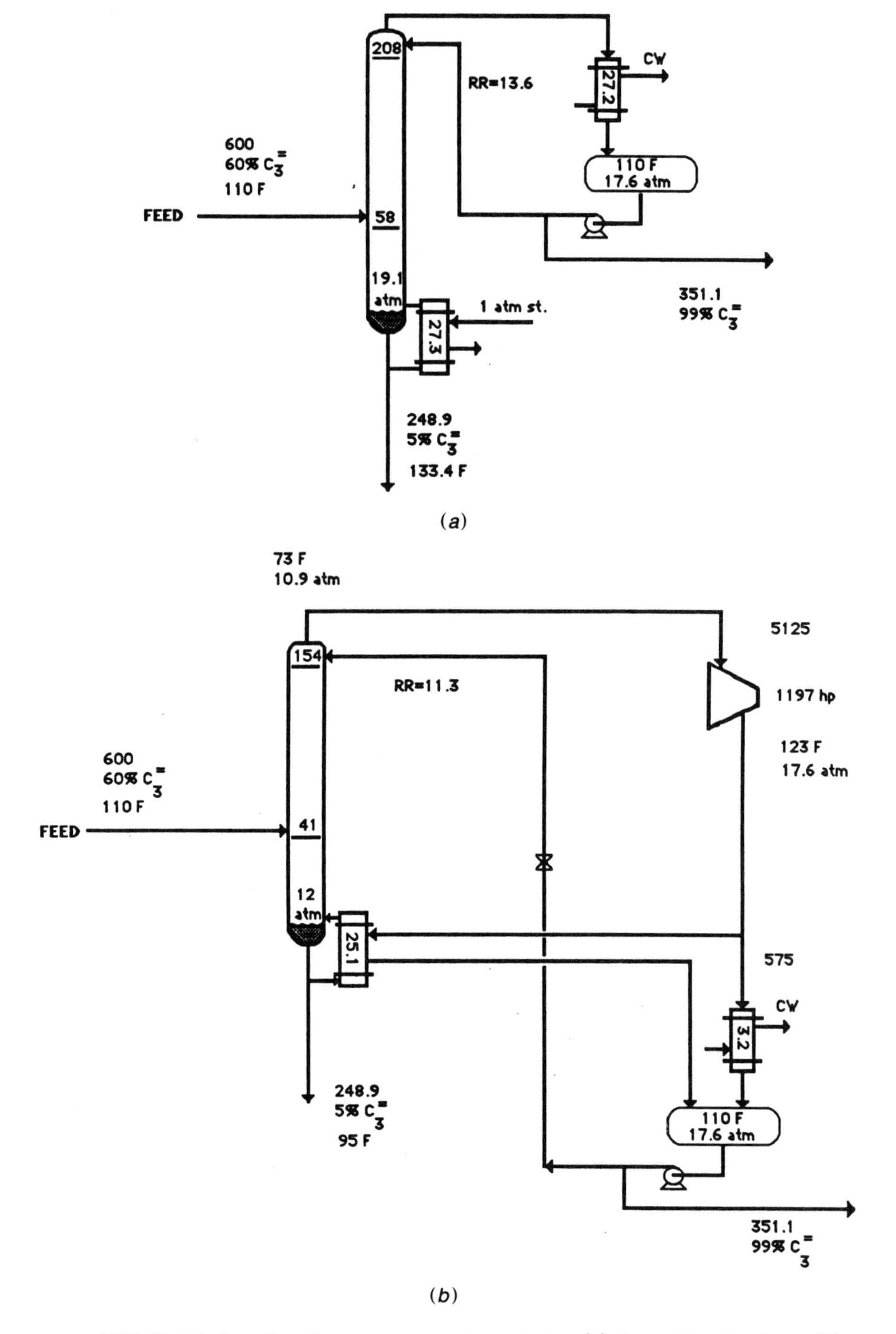

FIGURE 23-12. Propylene–propane column design. (*a*) Conventional column. (*b*) Vapor recompression column (numbers inside heat exchangers are 10^6 Btu/h; numbers without units along lines are flow rates in pound-moles per hour; composition are in mole percent).

The vapor recompression design for the same separation is presented in Figure 23-11*b*. The column itself is identical to the conventional one because the operating pressure and reflux ratio are the same. The energy consumption is 8000 hp or 20.4×10^6 Btu/h. Because the top temperature is high enough to use water to condense the excess vapor (the vapor not required to match reboiler duty), it is possible to use a low-pressure reflux drum scheme.

The capital investments for the conventional and vapor recompression designs are estimated to be \$1.3 and $\$5.5 \times 10^6$. Using utility costs of \$5 per 1000 pounds of steam and \$0.07 per kilowatt hour of electricity, the annual energy costs are \$5.6 and 3.8×10^6. These numbers give a 2.3 yr payout time for the incremental investment in going from conventional to vapor recompression systems. The very large capital cost of the vapor recompression system is mostly due to the cost of the compressor. Compressor cost is a function of the compression work required and the volume of vapor that must be handled. For this chemical system, the relatively large temperature difference (62°F) between the top and bottom of the column indicates the need for a large compression ratio. For systems with smaller temperature differences, less energy is required and capital cost and payout time decrease (as in the propylene–propane system).

Propylene–Propane System

The base case for the conventional propylene–propane distillation is shown in Figure 23-12*a*. It is necessary to operate at 17.6 atm (244 psig) to use water in the condenser. This pressure is based on a design overhead vapor temperature of 110°F. This high pressure reduces relative volatility, resulting in a very large column (208 trays, 100% efficiency) and high reflux ratio (13.6). It is not possible to preheat the feed using a column product because the temperatures of these streams are close together.

If colder cooling water were available, allowing condensation at 90°F, the pressure could be reduced to 15.6 atm, resulting in a 4.4% reduction in energy consumption.

The same separation using vapor recompression is presented in Figure 23-12*b*. In this case, it is possible to decrease the operating pressure and hence increase the relative volatility. A high-pressure reflux drum scheme is used because of the low top temperature. The temperature in the reflux drum is fixed to provide a 15°F temperature difference in the reboiler-condenser.

This design has a column with 54 fewer trays and an operating pressure 7 atm lower than the conventional column. The energy consumption drops from 27.3×10^6 Btu/h (steam) to 3.0×10^6 Btu/h (electricity).

The base case design of the conventional column has a capital cost of $\$1.5 \times 10^6$ and an annual energy cost of $\$0.56 \times 10^6$. The payout time for the incremental investment in going from conventional to vapor recompression in this system is only 0.25 yr. This is because of the small column ΔT and high pressure.

23-5-2 Dynamic Models

The distillation columns are simulated by using standard column modelling equations. For ethanol–water, the model includes two dynamic component balances and one energy balance per tray. For propylene–propane, equimolal overflow was assumed, eliminating the energy balances. The liquid flow rate from each tray is calculated from the Francis weir formula. Details of vapor–liquid equilibrium relationships, physical properties, and column dimensions are presented by Muhrer (1989).

To model the variable pressure in the column and in the compressor piping, the system is divided into three parts: suction piping, discharge piping, and the compressor proper. Mass balance equations are written for the suction and discharge volumes (Davis and Corripio, 1974). These differential equations give the suction and discharge pressure of the compressor at any point in time:

$$\frac{dP_s}{dt} = (F_s + F_g - F_c)\frac{z_s RT_s}{V_s} \quad (23\text{-}11)$$

$$\frac{dP_D}{dt} = (F_c - F_g - F_d)\frac{z_d RT_d}{V_d} \quad (23\text{-}12)$$

where F = molar flow rate
P = pressure
R = ideal gas constant
z = compressibility
T = temperature
V = volume

For the subscripts, c is the compressor, d is the discharge, g is the surge, and s is suction. The suction volume includes the effective volume of the column (Choe and Luyben, 1987).

The flows are calculated from algebraic equations. The compressor curves are used to determine F_c from the given speed and pressure difference. F_s is the sum of the vapor leaving the top tray in the column (from the energy equation) and the vapor that is flashed from the hot reboiler condensate. The flow rate F_d is what is being condensed in the reboiler and trim condenser. The bypass flow F_g is used, if necessary, to avoid surge in the compressor.

Ethanol–Water System

The column operating pressure is atmospheric in both the conventional and vapor recompression cases, so vapor holdup is a small fraction of the total holdup (2%) and is neglected. Other specific assumptions are:

1. Constant pressure drop (5.3 mmHg per tray).
2. Temperature measurement lag of 15 s, composition measurement deadtime of 1 min, compressor speed lag of 30 s, heat input lag of 30 s.
3. Compressor polytropic efficiency of 70%.

Propylene–Propane

Because of the very large number of trays and the long time constants of this system, it is necessary to use as simple a dynamic model as possible. Several simplifying assumptions are made.

Due to the high operating pressure, vapor holdup represents about 30% of the total holdup and needs to be considered. As a good approximation, additional liquid holdup is added to simulate the impact of the vapor holdup on the column dynamics. This model gives dynamic responses to all disturbances that are essentially the same as the rigorous vapor holdup model proposed by Choe and Luyben (1987).

An equimolal overflow model, which removes the need for energy equations, gives the same dynamic responses as a rigorous model for all disturbances except pressure changes. Pressure effects are not seen because vapor flows are not calculated from the energy balance of each tray, but are instead assumed constant throughout the column. Because relatively small changes in pressure are seen in the dynamic tests of the column, an equimolal overflow model is used for propylene–propane.

The following additional assumptions are made:

1. Constant pressure drop (5.3 mmHg per tray).
2. Composition measurement deadtime of 5 min, compressor speed lag of 30 s, heat input lag of 30 s.
3. Compressor polytropic efficiency of 70%.

23-5-3 Control System Design

An effective control system is essential for the efficient operation of any process. For these applications, multiloop, single-input–single-output controllers (diagonal structure) perform well, so multivariable controllers are not considered.

The controller design procedure employed is similar to that presented by Yu and Luyben (1986). Steady-state "rating" programs are developed for all columns. They are used to determine incentives for dual composition control by plotting the steady-state changes required in all manipulated variables to hold the controlled variables constant for disturbances in feed composition (Luyben, 1975). Disturbances in feed flow rate do not have to be considered because they can be handled with ratio schemes. Figure 23-13 gives these results for both systems.

The curves for ethanol–water show that both vapor boilup and reflux flow must be adjusted to compensate for feed composition disturbances, but the reflux ratio remains almost constant over the range considered. This means that single-end composition control could be used with very little penalty in energy consumption. Singular value decomposition (Moore, 1986) recommends the use of tray 6 temperature. The recommended control system has heat input controlling tray 6 temperature and reflux

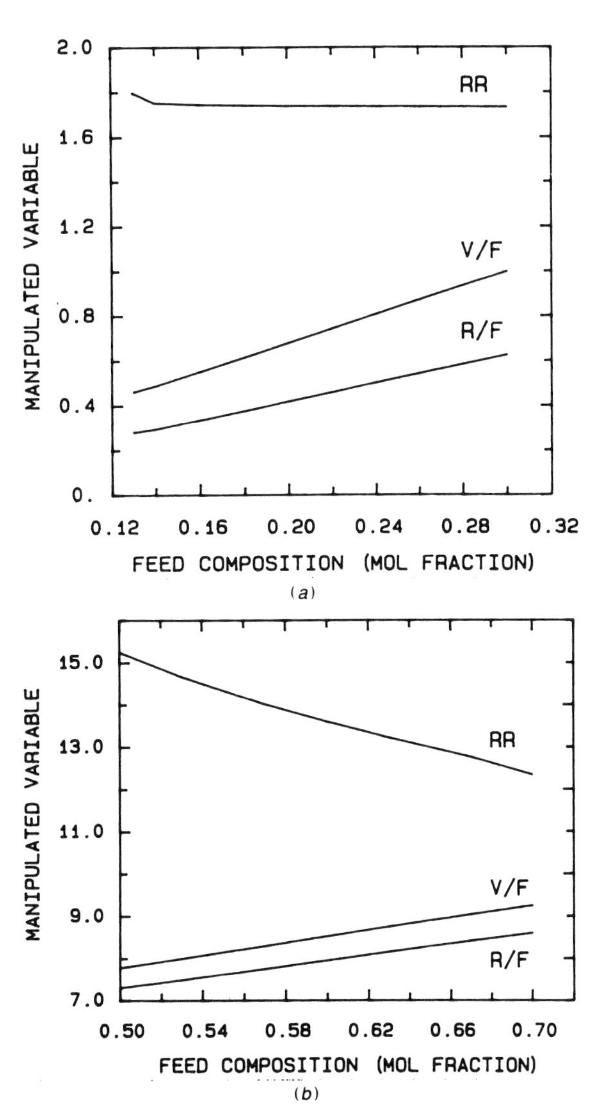

FIGURE 23-13. Steady-state changes in manipulated variables (*a*) Ethanol–water; (*b*) propylene–propane.

ratio fixed. Although dual composition is not required for the ethanol–water system, it is examined in the interest of increasing our understanding of the generic behavior of vapor recompression columns. An analyzer is used to measure the top composition instead of inferring it from the temperature because of the poor sensitivity of the temperature on the top trays. Figure 23-14 shows the control system for the ethanol–water separation with vapor recompression.

The curves in Figure 23-13*b* for propylene–propane show that all manipulated variables change significantly as feed composition changes, so dual composition control is required. Several choices of manipulated variables are explored. Finco, Luyben, and Polleck (1989) studied the control of conventional propylene–propane columns and recommended an unorthodox *D-B* control structure: distillate flow controls distillate composition, bottoms flow controls bottoms composition, reflux flow controls reflux drum level, and heat input controls bottoms level. Unfortunately, this structure has poor integrity and cannot be analyzed by steady-state methods. The *RR-BR* structure (reflux ratio controls distillate composition and boilup ratio controls bottoms composition) is equivalent to the *D-B* structure assuming tight level control. It requires more complex instrumentation and is more difficult to operate (Waller et al., 1988).

The *D-B*, *RR-BR*, and the more conventional $D\text{-}Q_R$ structures are all studied. Con-

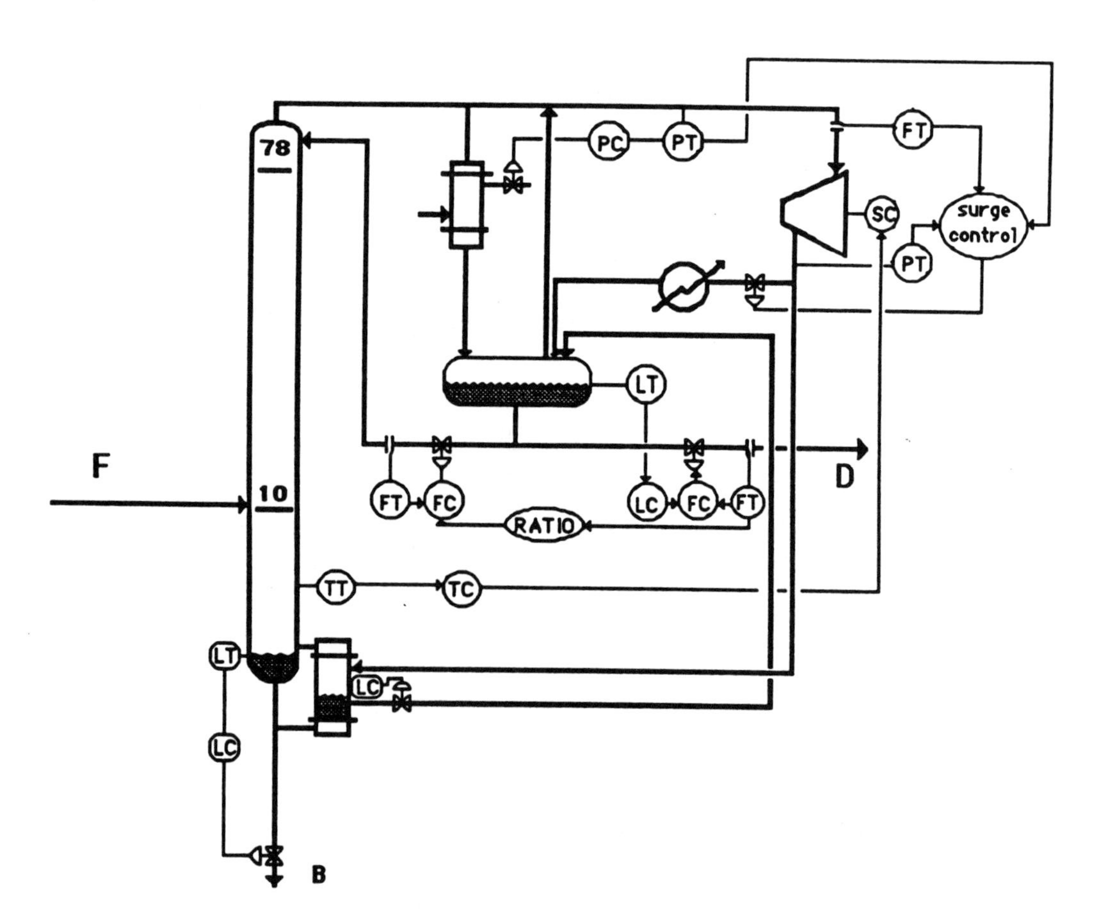

FIGURE 23-14. Control system for ethanol–water vapor recompression column.

trollers are designed for each case. Figure 23-15 shows that all control schemes perform similarly. Therefore, the widely used D-Q_R structure is chosen for the conventional column. The corresponding control structure is used for the vapor recompression column, distillate flow and compressor speed (D-S_p). Figure 23-16 shows the control system with vapor recompression.

The steady-state rating programs are also used to calculate steady-state gains. These are used in obtaining transfer functions so that manipulated variables can be selected and unworkable pairings can be eliminated.

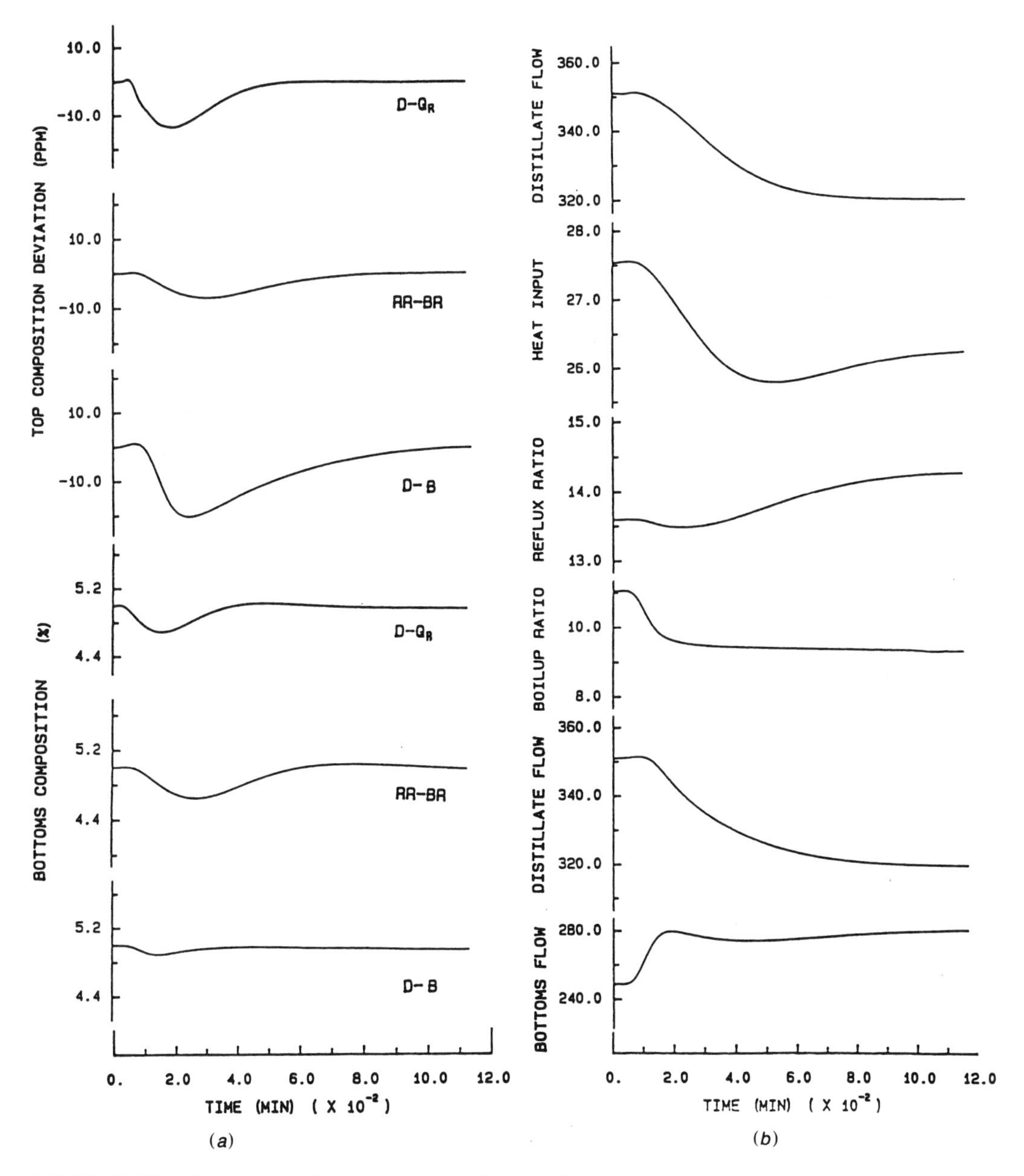

FIGURE 23-15. Comparison of control schemes for propylene–propane conventional column. (a) Controlled variables; (b) manipulated variables.

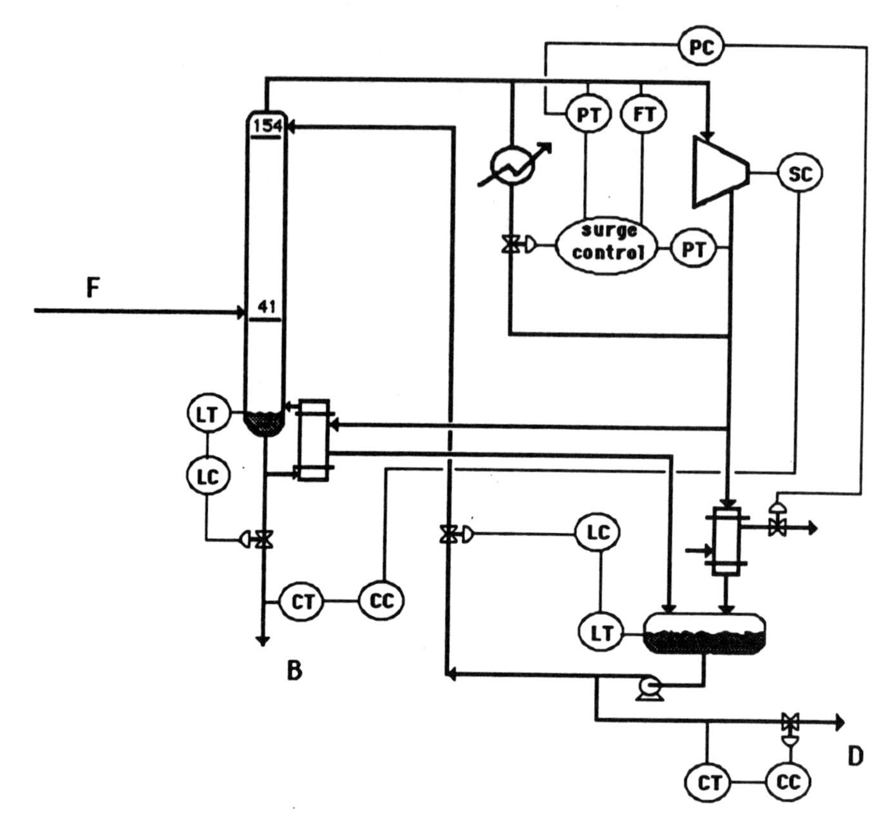

FIGURE 23-16. Control system for propylene–propane vapor recompression column.

Transfer functions of the processes are identified by using the auto-tune (ATV) identification procedure (Luyben, 1987). Table 23-1 gives the transfer functions for the conventional and vapor recompression columns for both chemical systems.

Note that the transfer functions obtained between manipulated variables and pressure have time constants that are much smaller than those for the compositions (see the last entry row of Table 23-1 for propylene–propane). Note also that compressor speed has little effect on pressure; it mostly affects compositions. The Niederlinski index is +3.6 for the pairings x_D-D, x_B-S_p, and P-F_w. The RGA for this system shows that the only pairings that do not give negative RGA elements on the diagonal are those that pair cooling water with pressure:

$$\text{RGA} = \begin{bmatrix} 0.240 & 0.893 & -0.137 \\ 0.754 & 0.255 & -0.010 \\ 0.001 & -0.148 & 1.15 \end{bmatrix}$$

It is interesting to note that the Niederlinski index does not show that pressure must be paired with cooling water. It gives a positive index of 36.3 for the pairings x_D-S_p, x_B-F_w, and P-D.

The multiple SISO controllers are tuned by using the BLT technique (Luyben, 1986). Table 23-2 gives tuning parameters.

TABLE 23-1 Transfer Functions

Ethanol–Water

Conventional column

$$\begin{bmatrix} x_D \\ T_6 \end{bmatrix} = \begin{bmatrix} \dfrac{3.7e^{-1.2s}}{(6.9s+1)(162s+1)} & \dfrac{-5.8e^{-1.3s}}{(1.8s+1)(1190s+1)} \\ \dfrac{-170e^{-0.2s}}{(0.6s+1)(127s+1)^2} & \dfrac{323e^{-0.2s}}{(1.3s+1)(400s+1)} \end{bmatrix} \begin{bmatrix} RR \\ Q_R \end{bmatrix}$$

Vapor recompression column

$$\begin{bmatrix} x_D \\ T_6 \end{bmatrix} = \begin{bmatrix} \dfrac{1.6e^{-s}}{(0.02s+1)(19.3s+1)^2} & \dfrac{-26.6e^{-s}}{(0.8s+1)(1510s+1)} \\ \dfrac{168}{(0.6s+1)(140s+1)^2} & \dfrac{2130}{(1211s+1)(0.08s^2+0.6s+1)} \end{bmatrix} \begin{bmatrix} RR \\ S_p \end{bmatrix}$$

Propylene–Propane

Conventional column

$$\begin{bmatrix} x_D \\ x_B \end{bmatrix} = \begin{bmatrix} \dfrac{-6.84e^{-5s}}{(1.55s+1)(4310s+1)} & \dfrac{5.4e^{-5s}}{(0.8s+1)(175s+1)^2} \\ \dfrac{-10.8e^{-10s}}{(46s+1)(472s+1)} & \dfrac{-1.91e^{-5s}}{(3.2s+1)(7.2s+1)^2} \end{bmatrix} \begin{bmatrix} D \\ Q_R \end{bmatrix}$$

Vapor recompression column

$$\begin{bmatrix} x_D \\ x_B \\ P \end{bmatrix} = \begin{bmatrix} \dfrac{-9.2e^{-5s}}{(1.5s+1)(4800s+1)} & \dfrac{7.9e^{-5s}}{(7s+1)(141s+1)^2} & \dfrac{-0.27e^{-5s}}{(178s+1)(24150s^2+252s+1)} \\ \dfrac{-10e^{-5s}}{(3s+1)(115s+1)^2} & \dfrac{-2.8e^{-5s}}{(4.5s+1)(14s^2+6s+1)} & \dfrac{0.1e^{-5s}}{(11s+1)(78s^2+10.5s+1)} \\ \dfrac{-0.88}{(0.02s+1)(92s+1)} & \dfrac{1}{(4.2s+1)(11.5s^2+3s+1)} & \dfrac{-0.275}{(0.4s+1)(12s+1)} \end{bmatrix} \begin{bmatrix} D \\ S_p \\ F_w \end{bmatrix}$$

TABLE 23-2 Controller Tuning[a]

System	Loop	Conventional Column K_c	Conventional Column τ_I	Vapor Recompression Column K_c	Vapor Recompression Column τ_I
Ethanol–water	x_D-RR	4.8	57	5.0	37
	T_6-$Q_R(S_p)$	0.9	9.6	0.8	3.4
Propylene–propane	x_D-D	−36	44	−36	37
	x_B-$Q_R(S_p)$	−0.385	60	−0.18	42

[a]Controller gains are dimensionless. Flow spans are twice the steady-state value and transmitter spans are: for ethanol–water, $x_D = 0.20$, tray 6 100°F; for propylene–propane, $x_D = 0.05$, $x_B = 0.20$.

23-5-4 Results for Specific Systems

Ethanol–Water

The performance is evaluated with the nonlinear model by introducing step disturbances in the feed composition. The results are shown in Figure 23-17*a* for the conventional column and in Figure 23-17*b* for the vapor recompression column. Constant pressure is assumed in the conventional column because it would typically be vented to the atmosphere through a vent condenser. In the vapor recompression column, the pressure loop is quite fast. Therefore, it is tuned independently from the composition loops.

The control strategy used on the conventional column works quite well on the vapor recompression column. Controller settings are somewhat different because the open-loop behavior of the vapor recompression system is slightly faster than the conventional column (smaller deadtimes and time constants). This behavior could be due to the effect of pressure transients.

These results indicate that the control structure used in the conventional column can also operate successfully in the vapor recompression column if compressor control is substituted for steam flow.

Propylene–Propane

The performance of the conventional propylene–propane column has already been shown in Figure 23-15. The same basic structure is used on the vapor recompression column. Compressor speed is used instead of steam flow. Controllers are retuned because the dynamics of the vapor recom-

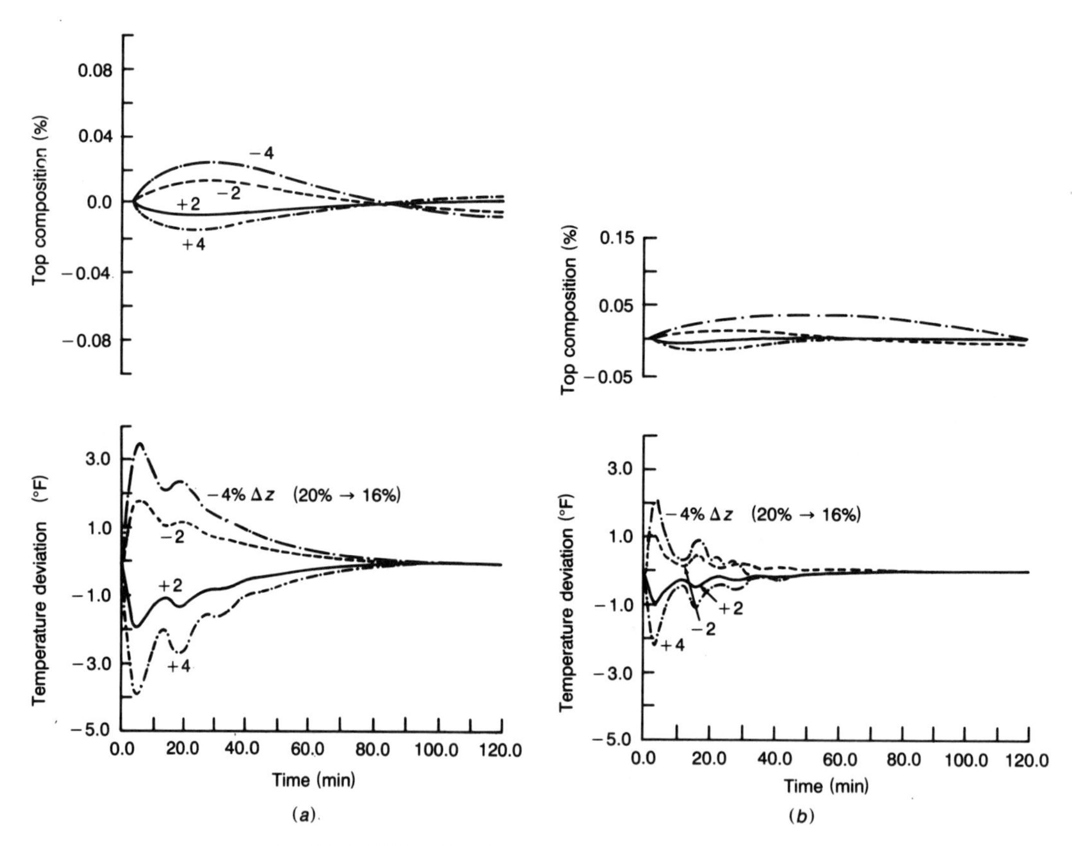

FIGURE 23-17. Feed composition disturbances, ethanol–water column. (*a*) Conventional; (*b*) vapor recompression.

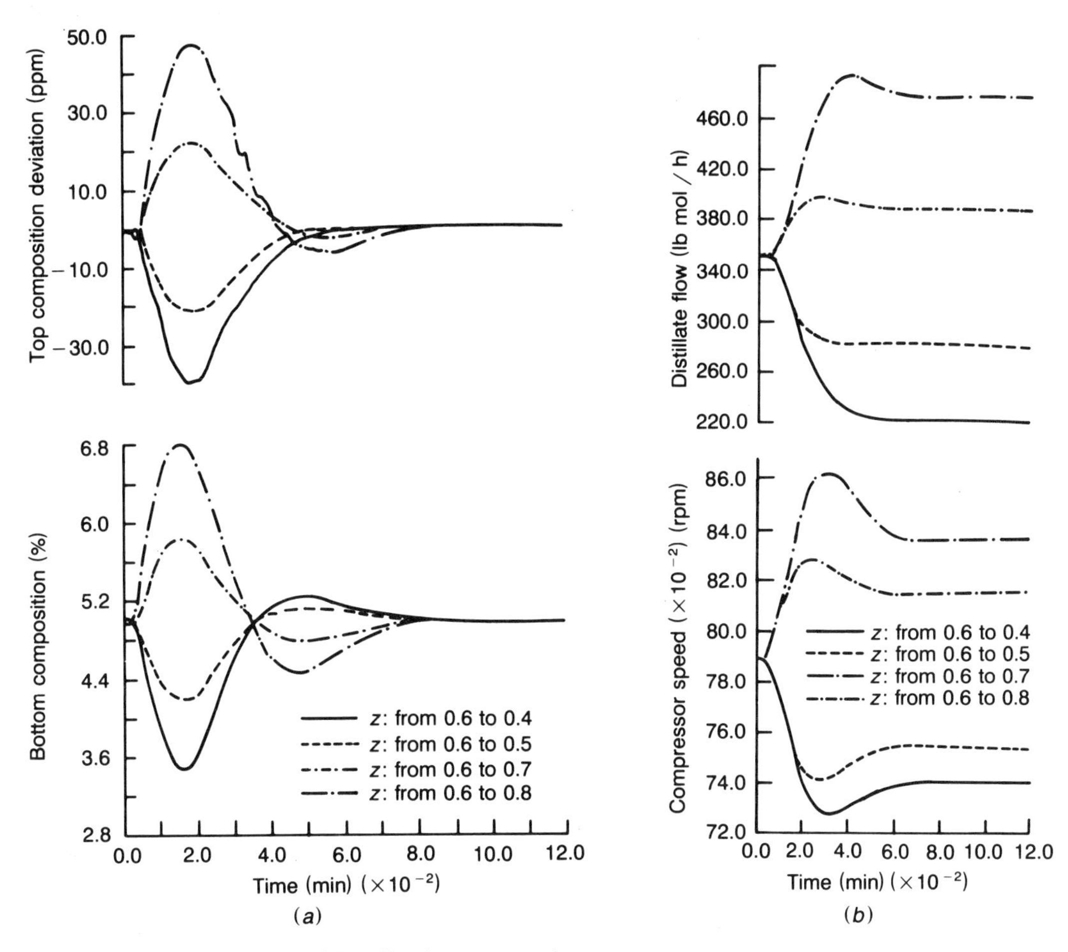

FIGURE 23-18. Feed composition disturbances, propylene–propane vapor recompression column. (*a*) Controlled variables; (*b*) manipulated variables.

pression column are slightly faster than the conventional. This can be explained by the lower operating pressure, higher relative volatility, and fewer trays.

Cooling water flow rate to the trim condenser is used to control the pressure. As shown in Table 23-2, the time constants of the pressure-to-cooling water loop are much faster than the composition dynamics. Therefore, the pressure loop is tuned tightly and the composition loops are treated as a 2×2 system in the BLT tuning method.

The performance of the vapor recompression system for step changes in feed composition is given in Figure 23-18. The response is slightly faster than the conventional column. The manipulated variable movements indicate that the range of controllability could be wider (bigger disturbances could be handled) without running into valve saturation.

23-6 CONCLUSION

Both of the separations studied yielded the same conclusion: The control system used on the conventional column can be applied to the vapor recompression column with the simple substitution of compressor speed for steam flow. The effect of a change in compressor speed is primarily to change the vapor rate through the column, not to make large changes in suction or discharge pressures. The discharge pressure must change somewhat to produce the required change in the temperature differential driving force in the reboiler.

23-7 NOMENCLATURE

A heat-transfer area in reboiler-condenser
b intercept of straight-line plant characteristic curve
D distillate flow rate
F_c flow rate through compressor
F_d flow rate to reboiler and trim condenser
F_g flow rate through bypass
F_s flow rate into suction volume from column and reflux flashing
F_w flow rate of cooling water to condenser
K_c controller gain
m slope of straight-line plant characteristic curve
P column pressure
P_D discharge pressure
P_s suction pressure
P^s vapor pressure of distillate product
Q_R heat-transfer rate in reboiler-condenser
R perfect gas law constant
RR reflux ratio
S_p compressor speed
T_B temperature in column base
T_d temperature at compressor discharge
T_R temperature on hot side of reboiler-condenser
T_s temperature at compressor suction
T_6 tray 6 temperature
U overall heat-transfer coefficient in reboiler-condenser
V vapor flow rate
V_d volume compressor discharge
V_s volume compressor suction
x_D distillate composition
x_B bottoms composition
z compressibility

Greek Symbols

α constant in vapor pressure equation
θ constant in vapor pressure equation
ΔP pressure differential across compressor
λ heat of condensation of vapor in reboiler
ρ vapor density
τ_I controller integral time

References

Brousse, E., Claudel, B., and Jallut, C. (1985). Modeling and optimization of the steady-state operation of a vapor recompression distillation column. *Chem. Eng. Sci.* 40, 2073–2078.

Choe, Y. S. and Luyben, W. L. (1987). Rigorous dynamic models of distillation columns. *Ind. Eng. Chem. Res.* 26, 2158–2161.

Collura, M. A. and Luyben, W. L. (1988). Energy-saving distillation designs in ethanol production. *Ind. Eng. Chem. Res.* 27, 1686–1696.

Davis, R. T. and Corripio, A. B. (1974). Dynamic simulation of variable speed centrifugal compressors. In *Advances in Instrumentation*. Research Triangle Park, NC: Instrument Society of America.

Ferre, J. A., Castells, F., and Flores, J. (1985). Optimization of a distillation column with a direct vapor recompression heat pump. *Ind. Eng. Chem. Process Des. Dev.* 24, 128–132.

Finco, M. V., Luyben, W. L., and Polleck, R. E. (1989). Control of low relative-volatility distillation columns. *Ind. Eng. Chem. Res.* 25, 75–83.

Luyben, W. L. (1975). Steady state energy conservation aspects of distillation column control systems design. *Ind. Eng. Chem. Fundam.* 14, 321–325.

Luyben, W. L. (1986). Simple method for tuning SISO controllers in multivariable systems. *Ind. Eng. Chem. Process Des. Dev.* 25, 654–660.

Luyben, W. L. (1987). Derivation of transfer functions for highly nonlinear distillation columns. *Ind. Eng. Chem. Res.* 26, 2490–2495.

Meili, A. (1987). Experience with heat pump system for energy saving in distillation column. In *Distillation and Absorption 1987*. London: European Federation of Chemical Engineering.

Meili, A. and Stuecheli, A. (1987). Distillation column with direct vapor recompression. *Chem. Eng.* 94, 133–143.

Moore, C. (1986). Application of singular value decomposition to the design analysis and control of industrial processes. Presented at the American Control Conference, Seattle, WA.

Mosler, H. A. (1974). Control of sidestream and energy conservation distillation towers. Presented at AIChE Continuing Education Meeting on Industrial Process Control, Tampa, FL.

Mostafa, H. A. (1981). Thermodynamic availability analysis of fractional distillation with vapor recompression. *Can. J. Chem. Eng.* 59, 487–491.

Muhrer, C. A. (1989). Study of the dynamics and control of vapor recompression columns. Ph.D. Thesis, Lehigh University, Bethlehem, PA.

Muhrer, C. A., Collura, M. A., and Luyben, W. L. (1990). Control of vapor recompression distillation columns. *Ind. Eng. Chem. Res.* 29, 59–71.

Nielsen, C. S., Andersen, H. W., Branbrand, H., and Jorgensen, S. B. (1988). Adaptive dual composition control of a binary distillation column with a heat pump. Presented at IFAC Symposium on Adaptive Control of Chemical Processes, ADCHEM-88, Lyngby, Denmark.

Null, H. R. (1976). Heat pumps in distillation. *Chem. Eng. Prog.* 73, 58–64.

Quadri, G. P. (1981). Use of heat pump in *P-P* splitter. Part 1: Process Design. Part 2: Process Optimization. *Hydrocarbon Process.* 60, 119–126, 147–151.

Robinson, C. S. and Gilliland, E. R. (1950). *Elements of Fractional Distillation*. New York: McGraw-Hill.

Shinskey, F. G. (1979). *Process Control Systems*. New York: McGraw-Hill.

Waller, K. V., Finnerman, D. H., Sandelin, P. M., Haggblom, K. E., and Gustafsson, S. E. (1988). An experimental comparison of four control structures for two-point control of distillation. *Ind. Eng. Chem. Res.* 27, 624.

Yu, C. C. and Luyben, W. L. (1986). Design of multiloop SISO controllers in multivariable processes. *Ind. Eng. Chem. Process Des. Dev.* 25, 498–503.

24

Heat-Integrated Columns

William L. Luyben
Lehigh University

24-1 INTRODUCTION

The rapid increase in energy prices in the 1970s spurred a flurry of activity in energy conservation programs in the chemical industry. Because distillation columns are major energy consumers in many chemical plants, more energy-efficient designs for distillation systems were actively studied.

One of the most successful energy conservation measures for distillation columns is the use of heat integration: the overhead vapor from one column is used to provide heat to a second column. The idea is similar to the use of multiple effects in evaporation.

Major reductions in energy consumption (about 30 to 40%) can be achieved in some separation systems. Therefore the economic incentives can be very large (payback periods of a few months). In this chapter we will discuss some of the steady-state design aspects of heat-integrated distillation columns and will explore methods for developing effective control systems.

24-2 TYPES OF SYSTEMS

There are a large variety of heat integration schemes. Some of the more common ones are discussed in a fair amount of detail in the following text. All of these schemes feature multiple columns and interconnected energy systems. This markedly increases the complexity of the process, produces higher-order systems, and introduces significant interactions. Thus, these heat-integrated columns pose some challenging control problems.

Heat integration is used in distillation systems in two basic forms:

1. Energy integration only (columns that are separating different mixtures and making four or more different products).
2. Energy and material integration (multiple heat-integrated columns are used in the separation of a single feedstream into two or more products).

24-2-1 Energy Integration Only

Two Binary Feeds

Figure 24-1 illustrates the case where there are two feedstreams, F_1 and F_2, that contain different chemical components. The first column separates components A and B; the second column separates components C and D. Figure 24-1a shows the conventional non-heat-integrated case where the two columns operate completely independently, with each having its own source of energy and cooling. Because each uses cooling water in its condenser, the reflux-drum temperatures are about 110°F. Operating pressures are the bubble-point pressure of the overhead product at 110°F. Base temperatures are the bubble-point temperatures of the bottoms product at the base pressure (condenser pressure plus tray pressure drop). The figure shows some hypothetical but typical temperatures and pressures. Some specific real examples will be given later in this chapter.

Figure 24-1*b* shows the case where these two columns have now been heat-

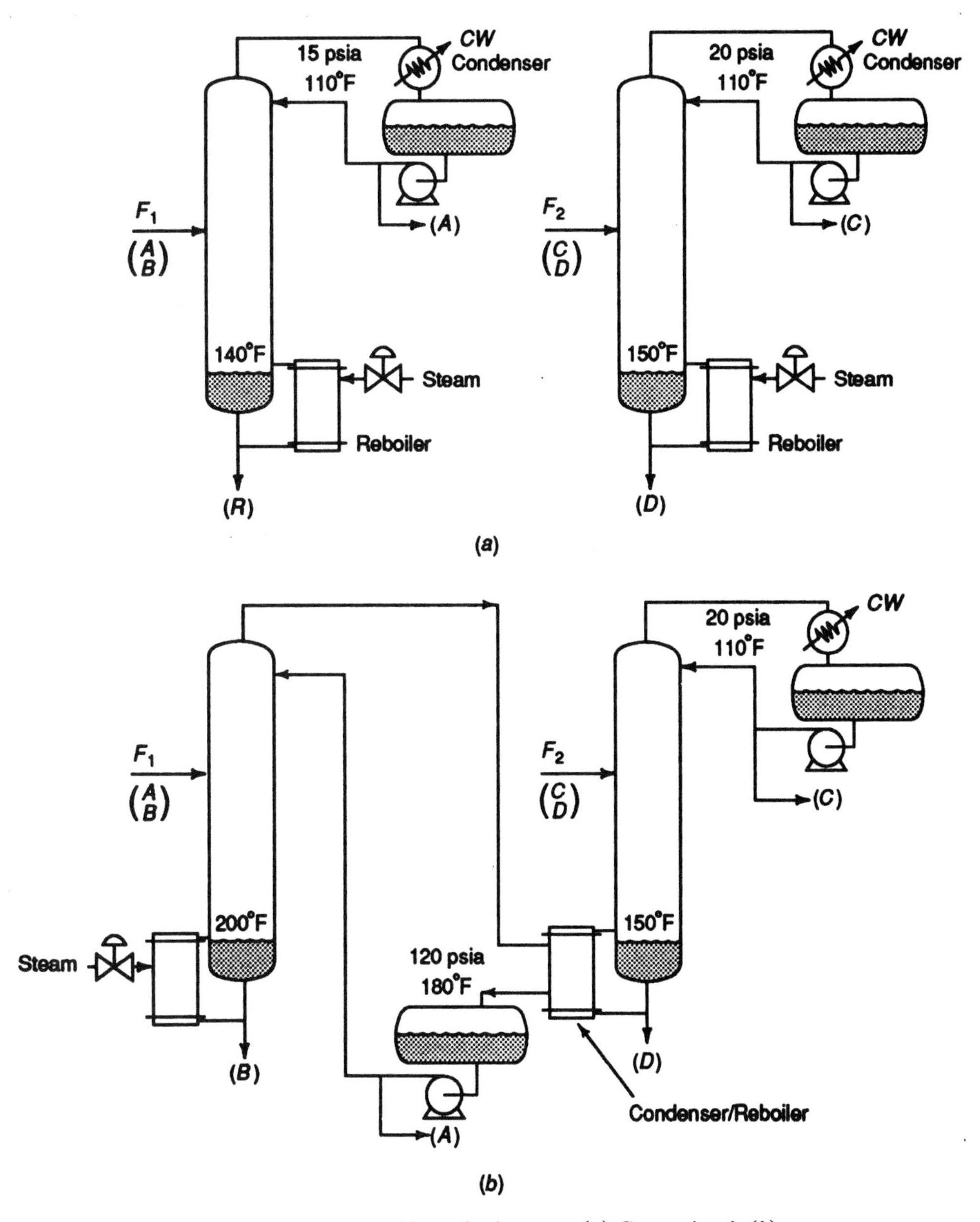

FIGURE 24-1. Energy integration, two binary feedstreams. (*a*) Conventional; (*b*) heat-integrated.

integrated. The operating pressure in the first column is raised from 15 to 120 psia so that the temperature in its reflux drum is raised from 110 to 180°F. Because this temperature is now higher than the base temperature of the second column, the overhead vapor from the first column can be used to reboil the second column. The energy that is put into the reboiler on the first column is being used twice, so very significant reductions in energy consumption are realized.

Note that the temperatures and pressure in the second column are unchanged, but the temperatures and the pressure in the first column are higher. This has two results:

1. If the relative volatility between components A and B decreases significantly as temperature is increased, the column in the heat-integrated system will require more trays and/or a higher reflux ratio.
2. The base temperature of the reboiler in the first column is increased, so higher-pressure steam or other higher temperature heat sources must be used. Thus we end up using less energy (fewer British thermal units per hour or pounds per hour of steam) but requiring a higher temperature source of heat (higher-pressure steam).

Single Ternary Feed

Another variation of this type of system is when there is a single feedstream with three components (A, B, and C), and two columns are used to make the separation into three products. As shown in Figure 24-2a, the conventional non-heat-integrated separation scheme may consist of two columns, each with its own sources of both cooling and heating. The lightest component (A) would be taken overhead in the first column, and the mixture of components B and C would be taken out the bottom of the first column. This bottoms stream is fed into the second column, which separates B and C.

In Figure 24-2b the columns have been heat-integrated by raising the pressure in the *first* column high enough to make its reflux-drum temperature higher than the bottoms temperature of the second column. Figure 24-2c shows another alternative where the columns have been heat-integrated by raising the pressure in the *second* column high enough to make its reflux-drum temperature higher than the bottoms temperature of the first column. We call the first type *forward* integration because energy flows in the same direction as the process flows (feed and energy are both introduced into the first column). We call the second type *reverse* integration because the flow of energy is in the opposite direction to the process flows (energy is introduced into the second column while the fresh feed is introduced in the first column). These configurations are somewhat analogous to multiple-effect evaporators with forward and reverse feeds.

Which of these alternatives is better depends on the chemical system (temperature dependence of vapor pressures, critical temperatures and pressures, effect of pressure on relative volatilities, thermal sensitivity of components, corrosiveness, etc.) and on the specific plant and process environment (steam pressures available, existence of waste energy at some fixed temperature level, safety consequences of leaks in the condenser–reboiler, etc.).

Many other alternatives are also possible. For example, we could take component C out the bottom of the first column and take the A/B mixture out the top. This distillate would be fed into the second column where the A/B separation would be achieved. Heat integration could be in either the reverse or the forward directions. Figure 24-2d shows an interesting possibility in which the second column is a complex configuration (two feeds and a sidestream). The first column is a "prefractionator," which produces an overhead that contains almost all of the lightest component A, about half of the intermediate component B, and very little of the heaviest component C. The bottoms from the prefractionator contains very little A, about half of the B, and almost all of the C. These two streams are fed into the second column at different feed tray locations. A sidestream product of mostly component B is removed between the feed trays. Overhead product is A and bottoms product is C. A reverse heat integration scheme is shown in Figure 24-2d, but forward heat integration can be used in some systems.

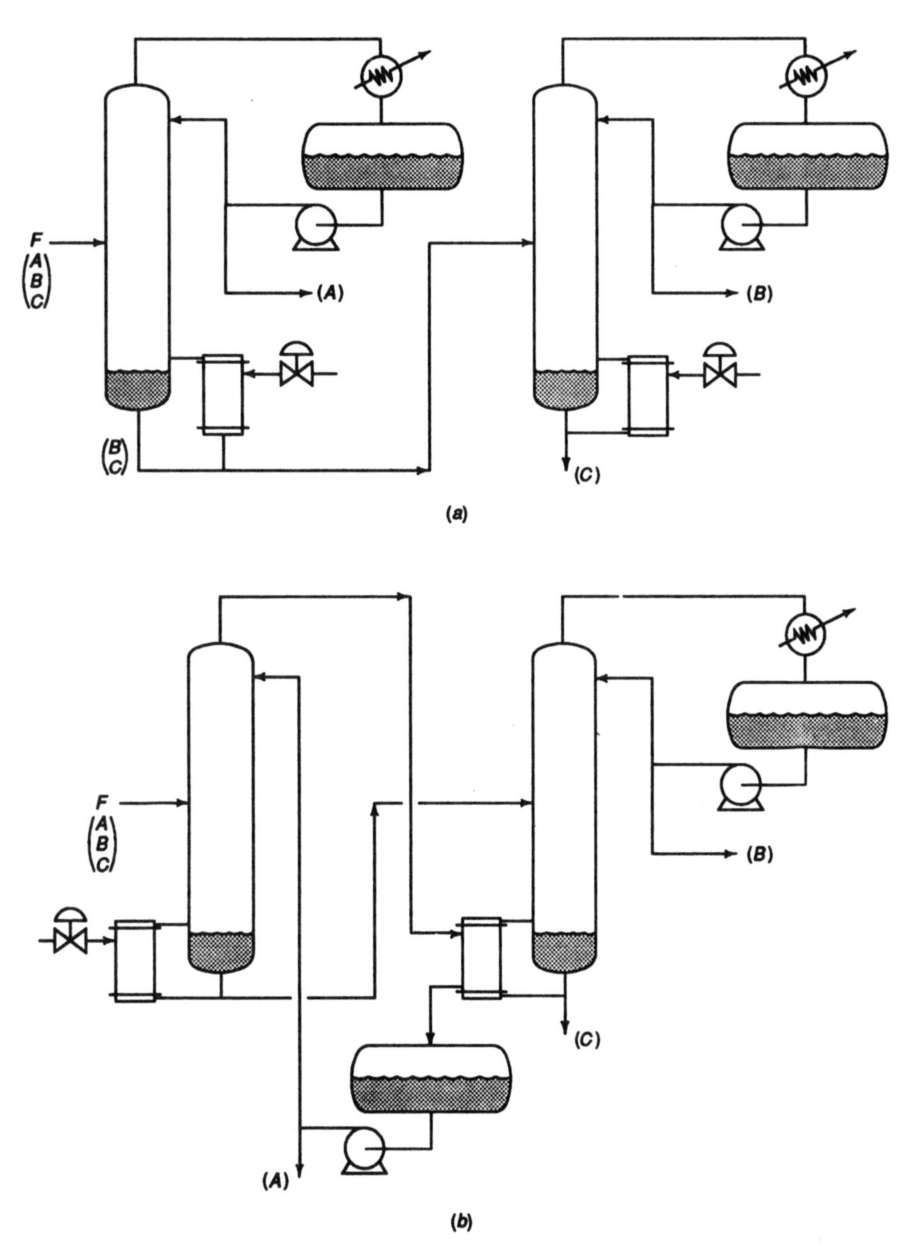

FIGURE 24-2. Energy integration, one ternary feedstreams. (a) Conventional; (b) heat-integrated (forward); (c) heat-integrated (reverse); (d) prefractionator heat-integrated reverse.

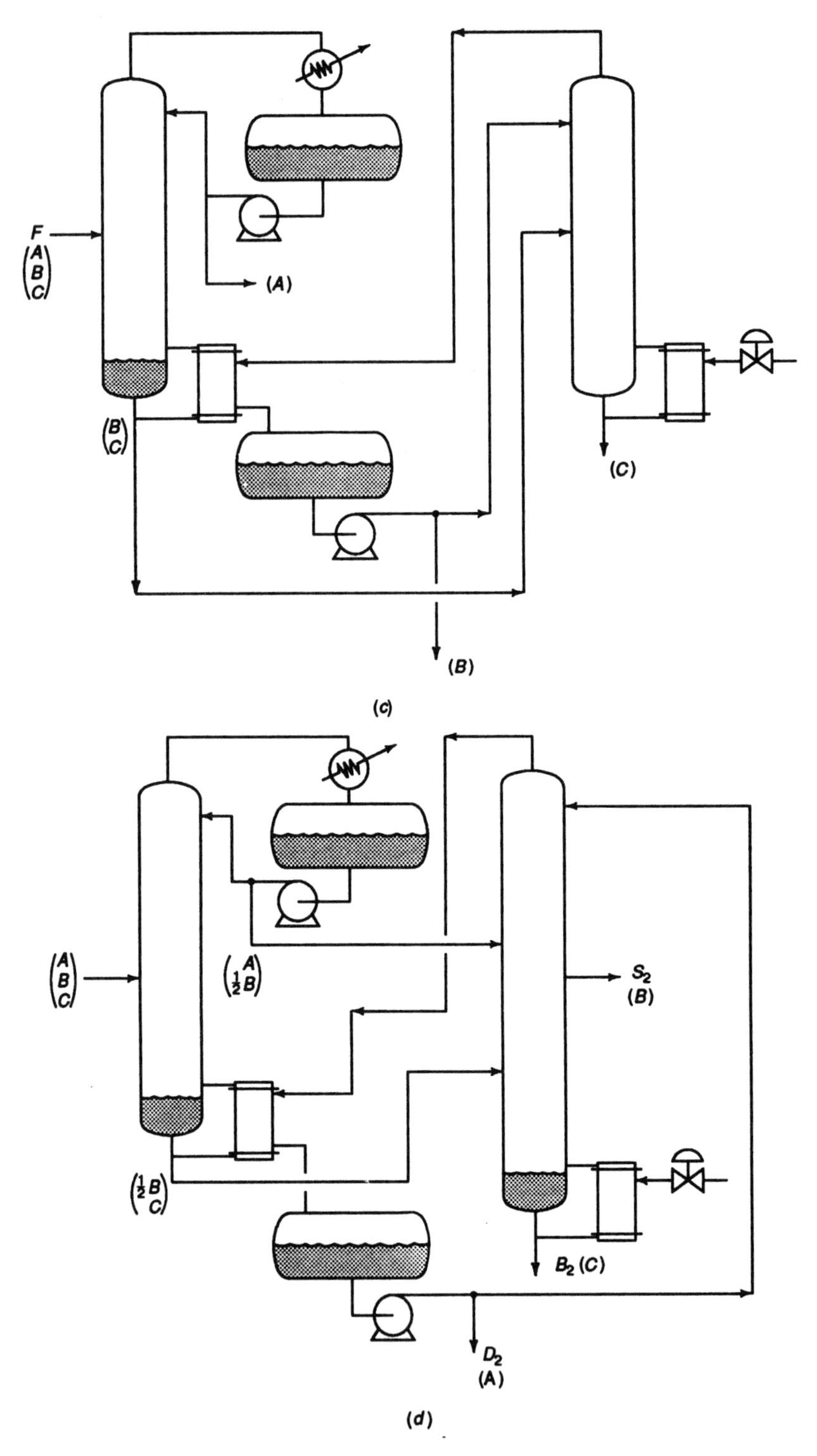

FIGURE 24-2. *Continued.*

In almost all situations there will be an imbalance between the reboiler heat-input requirements of the second column and the condenser heat-removal requirements of the first column. Therefore an auxiliary reboiler or an auxiliary condenser will almost always be used, even under steady-state conditions. These auxiliary heat exchangers can then be used dynamically to control the energy input to the second column. Figure 24-3 shows the two cases, one involving an auxiliary reboiler (when the first column does not

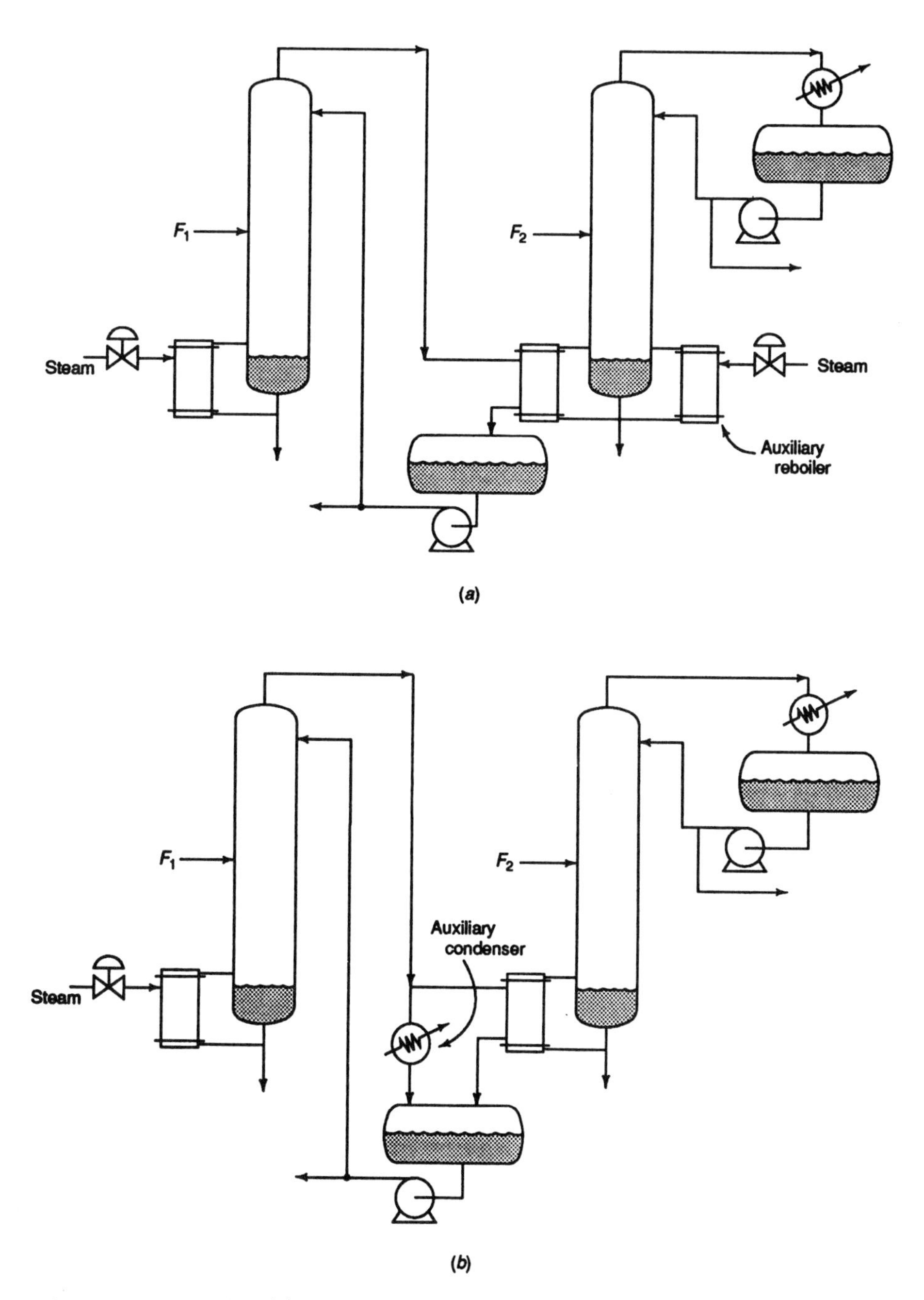

FIGURE 24-3. Auxiliary (*a*) reboiler or (*b*) condenser.

produce enough vapor to provide heat for the reboiler of the second column), and the second involving an auxiliary condenser (when the second column does not require all the vapor from the first column overhead). In the event that the columns are almost exactly balanced in terms of heat requirements (condenser heat removal requirements in the high-pressure column equal the reboiler heat-input requirement in the low-pressure column), both an auxiliary condenser and an auxiliary reboiler must be used to handle dynamic transients. We will discuss this in more detail later. Both an auxiliary condenser and an auxiliary reboiler may be needed for start-up or so that each column can be operated independently under certain abnormal conditions.

24-2-2 Energy and Process Integration

In the configurations discussed in the preceding text, each column had different feeds and produced different products. Another very common type of application of heat integration is the situation where we have a single binary feedstream. Conventionally we would use one column to separate the two components (see Figure 24-4*a*) with a steam-heated reboiler and a water-cooled condenser.

Instead of using just one column, it is sometimes economical to use two (or more) columns that are heat-integrated. There are many alternative configurations. Two of the more common are shown in Figures 24-4*b* and *c*.

1. *Split feed* (FS). The total feed is split between a high-pressure column and a low-pressure column. Each column makes specification distillate product (D_1 and D_2) and bottoms product (B_1 and B_2). Thus there are four product streams and there are potentially four compositions to be controlled.
2. *Light split with forward integration.* All the fresh feed is fed into the first column, which operates at higher pressure. About half of the lighter component is removed as the distillate product (D_1). The bottoms B_1 is a mixture of both components and is fed into the second column, which operates at lower pressure. The other half of the lighter component is removed in D_2 and almost all of the heavier component leaves in the bottoms product B_2. There are only three product streams, so there are only three compositions to be controlled. Steam is used in only the first column.

Many other alternatives exist such as using reverse heat integration in the light-split case (operate the second column as the high-pressure column), taking about half of the heavier component out the bottom of the first column and feeding the distillate A/B mixture to the second column (heavy split), and so forth.

24-3 ECONOMIC INCENTIVES

Table 24-1 gives a fairly detailed comparison of the steady-state designs of several alternative configurations for the separation of a specific binary system: methanol–water. The single conventional column consumes 369,000 kcal/min of energy. The feed-split (FS) heat-integrated configuration consumes only 227,000 kcal/min. This is a 38% reduction in energy consumption! The light-split forward (LSF) and the light-split reverse (LSR) configurations consume about the same amount of energy as the feed-split configuration: 241,000 and 226,000 kcal/min, respectively, so all of these heat integration schemes offer very significant savings in energy.

Building two columns will cost more than building one column, even though the diameters of the two columns are smaller than the diameter of the single column. Typically any of the heat-integrated schemes will require higher capital investment. But the payback periods for the incremental investment are usually quite short.

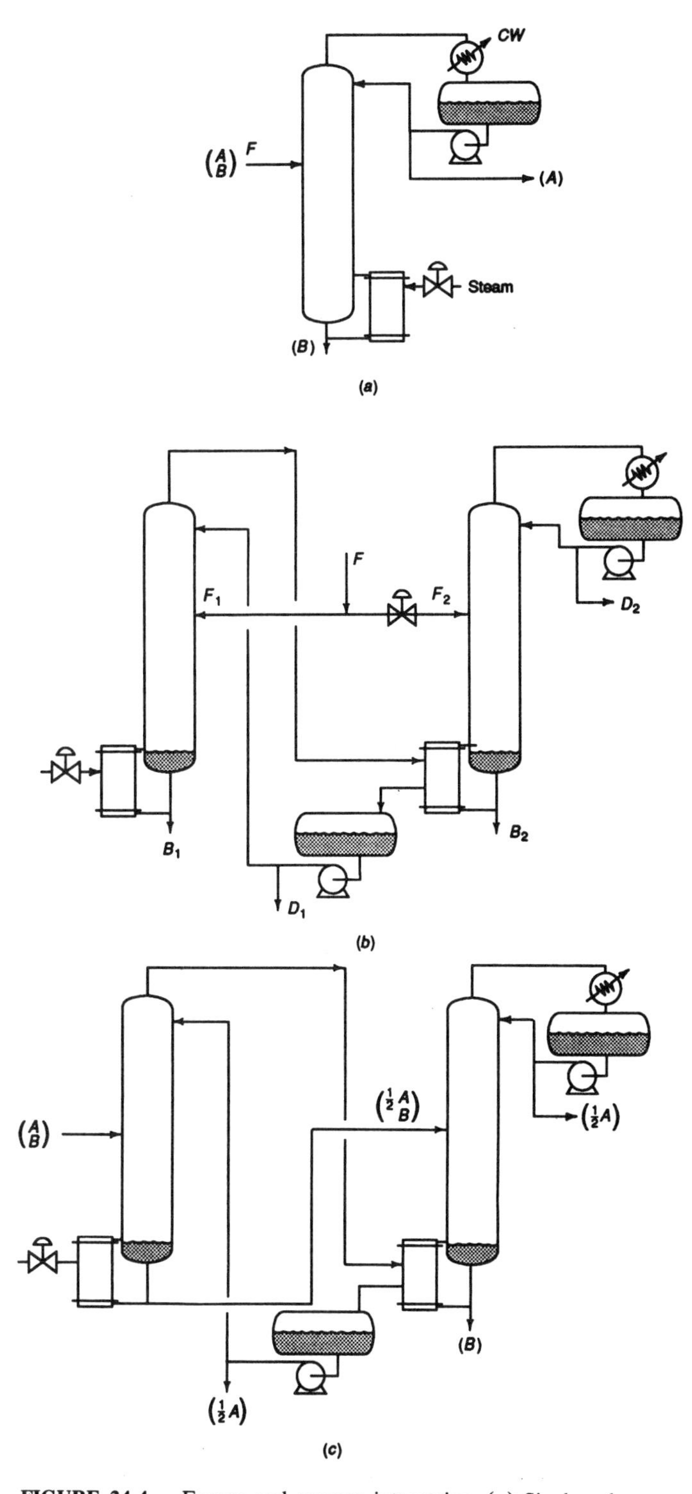

FIGURE 24-4. Energy and process integration. (*a*) Single column; (*b*) feed split (FS); (*c*) light split with forward integration.

TABLE 24-1 Design Specifications for the Binary Separation of Methanol–Water: Conventional Single Column and Several Heat-Integrated Configurations

Single Column	
Feed flow rate, kg-mol/min	45
Feed composition, m.f. methanol	0.5
Distillate flow rate, kg-mol/min	22.3
Distillate composition, m.f. methanol	0.96
Bottoms flow rate, kg-mol/min	22.5
Bottoms composition, m.f. methanol	0.04
Pressure, mm Hg	760
Total trays	12
Feed tray	6
Reflux ratio	0.831
External reboiler heat, kcal/min	369,000
Base temperature, °C	94.5
Reflux drum temperature, °C	64.9
Column diameter, m	3.2

Feed Split	Column 1	Column 2
Feed flow rate, kg-mol/min	22.506	22.494
Feed composition, m.f. methanol	0.5	0.5
Distillate flow rate, kg-mol/min	11.253	11.247
Distillate composition, m.f. methanol	0.96	0.96
Bottoms flow rate, kg-mol/min	11.253	11.247
Bottoms composition, m.f. methanol	0.04	0.04
Pressure, mm Hg	3900	760
Total trays	15	12
Feed tray	5	5
Reflux ratio	1.173	0.831
External reboiler heat, kcal/min	227,000	0
Base temperature, °C	146.7	94.5
Reflux drum temperature, °C	112.7	64.9
Column diameter, m	1.8	2.3

Light-Split Forward	Column 1	Column 2
Feed flow rate, kg-mol/min	45	34.263
Feed composition, m.f. methanol	0.5	0.356
Distillate flow rate, kg-mol/min	10.737	11.763
Distillate composition, m.f. methanol	0.96	0.96
Bottoms flow rate, kg-mol/min	34.263	22.5
Bottoms composition, m.f. methanol	0.356	0.04
Pressure, mm Hg	3900	760
Total trays	15	12
Feed tray	4	3
Reflux ratio	1.1	1.0
External reboiler heat, kcal/min	241,000	0
Base temperature, °C	125.6	94.5
Reflux drum temperature, °C	112.7	64.9
Column diameter, m	2.0	2.4

TABLE 24-1 *Continued*

Light-Split Reverse

	Column 1	Column 2
Feed flow rate, kg-mol/min	45	33.388
Feed composition, m.f. methanol	0.5	0.34
Distillate flow rate, kg-mol/min	11.612	10.888
Distillate composition, m.f. methanol	0.96	0.96
Bottoms flow rate, kg-mol/min	33.388	22.5
Bottoms composition, m.f. methanol	0.34	0.04
Pressure, mm Hg	760	2800
Total trays	12	15
Feed tray	5	5
Reflux ratio	0.772	1.24
External reboiler heat, kcal/min	0	22,600
Base temperature, °C	76.4	128.1
Reflux drum temperature, °C	64.9	96.0
Column diameter, m	2.3	2.0

Note that pressures in the high-pressure columns of the FS and the LSF configurations are the same (3900 mm Hg), but the base temperature in the high-pressure column of the LSF configuration is lower (125.6°C versus 146.7°C). This occurs because the material in the base of the column in the FS configuration is mostly water, whereas in the LSF it is a mixture of methanol and water. The lower base temperature means that lower-pressure steam can be used in the LSF configuration. The single column has the lowest base temperature (94.5°C) but consumes much more energy.

Note also that the pressure in the high-pressure column of the LSR is the lowest of all the heat-integrated configurations (2280 mm Hg). This configuration should be the most economical in those chemical systems where relative volatilities decrease more quickly with increasing pressure.

24-4 LIMITATIONS

Heat integration is quite attractive in many systems, but it has some limitations. It is seldom economical to lower pressures below the point where cooling water can be used in the condenser. It is also seldom economical to lower pressures too far into the vacuum range (except in the case of thermally sensitive material) because the low vapor densities increase column, condenser, and vapor-piping sizes quite significantly.

There is a limit to how high pressures can be raised to achieve heat integration. These limits can be imposed by several factors:

1. *Temperature sensitivity.* The temperature in the reboiler of the high-pressure column may be limited to some maximum value because of several possible factors: thermal degradation, polymerization, decomposition, fouling, corrosion, and so forth.
2. *Effect of pressure on vapor–liquid equilibrium.* Raising pressure reduces relative volatilities in some systems. This requires more trays and higher reflux ratios. Therefore pressures cannot be raised so high that the separation becomes uneconomically difficult. Note that some chemical systems show increases in relative volatility as pressures are increased, so the energy efficiency of these systems would be improved by raising pressures.

Knapp and Doherty (1990) discuss this problem for azeotropic separations.

3. *Critical temperatures or pressures.* The pressure must be kept well below the critical temperatures and pressures of the principal components so that the densities of the vapor and liquid phases are appreciably different. The hydraulics of the column (liquid flowing downward against a positive pressure gradient) would not function properly if the densities of the phases became too close.
4. *Temperature level of heat.* The temperature of the heating medium may be limited. For example, steam in the plant may only be available at pressures up to 150 psig (saturation temperature 366°F). Assuming a 40°F temperature difference in the high-pressure column reboiler, base temperatures above 326°F could not be used.

Heat integration is seldom used in multicomponent systems. If lighter-than-light key components exist in the feed, they will require a higher pressure for the same reflux-drum temperature. This will raise the base temperature in the low-pressure column and require a much higher pressure in the high-pressure column. If heavier-than-heavy key components exist in the feed, they will raise the base temperatures. This will also increase the required pressure in the high-pressure column. Thus most heat integration applications are in binary or ternary systems with components that do not have extremely different boiling points.

24-5 CONTROL PROBLEM

Heat-integrated systems involve two columns that are interconnected by energy flows and sometimes by process flows. They are usually multivariable.

Energy Integration Only

The control problem for those systems that have only energy integration is usually not severe because auxiliary reboilers and/or auxiliary condensers can be used to isolate the energy inputs for each column. As shown in Figure 24-3*a* the auxiliary reboiler on the second column would provide the additional energy that the second column requires. As shown in Figure 24-3*b* the auxiliary condenser removes the additional energy that the second column does not need.

Energy and Process Integration

The control problem for those systems that have both energy and process integration is more difficult. These systems feature higher-order multivariable dynamics and more severe interaction.

The feed-split configuration shown in Figure 24-7 is a 4 × 4 multivariable process. The four controlled variables are the purities of the two distillate products and the two bottoms products. The four manipulated variables are the heat input to the reboiler in the high-pressure column, the two reflux flows, and the feed split (what fraction of the total feed is fed into each column). In this multivariable environment, the methods for diagonal controllers discussed in Chapter 11 or multivariable controllers should be used.

The light-split forward configuration (Figure 24-4*c*) and the light-split reverse configuration have only three final products to control: compositions of two distillate products and one bottoms product. The three manipulated variables are heat input to the reboiler of the high-pressure column and the two reflux flow rates. This is a 3 × 3 multivariable process.

24-6 TOTAL HEAT-INPUT CONTROL

The use of a "total heat-input" controller is recommended in those heat-integrated systems that contain auxiliary reboiler and/or condensers. Figure 24-5 illustrates the idea.

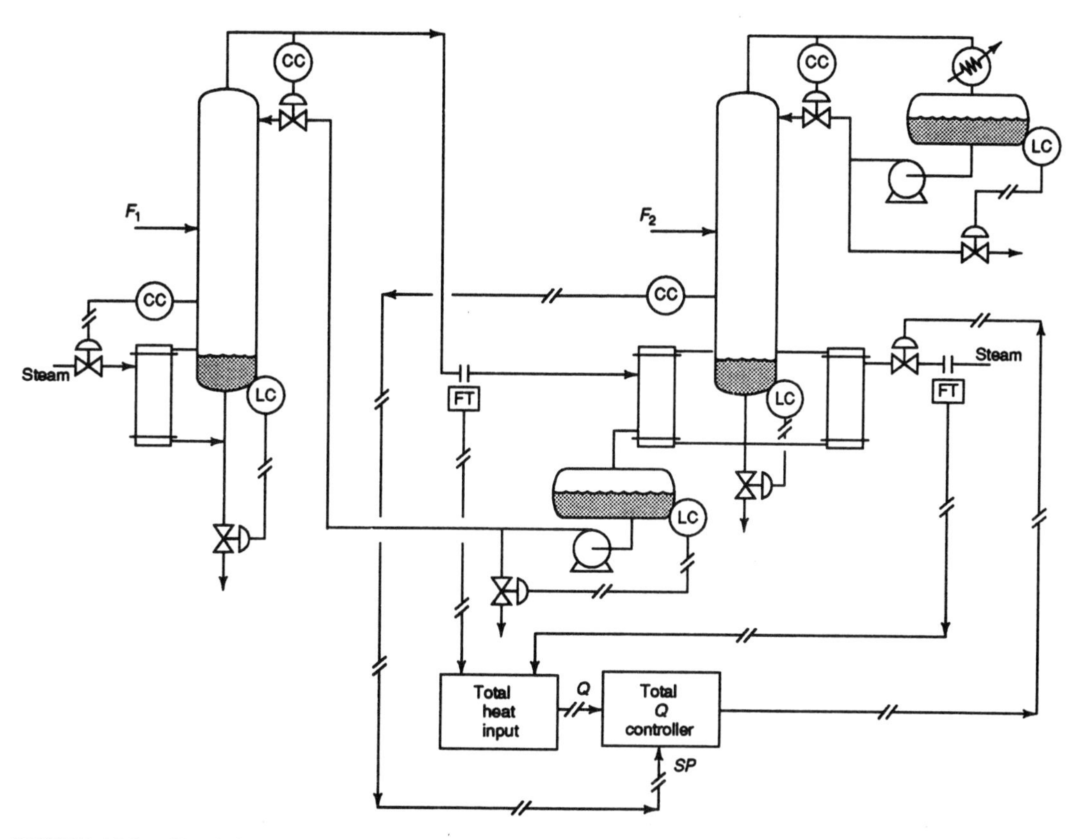

FIGURE 24-5. Total Q control with auxiliary reboiler.

The first high-pressure column is controlled in a conventional way: a composition controller (CC) adjusts reflux to control distillate purity, and a composition controller adjusts steam to the reboiler to control bottoms purity. In the low-pressure column the flow rates of the steam to the auxiliary reboiler and the vapor from the high-pressure column to the condenser–reboiler are both measured. These flow rates are multiplied by their appropriate heats of condensation and the results are added together to obtain the total heat removed from both streams as they condense in the reboiler of the low-pressure column. This signal is fed into a "total Q" controller that adjusts the steam valve on the auxiliary reboiler to keep the total heat input at its setpoint. The setpoint of the total Q controller is the output signal from the bottoms composition controller.

Note that the pressure in the high-pressure column is not controlled in the scheme shown. It simply floats up and down as dictated by the required differential temperature for heat transfer in the reboiler. Thus at low throughput rates the pressure in the high-pressure column will be lower than at high throughputs because a smaller ΔT is needed to transfer the smaller amount of energy. The pressure in the high-pressure column will also change if the compositions change in either the base of the low-pressure column or the reflux drum of the high-pressure column. The variable pressure

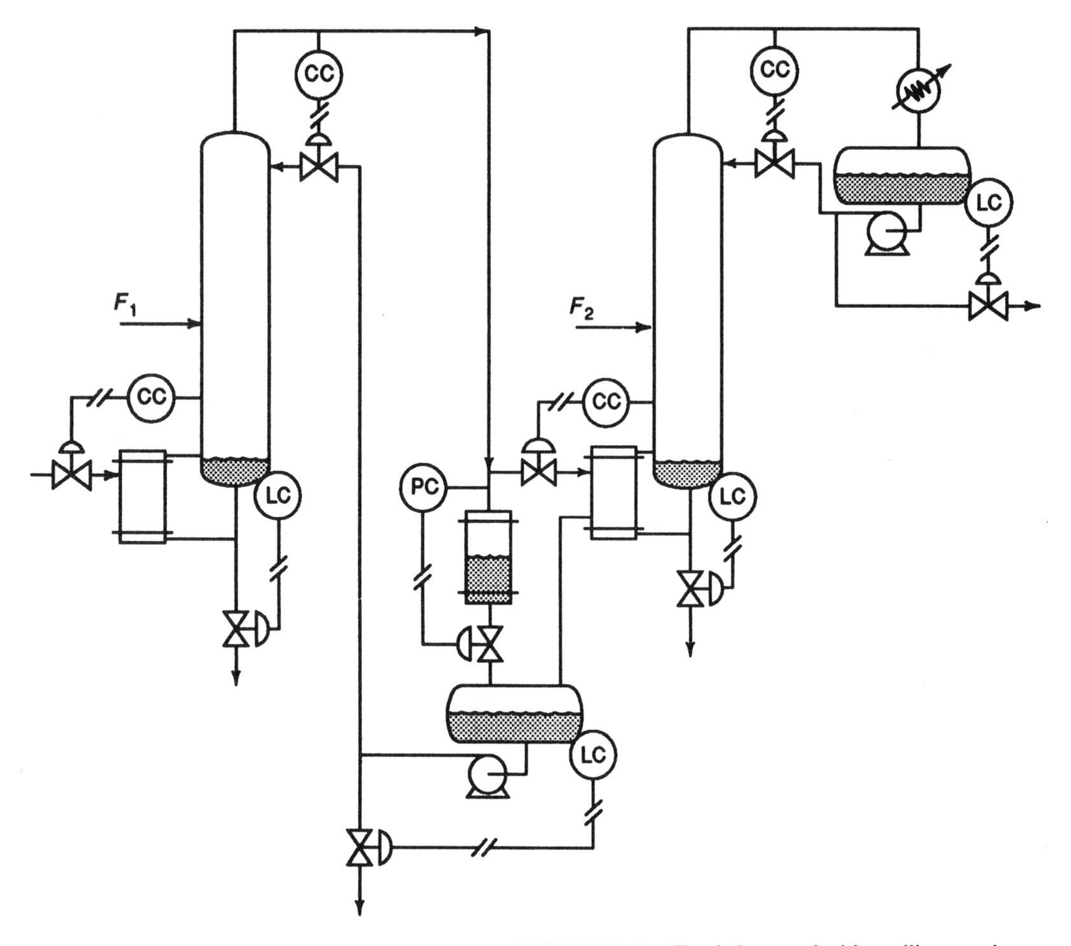

FIGURE 24-6. Total Q control with auxiliary condenser.

makes it vital to use pressure compensation of temperature signals if they are being used to infer composition.

If an auxiliary condenser is being used, the heat removed in the condenser (cooling water flow rate times the temperature increase of the cooling water) can be *subtracted* from the total heat available in the vapor from the high-pressure column to determine the "net" heat input to the reboiler. A simple and effective alternative is to place a control valve in the vapor line to the condenser–reboiler and use it to control the heat input to the second column. The heat removal rate in the auxiliary condenser could be used to control the pressure in the high-pressure column. As sketched in Figure 24-6, the auxiliary condenser heat removal is adjusted by the control valve on the liquid line from the bottom of the condenser. The area for condensation is varied by covering or exposing more area for heat transfer.

This total heat-input controller decouples the heat inputs for the two columns and eliminates some of the interactions present in the system.

24-7 INCENTIVES FOR COMPOSITION CONTROL OF ALL PRODUCTS

We mentioned earlier in this book that for a single distillation column the minimum en-

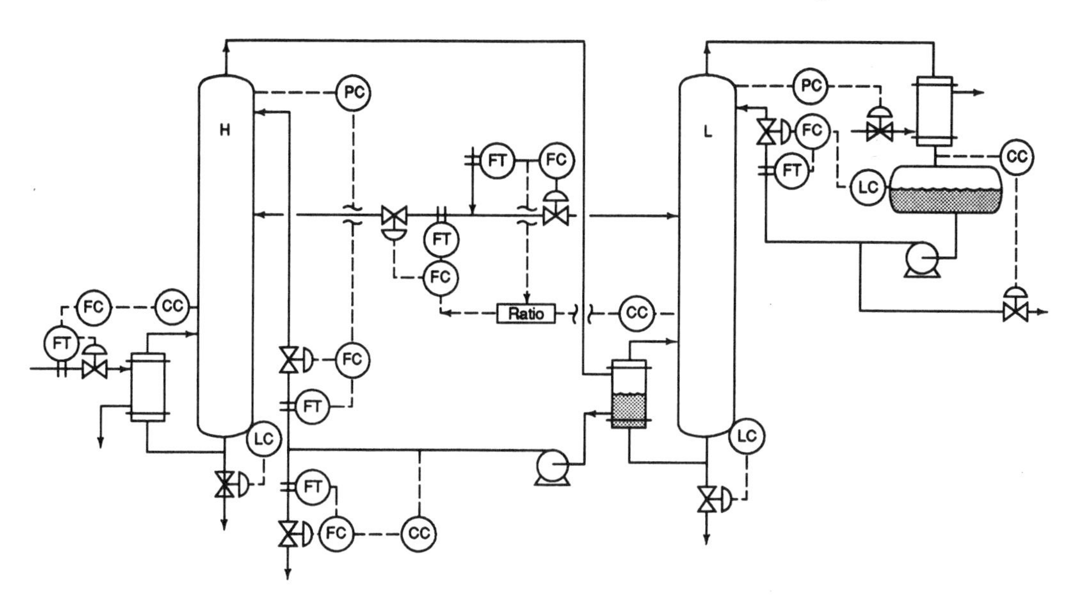

FIGURE 24-7. 2-2 scheme (FS).

ergy is consumed by controlling both products at their specified purity levels. This "dual composition" control strategy is, however, frequently not economically justified because the energy savings compared to single-end control are often fairly small in many binary separations. The control engineer needs to quantify the economic incentives for dual composition control.

The same question needs to be answered in heat-integrated systems. For binary systems is it often possible to simplify the control problem by not controlling the compositions of all product streams. Figure 24-7 shows the feed-split configuration with the compositions of both products in both columns controlled (called the 2-2 scheme). This is a full-blown 4 × 4 multivariable control problem.

Figure 24-8 illustrates some typical steady-state results for many binary systems. The specific chemical system is isobutane–*n*-butane. Feed composition is varied over the expected range and the compositions of all four products are held constant. The required steady-state changes in several of the manipulated variables are shown. Because the curves of reflux-to-feed in the high- and low-pressure columns and the curve of the feed split (F_H/F_L) are fairly flat, there may be very little energy saved by using the 2-2 scheme (4 × 4 multivariable).

The control problem can be reduced to 3 × 3 by not controlling bottoms composition in the low-pressure column and simply fixing the feed split (see Figure 24-9*a*; called the 2-1 scheme). The first column is just a 2 × 2 system. The second column is disturbed by changes in the vapor entering its reboiler from the first column, but the control problem in the second column is just single-input–single-output (SISO). The interaction between the columns has been broken.

Figure 24-9*b* shows an even further simplification (the 1-1 scheme) in which only the distillate compositions are controlled. The feed-split ratio and the reflux to the high-pressure column are fixed. Heat input in the high-pressure column and distillate flow rate in the low-pressure column are the manipulated variables. Each column is a simple SISO control problem.

These simpler approaches will not be economically attractive in all cases, particularly multicomponent systems and systems with low relative volatilities because too much energy will be wasted.

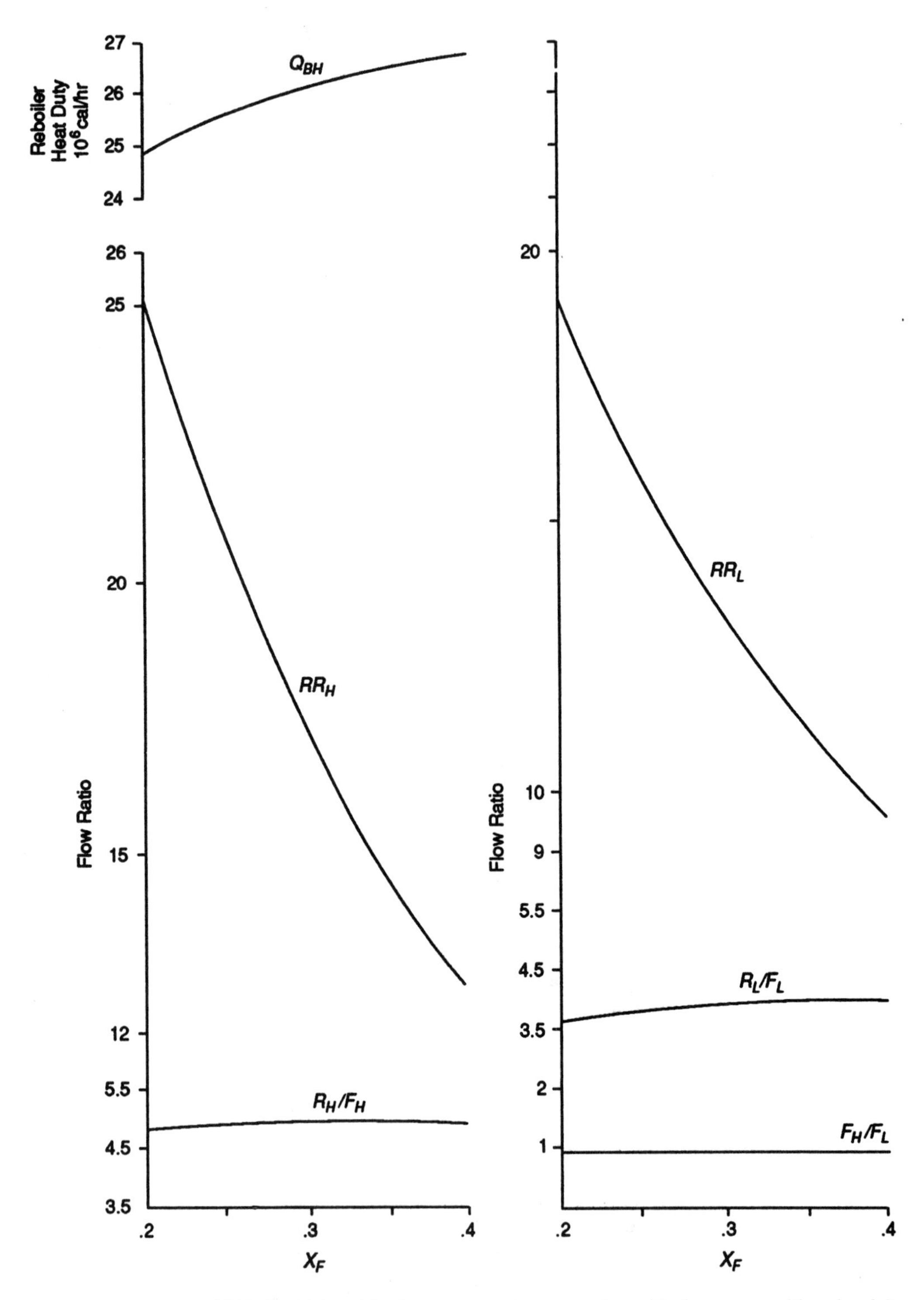

FIGURE 24-8. FS changes in system variables with feed composition for 2-2 scheme.

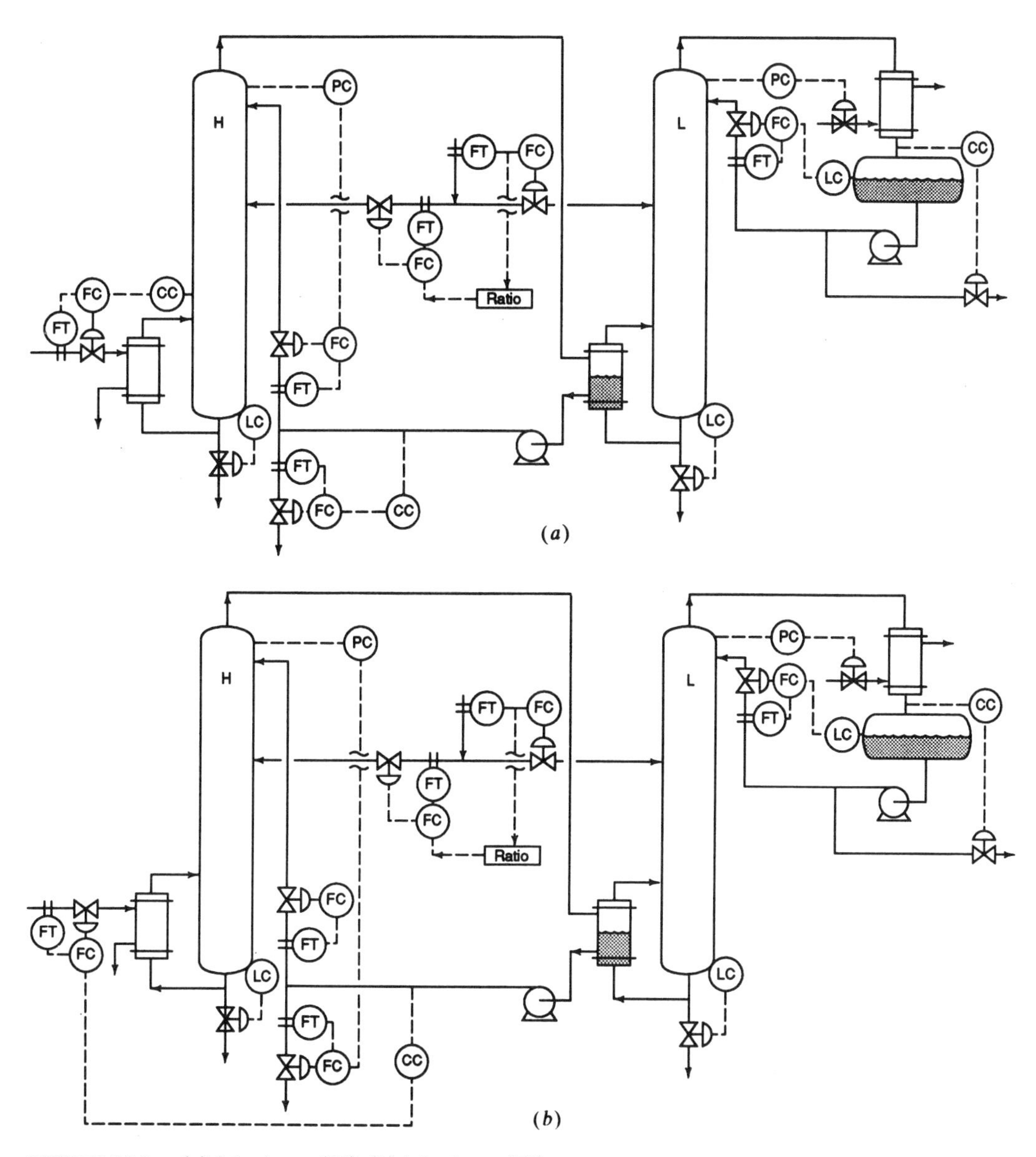

FIGURE 24-9. (*a*) 2-1 scheme (FS). (*b*) 1-1 scheme (FS).

24-8 CONCLUSION

This chapter has discussed some of the many possible heat integration configurations and illustrated their very significant savings in energy consumption. The control of these systems is more difficult than single columns because of the interactions of both energy and material flows.

Conventional diagonal control systems have been shown to provide effective control of these complex systems, but controller tuning must take the interactions into account.

Reference

Knapp, J. P., and Doherty, M. F. (1990). Thermal integration of homogeneous azeotropic distillation sequences. *AIChE J.* 36, 969–984.

25

Batch Distillation

Cristian A. Muhrer
Air Products and Chemicals, Inc.

William L. Luyben
Lehigh University

25-1 INTRODUCTION

Up to this point in this book, almost all of the distillation columns considered were operated continuously. These columns are fed continuously and produce product streams continuously.

There is another type of distillation column that operates in a batch mode. Batch distillation columns have always been important in the specialty chemical industry with its small-volume, high-value products. Batch distillation has the advantage of being able to produce a number of products from a single column that can handle a wide range of feed compositions, numbers of components, and degrees of difficulty of separation (wide ranges of relative volatilities and product purities). Even though batch distillation typically consumes more energy than continuous distillation, it provides more flexibility and involves less capital investment. Because energy costs are not too significant in many small-volume, high-value chemical processes, batch distillation is often attractive for this class of products.

The purpose of this chapter is to present a brief overview of the design, operation, and control of batch distillation columns. This area has been the subject of quite a bit of work in recent years, and a comprehensive treatment would be too lengthy for a single chapter. Therefore, we will limit our discussion to some of the important concepts and problems. A fairly extensive list of references is provided for the reader who wishes to dig into this subject in more depth.

25-2 BASIC OPERATIONS

25-2-1 Process

In conventional batch distillation, a liquid mixture is charged to a vessel and heat is added to produce vapor that is fed into a rectifying column. Vapor from the top of the column is condensed and some of the liquid condensate is returned to the top of the column as reflux. The rest is removed as products that are collected according to their boiling points. This scheme is referred to as a conventional or "direct" batch distillation.

In contrast, when the liquid from the vessel is fed into the top of a stripping column and products are withdrawn from the bottom, this scheme is called an *inverted batch distillation* process (Robinson and Gilliland, 1950). Bernot, Doherty, and Malone (1991) discuss the use of a sequence of these processes to separate an azeotropic mixture into its pure components.

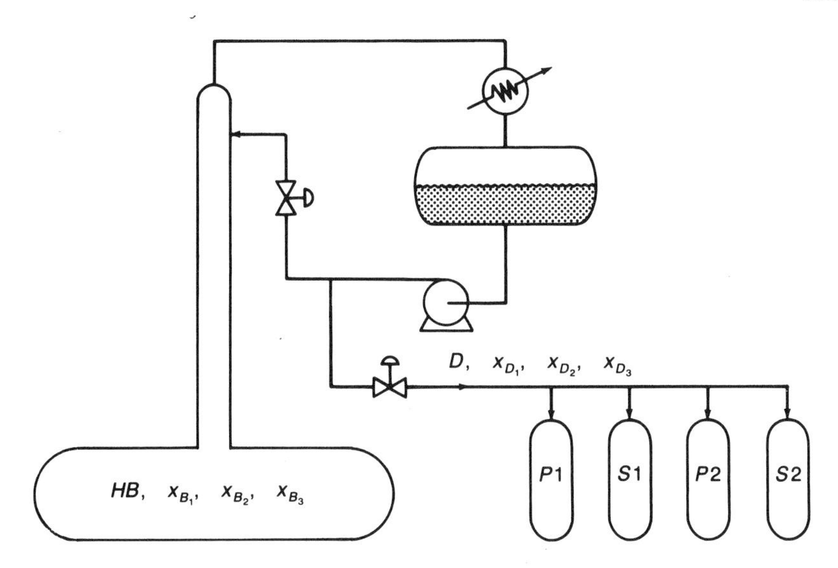

FIGURE 25-1. Ternary batch distillation.

In binary separations, there are two products and one slop cut (the material removed as distillate during the period when the distillate contains too much heavy component to include in the light product and the contents of the still pot and column contain too much light component). In ternary separations, there are three products and two slop cuts; in general we obtain at most NC − 1 slop cuts and NC products (where NC is the number of components). Figure 25-1 shows the configuration and nomenclature for a ternary batch distillation.

After the fresh feed and any recycled slop cuts have been charged to the still pot, there is an initial start-up period during which the column operates at total reflux until a distillate of the desired purity in the most volatile component can be withdrawn. Batch time is established by the time it takes to produce NC − 1 distillate products and to purify the heavy product left in the still pot.

25-2-2 Composition Profiles

In batch distillation the compositions at all points in the column are continually changing. Figure 25-2 shows some typical changes in distillate composition (x_D) and bottoms composition (x_B) for a binary system. During the start-up period when the column is on total reflux, the top composition gets richer and richer in the more volatile component. When the purity of the material in the column overhead reaches its specified level, distillate flow is begun at some appropriate flow rate (which reduces the reflux ratio from infinity to some finite value). There are a number of ways to adjust the

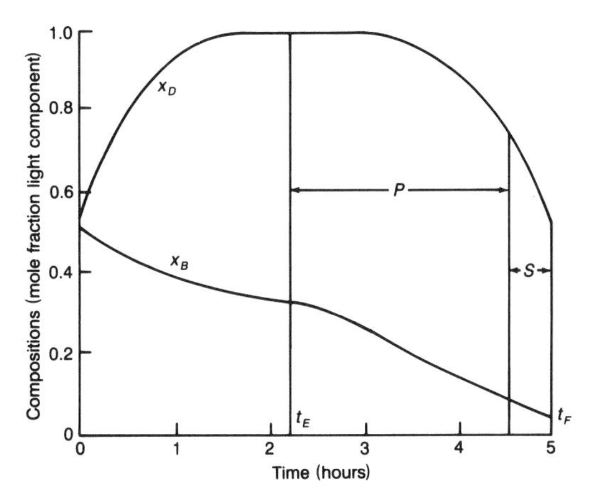

FIGURE 25-2. Typical composition profiles for binary batch distillation.

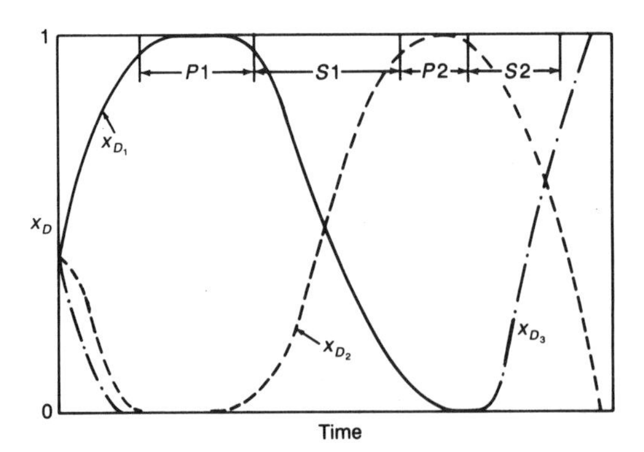

FIGURE 25-3. Typical distillate composition profiles for ternary batch distillation.

reflux ratio during the course of the batch. The easiest and most commonly used method is to simply hold the reflux ratio constant (at some optimum constant value). This policy produces distillate whose composition varies during the batch.

The distillate is stored in a product tank. When the average composition of the material in the tank drops to the specified purity level x_D^{spec}, the distillate stream is diverted to another tank and the slop cut S is produced. When the material in the still pot and in the column (not including the reflux drum because this is added to the slop cut) meets its specification for the heavy product, x_B^{spec}, the binary batch cycle is stopped.

In a ternary system (see Figure 25-3) the lightest component is removed in the first distillate product ($P1$). Then the first slop cut ($S1$) is produced until the distillate purity (in terms of the intermediate component) reaches the specification value. The second distillate product ($P2$) is produced until its average purity drops to its specified value. At that point the purity of the last product ($P3$) may be better than specified. If not, a second slop cut ($S2$) is produced as required to purify the heavy product.

25-2-3 Slop Cuts

As already mentioned, in a binary distillation the slop cut is the distillate that is removed during the period when the overhead contains too much heavy component to be used in the light product and the material left in the still pot and column still contains too much light component. Note that it is assumed that the liquid on the trays (or packing) in the column drains into the still pot. This slop cut will contain both components. It is usually recycled back to the next charge to the still pot.

For ternary systems, there can be two slop cuts. The first will contain mostly the light component and the intermediate component. The second slop cut will contain mostly intermediate component and the heavy component. The size of these cuts depends on the composition of the initial charge, the number of trays in the column, the reflux ratio used, and the degree of difficulty of the different separations (relative volatilities and product purities).

25-3 ASSESSMENT OF PERFORMANCE: CAPACITY FACTOR

In a batch column we can have many objectives, such as minimize batch time, maximize production in a fix period of time, and so forth. One quantitative measure of performance is the capacity factor defined by Luyben (1971). The capacity factor (CAP) of the batch still is defined as the total specification products produced divided by the total time of the batch. Extending the original equation to a multicomponent system and assuming 0.5 h to empty and recharge the still pot

$$\text{CAP} = \frac{\sum_{j=1}^{NC} P_j}{t_F + 0.5} \tag{25-1}$$

In this definition it is assumed that all products have the same economic value. A weight factor can be easily included to account for differences in prices. The total batch time t_F (hours) includes the time at total reflux and the time producing the products and slop cuts. Note that the total products produced also will be equal to the net fresh feed charged to the still pot, which is equal to the total still pot charge (H_{B0}) minus the sum of the slop cuts:

$$\sum_{j=1}^{NC} P_j = H_{B0} - \sum_{j=1}^{NC-1} S_j \tag{25-2}$$

The capacity factor can be used to determine the optimum column design. As one would intuitively expect, CAP increases as more trays are added to the column for the same vapor rate in the column. This corresponds to the same energy consumption and the same column diameter. Thus, more feed can be processed for the same energy cost. Alternatively, a smaller diameter column and less energy could be used to handle the same amount of material per unit time. Of course, the capital cost of the column increases as more trays are added. Therefore, the capacity factor can be used to perform the necessary economic trade-off calculations between capital and energy cost to determine the optimum number of trays and optimum reflux ratio.

25-4 MODELS

Since the early 1960s when large digital computers become available, rigorous computer-based calculation procedures have been generated for binary and multicomponent batch distillation. Meadows (1963) presented a multicomponent batch distillation model, Distefano (1968a, b) extended the model that was used to simulate several commercial batch distillation columns, and Boston et al. (1981) further extended the model to efficiently handle the nonlinear nature of the problem. The stiff nature of multicomponent batch distillation equations sometimes requires the use of special numerical methods. Distefano (1968b) presented the cause of instabilities that arise during the numerical solutions of ordinary equations, showed it in the solution of unsteady-state distillation, investigated theoretically the stability characteristics of several numerical methods, and predicted the maximum increment size to be used in each one. Holland and Liapis (1983) solved the equations for multicomponent batch distillation using the two-point implicit method and the theta method convergence, as well as a combination of the two implicit method and the $2N$ Newton–Raphson method. Azeotropic multicomponent batch distillation was treated by Van Dongen and Doherty (1985), who developed a model for ternary azeotropic columns operating at high reflux ratios and with a large number of plates. The model agreed well with experimental results. Bernot, Doherty, and Malone (1991) predicted the possible set of fractions that can be obtained for any initial composition of an azeotropic mixture.

25-4-1 Differential Distillation

The simplest type of batch distillation is called *differential distillation*, shown in Figure 25-4. The vapor is generated by boiling the liquid in the still pot. No rectifying column is used. Rayleigh developed an equation for differential distillation. Consider a binary batch of H_0 moles of liquid. At any instant during the distillation there are H moles of liquid left in the still pot. At that instant, the mole fraction of the more

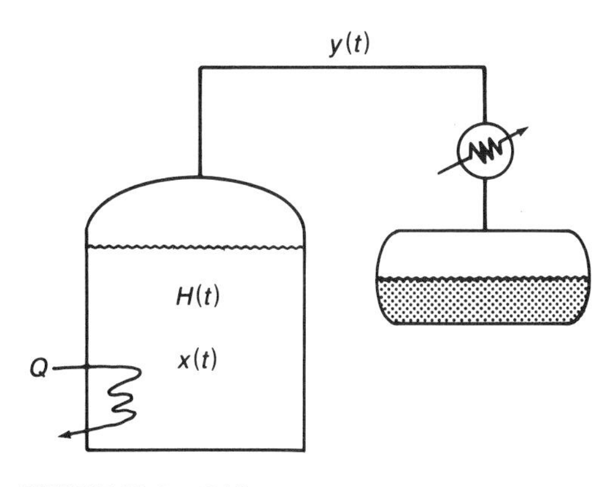

FIGURE 25-4. Differential distillation.

volatile component in the liquid is x and its corresponding vapor composition is y. The total amount of light component in the liquid is given by Hx. If a differential amount of liquid dH is vaporized, the liquid composition will decrease from x to $x - dx$, and the amount of liquid will decrease from H to $H - dH$ moles. There will be left in the still $(x - dx)(H - dH)$ moles of light component, whereas the amount of light component removed is $y\,dH$. The material balance for the light component is

$$xH = (x - dx)(H - dH) + y\,dH \quad (25\text{-}3)$$

Expanding and neglecting the second-order differential gives

$$\frac{dH}{H} = \frac{dx}{y - x} \quad (25\text{-}4)$$

By applying the proper limits, we obtain the well-known Rayleigh equation (Coates and Pressburg, 1961; Block, 1961):

$$\ln\left(\frac{H_1}{H_0}\right) = \int_{t=0}^{t_1} \frac{1}{y - x}\,dx \quad (25\text{-}5)$$

This simple example illustrates the "dynamic" nature of the batch distillation process. As a result, a steady-state analysis of the type employed for continuous distillation cannot be made of the column behavior.

25-4-2 Pseudo-Steady-State Models

Several attempts to study the process as a variation of a continuous column have been made. Galindez and Fredenslund (1988) modelled the unsteady-state process by simulating a succession of short periods of time in which a continuous, steady-state distillation process is assumed. To illustrate the procedure let us assume a binary system with equimolal overflow where we can use the McCabe–Thiele diagram to relate compositions. Figure 25-5 shows the McCabe–Thiele analysis for a two tray column that is operated at constant reflux ratio. The operating lines at different times are a series of parallel lines, the slopes being the same because L/V is constant. As the light component is depleted from the system, the composition of the heavy component increases on all trays and in the still pot. The dashed lines give the construction for a later time when the overhead composition is x_{D_2}.

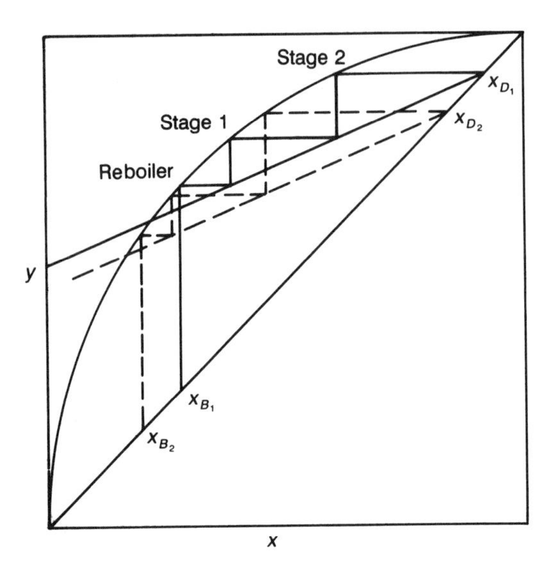

FIGURE 25-5. McCabe–Thiele diagram of steady-state batch distillation.

This type of pseudo-steady-state model can only be used in those situations where the liquid holdups on the trays in the column and in the reflux drum are negligible compared to the volume of the still pot. Chiotti and Iribarren (1991) used this type of steady-state model for optimization studies.

25-4-3 Rigorous Dynamic Models

To describe batch distillation columns accurately, we must use a dynamic model that includes tray and reflux-drum holdup dynamics. Based on the assumptions of constant tray and reflux drum holdups, constant relative volatilities, and equimolal overflow, the equations describing a multicomponent

batch distillation system are
Still pot:

$$\frac{dM_B}{dt} = -D \tag{25-6}$$

$$\frac{d[M_B x_{Bj}]}{dt} = Rx_{1j} - Vy_{Bj} \tag{25-7}$$

$$y_{Bj} = \frac{\alpha_j x_{Bj}}{\sum_{k=1}^{NC} \alpha_k x_{Bk}} \tag{25-8}$$

Tray n:

$$M_n \frac{dx_{nj}}{dt} = R[x_{n+1,j} - x_{nj}] + V[y_{n-1,j} - y_{nj}] \tag{25-9}$$

$$y_{nj} = \frac{\alpha_j x_{nj}}{\sum_{k=1}^{NC} \alpha_k x_{nk}} \tag{25-10}$$

Tray N_T (top tray):

$$M_{N_T} \frac{dx_{N_T,j}}{dt} = R[x_{Dj} - x_{N_T,j}] + V[y_{N_T-1,j} - y_{N_T,j}] \tag{25-11}$$

$$y_{N_T,j} = \frac{\alpha_j x_{N_T,j}}{\sum_{k=1}^{NC} \alpha_k x_{N_T,k}} \tag{25-12}$$

Reflux drum:

$$M_D \frac{dx_{Dj}}{dt} = Vy_{N_T,j} - [R + D]x_{Dj} \tag{25-13}$$

$$R = V - D \tag{25-14}$$

These equations can be solved using the numerical integration methods discussed in Chapter 3.

25-4-4 Fitting Models to Experimental Batch Distillation Data

Muhrer, Grassi, and Silowka (1990) presented an industrial case study that illustrated how a dynamic model can be used in conjunction with plant data to increase process productivity. It was assumed that the vapor–liquid equilibrium was reasonably well known. Simulations using the model were run and tray efficiency and holdup were adjusted until the model matched the plant data. Figure 25-6 gives a comparison of plant data and model predictions.

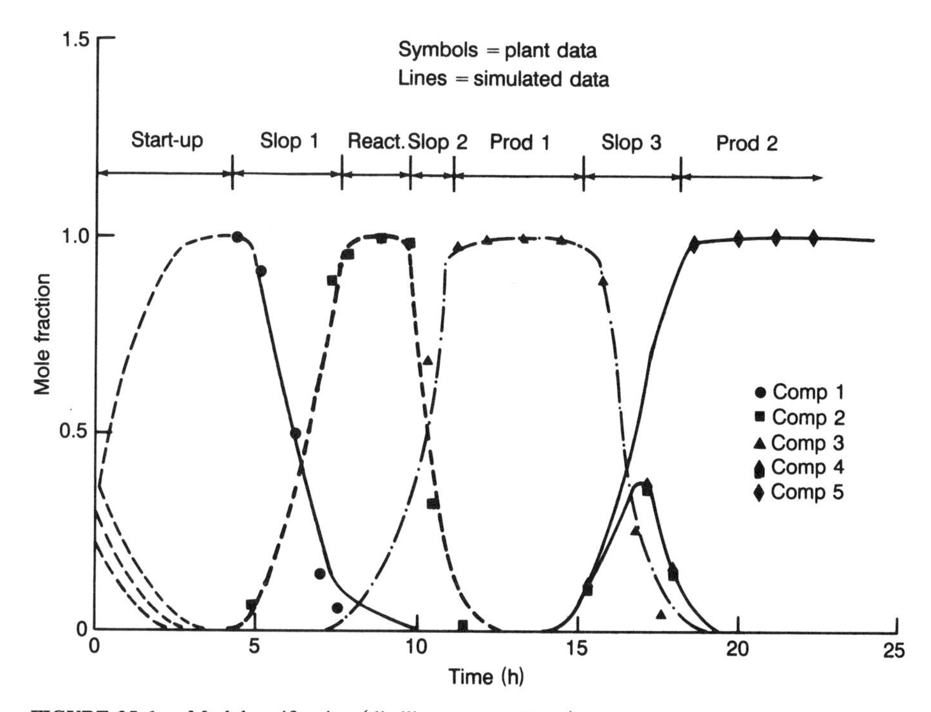

FIGURE 25-6. Model verification (distillate compositions).

The model was then used to test various control schemes, investigate the effect of operating procedures on batch cycle, and determine the necessary measurements needed to infer product and slop cuts. In one typical example process, the amount of slop cuts and the cycle time for the batch were reduced significantly. Cycle time was reduced from 21 to 16 h by using model analysis.

The model was also used to design new columns by applying first principles and augmenting the analysis with engineering experience obtained from working systems at similar plants.

25-5 COMPARISON WITH CONTINUOUS DISTILLATION

Continuous distillation is normally more energy efficient than batch distillation, but batch distillation requires only one distillation column, even for the separation of a multicomponent mixture, whereas in continuous distillation a sequence of columns is usually required. To illustrate the differences let us compare the two alternative processes for a specific ternary separation: relative volatilities of 9/3/1, product purities of 95 mol%, and feed composition 30 mol% light, 30 mol% intermediate, and 40 mol% heavy. By use of standard short-cut design procedures for continuous distillation with reflux ratios 1.1 times the minimum, a total of 100 mol/h of vapor boilup in two continuous columns would correspond to a feed rate (or capacity) of about 80 mol/h. Each column would require 12 trays.

Making the same separation in a 20-tray batch distillation with the same energy consumption (100 mol/h of vapor boilup) yields a capacity of about 37 mol/h. If 40 trays were used in the batch column, the capacity would be about 45 mol/h. If 80 trays were used, capacity would increase to about 55 mol/h.

25-6 REFLUX RATIO TRAJECTORIES

One of the most interesting aspects of the operation of a batch distillation column is the question of how to best set the reflux ratio during the course of the batch cycle. There have been a variety of approaches. The optimum theoretical solution to the problem involves the use of the calculus of variations to solve for the optimum time trajectory of reflux ratio that extremizes some objective function (Coward, 1967). The mathematical complexity of this elegant approach makes it very difficult to apply in realistically complex industrial batch distillation columns.

The two most frequently used approaches are to operate at a constant reflux ratio (with distillate composition varying) or to operate with distillate composition (or overhead temperature) held constant (with reflux ratio varying). The former strategy is the more commonly used because of its simplicity. Several investigators have studied the differences between these two approaches. The results reported are conflicting: Some researchers claim significant differences and others have found only slight differences. This remains an open question. Results are probably strongly dependent on the size of the column, the difficulty of separation (relative volatility), and the product purities.

As one specific example, Luyben (1988) studied a ternary separation with relative volatilities of 9/3/1 in two different columns (20 trays and 40 trays). Product purities were 95 mol%. Capacity factors for the optimum fixed reflux operation were 38.6 and 45.4 for the 20 and 40 tray columns. Capacity factors for a variable reflux operation were 33.8 and 44.4. These results indicate no major differences.

As a conflicting example, Farhat et al. (1990) studied fixed, linear, and exponential reflux trajectories. They claim a 10% improvement, which they felt was significant. However, they used a very small column (nine trays) and a very simplified model (neglected tray and reflux-drum holdups).

More work is needed to clear up some of these questions. As of now, a reasonable mode of operation would appear to be the use of optimal fixed reflux ratios, which could be different for each product cut if relative volatilities are different between the various components.

25-7 PRESSURE TRAJECTORIES

There has been very little investigation of how column pressure should be varied over the course of the batch. Some of the possible strategies are now briefly discussed.

25-7-1 Constant Pressure

If pressure is held constant, overhead temperature will increase as the components leaving in the distillate become less volatile. Still-pot temperature will also increase. This may be undesirable if temperature-sensitive components are present.

25-7-2 Constant Reflux-Drum Temperature

If the reflux drum temperature is held constant during the batch, the pressure in the column will decrease as the heavier components come overhead. This may be desirable from the standpoint of temperature sensitive material. It also may make the separation easier if relative volatilities increase as pressure decreases.

However, there is a question of capacity. Lower pressure results in lower vapor densities and higher vapor velocities. If the column is vapor-flooding limited, the throughput of the column may be reduced if pressure is reduced. In this situation, there is probably an optimum operating strategy that varies both heat input and column pressure during the batch cycle to maximize capacity.

25-8 COLUMN DESIGN

Most studies of batch distillation start with a given column (i.e., the number of trays is fixed) and only consider the optimum operation using this fixed column. There is, of course, the very important and fundamental design question of how many trays should be used.

Using more trays in a batch distillation column increases the capacity factor. This means that energy costs are decreased for a given separation job. Alternatively it means that capacity is maximized for a given energy consumption. These statements assume that there are no hydraulic problems with

TABLE 25-1 Cases for Engineering Economic Analysis

Case Number	Material of Construction	Product Purities (mol%)	Energy Cost ($/10^6 Btu)
1	Carbon steel	95	2.50
2			5.00
3		99	2.50
4			5.00
5	Stainless steel	95	2.50
6			5.00
7		99	2.50
8			5.00
9	Monel	95	2.50
10			5.00
11		99	2.50
12			5.00

adding additional trays. This is not always the case, particularly in vacuum distillation systems because of the increased pressure drop.

However, using more trays also increases the height of the column and therefore its capital cost. So the optimum design of a batch distillation column must balance the economic and operational aspects of all these considerations. Al-Tuwaim and Luyben (1991) gives some simple guidelines to aid in the shortcut design of batch distillation columns for different cost of energy, materials of construction, relative volatilities, and product purities.

For example, Table 25-1 defines several cases with typical materials of construction, product purities, and energy costs. For each of these cases, Figure 25-6 gives the number of trays that gives the minimum cost (capital plus energy) as a function of relative volatility.

For a specific example, take a batch separation of a ternary mixture with 95% product purities, Monel materials, and $2.50/10^6$ Btu energy cost (case 9 in Table 25-1). Figure 27-7 indicates that 70 trays should be used if relative volatilities are 4/2/1, but 30 trays should be used if relative volatilities are 9/3/1. Figure 25-8 gives the reflux ratio that gives the maximum capacity factor. The optimum reflux ratios for the two cases considered above are 2.2 and 1.5, respectively.

25-9 SLOP CUT PROCESSING

As discussed earlier, there can be up to NC − 1 slop cuts produced when a mixture of NC components is separated in a batch distillation column. The best way to reprocess these slop cuts is an interesting problem. There are a host of alternative procedures.

The simplest is to recycle all the slop cuts to the next batch. The initial charge to the still pot is then all the slop cuts plus fresh feed. However this simple strategy makes little sense from a thermodynamic point of view. We are mixing together streams that have already been partially separated. Each slop cut is approximately a binary mixture.

Quintero-Marmol and Luyben (1990) studied several slop-handling strategies. Some typical results are summarized in the following text for two of the most interesting alternatives and are compared with the total slop recycle strategy.

25-9-1 Alternatives

Three operating strategies are considered:

1. *Total slop recycle.* All slop cuts are mixed with fresh feed to make the initial charge to the still pot for the next batch. The capacity of the column is total product streams produced, which is equal to the total still-pot charge minus the slop cuts produced. This is the most simple procedure and the most commonly used. However, its thermodynamic efficiency is low.
2. *Multicomponent-binary batches.* A number of batch distillations are run producing the NC product streams and the NC − 1 slop cuts. All slop cuts are stored in their own segregated tankage until enough material has been accumulated of each slop cut to run a batch distillation of each slop cuts. Because these slop cuts are essentially binary mixtures, these batch distillations are binary. The slop cut from each binary distillation is recycled to its own slop cut tank. Products of the binary distillations are mixed with the products of the multicomponent distillations.

   This procedure is the most efficient, but it requires more tankage and more complex operating procedures.
3. *Fed batch.* The charge to the still pot is all fresh feed. During the course of the batch distillation each slop cut is fed into the middle of the column when the composition on the feed tray is approximately equal to the composition of the slop cut. This procedure requires less

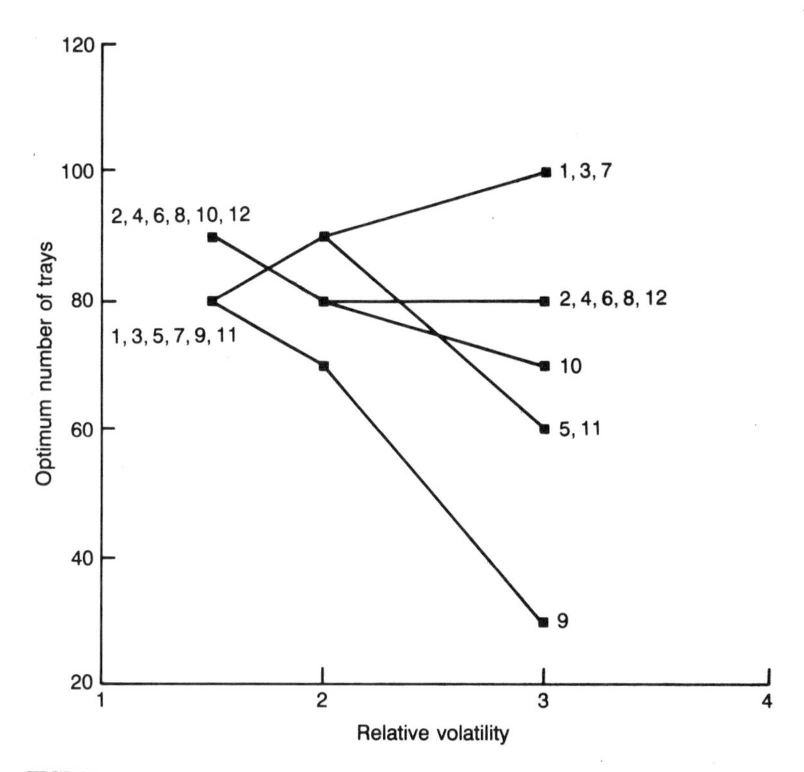

FIGURE 25-7. Cases for economic design.

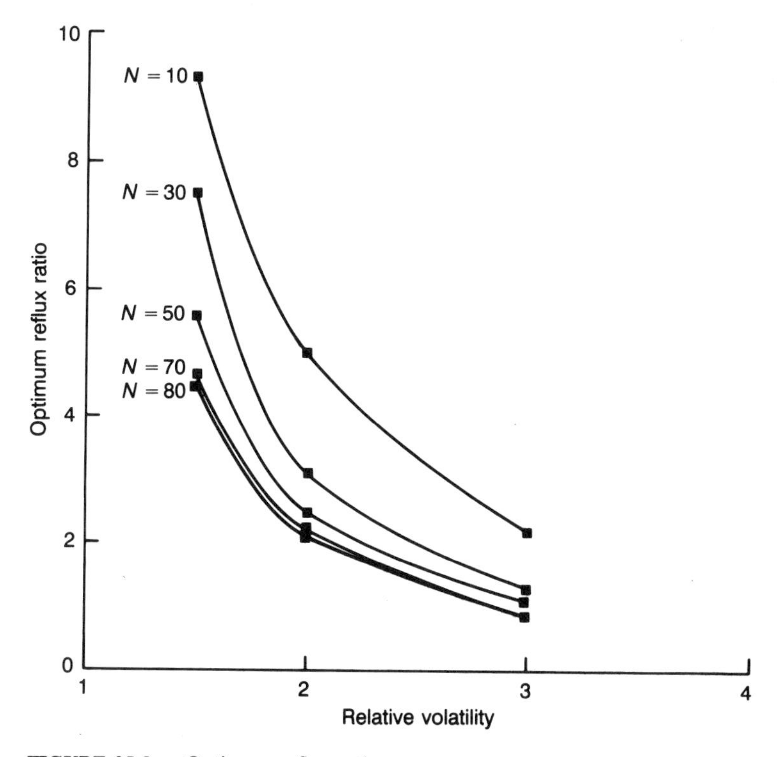

FIGURE 25-8. Optimum reflux ratios.

TABLE 25-2 Capacity Factors for Alternative Slop-Handling Strategies[a]

Relative volatilities	9/3/1	4/2/1
Product purities = 95%		
Slop recycle	46.2	25.4
Multicomponent binary	59.2	33.2
Fed batch	50.6	27.9
Product purities = 99%		
Slop recycle	31.4	15.7
Multicomponent binary	40.8	21.7
Fed batch	32.5	18.0

[a]Basis: 40 tray column, ternary mixture, fresh feed 30/30/40 mol%.

tankage and is more efficient than slop recycle. However there are a large number of parameters to optimize: slop feed time and flow rate for each slop cut, feed tray location, reflux ratio, and so forth.

25-9-2 Results

Table 25-2 gives typical results (in terms of capacity factors) for four different systems (two product purities and two relative volatilities). The multicomponent-binary slop-handling strategy gives the highest capacity factor. Figure 25-9 compares the two slop-handling strategies for different product purities and different numbers of trays.

The increases in capacity factors in going from the easy slop recycle operation to the more complex binary-multicomponent operation are on the order of 30 to 50%. This

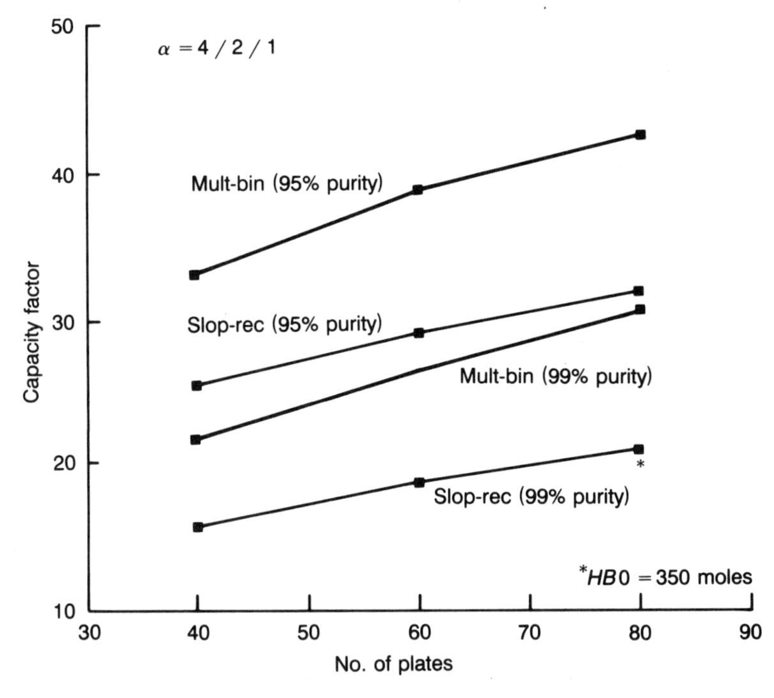

FIGURE 25-9. Comparison of two slop-handling strategies.

could be very significant in some batch distillation applications. However, in some plants the increased complexity may not be justified by the economic gain.

25-10 INFERENTIAL CONTROL OF BATCH DISTILLATION

25-10-1 Problem

In theory, the control of batch distillation columns should be fairly simple. If a composition analyzer is available that gives the instantaneous distillate compositions during the course of the batch cycle, one could simply switch from product to slop cuts at the appropriate times while following the optimal reflux ratio and pressure trajectories. However, composition analyzers are often not available and seldom instantaneous.

As an alternative, a mathematical model could be used to predict how compositions will change with time and could predict switching times. However, two requirements would have to be met if this "open loop" model is to be successfully used:

1. The model must be perfect. It must very accurately describe the system, including such parameters as the vapor–liquid equilibrium, tray efficiencies, heat losses, and so on.
2. The initial conditions in the still pot and in the column and reflux drum must be accurately known. If the model starts integrating from the incorrect initial conditions, the entire trajectory will be incorrect.

Almost never will these two requirements be satisfied. Therefore, if a mathematical model is to be used to predict compositions, some type of feedback correction must be used. Temperature information is usually available from a number of locations in the column during the entire batch. These actual measured temperatures can be compared with the temperatures predicted by the mathematical model, and any differences can be used to correct the model predictions.

This procedure in the control field is called constructing a *state estimator* or *observers*. A *Kalman filter* is one example of such a state estimator. It can be used when signals are noisy (contain a lot of uncertainty). Some very elegant statistical mathematics and optimal control theory are used in Kalman filter design.

A more simple observer is a Luenberger observer that uses deterministic models. Other types of observers (models) have also been used. Quintero-Marmol and Luyben (1992) and Quintero-Marmol, Luyben, and Georgakis (1991) explored the application of several observers to the problem of the control of batch distillations. Some of the highlights of this work are discussed in the following text.

25-10-2 Basic Insight

One of the most interesting and important findings of the work on inferential control of batch distillation is the way the composition fronts move through the column during the course of the batch cycle. Figure 25-10 illustrates this phenomenon for a ternary system. At the beginning of the batch (time = 0) the compositions on all trays and in the still pot are assumed to be the same (flat composition profiles). While the column is operated under total reflux conditions, the concentration of the lightest component (x_1) builds up on the upper trays in the column and the concentrations of the intermediate component (x_2) and the heaviest component (x_3) decrease in the top of the column but increase in the still pot. At the time when withdrawal of the first product ($P1$) is begun (time = 1.5) there is a composition front located in the lower part of the column that separates the lightest

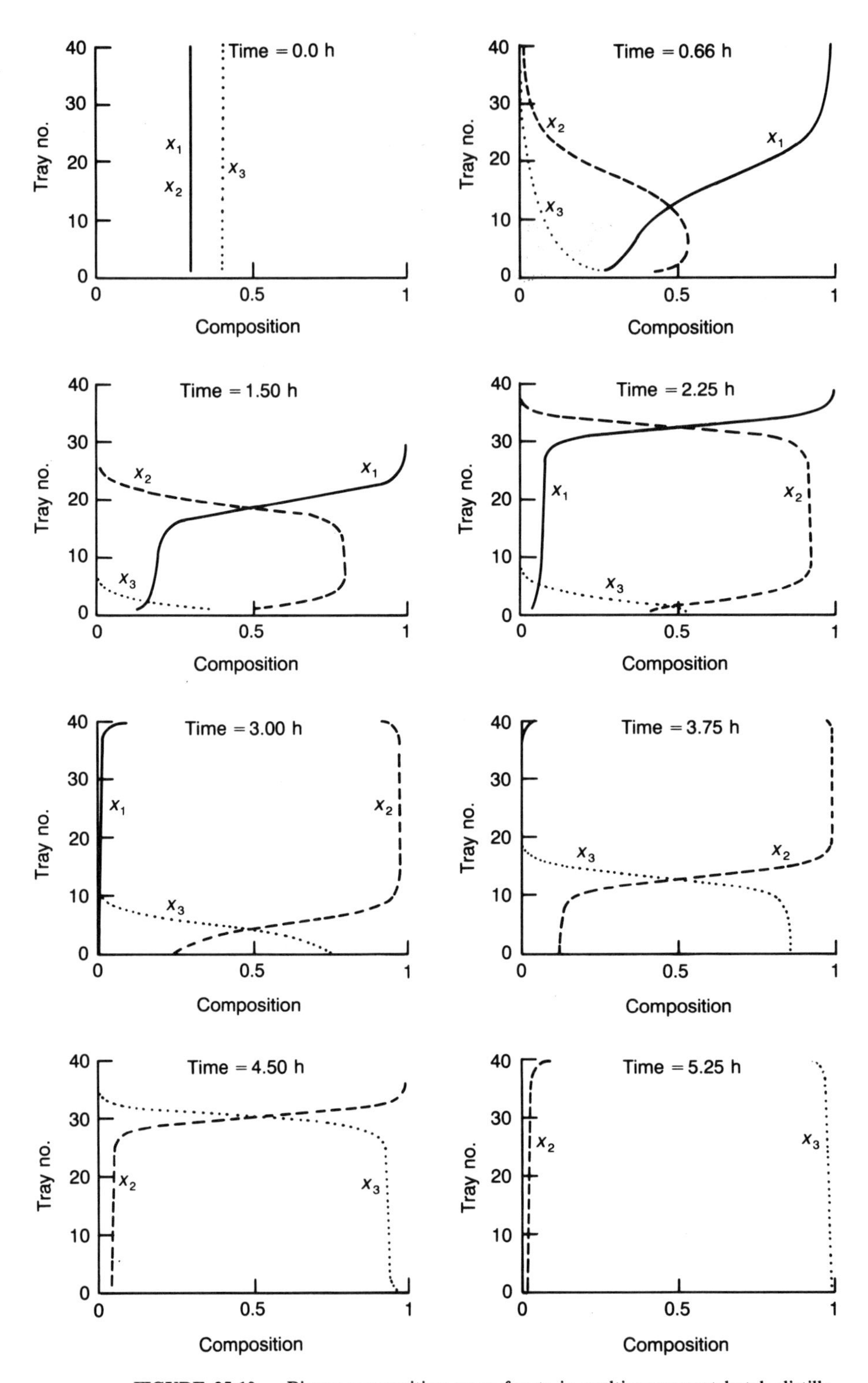

FIGURE 25-10. Binary composition wave fronts in multicomponent batch distillation.

and intermediate components; this front moves up the column (time = 2.25) as light product is removed.

When this front nears the top of the column, the first slop cut ($S1$) is begun. It is essentially a binary mixture of components 1 and 2. When the purity of component 2 reaches its specification, the withdrawal of the intermediate product ($P2$) is begun.

Now a second composition front forms near the base of the column, which represents the interface between components 2 and 3. This front sweeps up the column as intermediate component 2 is removed in the distillate. The shapes of these fronts depend on the reflux ratio and the relative volatilities.

Thus, what is in the upper part of the batch distillation column at any one time during most of the batch cycle is some binary mixture. Initially it is the binary mixture of components 1 and 2. Then it becomes components 2 and 3. This continues for as many components as there are in the initial feed to the unit.

The significance of this behavior is that simple temperature and pressure measurements can be used to estimate compositions in these binary systems. The components change during the batch cycle, but for any specific binary mixture, compositions can be inferred from temperature and pressure measurements.

25-10-3 Quasidynamic Model

This binary wave-like behavior of batch distillation columns is the basis for the "quasidynamic" ad hoc estimator studied by Quintero-Marmol and Luyben (1992). A temperature is measured in the middle section of the column. The composition of the vapor phase of the appropriate binary mixture at that point in time and at that location is calculated from the measured temperature and pressure and the vapor–liquid equilibrium relationships.

As illustrated in Figure 25-11, this calculated *binary* vapor is fed into a dynamic

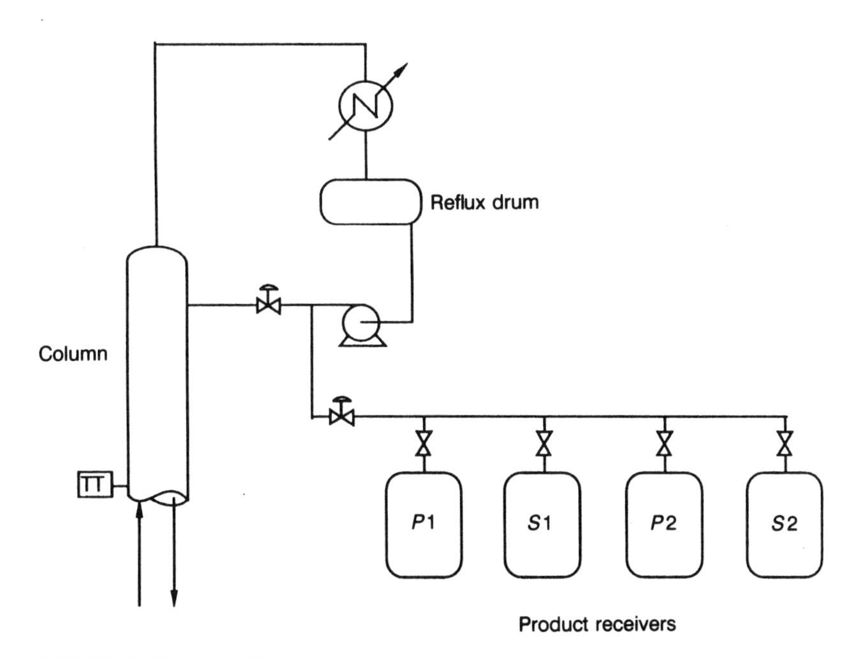

FIGURE 25-11. Quasidynamic model.

mathematical model of the section of the column above this temperature measurement tray. The model is a *ternary* mixture, not a binary mixture. The vapor feeding it is always binary, but changes from a binary mixture of components 1 and 2 early in the batch to a binary mixture of components 2 and 3 later in the batch.

We make an educated guess to establish values for initial conditions of the model (the compositions on all the trays in the upper section) because we usually have only

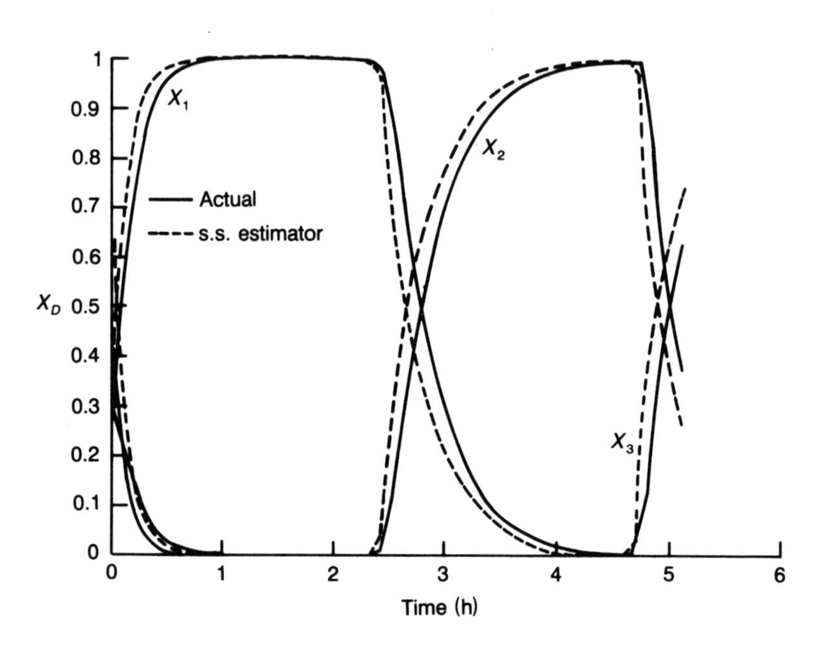

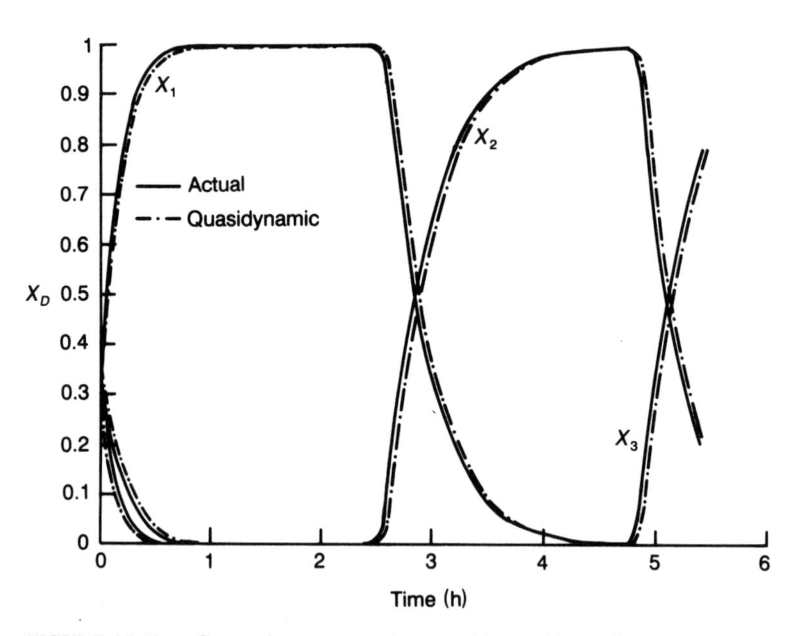

FIGURE 25-12. Comparison on actual compositions with predictions of steady-state and quasidynamic estimators.

an approximate idea of the composition of the initial charge to the still pot. The distillate compositions predicted by the model converge fairly quickly to the actual distillate compositions during the period of total reflux operation.

Figure 25-12 gives some typical results. In the top part of the figure actual distillate compositions are compared with the predictions of a simple steady-state estimator (holdups on trays and in the reflux drum are neglected). The steady-state estimator lags the real column by 5 to 10 min. In the bottom half of the figure actual distillate compositions are compared to the predictions of the quasidynamic estimator. The predictions are quite good.

25-10-4 Extended Luenberger Observer

Most of the work on observers in the control literature has been limited to low-order systems. Our batch distillation processes is quite high order (126 differential equations

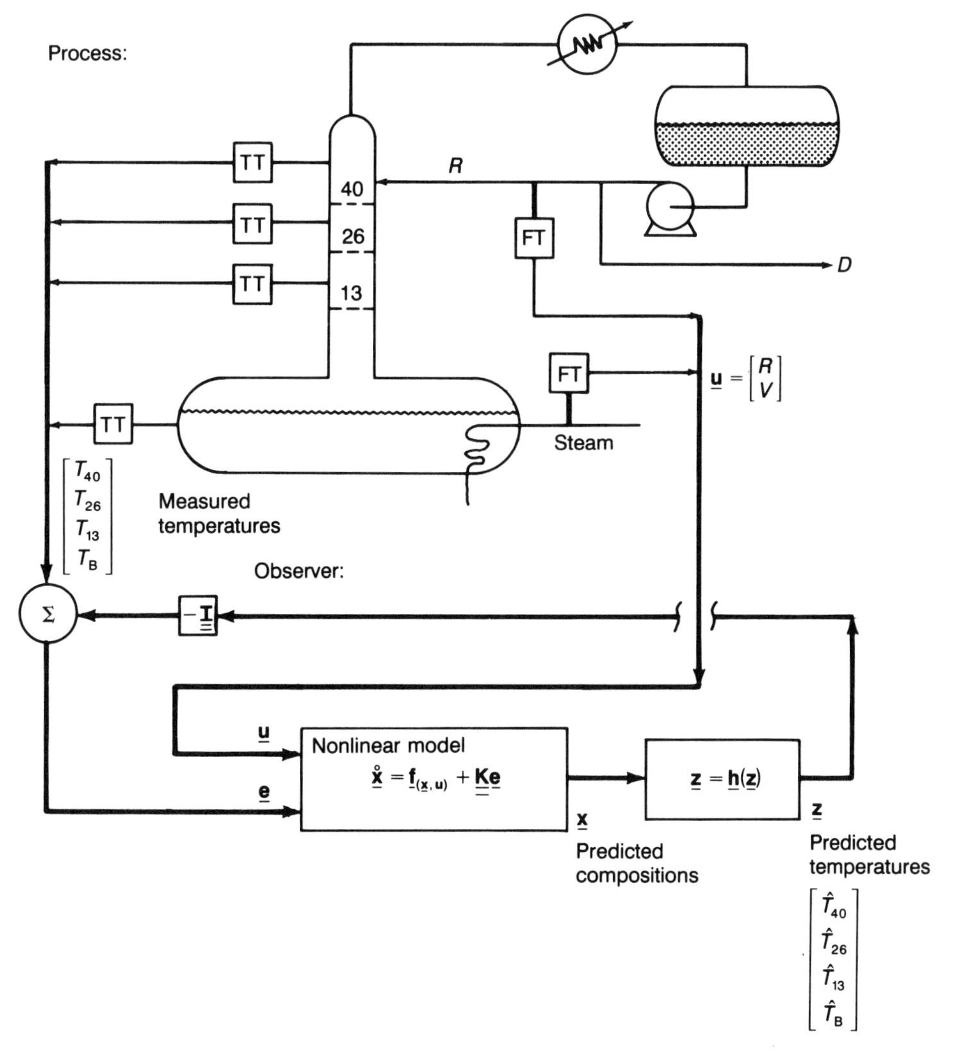

FIGURE 25-13. Extended Luenberger observer.

for a 40 tray ternary column). There have been very few studies of the application of observers to complex, nonlinear chemical engineering systems.

A Luenberger observer uses a linear model of the process to predict its states. An extended Luenberger observer uses a nonlinear model of the process. Because this is a batch process, compositions vary drastically during the course of the batch. Therefore, a linear model gives very poor predictions of the actual behavior and a rigorous nonlinear model must be used.

The nonlinear dynamic model has the form

$$\dot{\underline{x}} = \underline{f}_{(\underline{x}, \underline{u})} \tag{25-15}$$

$$\underline{z} = \underline{h}_{(\underline{x})} \tag{25-16}$$

where $\underline{x}$ = states of the process (compositions on all trays, reflux drum, and still pot; for a 40-tray ternary column there are 126 states)

$\underline{z}$ = measurements (temperatures on several trays in the column, for example, four temperature measurements)

The structure of the extended Luenberger observer is sketched in Figure 25-13. At the top of the figure the process is shown with the measured inputs R and V (reflux flow rate and steam flow rate) and the measured temperatures (in the still pot and on trays 40, 26, and 13). These process data are fed into the observer. The differences between the measured temperatures and the temperatures predicted by the model are used in a feedback loop to change all of the states. The $\underline{\underline{K}}$ is a matrix of gains that multiplies the differences between the measured and predicted temperatures (the error $\underline{e}$). This error drives the model to change the predicted states (and temperatures). If the feedback loop in the observer is stable, the observer will converge the predicted temperatures toward the measured temperatures. Then the distillate composition predicted by the model can be used to control the batch cycle.

The rate of convergence of the observer depends on the gains $\underline{\underline{K}}$ used. Making these gains large produces small closed-loop time constants (large negative eigenvalues) and rapid convergence. However, if the gains are too large, the observer can become too underdamped or even closed-loop unstable. Therefore, the design of the $\underline{\underline{K}}$ matrix is crucial to the successful operation of the observer. "Pole-placement" methods are usually employed using a linear model of the system for the local operating point.

The observer will only be useful in the batch distillation process if it is able to converge to the correct distillate compositions (starting from some assumed initial conditions because we seldom know the initial compositions in the charge to the still pot) before the end of the total reflux operating period (before the withdrawal of the first product $P1$ has begun). Therefore, the observer must converge fairly rapidly, but it must be stable.

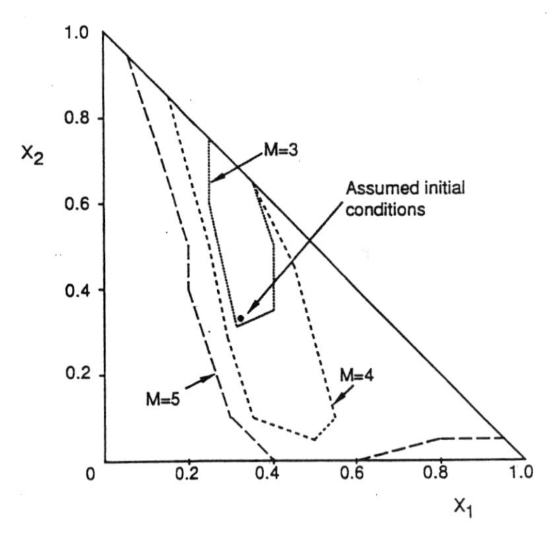

FIGURE 25-14. Regions of convergence for different numbers of temperature measurements.

There are a number of important practical design questions to be answered about the design of the extended Luenberger observer:

1. How many temperature measurements are needed?
2. Where should they be located in the column?
3. Where should the closed-loop poles be located?
4. Does the $\underline{\underline{\mathbf{K}}}$ matrix of gains have to be changed as the nonlinear process changes?
5. How close must the guessed initial conditions be to the actual initial conditions in order to get successful convergence of the observer?

Quintero-Marmol et al. (1991) studied most of these questions and proposed the following design procedure:

1. Guess the number of temperature measurements (M).
2. Locate these temperature measurements at the top of the column, in the still pot, and equally spaced up the column.
3. Use a linear model (linearized around the guessed initial conditions) and a robust pole-placement algorithm (for example, the MATLAB package, which uses the work of Kautsky, Nichols, and van Dooren, 1985). Specify the closed-loop poles to be equal to the open-loop poles with the smallest of the open-loop poles moved to the left in the s plane. These should be moved to the left until the closed-loop response of the observer is as fast as possible but still has a reasonable damping coefficient (greater than 0.3).
4. If the observer does not converge quickly enough, increase the number of measurements and repeat steps 2 and 3.
5. Test the sensitivity of convergence to the error in the guessed initial conditions. If the region of convergence is small, increase the number of measurements.

Quintero-Marmol et al. (1991) found that it was not necessary to update the $\underline{\underline{\mathbf{K}}}$ matrix of gains during the batch. The gains were fairly insensitive to the composition profile in the column, and once convergence was obtained during the initial total reflux operation, it did not matter much what gains were used.

A typical number of temperature measurements was four to five. Figure 25-14 illustrates how the region of convergence increases as the number of temperature measurements (M) is increased.

Figure 25-15 illustrates how distillate and bottoms compositions (actual and predicted by the observer) vary with time. The specific numerical case is a ternary 20-tray column with five temperature measurements (base and trays 5, 10, 15, and 20). The actual initial conditions were 0.25/0.65/0.15, whereas the initial conditions guessed were 0.333/0.333/0.334. Note that the observer converges in about 15 min, which is well before the $P1$ withdrawal begins. Figure 25-16 shows what happens when only three temperatures were measured (base and trays 10 and 20). The response is slower and much more oscillatory, particularly in the reboiler where they last for almost 2 h.

The extended Luenberger observer was found to be somewhat better than the quasidynamic, but it is more difficult to design. Most batch distillation applications can probably use the quasidynamic observer, but others may require a more complex observer.

25-11 CONCLUSION

This chapter has attempted to discuss briefly some of the many important and interesting aspects of the design and operation of batch distillation columns. Much more study in this area is needed.

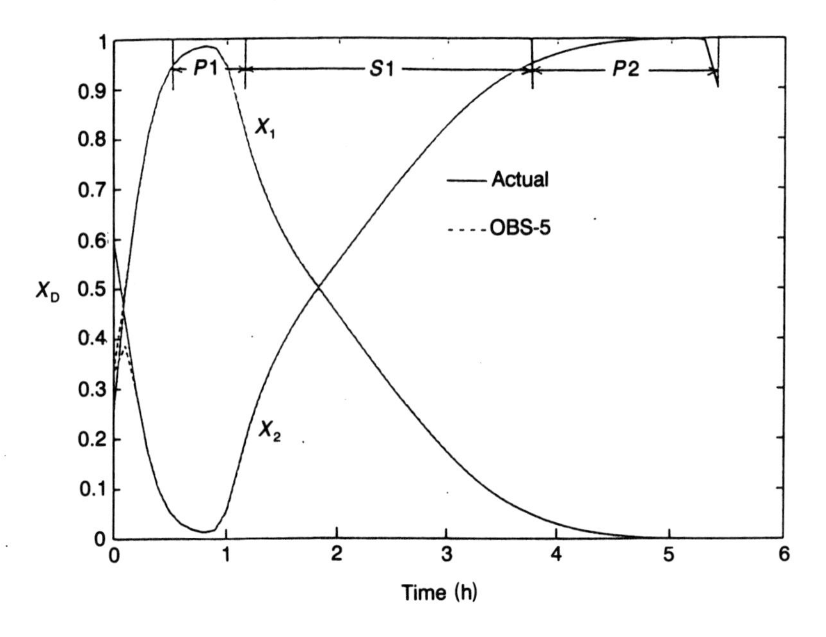

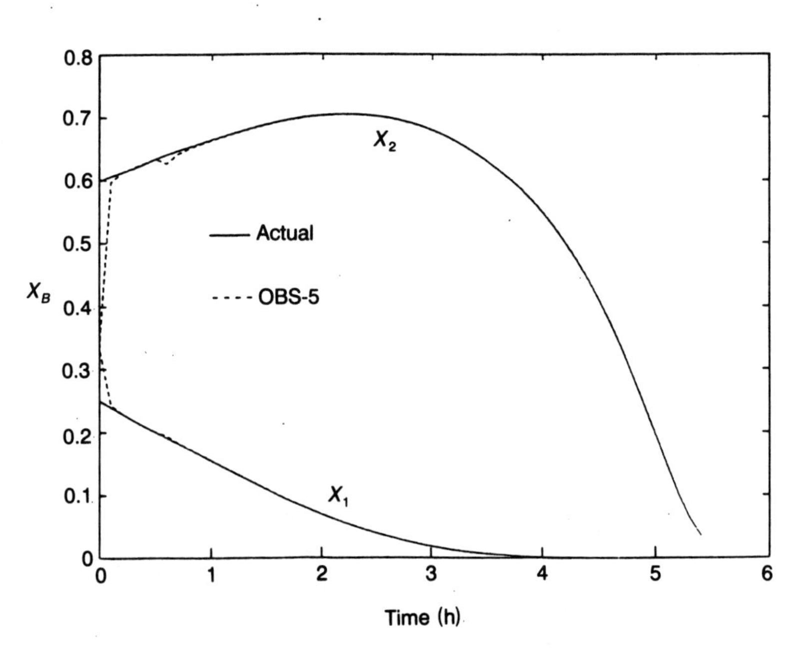

FIGURE 25-15. Observer with five temperature measurements.

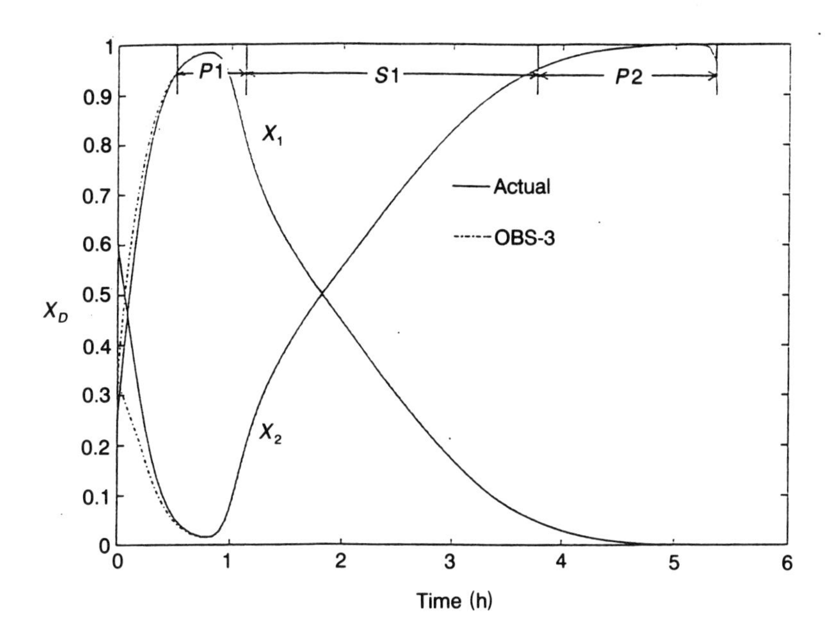

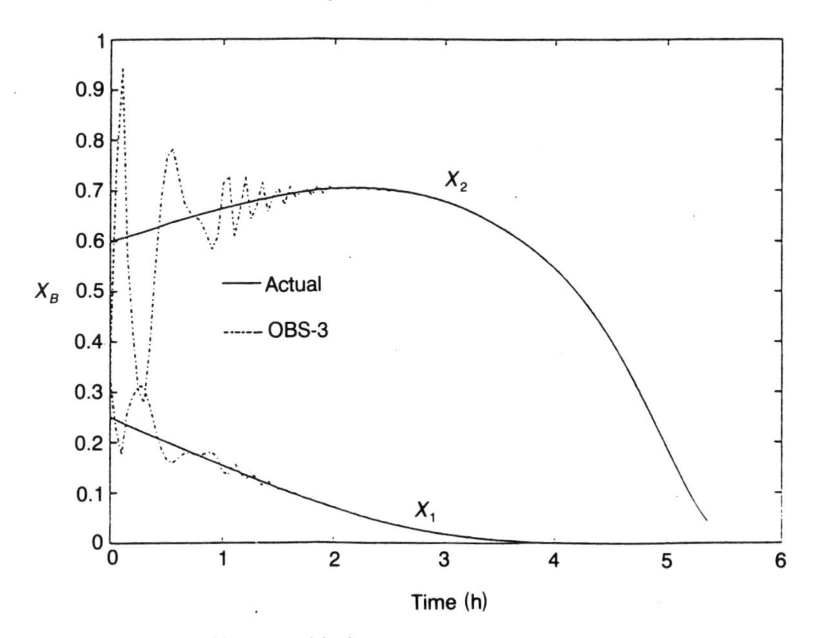

FIGURE 25-16. Observer with three-temperature measurements.

References

Al-Tuwaim, M. S. and Luyben, W. L. (1991). Multicomponent batch distillation. 3. Shortcut design of batch distillation columns. *Ind. Eng. Chem. Res.* 29, 507–516.

Bernot, C., Doherty, M. F., and Malone, M. F. (1991). Feasibility and separation sequencing in multicomponent batch distillation. *Chem. Eng. Sci.* 46, 1311–1326.

Block, B. (1961). Batch distillation of binary mixtures provides versatile process operation. *Chem. Eng.* 68, 87–93.

Boston, J. F., Britt, H. I., Jirapongphan, S., and Shah, V. B. (1981). Foundations of computer aided chemical process design. New York: Engineering Foundation.

Chiotti, O. J. and Iribarren, O. A. (1991). Simplified models for binary batch distillation. *Comput. Chem. Eng.* 15, 1–5.

Coates, J. and Pressburg, B. S. (1961). How to analyse the calculations for batch rectifications in tray columns. *Chem. Eng.* 68, 131–136.

Coward, I. (1967). The time-optimal problem in binary batch distillation. *Chem. Eng. Sci.* 22, 503–516.

Distefano, G. P. (1968a). Mathematical modeling and numerical integration of multicomponent batch distillation equations. *AIChE. J.* 14, 190–199.

Distefano, G. P. (1968b). Stability of numerical integration techniques. *AIChE. J.* 14, 946–955.

Farhat, S., Czernicki, M., Pibouleau, L., and Domenech, S. (1990). Optimization of multiple-fraction batch distillation by nonlinear programming. *AIChE J.* 36, 1349–1360.

Galindez, H. and Fredenslund, A. (1988). Simulation of multicomponent batch distillation processes. *Comput. Chem. Eng.* 12, 281–288.

Holland, C. D. and Liapis, A. I. (1983). Computer methods for solving dynamic separation problems. New York: McGraw-Hill.

Kautsky, J., Nichols, N. K., and van Dooren, P. (1985). Robust pole assignment in linear state feedback. *Int. J. Control* 41, 1129–1155.

Luyben, W. L. (1971). Some practical aspects of optimal batch distillation design. *Ind. Eng. Chem. Proc. Des. Dev.* 10, 54–59.

Luyben, W. L. (1988). Multicomponent batch distillation. 1. Ternary systems with slop recycle. *Ind. Eng. Chem. Research* 27, 642–647.

Meadows, E. L. (1963). Multicomponent batch distillation calculations on a digital computer. *Chem. Eng. Prog. Symp. Ser.* 45, 59, 48. AIChE, New York.

Muhrer, C. A., Grassi, V. G., and Silowka, W. (1990). Process improvements in batch distillation through dynamic simulation. AIChE Annual Meeting, Chicago.AIChE, New York.

Quintero-Marmol, E. and Luyben, W. L. (1990). Multicomponent batch distillation. 2. Comparison of alternative slop handling and operation strategies. *Ind. Eng. Chem. Res.* 29, 1915–1921.

Quintero-Marmol, E. and Luyben, W. L. (1992). Inferential model-based control of multicomponent batch distillation. *Chem. Eng. Sci.* 47(4), 887–898.

Quintero-Marmol, E., Luyben, W. L., and Georgakis, C. (1991). Application of an extended Luenberger observer to the control of multicomponent batch distillation. *Ind. Eng. Chem. Res.* 30, 1870–1880.

Robinson, C. S. and Gilliland, E. R. (1950). Elements of fractional distillation. New York: McGraw-Hill.

Van Dongen, D. B. and Doherty, M. F. (1985). On the dynamics of distillation processes IV—Batch distillation. *Chem. Eng. Sci.* 11, 2087–2093.

Index